Finite Element Analysis for Civil Engineering with DIANA Software

Shun Chai

Finite Element Analysis for Civil Engineering with DIANA Software

南京大學出版社
NANJING UNIVERSITY PRESS

Springer

Shun Chai
Department of Civil Engineering
Southeast University
Nanjing, Jiangsu, China

ISBN 978-981-15-2944-3 ISBN 978-981-15-2945-0 (eBook)
https://doi.org/10.1007/978-981-15-2945-0

Jointly published with Nanjing University Press
The print edition is not for sale in China Mainland. Customers from China Mainland please order the print book from: Nanjing University Press.

This Springer imprint is published by the registered company Springer Nature Singapore Pte Ltd.
The registered company address is: 152 Beach Road, #21-01/04 Gateway East, Singapore 189721, Singapore

Preface

DIANA (Displacement ANAlyzer), which is also named as Diana, is a brilliant type of structural finite element nonlinear analysis software applicable for both engineering design institutions as well as scientific research institutions in civil engineering. In the past two decades, DIANA has been upgraded from Release 8.1 to 10.3, experiencing constantly tremendous improvement in graphical user interface manipulations, command console syntax simplification, enrichment of element and material library. Compared with other kinds of finite-element software, it has received vast attention owing to the simulation advantages of concrete structure cracking, hydration heat simulation, sand liquefaction, random field prediction, concrete time-dependent performance and earthquake resistance of building structure in the nonlinear finite-element analysis of reinforced concrete structure over other kinds of finite-element software.

However, there has been no such related academic works to introduce this kind of software so far. Moreover, due to the language obstacle, the access to referring to the English manual of this software may have become a bottleneck for some beginners to learn and understand this kind of software. In view of the complexity of such an issue, the author expects to fill in the vacancy in the current academic field via the comprehensive and systematic introduction in this book. Through both theoretical introduction and abundant numerical cases of this excellent civil engineering finite-element software as well as based on years of experience of the author, the university researchers and engineering designers all over the world can have a targeted view when studying and mastering the basic manipulations of this software as soon as possible.

The main feature of this book is easy-to-interpret. Vast amount of complicated, highly difficult and hard-to-interpret basic theoretical knowledge of finite-element method is simplified and replaced by the plain and understandable words, which help beginners in mastering the basic modeling skills.

The other edge of this academic work lies in its manipulation diversity. Manipulations in this book are displayed not only according to the graphical user interface visual operation mode but also the command console in Python language. In order to facilitate the reader's study, DIANA command console in python language for every numerical case is listed at the end of each part and uploaded as attachment in the corresponding given official website.

The third advantage of this book lies in that it has abundant numerical cases concerning emerging material and structures in a wide range of sources to satisfy current engineering requirement. For example, numerical cases are compiled in Chap. 5 focused on the current emerging precast segmental structures, including direct shear, long-term analysis and cracking propagation prediction via random field. Degenerated long-term performance under mutual time-dependent variables concerning creep, shrinkage and relaxation, ultra-high performance concrete (UHPC) beam under flexural bending capacity and cracking process, hysteresis analysis of shear wall, nonlinear dynamic analysis for reinforced concrete, phase analysis for box-girder bridge as well as time-history analysis are all displayed and illustrated in Chap. 3.

This book is only written for the related fields that the editors of DIANA model are familiar with in civil engineering. In fact, it is a kind of software suitable for many fields and many directions. It has a very broad application prospect, not only limited to the application of structural direction. Since the author is experienced in the structural and bridge engineering, this book mainly focuses on the structural direction of civil engineering, and tends to put emphasis on nonlinear analysis and calculation based on iterative methods.

This book also has some reference value for the following future academic research, and the author wishes that more experts and masters may write more theoretical-deepening and high-quality works about DIANA.

Nanjing, China Shun Chai

Contents

About the Author

Shun Chai a Ph.D. of Civil Engineering Department in Southeast University situated in Nanjing, writes this academic work. The research field of the author is mainly focused on evaluation of structural state, including the long-term assessment of precast segmental bridges, resilient design of bridges, stochastic finite-element method of polynomial chaos extension on structural reliability research, as well as the current research on UHPC bridges.

Dr. Chai participated in many experiments, including long-term performance experiment of precast segmental concrete (PCS) bridges and the flexural research of precast segmental concrete (PCS) bridges, and he has published three academic papers in the year of 2016 and 2019 respectively. During the process of researching stochastic finite-element method, he has gained vast expertise as well as rich experience in all versions of DIANA. In 2018, the academic work named *Finite Element Analysis of DIANA 10.1 for Civil Engineering* in Chinese language was launched by him and published in Nanjing University Press (ISBN: 978-7-305-20282-7).

Chapter 1
Introduction of DIANA

Abstract As an initial chapter of this academic work, the background and application scope of DIANA (Displacement ANAlyzer, also named as Diana) software is introduced in brief. Besides, functions and installation are illustrated in the second part. The most important part lies in the introduction of element types, where the shapes, interpolation orders as well as integration schemes of truss elements, beam elements, plane stress and strain elements, plate bending elements, axisymmetric elements, flat and curved shell elements, solid elements, reinforcement elements, interface elements, contact elements as well as spring elements are introduced in detail one by one. In order to render more convenience to beginners, Sect. 1.4 focuses on file system and opening paths, and the working window of DianaIE is also presented in Sect. 1.5. The two key methods for DIANA preprocessing modeling procedure—the graphical user interface manipulation in DianaIE and the editing command console syntaxes in Python language—are also explicated in Sects. 1.6 and 1.7, respectively. Moreover, unit systems in DIANA are also illustrated in this chapter.

1.1 Background

DIANA (Displacement ANAlyzer, also named as Diana) was established in 1970 in Holland, which is an outstanding structural finite-element software developed by TNO DIANA company, applicable to all structural fields in civil engineering. Recently, it has been widely applied in structural engineering, bridge engineering, geotechnical engineering, tunnel, underground structural engineering, pile engineering and their like. In the past two decades, DIANA has gone through continuously tremendous development and improvement in GUI manipulation, command console syntax simplification enrichment of element and material library. Meanwhile, it has

---

Electronic supplementary material The online version of this chapter (https://doi.org/10.1007/978-981-15-2945-0_1) contains supplementary material, which is available to authorized users.

S. Chai, *Finite Element Analysis for Civil Engineering with DIANA Software*,
https://doi.org/10.1007/978-981-15-2945-0_1

received vast attention and application from scientific researchers and structural designers across the globe, owing to its extraordinary modeling effects of structural nonlinear analysis in concrete cracking simulation, hydration reaction, liquefaction, random field prediction as well as time-dependent performance of structures. However, no such academic work has systematically introduced its manipulation so far. Besides, consulting and studying manual accompanied by the software has become an obstacle for many users and beginners. In view of such issues, the author hopes the gap in this field can be filled via systematically introducing this structural software in this academic work so that scientific researchers and engineering designers can have a definite study aim when using this book coupled with manual, and they can rapidly grasp the basic operations of this structural software.

The main feature of this book is easy-to-interpret. Contrary to traditional comprehensive explication and excellent theoretical analysis, based on the platform of DIANA 10.1 release and long-term experience of applications of both old and new versions, vast amount of complicated, highly difficult and hard-to-interpret basic theoretical knowledge of finite-element method is simplified and replaced by the plain and understandable words, which help beginners in mastering the basic modeling skills. Besides, relevant features in the application characteristics of the software are illustrated, and engineering cases are made according to the experience of the author. Constitutive relationship setting and specific manipulation procedures are also demonstrated as real examples to aid readers get started quickly. These engineering cases are combined with both DIANA 9.4 and DIANA 10.1 and many of them derive from simplified engineering models in construction site and domestic relevant hottest researching points in recent years. Meanwhile, this book provides attachments on command console syntax in Python language, and the author believes that these attachments also play a key role in making users easier to interpret and grasp complicated manipulation, consuming the shortest time. DianaIE modules are split in this book in order to introduce common functions one by one, which is the main stating thought. The details and highlights are emphasized so that the readers grasp DIANA from macro perspective and this book ensures that the key parts are introduced and illustrated in detail. The second priority lies in relevant simulation cases being integrated with the experience of numerical simulation, experiment as well as engineering background, as illustrated in Chaps. 3–5, respectively.

Another characteristics of this book lies in that other than vast amount of pictures and tables introducing DianaIE module and software operation, command console syntaxes in Python language are attached in most numerical examples in order to assist readers to compare two distinguished modeling methods so that they can adapt to a series of basic operations such as opening, running, editing and saving files with different suffix names. The third feature is the learning essentials being added to the examples in the following three chapters. That is to say, four to five representative classical manipulations are extracted purposefully and placed at the beginning of each section. These learning essentials are displayed in simple and complicated ways so that DIANA users have a step-by-step and targeted learning process, as well as clear-headed state when studying this book. It is worth to note that the data in examples are dependent on author's engineering experience and numerical assumption and no such sufficient civil engineering experiments validation is performed; thus there may be deviation between cases and real engineering states. Therefore, readers should have a scrutiny and critical attitude to this academic work. On the one hand, the focus of the reader should be concentrated on studying the essentials of software manipulation. The calculation and analysis given in the book can be critically absorbed and the modeling operation methods mentioned by the author in the book can be mastered by means of analogy. On the other hand, context needs further improvement, so the reader should consider and pay attention whether there are better constitutive models, faster modeling thoughts and ways when learning these specific manipulations. Under such self-considerations above, one can handle the operation and functions in a better and more precise way.

The following several chapters forge this academic work. This chapter mainly introduces relative characteristics about DIANA, ranging from main functions, installation and operation, command element classes, system of files and GUI operation to common command console. The focus of this book is on GUI operation in order to provide a favorable access to mastering DIANA for beginners. Chapter 2 mainly puts focus on the following three aspects: (1) constitutive material properties of concrete, steel and reinforcements, (2) cracking models for concrete and (3) codes for design of concrete and steel in the world concerning DIANA.

Chapters 3–5 are examples of the modeling operation. The third chapter includes many traditional fields of civil engineering, such as cracking analysis of simply supported beam, bond-slip material model for reinforcement of box-girder, long-term nonlinear analysis for mutual time-dependent variables concerning creep, shrinkage and relaxation, hysteretic analysis about shear wall so that readers can have a comprehensive understanding about nonlinear analysis through studying the examples above. Chapter 4 introduces one of the unique features of DIANA-transient heat flow analysis of concrete in large volume according to the current hydration reaction in the construction phase. Chapter 5 is focused on the current emerging precast segmental structures, degenerated long-term performance under mutual time-dependent variables concerning creep, shrinkage and relaxation and random field in demonstrating modeling operation on these primitive examples of precast segmental structures. Considering the complexity of structural behaviors and that the characters of this type have high demand on geometric modeling, interface element modeling as well as nonlinear iterative calculation, this chapter is one of the core chapter and also the flashpoint of this book. The last chapter, Chap. 6, provides the feedbacks and suggestions for improving DIANA according to the author's daily modeling experience.

1.2 Main Functions, Installation and Operation

DIANA is not only extraordinary structural finite-element software with enriched constitutive model and strong nonlinear analysis function but also a nonlinear widely used tool for civil engineering, bridge engineering, geotechnical engineering, tunnel, underground structural engineering, irrigation works, municipal engineering and fire engineering. Here nonlinear means that there are large deformations and displacements in the structures under the load while such deformation has high impact on equilibrium. Thus the deformation compatibility equations are established on the post-deformation state. For nonlinear calculation, iterative calculation is always selected. In any versions of DIANA, the whole failure process of concrete structures-from initial state, cracking get started, cracking propagation to the ultimate collapse can be simulated precisely. Structural real geometric characters, material models and their like are all taken into account in this type of simulation, which also has high precise on the coupling between concrete elements and embedded reinforcements (bar elements as well as grid elements). Moreover, three bonding types—fully bonded, non-bonded as well as bond-slip between reinforcement and concrete—can be simulated precisely, which is also a unique technique in DIANA compared with other kinds of finite-element analysis software. DIANA not only provides structural linear dynamic analysis functions but also considers nonlinear analysis module under the circumstance of cyclic loading or action of seismic wave when it comes to the seismic design. Excellent functions are also displayed in the analysis of geotechnical excavation and dam analysis such as construction phase analysis, soil-structure coordination analysis, fluid-structure coupling analysis, user-specified constitutive model, multiple interface elements, large deformation and strain analysis, nonlinear material analysis, time-dependent and ambient analysis and nonlinear dynamic analysis. DIANA also has wide applications in the tunnel engineering, and the traditional common analysis and design of tunnel are mainly concentrated on the stress analysis of tunnel excavation and lining. It is also worth to mention that the analysis of structural performance under the influence of temperature field such as structures in the fire and hydration heat reaction of mass concrete are widely applied in the emerging module of DIANA software.

DIANA 10.1 consists of two major moduli. One is a newly developed graphical user interface (GUI) DianaIE, which was developed by the DIANA developing institution, while the other is the former preprocessing module belonging to the original DIANA 9.6. Both old and new users not only manipulate DianaIE interactive environment directly but also use preprocessing interface to solve the analysis and calculation or they can even import traditional edited binary files in the suffix name of .bat to create the numerical model according to their extent of expertise. Moreover, DianaIE can also import CAD files such as IGES and STEP, and the following procedure of modeling work can be conducted based on this import. Mature applications of DIANA software in civil engineering will be introduced in the following part.

The main application features of DIANA are listed as follows:

(1) Reinforced concrete cracking analysis
(2) Hydration analysis for mass concrete
(3) Time-dependent nonlinear analysis for reinforced concrete
(4) Phased analysis
(5) Seismic analysis for concrete and masonry

(6) Passive and active strengthening measures
(7) Cracking prediction via random field
(8) Hysteresis analysis for reinforced concrete under low-period test and cyclic loading conditions.

On comparing with the former old DIANA release 9.4 version, users are required to purchase this kind of DIANA software and obtain permission before application. Contrary to the 9.4 release version that supports XP system, 32-bit Windows as well as 64-bit Windows, this new version DIANA is only applicable for the 64-bit Windows. Meanwhile, the whole procedure of installation must be under computer networking state. In order to ensure success of following installation, all kinds of anti-virus software should be closed and installation files (i.e., Setup.exe and Dianahasprus.exe) must be added into the trusted zone. Encryption mode for DIANA is software key encryption coupled with computer physical binding type. Users ought to send the generated **c2v** file to an agent company via generated by **Dianahasprus.exe** before activation of this software. After activation, users can log on to the DIANA official website (http://tnodiana.com/Diana-downloads) and can download the installation moduli of DIANA software of various versions.

On comparing the starting file Setup.exe, software key encryption HASPUser Setup.exe and MIDAS moduli with former old version of DIANA 9.4, we found installations of DIANA 10.1 and 10.2 mainly incorporates the following three aspects:

(1) Installation of **DianaIE**

Installation of **DianaIE** is completed via downloading **Setup.exe** file. DIANA document after download contains bin starting folder, binseg folder, lib folder to perform registration function, python folder applied for editing Python language and PDF of DIANA manual dominated by current 9.6 release version (see Fig. 1.1).

Fig. 1.1 Folders of DIANA software

(2) Installation of software key encryption **Dianahasprus.exe**

Before the installation of **Dianahasprus.exe,** users should purchase DIANA software and apply to general company or to local agent service companies for c2v file key activation and license status activation. After activation by general company and update of license information, installation can be completed.

Ownerships of key number information, activated c2v file and procedure information of updated license information belong to general company, so users are not allowed to disclose without authorization.

(3) Installation of **iDiana**

In view of preprocessing features of DIANA release version 9.6 retained in iDiana, installation type is the same as former old version.

After license and update of **Dianahasprus.exe** are completed, click **Setup.exe** then the ejected initial installation interface is displayed as shown in Fig. 1.2.

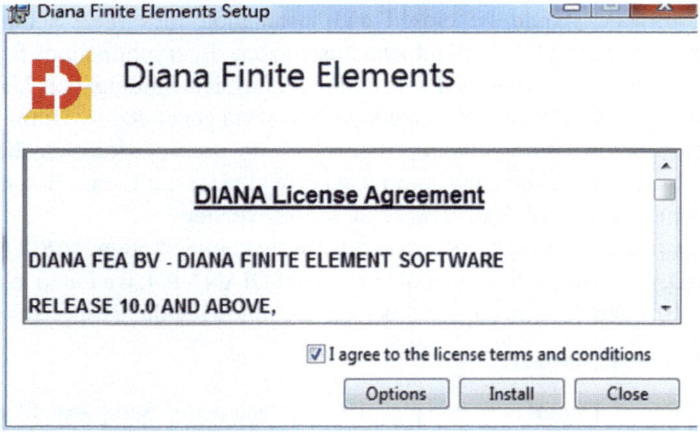

Fig. 1.2 Installation interface after clicking **Setup.exe**

Clicking the **Install** button; initialized program interface appears and the procedure automatically enters the installation process (see Fig. 1.3).

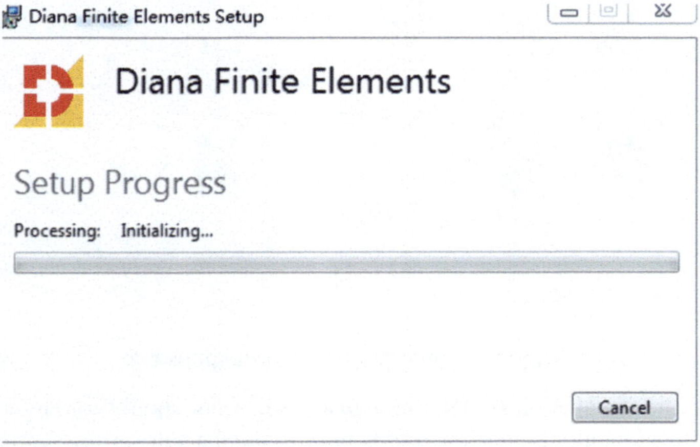

Fig. 1.3 Initializing process in installation procedure

After the installation, click OK button; **DianaIE** interface pops up, representing successful installation of DianaIE (see Fig. 1.4).

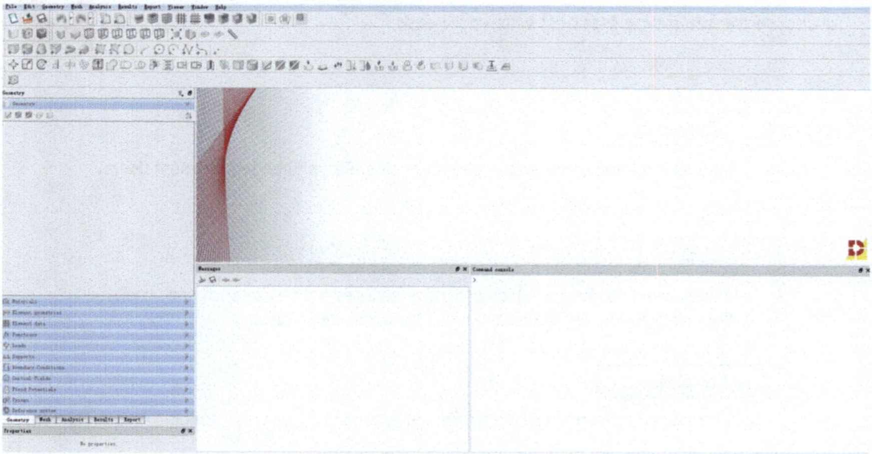

Fig. 1.4 DianaIE interface

Taking the installation of emerging new DIANA 10.2 release version for instance, installation procedure is displayed as shown in the figures. When the starting interface ejects, we click the **Next** button to enter the **Choose Setup type interface**, where **Complete** installation selection is chosen in order to install complete full set of functions (see Figs. 1.5 and 1.6).

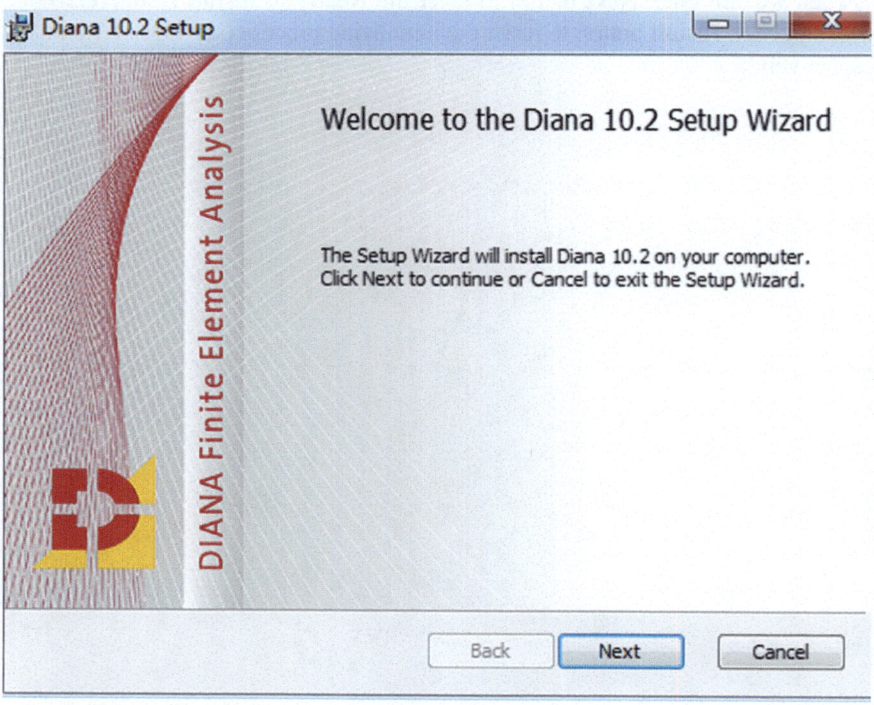

Fig. 1.5 Starting and initializing interface of DIANA10.2

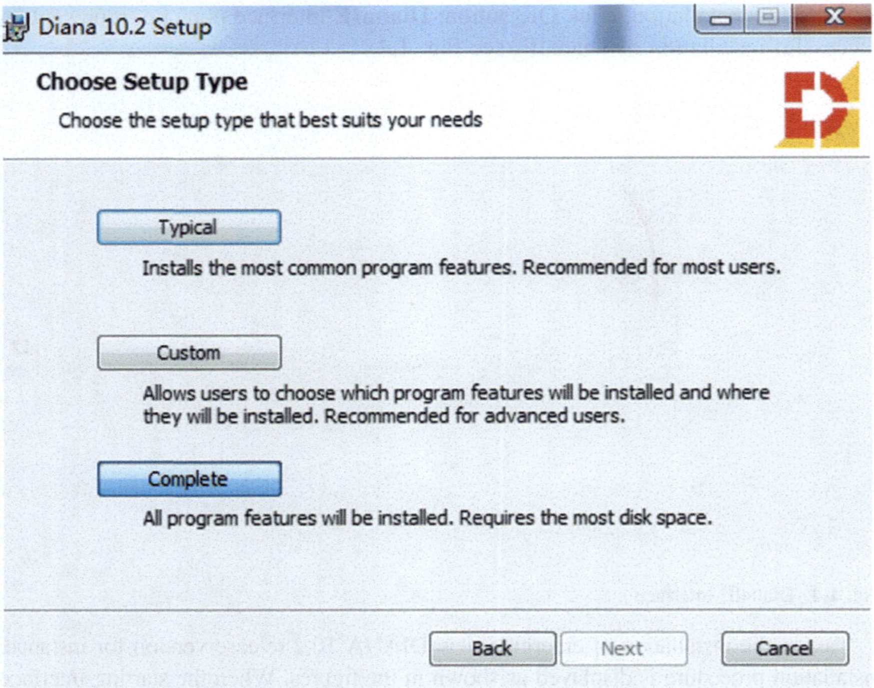

Fig. 1.6 Selection of complete installation

Then we still click **Next** button to enter the **Ready to install Diana10.2** interface, and click **Install** button to resist the initializing process (see Figs. 1.7 and 1.8, respectively).

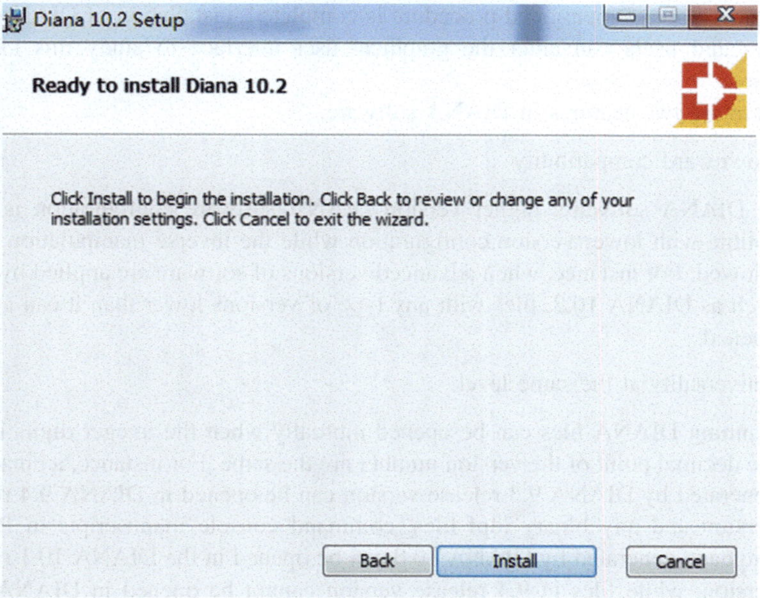

Fig. 1.7 Ready to install Diana10.2 interface

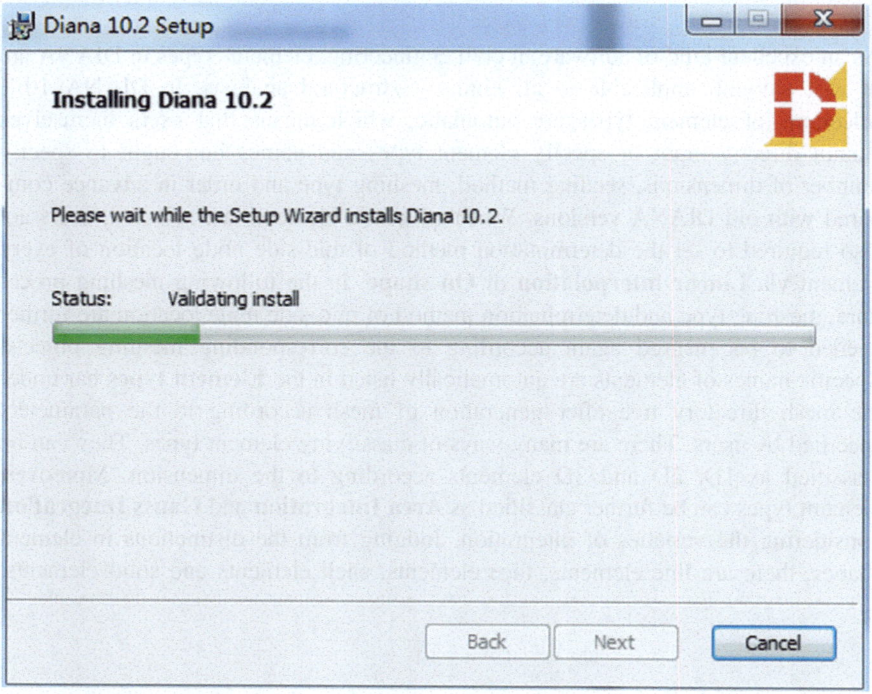

Fig. 1.8 Initializing interface

After the above-mentioned procedure is completed, installation of DIANA 10.2 finishes and users can enter the graphical user interface to study this kind of software.

There are two features in DIANA software:

(1) Downward compatibility

In DIANA software, higher version DIANA software configuration is compatible with lower version configuration while the inverse manipulation is not allowed. For instance, when advanced versions of software are applied by users such as DIANA 10.2, files with any type of versions lower than it can also be opened.

(2) Universality at the same level

Running DIANA files can be opened mutually when the integer digits before the decimal point of the version number are the same. For instance, a binary file generated by DIANA 9.3 release version can be opened in DIANA 9.4 release version and any binary .dpf files, command console manuscripts in Python language generated by DIANA 10.0 can be opened in the DIANA 10.1 release version, while files in 9.4 release version cannot be opened in DIANA 10.1 release version.

1.3 Typical Element Types in DIANA

As an excellent kind of software in civil engineering, elements types in DIANA are in vast amount, applicable to all kinds of structural analysis. In DIANA 10.1, selections of element types are automatic, which means that users themselves cannot directly input or specify element types and names but ought to specify number of dimensions, seeding method, meshing type and order in advance compared with old DIANA versions. When quadratic elements are selected, users are also required to set the determination method of mid-side node location of every element via **Linear interpolation** or **On shape**. In the following meshing procedure, meshing type and determination method of mid-side node location are further needed to be ensured again according to the corresponding meshing objects. Specific names of elements are automatically listed in the **Element types** bar under the mesh directory tree after generation of mesh according to the parameters specified by users. There are many ways of classifying element types. They can be classified as 1D, 2D and 3D elements according to the dimension. Moreover, element types can be further classified as **Area Integration** and **Gauss Integration** considering the varieties of integration. Judging from the distinctions in element shapes, there are line elements, face elements, shell elements and solid elements, where in face elements are further classified as triangle elements and quadrilateral elements, while solid elements are also further classified as pyramid elements, wedge elements and brick elements. Additionally, according to whether the order of the shape function in the coordinate transformation and displacement interpolation function of the element are equal, elements can be divided into isoparametric and non-isoparametric elements. According to the different mechanic behaviors, elements can be further classified as truss elements, beam elements, plane stress elements, plane strain elements, plate bending elements, axisymmetric elements, flat shell elements, curved shell elements, solid elements interface elements, contact elements, spring elements, composite elements and other structural-applicable elements, which will be highlighted in the following part. Structural elements are introduced in this chapter while elements for heat flow or other fields are not highlighted here.

1.3.1 Truss Elements

Truss elements can be classified as 2D truss elements and 3D elements. Besides, according to the differences in displacement variables, they can be further sorted as **Regular elements** and **Enhanced elements** [1]. The main feature of truss elements lies in that the diameter perpendicular to the length of the element is negligible relative to the length of the element. Deformation variable of this element is only axial elongation along the direction of length without any bending or shear deformations. On the basis of number of dimensions, nodes and degrees of freedom, truss elements are divided into L2TRU, L4TRU and L6TRU.

L2TRU belongs to regular elements, which is composed of two nodes, applicable for simulating mechanic behaviors of truss, springs or prestress tendons, where L represents the shape of line, 2 represents element degrees of freedom and TRU represents truss. Each node has only one axial elongation translation displacement in uniaxial X, Y or Z directions (see Fig. 1.9). Displacement interpolation function is linear. This type of element can only bear compression but not bending moment [1]. There is mass distribution along the local x direction. However, mass cannot be distributed in X, Y and Z directions under global coordinate system. Therefore, this kind of truss element may not be applied to dynamic analysis issues. The characteristics of regular truss element are displayed in Table 1.1.

Fig. 1.9 Variables of translational displacement in L2TRU

Table 1.1 Parameters of L2TRU

Element features	Parameters
Name of element	L2TRU
Number of nodes	2
Number of dimensions	2
Nodal degrees of freedom per node	1
Total degrees of freedom	2
Displacement interpolation function	$u_x(\xi) = a_0 + a_1\xi$
Displacement variables	U_X
Geometric parameters	Cross-section area
Scope of application	Uniaxial elongation of 2D truss structures cables, springs and prestress tendons

L4TRU, a two-node truss element, with 2 degrees of freedom on each node in X and Y directions, is in line shape, where 4 represents total degrees of freedom. There are translational displacements in both X and Y directions of each node, which can translate along the axial directions. Similar to L2TRU, this kind of truss element can only bear tension and compression without the mechanic behavior of bending. Parameters of this kind of element are displayed in Table 1.2.

Table 1.2 Parameters of L4TRU

Element features	Parameters
Name of element	L4TRU
Number of nodes	2
Number of dimensions	2
Nodal degrees of freedom per node	2
Total degrees of freedom	4
Displacement interpolation function	$u_x(\xi) = a_0 + a_1\xi$
Displacement variables	U_X, U_Y
Geometric parameters	Cross-section area
Scope of application	2D truss structures, cables, springs and prestress tendons

Truss element of L6TRU is also constituted by two nodes, and is in line shape, where 6 in the name of the element represents total degrees of freedom, and every node has three translational displacements in X, Y and Z directions, respectively. Similarly, this type of element also bears tensile and compressive behaviors instead of bending moment. Parameters are displayed in Table 1.3.

Table 1.3 Parameters of L6TRU

Element features	Parameters
Name of element	L6TRU
Number of nodes	2
Number of dimensions	3
Nodal degrees of freedom per node	2
Total degrees of freedom	6
Displacement interpolation function	$u_x(\xi) = a_0 + a_1\xi$
Displacement variables	U_X, U_Y, U_Z
Geometric parameters	Cross-section area
Scope of application	3D truss structures, springs and prestress tendons

Cable elements are all in cable shapes. Compared with truss elements, they have more degrees of freedom and are applicable for geometric nonlinear analysis of large deformation for suspension structures and suitable for simulating single curved prestress tendon or curved reinforcement in reinforced concrete structures. Cable elements can be classified as CL6TR (2 dimensions, 6 degrees of freedom), CL8TR (2 dimensions, 8 degrees of freedom), CL10TR (2 dimensions, 10 degrees of freedom), CL9T (3 dimensions, 9 degrees of freedom), CL12T (3 dimensions, 12 degrees of freedom), CL15T (3 dimensions, 15 degrees of freedom) according to the variances in number of dimensions and total degrees of freedom.

1.3.2 Beam Elements

As background applications of DIANA are wide, especially in simulating reinforced concrete beam and long-span prestress concrete bridges, there may be axial deformation Δl, shear deformation γ, curvature κ and torsion, which correspondingly describe axial force, shear force and moment in-plane and out of the plane, respectively. According to the distinctions in spatial dimensions as well as total degrees of freedom, beam element can be further divided into L6BEN, L7BEN, CL19BE, CL12B and CL15B, where letter C at the beginning is the abbreviation of **CURVED**, representing that the shape of this element kind is curve. Besides, beam elements can be further divided into elements in line shape (starting with letter L) and curve shape (starting with letter C) based on the distinctions of element shape. Moreover, judging from the distinctions in application theories and mechanic behaviors, beam elements can also be classified as Class-I beam, Class-II beam and Class-III beam, which are the core conceptions in the numerical simulation of DIANA software for beam element.

Class-I beam

Class-I beam is mainly composed of beam elements in line shape, which are based on the **Plane cross-section assumption** as well as **Euler–Bernoulli beam theory**, where the shear deformation is not considered in the analysis. This kind of beam element is applicable for linear and geometric nonlinear analysis. During the modeling process, various cross-section geometric properties in different shapes (such as rectangular shape, T shape, I shape and box shapes) and required sizes are assigned after material assignment according to the need of the users. Major strain variables are longitudinal elongation, bending strain and torsional deformation out of the plane, especially for 3D beam element. Stress is composed of normal stress and moment. In this type of beam element, it is deemed that displacements and rotations are independent variables. Therefore, curvature is usually expressed by the second-order derivative of the element in y direction. **Timoshenko beam** with shear-locking characteristic also belongs to this type of element, where the beam element in linear shape such as L6PE is sensitive for shear locking. **Class-I beam** is not only applicable for analysis of concrete structures modeled by beam element but also suits for solving the coupling issue of single discrete reinforcement or prestress tendon elements with fully bonded, non-bonded and bond-slip mechanic behaviors in reinforced concrete. Conversion between **Class-I beam** and **Timoshenko beam** with shear deformation, shear-locking behaviors as well as moments of inertia is realized via specification of shape factor.

Compared with **Class-I beam**, **Class-II beam** is also based on the **Plane cross-section assumption** as well as **Euler–Bernoulli beam theory**. Shear deformation is also omitted in this type of element. Contrary to first class, axial relative deformation of beam element is taken into account. Since numerical integration of interpolation type is along the axial bar direction as well as in the area of cross-section, besides linear and geometric nonlinear analysis, physical nonlinear analysis is also allowed in this type of beam element. Common star element L7BEN belongs to this type of element.

Class-III beam elements are mostly in curved geometric shape. Similar to second class, numerical integration of interpolation type is also along the axial bar direction and in the area of cross-section. In the finite-element analysis of DIANA, independent variable of shear deformation is taken into consideration and the displacements and rotations are also individually independent interpolations, meaning that nodal normal displacements and rotations are individually and independently interpolated. Owing to more nodes on the element, the shapes of this kind of elements are curved, where developed **Class-III beam** elements before release 9.6 version are all in curved shape. Meanwhile, displacement functions are usually in high orders, thus better displacement compatibility and element boundary adaptability are demonstrated when connected with other kind of structural elements.

Features of the three kinds of beam elements are listed in Table 1.4.

Table 1.4 Features of the three kinds of beam elements

Element class	Name	Dimensions	Nodal number	Total degrees of freedom	Mechanic behaviors	Displacement variables
Class-I beam	L6BEN	2	2	6	Omitting shear deformation and axial relative deformation	U_x, U_y, ϕ_z
	L12BE	3	2	12	Omitting shear deformation and axial relative deformation	U_x, U_y, U_z, $\phi_x \phi_y \phi_z$
Class-II beam	L7BEN	2	2	7	Omitting shear deformation but considering axial relative deformation	U_x, U_y, ϕ_z Δux
	L13BE	3	2	13	Omitting shear deformation but considering axial relative deformation	U_x, U_y, U_z $\phi_x \phi_y \phi_z \Delta$ux
Class-III beam	CL9BE	2	3	19	Considering shear deformation	U_x, U_y, ϕ_z
	CL12B	2	4	12	Considering shear deformation	U_x, U_y, ϕ_z
	CL15B	2	5	15	Considering shear deformation	U_x, U_y, ϕ_z
	CL18B	3	3	18	Considering shear deformation	U_x, U_y, U_z, $\phi_x \phi_y \phi_z$
	CL24B	3	4	24	Considering shear deformation	U_x, U_y, U_z, $\phi_x \phi_y \phi_z$
	CL30B	3	5	30	Considering shear deformation	U_x, U_y, U_z, $\phi_x \phi_y \phi_z$

Note Self-weight and distributed load are not taken into consideration in the **Class-II beam** when calculating initial strain and initial stress [1]

1.3.3 Plane Elements

Plane elements are dominated by two-dimensional elements, where a model is created on the geometric neutral surface. When the element type is determined, plane elements are generated via assignment of cross-section geometric properties. The features and mechanic behaviors of 3D plane elements are almost the same as shell elements. According to the different mechanic behaviors, these kinds of element can be further classified into **Plane stress elements** and **Plane strain elements**.

Plane stress elements consist of elements in flat plate shape, which are also called membrane elements, where all the nodes must be in-plane. Two important features of plane stress elements are as follows: (1) Size in the thickness direction must be small relative to the dimensions of element length and width. (2) Stress components perpendicular to the face are zero; that is to say, local stress perpendicular to plane along the thickness direction $\sigma_{zz} = 0$. DIANA software caters to the demand of users, by providing them both 2D and 3D plane stress elements, where 3D plane stress elements exist in non-flat geometry or solid elements when they are connected with other elements with stiffness in the transverse direction. Usually, 3D plane stress elements are also called 3D membrane elements.

There are usually two degrees of freedom of every node in 2D plane stress elements, which are translational degrees of freedom in x and y directions, respectively. Displacement variables of 3D plane stress elements are translational displacement variables U_x, U_y and U_z along the axial bars, respectively. Strains are composed of normal strains in x, y and z directions ε_{xx}, ε_{yy} and ε_{zz} as well as shear strain γ_{xy} with the corresponding stress σ_{xx}, σ_{yy}, σ_{zz} and τ_{xy}, where stress corresponding to z direction satisfies $\sigma_{zz} = 0$. Normal and shear stress can be automatically calculated via thickness integration, that is to say, $\{n_{xx}, n_{yy}, n_{xy}, n_{yx}\}^T$, where shear stress satisfies reciprocal theorem $n_{xy} = n_{yx}$ [1]. According to the differences in basic displacement variables, besides 2D and 3D plane stress elements, there is another special element called **Elements with Drilling Rotations.** The latter not only has the same translational displacements as the conventional plane stress element along the coordinate axes under the global coordinate system, but also a rotation variable rotating around Z-axis ϕ_Z. Additionally, there is another single special plane stress element for wrinkling, where displacement variables are only translational displacements along three coordinate directions, while the stress vector includes normal and shear stress along three coordinate axes.

Thickness specification for plane stress elements in DIANA is special. For isotropic elements, assignment of thickness has nothing to do with directions of cross-section, while orthotropic thickness may be feasible for some special ones. There are two types of thickness specification in DIANA. One is uniform-thickness assignment and the other is non-uniform-thickness assignment. For uniform-thickness assignment, a uniform value of thickness is input into module of cross-section geometric properties after element type and shape are defined, where the uniform value represents the thickness of all the nodes to be exactly of same value. For elements with non-uniform thickness, values of every node are required to be input one by one in order to generate ultimate elements after element type and geometric shape are determined. Uniform-thickness assignment and non-uniform-thickness assignment are displayed in Figs. 1.10 and 1.11, respectively. Under the conditions of uniform-thickness assignment, number of thickness values needed to be input for plane stress elements is related with selected element types and the number of nodes. For instance, number of thickness values needed to be input for rectangular elements with 4 nodes is 4, or 8 different thickness values are required for eight-node isoparametric elements. Moreover, there may be three or six different thickness values for triangular elements according to the type of element and the number of interpolation nodes.

Fig. 1.10 Uniform-thickness assignment

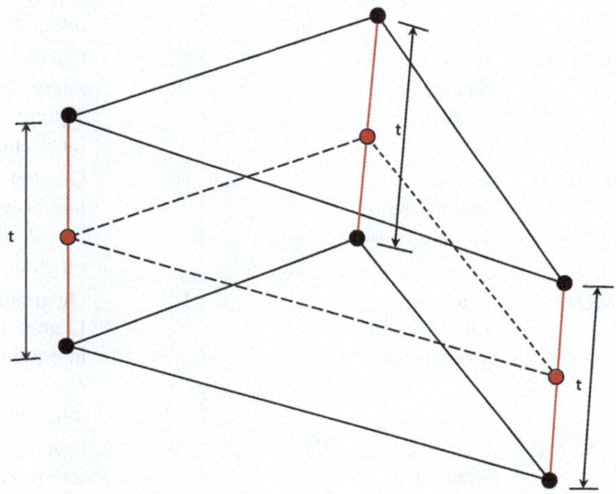

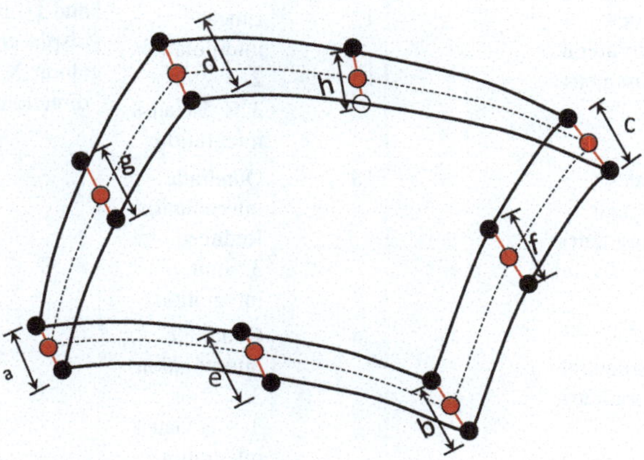

Fig. 1.11 Non-uniform-thickness assignment

Element characteristics for plane stress elements are displayed in Table 1.5.

Table 1.5 Characteristics for plane stress elements

Name	Shape	Dimensions	Total degrees of freedom	Numerical integration	Characteristics
T6MEM	3 nodes, triangular isoparametric	2D	6	Linear interpolation, 1-point area integration	(1) These kinds of elements are typical for 2D models. (2) There are U_x and U_y translational displacement variables along X and Y directions
Q8MEM	4 nodes, quadrilateral isoparametric		8	Linear interpolation, 2 × 2 Gauss integration	
CT12M	6 nodes, triangular isoparametric		12	Quadratic interpolation, 3-point area integration	
CQ16M	8 nodes, quadrilateral isoparametric		16	Quadratic interpolation, 2 × 2 Gauss integration	
CQ18M	9 nodes, quadrilateral isoparametric		18	Quadratic Lagrange interpolation, 3 × 3 Gauss integration	
T9GME	3 nodes, triangular isoparametric	3D	9	Linear interpolation, 3-point area integration	(1) These kinds of elements are typical for 3D models (2) There are U_x, U_y and U_z translational displacement variables along X, Y and Z directions
Q12GME	4 nodes, quadrilateral isoparametric		12	Linear interpolation, 2 × 2 or 3 × 3 Gauss integration	
CT18GM	6 nodes, triangular isoparametric		18	Quadratic interpolation, Reduced 3-point integration	
CQ24GM	8 nodes, quadrilateral isoparametric		24	Quadratic interpolation, 2 × 2 or 3 × 3 Gauss integration	

(continued)

Table 1.5 (continued)

Name	Shape	Dimensions	Total degrees of freedom	Numerical integration	Characteristics
T9MEM	3 nodes, triangular isoparametric	2D	9	Linear interpolation, 3-point area integration	(1) Elements with drilling rotation, (2) Besides translations U_x, U_y and U_z, another basic drilling rotation is ϕ_z
Q12ME	4 nodes, quadrilateral isoparametric		12	Linear interpolation 2×2 area integration.	
T6OME	triangular isoparametric	2D	6	Linear interpolation 1-point area integration	(1) Elements with orthotropic thickness, thickness values ought to be assigned respectively. (2) Linear interpolation of displacement function (3) Displacement variables are U_x and U_y
Q8OME	4 nodes, quadrilateral isoparametric		8	Linear interpolation, 2×2 Gauss integration	
CT12O	6 nodes, triangular isoparametric		12	quadratic interpolation, 3-point or 4-point area integration	
CT16O	8 nodes, quadrilateral isoparametric		16	quadratic interpolation, 2×2 Gauss integration	
T9MWE	3 nodes, triangular isoparametric	3D	9	linear interpolation, 1-point area integration	(1) Applicable for wrinkling analysis, (2) There are U_x, U_y and U_z translational displacement variables along X, Y and Z directions while stress vector contains normal stress along X, Y and Z directions and shear stress

Typical plane stress element is CQ16M (Fig. 1.12), which is eight-node quadrilateral isoparametric element based on quadratic interpolation and **Gauss integration**. There are two degrees of freedom along x and y directions in every point, applicable for simulating 2D in-plane concrete models such as beam and floor slab. Additionally, steel constitutive material properties can also be solely assigned for open-hole steel plate structures and fatigue mechanic characteristics. This type of element is especially suitable for smeared cracking models, and has a good coupling performance with embedded reinforcement bar and grid elements to simulate longitudinal steel bars, stirrups and prestress tendons. Moreover, it also has good coordination and convergence for nonlinear calculation.

Fig. 1.12 CQ16M

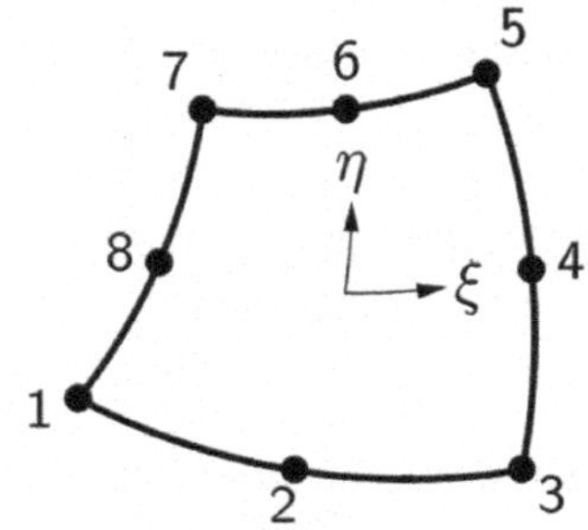

Nodes of plane strain elements are located in XOY coordinate area zone. Analogical to plane stress elements, element nodes under coordinate global system in Z direction and strain components perpendicular to the face under Z coordinate axis are zero. Similar to plane stress elements, load must be positioned in the model XOY plane. For the strain variables of plane strain elements, it consists of three types of normal strains and shear strain, which are ε_{xx}, ε_{yy} and ε_{zz}, coupled with shear strain γ_{xy}. Contrary to the expression of plane stress elements, normal strain in z direction is $\varepsilon_{zz} = 0$. Displacement variables of such elements are translational variables U_x and U_y in X and Y directions, respectively. Similar to plane stress elements, there are also elements with drilling rotations in-plane strain elements. Meanwhile, corresponding stress variables in x, y and z directions are σ_{xx}, σ_{yy}, σ_{zz} and τ_{xy}, respectively.

Besides bearing distributed load, temperature effect can also be taken into account in plane strain elements [1]. All the line or surface distributed load is applied on the node. Analogical to thickness assignment in plane stress elements, loading assignment for plane strain elements are completed via determining values and directions of every node. If there are no specifications and there is only one value assigned, it is determined, by default, in DIANA that all the nodes have the same load value out of the plane, where the directions can be along the coordinate system (such as X, Y and Z) and via normal or shear way to determine load values of each nodes. Taking Fig. 1.13 for example, it represents first to third nodes that sustained normal load in an eight-node plane strain elements with the values: $F_1 = 300$ N, $F_2 = 400$ N and $F_3 = 500$ N, respectively. In the geometric and material file with the suffix name **.dat**, they can be expressed as follows:

'LOADS'
CASE 1
ELEMEN
1 EDGE L1
FORCE 300 400 500
DIRELM NORMAL

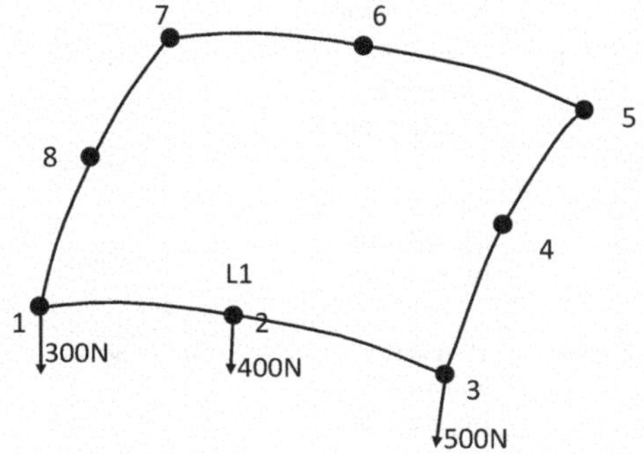

Fig. 1.13 Sustained normal load in an eight-node plane strain elements

According to practical modeling requirements, values and directions of every node can be adapted or modified by users.

In addition to regular plane strain elements, there is another 3D **Complete Plane Strain Elements** specific for hypothesis analysis based on plane strain. Compared with the former, besides eliminating displacements out of plane, components of stress and strain vectors, translational displacement in Z-axis direction is added while normal strain variables are along out of the plane Z-axis direction ε_{zz}, and shear strain γ_{yz}, γ_{xz}. Meanwhile, corresponding shear stress σ_{yz} and σ_{xz} are added as stress components in out of the plane shear direction. Moreover, a type of finite element with shape similar to beam elements is emerging as the time requires. Strictly speaking, it belongs to a special plane strain elements between plane stress elements and shell elements. There are only three degrees of freedom—translational variables u_x, u_y and rotational variables ϕ_z around Z-axis. This kind of element is sensitive to shear-locking issues. When the geometric shape is line, normal strain ε along local coordinate x direction is constant, while it takes on the linear alteration when the shape of element is curve. Integration points and types in ζ direction are distinctive with those in x direction: 2-point Gauss integration in thickness ζ direction is adopted when physical linear analysis is conducted, while Simpson integration is adopted when physical nonlinear analysis is required, where the number of integrations is high related with the extent of physical nonlinear analysis. In the DIANA library plane strain elements, there is a kind of special plane strain elements—rubber elements, which is applicable for simulating hyper-elastic structures and components rubber mechanic behaviors under nonlinear conditions. Furthermore, seismic isolation structure or bearing supports may be specifically simulated when integrating spring elements and dashpot material constitutive properties, such as rubber seismic isolation pads and damping dashpots.

All kinds of plane strain elements are listed in Table 1.6.

Table 1.6 Plane strain elements

Name	Shape	Dimensions	Total degree of freedom	Numerical integration	Characteristics
T6EPS	3 nodes triangular isoparametric	2D	6	Linear interpolation, 1-point area integration	Displacement variables of every node are U_x and U_y along X and Y directions
Q8EPS	4 nodes, quadrilateral isoparametric		8	Linear interpolation, 2×2 Gauss integration	
CT12E	6-node triangular isoparametric		12	Quadratic interpolation 3-point area integration	
CQ16E	8 nodes, quadrilateral isoparametric		16	Quadratic interpolation 2 × 2 Gauss integration	
CT30E	15 nodes triangular isoparametric		30	Fourth-order interpolation 12-point area integration	Besides features of regular plane stress elements mentioned above, this kind of element specifically applicable for nonlinear analysis in Geotechnical field
CQ20E	8 nodes, quadrilateral isoparametric	2D	20	Quadratic interpolation 3 × 3 Gauss integration	Both are quadratic and quadrilateral elements with Gauss integration. Displacement variables of every node are translational displacements U_x and U_y in x and y directions
CQ22E	9 nodes, quadrilateral isoparametric		22	Quadratic interpolation 3 × 3 Gauss integration	
CT18GE	6-node triangular isoparametric	3D	18	Quadratic interpolation 3-point area integration	(1) Displacement variables of every node are U_x, U_y, and U_z along the X, Y and Z directions respectively (2) For coordinate axis pointing out of the plane, stress and strain vales are mutually independent (3) In addition to the default settings, there are other

(continued)

Table 1.6 (continued)

Name	Shape	Dimensions	Total degree of freedom	Numerical integration	Characteristics
					alternative number of suitable integral points in every element. However, once the upper limit number of integral points is exceeded, this element is unavailable
CQ24GE	8 nodes, quadrilateral isoparametric		24	Quadratic interpolation 2 × 2 Gauss integration	
CT27GE	9 nodes triangular isoparametric		27	Cubic interpolation 7-point area integration	
CQ36GE	12 nodes, quadrilateral isoparametric		36	Cubic interpolation 3 × 3 Gauss integration	
L6PE	2 nodes, linear	3D	6	Linear interpolation, 1·2 integration	(1) Element is in linear shape and x direction under local coordinate system is in tangent direction (2) Normal stress in Y direction is zero, which is sensitive to shear-locking issues (3) Displacement variables are translational variables u_x and u_y rotational variables ϕ_z around Z axis of each node (4) 2-point Gauss integration in thicknessζdirection when physical linear analysis is conducted while Simpson integration is adopted when physical nonlinear analysis is required
CL9PE	3 nodes, curved		9	Second-order interpolation, 2-point Gauss integration	

1.3.4 Plate Bending Elements

When it comes to the geometric size, plate bending elements are like plane stress elements; that is to say, all the coordinate values of element nodes must be located in the same flat plate elements. Furthermore, element thickness relative to size of width can be omitted. For mechanic behaviors, if load applied on the element is merely longitudinal load parallel to element surface, then these kinds of elements are named as plate stress elements, or else if transverse load perpendicular to element plane is applied on the element, these elements are named as plate bending elements. Load must be perpendicular to element surface, and the stress perpendicular to element surface along the direction thickness satisfies $\sigma_{zz} = 0$. Different from plane stress elements, besides loading force, moment in-plane can also be acted on the plate bending elements, where the direction is rotating around a local axis [1]. Plate elements must satisfy both deformation compatibility as well as equilibrium conditions. Plane cross-section assumption is satisfied before and after element deformation with load types as follows: point load, edge load, face load, temperature, concentration load and initial stress.

Displacement variables are in vast variety compared with plane stress elements. Above all, there are two rotation variables in the element plane rotating around positive x and negative y coordinate axes, respectively, in plate bending elements with regard to bending moment in-plane. Meanwhile, there is translational displacement Uz along Z direction. What is also different is that the number of strain variables is only five owing to moment in-plane, where there is no normal strain, but there are curvature strains in x, y directions and xoy plane κ_{xx}, κ_{yy}, κ_{xy} as well as torsional curvature Ψ_{yz} and Ψ_{zx}. Judging from the mechanic behaviors of the whole element type, plate bending elements can be approximately regarded as a series of transitional elements between plane stress elements and regular curved elements. Stress variables are complex where, in DIANA, there are two types of stress output forms: one is the stress form of bending plate element output via bending stress and load concentration while the other is output by Cauchy stress. The former is constituted by moment stress m_{xx}, m_{yy} and m_{zz} and tangential concentration stress qyz and qxz, while Cauchy stress of the latter is composed of normal stress in three directions σ_{xx}, σ_{yy} and σ_{zz} coupled with tangential stress τ_{xy}, τ_{yx}, τ_{xz}, τ_{zx}, τ_{yz} and τ_{zy}, where normal stress in Z direction is zero.

Thickness assignment for plate bending elements is similar to plane stress elements with uniform-thickness assignment as well as non-uniform-thickness assignment, which is introduced in the former part and thus it is not repeated here.

Another two kinds of plate bending elements are listed as follows. One is based on discrete Kirchhoff line method, where discrete Kirchhoff bending plate is retained in element or at the edge of the element. The other is based on the Mindlin plate principle, where lines perpendicular to neutral surface still keep linear shape in this kind of element but not necessarily perpendicular to deformed neutral surface (Table 1.7).

Table 1.7 List of plate bending elements

Name	Shape	Total degrees of freedom	Numerical integration	Characteristics
T9PLA	3 nodes, triangular isoparametric	9	Linear interpolation, 3-point integration scheme, Second-order polynomial for the rotations	(1) Based on **Discrete Kirchhoff Line Theory**, (2) Impact of shear deformation is taken into account
Q12PL	4 nodes, quadrilateral isoparametric	12	Linear interpolation, polynomials for the displacements u_z and rotations ϕ_x and ϕ_y are both linear	Based on the **Mindlin plate principle**, where lines perpendicular to neutral surface still keep linear shape
CT18P	6 nodes, triangular isoparametric	18	Quadratic interpolation, 3-pointor1-point area integration polynomials for the displacements u_z and rotations ϕ_x and ϕ_y are both quadratic	
CQ24P	8 nodes, quadrilateral isoparametric	24	Quadratic interpolation, polynomials for the displacements u_z and rotations ϕ_x and ϕ_y are both linear	

1.3.5 Axisymmetric Elements

Since structures are often symmetrical ones in engineering reality, that is to say, they are generated by rotating an axis or some axes to form a symmetric geometric figure. Moreover, owing to another reason that structures are often in large scale, which may take up CPU or computer storage space and consume a vast quantity of time, numerical models of such 1/4 structures or semi-structures are taken in order to shorten calculation time and improve calculation efficiency. Meanwhile, corresponding semi-structure constraints should be rightly attached before nonlinear calculation. In view of this, some axisymmetric elements developed by DIANA can directly be applied to replace semi-structures under some circumstances.

There are two kinds of axisymmetric elements in DIANA. One is variables only with basic displacements such as U_x and U_y, including triangular or quadrilateral or solid ring elements, where they can be further classified as **Regular Solid Rings** and **Rubber Solid Rings** according to the characteristics of elements [1]. This kind of elements is based on simple principle, with simplified calculation procedure as well as high efficiency, and thus they are applied universally. The other element type is **Shells of Revolution**, which are line-shaped elements [1]. Displacement variables are translational displacements along the directions of X and Y axes and rotational degree of freedom around Z axis. Similar to regular flat shell and curved shell elements, a thickness value is required to be assigned, where there is also uniform-thickness assignment and non-uniform-thickness assignment, and the size of thickness relative to length can be omitted. The characteristics of symmetrical elements are displayed in Table 1.8.

Table 1.8 Characteristics of symmetrical elements

Name	Shape	Total degrees of freedom	Numerical integration	Characteristics
T6AXI	3-node axisymmetric isoparametric, triangular cross-section	6	Linear interpolation 1-point area integration	(1) Axisymmetric solid rings, cross-sections are composed of a series of triangular and quadrilateral elements, (2) Displacement variables of every node are translational variables U_X and U_Y
Q8AXI	4-node axisymmetric isoparametric, quadrilateral cross-section.	8	Linear interpolation, 2 × 2 Gauss integration	
CT12A	6-node axisymmetric isoparametric, triangular cross-section	12	quadratic interpolation 1-point, 3-point or 4-point area integration	
CQ16A	8-node axisymmetric isoparametric, quadrilateral cross-section.	16	quadratic interpolation 2 × 2 Gauss integration	
CT30A	15-node axisymmetric isoparametric, triangular cross-section	30	fourth-order interpolation 12-point area integration	
CQ20A	8-node axisymmetric isoparametric, quadrilateral cross-section.	20	quadratic interpolation for displacements and linear interpolation for pressures, 2 × 2 or 3 × 3 Gauss integration	(1) Quadrilateral cross-sections for rubber solid rings (2) Lagrange interpolation, orders of interpolations for displacement and pressure are different with incongruity (3) Specifically for nonlinear analysis with hyper-elasticity in rubber structures

(continued)

Table 1.8 (continued)

<table>
<tr><td>CQ22A</td><td>9-node axisymmetric isoparametric, quadrilateral cross-section.</td><td>22</td><td>interpolation for the cross-section displacements is quadratic Lagrange, while for pressure is linear Lagrange
2 × 2 or 3 × 3 Gauss integration</td><td></td></tr>
<tr><td>L6AXI</td><td>2-node straight, axisymmetric shell of revolution element</td><td>6</td><td>Linear interpolation polynomial for translational displacement,
1·2 Gauss integration in thickness ζ direction</td><td rowspan="2">(1) Element shape is similar to 3D beam element, while mechanic behaviors are similar to flat and curved shell elements, which can be generated by degenerated solid element
(2) Elements are generated via thickness assignment and thickness relative to length can be omitted
(3) Displacement variables of every node are translational variables U_x, U_y rotational variables ϕ_z around Z axis of each node
(4) 2-point Gauss integration in thickness ζ direction is adopted when physical linear analysis is conducted while Simpson integration is adopted when physical nonlinear analysis is required</td></tr>
<tr><td>CL9AX</td><td>3-node curved, axisymmetric shell of revolution element</td><td>9</td><td>Linear interpolation polynomial for translational displacement,
2·2 Gauss integration in thickness ζ direction</td></tr>
</table>

1.3.6 Shell Elements

Shell elements are divided into flat shell elements as well as curved shell elements according to the distinctions of element shape. They are based on a combination of plane stress elements and plate bending elements. Load applications are numerous, where they can be not only applied perpendicular to the surface of element but also acted on the shell plane [1]. There are two kinds of elements for shell elements; one is **Regular elements**, and the other is **Element with Drilling Rotations**. The main feature of regular shell elements is combination of plane stress element as well as plate bending elements. Thickness size relative to width is negligible. Flat shell element conforms to **Mindlin plate principle,** that is to say, central lines perpendicular to element surface still keep linear shape after deformation. Basic displacement variables of regular plane stress elements are translational displacements U_x, U_y and U_z along x, y and z under local coordinate system and rotational variables ϕ_x and ϕ_y rotating around x and y. Compared with regular flat shell elements, in order to display the characteristic of drilling rotation, additional rotational variable ϕ_z rotating around z axis in every node under local system is included based on the original displacement variables. Therefore, this element type is also called as second flat elements, which can avoid ill-condition of the global stiffness matrix [1]. In the procedure of 3D modeling, simulation effect of flat shell element is better than the 3D plane stress element as well as plate bending elements.

Stress variables of flat shell elements are also classified as two types: Cauchy stress, generalized moment and forces. Stress variables of former is totally the same as plate bending elements, containing normal stress σ_{xx}, σ_{yy} and σ_{zz} and shear stress $\tau_{xy}, \tau_{yx}, \tau_{xz}, \tau_{zx}, \tau_{yz}, \tau_{zy}$,, where except moment variables keeping the same as plate bending elements, normal concentration variables n_{xx},n_{yy} along x and y direction under local coordinate system, n_{xy} perpendicular to x axis and parallel to y axis and n_{yx} perpendicular to y axis and parallel to x axis are added.

There is another **Spline Elements** type in flat shell elements, where the element shape is rectangle in segments. Mechanic behaviors are similar to regular plane stress elements, where the thickness value relative to length and width can be ignored. Contrary to regular plate bending elements, spline elements are rectangular elements divided into segments, and the width and thickness are the same for nodes in the same segment. However, they are classified into several segments with various lengths in longitudinal direction. Compared with other plate bending elements, transverse shear deformation is also taken into account in the spline elements [1]. The shape is displayed in Fig. 1.14 (Table 1.9).

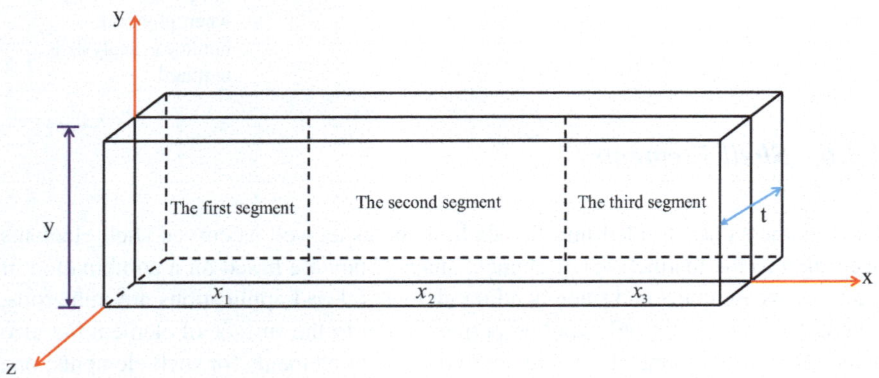

Fig. 1.14 Spline element

Table 1.9 Characteristics of flat shell element

Name	Shape	Total degrees of freedom	Numerical integration	Characteristics
T15SF	3 nodes, triangular isoparametric	15	Bi-linear interpolation for both geometry and displacements 1-point integration	Based on Mindlin Reissner theory
Q20SF	4 nodes, quadrilateral isoparametric	20	Bi-linear interpolation for both geometry and displacements 2 × 2 integration	
CT30F	6 nodes, triangular isoparametric	30	Quadratic interpolation, 3-point integration	
CQ40F	8 nodes, quadrilateral isoparametric	40	bi-quadratic interpolation for both geometry and displacements 2 × 2 integration	
T18SF	3 nodes, triangular isoparametric	18	Linear interpolation for geometry function Linear and hierarchical quadratic interpolations for displacements 3-point integration	1. Compared with regular shell elements, there is an additional rotational variables rotating around z-axis ϕ_z 2. The highest order of geometric function and displacement interpolation function of some elements are various, indicating coordination between displacement and geometry
T18FSH	3 nodes, triangular isoparametric	18	1. Applicable for postbucking and nonlinear vibration 2. Analysis element output is only available at the element center and nodes 3. Element analysis is analytical synthesis	
Q24SF	4 nodes, quadrilateral isoparametric	24	Linear interpolation for geometry function bi-linear and hierarchical bi-quadratic for translational displacements in x and y directions bi-linear interpolation for normal translational displacement in z direction and drilling rotation ϕ_z 2 × 2 integration	
CT36F	6 nodes, triangular isoparametric	36	Quadratic interpolation for both geometry and displacements	

(continued)

Table 1.9 (continued)

Name	Shape	Total degrees of freedom	Numerical integration	Characteristics
			Geometric and displacement interpolation are both quadratic Only 3-point integration is allowed	
CQ48F	8 nodes, quadrilateral isoparametric	48	Bi-quadratic interpolation for both geometry and displacements 2 × 2 integration	
Q48SPL	8 nodes, quadrilateral 3 sections	48	Spline interpolation in longitudinal x direction bi-linear interpolation in y direction 2·2·2 Gauss integration	1. Element shapes are rectangles in segments 2. Linear interpolation 3. Width and thickness in each segment are the same but longitudinal length in different segments can vary
Q56SPL	10 nodes, quadrilateral 4 sections	56	Spline interpolation in longitudinal x direction bi-linear interpolation in y direction 2·2·2 Gauss integration	

Curved shell elements in DIANA are based on the isotropic composite degenerated solid elements with the same mechanic behaviors as flat shell elements. They can be further divided into T15SH, Q20SH, CT30S, CQ40S, CT45S as well as CQ60S according to the element types and degrees of freedom. They are also classified as triangular and quadrilateral elements according to the element shapes. Moreover, according to the assignment in thickness ζ direction, curved shell elements are also sorted as regular curved shell elements as well as layered curved shell elements. Furthermore, on the basis of whether there is additional drilling rotation, they can be further divided into curved shell elements as well as curved shell elements with drilling rotations. Finite-element models of the structures shall be established on the center line or the neutral surface when shell elements are applied, and the thickness should be assigned after the element type is determined. Since the edge shapes of curved shell elements are mostly quadratic and cubic curves and the nodes are in vast amount, they have excellent boundary adaptability, better element compatibility and higher calculation convergence so that they are widely applied in the 3D thin-walled structures, where they are the best choice especially for thin-walled box bridges in the nonlinear analysis. All the curved shell elements are listed in Table 1.10.

Table 1.10 All the curved shell elements

Name	Shape	Total degrees of freedom	Numerical integration	Characteristics
T15SH	3nodes, triangular isoparametric	15	Linear interpolation, 3-point area integration 3-point Simpson integration in thickness ζ direction by default	1. Displacement variables of every node are translational variables U_X, U_Y and U_Z as well as rotational variables RotX and RotY 2. Material properties and thickness are uniform in the thickness ζ direction. That is to say, once material properties and thickness of certain node is determined, they are the same in the thickness ζ direction 3. 3-point Simpson integration in thickness ζ direction by default while 2-point Gauss is a suitable option Schemes higher than 3-point in ζ direction are only useful in case of nonlinear analysis
Q20SH	4 nodes, quadrilateral isoparametric	20	Linear interpolation, 2 × 2 Gauss integration in element $\xi\eta$ area, 3-point Simpson integration in thickness ζ direction by default	
CT30S	6 nodes, triangular isoparametric	30	Quadratic interpolation, 3-point reduced area integration 3-point Simpson integration in thickness ζ direction by default.	
CQ40S	8 nodes, quadrilateral isoparametric	40	Quadratic interpolation, 2 × 2 Gauss integration in element $\xi\eta$ area, 3-point Simpson integration in thickness ζ direction by default	
CT45S	9 nodes, triangular isoparametric	45	Cubic interpolation, 7-point area integration 3-point Simpson integration in thickness ζ direction by default	
CQ60S	12 nodes, quadrilateral isoparametric	60	Cubic interpolation, 3 × 3 Gauss integration in element $\xi\eta$ area, 3-point Simpson integration in thickness ζ direction by default	

(continued)

Table 1.10 (continued)

Name	Shape	Total degrees of freedom	Numerical integration	Characteristics
T18SH	3 nodes, Triangular isoparametric	12	Linear interpolation 3-point area integration per layer, 3-point Simpson integration in thickness ζ direction	Besides basic displacement variables, an additional rotation Rotz ϕ_z is added An ill-condition of the assembled global stiffness matrix can be avoided when these elements are applied
Q24SH	4 nodes, Quadrilateral isoparametric	24	Linear interpolation 2×2 Gauss integration in $\xi\eta$ element area of each layer, Default scheme in ζ direction (layer thickness) is 3-point Simpson integration	
CT36S	6 nodes, Triangular isoparametric	36	Quadratic interpolation 3-point reduced area integration per layer, 3-point Simpson integration in thickness ζ direction	
CQ48S	8 nodes, quadrilateral isoparametric	48	Linear interpolation 2×2 Gauss integration in $\xi\eta$ element area per layer, 3-point Simpson integration in thickness ζ direction	
T15LA	3 nodes, triangular isoparametric	15	Linear interpolation 3-point area integration in element $\xi\eta$ area of per layer 3-point Simpson integration in thickness ζ direction per layer	1. Displacement variables of every node are translational variables U_X, U_Y and U_Z as well as translational variables RotX and RotY 2. Thickness is subdivided into number of layers. Each layer has its own material properties and is numerically integrated separately 3. The default in ζ direction (layer thickness) is 2-point, 3-point is a suitable option Schemes higher than 3-point in ζ direction are only useful in case of nonlinear analysis
Q20LA	4 nodes, quadrilateral isoparametric	20	Linear interpolation 2×2 Gauss integration in $\xi\eta$ element area of each layer, Default scheme in ζ direction (layer thickness) is 3-point Simpson integration	
CT30L	6 nodes, triangular isoparametric	30	Quadratic interpolation, 3-point reduced area integration in element $\xi\eta$ area of each layer Default scheme in ζ direction (layer thickness) is 3-point Simpson integration	

(continued)

Table 1.10 (continued)

Name	Shape	Total degrees of freedom	Numerical integration	Characteristics
CA40L	8 nodes, quadrilateral isoparametric	40	Quadratic interpolation, 2 × 2 reduced Gauss integration in $\xi\eta$ element area per layer, Default scheme in ζ direction (layer thickness) is 2-point	
T18LA	3 nodes, triangular isoparametric	18	Linear interpolation 3-point area integration in $\xi\eta$ element area per layer 3-point Simpson integration in thickness ζ direction per layer	Layered elements with drilling rotation; an additional rotation Rotz ϕ_z is added to avoid ill-condition of the assembled global stiffness matrix
Q24LA	4 nodes, quadrilateral isoparametric	24	linear interpolation 2 × 2 Gauss integration in $\xi\eta$ element area 3-point Simpson integration in thickness ζ direction per layer	
CT36L	6 nodes, triangular isoparametric	36	Quadratic interpolation 3-point area integration in $\xi\eta$ element area of each layer 3-point Simpson integration in thickness ζ direction per layer	
CQ48L	8 nodes, Quadrilateral, isoparametric	48	Quadratic interpolation 2 × 2 Gauss integration in $\xi\eta$ element area 3-point Simpson integration in thickness ζ direction per layer	

Basic displacement variables for this kind of element are translational displacement variables U_X, U_Y and U_Z in the global coordinate system while rotational variables are Rotx and Roty rotating around local x and y axes with corresponding stress variables ε_x, ε_y and ε_z as well as shear strain γ_{xy}, γ_{xz} and γ_{yz}. However, a series of new elements with an additional drilling rotational variable Rotz is emerging after the release of DIANA10.0 version, which are called curved shell elements with drilling rotations. There are both global and local strains in the curved shell elements at the same time, where the conversion between the local strain matrix and global strain matrix can be achieved via coordinate transformation matrix [1]. Similar to plane stress elements, stress in z direction satisfies $\sigma_{zz} = 0$ according to **Plate hypothesis** as well as **Kirchhoff's law**.

Stress of curved shell elements is similar to flat shell elements and there are two kinds of stress at the same time: **Cauchy stress** as well as **Generalized Moments and Forces**.

Thickness assignment for curved shell elements is the same as plane stress elements, where geometric sheet models are created on the neutral surface and then the element class as well as shape of cross-section should be specified. After that, thickness value is assigned to the cross-section geometric properties. Similarly, there are also uniform-thickness and non-uniform-thickness assignments. If there is a regular curved shell element and a single solo thickness value is specified, it is reckoned, by default, in DIANA that thickness values for all the nodes in this element are the same and is in thickness ζ direction. However, when nodal thickness values are different or there is a layered element, then thickness values should be specified node by node or layer by layer.

It is commonly acknowledged that a curved shell element is the superposition of a plane stress element and a plate bending element. It has been said that curved shell elements are sometimes deemed as pseudo 3D elements, where shell elements can be viewed as a transitional element between planar elements and solid elements and it seems that curved shell elements are like solving 2D issues under 3D coordinate systems. The reason is that thickness assignment ways are the same although geometric models are created on the neutral surface under 3D coordinate system, where cross-section geometric properties are generated by assigning parameters to generate the ultimate curved shell elements. Therefore, it is in essence that curved shell elements are the combinations of plane stress elements and plate bending elements created in the 3D coordinate system, which is also a unique difference in curved shell elements and plane stress elements.

Next, we focus on a typical eight-node composite degenerated regular curved shell element—CQ40S, displayed as in Fig. 1.15, which will be introduced in detail, and also it is a frequently applied element in the following chapters. It is surrounded by upper and lower surfaces as well as the surface formed by the thickness of the shell in the direction of the generatrix, where the first letter C in the name represents the curved element shape and Q refers to the quadrilateral shape of this element; 40 is the total degrees of freedom and S is representing that this element belongs to shell elements. Thickness of this element is uniform and it has only a single layer. This element is applicable to simulating thin-walled reinforced or prestress concrete structures, and is especially suitable for thin-walled box girders. Meanwhile, it has extraordinary simulation and is compatible with all the embedded reinforcement bar and grid elements in it. Displacement variables of every node are: three translational variables U_x, U_y and U_z as well as two rotational variables ϕ_x and ϕ_y. In order to avoid stiffness matrix distortion resulting from element stiffness that is too large or membrane and shear locking, a reduced 2×2 Gauss integration scheme over the $\xi\eta$ element area and 3-point Simpson integration in thickness ζ direction are applied by default.

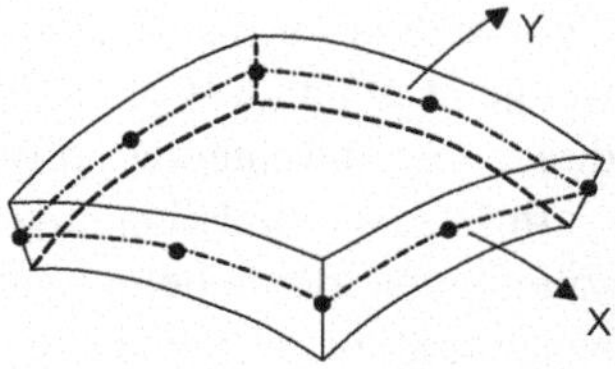

Fig. 1.15 Curved shell elements CQ40S

The characteristics of CQ40S are displayed in Table 1.11.

Table 1.11 Characteristics of CQ40S

Element features	Parameters
Name	CQ40S
Number of nodes	8
Dimensions	3D
Degrees of freedom for single node	5
Total degrees of freedom	40
Characteristics of interpolation	2 × 2 Gauss integration over the ^ij area 3-point Simpson integration in the ζ thickness direction
Displacement variables of every node	U_X, U_Y, U_Z, Rot_X, Rot_Y
Geometry parameters	Thickness of shell elements; Local element directions corresponding to global coordinate system
Application scope	Thin-walled box girders as well as thin-walled reinforced concrete structures

1.3.7 Solid Elements

Different from earlier kinds of elements, solid elements are the combinations of quadrilateral and quadrilateral, quadrilateral and triangle, and triangle and triangle to form corresponding brick, wedge and pyramid element shapes. Therefore, integration schemes can be classified as two types according to the feature where surfaces constructing the volume shape are the same or two different kinds of shape planes. For the pyramid and brick elements with uniform shapes (triangular or quadrilateral), regulation of integration scheme is that when number of nodes in an element is more and interpolation order is higher, integration points in the element are more. For solid element such as wedge, where shapes of top and side surfaces are different, integration scheme in both $\xi\eta$ and thickness ζ directions should be taken into account, respectively. That is to say, the number of integration points in the triangular or quadrilateral domain at the top and bottom of the element such as wedge or pyramid is often different from that along the direction of the generatrix ζ directions, and integration types are also the combinations in these two directions.

In DIANA modeling, a volume can be realized not only through directly creating solid geometric modeling but also via extrusion or sweep to achieve a volume after generation of plane sheets. Different from other kinds of structural elements, cross-section geometric properties are not needed more to assign to solid elements. One reason is that the generated points in solid elements are the geometric points forming the geometric volume at the same time. Compared with other elements such as plane stress elements and shell elements, cross-section shape as well as size is determined at the beginning phase of modeling, and they are present in the volume characteristics. The other reason is that the shapes of solid elements are all in volume, and cross-section geometric parameters such as thickness or area are no more required.

Analogical to plane stress elements, basic displacement variables of solid elements are all translational displacement variables in *X*, *Y* and *Z* directions without considering any rotations. Strain variables contain normal strains in three directions ε_x, ε_y and ε_z as well as shear strain γ_{xy}, γ_{xz} and γ_{yz}. The corresponding stress variables are normal stress in three directions as well as shear stress in three directions.

There is another special kind called composite solid element constituted by reference surface, base elements as well as matching composed elements. Unlike regular solid elements, there is only a translational variable in thickness z direction of every node. The reason is that every composed solid element has a corresponding base element, which is regular non-layered solid element and there is only single one-layer numerical integration along thickness direction in these elements, which are all composed of prementioned regular solid elements. Composed elements do not have mechanical properties such as stiffness or mass, and thus they have no influence on the behavior of the finite-element model. Table 1.12 lists all kinds of solid elements in DIANA.

Table 1.12 Lists of solid elements

Name	Shape	Types	Total degrees of freedom	Numerical integration	Characteristics
TE12L	4 nodes, pyramid, isoparametric	Regular solid elements	12	Linear interpolation, 1-point integration in element volume by default	1. All the elements are 3D 2. Displacement variables of every node are translational variables U_X, U_Y and U_Z 3. Number of integration points can be decreased via reduced integration 4. There is no need to further add cross-section geometric properties once solid elements are specified 5. Integration scheme is dominated by the number of integration points and integration schemes at the top surface and the side 6. Besides integration scheme, by default, there are other suitable alternative integration schemes in every kind of element However, the number of integration points has upper limit; once this upper limit is exceeded, these elements are no longer available
TP18L	6 nodes, wedge, isoparametric		18	Linear interpolation in triangular domain $\xi\eta$ direction, Linear isoparametric interpolation in ζ direction, 1-point integration in triangular domain $\xi\eta$ direction, 2-point in ζ direction	
HX24L	8 nodes, brick, isoparametric		24	Linear interpolation, $2 \times 2 \times 2$ Gauss integration in volume	
CTE30	10 nodes, pyramid, isoparametric		30	Quadratic interpolation, 4-point integration in the volume	
CPY39	13 nodes, pyramid, isoparametric		39	Quadratic interpolation, 13-point integration in the volume	
CTP45	15 nodes, wedge, isoparametric		45	Quadratic interpolation, 4-point integration in triangular domain in $\xi\eta$ direction, 2-point in ζ direction	
CHX60	20 nodes, brick, isoparametric		60	Quadratic interpolation, $3 \times 3 \times 3$ Gauss integration in volume	
CTE48	16 nodes, pyramid, isoparametric		48	Cubic/three-order interpolation, 27-point integration in the volume	

(continued)

Table 1.12 (continued)

Name	Shape	Types	Total degrees of freedom	Numerical integration	Characteristics
CTP72	24 nodes, wedge, isoparametric		72	Cubic/three-order interpolation, 9-point integration in triangular domain in $\xi\eta$ direction 4-point in ζ direction	
CHX96	32 nodes, brick, isoparametric		96	Cubic/three-order interpolation, $4 \times 4 \times 4$ Gauss integration in volume	
HX25L	8 nodes, brick, isoparametric	Rubber elements	25	Linear interpolation, $2 \times 2 \times 2$ Gauss integration in volume	1. Composed of brick elements, 2. Integration schemes are Gauss integration 3. Suitable for nonlinear analysis with hyper elasticity
CHX64	20 nodes, brick, isoparametric		64	Quadratic interpolation, $3 \times 3 \times 3$ Gauss integration in volume	
Composed solid elements	Shape	Characteristics	Total degrees of freedom	Base elements	
T3CMP	3-node triangular curved base element	1. Compromised by reference surface, base elements as well as composed elements 2. There is only a translational displacement degree of freedom in ζ direction of every element 3. Since composed elements do not have mechanical properties such as stiffness or mass, they have no influence on the behavior of the finite element model	3	TP18L	
CT6CM	6-node triangular curved base element		6	CTP45	
CT9CM	9-node triangular curved base element		9	CTP72	
Q4CMP	4-node quadrilateral curved base element		4	HX24L	
CQ8CM	8-node quadrilateral curved base element		8	CHX60	
CQ12C	12-node quadrilateral curved base element		12	CHX96	

This part focuses on a common regular solid element—CHX60, which is displayed in Fig. 1.16. This kind of solid element is formed by 20 nodes and the highest interpolation order for displacement function is quadratic, where 3 × 3 × 3 Gauss integration is adopted in volume. This kind of solid element is applicable for simulating the large volume structure of the dam, slope or pipe gallery and also widely applied in the durability and hydration analysis of mass concrete and square piles. Owing to the reason that mass concrete is often in brick volume shape while the shape of this kind is exactly in this shape, thus it displays better boundary adaptability and convergence. Surface to surface connected interface elements can also be added between the contact surfaces of these elements to simulate strengthening case such as bonded steel or CFRP sheet, which has wide application in the calculation of structural reliability taking solid element modeling method.

Fig. 1.16 CHX60 solid element

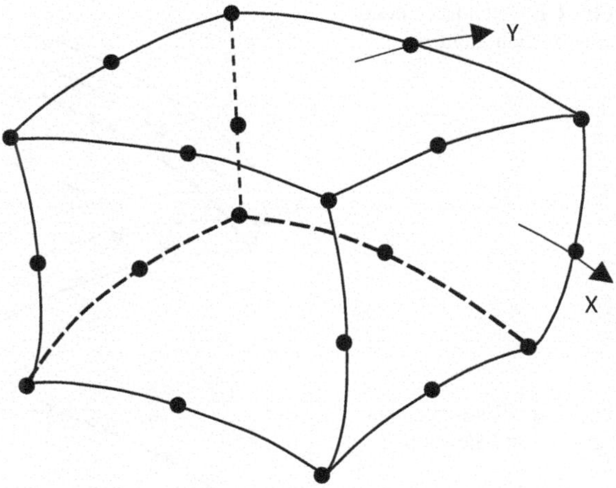

1.3.8 Reinforcements Elements

It is typical in DIANA to simulate longitudinal bars, stirrups, erection reinforcement as well as prestress tendons with automatically embedded reinforcement type with reinforcement bars and grids, which can be embedded into all structural element types. The so-called embedding conception is that reinforcement elements can automatically enter and couple with any kind of surrounding mother concrete elements without users manually establishing the bond interface elements between reinforcement and concrete, which also means that stiffness can only be contributed by embedded reinforcement as well as mother concrete elements. This design is also a unique prior technique in DIANA compared with other kinds of general finite-element software. There are two main reinforcement element types. One is reinforcement bar element applicable for simulating longitudinal steel bars as well as prestress tendons. The other is reinforcement grid elements simulating distributed reinforcement as well as reinforcement grid. Reinforcement bars in shell elements as well as solid elements are displayed in Figs. 1.17 and 1.18, respectively. For reinforcement bar elements, the users not only need to specify material constitutive model, class and cross-section area but also need to determine shape of bars according to geometric modeling points and integration points. Grid elements are often applied to simulate embedded reinforcement grids for all kinds of elements, where the specifications for them are further classified into two types: spacing and diameter as well as equivalent thickness. Discretization method of reinforcement bar elements are of two categories: **Section wise** as well as **Element by element**. Convergence of the former tends to be better than the latter in the structural nonlinear analysis. Reinforcement grids in solid, plane stress and shell elements are displayed in Figs. 1.19, 1.20 and 1.21, respectively.

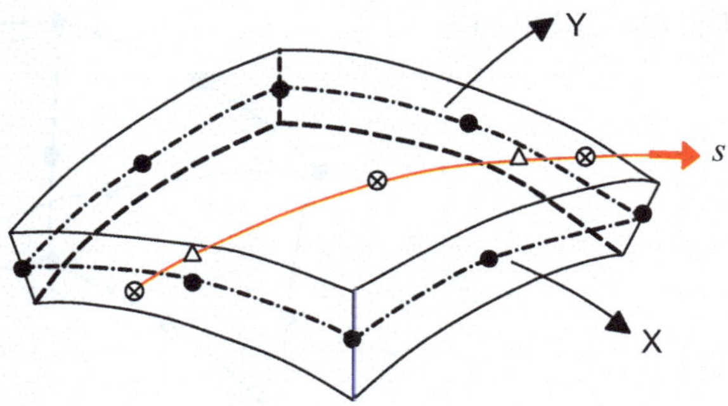

Fig. 1.17 Reinforcement bars in shell elements

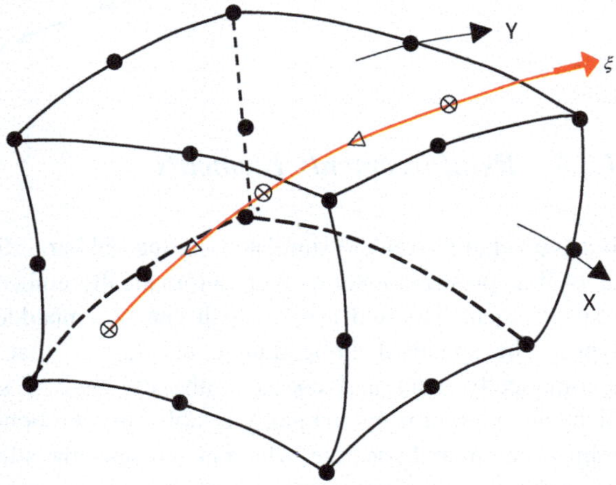

Fig. 1.18 Reinforcement bars in solid elements

In prestress concrete structures, shapes of prestress tendons are often required to match with the form of external load in order to store the prestressing stress to offset the influence of external load applied on the structures. For instance, tendons in straight line are required to be arranged in the pure bending zone while tendons in polyline line are required to be placed in the structures under concentrated load. Moreover, when distributed load is attached to the structures, it is common to arrange parabolic curved tendons to act as equivalent prestress load, and tendons in harp shape are feasible for symmetric concentration load under the condition of four-point loading. The instances mentioned above will be further introduced via numerical cases in the following chapters. When there is an eccentricity from neutral axis, two ways are proposed to deal with this. One way is to directly determine the location of a prestressed tendon by establishing the coordinate value of every point on the bar reinforcement elements, which is suitable for the case of a single prestressed reinforcement bar element with lower order.

The other way is to use the eccentricity definition function provided by DIANA software to realize the more accurate simulation of the attachment position of every point on the curved tendons with higher order or prestressed tendons with steering block.

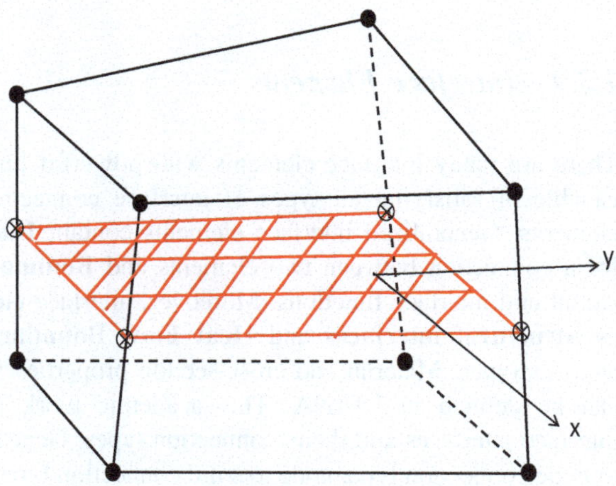

Fig. 1.19 Reinforcement grid in solid element

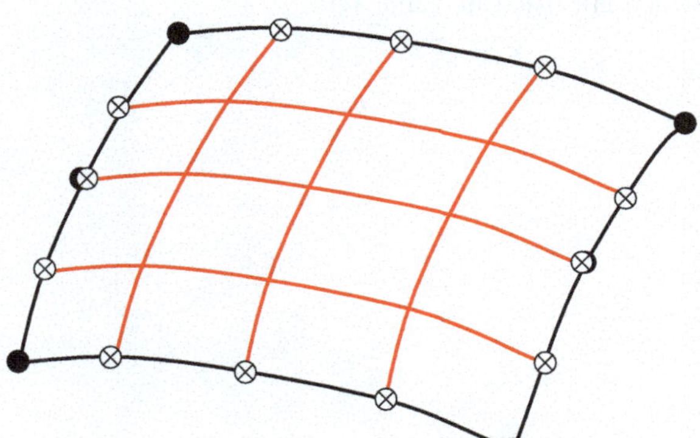

Fig. 1.20 Reinforcement grid in plane stress element

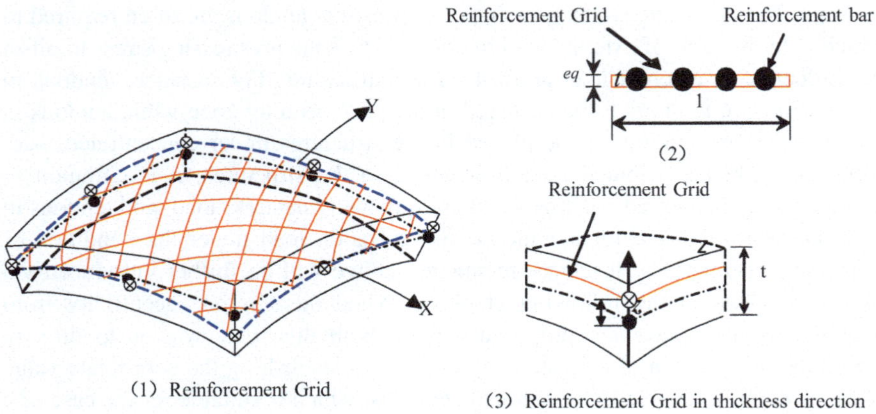

Fig. 1.21 Reinforcement grid in shell elements

Material properties of reinforcement elements will be introduced in detail in Chap. 2.

1.3.9 Interface Elements

There are many interface elements with powerful functions in DIANA, which is capable of satisfying all types of interface connection in all types of structural elements. Generalized interface elements contain **Interface elements** responsible for a connection between two elements and **Boundary interface** with both constraint and interface functions. Moreover, interface elements can also be classified as **Structural interfaces** and **Heat Flow Boundary** according to the specific analysis types. Material and cross-section properties for all the interface elements can be defined in DIANA. This academic work mainly introduces structural interface interfaces and theirs connection types. Generally speaking, there are node to node connection type, node to solid connection type, line to line connection type, line to solid connection type and surface to surface connection type in DIANA, which are listed in Table 1.13.

Table 1.13 Structural interface elements in DIANA

Connection type	Name	Connection characteristics	Type of mother elements matching interface
Node to node connected interface elements	N4IF	2D, 1-1 nodes	2D beam element Plane stress elements Plane strain elements
	N6IF	3D, 1-1 nodes	3D beam element, shell element
Line to line connected interface elements	CL12I	2D, 3-3 nodes	Plane stress elements Plane strain elements
	CL20I	2D, 5-5 nodes	Plane stress elements Plane strain elements
	CL24I	3D, 3-3 nodes	Curved shell elements
	CL32I	3D, 4-4 nodes	Curved shell elements
	L8IF	2D, 2-2 nodes	Plane stress elements Plane strain elements
	L16IF	3D, 2-2 nodes	Curved shell elements
	L20IF	3D, 3-2 nodes	Curved shell elements
Node to solid connected interface elements	TE15IF	3D, 1-4 nodes	Linear pyramid solid elements
	TP21IF	3D, 1-6 nodes	Linear wedge solid elements
	HX27IF	3D, 1-8 nodes	Linear brick solid elements
	CTE33I	3D, 1-10 nodes	Quadratic pyramid solid elements
	CTP48I	3D, 1-15 nodes	Quadratic wedge solid elements
	CHX63I	3D, 1-20 nodes	Quadratic brick solid elements
Line to solid connected interface elements	L12IF	3D, 2-2 nodes	Plane stress elements Plane strain elements
	CL18I	3D, 3-3 nodes	Solid elements
	TE18IF	3D, 2-4 nodes	Linear pyramid solid elements
	TP24IF	3D, 2-6 nodes	Linear wedge solid elements
	HX30IF	3D, 2-8 nodes	Linear brick solid elements (Connect with linear straight line)
	CTE39I	3D, 3-10 nodes	Quadratic pyramid solid elements (connect with quadratic curve)
	CTP54I	3D, 3-15 nodes	Quadratic wedge solid elements (connect with quadratic curve)
	CHX69I	3D, 3-20 nodes	Quadratic brick solid elements (connect with quadratic curve)
Surface to surface connected interface elements	T18IF	3D, 3-3 nodes	Solid elements
	Q24IF	3D, 4-4 nodes	Solid elements
	CT36I	3D 6-6 nodes	Solid elements (triangular interface)
	CQ48I	3D, 8-8 nodes	Solid elements (quadrilateral interface)

Since there are many transitional elements and in vast amount, three typical and frequently applied interface elements are introduced in this part.

1. N4IF

N4IF is a 2D interface element in all versions of DIANA, displayed in Fig. 1.22. Connection type of this interface element is node to node, and mechanic behaviors such as spring elements and one-way stressed truss can be realized by controlling material stiffness. Moreover, various material stiffness curves can be generated via specifications for stress-relative strain curves. This interface element is frequently applied to simulate 2D new seismic energy dissipation members such as coupling beams in shear walls, mew material for energy dissipation embrace bar in 2D structures and it is applicable for truss elements, 2D beam elements, plane stress elements as well as plane strain elements.

Connection type is node to node, and the local x axis of this interface element is along the connected line direction of two points while the local y direction is perpendicular to connection direction. Displacement variables of contact points are translational in *X* and *Y* directions. Material constitutive properties are various. Common types include **Linear elasticity**, **Nonlinear elasticity** and **Coulomb friction**. Normal and shear stiffness of per material properties are required to be specified. According to the simulation objects, there are **No tension with constant shear stiffness, No tension with reduced shear stiffness** as well as **user-specified**, which is specific for shear stiffness and tensile stiffness. Higher precision can be achieved through editing .dat files in the command console in Python language or specifying relationship in the graphical user interface operation to generate normal, shear as well as tensile curves. However, convergence of user-specified mode tends to be lower than the former two options. Normal and shear stiffness of cross-section materials are independent in **Nonlinear elasticity** model while the normal and shear stiffness are dependent in **Coulomb friction** constitutive model. The relationship between the normal and shear stiffness of interface element materials is carried on through **Coulomb friction angle**. Additionally, this element is a perfect combination with Class I beam elements as well as some Class II elements such as L7BEN, which will be introduced in the numerical cases in the following Chap. 3.

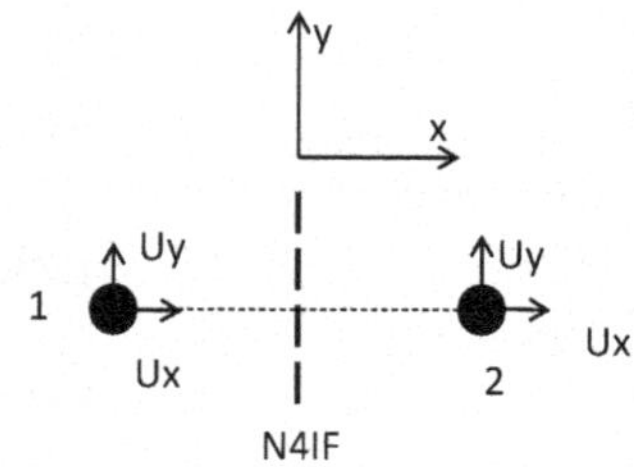

Fig. 1.22 N4IF

The characteristics of N4IF mentioned above are not only applicable for itself but also can be extended to all kinds of node to node interface elements through analogy.

2. Cl24I

CL24I is a line to line connected interface element specific for curved shell elements. Interpolation order of this interface element is quadratic. It is generated by constructing three pairs of nodes on the edges between shell elements. Default integration scheme is 3-point Newton-Cotes integration scheme in the longitudinal ξ direction while 3-point Simpson scheme is in thickness ζ direction [1]. Element x direction in the local coordinate system is the first node pointing to the second one on the edges of contacted curved shell elements and local y axis is perpendicular to the x axis. Element z direction is the outer plane direction. Variables are translational U_x, U_y and U_z along the local x, y and z direction and rotational variables ϕ_x rotating around x-axis. The orientation of the local coordinate system of the interface element conforms to the right-hand system law. Interface element figures, basic variables as well as mechanic behaviors are displayed in Figs. 1.23, 1.24, 1.25 and 1.26.

Fig. 1.23 CL24I between curved shells

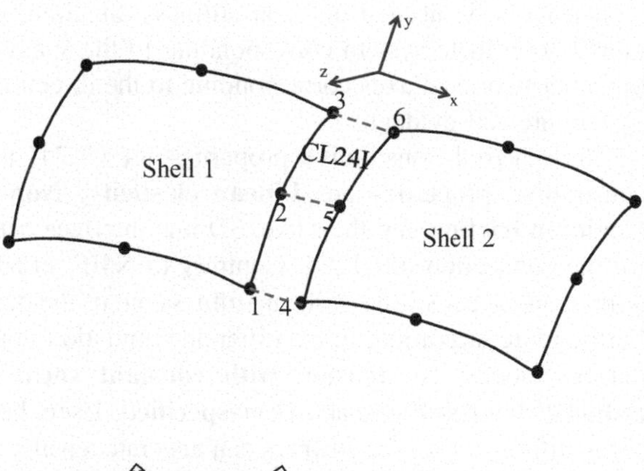

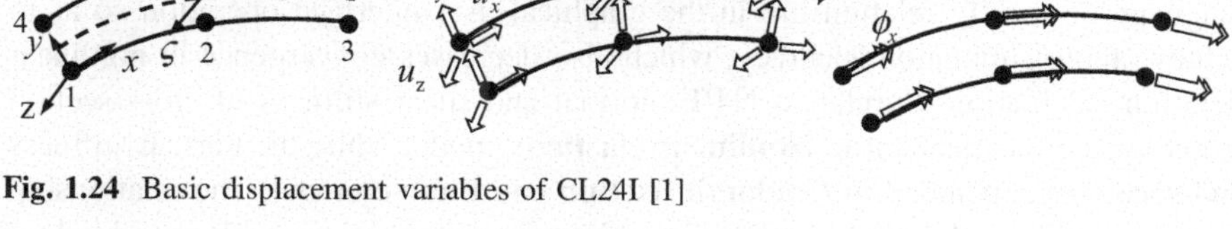

Fig. 1.24 Basic displacement variables of CL24I [1]

Fig. 1.25 Stiffness and nodes of CL24I

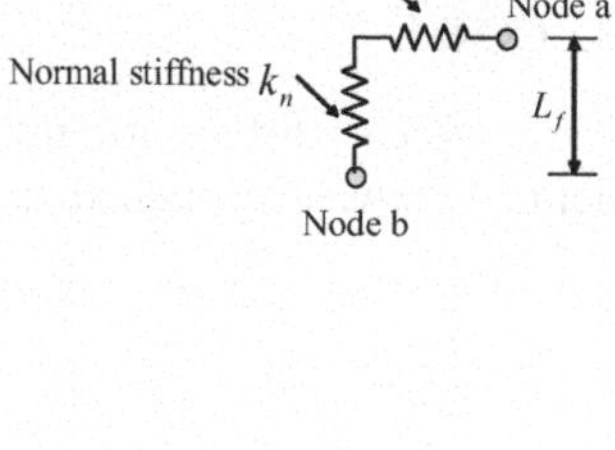

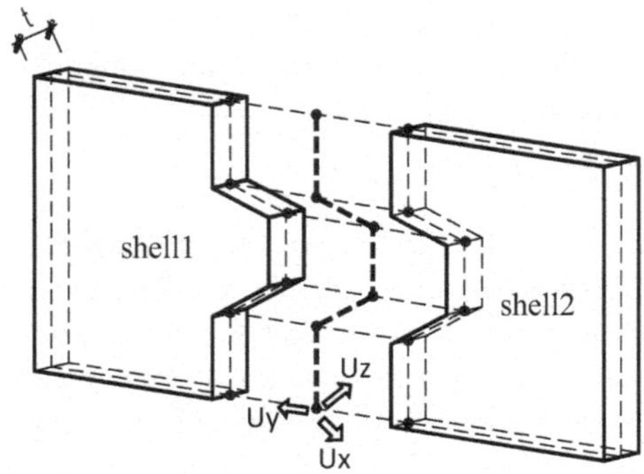

Fig. 1.26 Mechanic behaviors of CL24I

It is worth to note that in the new DIANA versions, there are two types for specifying line to line connected interface elements. One is **Direction vector parallel to shell plane** and the other is **Direction vector normal to shell plane**. The former is required to specify element directions corresponding to the directions in the global coordinate system while the latter is required to local coordinate z-axis corresponding to direction under global coordinate system. Element y direction represents basic normal or shear stiffness direction, and once the direction of the global coordinate system corresponding to the y-axis is determined, properties of the other two local axes corresponding to the direction under the global coordinate system are self-evident.

The material constitutive properties for CL24I are also many. The common constitutive properties are **Linear elasticity, Nonlinear elasticity, Bond-slip, Coulomb friction** and their like. 3D line interface between shells is always typical for the connection of CL24I. Contrary to N4IF, in addition to specifying in-plane normal and shear stiffness, shear stiffness out of the plane is also required to specify. Furthermore, according to the different simulation objects, there are three types for tension modes: **No tension with constant shear stiffness, No tension with reduced shear stiffness** and **User-specified**. Users having specific requirement for shear stiffness or tensile stiffness can generate tensile, compressive and shear curves and mechanic properties via editing **.dat** files in the command console in Python language or specify relationship in the graphical user interface operation so as to achieve more simulation accuracy, which also decreases convergence in nonlinear iteration calculation. Similar to N4IF, normal and shear stiffness of cross-section materials are independent in **Nonlinear elasticity** model while the normal stiffness and shear are dependent in **Coulomb friction** constitutive model. The relationship between the normal and shear stiffness of interface element materials is carried on through **Coulomb friction angle**. The most suitable combination couple for CL24I is CQ40S owing to their common interpolation order.

3. CQ48I

As a typical surface to surface connected interface element, interface surface is defined between touched surfaces of two solids (Fig. 1.27). Normal and shear stiffness values are assigned to these surfaces. Element local x-axis in the local coordinate system is the first node pointing to the second one on the surfaces of contacted solid elements and y-axis is in the normal direction. Element z-axis is perpendicular to plane co-determined by x and y coordinate axes, which is also the outer-plane direction perpendicular to element plane.

The main material constitutive models for CQ48I are **Linear elasticity**, **Nonlinear elasticity** as well as **Coulomb friction**. Connection type for CQ48I is 3D surface interface. Besides specifying in-plane normal and shear stiffness, an outer-plane shear stiffness value is also required to be specified. Moreover, according to the different simulation objects, there are also three types for tension modes: **No tension with constant shear stiffness**, **No tension with reduced shear stiffness** and **User-specified**, which are the same as CL24I and hence is not repeated here.

It is evident to find that this element is a perfect combination with solid element CHX60 integrated with introduction mentioned above. Reasons are as follows. Above all, both the elements are quadratic elements, indicating better boundary compatibility and convergence. Besides, they are both in quadrilateral shape with 8 nodes, which contributes to mesh and follows nonlinear calculation. This interface element also has excellent connection effect, thus has higher convergence in iteration calculation, which is especially suitable for simulating strengthening structures such as CFRP sheet as well as bonded steel plates. Meanwhile, CQ48I interface element is also applicable for surface to surface issues between sheets of two curved shell elements.

Fig. 1.27 CQ48I between two solid elements

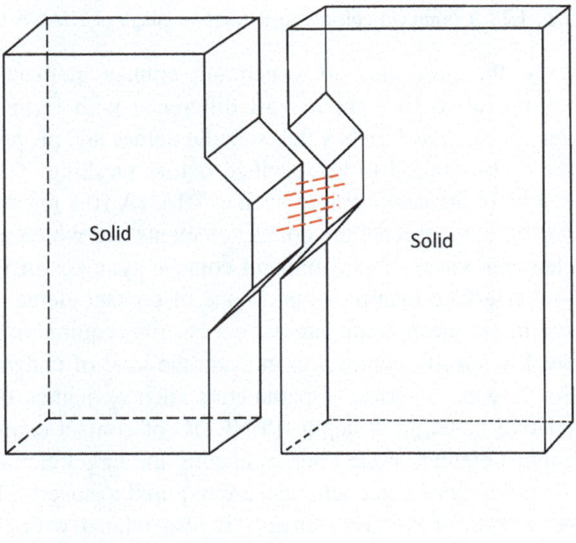

1.3.10 Contact Elements

Contact element is a kind of special element, where contact surface must include two sections. One is contacters and the other is targets. Similar to interface elements, contact types can be node to node contact, line to line contact and line to surface contact, and there are not only 2D elements but also elements under 3D conditions. Displacement variables in 2D are only translational variables U_X and U_Y in X and Y directions, while an additional translational variable U_Z is added in Z direction. The corresponding stress variables are normal stress without shear stress. Command console in Python language of DIANA 9.4 is displayed in Fig. 1.28.

文件(F) 编辑(E) 格式(O) 查看(V) 帮助(H)

```
PROPERTY ATTACH S19 MACON
PROPERTY ATTACH S20 MACON
PROPERTY ATTACH S21 MACON
PROPERTY ATTACH S22 MACON
PROPERTY ATTACH S23 MACON

PROPERTY ATTACH S12 MATARG
PROPERTY ATTACH S13 MATARG
PROPERTY ATTACH S14 MATARG
PROPERTY ATTACH S29 MATARG
PROPERTY ATTACH S30 MATARG
```

```
PROPERTY MATERIAL MACON STATNONL CONTACT CONTAC
PROPERTY MATERIAL MATARG STATNONL CONTACT TARGET 1.00 0.001 0.0000001 0.0 0.0
```

Fig. 1.28 Command console in Python language of DIANA 9.4

In the procedure of generating contact elements, manipulations in DIANA version after 10.1 are in vast difference with former 9.4 version. In the former version such as DIANA 9.4, specific names and properties of targets and contacters are both required to be specified before meshing. Then contact elements are generated. In the new versions such as DIANA 10.1 release, users should specify which points, lines or faces are contact elements and which points, lines or faces are target elements via texting command console syntaxes in Python language or graphical user interface manipulation. Shape of contact elements should be specified before starting to mesh. Users are not necessarily required to input specific parameters, but need to specify contacts to indicate the kind of material used for contact elements. For targets, a series of parameters such as names of target elements, maximum relative penetration depth (PENETR) of contact elements, relative distance of the contact element node after contacting the target element surface, Coulomb friction as well as cohesion coefficient are required to specify. In the real contacting process, generation of contact elements is also related with axial orientation. For 2D elements, positive y direction in the local coordinate system must point to the outer direction of target elements, while the positive z direction must point to the external normal direction of the plane forming target elements. A couple of contact elements are displayed in Fig. 1.29.

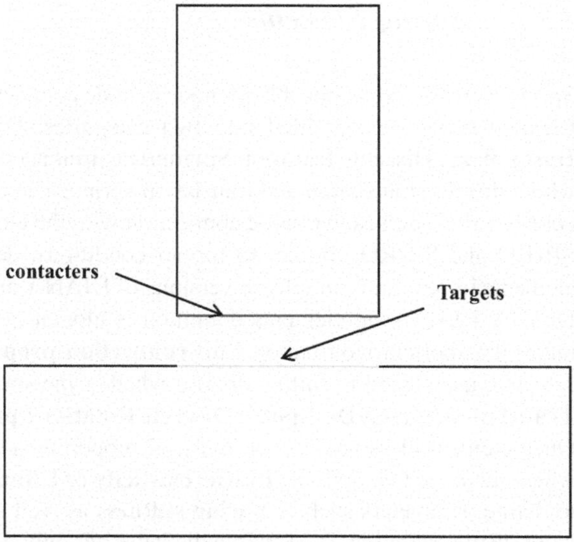

Fig. 1.29 Contact element

All the contact elements in DIANA are listed in Table 1.14.

Table 1.14 Basic contact elements

Name	Dimensions	Shape	Total degrees of freedom	Numerical integration	Characteristics
L4CT	2D	2-node Straight line	4	Linear interpolation	1. There are only two translational displacement variables u_X and u_Y in X and Y directions of every node 2. All the nodes must be in the XOY plane
CL6CT		3-node Quadratic curved line	6	Quadratic interpolation	
T9CT	3D	3-node triangular	9	Linear interpolation	1. There are only three translational displacement variables u_X , u_Y and u_Z in X, Y and Z directions 2. Positive direction of the x-axis is the direction from the first node of the element pointing to the second node under coordinate local system
Q12CT		4-node, quadrilateral	12	Linear interpolation	
CT18C		3-node triangular	18	Quadratic interpolation	
CQ24C		8-node, quadrilateral	24	Quadratic interpolation	

1.3.11 Spring Elements

Spring elements are applicable for node to node connection in a finite-element model. Element types are classified into two categories: **Discrete Translation Spring/Dashpot** and **Discrete Rotation Spring/Dashpot** according to displacement modes, where discrete translation and translation springs can also be classified as one-node connection as well as two-node connection with the element names SP1TR, SP2TR, SP1RO and SP2RO. Similar to former conditions, definition operations of spring elements in new and old release versions of DIANA are also many. For example, in DIANA 10.1, spring elements definition is hidden in the **Connection type** option under the shortcut icon button **Edit connection property**. When **Spring** option is selected, users need to further choose whether the spring element class is **Discrete Translation Spring/Dashpot** or **Discrete Rotation Spring/Dashpot** (see Fig. 1.30). Once element class is selected, material properties are also required to be chosen, where there are two options: **Linear elasticity** or **Ultimate forces**. The former needs to define parameters such as **Spring stiffness** as well as **Constant damping coefficient** displayed in Fig. 1.31, while the latter not only contains definitions mentioned above but also needs to further specify minimum and maximum force (translational spring element) or moment (rotational spring element) values displayed in Fig. 1.32.

Fig. 1.30 Spring/Dashpot elements in DIANA 10.1

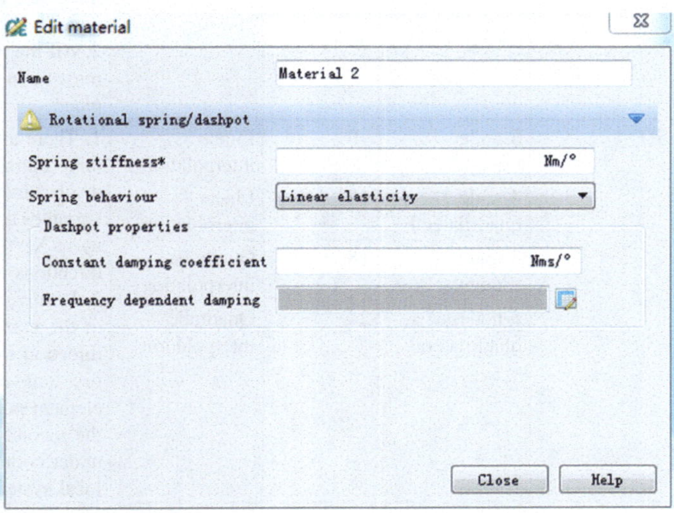

Fig. 1.31 Linear elastic material constitutive properties for rotational spring elements

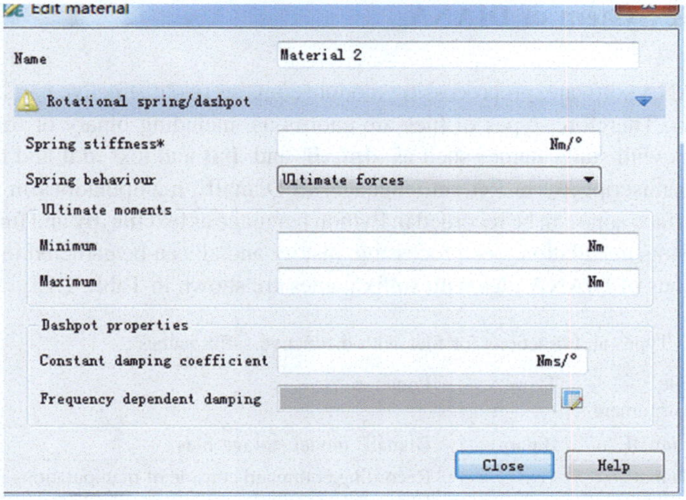

Fig. 1.32 Material constitutive specifications for rotational spring elements under ultimate forces model in DIANA 10.1

Constitutive properties of translational spring elements are described via load–displacement relationship, while constitutive characteristics of rotational spring elements SPARO are described via moment–rotation relationship. Rotational spring elements are often applied to simulate shear deformation in joint core zone of frame-shear wall structures. Therefore, it is not only essential to simplify simulation joints in geometry but also require force equivalence (Fig. 1.33).

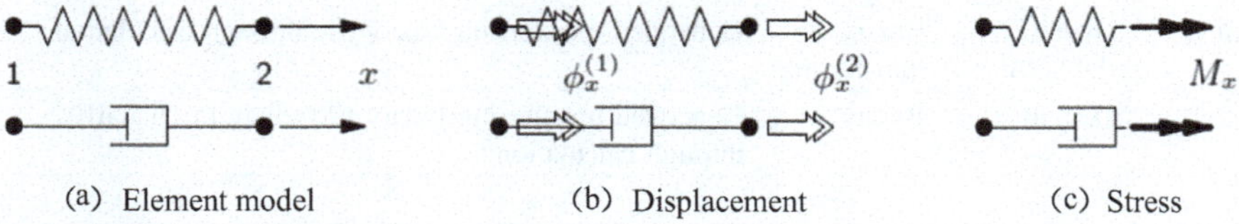

Fig. 1.33 Rotational spring element in two-node connection(SP2RO) [1]

In the specific modeling process, beam and column elements at the joint are coupled by degrees of freedom, including the horizontal and the vertical directions. The two joints are connected by SP2RO element so as to achieve the effect of transmitting bending moment. In order to simulate the shear transfer, the vertical degree of freedom coupling can be directly conducted in the corresponding parts where the spring element needs to be added.

Spring/Dashpot elements are applicable for simulating isolation damping devices as well as isolation bearing models with high damping coefficients. Moreover, this kind of element can also be applied in seismic energy dissipation structures.

1.4 File System of DIANA

In the DIANA software, preprocessing module contains two distinctive parts: DianaIE and iDiana. Therefore, types of files are enormous, including binary or transferring binary files with suffix names such as **.dpf**, **.ff**, and **.bat** and text **.dat** and command console manuscript **.py** in Python language. In DianaIE, manipulations in graphical user's interface zone can be recorded in Python language as text file **.py** and functions of preprocessing, calculation, postprocessing, history and all can be gathered in it. Types and functions of DIANA files with suffix names are shown in Table 1.15.

Table 1.15 Types and functions for files with distinctive suffix names

Suffix name	File assignment	Type	Function
.dpf	DianaIE	Binary	DianaIE model storage files
.py	DianaIE	Text	Recording command console of manipulations in Python language
.dat	DIANAIE, iDIANA	Text	Recording parameters such as nodes, elements, material properties, load cases, number of elements and boundary conditions in model file
.bat	iDIANA	Binary	1. Binary model parameter storage file in iDiana 2. Text file inputting material property parameters
.dcf	iDIANA	Text	Adding calculation commands
.out	DianaIE, iDIANA	Text	Recording result for every load step and calculation outputs
.G72	iDIANA	Binary	Binary files generated in preprocessing procedure.
.V72	iDIANA	Binary	Binary files generated in postprocessing procedure
.his	iDIANA	Text	History file recording every manipulation in modeling procedure
.ff	DianaIE, iDIANA	Binary transferring	Generated transferring binary file through calculation
.dnb	DianaIE	Binary	Generated transferring binary file solely in DianaIE through calculation

Macro logic of the relations between various files in DianaIE as well as iDiana is shown in Fig. 1.34.

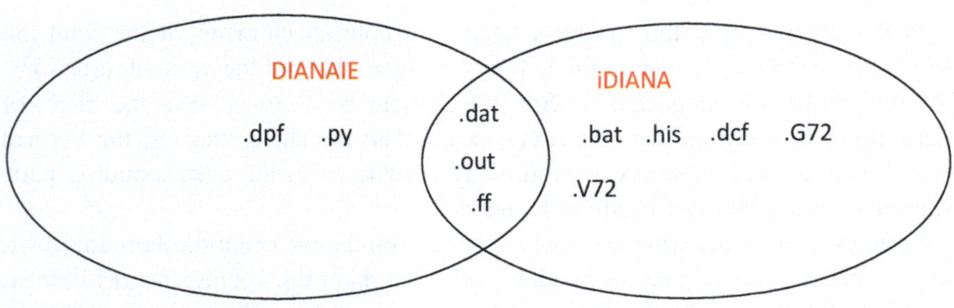

Fig. 1.34 Macro logic of the relations between various files in DianaIE as well as iDiana

Transformations of various files with distinctive suffix names in DianaIE are displayed in Fig. 1.35.

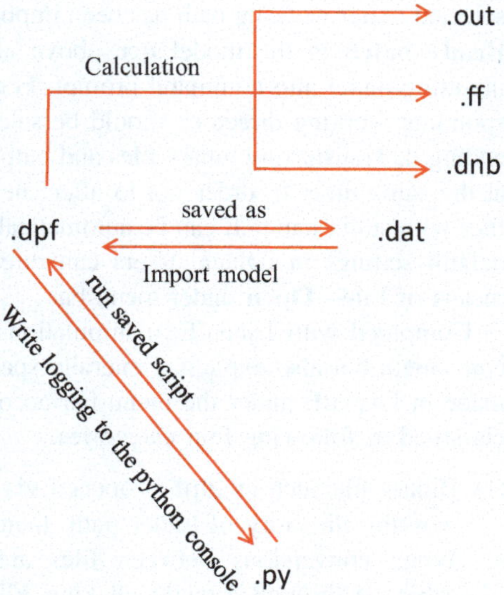

Fig. 1.35 Transformations of various files with distinctive suffix names in DianaIE

Transformations of various files with distinctive suffix names in iDiana are displayed in Fig. 1.36.

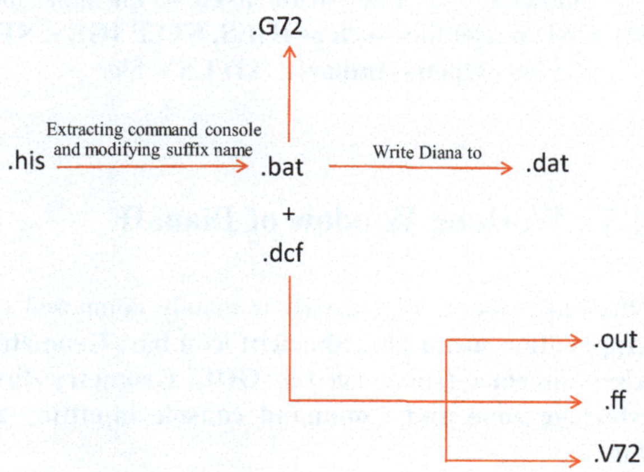

Fig. 1.36 Transformations of various files with distinctive suffix names in iDiana

DianaIE and iDIANA modules in DIANA software have their own way of opening, editing, saving and closing files. For example, whatever for DianaIE or iDiana files, users can open files by using **File**-related operations under the menu bar, or creating, opening, closing, saving or running files directly through series of **Ctrl** operations on the keyboard. For older release version 9.4, when user's file is a binary model database file, where it is a binary preprocessing modeling file with suffix name **.bat**, users should not directly click **File—Open** to open the file. Instead, working directory of folder path where the **.bat** file resides should be selected. After working path has been imported into the iDiana, searching **Utility—Read—batch** in the model tree above all, then the file can be opened before inputting name into command prompt. For .dcf calculation controlling file, corresponding working directory should be selected in the module of **Diana_w** before adding it. Transferring binary files and output files such as .ff and .out are generated at the same time. In order not to affect next calculation, these transferring binary files with suffix name **.ff** can be automatically deleted by system itself via adjusting default settings in iDiana. Users can directly open **.dat** model database file by means of **File—Open** under menu bar.

Compared with DianaIE, manipulations and opening types are not only more convenient but also explicit. Generally speaking, different opening regulations are made in DianaIE under the menu bar according to different kinds of files, mainly classified as following four categories.

(1) Binary file such as **.dpf** is opened via **File—Open** to determine **.dpf** file and working directory of folder path, then double-clicking to open this **.dpf** file. When conversions between files are required, these conversions can be achieved through directly clicking **File—Open—Discard-**clicking next **.dpf** file.
(2) Generated model database **.dat** files can be opened via **File—Import model** under menu bar.
(3) Command console manuscript **.py** files in Python language is opened in the following way: **File—Run saved script** under menu bar.
(4) CAD model files such as **IGES, STEP IGES, STEP** are opened by the means of **File—Open—Import CAD/CSV file**.

1.5 Working Window of DianaIE

Working window of DianaIE is mainly composed of following interface plates: **Application menu bar, Shortcut icon bar, Geometry directory tree, Graphical User interface** (abbreviated as **GUI**), **Geometry directory tree, Property zone, Message zone and Command console inputting zone**, which is displayed in Fig. 1.37.

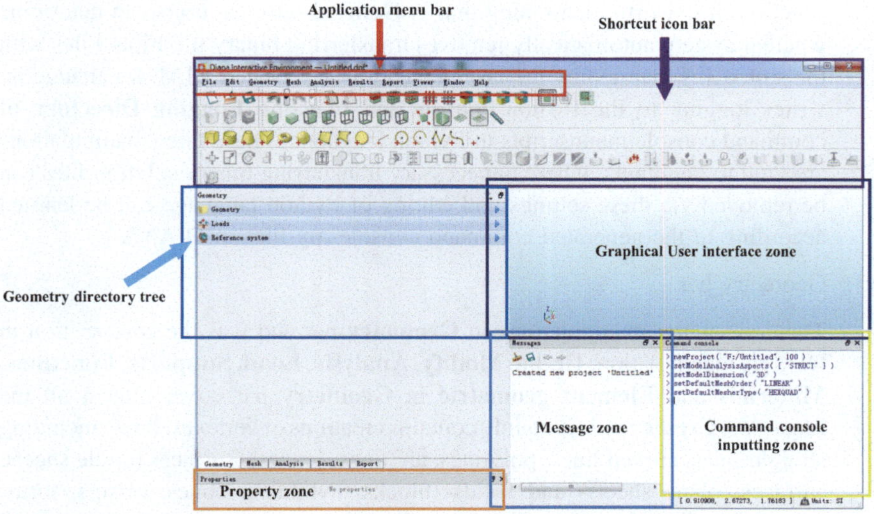

Fig. 1.37 Working window and interface plates of DianaIE

Application menu bar, often abbreviated as menu bar in the following chapters, includes File, Edit, Geometry, Mesh, Analysis, Report, Results, Viewer and Help bars. Several bars in common use are introduced now.

(1) File bar

Function of file bar is document operating bar, including open, generation, creation, save and modeling of files. Main manipulation ways for file are creating binary **.dpf** files (**File-New**), opening **.dpf** files (**Open**), saving .dpf files (**Save/ Save as**), importing **.dat** files to convert them into **.dpf** ones (**Import model**), reading text files with **.py** in Python language to convert them into **.dpf** files (run saved scripts). Files with distinctive suffix names and formats can be not only opened, imported and mutually opened in menu bar, but also allow files in CAD formats to convert into graphical user interface (Run CAD/ CAS file). Since conversions of different files are introduced in former part, it is not repeated here.

(2) Edit bar

Edit bar mainly focus on settings for DIANA manipulations, which can be further classified as eight types: **Undo**, **Redo**, **Preferences**, **Project settings**, **Lighting preferences**, **Move working plane**, **Working plane Grid**, **Distance between points**, where **Preferences** and **Project settings** are two main functions. Users can adapt series of settings such as colors of DianaIE, background colors of graphical user interface, colors of nodes, width of line and sizes of nodes and their like via **Preferences**. Meanwhile, corresponding settings of geometric colors and sizes of models are specified under this function. It is worth mentioning that in **General** settings, users can determine whether system automatically removes transferring binary file **Filos Files** with the suffix **.ff** generated by the last calculation in repeated DIANA calculation, writes logging to the Python console and selecting **Logging Directory** of command console manuscripts in Python language. These three manipulations are vital to beginners, where unnecessary transferring binary solution files can be removed via these settings and editing of Python language can be learned according to the generated command console **.py** files in DIANA.

(3) Geometry bar

There are many manipulations in **Geometry** bar and it is the core section in DIANA manipulation. **Create Modify**, **Analysis**, **Load**, **Supports**, **Functions**, **Materials** and **Element geometric** in **Geometry** bar cover almost all the functions. **Create** module mainly contains creations of vertexes, lines (including straight lines, curved lines, polylines, arc lines, circles), surfaces (circle sheets, surfaces, plane sheets) and solids (blocks, cylinders, cones, prisms, torus, spheres). In DIANA, not only straight lines but also curved lines are created. Moreover, polylines, closed lines and curved sheets can also be established. **Modify** module is concerned with graphic transformation manipulations such as **Move shape** (translational movements of geometric shapes), **Scale shape, Extrude** (from **2D** surface into a 3D volume through a longitudinal extrusion)**, Sweep** (from **2D** surface into a 3D volume through a longitudinal sweep), **Mirror shape** (making graphics axisymmetric according to a certain symmetry), **Array copy** (Duplications and movements of a geometric objects). **Load** module is responsible for defining and attaching all kinds of loads and acts, including gravity, distributed force, concentrated force, post-tensioning load, pressure, prescribed deformation, temperature and their like. **Support** module mainly refers to attaching constraints or boundary conditions. **Material** and **Element geometries** are assigning material properties and cross-section geometric properties for numerical models, corresponding to shortcut icons of Property assignments in yellow, blue and red, respectively.

(4) Mesh bar

Mesh bar is mainly responsible for mesh properties settings and generation after the preprocessing procedures of geometric modeling, assignments for material properties and attachments of loads and supports. There are also items of **Load, Support**, **Material**, **Element geometries** and **Functions** under **Mesh bar**. However, these items are nothing but the modifications or additions for meshed model under **Geometry directory tree**.

(5) Results bar

It is in the light of output results after calculation. Users can not only check results under certain perspective, such as **Normalized deformed results**, but can also check deformed results under full perspective, such as Absolute deformed results. In addition, meshed results before and after deformation can also be checked, such as Deformed mesh feature edges.

(6) Viewer bar

Viewer bar is mainly responsible for checking selective output results. In viewer bar, users can not only select to check geometric model before meshed (**View geometry**) but also check results after calculation (**View results**). Moreover, all kinds of geometric perspectives can be mutually converted (such as **Fit all**, **perspective projection**, **show mesh seedings**). After accomplishment of solution, all primary outputs of certain part or shape can be checked in **Viewer** bar (Node/Shape/Face/vertex selection), element information (**Element selection**) as well as reinforcement information (Reinforcement selection).

(7) Window bar

It contains **Windows panes**, **Tool bars**, **Model sections** as well as **Mesh sections**, which include the majority of layouts and display of **Geometry directory tree**. Users can select which functions should be displayed or omitted as their like.

(8) Help bar

It consists of **DIANA Manual**, **Release notes**, **Activate new DIANA license**, **update DIANA license** and **About DIANA Interactive Environment.** Users should apply for reactivation of this software in order to continue their use when the time exceeds the period of **DIANA Interactive Environment**.

(9) Report bar

This portion is mainly concerned with the report of output after solution. A report can be added, removed or run in this part. Besides, any section of report can also be added and converted into a chapter.

(10) Analysis bar

Functions of this portion correspond with **Analysis** module and the functions are equal, including add, remove, duplicate or run an analysis. Additionally, analysis command can also be added, loaded and saved. Phased analysis is added through this module.

There are many shortcut buttons in shortcut icon bar. Through clicking these buttons, users can rapidly create vertex, line, surface, solids or blocks and realize manipulations such as sweep, extrusion and Boolean logic operations. Meanwhile, operations such as concrete and reinforcement assignment as well as generation of interface elements are also conducted via shortcut icon buttons. Icons of shortcut button are common-in-use ones of graphical user interface.

Geometry directory tree, listing all the steps as well as items in the preprocessing procedure and viewed as main displayed body of graphical user interface, is required to coordinate with application menu bar as well as shortcut icons in shortcut icon zone. When any manipulation is generated via two processing ways mentioned above, corresponding results and sets are displayed in the **Geometry directory tree**. Sets rename and properties assignment can be completed via right-clicking in **Geometry directory tree**. In the **Analysis directory tree**, analysis type can be directly added to execute settings of Load block.

As an information generated bar in interface manipulation, **Properties** bar is a record as well as reflection of former manipulation step. Parameter controlling method is applied in DainaIE **Properties** window. Tables in the **Properties** bar are applied for recording parameters and characteristics of successfully defined properties. **Geometry, Mesh, Analysis, Results and Export** module all have their own lists of properties, where results of postprocessing settings are specified in **Results** bar according to user's personal preferences.

Command console zone records and inputs command console information of every manipulation step. After ticking **Write loggings to python**, any modeling step manipulated by the users in graphical user interface zone can be recorded in Python language in this zone. Meanwhile, users can also run the **.py** files by means of editing command console in a .py file, and then copying and pasting it into the command console zone. Moreover, owing to the diversity and universality of Python manuscripts, users can also use interface manipulation and editing Python language in a mixed way. For example, some part of pre-edited command console in Python language can be first copied and pasted into command console zone to run the model. When the graphical user interface appears, they can directly take interface manipulations in the graphical user interface zone, which can effortlessly realize the conversion between command console and interface manipulation. When creating numerical model with command console, users can not only click **File-Run saved script** to open the **.py** files, but also can copy and paste the command console into the corresponding zone, where the latter method is more prone to check and modify errors in command console sentence by sentence according to author's experience. When confronting severe errors, running process and iterative calculation abort, and tips in red appear in the **Message** zone (see Fig. 1.38). In view of this condition, users can extract command console section by section or line by line to input into command console zone in order to search locations and reasons of errors.

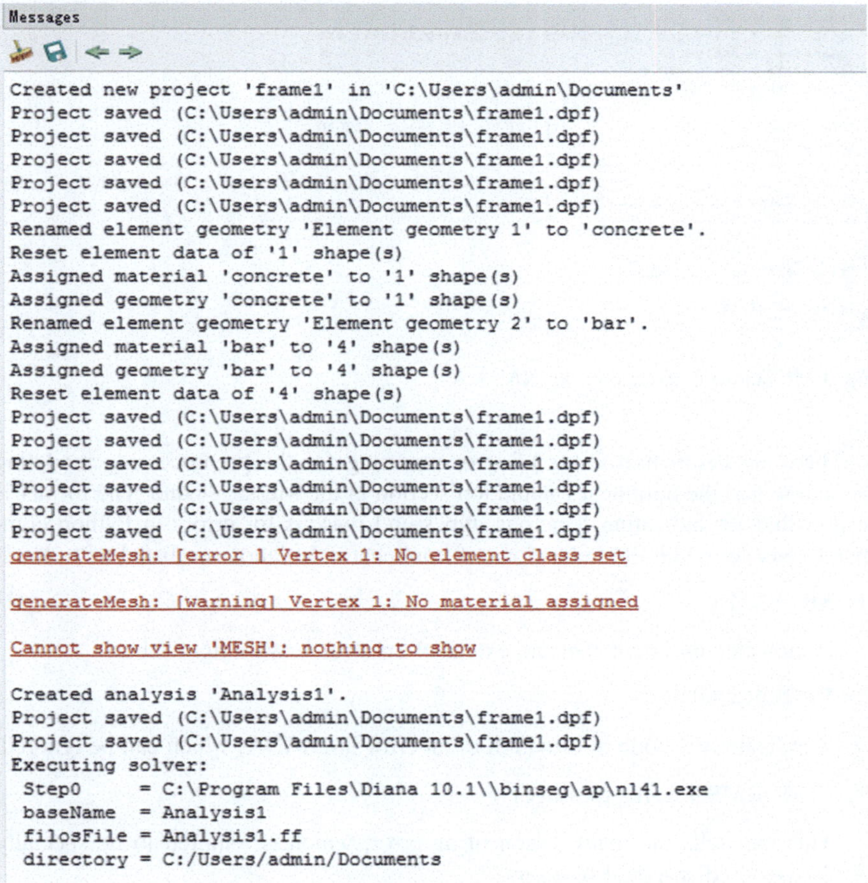

Fig. 1.38 Tips of errors in **Message** zone

All the errors, regardless of the extents and ranks, can be found in **Message** zone, which contains two main functions. One is reminding users of phased states (open, import, save and close) for all kinds of files. When the manipulation errors under current state appear, tips in red appear immediately in **Message** indicating zone in DIANA so as to render convenience for users to search. The other one is to point out errors that appear in modeling and calculation procedure, where two distinctive errors in various colors take on the **Message** indicating zone, one is SEVERITY:WARNING in yellow extent while the other is SEVERITY:ABORT in red. If the former warning happens, it means that this kind of error does not have influence on calculation running smoothly but may affect the precision of results and further improvement is expected to be conducted. However, when errors in red appear, it indicates that the error extent is severe (see Fig. 1.39), then the DIANA will abort the calculation or solution in Abort format, and the finite-element analysis is accordingly terminated. Most frequent severely aborted error confronting in DIANA is disconvergence in nonlinear iterative calculation.

```
Executing solver:
 Step0     = C:\Program Files\Diana 10.1\\binseg\ap\nl41.exe
 baseName  = Analysis3
 filosFile = Analysis3.ff
 directory = G:/例题

/DIANA/AP/NL41    23:14:14      0.03-CPU     0.06-IO    BEGIN
SEVERITY : ABORT
ERROR CODE: /DIANA/LB/AU32/0043
ERRORMSG.A: All licensed seats for module NONLIN are in use. See: http://localhost:1947
Extend your license or try again later.
DIANA-JOB ABORTED


Diana abort routine called

Trying to close things.
```

Fig. 1.39 Common severe error in DIANA

There are distinctive forms of error reporting for the preprocessing modeling procedure and the nonlinear calculation section in the Message zone. The former is responsible for indicating locations, types and reasons for errors in Python command console, while the indications of nonlinear calculation are listed as follows:

(1) SEVERITY

It indicates users whether the extent of an error is warning or abort

(2) ERROR CODE

It indicates the code information of an error in DIANA, which can be ignored.

(3) ERRORMSG (error message)

This part tells the main reason of an error, which is required to be specially focused and searched by users.

(4) Error suggestions

It renders proposals for users to correct this error.

Graphical User interface (abbreviated as GUI) is a major module in DianaIE, rendering attachments of supports, load, meshing, checking postprocessing results as well as contour plots. Compared with former old release version 9.4 and 9.6 with black interface style, vast improvement happens in new DIANA release versions. Users not only use default background color of GUI but also individually alter GUI background color as their preferences via Edit—Preferences settings in the menu bar. For the same reason, colors of concrete elements, reinforcements, or colors of generated meshed elements, are all can be changed under this working directory. Every manipulation results such as creating models, modifying, adding load cases and supports are displayed in GUI zone. After generation of meshed elements, numerical models of meshed structures will be taken in this zone too. Furthermore, deformed shapes will also be displayed in GUI zone after nonlinear calculation. When output of the results is specified, calculation results are displayed in the format of contour plots. For the conditions with multiple load cases as well as load steps, checking contour plots instead of tabular **.out** files provides users (especially beginners) more convenience and help.

1.6 Finite-Element Analysis Procedure for DIANA

Procedure for finite-element analysis can be classified as follows:

(1) Creation of geometric numerical model

This step can be completed via clicking items in the menu bar or shortcut icon button in the shortcut icon zone. Besides, geometric numerical model can also be created via importing edited command console in Python language. Geometric forms constituting numerical models are vertex, line, surface and solid. Contrary to DIANA 9.4 release version, which is to create vertexes above all then connect them by lines, surfaces and solids, geometric model in new release DIANA, especially after release version 10.0, can be created directly through establishing coordinate points rather than creating coordinate vertexes in solo. It is also worth to mention that when inputting coordinate values to construct surfaces or solids, coordinate points ought to be constructed in clockwise or counterclockwise direction and intersect points are not allowed in DIANA.

(2) Assignment for material and cross-section geometric properties

This step can be accomplished via directly clicking shortcut icons in shortcut icon region or selecting numerical model to right-click it. Three following icons can be displayed as in Fig. 1.40.

Fig. 1.40 Shortcut icons for geometric properties of various materials

(3) Attachment of load and boundary constraints

Load types are various in DIANA, which are mainly **Force**, **Distributed force**, **Posttensioning load**, **Point load**. Contrary to other software, prestress load can be directly attached as load type, which can effectively simulate post-tensioning load in post-tensioning construction technology.

(4) Mesh

There are two meshing types: **Division** and **Element size**. After determination of meshing types, users not only mesh numerical model via **Face** but also through **Edge**. In seeding method, DIANA allows users to choose either the **Division** method specifying number of divisions or **Element sizes** method specifying meshed element sizes. Meshing type is intelligent. Based on the convergence of nonlinear calculation, quadrilateral (2 dimensions) or hexahedron (3 dimensions) elements (**Hexa/Quad**) are taken into account as priority. When meshing for edges in geometric model, meshing divisions should be in proportion to the length of every side. Meshing step can be conducted in any preprocessing phase in former old release versions such as DIANA 9.4, while it must be the last step for preprocessing procedure in release versions after DIANA 10.0. However, names of required element type can be directly specified before generation of meshed shape, while it can only be indirectly and passively determined through specifying meshing shape or meshing order after DIANA10.1 release version. Therefore, ultimate names of element types are checked only after successful completion of mesh. In DIANA, meshing step can be completed both in editing command console way as well as in graphical users interface region under geometry tree directory. There are several steps for meshing.

1. Specifying element type (such as edges, sheets or solids)
2. Seeding method (element size or divisions)
3. Specifying element shape (usually selecting quadrilateral/ hexahedron or triangle/pyramid)
4. Determination of mid-side node location (on shape or linear interpolation)
5. Element sizes
6. Generation of meshed elements.

(5) Attaching load

There are three blocks for attaching load cases. The main load types are Point, Line, Face, Solid, Temperature, Prescribed Deformation, Posttensioning load and their like. Whatever attaching which load type, users are required to specify the geometric characteristics: load target type (point, line, edge, face or solid), load type (point load, line distributed load, surface distributed load or solid distributed load), loading value, direction as well as attaching objects. Attaching points can be ticked through mouse click. For post-tensioning load, selections of **Tension type** (one or both end), **Anchor point**, **Nodal anchor force**, **Anchor retention length** as well as **Post-tensioning schemes**, **Coulomb friction coefficient** as well as **Wobble factor** should be specified, where **Anchor retention length** and **Wobble factor** have great influence on descending amplitude of priestess tendons with time. When nodal anchor force is attached, priestess loss of anchor retention length and Coulomb friction coefficient can be automatically deducted. According to author's experience, the larger the anchor retention length and Coulomb friction coefficient are, the more reduction long-term priestess loss is. Load cases are added into the structural analysis by means of geometry load combinations.

(6) Adding analysis block and running calculation

In the structural nonlinear analysis module, iterative methods, maximum iterations, convergence norm, convergence tolerance and abort criterion are required to be specified, where tolerance by default is 0.01. When disconvergence occurs, load steps, iterative method as well as convergence in structural nonlinear analysis are all required to be modified. In the convergence norm, both force and displacement are selected at the same time, which is much faster than single convergence criterion such as displacement, force or energy convergence norm. The reason is that the DIANA system must judge the results of force and displacement calculation at the same time under practical nonlinear iteration calculation for multiple load steps. As long as either of them reaches predefined convergence tolerance, iteration of this load step is deemed as convergence.

(7) Checking output results

Users often check contour plots of all primaries such as strain, displacement in certain direction as well as local stress and their like. Besides, some specific calculation results in **OUTPUT** module can be displayed via user selection function before calculation and after additions of load cases. Furthermore, output mode is displayed with DIANA **native** output binary format or **Tabulated** output model with suffix name of .tb files to view the results in the **Device** option of the OUTPUT module.

1.7 Command Console of DIANA in Python Language

In DIANA, users can complete numerical model through clicking menu bar or shortcut icon buttons or they can also edit command console in Python language in **.py** files first, and then import or copy and paste console into command console zone to accomplish modeling procedure. On comparing manipulations with GUI, this way has high universality, especially for finite-element analysis in civil engineering. When encountering similar parameter modeling issues, users do not need to create new files and repeat the modeling manipulations, rather they can realize batch-oriented numerical models rapidly by simply copying, editing or modifying command console in **.py** files. Compared with complicated and obscure reduced command console in binary files **.bat** and calculation controlling files with suffix name **.dcf** in iDiana9.4, Python syntaxes of command console in DianaIE are more casual and random in grammar and format. Users are even not bothered to consult syntax manual so that they could interpret and edit command console (Table 1.16).

Table 1.16 Python language and corresponding explanation

Python language	Explanation
New Project("working directory A", model size)	Creating a new file in the working directory of A and setting maximum scale for model
Set Model Analysis Aspects	Specification for analysis type
Set Model Dimension	Determination on model dimension
Set Default Mesh Order	Setting Default mesh order
Set Default Mesher Type	Settings for mesher shape
set DefaultMidSideNode Location	Determination of Mid-site node location
Createveretex/line/Sheet/Solid	Creating vertex, line, surface, solid
saveProject	Saving files
setParameter	Specifying material and geometric parameters for numerical objects
addMaterial	Adding material properties
addGeometry	Adding geometric properties
createLineConnection	Generation of line to line connected interface element
attachTo	Attachment on objects
setElementClassType	Specifying element type
assignMaterial/Geometry	Assignment for Material and geometric properties
rename	Rename a set
addGeometryLoadCombination	Adding geometry load combinations
Set Geometry Load Combination Factor	Specifying user specified load factor for every load sets or combinations
Set MesherType	The same as "Set Default Mesher Type"
Set Mid Side Node Location	The same as "setDefaultMidSideNodeLocation"
Set Element Size	mesh settings for element size
generateMesh	Mesh generation
setActivePhase	Activation on phased analysis
addAnalysisCommand	Adding analysis module for load case
runSolver	Running solution
setResultPlot	Checking results for stress contour plots
setResultCase	Checking calculation results for corresponding load steps
arrayCopy	Copy and translate objects

All the manipulations generated in graphical user interface zone under the DIANA interactive environment (abbreviated as DIE) are recorded and saved in **.py** files in Python language. There are three ways for users to generate graphical user interface: (1) direct manipulations in graphical user interface zone; (2) inputting every syntax in Python language into command console zone; (3) pre-editing command console in a manuscript file with suffix name **.py** and then importing it or copying command console and pasting it into command console zone; (4) mixed applications mentioned above, which is a typical way for the majority of users. In DianaIE, Python syntaxes are random like draft. Redundant syntaxes in command console can be removed so that standard python syntaxes are established, which also guarantees parameter modeling and graphical features applicable in DIANA. When command console syntaxes in brackets are a statement that needs user to specify, the affiliation between the modules corresponding to the graphical user interface manipulation is expressed by/in brackets.

1.8 Units in DIANA

Whether in DianaIE or retained former iDiana module, international system of units (SI) is defined as solo in DIANA. Contrary to other general finite-element software with no specific units such as ABAQUS, units in DIANA are standard, explicit and various, which is a priority for DIANA. When the calculation results are achieved, users can identify units and orders of magnitude in order to help them evaluate or judge the correctness and precision at first glance, which renders valuable simulation reference in real construction. Units in DIANA concerned with civil engineering are length, force, mass, time, temperature and angle, and these types of units and conversion between different magnitude orders under the same unit will be introduced in detail. Once the unit system is changed, one of the three units—length, force and mass—will be automatically displayed in the derived form.

Quantity Unit under **Properties** zone in DianaIE is set to define the unit system of model files. In DianaIE, the default unit system is the international unit system, that is, meter, kilogram, Newton, second, Kelvin, radian, Unit, as shown in Fig. 1.41. In DIANAIE software operation module, the unit system of other units can be guided by the above units. Default standard units in DianaIE are meter, kilogram, Newton, second, Kelvin, radian. Other units can be derived from the above units.

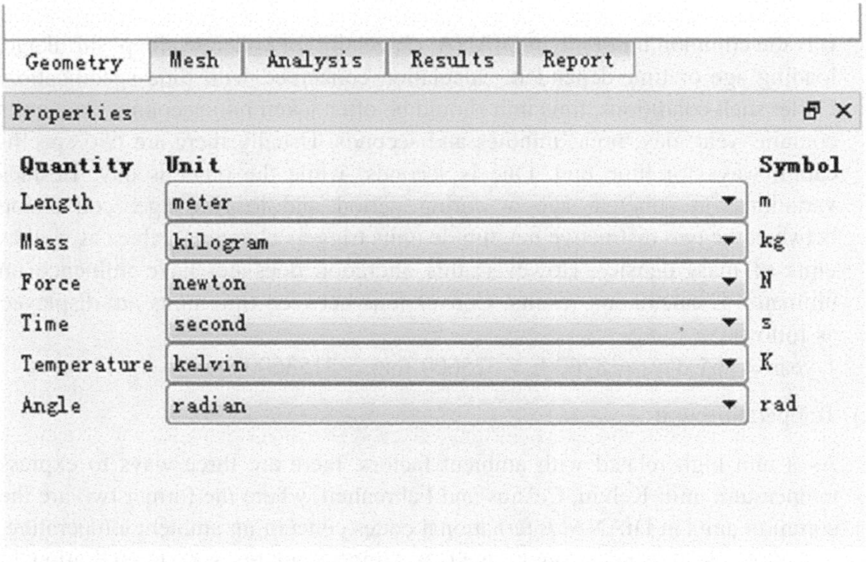

Fig. 1.41 Default standard units

(1) Length unit

Units representing length are of four types: meter (m), decimeter (dm), centimeter (cm) and millimeter (mm), while units in unit system in Europe and America are inch, feet, yard and mile, where m and mm are the common-in-use units.

(2) Force unit

Force units common-in-use in DIANA are N and kN, where combination distinctions in force and length may affect the output forms of elastic modulus, mass density as well as intensity of pressure. Once units are converted, corresponding values are changed with the alteration of units. For example, when length unit is m and force unit is N, unit of elastic modulus is N/m^2. However, as length unit is altered as mm while force unit retains unchanged, unit of elastic modulus is automatically converted into kg/mms^2. When force unit is kN and the length unit is converted into mm, unit of elastic modulus is kNs^2/mm^4. In order to unify units in this academic work as well as for convenience, force and length units in all the following numerical cases are all specified as m and N, respectively.

(3) Mass unit

There are many unit systems in mass unit, including kilogram, gram, ton, ounce, pound, kilo-pound, where kilogram and ton are main units applied in DianaIE. Similar to length and force unit, when mass unit is changed, concrete elastic modulus of constitutive model, values and unit of mass density are altered at the same time, and either of length or force unit is exported as **derived.**

(4) Time unit

It is the common unit type in DIANA, especially for concrete creep, shrinkage, loading age or time-dependent calculation concerned with time specification. Under such conditions, time unit should be often taken into account. Time units contains year, day, hour, minutes and seconds. Usually there are two specification ways for time unit. One is seconds, while the other is day. Besides variations on concrete age at curing period and loading age, conversion between the two distinctive magnitude units triggers changing values as well as units of mass density. However, this alteration does not have influence on ultimate FE calculation results. Conversions between time units are displayed as follows:
1 year = 365 days = 8760 h = 525600 min = 31536000 s.

(5) Temperature unit

As a unit high related with ambient factors, there are three ways to express temperature unit: Kelvin, Celsius and Fahrenheit, where the former two are the common units in DIANA. International codes concerning ambient temperature, heat flow, heat conduction as well as temperature reaction in hydration in DIANA is by default 293.15 K, which is equal to 20 °C.

(6) Angle unit

Angle is also a common unit in civil engineering, geotechnical engineering as well as underground engineering, especially for Coulomb friction constitutive model in DIANA. Main units in angle are radian and angle and the conversion between them is 1 rad = 57.3°

Reference

1. DIANA user's manual-element library, release 9.3. (2008). TNO Building and Construction Research, Holland

Chapter 2
DIANA Material Constitutive Models and International Codes

Abstract One of the main features of DIANA compared with other types of finite-element software lies in its abundant material constitutive model and international design codes. This chapter mainly focuses on illustrating concrete material constitutive model, steel material model and famous international concrete and steel design codes around the world. Besides, dozens of cracking models under various mechanic behavior conditions in DIANA are illustrated in turn to display the powerful functions of material specifications. Meanwhile, based on the numerical experience, modules of long-term performance concerning creep, shrinkage, heat flow and Rayleigh damping are also introduced in this part.

2.1 Introduction of Material Constitutive Models

There are abundant and powerful material constitutive model library in DIANA compared with other kinds of finite-element software, including some kinds of emerging material concrete model such as fiber-reinforced concrete concerning ultra-high performance concrete (abbreviated as UHPC). The priority lies in abundant constitutive model library in that many unnecessary redundant secondary development processes are omitted and the efficiency of modeling in DIANA numerical simulation is increased. Meanwhile, many famous international codes are included in the DIANA, such as European CEB-FIP Model Code 1990, fib Model Code for Concrete Structures 2010, AASHTO LRFD Highway Bridge Design Specifications, American ACI 209R-92, Eurocode 2 EN 1992-1-1 model, Japanese JSCE code and Dutch NEN 6720/A4 model code and their like [1], which are functional to simulate all kinds of conditions in concrete such as linear analysis, nonlinear analysis, large geometric deformation so that the secondary development is often omitted in DIANA.

There are many material constitutive models in DIANA and the main types are listed as follows:

S. Chai, *Finite Element Analysis for Civil Engineering with DIANA Software*,
https://doi.org/10.1007/978-981-15-2945-0_2

(1) Concrete and masonry

This type of material constitutive model is applicable to simulating short-term performance of structures, shrinkage in early age and the whole propagation process of cracking under short-term state, and it has favorable simulation results and cracking distribution. However, the drawback lies in that it lacks considerations on long-term parameters such as creep, concrete element age, and thus it is unable to simulate time-dependent issue as international codes do. Generally speaking, there are many material constitutive types, such as linear elastic isotropic model, linear elastic orthotropic model, **Total Strain-Based Crack Models**, **Multi-Directional Fixed Crack Models**, crack and plasticity model and Rankine principal stress model suitable for multi-axial stress condition. It is worth to mention that the curve type of tension cut-off is constant or linear in the Rankine principal stress model, which can be further classified as Rankine plasticity model, Rankine von Mises plasticity model and Rankine/Drucker-Prager plasticity model based on the difference in the multi-axial stress condition. Besides, there are many other kinds of material model under this type: Mohr-Coulomb and Drucker-Prager model concerning friction angle, dilatancy angle and cohesion coefficient, Maekawa-Fukuura concrete model proposed by Japanese especially for quasi-static and hysteretic analysis. To overcome concrete material anisotropy issue, Rankine Hill anisotropy model can be applied. All the concrete and masonry material constitutive models in DIANA 10.1 are shown in Fig. 2.1.

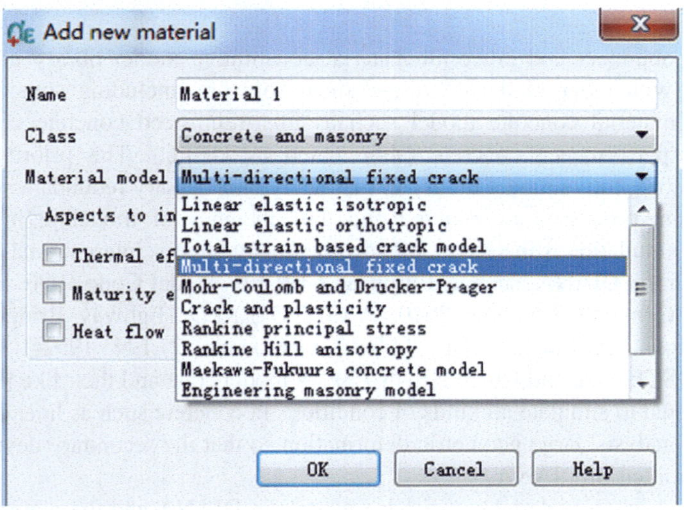

Fig. 2.1 Concrete and masonry material constitutive model in DIANA 10.1

For linear elastic isotropic model, parameters such as elastic module, Poisson's ratio and density are needed to be specified. When the parameters mentioned above are given as input, parameter performance indicators in all directions are the same, as shown in Fig. 2.2.

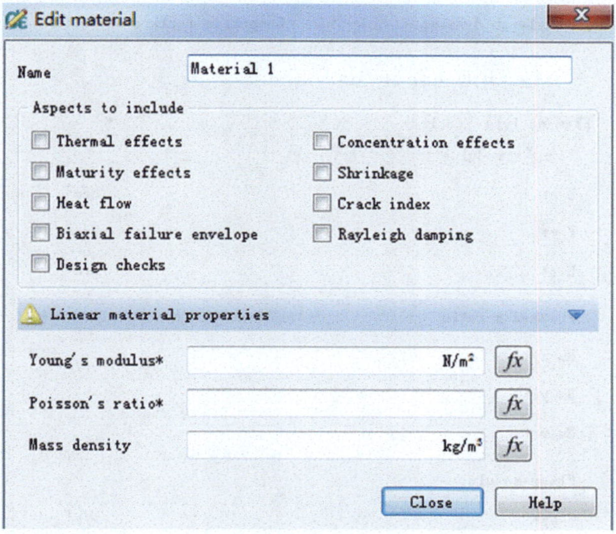

Fig. 2.2 Constitutive parameters of linear elastic isotropic model

The principle of orthotropic model is that the parameter performance indicators such as elastic modulus, Poisson's ratio, mass density and new supplementary shear modulus are different in the two orthotropic directions, thus it is essential to input these values in turn in distinctive directions. Specification of parameters in orthotropic model manipulation interface is shown in Fig. 2.3.

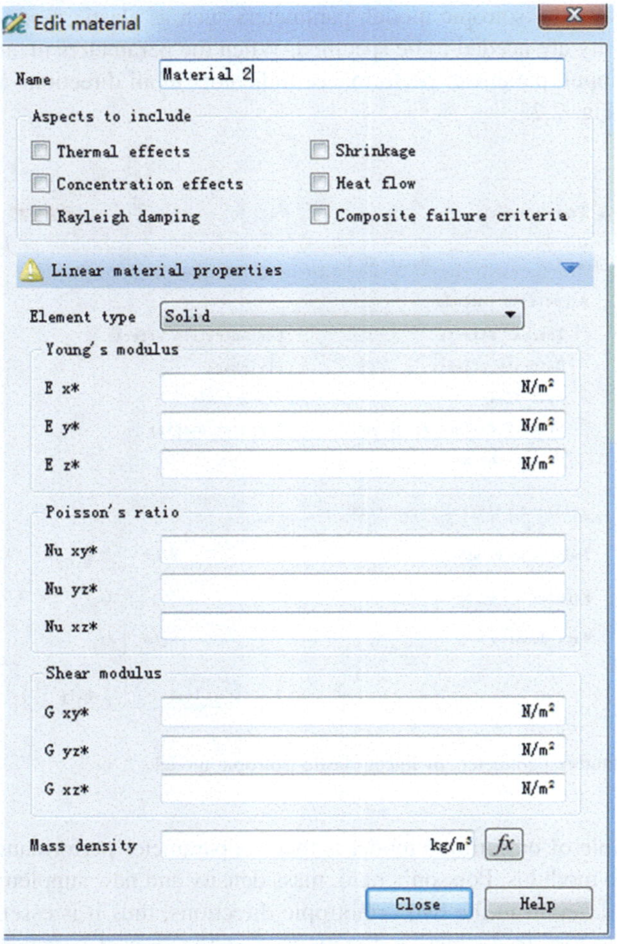

Fig. 2.3 Constitutive parameters of linear elastic orthotropic model (taking solid element for example)

There are two important and frequently applied material constitutive models in concrete and masonry aspect: **Multi-Directional Fixed Crack Models** and **Total Strain-Based Crack Models**. These are mainly applied to simulate structural cracking features under smeared cracking, and further details are introduced in the following section. Generally speaking, compared with the linear elastic model, not only parameters such as elastic modulus, Poisson's ratio and density but also the cracking features such as tensile behavior, shear behavior and compressive behavior are needed as input under **Total Strain-Based Crack Models**, as shown in Fig. 2.4. The cracking features mainly include crack orientation option (Fixed/Rotating/Rotating to fixed) and crack bandwidth specification (User specified/Rots/Govindjee). Tensile behavior is mainly concerned with the specifications of tensile curve, tensile strength, ultimate strain and residual tensile strength after cracking. In the Poisson's ratio reduction option, reduction model is required to ensure that whether **No reduction** or **Damaged based** is selected. Besides, it is worth to note that in nonlinear analysis and nonlinear tensile constitutive model curve such as exponential or Hordijk type, Model I tensile fracture energy (per height) is often required to be specified to simulate the nonlinear cracking features described by these curves (Fig. 2.5).

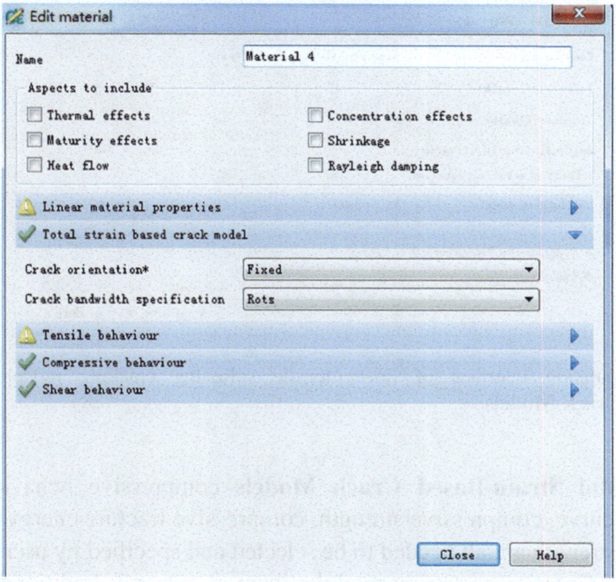

Fig. 2.4 Manipulation interface of **Total Strain-Based Crack Models**

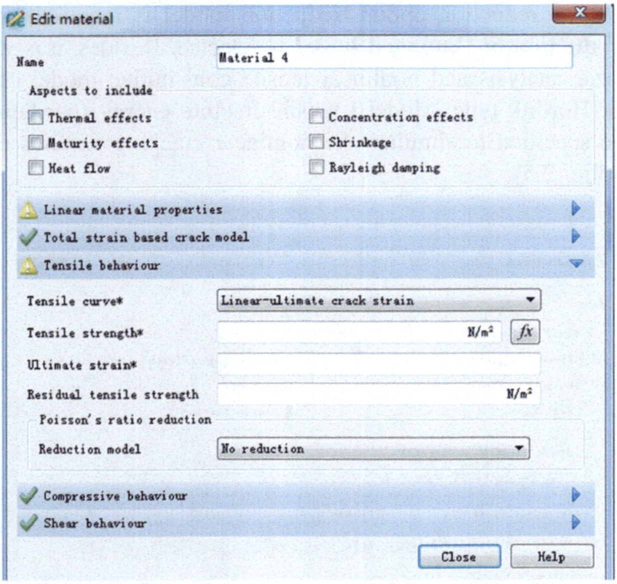

Fig. 2.5 Manipulation interface of linear tensile softening cracking model under **Total Strain-Based Crack Models**

In the **Total Strain-Based Crack Models** compressive behaviors such as compression curve, compressive strength, compressive fracture energy and residual compressive strength are all needed to be selected and specified by users themselves (see Fig. 2.6). Contrary to tensile module, there is not only traditional hypothesis model under residual compressive strength due to lateral cracking, but also newly proposed models according to the research results (Multi-linear/Vecchio and Collins 1986/1993) and international codes (JSCE 2012 Fig. 2.2.5), not only residual compressive strength due to lateral direction cracking, but also stress confinement required to be considered in some conditions. Users are also needed to specify lower bound reduction curve in the Multi-linear and Vecchio and Collins 1993 model. Similarly, No increase/Selby and Vecchio/Multi-linear options are required to be chosen in the stress confinement aspect. Additionally, for the Multi-linear model, factor-strain curve of confinement diagram is required to be specified in the DIANA.

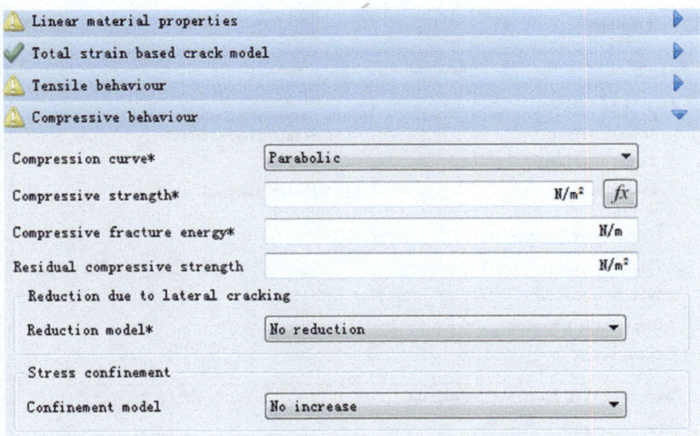

Fig. 2.6 Manipulation interface of nonlinear compressive model in **Total Strain-Based Crack Models**

Compared with the **Total Strain-Based Crack Models**, compressive behavior cannot be simulated in **Multi-Directional Fixed Crack Models**. Thus the major focus is concentrated on the tensile softening behavior. The main parameters used as input are tension cut-off, tensile strength, tension softening type, ultimate strain and their like. Besides, fracture energy and Hordijk factors are also needed except that the selection of crack bandwidth specification is the same in **Total Strain-Based Crack Models** when tension softening curve is nonlinear. Meanwhile, there are two types of selections for shear retention. One is full shear retention without value, while the other is constant shear retention. Under the option of constant shear retention, factor beta β needs to be specified via manual input, where the default value is 0.01 (see Fig. 2.7). It should also be noted that c1 and c2 representing the ratio of stress to ultimate tensile strength are required as input with the default value 3 and 6.93, respectively, under the Hordijk curve model, as displayed in Fig. 2.8. The default parameters of shear retention in DIANA will be introduced later, so it is not repeated here.

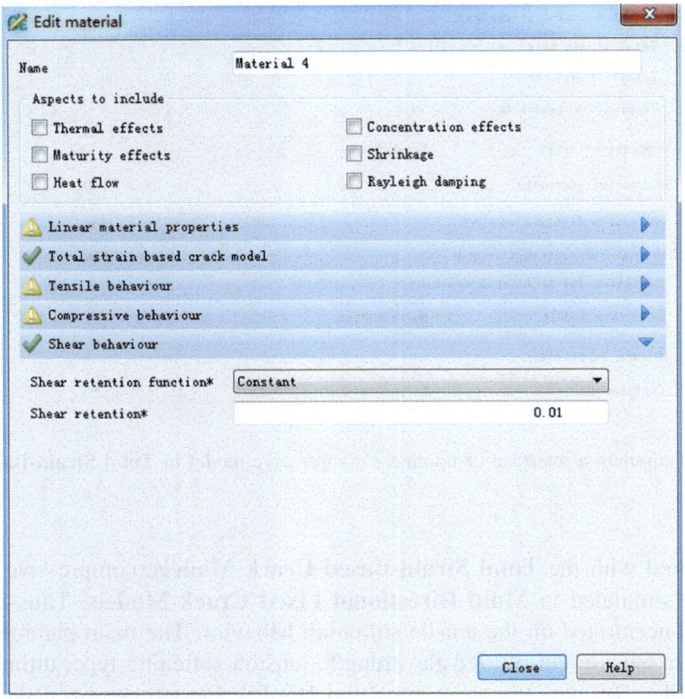

Fig. 2.7 Shear behavior of **Total Strain-Based Crack Models**

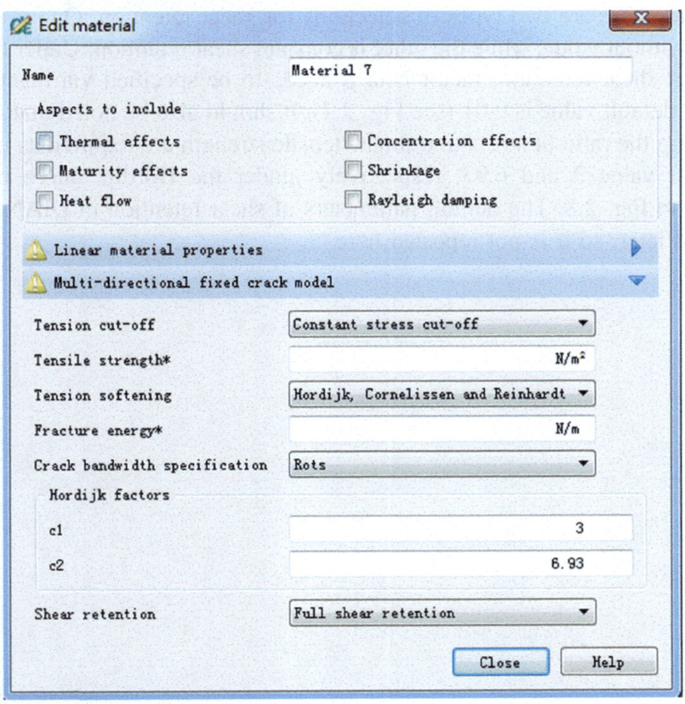

Fig. 2.8 Manipulation interface of Hordijk tension softening model under **Multi-Directional Fixed Crack Models**

(2) Soil and rock

Soil and rock are usually considered as semi-brittle structural material with orthotropic features. Compared with calculations on concrete under linear elastic, elastoplastic and plastic states, plastic analysis and calculation are required on them. In DIANA software, material constitutive models of soil and rock originated from many mature and feasible theoretical research results. Besides, Rayleigh damping analysis, heat flow analysis and heat effect analysis for concrete are also conducted. What is different from the old 9.4 release version is that initial stress model should sometimes be taken into account in this type of material constitutive model.

Constitutive model for soil and rock is of the following four types: Orthotropic elasticity, Duncan-Chang hyperbolic model, Geotexile-simple stress models and user-supplied, where the common and frequently applied type is Orthotropic elasticity, as shown in Fig. 2.9. Similar to the specifications of Orthotropic elasticity in concrete and masonry model, the fundamental parameter values such as elastic modulus, Poisson's ratio and shear stiffness vary in orthogonal directions, thus these values are needed to be specified in every direction (see Fig. 2.10). Additionally, definition of initial stress is also the consideration for material constitutive of orthotropic elasticity, where lateral pressure ratio, effective stress isotropic or effective stress orthotropic or the selection for total stress isotropic/ orthotropic is required.

Whichever stress option is selected, stress coefficient KO should be found for the orthotropic material, as displayed in Fig. 2.11. However, for orthotropic material, maximum and minimum values of lateral pressure ratio, KTOmax and KTOmin, respectively, and direction of maximum pressure ratio should be specified.

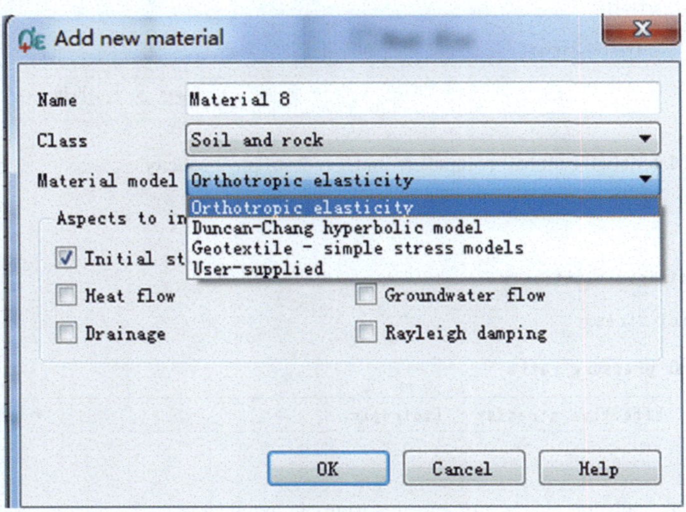

Fig. 2.9 Soil and rock material constitutive model

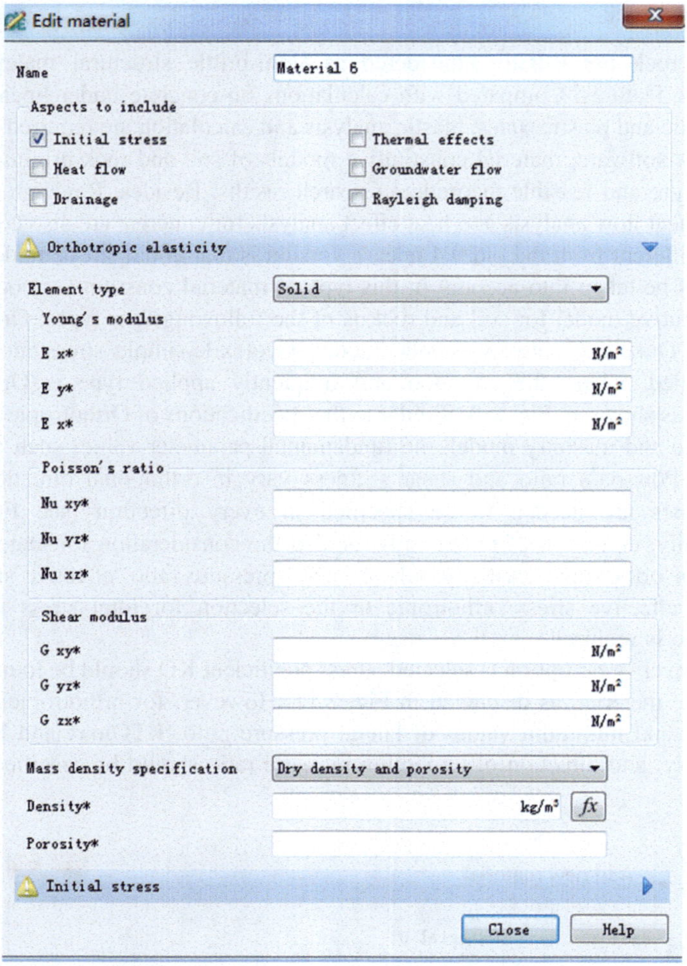

Fig. 2.10 Material constitutive parameters under orthotropic elasticity

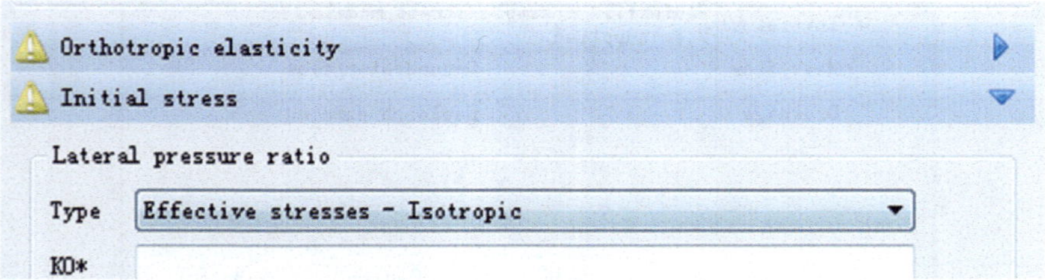

Fig. 2.11 Manipulation interface of effective isotropic initial stress

(3) Composite and rubber

For composite and rubber material type, not only linear elastic isotropic and orthotropic material but also anisotropic constitutive model can be applied. For anisotropic constitutive model, not only linear elastic property but also normal and shear stress values are needed (see Fig. 2.12).

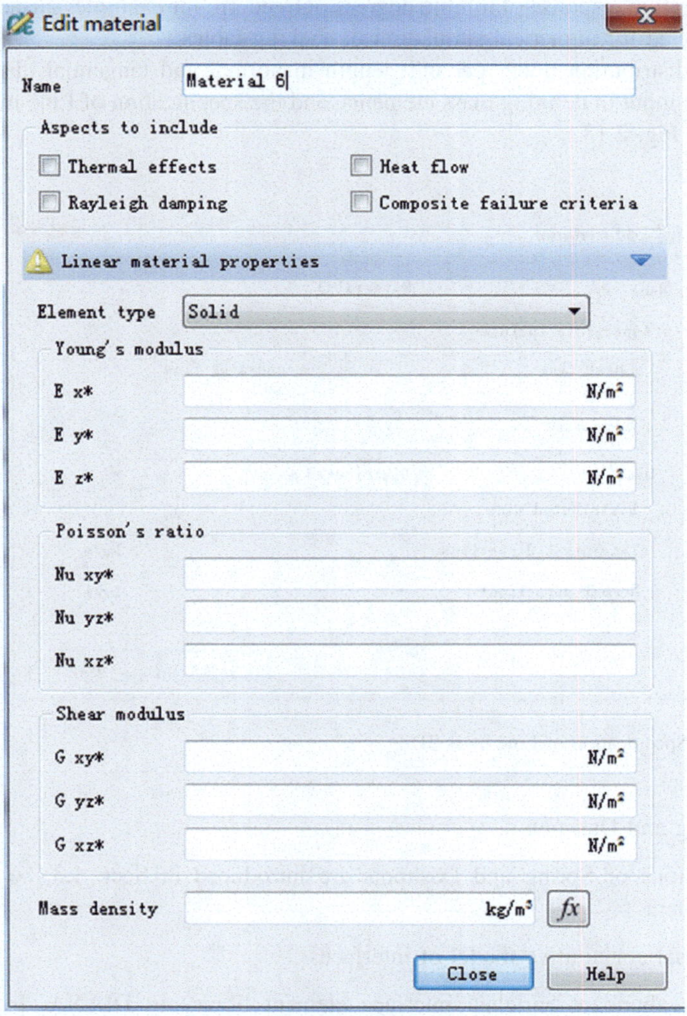

Fig. 2.12 Specification of composite and rubber material parameters

(4) Mass element

Mass element is feasible for simulating response issues in structural dynamics where there are multiple various constitutive types: Point mass, Line mass 2D, Line mass 3D and Surface mass. Point mass is applied to structural response issues concerning single and multiple degrees of freedom, while Line mass 2D and 3D are proper to simulate member response issues concerning infinite degrees in two dimensions and three dimensions, respectively. Surface mass is used for simulating plane vibration response of infinite degrees of freedom; for example, surface mass is often applied for numerical simulation of shaking tables.

Mass distribution under per unit length in normal and tangential directions is needed as input in defining mass elements, and the specification of Line mass 3D is shown in Fig. 2.13.

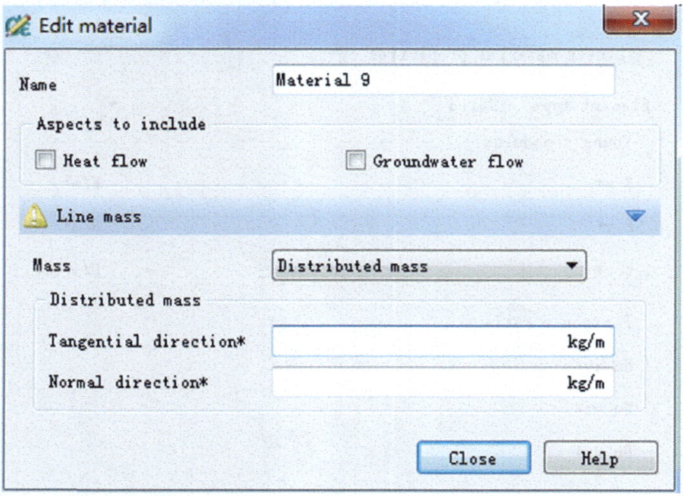

Fig. 2.13 Specification of Line mass 3D

(5) Spring and Dashpots

Specifications of Spring and Dashpots are introduced in Sect. 1.3, so it is not repeated here.

(6) Material constitutive model of interface

There are abundant enriched interface element library in DIANA. In DIANA interface element library, there are not only linear elasticity and nonlinear elasticity but also discrete cracking model specifically applied to simulate cracking behavior at known fixed positions, which will be introduced in detail in the following chapter. It is common sense that failure of structures are concerned with dilatancy angle or friction factor, thus corresponding interface elements are required to be specified. In view of such issues, two interface element models relative with friction constitutive-Coulomb friction and nonlinear elastic friction are added since the release version of DIANA 10.1. Besides, considering that prestress tendon cannot be embedded into the mother concrete elements automatically when truss or beam elements are applied to simulate single strand and there may be bond-slip for structural strengthening or issue of grouting compactness occurring in the prestress structures, a new type of bond-slip material constitutive model is proposed in DIANA software, which is also a prominent edge compared with other types of software and the specifying interface is shown in Fig. 2.14. Truss and beam elements are taken as bond-slip interface elements to specify the bond-slip interface between the steel and concrete; then the discrete material model is inverted into bond-slip state via integral conversion options under the mother option of INTERF belonging to the DATA aspect. It is worth to mention that this type of interface element material constitutive model is also often applied for bonded steel strengthening method via specifying bond-slip shear traction–displacement relationship, and the numerical case will be further introduced in the following chapter.

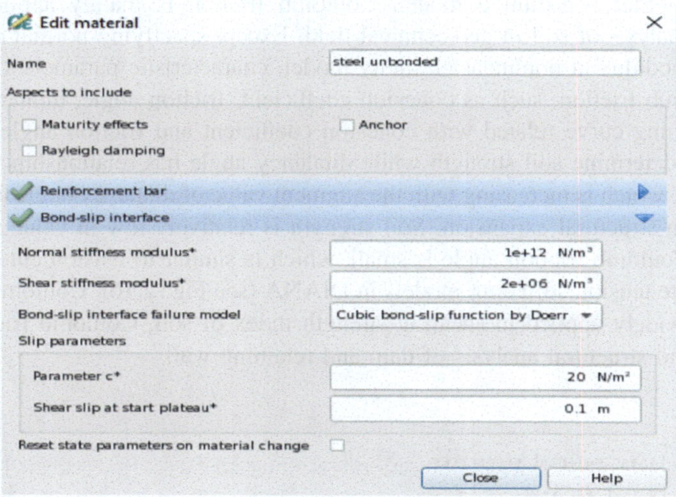

Fig. 2.14 Specification of bond-slip material constitutive model

Now, two common typical material constitutive models of interface elements are introduced: Nonlinear elasticity and Coulomb friction. In specifying linear material properties of nonlinear elasticity, **Type** option is selected. Meanwhile, according to the variety in dimension, normal stiffness modulus-z and shear stiffness modulus-y are specified respectively. Additionally, for the constitutive model of 3D structural interface elements, shear stiffness modulus-y out of the plane is also required. Manipulation interfaces for specifying material stiffness of nonlinear elasticity are shown in Fig. 2.15.

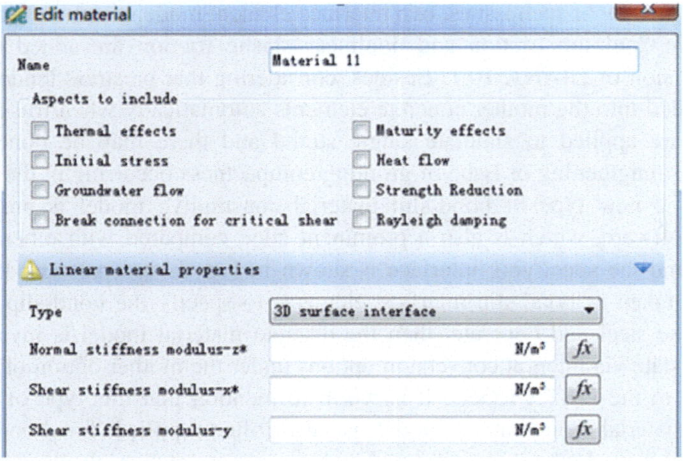

Fig. 2.15 Manipulation interfaces for specifying material stiffness of nonlinear elasticity

The material constitutive model, Coulomb friction is mainly applicable for strength analysis of soil in geotechnical field. Except specifying normal and shear stiffness modulus in nonlinear elasticity model, characteristic parameters concerning Coulomb friction, such as cohesion coefficient, friction angle, dilatancy angle and hardening curve related with cohesion coefficient and friction angle, are key factors to determine soil strength while dilatancy angle has relationship with volume strain, which is increasing with the augment value of dilatancy angle ultimately resulting in structural expansion. Soil strength is relatively low in value when the value of Coulomb friction angle is small, which is similar to tension cut-off under the concrete tension softening models in DIANA (see Fig. 2.16). Coulomb friction model is widely applied in shearing strength index of soil, Coulomb friction calculation and structural analysis of dam and retaining wall.

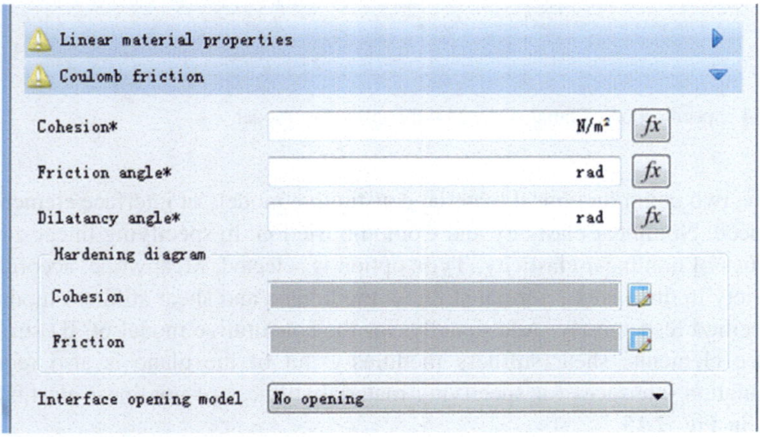

Fig. 2.16 Specification of Coulomb friction model

(7) Steel

There are many ways of simulating steel material with various constitutive aspects. For steel, its material properties can not only be assigned via pure steel material model but also through international famous steel design codes in alternative, which will be further introduced in detail in the following chapter. Material constitutive model of steel in DIANA mainly contains **Linear elastic isotropic** model, **Linear elastic orthotropic** model, **Von Mises and Tresca plasticity** model, **uniaxial nonlinear elasticity** model, **Modified two-surface model**, **Boundary elements** model and **Direct stiffness matrix for flat shells** model, where **Linear elastic isotropic** and **Von Mises and Tresca plasticity** models are the two commonly applied constitutive ones applicable for simulating issues of reinforced concrete and steel under linear state and fatigue state after cracking, respectively. In the linear elastic model, only linear elastic material features such as elastic modulus, Poisson's ratio and mass density are required to be specified, while such parameters mentioned above in various directions are needed to be input under linear elastic orthotropic model. In the **Von Mises and Tresca plasticity** model, not only basic parameters are required to be input but also the type of plastic model, hardening type and yield stress are needed to be specified (Fig. 2.17).

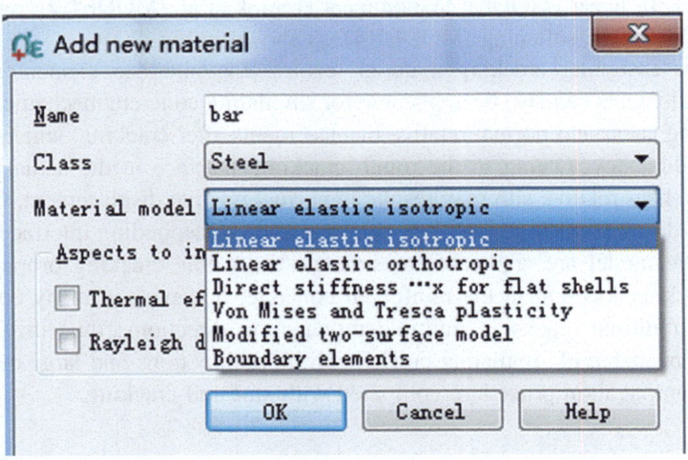

Fig. 2.17 Steel material constitutive models

2.2 Concrete Cracking Model in DIANA

Cracking models in DIANA are mainly classified as two distinctive types: discrete cracking and smeared cracking, where the former is the mainstream in simulating cracks in the reinforced concrete structures. In DIANA software, smeared cracks are simulated via user specifying material constitutive features such as elastic modulus, Poisson's ratio, tension cut-off, cracking mechanic behaviors (tensile, compressive and shear) to achieve ultimate cracking effects while discrete cracks are simulated through geometric modeling and specific interface elements. Besides, **Tension softening** relationships and **Shear retention** are also required to be selected and specified in DIANA **Multi-Directional Fixed Crack Models**. The essence of discrete cracking is a series combination of separate numerical structural components and interface elements and cracking behaviors can be simulated via relative displacement relationship between elements. That is to say, two segments at the location of cracks are regarded as two independent parts and modeled separately when discrete cracking model is applied in DIANA. Then line to line connected or surface to surface connected interface elements are added and material constitutive parameters of these interface elements are specified according to the mechanic behaviors and properties of cracks, thus the process of discrete cracking numerical simulation is completed. In the numerical model of discrete cracking, there is a principal crack and it is recognized that cracks exist in concrete when the normal stress of interface elements reaches tensile strength, thus relative displacement has appeared between main elements (see Fig. 2.18). Discrete cracking model is suitable for checking actual distribution figures after structural nonlinear calculation and conditions of local stress. After discrete cracking is completed, there are relative displacement and relative slip angle between concrete elements. Five tension softening types, also named as Mode-I tension softening, consist of discrete cracking tension softening model: Brittle (MODEL1 0), linear (MODE1 1), nonlinear Hordijk et al. (MODE1 2), multi-linear (MODE1 3), JSCE softening (MODE1 4), as shown in Fig. 2.19.

Besides assigning cracking material features via interface elements, discrete interface elements can also be applicable for simulating concrete mechanic behavior of coupling shear and normal relative displacements after cracking, which is called as crack dilatancy. Owing to the rough cracking interface in the actual cracking structure, shear relative slip may results in normal relative displacement. Generally speaking, discrete cracking interface elements and corresponding interface material constitutive model are applicable to simulate the whole cracking propagation at given cracking location in the reinforced concrete; however, in many conditions, there are multiple cracks in multiple propagation directions; thus this cracking model is incapable of simulating cracks with more directions and large quantity in cracking propagation procedure compared with smeared cracking.

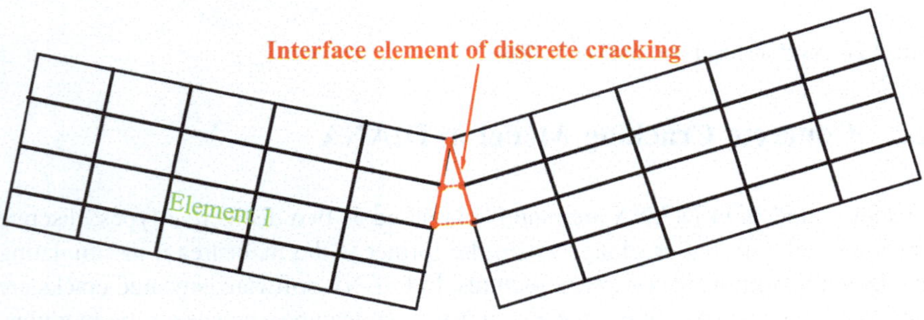

Fig. 2.18 Simplified diagram of discrete cracking model

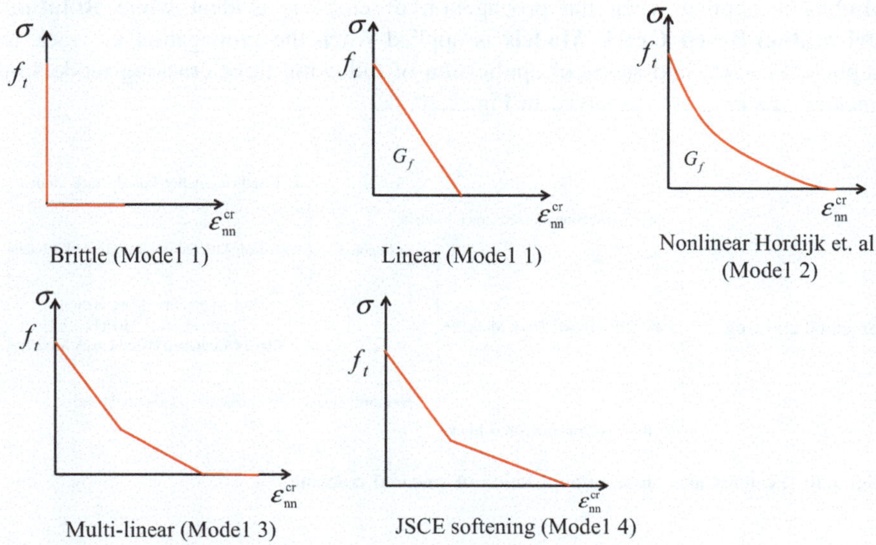

Fig. 2.19 Tension softening types of discrete cracking model

Cracking objects are taken as continuum of material anisotropy such as concrete in the smeared cracking modeling. Mechanic behaviors of crack and cracking mechanism are realized via the reduction of elastic modulus, compressive and tensile strength as well as value of strain, which is only applicable for observing structural macroscopic features after finite element calculation, such as total displacement, displacement in the middle site of the span or corresponding load–displacement curve and cracking behaviors are defined by stress–strain relationship. Unlike discrete cracking model, cracking feature of smeared cracking is that cracks are mainly distributed in the integral point locations in elements and the size of crack is small without principal crack compared with the discrete cracking. Another smeared cracking characteristic is that cracking propagation is along the direction of principal stress, presenting orthogonal relations and there are at most three cracks in every element [2]. Owing to the fracture, energy dissipated during smeared cracking may be affected by the size of element, which is also called as grid sensitivity, thus a new conceptual parameter is introduced into the DIANA-crack bandwidth to resolve the problem. Smeared cracking model is mainly classified as three types: **Multi-Directional Fixed Crack Models, Total Strain-Based Crack Models** and Rankine principal stress model, where the **Total Strain-Based Crack Models** is the major model for reinforced concrete cracking issue in DIANA. Furthermore, it can also be classified as **Fixed Orientation-Base Crack Models**, **Rotating Orientation-Based Crack Models** and **Rotating to Fixed Orientation-Based Crack Models** (also called as **Mixed Orientation-Based Crack Models**), where **Fixed Orientation-Based Crack**

Models is applied when the propagation of crack is evident while **Rotating Orientation-Based Crack Models** is applied when the propagation of crack is implicit. Features and scope of application of following three cracking models in smeared cracking are displayed in Fig. 2.20.

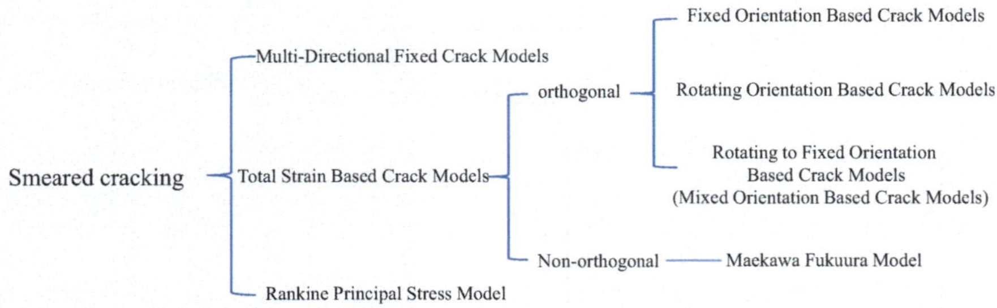

Fig. 2.20 Features and scope of application of smeared cracking

(3) Multi-Directional Fixed Crack Models

Multi-Directional Fixed Crack refers to the concept that there are multiple cracks in various directions in a unit simultaneously. The core of this model is to divide the total strain into two distinctive parts: elastic strain ε^{e} and cracking strain ε^{cr}, which means that the total strain is the sum of elastic strain and cracking strain [1]. In DIANA software, the characteristics of **Multi-Directional Fixed Crack Models** after cracking are determined via tension softening relationships and loading secant stiffness of loading and unloading curves; thus there are two features: tensile behavior and shear behavior. The former is for simulating concrete mechanic behaviors under tensile cracking and post cracking conditions. Owing to the curve of concrete compression section is not contained in this model, so the concrete compressive characteristic is omitted in this model, hence this model is suitable for cracking numerical circumstances dominated by tensile mechanic behaviors.

As the curves of tension softening under smeared cracking model are in vast amount, for the **Multi-Directional Fixed Crack Models**, there are eight types of tension softening models: Brittle (TENSIO 0), ultimate strain-based linear tension softening model (TENSIO 1), multi-linear tension softening model (TENSIO 2), nonlinear Moelands et al. (TENSIO 3), fracture energy-based linear tension softening model (TENSIO 4), nonlinear Hordijk et al. (TENSIO 5), JSCE softening (TENSIO 6) and JSCE stiffening (TENSIO 7) (see Fig. 2.21) [1].

The most widely applied tension softening curve among the eight types is the TENSIO 5, that is nonlinear Hordijk curve, where the descending segment of the curve ranging from ultimate tensile stress to the zero after cracking is taken into account. Meanwhile, crack bandwidth and fracture energy in the crack propagation process are also taken into consideration, thus the variation stress under cracking state can be simulated effectively.

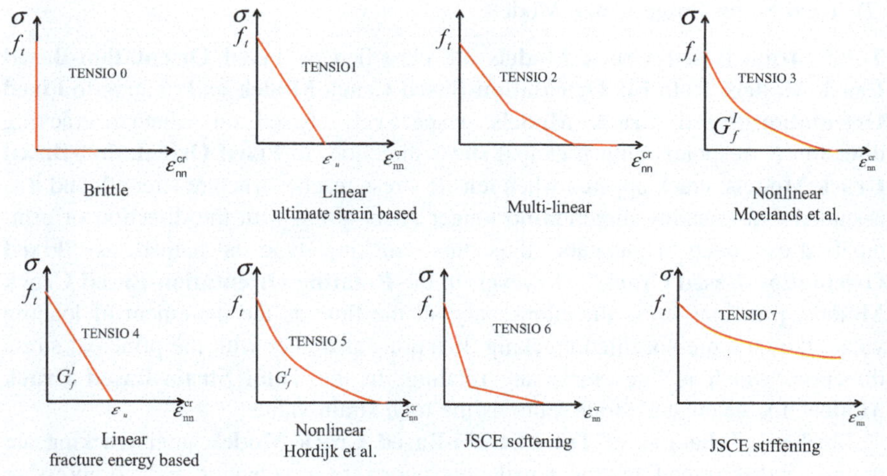

Fig. 2.21 Tension softening curves of **Multi-Directional Fixed Crack Models**

Another feature of **Multi-Directional Fixed Crack Models** is the shear retention, or illustrated as shear transferring characteristics, which only oriented for shear stiffness. When cracking in concrete occurs, material shear stiffness value at the location of cracks reduces compared with the initial ones but does not descend to zero directly. Then except specifications of tensile curve, tensile strength and fracture energy under tensile behavior module in DIANA, Poisson's ratio reduction and reduction model specifications are also required. Generally speaking, the option of **Full Shear Retention** or **Constant Shear Retention** is expressed by the symbol TAUCRI and its following numbers. For example, TAUCRI0 represents **Full Shear Retention Model** without specifying a factor β owing to its default value 1, while TAUCRI1 is **Constant Shear Retention Model** required to specify the factor β. In DIANA software, when constant shear retention model is selected, default value is 0.01, where the shear stiffness in the cracking process can be expressed as formula:

$$D = \frac{\beta}{1-\beta} G$$

where D represents the stiffness after cracking, β is the shear retention factor and G is the initial shear stiffness. The shear retention feature in **Multi-Directional Fixed Crack Models** is observed from the formula and that element material stiffness value at the site of cracks in shear direction is reducing.

Although **Multi-Directional Fixed Crack Models** can effectively simulate concrete tensile cracking in the nonlinear calculation, it is incapable of simulating nonlinear compressive calculation. So the most commonly applied and flexible cracking propagation models under short-term in reinforced concrete is **Total Strain-Based Crack Models**, which will be chiefly introduced in the following part.

(2) Total Strain-Based Crack Models

Total Strain-Based Crack Models are classified as **Fixed Orientation-Based Crack Models**, **Rotating Orientation-Based Crack Models** and rotating to **Fixed Orientation-Based Crack Models**, respectively, based on whether cracking direction corresponds with principal stress direction. In **Fixed Orientation-Based Crack Models**, crack appears when tensile stress reaches fracture strength and it is assumed that cracking direction no longer alters along with the direction of principal stress once it appears, thus this cracking type is named as "**Fixed Orientation-Based Crack**". However, in the **Rotating Orientation-Based Crack Models**, principal stress directions vary all the time as the increment of loading value, thus the pre-specified cracking directions also alter with the principal stress direction, which is like cracks are rotating. In the **Total Strain-Based Crack Models**, the calculated strain value is the total strain value.

Mechanic behaviors of **Total Strain-Based Crack Models** after cracking are mainly codetermined by the tensile behavior, shear behavior and compressive behavior. Compared with the **Multi-Directional Fixed Crack Models**, the phase from initial tension stress to ultimate tension stress is added into the tension softening curves. Moreover, there are more tension softening curves applicable for more widely emerging structural fields in **Total Strain-Based Crack Models**. For example, fiber-reinforced material constitutive model can be applied to simulate fiber-reinforced concrete and ultra-high performance concrete. Tension softening model contains 16 types of curve, including brittle, elastic, ideal model, ultimate strain-based linear model, fracture energy-based linear model, multi-linear total strain-based model, nonlinear Hordijk et al. exponential, CEB-FIP Model Code 1990, fib Model Code for concrete structures 2010, JSCE softening model, JSCE stiffening model, fiber-reinforced total strain-based model, fiber-reinforced total crack opening-based model and Cervenka tension softening model (see Fig. 2.22). Generally speaking, brittle and ultimate strain-based linear model is not suitable for nonlinear calculation.

Nonlinear Hordijk et al. model and exponential model are the commonly applied tension softening models, where not only the descending phase before reaching ultimate tensile stress is nonlinear curve better applicable for nonlinear calculation but also structural fracture and bandwidth are taken into consideration in the material constitutive curve (where the area at the cracking site enclosed by horizontal and vertical coordinate axes represents the fracture energy per width unit). In addition, international famous codes such as CEB-FIP Model Code 1990, fib Model Code for concrete structures 2010, JSCE softening model, and corresponding values in these codes are automatically embedded into the DIANA constitutive curves according to the specifications of these codes.

Fiber-reinforced total crack opening-based model is also one of the recently commonly applied tension softening model for simulating fiber-reinforced or ultra-high performance concrete. When material constitutive parameters of CMOD model are specified, besides tensile strength fL shown in Fig. 2.23a, it is also necessary to input coordinates of the residual tensile stress values at two key points representing the mechanical characteristic curve of fiber-reinforced concrete-residual strength fL, coordinates of crack mouth opening values corresponding to the stress state (ε_i, fR_i) and (ε_j, fR_j) as well as ultimate crack mouth opening value ε_u in the interface frame as shown in Fig. 2.23b.

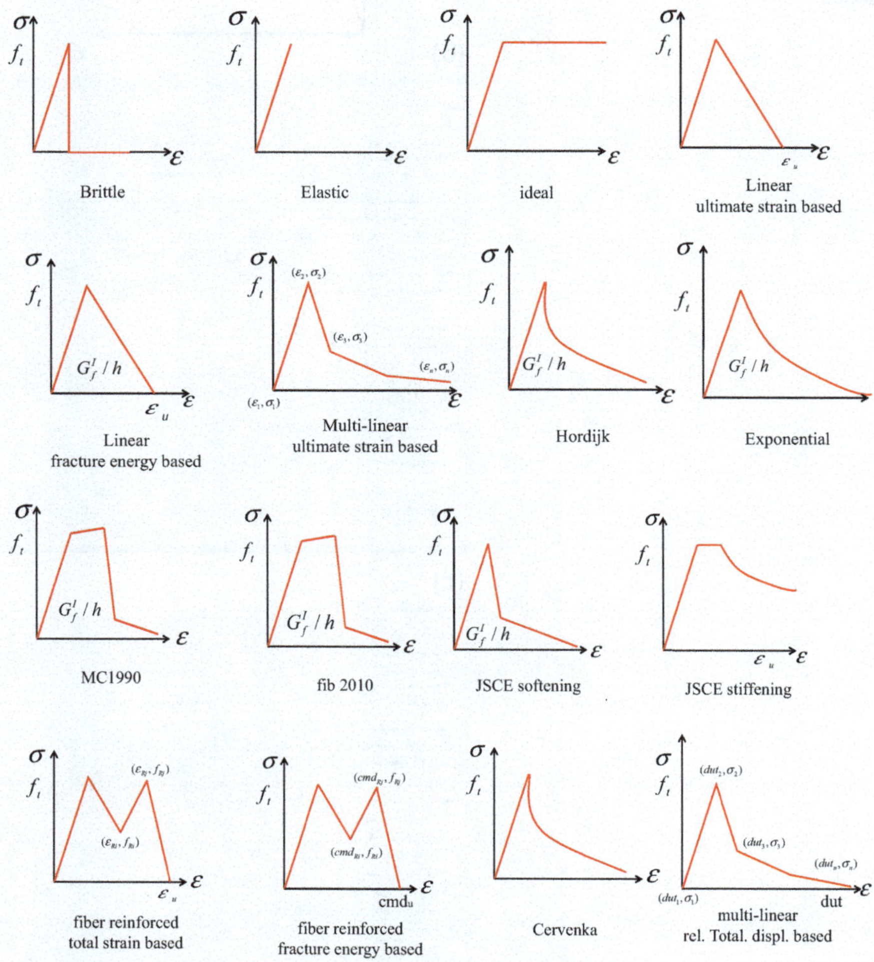

Fig. 2.22 Tension softening curves for **Total Strain-Based Crack Models**

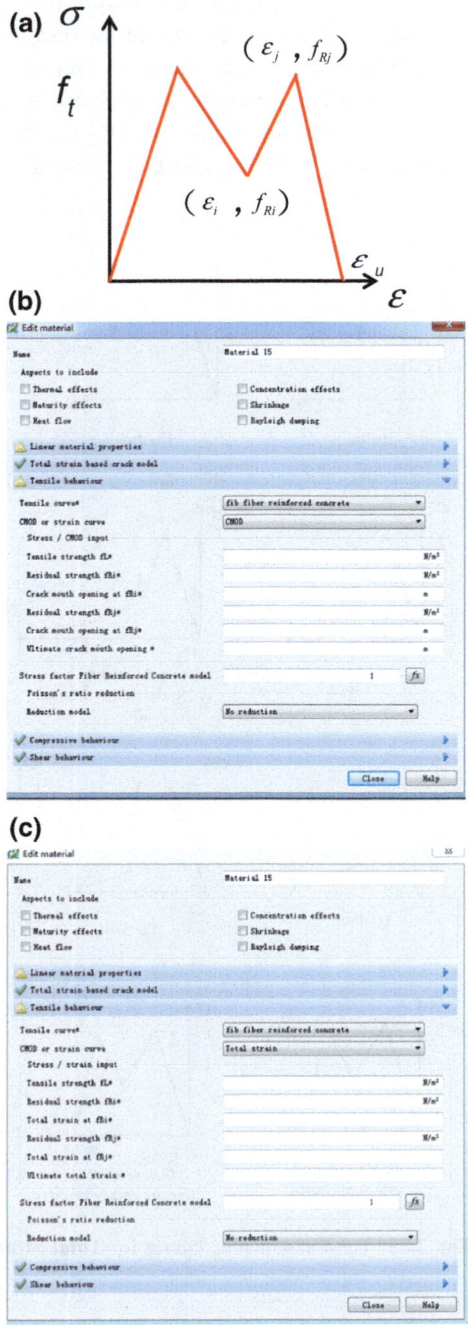

Fig. 2.23 **a** Fiber-reinforced total crack opening-based model (CMOD). **b** Specification interface of fiber-reinforced total crack opening-based model (CMOD). **c** Specification interface of fiber-reinforced **Total Strain-Based Crack Models**

When the fiber-reinforced total strain-based model is selected as tension softening model, except tensile strength as input shown in Fig. 2.23a, the residual tensile stress values at two key points representing the mechanical characteristic curve of fiber-reinforced concrete-residual strength fL, total strain values corresponding to the stress state (ε_i, fR_i) and (ε_j, fR_j) as well as ultimate total strain values ε_u in the interface frame, as shown in Fig. 2.23c, are also required as input.

Compressive behaviors of total strain-based model can simulate the whole state of concrete from compression to crushing, especially when there are abundant nonlinear compression models catering to nonlinear cracking calculation of large structures under the short-term loading. Besides classical elastic, ideal double-linear models and compressive stress–strain curve based on the Hognestad model, international codes such as CEB-FIP1990 and fib 2010 model are also added. In addition, parabolic model is also added in the **Total Strain-Based Crack Models** (Fig. 2.24). In DIANA software, when the international code is determined by the users (such as CEB-FIP1990 or EN1992), compressive behaviors are automatically specified by the DIANA without manual input while every compressive characteristic parameters are needed as manual input to realize simulation of stress–strain relationship when ordinary constitutive model is selected. Not only is the compressive constitutive law in the rising, as a process of reaching concrete ultimate compressive strength (peak stress) in simulation, but also the codes such as EN1992 and CEB-FIP1990 can continue simulating the complete stress–strain relationship for failure stage from concrete peak stress to ultimate compressive strain, where crack continues expansion and penetration, and the deterioration of structures is becoming increasingly serious. Meanwhile, inflection point and convergence point appear in the curve, thus the compressive strength of concrete is in rapid reduction.

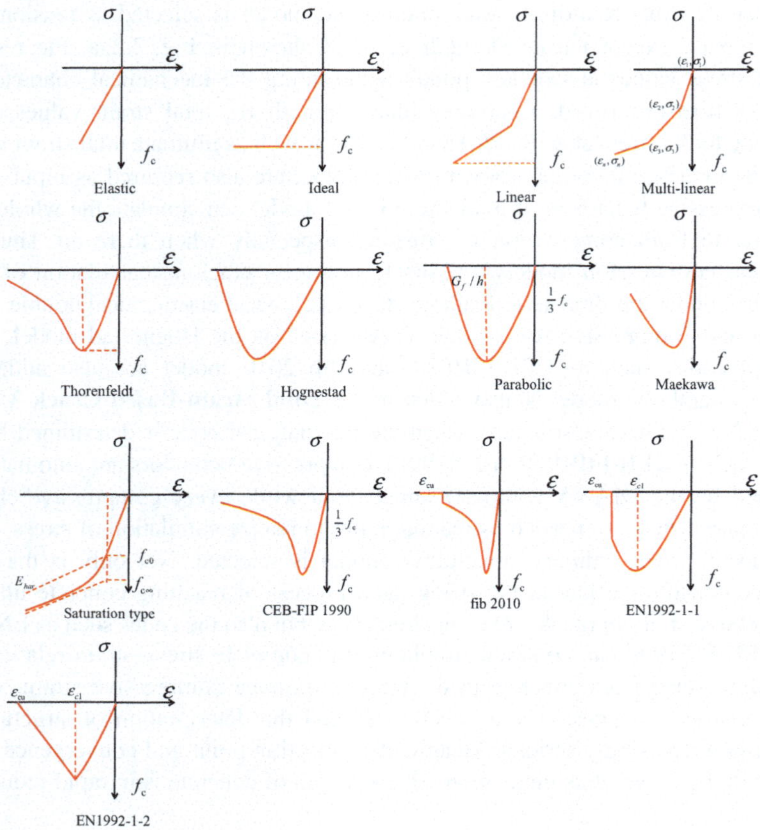

Fig. 2.24 Compressive behaviors of **Total Strain-Based Crack Models** [1]

Three key types of compressive behaviors in **Total Strain-Based Crack Models** are introduced—saturation type, EN1992 and fib2010. In the saturation type, there is a linear elastic phase and a nonlinear phase, where when the compressive strength reaches the maximum compressive strength corresponding to the proportional limit point of concrete linear elasticity, it enters the nonlinear phase. The slope of the curve in the nonlinear phase is the concrete hardening elastic modulus. Unloading of **Total Strain Crack-Based Model** is in the way of secant line, thus the dissipated energy value is relatively small on evaluation, and the results are usually insufficient to obtain accuracy.

EN1992 contains two types. One is the EN 1992-1-1, where the curves experience nonlinear rising and descending phase, respectively, coupled with a horizontal straight line limit to the ε_{cu}, and the other is EN 1992-1-2 where the constitutive relationship has only a nonlinear rising curve and a simplified descending downward slope line. Compared with other compressive stress–strain relationships, there is not only nonlinear rising section of stable crack propagation in EN1992 but also the descending phase where the continuous propagation and penetration of cracks result in structural damage and the compressive strength decreases to zero. It is worth to notice the contrast between EN 1992-1-1 and EN 1992-1-2. It is considered that the compressive stress descends linearly to zero in a constant rate while the EN 1992-1-1 insists that initial ultimate compressive strength descends in nonlinear type with an inflection point at certain value in different descending rate, then followed by a brittle sharp reduction above with a horizontal straight line limit to the ε_{cu}.

Based on the codes of CEB-FIP 1990, ultimate compressive strain ε_{cu} is added to the fib2010, and concrete compressive strength increases until the peak stress at first after cracking is achieved, then the strength in the curve takes on the nonlinear descending tendency. When the concrete strength descends to the point of crossing inflection and convergence and the value is small enough at the same time, it suddenly decreases to zero with the corresponding horizontal coordinate values ε_{cu}. After that, horizontal straight line on the coordinate axis means that the concrete always keeps the failure state of cohesion exhaustion, zero stress and rapidly sharp expansion of strain.

It is also essential to consider shear retention when **Fixed Orientation-Based Crack Models** is applied while it is not taken into account the **Rotating Orientation-Based Crack Models** as the cracking direction is always perpendicular to the direction of principal stress. Compared with the shear retention functions in **Multi-Directional Fixed Crack Models** with only full and constant shear retention options, there are abundant shear retention function models in **Total Strain Crack-Based Model**. Generally speaking, the shear retention function types in **Total Strain-Based Crack Models** are listed as follows [1]:

1. Constant shear retention
2. Variable shear retention
3. Damaged-based shear retention
4. Aggregate size-based shear retention
5. Normal crack strain-based shear retention
6. Maekawa shear retention curves
7. Al-Mahaidi shear retention function

Shear retention types mentioned above are applicable for all kinds of concrete cracking issues. For example, when the shear stiffness is under circumstance of damaged concrete plasticity, the damaged-based shear retention can be selected. Maekawa shear retention curves can be used to study the degeneration of shear stiffness under hysteretic analysis such as repeated loading and unloading action.

(3) Rankine principal stress model

Besides shear retention models mentioned above, there is another model called Rankine principal stress model, merely catering for simulating biaxial concrete compression model in two dimensions, where pressure in one direction has impact on the stress of the other direction. However, this type of model only suits for 2D structures and has limit on the crack propagation simulation under ambient atmospheric environment [2]. That is to say, the process of crack propagation is incapable of simulating structural deterioration via Rankine principal stress model under time-dependent factors such as creep and shrinkage and eroded by saline environment.

In addition, there is another special non-orthogonal crack to simulate elastic-plastic fracture model for hysteretic analysis of low cyclic loads under loading and unloading action—Maekawa-Fukuura model. It is worth to mention that this model is based on the elastoplastic damage model of concrete before cracking, while it attributes to **Total Strain-Based Crack Models** after cracking. Therefore, in this academic work, it belongs to one type of **Total Strain-Based Crack Models** based on post cracking features. In fact, Maekawa-Fukuura cracking model is non-orthogonal crack [2]. Unlike **Total Strain-Based Crack Models**, at most six non-orthogonal cracks are allowed in the Maekawa-Fukuura Model. Numerical case concerning this type of compressive model will be further illustrated in the following chapter.

All types of cracking models in smeared cracking are listed in Table 2.1.

Table 2.1 Categories of cracking model in DIANA

Smeared cracking		Applicable features
Multi-Directional Fixed Crack Models		1. Principal tensile stress exceeds tensile strength
		2. The direction of existing cracks and the current principal tensile stress are required to exceed the critical angle
Total Strain-Based Crack Models	**Fixed Orientation-Based Crack Models**	1. Crack does not alter with the changing direction of the principal stress crack, and the fixed direction is always kept unchanged once cracks appear
		2. Unable to be in combination with creep function in time-dependent analysis
	Rotating Orientation-Based Crack Models	1. Locations of cracks alter in accordance with relative rotation of cracking principal stress direction once cracks appear
		2. Unable to be in combination with creep function in time-dependent analysis
	Rotating to Fixed Orientation-Based Crack Models	1. **Rotating Orientation-Based Crack Models** is applied when the tendency of crack is not clear while it converts to **Fixed Orientation-Based Crack Models** when the locations and directions of cracks are evident
		2. Unable to be in combination with creep function in time-dependent analysis
	Maekawa-Fukuura Model	1. Elastoplastic damage model of concrete before cracking and **Total Strain-Based Crack Models** after cracking
		2. At most six non-orthogonal cracks are allowed in single integration point
		3. Applicable for hysteresis analysis under cyclic load and energy dissipation reflected precisely
Rankine principal stress model		1. Bi-axial concrete compression model,
		2. Pressure in one direction having impact on the stress of the other direction is taken into consideration
		3. Only applicable for 2D modeling
		4. Unable to be in combination with creep function in time-dependent analysis, incapable of simulating structural deterioration under time-dependent factors such as creep and shrinkage and eroded by saline environment [3]

Note: Whatever crack model mentioned above is selected, they can only be applied to the nonlinear analysis of concrete in short-term performance. It cannot be applied to the time-dependent or long-term analysis with shrinkage, creep and relaxation and the durability analysis related to environmental factors

2.3 Material Constitutive Model of Reinforcement

In DIANA software, reinforcement is a generalized concept where it contains both steel bars, steel grids and other new material with reinforcement function. There are two ways of modeling reinforcement: one is discrete reinforcement while the other is embedded reinforcement. Material constitutive specifications of embedded steel and corresponding reinforcement and pile foundations are highlighted in this part. As mentioned before, there are many steel models in DIANA software, mainly classified as reinforcement and pile foundations, steel material constitutive model and steel design codes, where every model can be further classified as linear elastic isotropic, linear elastic orthotropic, von Mises plasticity or von Mises and Tresca plasticity, and their like, which is displayed in Fig. 2.25. The aspect of steel design codes can be further classified as Eurocode 3 EN 1993-1-1, Dutch NEN 6770, Dutch NEN 6720 reinforcement steel and Dutch NEN 6720 prestress cable. In this part, reinforcement and pile foundations and steel material constitutive model are introduced, while the steel design codes are introduced in Sect. 2.5.

Discrete reinforcement is modeled by beam or truss elements and is discretized into separate truss or beam element via INTERF manipulation under DATA aspect in the DIANA software (see Fig. 2.26). Coupling effect is realized via merging the nodes of reinforcement bar element with concrete element for discrete reinforcement, which is applicable for analyzing bond-slip issue of single prestress tendon, where reinforcement bar element and surrounding mother concrete elements are connected via interface elements such as linear to linear, line to surface or line to solid connected type. In defining material constitutive models of discrete reinforcement bar element, beam or truss elements acting as reinforcement bars are required to be assigned with constitutive characteristics such as steel or strand, and the transformation from the truss and beam elements with specified steel bar attributes to embedded steel bar elements is realized via the module of DATA aspect in GUI operation interface. Generally speaking, discrete reinforcement modeling has higher accuracy but the modeling efficiency is relatively low at the same time.

Embedded reinforcement modeling is applicable for modeling reinforcement in large quantity, where the material constitutive model belongs to the aspect of reinforcement and pile foundations. Owing to embedded reinforcement without any degree of freedom, stiffness can only contribute to combining coordinate location points and integral points on the reinforcement element with concrete elements. Embedded reinforcement elements are further classified as reinforcement bar elements and grid elements based on the various modeling functions, shape and geometric properties, where the reinforcement bar elements are suitable for modeling post-tensioning tendons, longitudinal steel bars in vast quantity, while reinforcement grid element is applicable for stirrups or longitudinal steel bars in equidistant distribution. Material and geometric properties are directly selected and assigned sometimes. When embedded reinforcement is taken as prestress force in simulating post-tensioning construction, there are two bonding types in DIANA: bonded and non-bonded according to whether reinforcement is bonded to mother element before assigning material properties. Under bonded condition, embedded reinforcement is bonded to mother element to form as an integral, thus contributing to the whole common stiffness with mother element while non-bonded reinforcement has no contribution to the stiffness of mother element, thus stress and strain of reinforcement do not change with the deformation of mother element.

Contrary to the old version of DIANA such as DIANA 9.4, reinforcement bond-slip properties, which is applicable in the structural engineering cases of bonded steel strengthening method and issues of strand slip in the construction, can be directly applied based on the embedded reinforcement model and new bond-slip material model is no longer bothered to be built. This means that consuming time is saved, thus the modeling efficiency improves, providing convenience for beginners. In defining bond-slip model of reinforcement, not only bond-slip failure model is needed to be selected but also corresponding bond-slip parameters are required as input. The shapes of bond-slip reinforcement vary when selecting cross-section geometric properties. For example, truss bond-slip can be a better choice for mechanic behaviors of bond-slip model and beam bond-slip can also selected as a model when simulating prestress tendon. There are many types of cross-section shape in beam bond-slip model such as circle beam bond-slip, pipe beam bond-slip, rectangular beam bond-slip and box beam slip. Parameters in this beam bond-slip models vary according to the specific type, which will be further illustrated in the following chapter.

In addition, reinforcement bar element has good element adaptability, which is applicable for almost all the 2D and 3D elements. The shape and order of reinforcement bar element alter according to the variation in number of embedded location points. Material properties of embedded reinforcement can be assigned based on the parameters such as elastic modulus, Poisson's ratio, plastic hardening type and yield stress. User-specified model can also be applied to define the constitutive model according to the specific hardening type and yield stress–plastic strain relationship. When it comes to the specification of bond-slip model, besides parameters and specifications mentioned above, material properties of bond-slip interface are also needed as further input, including normal and shear stiffness modulus, bond-slip interface failure model, slip parameters containing cohesion and shear slip at start plateau. For nonlinear calculation, von Mises plasticity (or von Mises and Tresca plasticity) with plastic-yielding properties is often chosen as steel material constitutive model.

Sectional geometric properties are determined according to reinforcement type (embedded or other different types of bond-slip), cross-section input, cross-section area of bar or diameter. When embedded model is selected, cross-section area of bar or diameter is required as input according to the user's choice. However, once bond-slip model is chosen, not only the bond-slip anchor/tip surface area but also contact perimeter is required. Reinforcement types vary according to the specific shape and conditions.

Grid reinforcement is usually applied for simulating distributed reinforcement or net reinforcement. Similar to bar element, grid reinforcement can be embedded into all types of elements in 2D or 3D. Its material properties can be assigned via specifying elastic modulus, Poisson's ratio, plastic hardening type and yield stress as bar element while the sectional geometric properties can be determined by the parameters as diameters and spacing. For net reinforcement modeling, it can be modeled in turn one by one via bar element but can also be completed through grid element in a faster and more efficient way. It is also worth specifically emphasizing

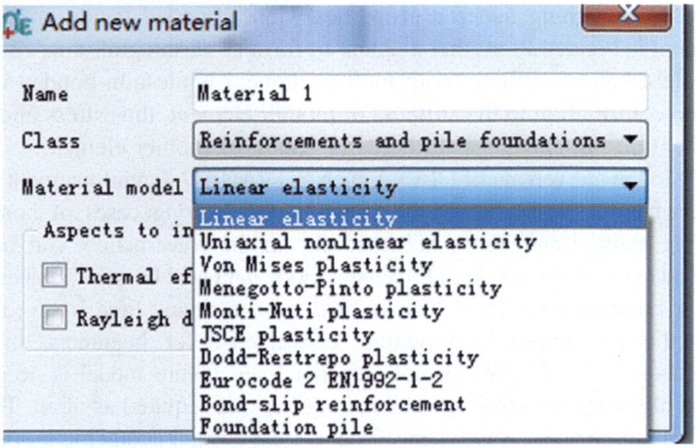

Fig. 2.25 Material constitutive aspects of reinforcement and pile foundations

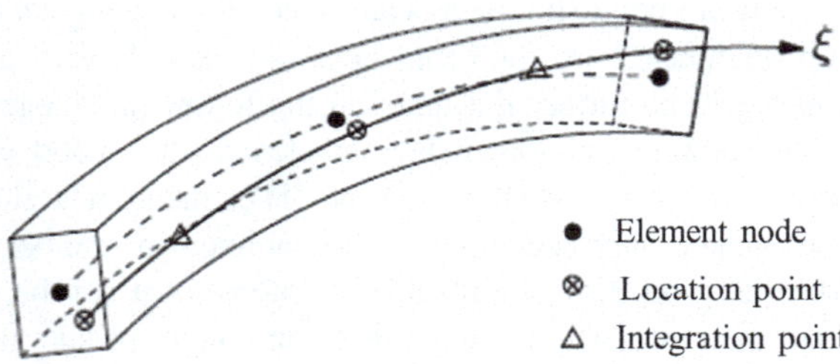

Fig. 2.26 Embedded reinforcement bar element in concrete beam element

that embedded reinforcement elements can also be coupled with beam and truss elements even though these reinforcement elements are beam or truss element themselves compared with other types of finite-element software.

Compared with bar element concerning cross-section geometric model, assignment of net reinforcement via grid element is relatively complex. Cross-section geometric definition of reinforcement grid can not only be completed through inputting the parameters of diameter and spacing values, but also can be specified via directly inputting equivalent thickness in two orthogonal directions under local coordinate system to represent reinforcement ratio, where the equivalent thickness is the area of the same stirrup in two orthogonal directions divided by the distance between the centers of the corresponding adjacent stirrups. Under these two types, users are not only required to define parameters mentioned above in local coordinate x and y directions, but also needed to determine the relationship between the x directions in the local and global coordinate systems.

Contrary to the conditions above, steel cables are usually exposed in the atmosphere and not bonded or touch with concrete, thus regular truss element such as L2TRU or L7BEN is taken as cables for numerical simulation. Prestress force is taken as cable prestress via geometric loading attachment in DIANA. Contrary to the post-tensioning loading unit, the unit of prestress load is N/m^2 or kN/m^2 instead of N or kN. The way of attachment is modeled according to truss or beam elements, then these elements are discretized under the DATA aspect. When the material constitutive model and cross-section geometric properties are specified, the steel properties can be assigned via steel, or direct attachment after specifying steel codes or even attaching load to reinforcement and pile foundations model. Prestress loading attachment is done after adding load cases in order to simulate mechanic behaviors of steel cables, which are often used for researching issues of corrosion, fatigue and durability in stay-cable bridges. Normally, prestress strands with large cross-section area in post-tensioning method are needed to attach prestress load while the stay-cables and steel tendons with small size are mainly attached by prestress tendons load such as prestress in DIANA. Besides, steel cables can also be assigned with FRP constitutive model to approximately simulate FRP mechanic behaviors.

Prestress force can be classified into two types: one is post-tensioning load and the other is ordinary prestress which is attached in the type of stress. Prestress option is selected as load case while solid is selected as load target type. Load type is reinforcement bar prestress and the manipulation interface of regular prestress load attached to the steel tendon is displayed in Fig. 2.27.

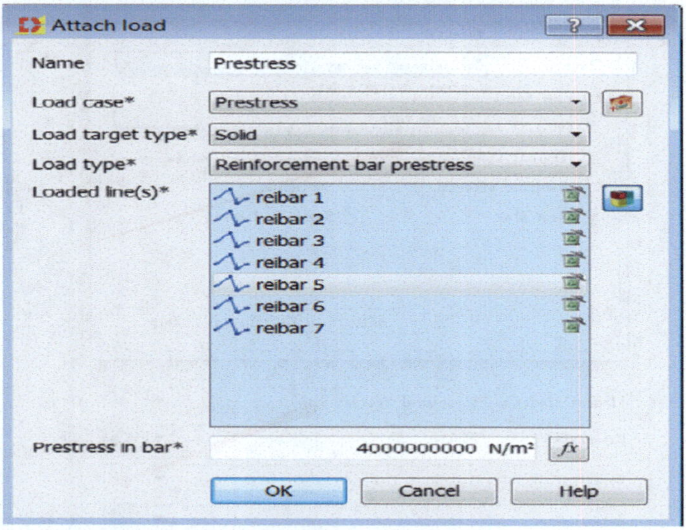

Fig. 2.27 Interface of attaching prestress load

Additionally, when structural stability issues, buckling issues, post buckling issues, steel fatigue issues or even steel fire issues are analyzed, material properties of main structures are usually assigned with steel. However, when steel constitutive aspect is selected, meshing for steel is required before the finite-element calculation is finished, which is contrary to the reinforcement and pile foundations model. In this circumstance, shortcut icon button in blue named as **Edit Reinforcement**

properties assignments should not be clicked for the reason that it only suits for specifying material properties of steel used as reinforcement in reinforced structures rather than structures with steel as the main body. Users should assign steel material properties via regular yellow shortcut icon **Edit property assignments** as assigning properties of concrete elements instead, then the steel should be meshed as concrete elements in the following meshing process. Taking solid element for instance, specification interface of steel material properties is shown in Fig. 2.28. Generally speaking, distinction between function of **Edit Reinforcement properties assignments** and **Edit property assignments** lies in that the former is incapable of selecting element type of steel once reinforcement properties are determined, which are displayed in Fig. 2.29.

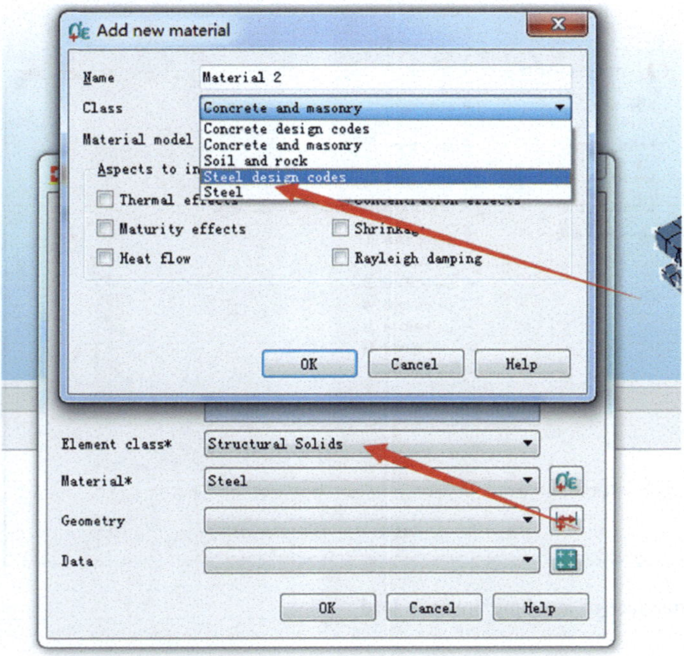

Fig. 2.28 Specifications for material properties with steel structure as the main body

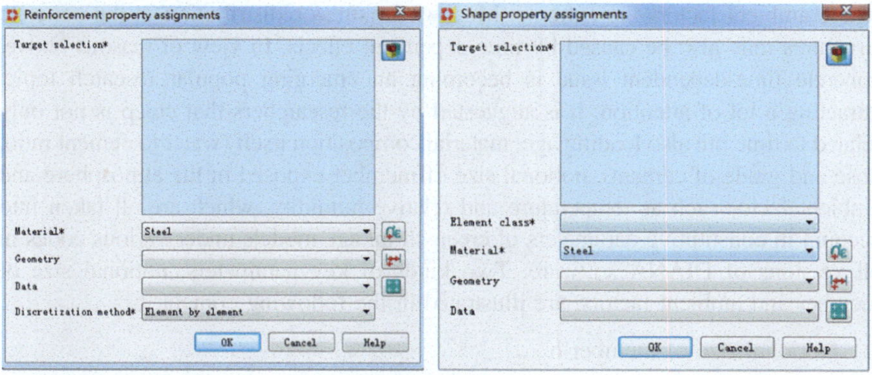

Fig. 2.29 Distinction of interface between **Reinforcement property assignments** and **Edit Property Assignments**

2.4 Time-Dependent Material Constitutive Model of DIANA

Time-dependent issue, or named as structural deteriorated long-term performance issue, refers to the prestress force loss and excessive deflection triggered by time-dependent multiple and mutual effects of creep, shrinkage and relaxation common in bridges under the factors of time and ambient influence, where creep and shrinkage are the common phenomenon that belong to inherent attributes of concrete material. Nowadays, codes considering constitutive model of creep and shrinkage around the world are listed as follows: (1) CEB-FIP Model Code 1990 (abbreviated as **CEB-FIP1990** model) proposed by the European Concrete Association (committee Euro-International du Beton). (2) fib Model Code for Concrete Structures 2010 (abbreviated as **fib2010)** proposed by the European Concrete Association in the year of 2010, (3) AASHTO LRFD Highway Bridge Design Specifications (abbreviated as **American AASHTO** model) proposed by the American Association steel Highway and Transportation officials, (4) American ACI 209R-92 proposed by the American Concrete Institute, (5) Eurocode 2 EN 1992-1-1 model (abbreviated as **EN 1992-1-1** model), (6) Korea Concrete Institute 2007 (abbreviated as **KCI** model), including two different standards: Korean KCI with civil standards and Korean KCI with industrial standards, and (7) Dutch NEN 6720/A4 model code proposed by the Netherlands [1]. It is pity that creep and shrinkage aspects are not included in the famous codes of Japanese codes: Japan Society of Civil Engineers (abbreviated as **JSCE**) and Japan Concrete Institute (abbreviated as **JCI**) in the DIANA software. The codes mentioned above are all incorporated in the DIANA software.

Deformations, deflections and prestress force loss are increased via the time-dependent effects of creep and shrinkage in civil engineering, especially resulting in potential hazards or even collapse for bridge structures in the zones where ambient factors are extreme. Meanwhile, stress redistribution in the concrete structures may also be caused by time-dependent effects. In view of reasons above, concrete time-dependent issue is becoming an emerging popular research topic, attracting a lot of attention. It is suggested by the researchers that creep is not only related to time but also loading age, material composition itself (water to cement ratio, dose and grade of cement), notional size of member exposed in the atmosphere and ambient factor such as temperature and relative humidity, which are all taken into account in constitutive parameters of creep-shrinkage models under various codes in all versions of DIANA software. Two kinds of key parameters, notional size of member and ambient factors, are illustrated in the following content.

(1) Notional size of member h

Notional size of member is a key inner parameter influencing creep and shrinkage, which is also the one concerning structural long-term nonlinear analysis in CEB-FIP model, which will be particularly discussed in the following chapter. The ratio of volume to surface area is reflected by notional size of member exposed in the atmosphere. When the surface area increases or volume decreases, the area of member exposed in the air is larger, which means that water loss accelerates and the effects of creep and shrinkage are relatively more prominent. In the DIANA software, notional size of member is defined as twice the cross-sectional area of the member divided by its perimeter in contact with the atmosphere.

(2) Ambient factor

The factors affecting creep and shrinkage are ambient temperature (°C) and relative ambient humidity (%), which have significantly prominent impact on the creep. Generally speaking, the higher the temperature, the higher is the creep, while the higher is the relative humidity, the lower is the creep. The reason is when temperature is higher or relative humidity is lower, the amount of water evaporation adsorbed in concrete is larger, thus the extent of hydration is higher and the creep is larger. Meanwhile, relative humidity also has influence on prestress force loss. Research indicates that prestress force loss is larger when ambient relative humidity is lower.

In the DIANA software, concrete creep, shrinkage and relaxation of steel all can be embedded into the DIANA via finite-element modeling. Time-dependent material model of concrete applied in DIANA is rate-dependent constitutive mode. Time-dependent factors of creep, shrinkage and relaxation mentioned above affecting long-term performance can be automatically reflected and considered. In the DIANA software, relaxation phenomenon can be automatically considered by the relaxation function, and selection of relaxation is also relative with the type of relaxation. The relaxation model that mainly focused on stress relaxation is described via both relaxation function and generalized Maxwell model, while relaxation model that mainly focused on creep and shrinkage is described via both creep function and generalized Kelvin model. Similar to series and parallel phenomena in circuits, a Maxwell element is constructed by an elastic component (usually spring) and a viscous one (usually dashpot) in series, while a Kelvin element is constructed by an elastic component and a viscous one in parallel, where a series of Maxwell elements is connected in parallel to construct a Maxwell chain model (see Fig. 2.30), while a series of Kelvin elements is connected in series (see Fig. 2.31) to construct a Kelvin chain model. That is to say, Maxwell chain is constructed by a series of multiple Maxwell elements, while Kelvin chain is constructed by multiple Maxwell elements in parallel. Considering Maxwell chain is based on concrete non-aging theory, while Kelvin chain is based on concrete aging theory, thus relaxation function and Maxwell chain model are usually applied to analyze time-dependent issue of reinforced concrete structures dominated by relaxation, while creep function and Kelvin chain are usually applied to analyze time-dependent issue of reinforced concrete structures dominated by creep and shrinkage. For the DIANA software, whether Maxwell or Kelvin chain is constructed, the number of chains is no less than 10. The results of relaxation dominated by Maxwell chain model without considering concrete aging theory are normally higher than the ones of creep dominated by Kelvin chain model considering concrete aging theory. The latter is more reliable to reach the default required number of chains. Meanwhile, owing to the feature that Maxwell chain model is relaxation dominated, the results of relaxation are slightly higher than the latter one. Thus it is essential that users should select model in caution based on the comparison between long-term concerned experimental outcomes, and numerical results are often essential. According to the author's experience, elastic modulus is one of the key parameters in Kelvin chain, which is relative with factors such as element age, ambient temperature and relative humidity. How to specify concrete compressive strength in 28 days is the key point in precise simulation via Kelvin numerical model.

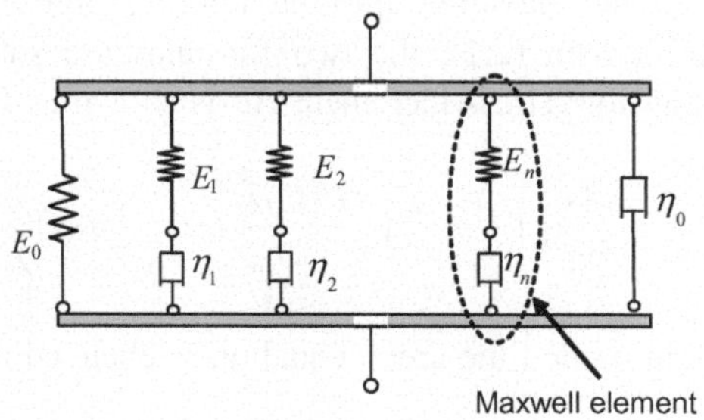

Fig. 2.30 Maxwell chain

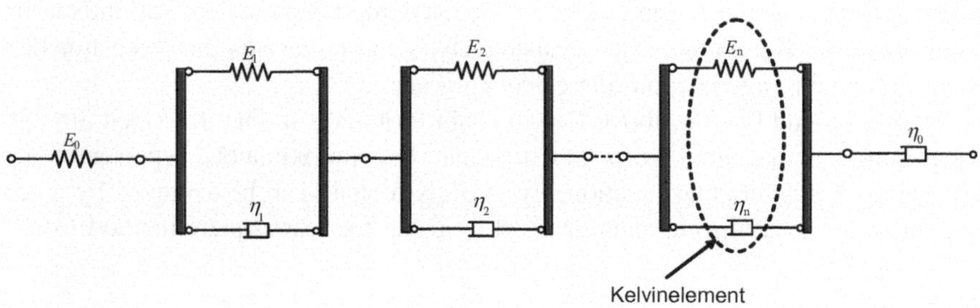

Fig. 2.31 Kelvin chain

In the rate-dependent model, relaxation and creep function can be approximately expressed via by expansion of the Dirichlet function, where the expansion of relaxation function is the product of exponential function and relaxation function of the ith element in the summation term of the expansion while the expansion of creep function model is an exponential function divided by the creep function of the ith element. In the generalized Maxwell model, stress–strain relationship of strand relaxation can be expressed as follows (2.1):

$$\sigma(t) = \int_{-\infty}^{t} E(t,\tau)\varepsilon d\tau \tag{2.1}$$

where $E(t,\tau)$ represents the relaxation function when loading age is τ, and calculation age is t. Relaxation function can be expanded via Dirichlet sequence as (2.2).

$$E(t,\tau) = \sum_{i=0}^{n} E_i(\tau) e^{-\frac{t-\tau}{\lambda_i}} \tag{2.2}$$

where $E_i(\tau)$ is the time-dependent stiffness under the Maxwell element and relaxation time is defined as

$$\lambda_t = \eta_t / E_i \tag{2.3}$$

Listing time-dependent relaxation formula under t and $t + \Delta t$, respectively, assigning $t^* = t + \Delta t/2$, then (2.1), the two formulas are substituted into (2.1) through Dirichlet expansion; stress increment $\Delta\sigma$ is shown as follows.

$$\Delta\sigma = \sum_{i=0}^{n} \left(1 - e^{-\frac{\Delta t}{\lambda_i}}\right)\left(\frac{E(t^*)\lambda_i}{\Delta t} \bar{D}\Delta\varepsilon - \sigma_i(t)\right) \tag{2.4}$$

For the Kelvin chain model, the creep equation is changed as follows [3]:

$$J(t,\tau) = \sum_{t=0}^{n} \frac{1}{E_i(\tau)}\left(1 - e^{-\frac{t-\tau}{\lambda_i}}\right) = \frac{1}{E_0(\tau)} + \sum_{t=1}^{n} \frac{1}{E_i(\tau)}\left(1 - e^{-\frac{t-\tau}{\lambda_i}}\right) \tag{2.5}$$

where $\lambda_i = \eta_i / E_i$, η_i and E_i represent the viscous damping coefficient and the elastic modulus of the Kelvin element i, respectively. $J(t,\tau)$ represents the creep function when the loading age is τ and the calculating age is t.

It can be found that the above Kelvin chain formula is in fact described through creep function and this creep function can be approximately represented by expansion of the Dirichlet function. Every Kelvin chain can be achieved by creep and shrinkage experimental data or fitted by using the least square method [3].

2.5 International Codes of DIANA

When manipulating DIANA software, users can not only select design codes introduced in the following part but also use the user-specified code, where European CEB-FIP Model Code 1990, fib Model Code for Concrete Structures 2010 and American AASHTO LRFD Highway Bridge Design Specifications are the most frequently used codes. In this part, European CEB-FIP Model Code 1990, fib Model Code for Concrete Structures 2010, American AASHTO LRFD Highway Bridge and Japanese JSCE code are introduced and the universality and features are compared in this part.

Concrete design codes in DIANA [1] are listed as follows:

(1) European CEB-FIP Model Code 1990 (abbreviated as CEB-FIP1990 in the following chapters)
(2) fib Model Code for Concrete Structures 2010 (abbreviated as fib 2010 in the following chapters)
(3) AASHTO LRFD Highway Bridge Design Specifications (abbreviated as AASHTO in the following chapters)
(4) American ACI 209R-92 (abbreviated as ACI in the following chapters)
(5) Japan Society of Civil Engineers (abbreviated as JSCE in the following chapters)
(6) Japan Concrete Institute (abbreviated as JCI in the following chapters)
(7) Eurocode 2 EN 1992-1-1 model (abbreviated as EN 1992-1-1 in the following chapters)
(8) Dutch NEN 6720/A4 model code
(9) Korea Concrete Institute 2007 (abbreviated as KCI model in the following chapters)
(10) Technical University of Denmark Joint Committee on Structural Safety (JCSS) Probabilistic Model Code (abbreviated as JCSS Probabilistic Model Code in the following chapters)

Steel design codes in DIANA are listed as follows:

(1) Eurocode 3 EN 1993-1-1 (abbreviated as EN 1993-1-1 in the following chapters)
(2) Dutch NEN 6770,
(3) Dutch NEN 6720 reinforcement steel
(4) Dutch NEN 6720 prestress cable

The steel design codes in DIANA are shown in Fig. 2.32.

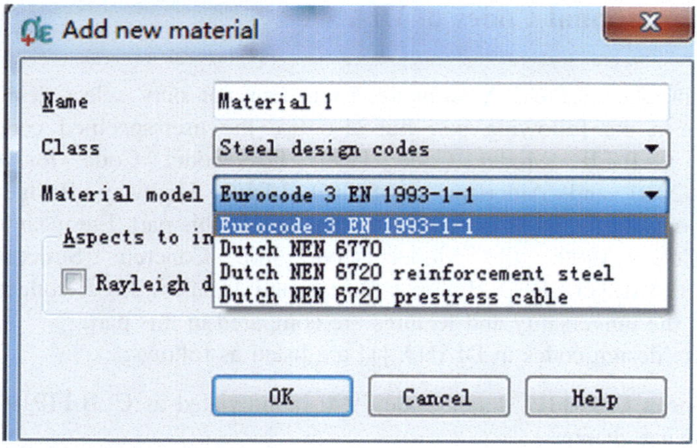

Fig. 2.32 Options of steel design codes in DIANA

CEB-FIP1990 design code is based on the CEB-FIP1978. In this design code, there are following modules required to be specified: Aspect to include, European CEB-FIP1990 containing specifications of concrete inner features, Direct input and time-dependent factors such as parameters of creep and shrinkage according to the various projects (creep curve type, concrete age at birth of element and concrete age at end of curing period), Heat flow and parameters concerning with damping coefficient (see Figs. 2.33 and 2.34, respectively).

The Aspect to include, Elasticity, Plasticity, Shrinkage, Creep and Heat flow aspects are selected where plastic calculation has high demand on convergence of model, while elasticity is not allowed to be selected with creep and shrinkage aspects at the same time. There are many parameters to be specified in the design codes. For example, CEB-FIP1990 not only contains the concrete basic parameters needed to be specified such as characteristic and mean cylinder compressive strength at 28 days, Young's modulus at 28 days, Poisson's ratio, mass density but also the ones related with concrete inner properties and atmosphere ambient factors such as cement type, notional size of member, aggregate type, ambient temperature and relative ambient humidity, which are key to simulating deterioration of structural long-term performance. Name, implications and default value of all the parameters in CEB-FIP 1990 are listed as Table 2.2, where the user-specified means that these values should be specified by users themselves according to the actual conditions.

Table 2.2 Parameters of CEB-FIP 1990

Name	Aspect	Default value
Concrete class	European CEB-FIP 1990	User specified
Cement type		Normal and rapidly hardening
Ambient temperature		20 °C/293.15 K
Notional size of member h		0.15 m
Relative ambient humidity RH in %		80%
Aggregate type		Quartzite
Young's modulus	Direct Input	User specified
Young's modulus at 28 days		User specified
Poisson's ratio		0.3
Thermal expansion coefficient		1×10^{-5}
Mass density		2500 kg/m^3
Characteristic strength at 28 days		User specified
Mean compressive strength at 28 days		User specified
Creep curve type	Creep	Non-aging
Concrete age at birth of element		2419200 s (28 day)
Concrete age at end of curing period	Shrinkage	86400 s (1 day)
Conductivity	Heat flow	User specified
Capacity		User specified
Conductivity/capacity function		No dependency
Heat of hydration method		Off
a-Factor for mass matrix	Rayleigh damping	User-specified
b-Factor for stiffness matrix		User-specified

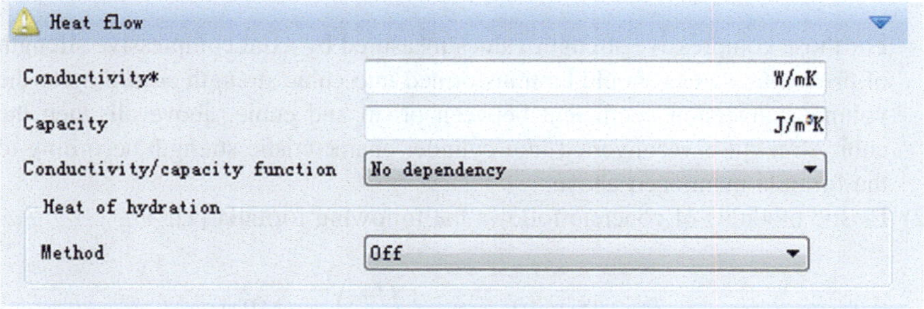

Fig. 2.33 Heat flow aspect of CEB-FIP1990

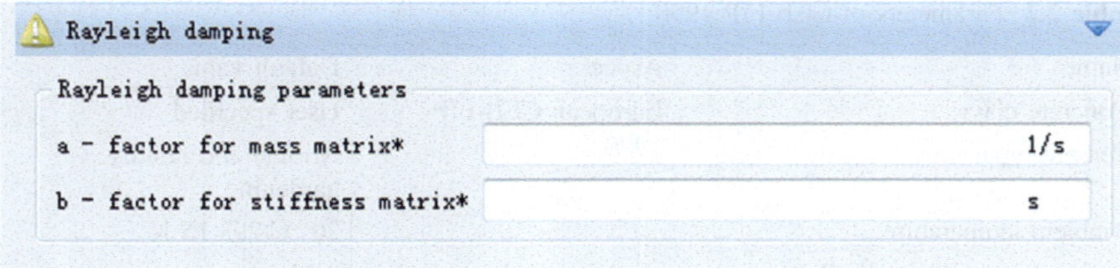

Fig. 2.34 Rayleigh damping aspect of CEB-FIP1990

The following points need special explanation:

(1) In European design codes, uniaxial compressive strength fck is deemed as characteristic strength determined by standard cylinder specimen with the diameter 152 mm and height 305 mm, where the measured compressive value corresponds to the number of concrete strength grade label, and this number is concrete characteristic strength at 28 days. For example, C30 means the cylinder characteristic strength is 30 N/m^2. Relationship between fck and mean compressive strength at 28 days fcm is defined as [1]:

$$\mathrm{f_{cm}} = \mathrm{f_{ck}} + 8$$

When cubic compressive strength is required to be measured in the material test condition, relationship between cylinder characteristic strength and cubic strength is shown as following piecewise formula:

$$f_{ck} = \begin{cases} 0.79 f_{cu,k}(<C60) \\ 0.833 f_{cu,k}(C60) \\ 0.857 f_{cu,k}(C70) \\ 0.875 f_{cu,k}(C80) \end{cases}$$

For those compressive strength values measured by axial compressive strength of prism, the values should be transformed into cubic strength according to the volume conversion coefficient between prism and cubic, above all, then the cubic strength is reconverted into cylinder characteristic strength according to the formula mentioned above.

(2) Elastic modulus of concrete follows the following formula [1]:

$$E = 2.15 \times 10^4 \times \alpha_E \times \left(\frac{f_{cm}}{10}\right)^{\frac{1}{3}} \quad \text{(MPa)}$$

where for the quartzite aggregate, aggregate type-dependent scaling factor α_E is 1; for the linear materials without any nonlinear material behavior, reduction coefficient of E is 0.85.

Tensile relationship is defined as [1]

$$f_{ctm} = 1.4 \times \left(\frac{f_{ck}}{10}\right)^{\frac{2}{3}} \quad \text{(MPa)}$$

fib Model Code for Concrete Structures 2010 is the newly added design code in DIANA release version 10.1, as Fig. 2.35 displays. Compared with CEB-FIP1990, the main priority of fib 2.10 lies in decreasing the number of parameters needed to be specified directly via manual in the aspect of Direct input. That is to say, parameters such as elastic modulus, Poisson's ratio, mass density, thermal expansion coefficient and mean compressive strength specified in CEB-FIP 1990 manual are all automatically determined once the grade of concrete is determined, saving vast of direct input time (see Fig. 2.36). However, owing to a certain degree of discreteness of material, values of size, elastic modulus and strength tend to fluctuate within a certain range. Therefore, constitutive parameters according to experimental results cannot be specified manually in DIANA when this design code is selected, which will trigger a certain deviation error. Generally speaking, it is a detrimental effect when numerical results needed to be compared with experimental ones under this circumstance.

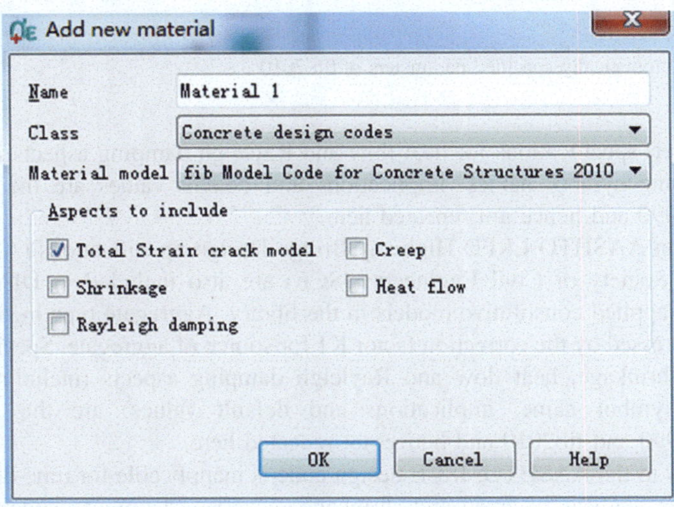

Fig. 2.35 Constitutive aspect of fib Model Code for Concrete Structures 2010

Another feature of fib 2010 lies in the notional size of member, and ambient factors are internal and external causes related to creep and shrinkage, respectively, and the mutual effects of creep, shrinkage, ambient temperature and relative humidity and notional size of member are considered together in the same module.

The third feature of fib 2010 is material safety factors in ultimate limit state (abbreviated as ULS), including Young's modulus, mean uniaxial tensile strength and mean compressive strength. Material safety factors are specified via considering variation in material elastic modulus, tensile and compressive strength, whose design value is material strength. Therefore, mechanic behaviors of reinforced concrete structures can be more comprehensively and accurately simulated under ultimate strength limit state by specifying the three parameters. Considering the material design value usually lower than the characteristic ones, the coefficient value of material safety factors should be less than 1.

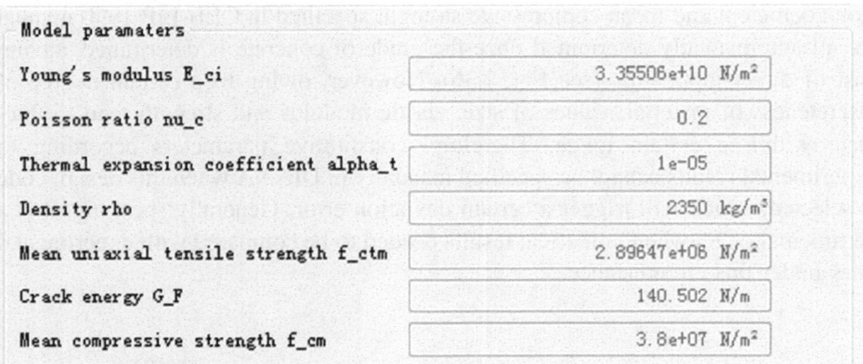

Fig. 2.36 Automatically specified parameters of fib 2010

Moreover, specifications for heat flow and Rayleigh damping aspects including corresponding symbol names, implications and default values are the same as CEB-FIP1990 and hence not repeated here.

American AASHTO LRFD Highway Bridge Design Specifications (AASHTO) and Japan Society of Civil Engineers (JSCE) are also included in DIANA and commonly applied constitutive models in the library. Aggregate type in AASHTO code is expressed by the correction factor K1 for source of aggregate. Specifications of creep, shrinkage, heat flow and Rayleigh damping aspects (including corresponding symbol name, implications and default values) are the same as CEB-FIP1990 and fib2010 and hence not repeated here.

Contrary to the AASHTO, JSCE design code is inapplicable for time-dependent or long-term analysis of creep and shrinkage. A series of time-dependent factors influencing creep and shrinkage such as concrete age at birth of element, notional size of member and concrete age at the end of curing period is not reflected in the JSCE module of DIANA software. Moreover, in this design code, strength of concrete is based on the standard cubic **Characteristic strength after 91 days** while the value of elastic modulus is **Modulus of elasticity at age of 91 days**. Figure 2.37 shows the interface of JSCE design code in DIANA.

The specifications of heat flow and Rayleigh damping in JSCE code are the same as other design codes (see Figs. 2.38 and 2.39).

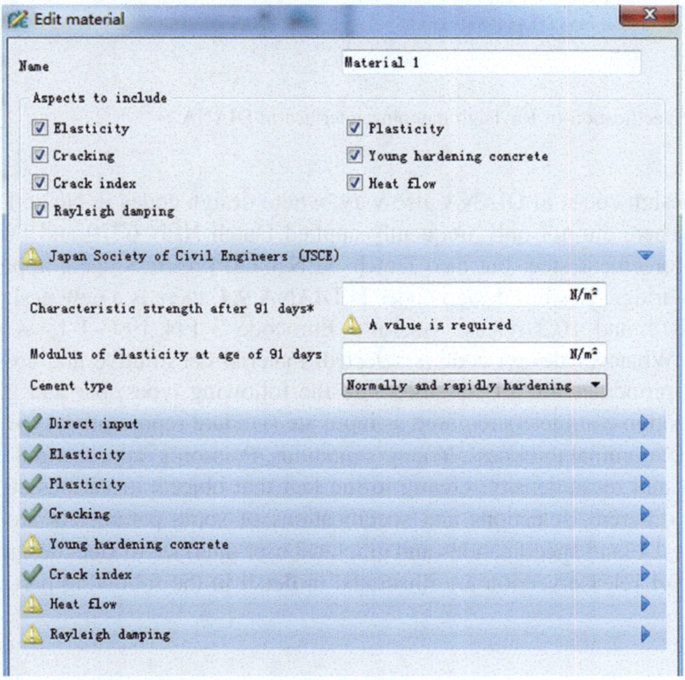

Fig. 2.37 Interface of JSCE design code in DIANA

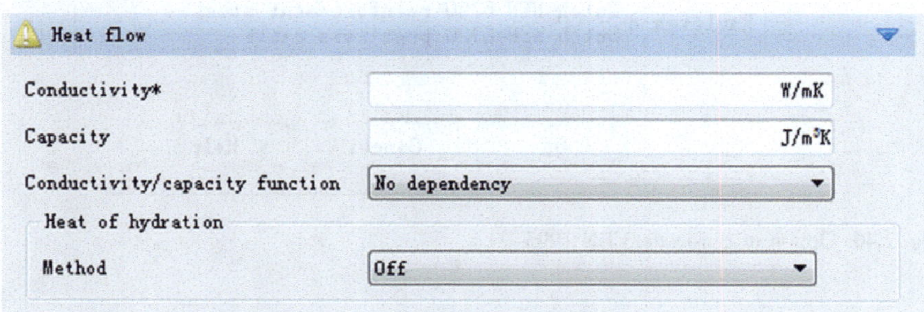

Fig. 2.38 Specification of heat flow interface in DIANA

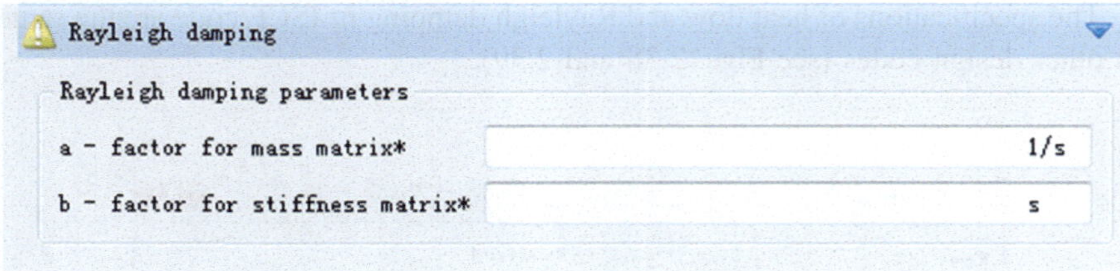

Fig. 2.39 Specification of Rayleigh damping interface in DIANA

Steel design codes in DIANA also vary, where design codes in Netherland are in majority. There are not only frequently applied Dutch NEN 6770 and Dutch NEN 6720 reinforcement steel but also Dutch NEN 6720 prestress cable applicable for stay-cable bridge. Besides design codes in DIANA 9.4, there is a new design code in DIANA 10.1 and 10.2 release versions: Eurocode 3 EN 1993-1-1, as shown in Fig. 2.40. Whatever design code is selected, material constitutive and cross-section geometric properties are all classified into the following types: bar and grid. Basic characterization parameters required as input are standard representing type of design code, grade, nominal thickness, Young's modulus, Poisson's ratio, thermal expansion coefficient and mass density. Owing to the fact that objects in various steel design codes are different, selections and specifications of some parameters are in subtle differences. Default specifications and graphical user interface of steel design codes in DIANA 10.1 release version are illustrated in detail in the following part.

Fig. 2.40 Option of Eurocode 3 EN 1993-1-1

Once steel geometric objects are created, selecting of required steel design code is done and the corresponding default values are generated. Meanwhile, these parameters can also be specified by users. Taking Eurocode 3 EN 1993-1-1 for example, the default parameters are shown in Fig. 2.41.

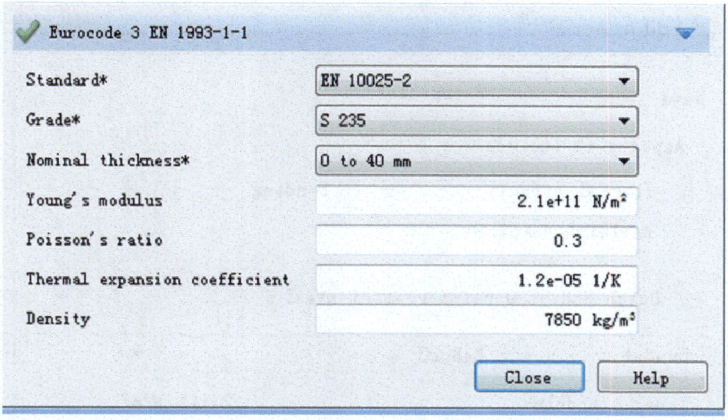

Fig. 2.41 Default values of Eurocode 3 EN 1993-1-1

Grade of steel strength, nominal thickness, elastic modulus, Poisson's ratio, thermal expansion coefficient and mass density are calculated based on Dutch NEN 6770 code, as Fig. 2.42 displays. Similar to the material constitutive model of **Reinforcement and pile foundations**, bonding types are required to be taken into account above all, which determines whether reinforcement is bonded to concrete mother element in the Dutch NEN 6720 reinforcement steel. Besides specifying grade of steel in FEB type, constitutive model of reinforcement is required to be selected as elasticity or ideal plasticity. Once the latter option is selected, yield stress is further needed to be specified. Default parameter values in ideal plasticity model are displayed in Fig. 2.43.

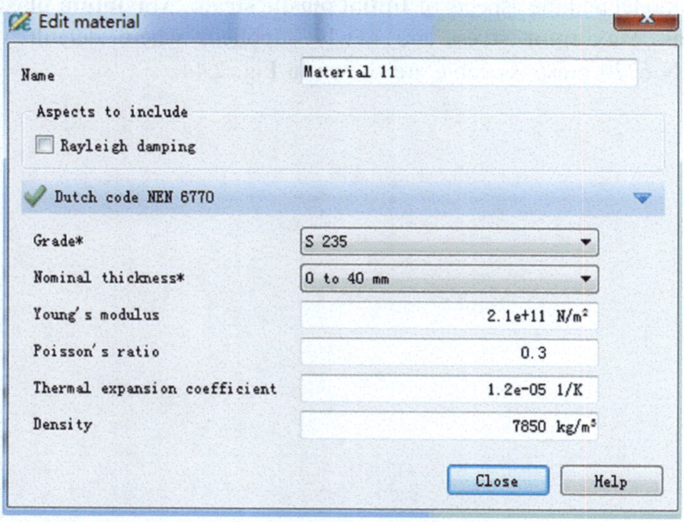

Fig. 2.42 Manipulation interface and default values of Dutch NEN 6770

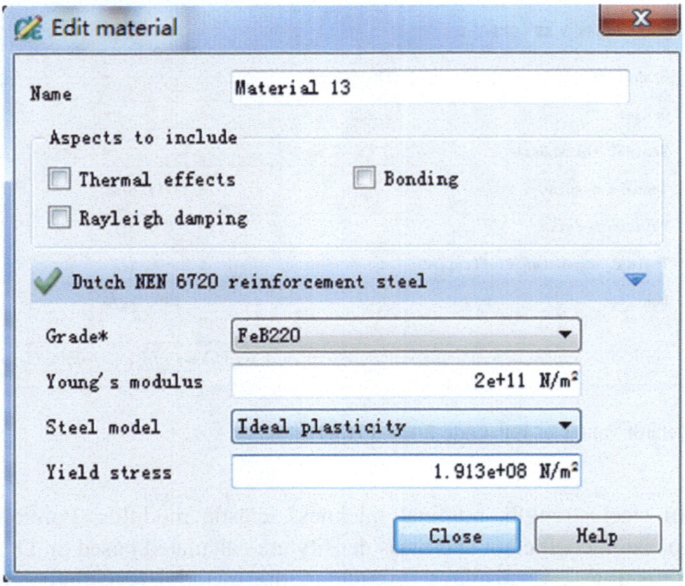

Fig. 2.43 Specifications of Dutch NEN 6720 reinforcement steel

Parameter specifications in Dutch NEN 6720 prestress cable are also similar to Dutch NEN 6720 reinforcement steel, where there are subtle variations in grade of strength and steel model in this model compared with the latter. Dutch NEN 6720 prestress cable is mainly applicable for simulating mechanic issues of steel cables in stay-cable bridges such as vibration, corrosion, fatigue and fracture, thus the strength is high. Moreover, **Ideal plasticity** model is replaced by the **Hardening plasticity** model and the aspects of **Initial plastic strain, Maximum plastic strain as well as Maximum stress** are newly supplied, where default values in Dutch NEN 6720 prestress cable are shown in Fig. 2.44.

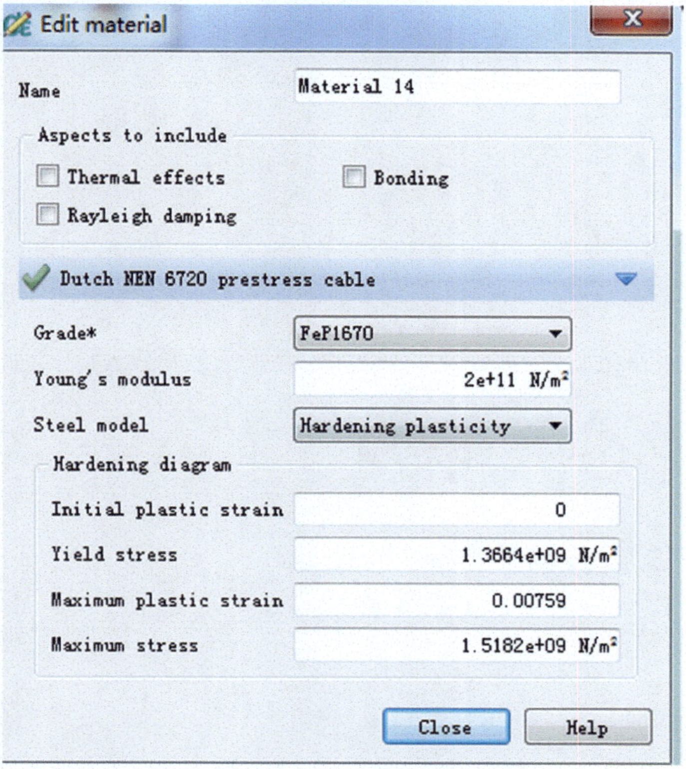

Fig. 2.44 Default values in Dutch NEN 6720 prestress cable

References

1. DIANA User's Manual-Material Library, Release 9.3. (2008). TNO Building and Construction Research, Holland
2. Jia MJ (2017) Wu fen zhong bang ni ren shi DIANA zhong de ge lei lie feng mo xing (五分钟帮你认识DIANA中的各类裂缝模型Mastering various crack models in DIANA in five minutes), Dunpu online training meeting, Shanghai. https://mp.weixin.qq.com/s?__biz=MzA3MzkyNzg0NQ==&mid=2247485165&idx=1&sn=f6b40888d5bc1e7863abda8c22db0067&chksm=9f06db27a871523196e560ea804f5205249420adceb1f0e03d22c08034f6f23872ae0b28995a&mpshare=1&scene=22&srcid=0904IF4JXkewJNl3P3MUPj1V#rd
3. Sun H, Ye LP, Ding JT (2004) Hun ning tu xu bian ji suan fen xi fang fa(混凝土徐变计算分析方法, Calculation and analysis method of concrete creep). Tsinghua University Symposium, Beijing

Chapter 3
Nonlinear Analysis of DIANA Modeling Cases

Abstract This chapter mainly focuses on exhibiting manipulation of DIANA modeling via numerical simulation. Traditional issues of concrete such as cracking analysis, time-dependent analysis, hysteresis analysis, phased analysis, dynamic analysis as well as time-history dynamic analysis concerning response modes as well as frequencies are simulated in this chapter. Moreover, the emerging tendency of ultra-high performance concrete (UHPC), stepwise loading as well as time-dependent analysis for UHPC structures are investigated and compared with ordinary C50 concrete based on the platform of solid elements and shell elements via both graphical user interface and editing command console in Python language. Phased analysis, a prominent function in DIANA, is also conducted in this chapter related with typical and commonly applied active and passive strengthening cases to validate and compare the effect of these two strengthening methods.

3.1 Structural Nonlinear for Prestress Frame

This model is a simple two-dimensional plane two-story concrete frame, where constraints between frame structure and foundation are fixed-based. There are reinforcement steel bars in both beam and column, with cross-section area 100 and 120 mm^2, respectively. Two-dimensional (2D) node to node connected interface elements are created between joints of beam and column. Each layer of frame has a single prestress tendon with elastic modulus, nominal diameter and strength as $1.95 \times 10^{11}\ N/m^2$, 15.24 mm and 1860 MPa, respectively. Concrete grade of both beam and column is C50. Cross-section size of frame beam and column is 500 mm × 500 mm and 350 mm × 400 mm, respectively. Height of lower layer frame is 4.6 m to the ground, while distance between upper and lower layers is 4 m coupled with the distance between columns is 6.1 m. Tensioning value of prestress tendons in each layer is 500 kN. Distributed load of 20 kN/m is attached to the upper layer of beam, while horizontal equivalent acceleration of $0.1g$ is applied to the frame. Beam elements of L7BEN are applied to simulate the frame. **Material constitutive model** is **Total strain based crack model**. Structural nonlinear

S. Chai, *Finite Element Analysis for Civil Engineering with DIANA Software*,
https://doi.org/10.1007/978-981-15-2945-0_3

analysis is conducted for the two-layer frame. Frame plane without longitudinal steel bars in beams and columns is shown in Fig. 3.1. Meanwhile, numerical model under equivalent acceleration is also demonstrated.

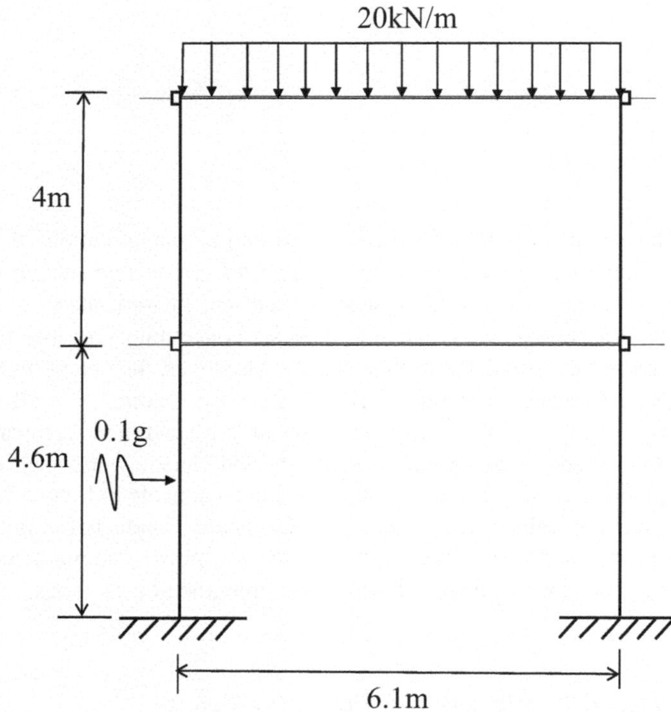

Fig. 3.1 Plane of frame

Note: It is worth to mention that in DIANA numerical cases, conditions of this book are all assumed values. The appropriateness of the analysis results is another matter.

Essentials of learning

(1) *Learning create line model*
(2) *Learning to specify cross-section geometric properties of beam elements*
(3) *Learning to specify material properties in total strain-based crack model for concrete*
(4) *Learning to establish node to node connected 2D interface element*
(5) *Learning to attach point load, distributed load, post-tensioning load and equivalent acceleration*
(6) *Learning to add geometry load combinations*
(7) *Learning to add load set and analysis block.*

Above all, clicking DianaIE, New project dialog ejects. Create a new file with suffix name **Frame.dpf**, which is stored in the working directory of G disk in computer. **Structural** analysis is selected as analysis type with the two-dimensional maximum **Model size** 100 m, indicating that the scope of the whole graphical user interface ranges from –50 to 50 m in both X and Y directions of the coordinate system. **Default mesher type** is **Hexa/Quad** and **Default mesh order** is **Linear** (see Fig. 3.2). On clicking OK button, generation of graphical user interface zone is completed instantaneously.

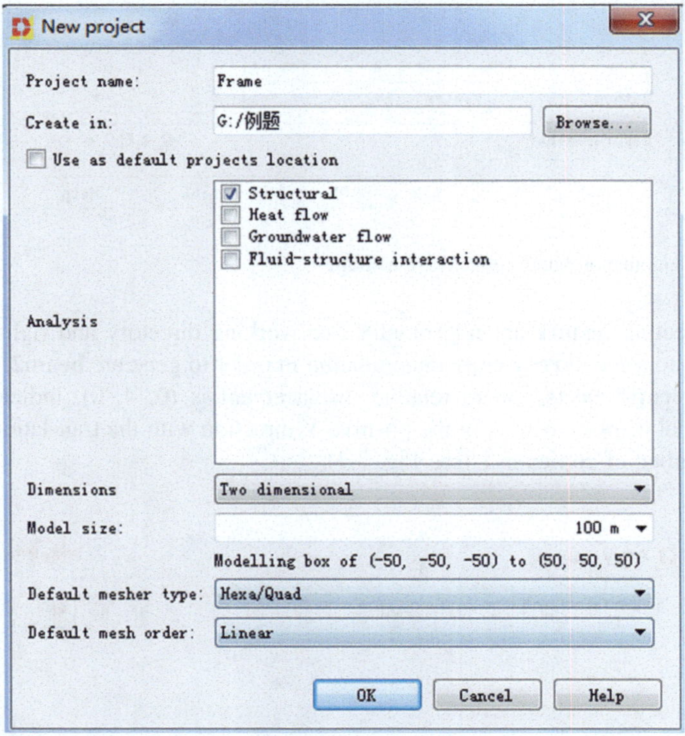

Fig. 3.2 Interface of new project

Now we start to create finite-element geometric model. Clicking shortcut icon **Adds a line** to add coordinate values (0, 0, 0) and (6.1, 0, 0) with the name of **beam1**, and click OK button to generate straight line. Then **beam1** under geometry tree directory is selected via right-clicking. We again right-click to select **Move a shape**, then the corresponding dialog box ejects, as shown in Fig. 3.3. Displacement in the positive Y direction is 4.6 m; manipulation interface of which is displayed in Fig. 3.3.

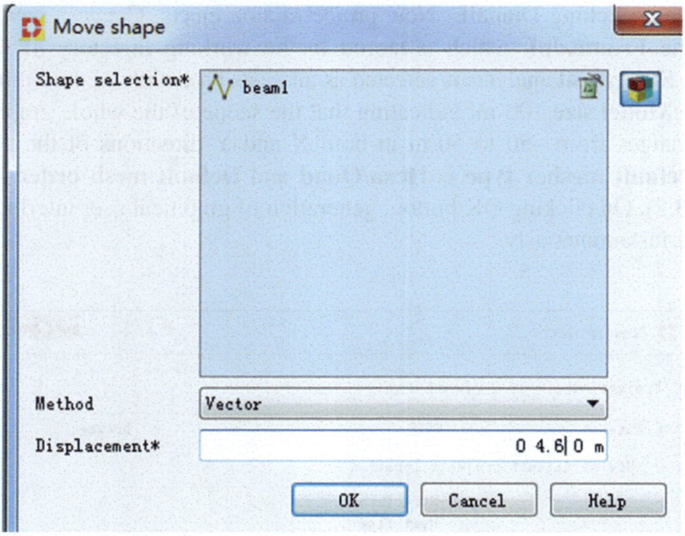

Fig. 3.3 Manipulation interface of **Move a shape**

By selecting **beam1** under geometry tree working directory and right-clicking the model to select **Array copy** manipulation in order to generate **beam2**, interface of **Array copy** ejects, where relative displacement is (0, 4, 0), indicating that translational displacement is in the positive Y direction with the translational value 4 m. **Number of copies** is 1 (see Fig. 3.4).

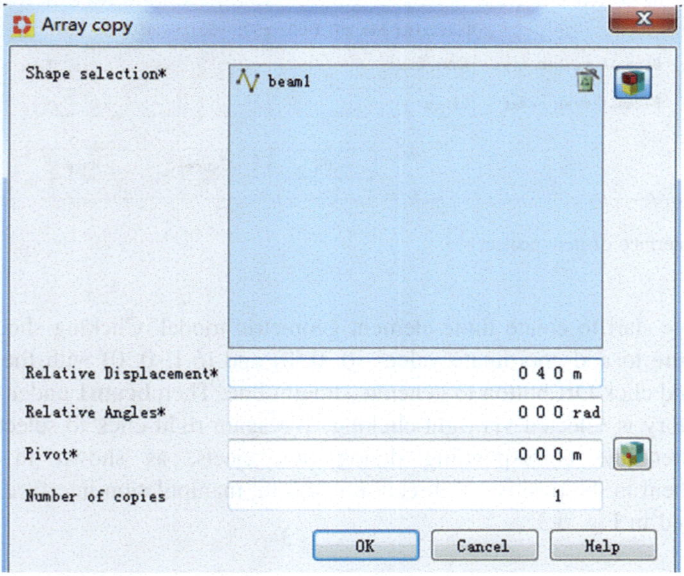

Fig. 3.4 Interface of Array copy

Click shortcut icon **Adds a polyline** to add a polyline (see Fig. 3.5). Untick Closed option to cancel multi-segment closed line. Coordinate values with corresponding locations at the left end of two beams (0, 8.6, 0) and (0, 4.6, 0) are ticked. Then we add (0, 0, 0) in manual displayed (Fig. 3.6). Click OK button; the first column with the name of **column1** is generated.

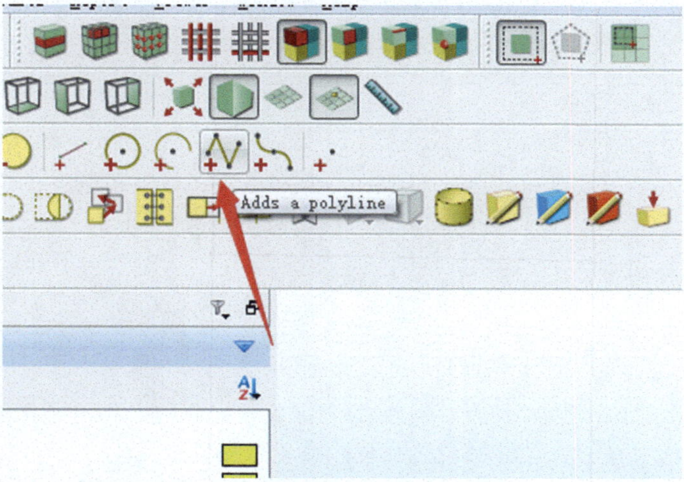

Fig. 3.5 Shortcut icon locations of **Adds a polylines**

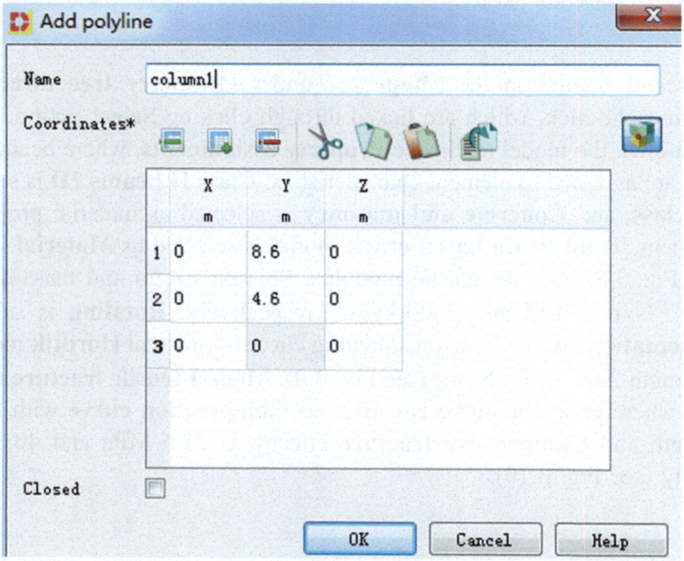

Fig. 3.6 Coordinate values of column

Applying the same method mentioned above to translate **column1** in the positive X direction to generate **column2** with the translational displacement of 6.1 m, integral numerical model is displayed (Fig. 3.7).

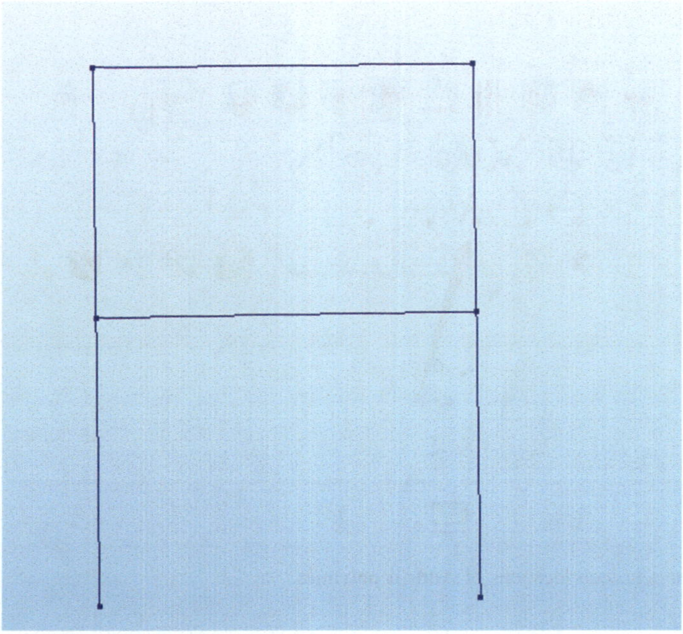

Fig. 3.7 Integral numerical model

Beam1 and **beam2** in the **Shape** bar under **Geometry tree directory** are selected via right-click, which are ticked through clicking **Select** option. Then we again right-click the model to choose **Property assignments**, where beam material properties are assigned to element characteristics. **Class II beams 2D** is selected as **Element class,** and **Concrete and masonry** is selected as material properties of beam element. **Total strain based crack model** is selected as **Material** model, as shown in Fig. 3.8, with the elastic modulus, Poisson's ratio and mass density as 3.45×10^{10} N/m^2, 0.15 and 2500 kg/m^3, respectively. **Rotating** is selected as **Crack orientation** option. Tension softening curve is classical **Hordijk** model with tensile strength 2.64×10^6 N/m^2 (see Fig. 3.9). **Mode-I tensile fracture energy** is 200 N/m, where parabolic curve is chosen as **Compression curve** with compressive strength and **Compressive fracture energy** as 32.5 MPa and 40,000 N/m, respectively (see Fig. 3.10).

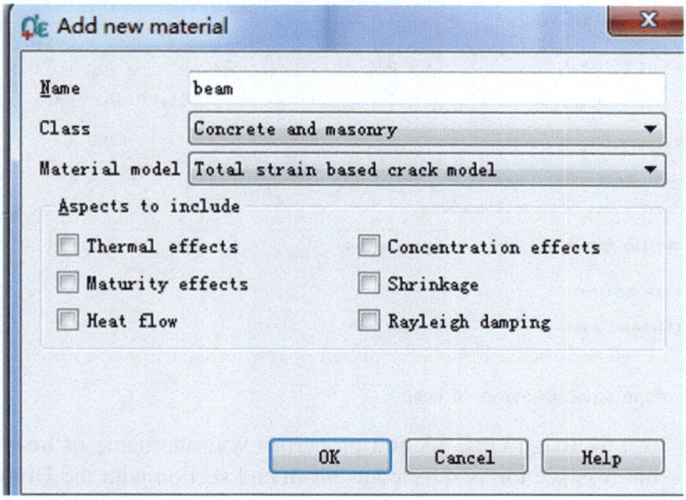

Fig. 3.8 Material model for beam

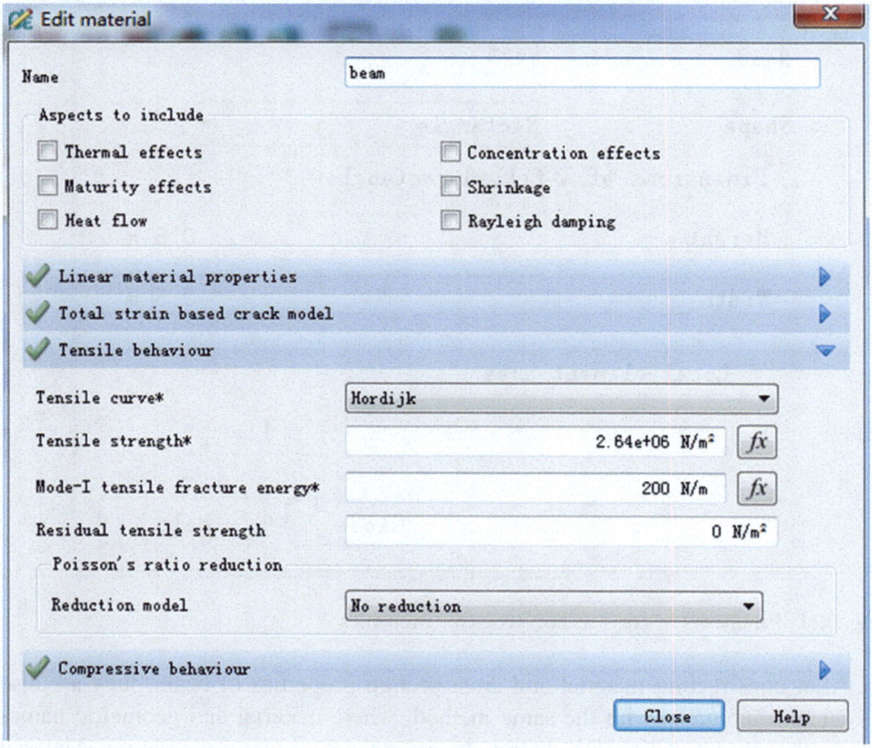

Fig. 3.9 Tensile behavior of Hordijk tension softening model

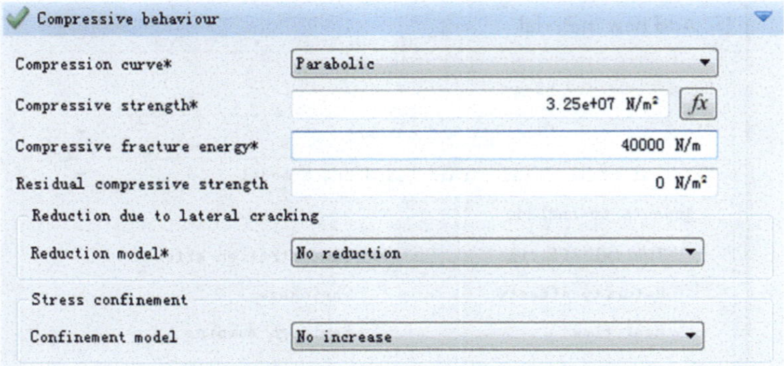

Fig. 3.10 Compressive behaviors of beam

Adding icon to assign cross-section properties with the name of **beam,** rectangular shape of cross-section is selected as beam end section with the **Dimension of a filled rectangle** as, both height and width, 0.5 m (see Fig. 3.11).

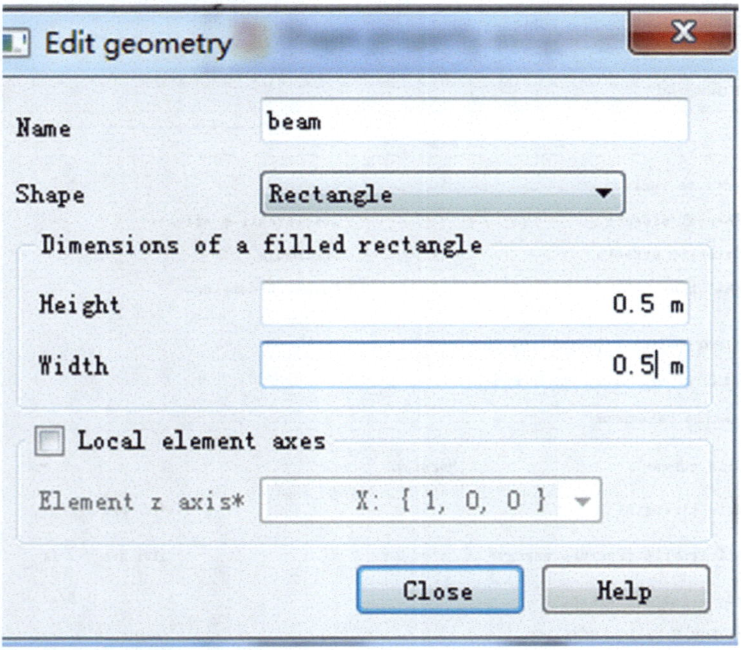

Fig. 3.11 Parameters of cross-section geometric properties

After constructing material and cross-section properties of beam, they are also assigned to columns with the same method, where material and geometric names are both **column**. Click OK button to generate properties of column. Material parameters of columns are the same as beams while the height and width of the cross-section geometric properties are 0.35 and 0.4 m, respectively.

Still clicking shortcut icon button **Adds a line** creates a line with the name of **bar1**, coordinate values (0, 4.5, 0) and (6.1, 4.5, 0) of prestress tendon in the first layer are created. **Bar1** in the graphical user interface region is selected. On right-clicking **Array copy**, prestress tendon in the first layer is duplicated and translated via manipulation of **Array copy** in the positive Y direction at a distance 4 m to generate prestress in the second layer with the name of **bar2** (see Fig. 3.12).

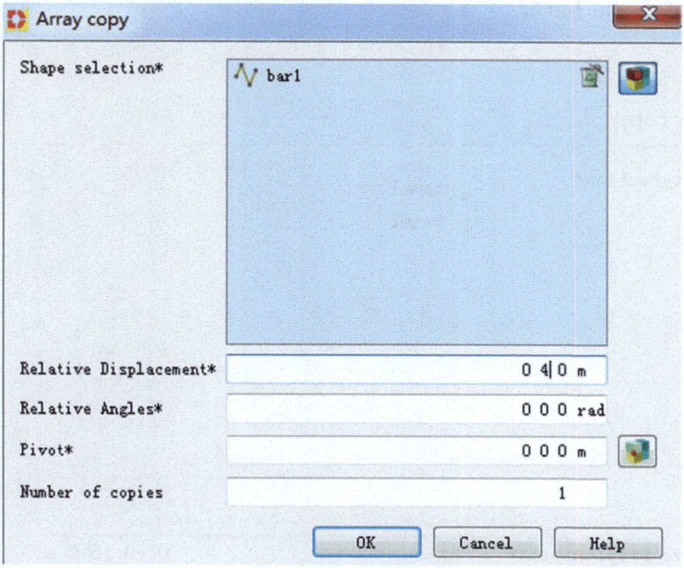

Fig. 3.12 Manipulation of **Array copy**

Both **bar1** and **bar2** are selected and material properties of prestress tendon are assigned to them. **Von Mises plasticity** model under material class **Reinforcements and Pile foundations** is chosen as **Material model**. Elastic modulus and yield stress are 1.95×10^{10} N/m^2 and 1860 MPa, respectively. In specifying cross-section geometric properties of prestress tendons, **Embedded** is selected as **Reinforcement type**. **Cross-section area of bar** is 1.4×10^{-4} m^2 (see Fig. 3.13). For better convergence in nonlinear calculation, **Section wise** is selected as **Discretization Method**.

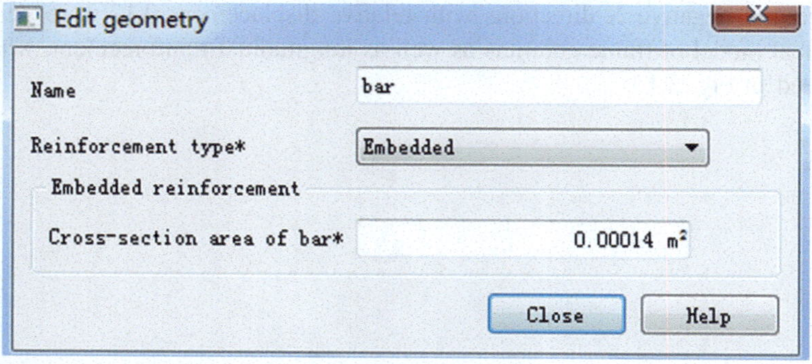

Fig. 3.13 Specification for cross-section geometric properties

Still selecting **beam1** and **beam2** and right-clicking to choose manipulation of **Array copy**, relation displacement is 0.15 m in the negative Y direction and the number of copies is 1, to generate longitudinal reinforcement steels in frame beam with the name of **bar3** and **bar4**, which is shown in Fig. 3.14. Then we apply the same manipulation again in the positive Y direction to translate 0.15 m in order to generate **bar5** and **bar6.**

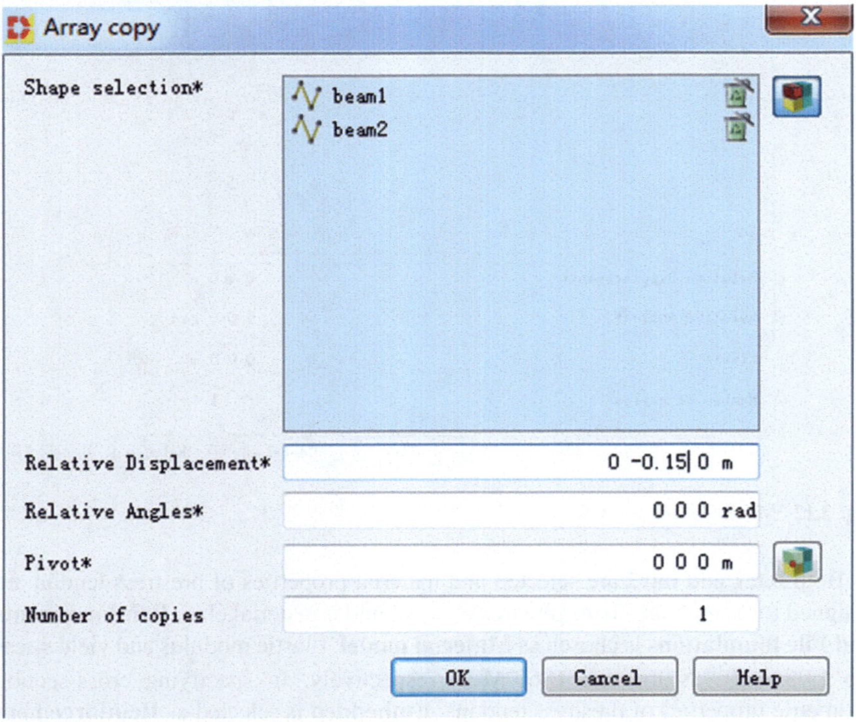

Fig. 3.14 Array copy of **bar1** and **bar2**

The same manipulation is still conducted for **column1** and **column2** in both positive and negative X directions with relative displacement 0.12 m to generate numerical model of frame columns as well as longitudinal reinforcement, which is displayed in Fig. 3.15.

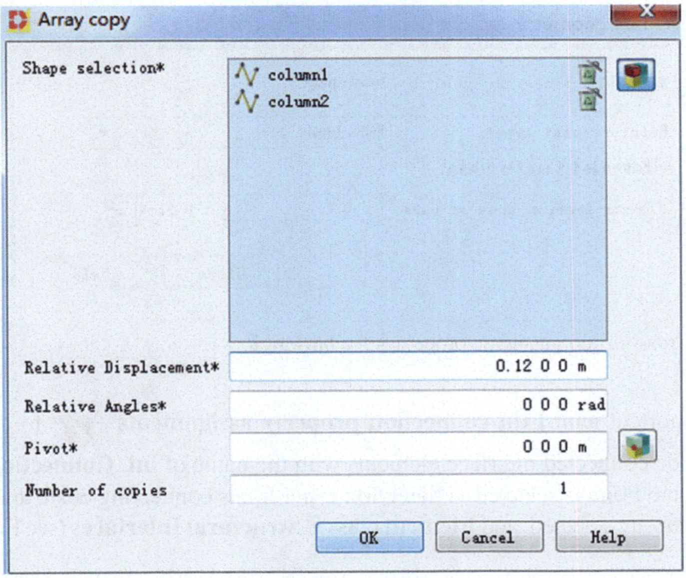

Fig. 3.15 Array copy for **column1** and **column2**

After generation of frame numerical model of beams and columns, material constitutive properties for longitudinal reinforcement steel bars in beams and columns are assigned where **Reinforcements and Pile foundations** is selected as **Class** and **Linear elasticity** is chosen as **Material model**. Names of dialog box are **barlong1** and **barlong2** with elastic modulus 2.1×10^{11} N/m^2. **Embedded** is selected as **Reinforcement type** and cross-section area of bars in beam and column is 1×10^{-4} m^2 and 1.2×10^{-4}m^2, respectively (see Figs. 3.16 and 3.17).

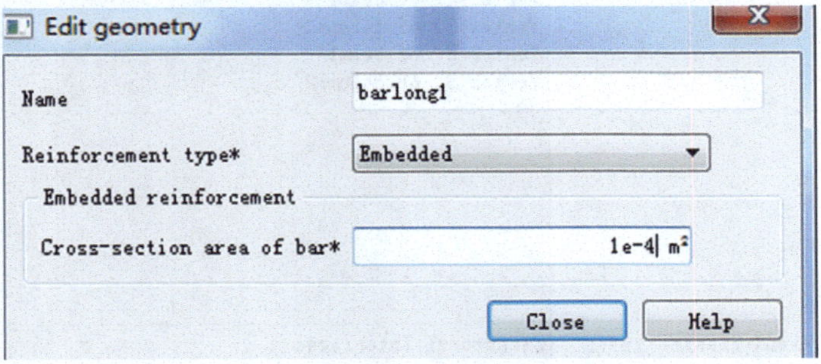

Fig. 3.16 Cross-section geometric properties for **barlong1**

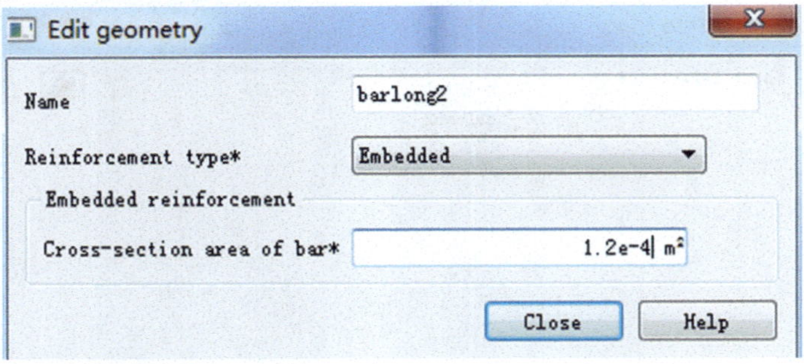

Fig. 3.17 Cross-section geometric properties for **barlong2**

Click shortcut icon **Edit connection property assignments**  to create 2D node to node connected interface elements with the name of **int**. **Connection type** is **Interface** and Point is selected as **Selection type**. Joints connecting beam and column in both layers are selected, and **Element class** is **Structural Interfaces** (see Fig. 3.18).

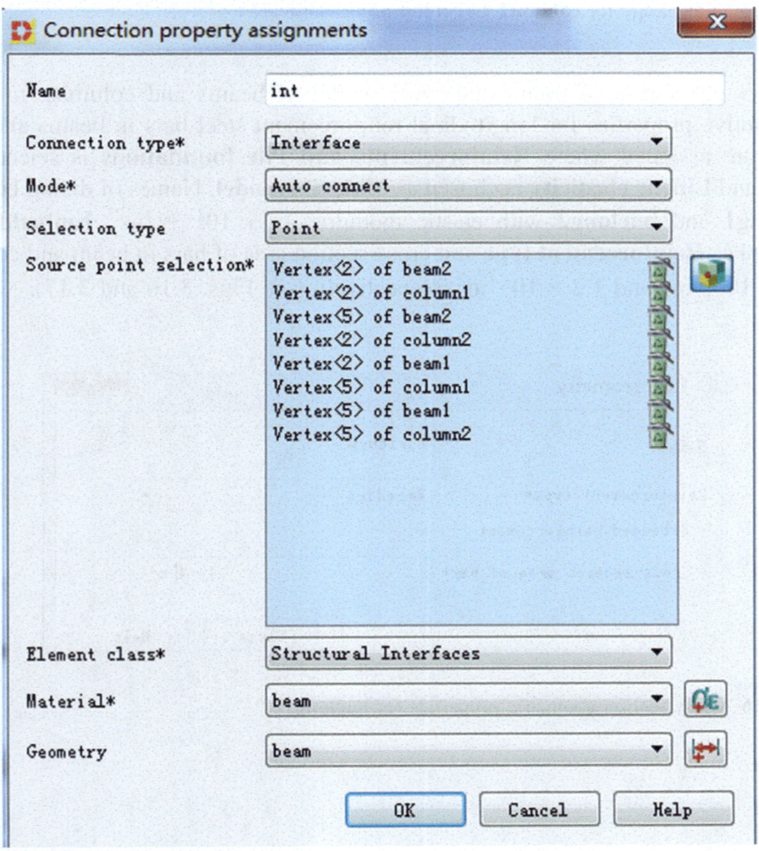

Fig. 3.18 Definition interface of **int**

Click icon [icon] in the **Material** option to specify material class and model. **Interface elements** is selected as **Class** while **Nonlinear elasticity** is chosen as **Material model** (see Fig. 3.19).

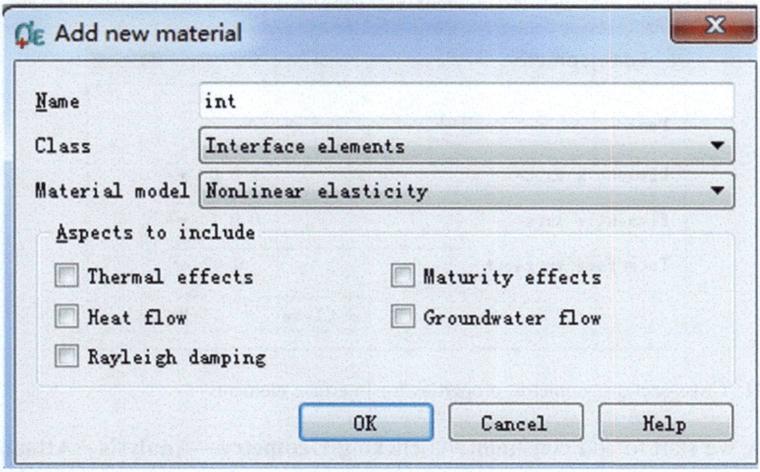

Fig. 3.19 Material class and model

On clicking OK button, material assignment dialog box ejects. **2D point interface** is chosen as **Type** option with **Normal stiffness modulus-x** and **Shear stiffness modulus-y** 3e16 and 3e12 N/m^3, respectively, which is displayed in Fig. 3.20.

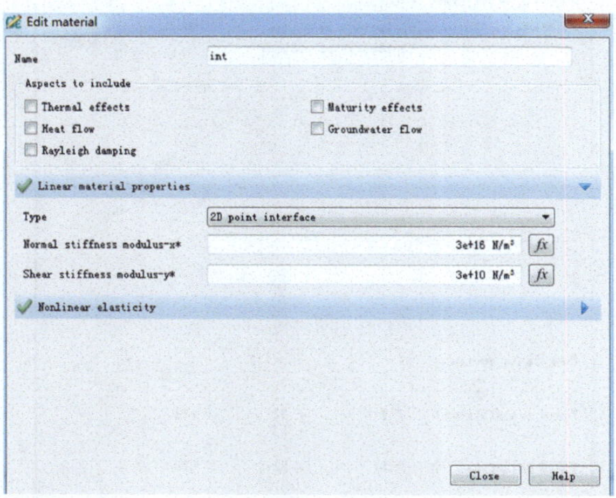

Fig. 3.20 Material properties assignments of **2D point interface**

Clicking shortcut icon  in the Geometry bar to specify cross-section geometric properties of 2D node to node connected interface elements with the name of **int**, Element x-axis under local coordinate system is X-axis under global coordinate system while Element z-axis under local coordinate system is Z-axis under global coordinate system, representing (1, 0, 0) and (0, 0, 1), respectively. Interface surface is 0.02 m^2 (see Fig. 3.21).

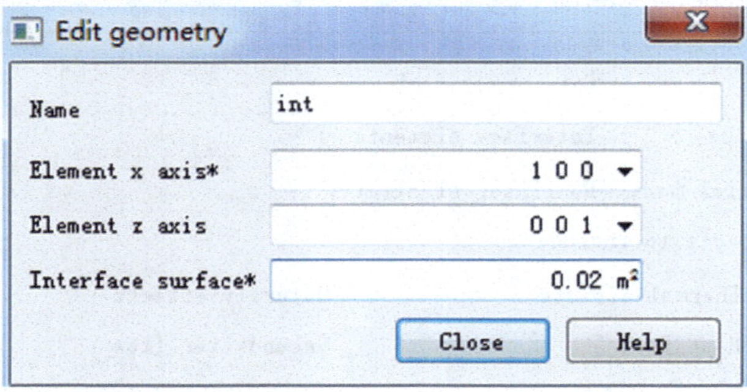

Fig. 3.21 Cross-section geometric properties for interface elements

Now we start to add constraint via clicking **Geometry—Analysis—Attach support** to create supports in the menu bar. Constraint **col** is constructed with **Point** selected as **Support Target type**, then translational and rotational constraints in X and Y directions are attached to the base of the columns, where definition and generation of constraints are displayed (Figs. 3.22 and 3.23).

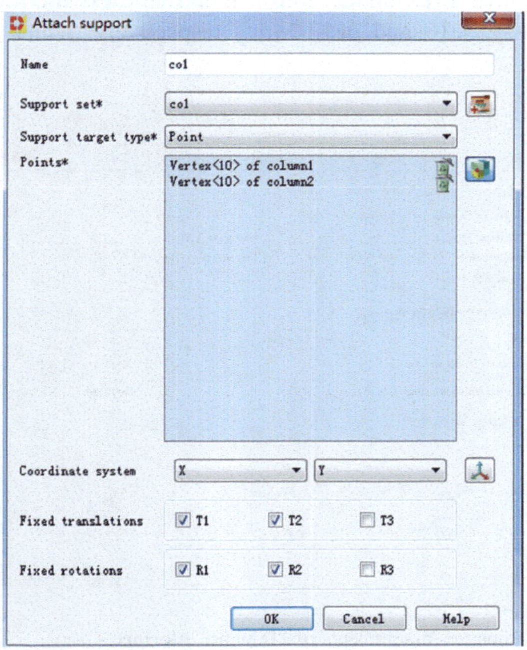

Fig. 3.22 Attachment of constraints

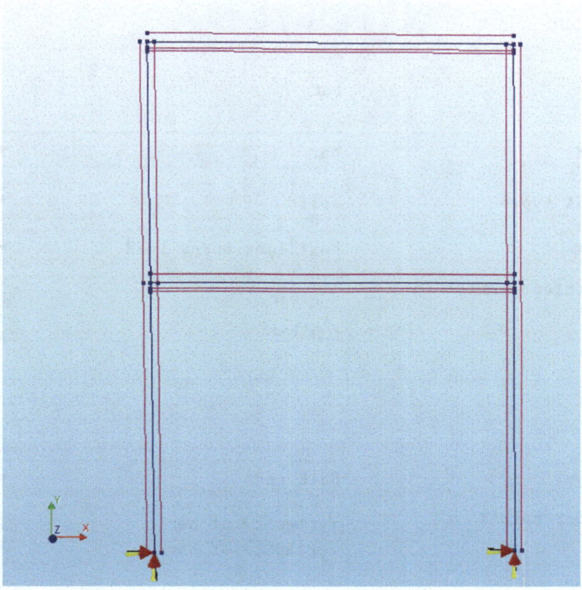

Fig. 3.23 Generation of constraints

The following step guides to attach load. Click **Define a global load** icon under **Load** bar (as displayed in Fig. 3.24); gravity is added with the name of **gravity**. **Dead load** is chosen as **Load type**.

Fig. 3.24 **Define a global load** icon

Load case of prestress tendon is added with the name of **bar**. **Load target type** is **Solid** while **Load Type** is **Post tensioning load**. Meanwhile, **Both ends** option is selected as **Tension type** and prestress value in both layers is 500 kN with **First&second anchor retention length** in the software both 0.01 m. **Coulomb friction coefficient** is 0.22 to simulate the friction between prestress tendons and pipeline wall of bellows. **Wobble factor** affecting curvature of tendon is 0.01/m (see Fig. 3.25).

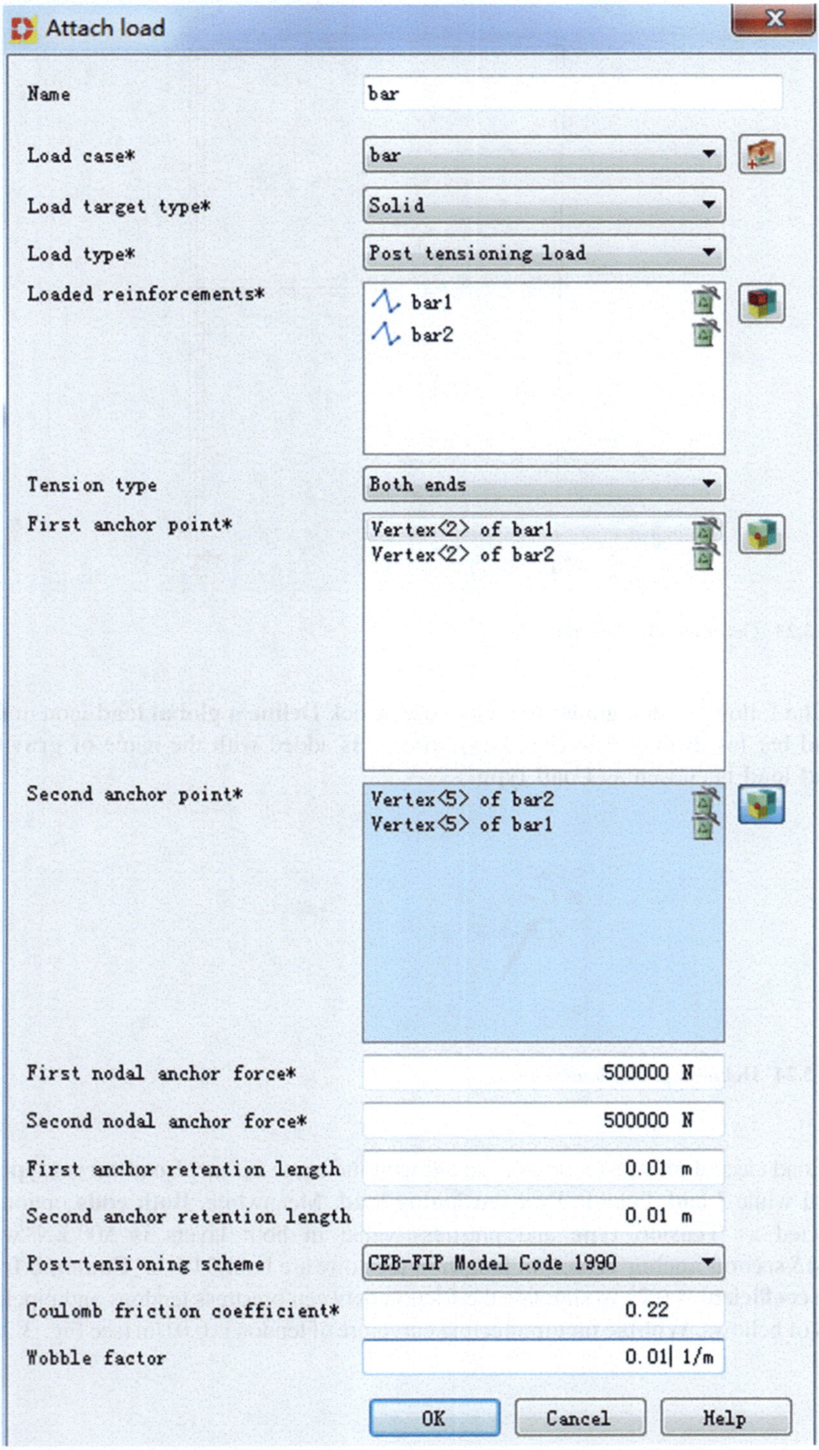

Fig. 3.25 Specification interface of attaching post-tensioning load

Note: Different from former procedure of assigning structural elements, solid in this part refers to geometric elements assigned with material and cross-section geometric properties owing to the fact that they are assigned to prestress tendons.

Another load case with the name of **pressure** is added where **Line** is selected as **Load target type**. **Load type** is **Distributed force**, which is attached in the negative Y position with the loading value of 20 kN/m (see Fig. 3.26).

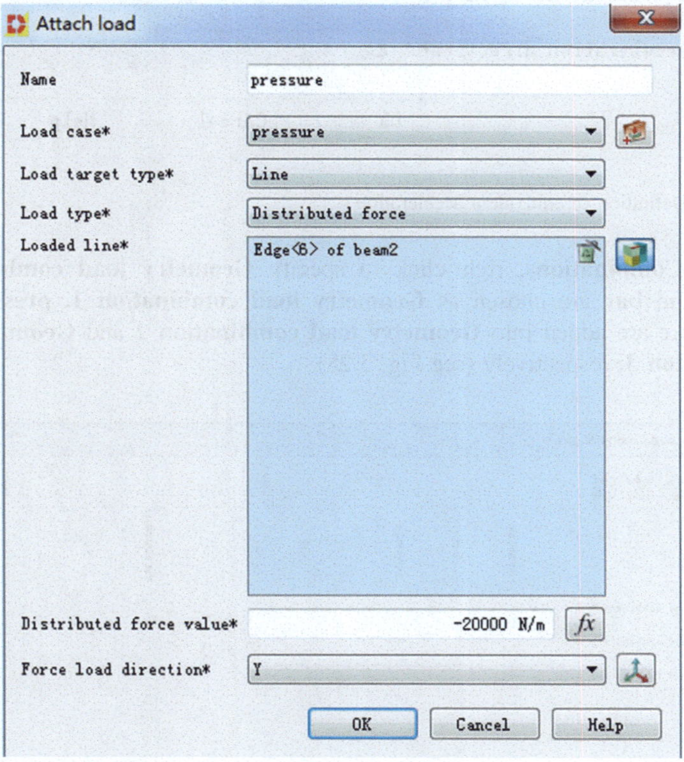

Fig. 3.26 Attaching interface of **pressure** load case

The last load case needed to attach is equivalent acceleration, where the icon is the same as gravity. However, contrary to the former definition of dead load, **Equivalent acceleration** is selected as global load with the name of load case **earthquake**, and the equivalent acceleration is taken as 0.1g in the negative X direction, which is –0.98 m/s^2 (see Fig. 3.27).

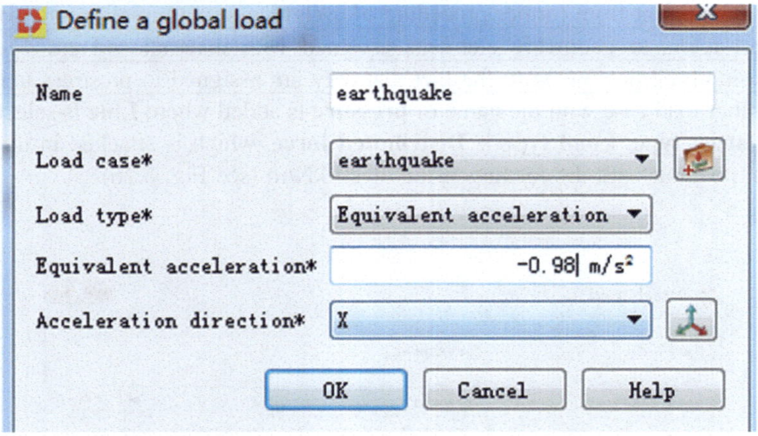

Fig. 3.27 Definition of equivalent acceleration

Click **Combinations**; right-click to specify **Geometry load combinations; gravity** and **bar** are chosen as **Geometry load combination 1, pressure** and **earthquake** are added into **Geometry load combination 2** and **Geometry load combination 3,** respectively (see Fig. 3.28).

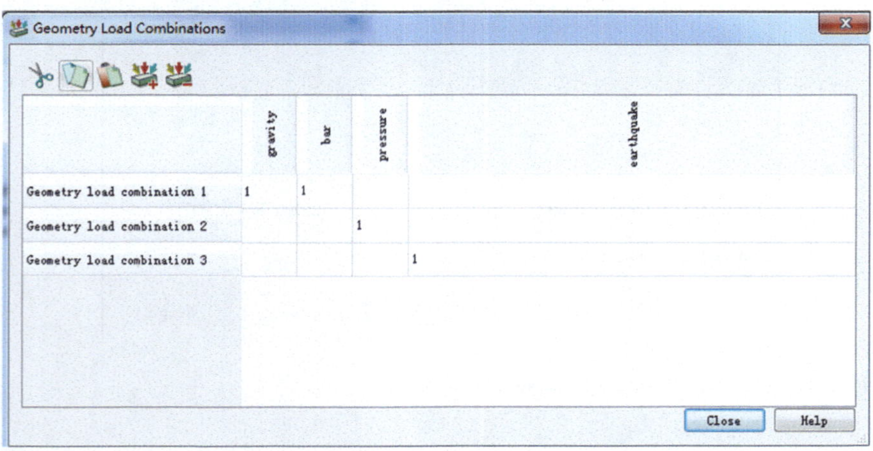

Fig. 3.28 Specification for geometry load combinations

Meshing is the last step for preprocessing procedure. Above all, the whole frame model is selected; then shortcut icon button **set mesh properties of a shape**  is clicked to specify meshing properties for numerical model shape. **Shape** is chosen as **Operation** while **Seeding method** is **Element size** with the **Desired size** 0.1 m (see Fig. 3.29). By clicking the shortcut icon button **Generate mesh of a shape**, meshed elements are generated. In order to further confirm element types are the ones we are desired for, **Element types** bar under the meshing tree directory is checked, and it is observed that Class II beam element type as well as 2D node to node connected interface elements are L7BEN and N4IF, respectively (Fig. 3.30), which also indicates the success of mesh.

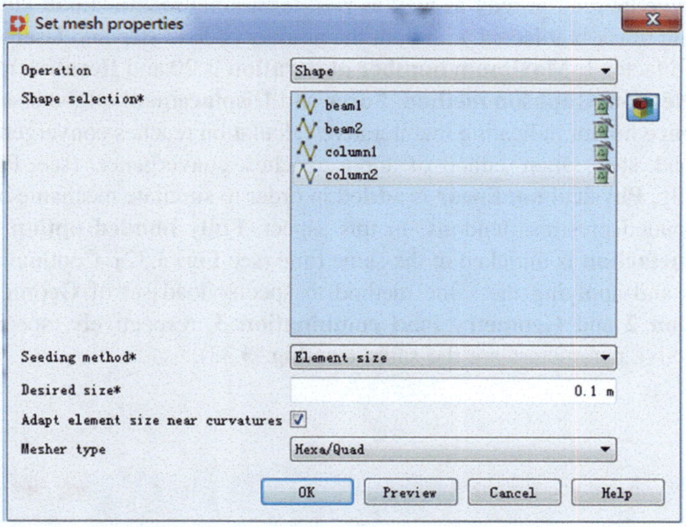

Fig. 3.29 Settings for mesh properties

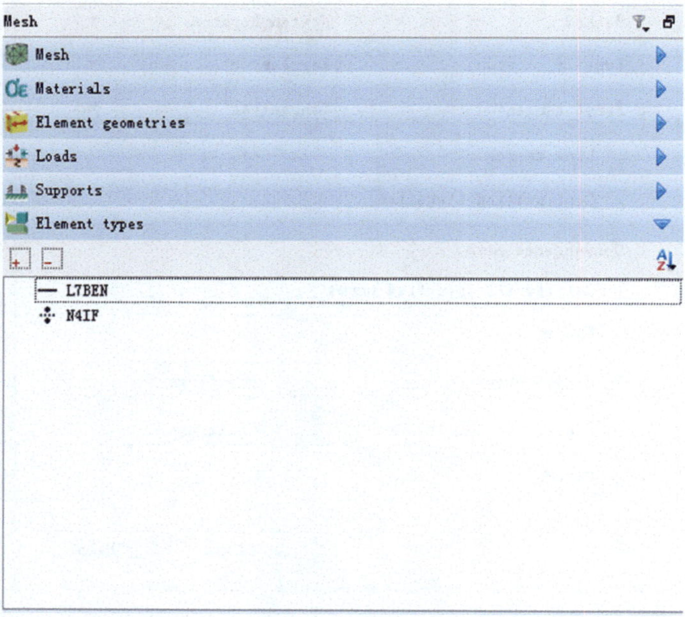

Fig. 3.30 Meshed element types

Create a new analysis module via clicking icon **Add an analysis** button. Then click icon **Add an analysis** , to create new analysis case. **Structural nonlinear** module under **Add command** is right-clicked, then again right-click **Structural**, click **Add-Execute steps-Load steps** to generate new execute block. **Load combination 1** containing gravity as well as post-tensioning load in **Load steps** under new execute block is selected with both the number of load step and **User specified size** of load factor 1. **Maximum number of iteration** is 20 and **Iterative method** is **Regular Newton–Raphson method**. **Force** and **Displacement** are both selected as **Convergence norm,** indicating that iterative calculation reaches convergence under current load step when either of them reaches convergence (see Fig. 3.31). Additionally, **Physical nonlinear** is **added** in order to simulate mechanic behaviors of fully bonded prestress tendons. In this aspect, **Fully bonded option** is ticked while **Liquefaction** is unticked at the same time (see Fig. 3.32). Continuing to add load steps and applying the same method to specify load set of **Geometry load combination 2** and **Geometry load combination 3**, respectively, specifications and interactive parameters are the same (see Fig. 3.33).

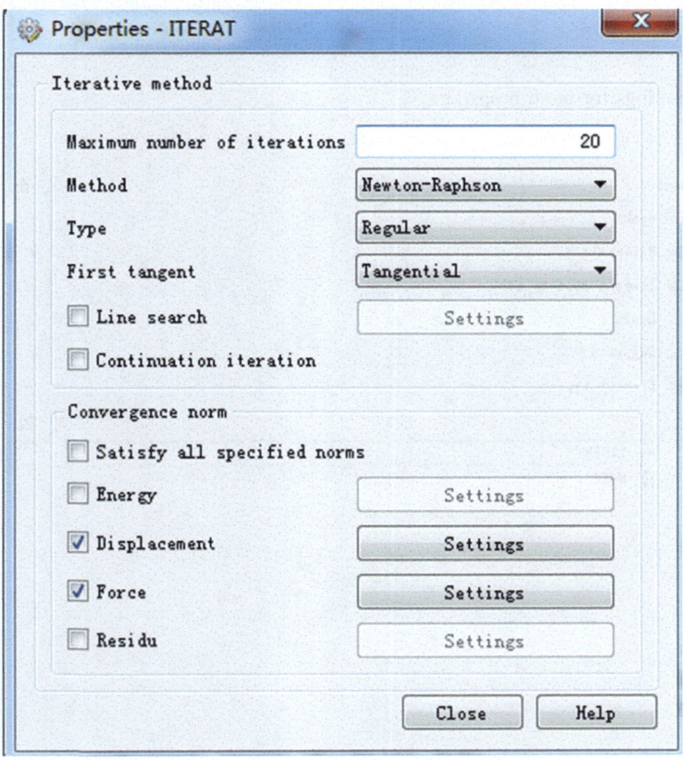

Fig. 3.31 Iterative parameters and specifications

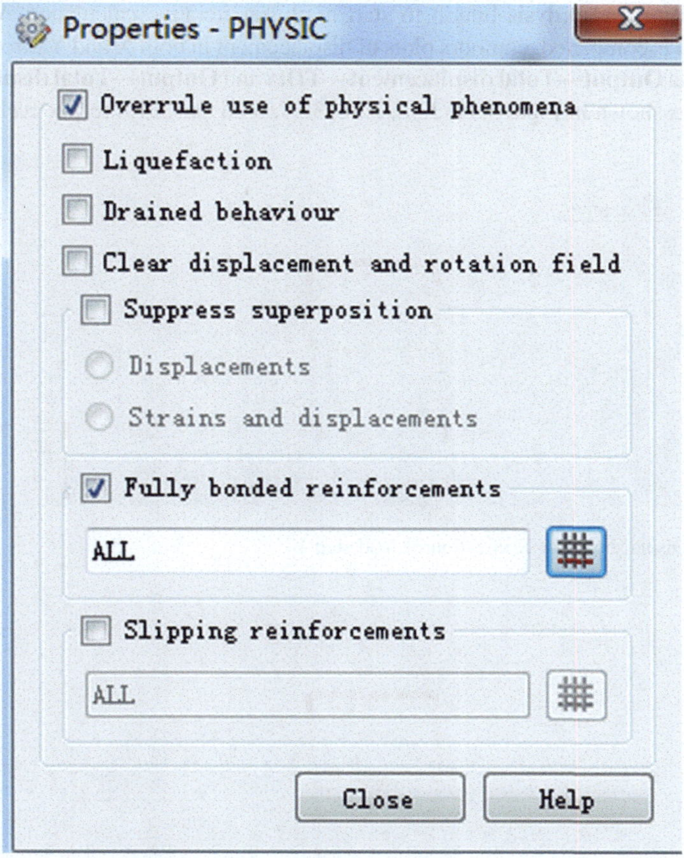

Fig. 3.32 Physical nonlinear aspect for fully bonded mechanic behavior

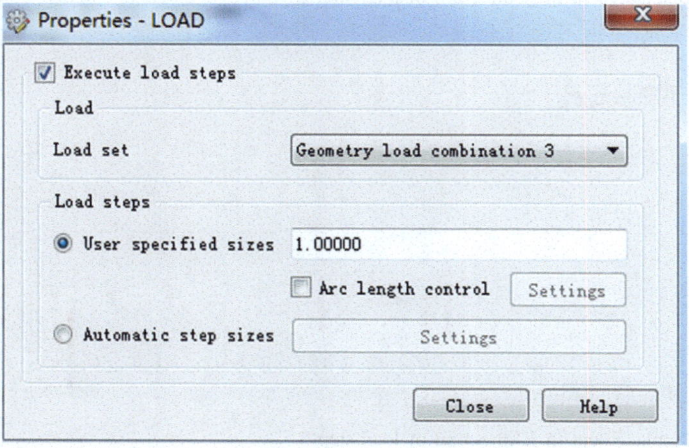

Fig. 3.33 Load set specifications for **Geometry load combination 3**

Click **Run an analysis** button to start nonlinear iterative calculation. When the calculation is completed, contours plots of displacement in both X and Y directions are checked via **Output—Total displacement—TDtx** and **Output—Total displacement—TDty**, as shown in Figs. 3.34, 3.35, 3.36, 3.37, 3.38 and 3.39, respectively.

Fig. 3.34 Displacement in X direction of load step 1

Fig. 3.35 Displacement in X direction of load step 2

Fig. 3.36 Displacement in X direction of load step 3

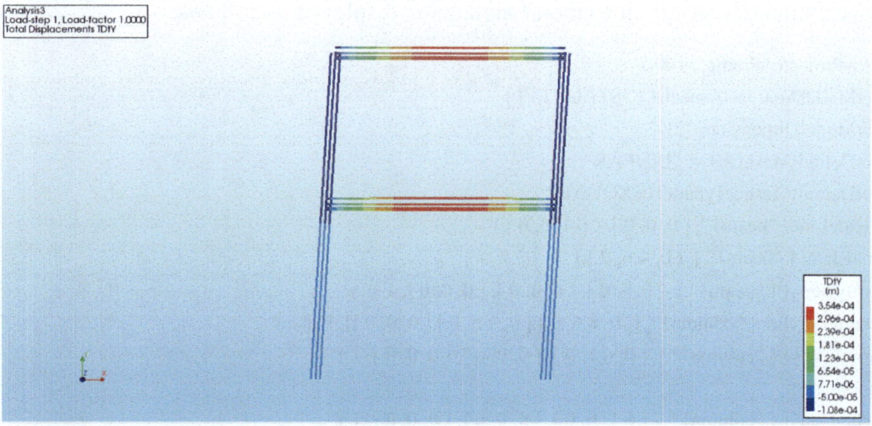

Fig. 3.37 Displacement in Y direction of load step 1

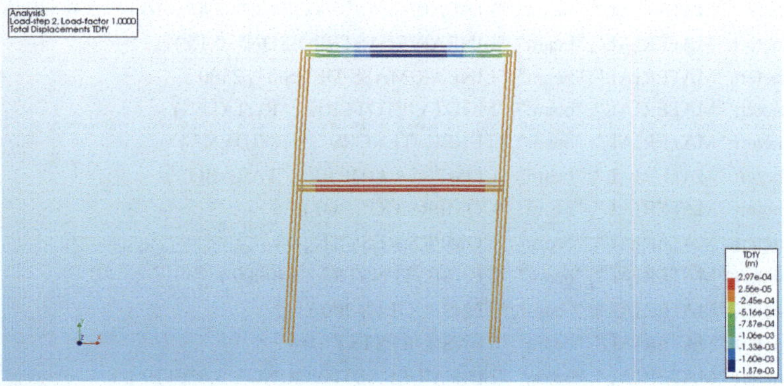

Fig. 3.38 Displacement in Y direction of load step 2

Fig. 3.39 Displacement in Y direction of load step 3

Command console in Python language is displayed as follows:

```
newProject( "Frame", 100 )
setModelAnalysisAspects( [ "STRUCT" ] )
setModelDimension( "2D" )
setDefaultMeshOrder( "LINEAR" )
setDefaultMesherType( "HEXQUAD" )
createLine( "beam1", [ 0, 0, 0 ], [ 6.1, 0, 0 ] )
translate( [ "beam1" ], [ 0, 4.6, 0 ] )
arrayCopy( [ "beam1" ], [ 0, 4, 0 ], [ 0, 0, 0 ], [ 0, 0, 0 ], 1 )
createPolyline( "column1", [[ 0, 8.6, 0 ],[ 0, 4.6, 0 ],[ 0, 0, 0 ]], False )
arrayCopy( [ "column1" ], [ 0, 6.1, 0 ], [ 0, 0, 0 ], [ 0, 0, 0 ], 1 )
undo( 1 )
arrayCopy( [ "column1" ], [ 6.1, 0, 0 ], [ 0, 0, 0 ], [ 0, 0, 0 ], 1 )
saveProject(  )
addMaterial( "beam", "CONCR", "TSCR", [] )
setParameter( "MATERIAL", "beam", "LINEAR/ELASTI/YOUNG", 3.45e+10 )
setParameter( "MATERIAL", "beam", "LINEAR/ELASTI/POISON", 0.15 )
setParameter( "MATERIAL", "beam", "LINEAR/MASS/DENSIT", 2500 )
setParameter( "MATERIAL", "beam", "MODTYP/TOTCRK", "ROTATE" )
setParameter( "MATERIAL", "beam", "TENSIL/TENCRV", "HORDYK" )
setParameter( "MATERIAL", "beam", "COMPRS/COMCRV", "PARABO" )
setParameter( "MATERIAL", "beam", "COMPRS/GC", 200 )
setParameter( "MATERIAL", "beam", "COMPRS/RESCST", 0 )
setParameter( "MATERIAL", "beam", "TENSIL/TENSTR", 2640000 )
setParameter( "MATERIAL", "beam", "TENSIL/GF1", 200 )
setParameter( "MATERIAL", "beam", "TENSIL/RESTST", 0 )
setParameter( "MATERIAL", "beam", "LINEAR/ELASTI/YOUNG", 3.45e+10 )
setParameter( "MATERIAL", "beam", "COMPRS/COMSTR", 32500000 )
setParameter( "MATERIAL", "beam", "COMPRS/GC", 200 )
setParameter( "MATERIAL", "beam", "COMPRS/GC", 40000 )
setParameter( "MATERIAL", "beam", "COMPRS/COMSTR", 32500000 )
addGeometry( "Element geometry 1", "LINE", "CLS2B2", [] )
rename( "GEOMET", "Element geometry 1", "beam" )
setParameter( "GEOMET", "beam", "SHAPE/RECTAN", [ 0.5, 0.5 ] )
setParameter( "GEOMET", "beam", "SHAPE/RECTAN", [ 0.5, 0.5 ] )
setParameter( "GEOMET", "beam", "SHAPE/RECTAN", [ 0.5, 0.5 ] )
clearReinforcementAspects( [ "beam1", "beam2" ] )
setElementClassType( "SHAPE", [ "beam1", "beam2" ], "CLS2B2" )
assignMaterial( "beam", "SHAPE", [ "beam1", "beam2" ] )
assignGeometry( "beam", "SHAPE", [ "beam1", "beam2" ] )
resetElementData( "SHAPE", [ "beam1", "beam2" ] )
saveProject(  )
```

```
addMaterial( "column", "CONCDC", "MC1990", [ "CRACKI", "ELASTI", "PLASTI" ] )
remove( "MATERIAL", "column" )
addMaterial( "column", "CONCR", "TSCR", [] )
setParameter( "MATERIAL", "column", "LINEAR/ELASTI/YOUNG", 3.45e+10 )
setParameter( "MATERIAL", "column", "LINEAR/ELASTI/POISON", 0.15 )
setParameter( "MATERIAL", "column", "LINEAR/MASS/DENSIT", 2500 )
setParameter( "MATERIAL", "column", "MODTYP/TOTCRK", "ROTFIX" )
setParameter( "MATERIAL", "column", "MODTYP/TOTCRK", "ROTATE" )
setParameter( "MATERIAL", "column", "TENSIL/TENCRV", "HORDYK" )
setParameter( "MATERIAL", "column", "TENSIL/TENSTR", 2640000 )
setParameter( "MATERIAL", "column", "TENSIL/GF1", 200 )
setParameter( "MATERIAL", "column", "TENSIL/RESTST", 0 )
setParameter( "MATERIAL", "column", "COMPRS/COMCRV", "PARABO" )
setParameter( "MATERIAL", "column", "COMPRS/COMSTR", 32500000 )
setParameter( "MATERIAL", "column", "COMPRS/GC", 40000 )
setParameter( "MATERIAL", "column", "COMPRS/RESCST", 0 )
setParameter( "MATERIAL", "column", "COMPRS/RESCST", 0 )
addGeometry( "Element geometry 2", "LINE", "CLS2B2", [] )
rename( "GEOMET", "Element geometry 2", "column" )
setParameter( "GEOMET", "column", "SHAPE/RECTAN", [ 0.35, 0.4 ] )
clearReinforcementAspects( [ "column1", "column2" ] )
setElementClassType( "SHAPE", [ "column1", "column2" ], "CLS2B2" )
assignMaterial( "column", "SHAPE", [ "column1", "column2" ] )
assignGeometry( "column", "SHAPE", [ "column1", "column2" ] )
resetElementData( "SHAPE", [ "column1", "column2" ] )
createLine( "bar1", [ 0, 4.5, 0 ], [ 6.1, 4.5, 0 ] )
saveProject(  )
arrayCopy( [ "bar1" ], [ 0, 4, 0 ], [ 0, 0, 0 ], [ 0, 0, 0 ], 1 )
saveProject(  )
addMaterial( "bar", "REINFO", "VMISES", [] )
setParameter( "MATERIAL", "bar", "LINEAR/ELASTI/YOUNG", 1.95e+11 )
setParameter( "MATERIAL", "bar", "LINEAR/ELASTI/YOUNG", 1.95e+11 )
setParameter( "MATERIAL", "bar", "PLASTI/HARDI1/YLDSTR", 1.86e+09 )
setMaterialAspects( "bar", [ "NOBOND" ] )
addGeometry( "Element geometry 3", "RELINE", "REBAR", [] )
rename( "GEOMET", "Element geometry 3", "bar" )
setParameter( "GEOMET", "bar", "REIEMB/CROSSE", 0.00014 )
setReinforcementAspects( [ "bar1", "bar2" ] )
assignMaterial( "bar", "SHAPE", [ "bar1", "bar2" ] )
assignGeometry( "bar", "SHAPE", [ "bar1", "bar2" ] )
resetElementData( "SHAPE", [ "bar1", "bar2" ] )
```

```
setReinforcementDiscretization( [ "bar1", "bar2" ], "SECTION" )
arrayCopy( [ "beam1", "beam2" ], [ 0, -0.15, 0 ], [ 0, 0, 0 ], [ 0, 0, 0 ], 1 )
arrayCopy( [ "beam1", "beam2" ], [ 0, 0.15, 0 ], [ 0, 0, 0 ], [ 0, 0, 0 ], 1 )
renameShape( "beam3", "bar3" )
renameShape( "beam4", "bar4" )
renameShape( "beam5", "bar5" )
renameShape( "beam6", "bar6" )
arrayCopy( [ "column1", "column2" ], [ 0.12, 0, 0 ], [ 0, 0, 0 ], [ 0, 0, 0 ], 1 )
renameShape( "column3", "bar7" )
renameShape( "column4", "bar8" )
arrayCopy( [ "column1", "column2" ], [ -0.12, 0, 0 ], [ 0, 0, 0 ], [ 0, 0, 0 ], 1 )
renameShape( "column3", "bar9" )
renameShape( "column4", "bar10" )
addMaterial( "barlong1", "REINFO", "LINEAR", [] )
setParameter( "MATERIAL", "barlong1", "LINEAR/ELASTI/YOUNG", 2.1e+11 )
addGeometry( "Element geometry 4", "RELINE", "REBAR", [] )
rename( "GEOMET", "Element geometry 4", "barlong1" )
setParameter( "GEOMET", "barlong1", "REIEMB/CROSSE", 0.0001 )
setReinforcementAspects( [ "bar3", "bar4", "bar5", "bar6" ] )
assignMaterial( "barlong1", "SHAPE", [ "bar3", "bar4", "bar5", "bar6" ] )
assignGeometry( "barlong1", "SHAPE", [ "bar3", "bar4", "bar5", "bar6" ] )
resetElementData( "SHAPE", [ "bar3", "bar4", "bar5", "bar6" ] )
setReinforcementDiscretization( [ "bar3", "bar4", "bar5", "bar6" ], "SECTION" )
saveProject(  )
addMaterial( "barlong2", "REINFO", "LINEAR", [] )
setParameter( "MATERIAL", "barlong2", "LINEAR/ELASTI/YOUNG", 2.1e+11 )
addGeometry( "Element geometry 5", "RELINE", "REBAR", [] )
rename( "GEOMET", "Element geometry 5", "barlong2" )
setParameter( "GEOMET", "barlong2", "REIEMB/CROSSE", 0.00012 )
setParameter( "GEOMET", "barlong2", "REIEMB/CROSSE", 0.00012 )
setReinforcementAspects( [ "bar7", "bar8", "bar9", "bar10" ] )
assignMaterial( "barlong2", "SHAPE", [ "bar7", "bar8", "bar9", "bar10" ] )
assignGeometry( "barlong2", "SHAPE", [ "bar7", "bar8", "bar9", "bar10" ] )
resetElementData( "SHAPE", [ "bar7", "bar8", "bar9", "bar10" ] )
setReinforcementDiscretization( [ "bar7", "bar8", "bar9", "bar10" ], "SECTION" )
saveProject(  )
saveProject(  )
addMaterial( "int", "INTERF", "NONLIF", [] )
setParameter( MATERIAL, "int", "LINEAR/IFTYP", "PNT2D" )
setParameter( MATERIAL, "int", "LINEAR/IFTYP", "PNT2D" )
setParameter( MATERIAL, "int", "LINEAR/ELAS1/DSNX", 3e+16 )
```

```
setParameter( MATERIAL, "int", "LINEAR/ELAS1/DSSY", 3e+10 )
setParameter( MATERIAL, "int", "NONLIN/IFNOTE", "NOTENS" )
addGeometry( "Element geometry 6", "POINT", "STRINT", [] )
rename( GEOMET, "Element geometry 6", "int" )
setParameter( GEOMET, "int", "SURFAC", 0.02 )
setParameter( GEOMET, "int", "SURFAC", 0.02 )
setParameter( GEOMET, "int", "SURFAC", 0.02 )
createPointConnection( "int" )
setParameter( GEOMETRYCONNECTION, "int", "CONTYP", "INTER" )
setParameter( GEOMETRYCONNECTION, "int", "MODE", "AUTO" )
attachTo( GEOMETRYCONNECTION, "int", "SOURCE", "beam1", [[ 0, 4.6, 0 ],[ 6.1, 4.6, 0 ]] )
attachTo( GEOMETRYCONNECTION, "int", "SOURCE", "beam2", [[ 0, 8.6, 0 ],[ 6.1, 8.6, 0 ]] )
attachTo( GEOMETRYCONNECTION, "int", "SOURCE", "column1", [[ 0, 8.6, 0 ],[ 0, 4.6, 0 ]] )
attachTo( GEOMETRYCONNECTION, "int", "SOURCE", "column2", [[ 6.1, 8.6, 0 ],[ 6.1, 4.6,
0 ]] )
setElementClassType( GEOMETRYCONNECTION, "int", "STRINT" )
assignMaterial( "int", GEOMETRYCONNECTION, "int" )
assignGeometry( "int", GEOMETRYCONNECTION, "int" )
saveProject(  )
addSet( "GEOMETRYSUPPORTSET", "co1" )
createPointSupport( "co1", "co1" )
setParameter( "GEOMETRYSUPPORT", "co1", "AXES", [ 1, 2 ] )
setParameter( "GEOMETRYSUPPORT", "co1", "TRANSL", [ 1, 1, 0 ] )
setParameter( "GEOMETRYSUPPORT", "co1", "ROTATI", [ 1, 1, 0 ] )
attach( "GEOMETRYSUPPORT", "co1", "column1", [[ 0, 0, 0 ]] )
attach( "GEOMETRYSUPPORT", "co1", "column2", [[ 6.1, 0, 0 ]] )
saveProject(  )
addSet( "GEOMETRYLOADSET", "gravity" )
createModelLoad( "gravity", "gravity" )
saveProject(  )
addSet( "GEOMETRYLOADSET", "Geometry load case 2" )
rename( "GEOMETRYLOADSET", "Geometry load case 2", "bar" )
createBodyLoad( "bar", "bar" )
setParameter( "GEOMETRYLOAD", "bar", "LODTYP", "POSTEN" )
setParameter( "GEOMETRYLOAD", "bar", "POSTEN/BOTHEN/FORCE1", 500000 )
setParameter( "GEOMETRYLOAD", "bar", "POSTEN/BOTHEN/FORCE2", 500000 )
setParameter( "GEOMETRYLOAD", "bar", "POSTEN/BOTHEN/RETLE1", 0.01 )
setParameter( "GEOMETRYLOAD", "bar", "POSTEN/BOTHEN/RETLE2", 0.01 )
setParameter( "GEOMETRYLOAD", "bar", "POSTEN/SHEAR", 0.22 )
setParameter( "GEOMETRYLOAD", "bar", "POSTEN/WOBBLE", 0.01 )
attach( "GEOMETRYLOAD", "bar", [ "bar1", "bar2" ] )
```

```
attachTo( "GEOMETRYLOAD", "bar", "POSTEN/BOTHEN/PNTS1", "bar1", [[ 0, 4.5, 0 ]] )
attachTo( "GEOMETRYLOAD", "bar", "POSTEN/BOTHEN/PNTS1", "bar2", [[ 0, 8.5, 0 ]] )
attachTo( "GEOMETRYLOAD", "bar", "POSTEN/BOTHEN/PNTS2", "bar1", [[ 6.1, 4.5, 0 ]] )
attachTo( "GEOMETRYLOAD", "bar", "POSTEN/BOTHEN/PNTS2", "bar2", [[ 6.1, 8.5, 0 ]] )
saveProject( )
addSet( "GEOMETRYLOADSET", "Geometry load case 3" )
addSet( "GEOMETRYLOADSET", "pressure" )
remove( "GEOMETRYLOADSET", [ "Geometry load case 3" ] )
createLineLoad( "pressure", "pressure" )
setParameter( "GEOMETRYLOAD", "pressure", "FORCE/VALUE", -20000 )
setParameter( "GEOMETRYLOAD", "pressure", "FORCE/DIRECT", 2 )
attach( "GEOMETRYLOAD", "pressure", "beam2", [[ 3.05, 8.6, 0 ]] )
saveProject( )
createModelLoad( "earthquake", "earthquake" )
setParameter( GEOMETRYLOAD, "earthquake", "LODTYP", "EQUIAC" )
setParameter( GEOMETRYLOAD, "earthquake", "EQUIAC/ACCELE", -0.98 )
setParameter( GEOMETRYLOAD, "earthquake", "EQUIAC/DIRECT", 1 )
saveProject( )
setDefaultGeometryLoadCombinations( )
setGeometryLoadCombinationFactor( "Geometry load combination 1", "gravity", 1 )
remove( GEOMETRYLOADCOMBINATION, "Geometry load combination 4" )
remove( GEOMETRYLOADCOMBINATION, "Geometry load combination 3" )
remove( GEOMETRYLOADCOMBINATION, "Geometry load combination 2" )
setGeometryLoadCombinationFactor( "Geometry load combination 1", "gravity", 1 )
setGeometryLoadCombinationFactor( "Geometry load combination 1", "bar", 1 )
addGeometryLoadCombination( "" )
setGeometryLoadCombinationFactor( "Geometry load combination 2", "pressure", 1 )
addGeometryLoadCombination( "" )
setGeometryLoadCombinationFactor( "Geometry load combination 3", "earthquake", 1 )
saveProject( )
setElementSize( "beam1", 1, [[ 3.05, 4.6, 0 ]], 0.1, 0, True )
setElementSize( "beam2", 1, [[ 3.05, 8.6, 0 ]], 0.1, 0, True )
setElementSize( "column1", 1, [[ 0, 6.6, 0 ],[ 0, 2.3, 0 ]], 0.1, 0, True )
setElementSize( "column2", 1, [[ 6.1, 6.6, 0 ],[ 6.1, 2.3, 0 ]], 0.1, 0, True )
saveProject( )
generateMesh( [] )
hideView( "GEOM" )
showView( "MESH" )
addAnalysis( "Analysis3" )
addAnalysisCommand( "Analysis3", "NONLIN", "Structural nonlinear" )
renameAnalysis( "Analysis3", "Analysis3" )
renameAnalysisCommand( "Analysis3", "Structural nonlinear", "Structural nonlinear" )
addAnalysisCommandDetail( "Analysis3", "Structural nonlinear",
```

```
"MODEL/EVALUA/REINFO/INTERF" )
setAnalysisCommandDetail( "Analysis3", "Structural nonlinear", "MODEL/EVALUA/REINFO/INTERF", True )
removeAnalysisCommandDetail( "Analysis3", "Structural nonlinear", "EXECUT(1)" )
setAnalysisCommandDetail( "Analysis3", "Structural nonlinear", "EXECUT/EXETYP", "LOAD" )
renameAnalysisCommandDetail( "Analysis3", "Structural nonlinear", "EXECUT(1)", "bar" )
setAnalysisCommandDetail( "Analysis3", "Structural nonlinear", "EXECUT(1)/ITERAT/MAXITE", 20 )
setAnalysisCommandDetail( "Analysis3", "Structural nonlinear", "EXECUT(1)/ITERAT/CONVER/DISPLA/NOCONV", "CONTIN" )
setAnalysisCommandDetail( "Analysis3", "Structural nonlinear", "EXECUT(1)/ITERAT/CONVER/FORCE/NOCONV", "CONTIN" )
setAnalysisCommandDetail( "Analysis3", "Structural nonlinear", "EXECUT/EXETYP", "LOAD" )
renameAnalysisCommandDetail( "Analysis3", "Structural nonlinear", "EXECUT(2)", "PRESSURE" )
setAnalysisCommandDetail( "Analysis3", "Structural nonlinear", "EXECUT(2)/LOAD/LOADNR", 2 )
setAnalysisCommandDetail( "Analysis3", "Structural nonlinear", "EXECUT(2)/ITERAT/MAXITE", 20 )
setAnalysisCommandDetail( "Analysis3", "Structural nonlinear", "EXECUT(2)/ITERAT/CONVER/DISPLA/NOCONV", "CONTIN" )
setAnalysisCommandDetail( "Analysis3", "Structural nonlinear", "EXECUT(2)/ITERAT/CONVER/FORCE/NOCONV", "CONTIN" )
setAnalysisCommandDetail( "Analysis3", "Structural nonlinear", "EXECUT/EXETYP", "LOAD" )
renameAnalysisCommandDetail( "Analysis3", "Structural nonlinear", "EXECUT(3)", "earthquake" )
setAnalysisCommandDetail( "Analysis3", "Structural nonlinear", "EXECUT(3)/LOAD/LOADNR", 2 )
setAnalysisCommandDetail( "Analysis3", "Structural nonlinear", "EXECUT(3)/ITERAT/MAXITE", 20 )
setAnalysisCommandDetail( "Analysis3", "Structural nonlinear", "EXECUT(3)/ITERAT/CONVER/DISPLA/NOCONV", "CONTIN" )
setAnalysisCommandDetail( "Analysis3", "Structural nonlinear", "EXECUT(3)/ITERAT/CONVER/FORCE/NOCONV", "CONTIN" )
saveProject( )
runSolver( "Analysis3" )
showView( "RESULT" )
setResultPlot( "contours", "Total Displacements/node", "TDtZ" )
setResultPlot( "contours", "Total Displacements/node", "TDtY" )
setResultPlot( "contours", "Total Displacements/node", "TDtX" )
```

3.2 Bonded Steel Strengthening Case of Box Girder

A simply supported concrete box girder is simulated in this case. Longitudinal length of the whole box girder is 10 m, while the width of top and bottom plate is 6 and 4 m, respectively, with the thickness of web 0.45 m, as displayed in Fig. 3.40. Concrete grade is C50, and CEB-FIP 1990 model for creep and shrinkage is applied in this numerical simulation. Additionally, the diameters of longitudinal steel bars at the top and bottom plate are both 8 mm with the spacing 1.2 m. Steel plate with the size of 10 m × 4 m × 0.01 m is pasted and bonded at the bottom of the girder. A quadratic solid hexahedral element CHX60 in DIANA 10.1 is modeled for concrete while the interface between plate and bottom concrete is simulated by a surface to surface connected interface element CQ48I with the bond-slip specified in material constitutive model of bonded contacted interface, as shown in Fig. 3.41. Vertical distributed load with the value of 50 kN/m^2 is applied to the top of the box girder in downward direction. Phased analysis is applied to simulate the whole-bonded steel strengthening process, where the long-term deterioration process of simply supported box girder is simulated in the first phase. Total time is set

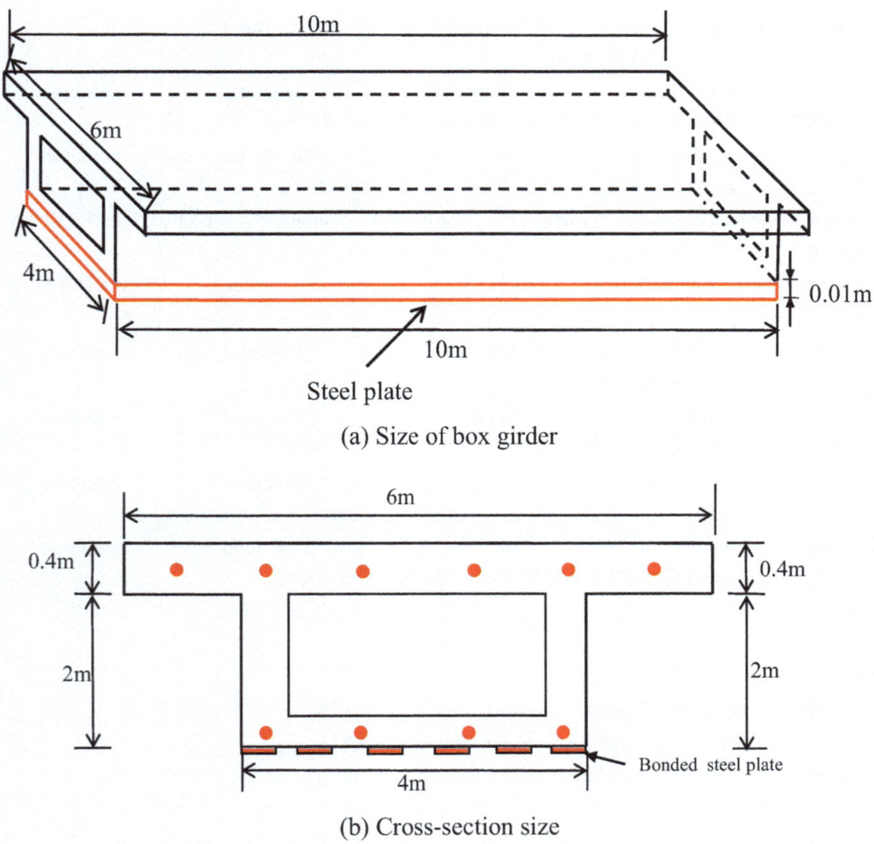

Fig. 3.40 Model size of the whole box girder

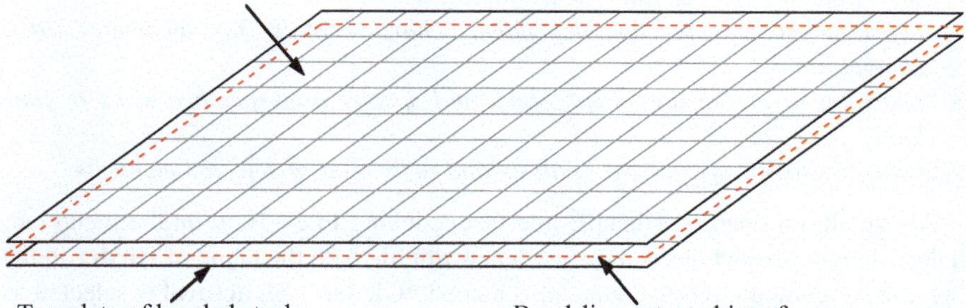

Fig. 3.41 Schematic diagram of contact surface

as 100 years (3.1536×10^9 s) and it is assumed in this case that deflection in the middle site of the span exceeds allowable controlling value when time reaches 100 years. In the second phase, steel plate element is activated and strengthening stage is started. After the nonlinear calculation, displacements before and after the paste in the middle of the site are compared to validate the strengthening effect of bonded steel plate. Moreover, extraction of .py files generated by graphical user interface and command console modification are also displayed in detail in this case and the command console in Python language is imported into DianaIE to achieve automatic calculation results.

Note: It is normally assumed that strengthening measures should immediately be taken when deflection exceeds allowable value. For instance, deflection to span of bridge is ranging from 1/200 to 1/250 under serviceability limit state. In order to better display the procedure of phase analysis for readers in this numerical case, 100 years is set as the initial strengthening point, and simulation of deflection controlling effect of steel plate under time-dependent effect after strengthening is omitted. Phase analysis of secondary loading strengthening method. Please forgive if there are any inconsistencies with the real engineering construction.

Essentials of learning

(1) *Learning Boolean addition and subtraction for constructing hollow model plane in geometric modeling.*
(2) *Learning to establish 3D volume model of box girder* via *the manipulation of* ***Extrude a shape***.
(3) *Learning to specify time-dependent parameters in European CEB-FIP 1990 model.*
(4) *Learning to define surface to surface connected interface elements between solid elements*
(5) *Learning to master a relationship of shear traction and relative displacement bi-linear constitutive model under multi-linear bond-slip material model in the surface to surface connected interface element.*

(6) *Learning the manipulations of adding phased analysis in the bonded steel strengthening process and stress element activation.*
(7) *Learning to create time-dependent characteristics for geometry load combinations*
(8) *Learning to create time steps block and specify user-specified sizes of time load.*
(9) *Learning to specify phased analysis and activation of selected elements.*

Above all, on opening DianaIE interface, clicking File—New on the menu bar, dialog of **New project** ejects, creating a new dpf file with the name of **Girder** in the directory of computer G-disk area, also named **"Girder"**. **Structural** is selected as **Analysis** type and the dimension is three (3D). **Model size** is 100 m, representing that the scope of the whole graphical visualization zone ranges from –50 to 50 m in X, Y and Z directions under global coordinate system. **Default mesher type** is Hexa/Quad, representing that the geometric element shapes are all quadrilateral (2 dimensions) or hexahedron (3 dimensions). **Quadratic** order is selected as default mesh order and the determination of **mid-side node location** is **Linear interpolation** (see Fig. 3.42).

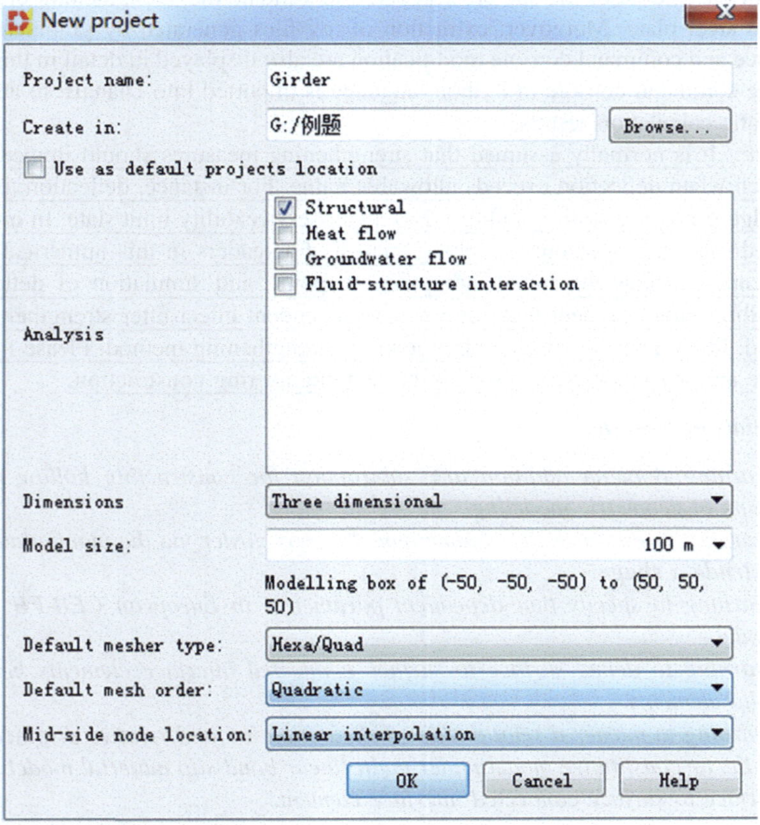

Fig. 3.42 New project

Click the shortcut icon under menu bar **Adds a sheet** to create **Sheet1**, as Fig. 3.43 displays. Input coordinate points to create outer profile of box girder cross-section, and the coordinate values of every point are shown in Table 3.1.

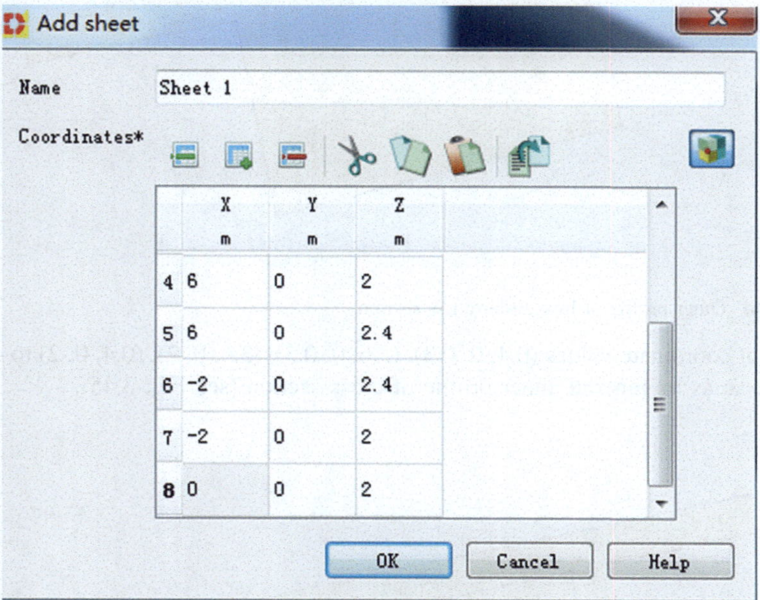

Fig. 3.43 Creation of **Sheet1**

Table 3.1 Outer profile of box girder cross-section

1	(0, 0, 0)
2	(4, 0, 0)
3	(4, 0, 2)
4	(6, 0, 2)
5	(6, 0, 2.4)
6	(–2, 0, 2.4)
7	(–2, 0, 2)
8	(0, 0, 2)

Clicking OK button generates outer profile of box girder cross-section, which is shown in Fig. 3.44.

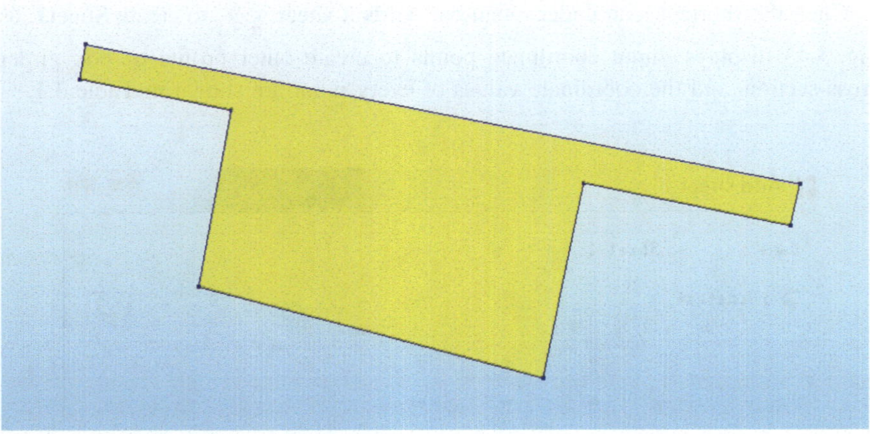

Fig. 3.44 Outer profile of box girder cross-section

Input coordinate values (0.4, 0, 0.3), (3.6, 0, 0.3), (3.6, 0, 2), (0.4, 0, 2) to create Sheet2 so as to generate inner profile of cross-section (see Fig. 3.45).

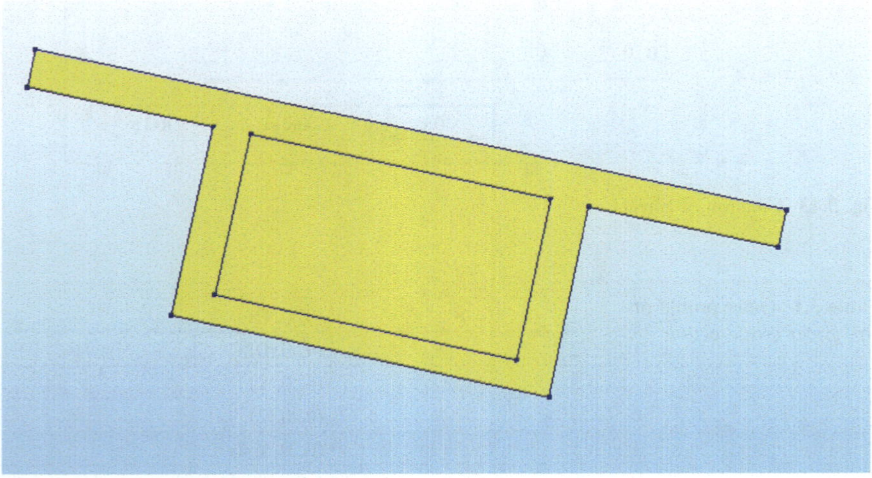

Fig. 3.45 Outer and inner profile of cross-section

Clicking to select interface figure with the left key of mouse, then clicking shortcut icon **Subtract two or more shapes**  to conduct Boolean logic subtraction operations, the site in the shortcut icon zone is displayed (Fig. 3.46). Then manipulation interface of **Subtract two or more shapes** ejects; **Sheet 1** is selected as **Target Selection**, while **Sheet2** is selected as **Tool selection**. Boolean logic subtraction operation for geometric figure is conducted and the manipulation interface is displayed (Fig. 3.47).

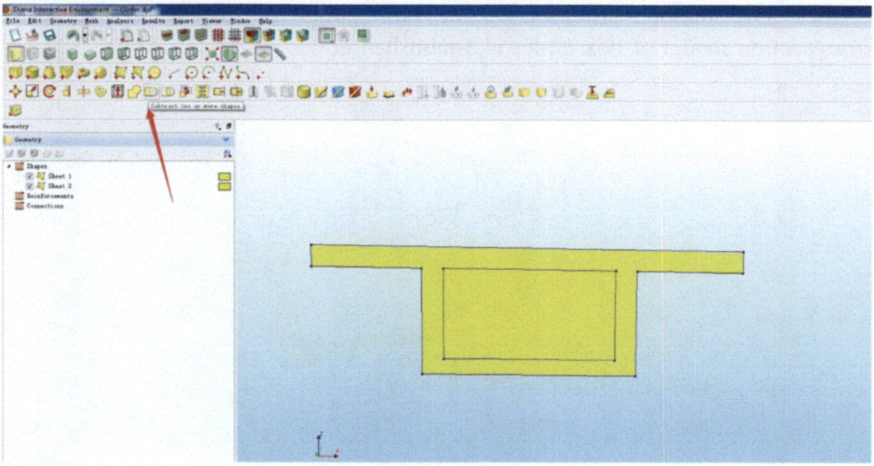

Fig. 3.46 The site of **Subtract two or more shapes** in shortcut icon zone

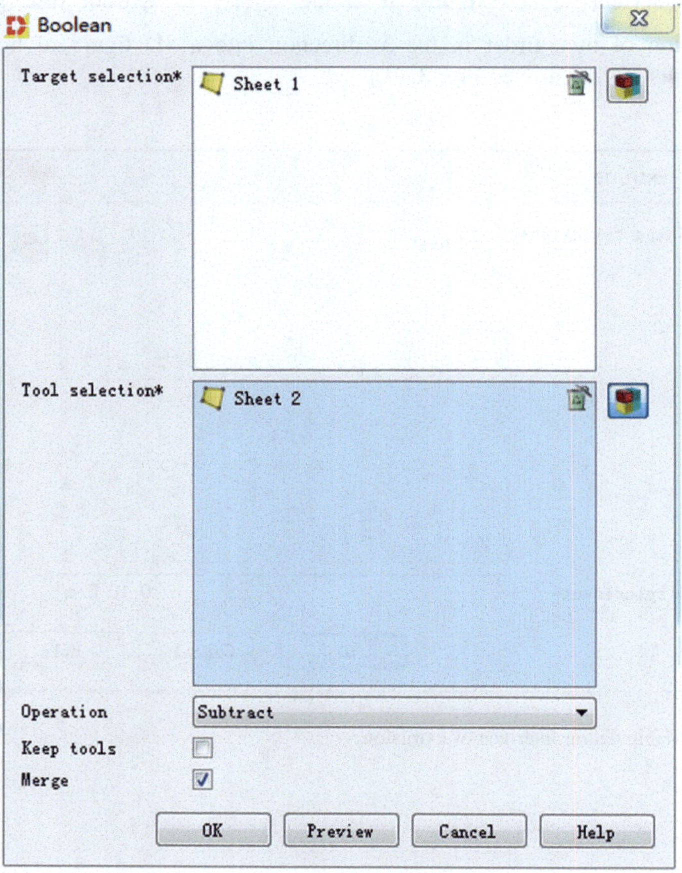

Fig. 3.47 Manipulation interface of Boolean logic subtraction operation

Click OK button, cross-section of box girder is displayed (Fig. 3.48). Cross-section model of box girder is established now.

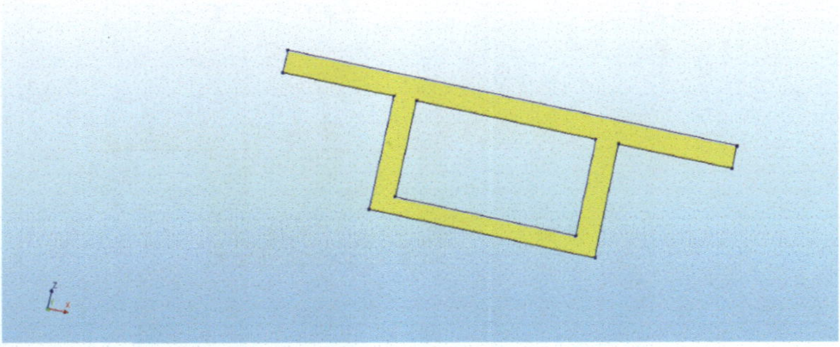

Fig. 3.48 Cross-section geometric model after Boolean logic subtraction operation

Click shortcut icon **Extrude a shape** to extrude the geometric cross-section of box girder in the Y direction into a 3D figure of box girder. **Displacement** is 10 m (see Fig. 3.49).

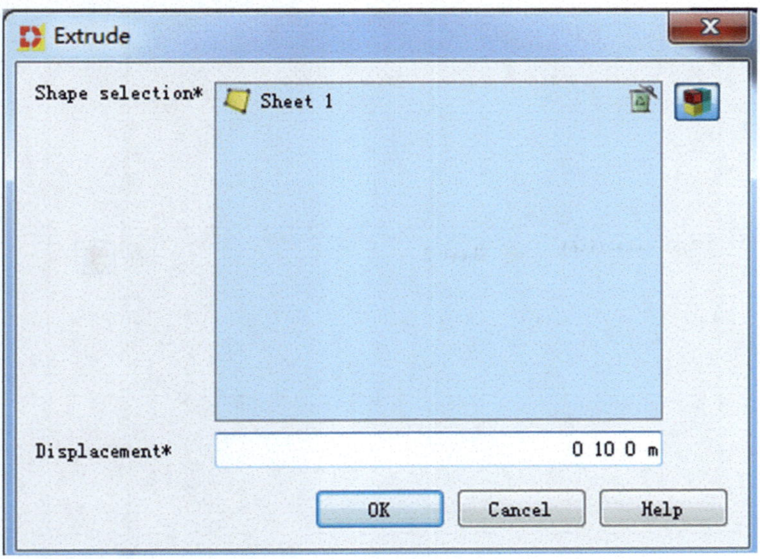

Fig. 3.49 Manipulation interface of extrusion

Clicking OK button generated 3D figure of box girder is displayed (Fig. 3.50).

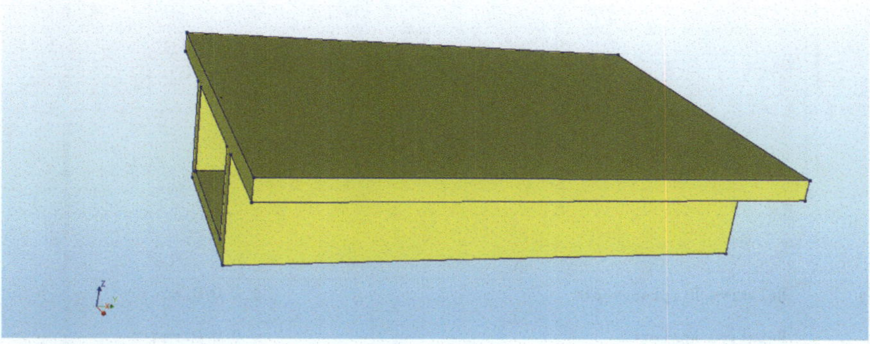

Fig. 3.50 Genearted 3D box girder

Inputting coordinate values of longitudinal steel bars in the bottom plate and clicking shortcut icon **Adds a line**  create geometric line. The coordinate values forming the first geometric longitudinal bars is (0.2, 0, 0.15) and (0, 2, 10, 0.15) with the name of **bar1** (see Fig. 3.51).

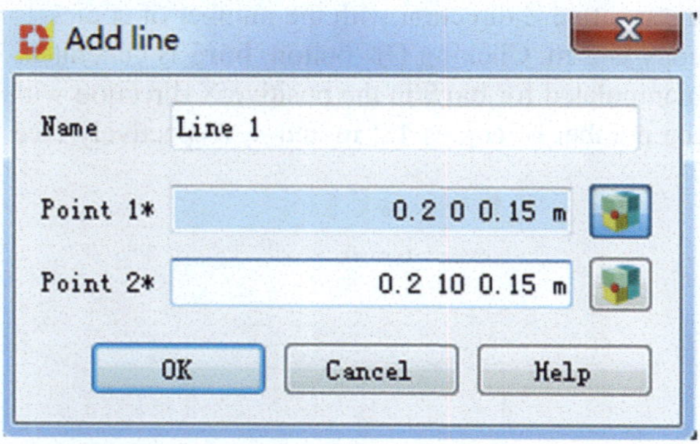

Fig. 3.51 Geometric coordinate values constructing longitudinal bar

Clicking **bar1** under **Geometry** bar then right-clicking to select the **select** function, **Array copy** is selected in the Graphical user interface; with further right-clicking, ejected manipulation interface is displayed (Fig. 3.52). The line of **bar1** is copied and translated in positive X direction with the **Number of copies** 3. **Relative Displacement** is **1.2 m**. Clicking OK button, **bar2**, **bar3** and **bar4** are generated.

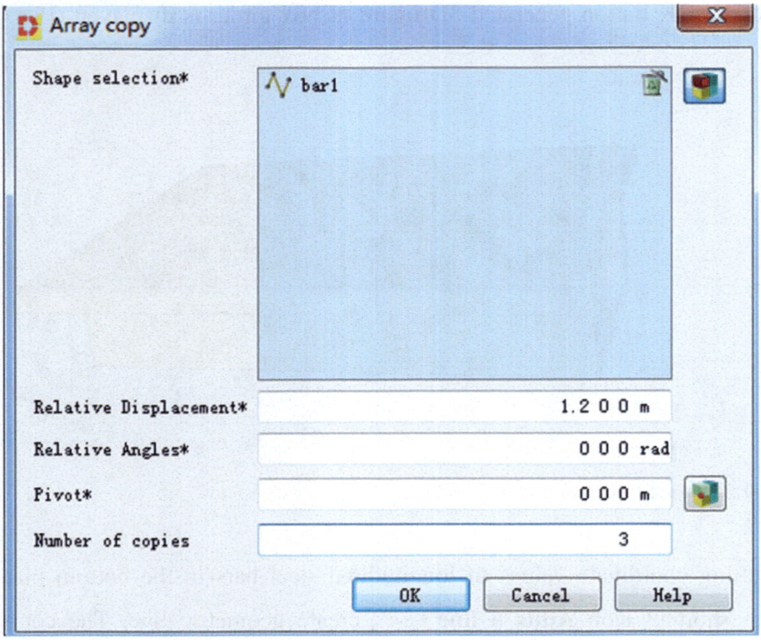

Fig. 3.52 Manipulation of Array copy

Applying the same manipulation mentioned above to copy and translate **bar1** in the positive Z direction with the number of copies 1, relative displacement of array copy is 2 m. Clicking OK button, **bar5** is generated. Then the same method is still manipulated for **bar5** in the positive X direction with the relative displacement and the number of copies 1.2 m and 4, respectively (see Fig. 3.53).

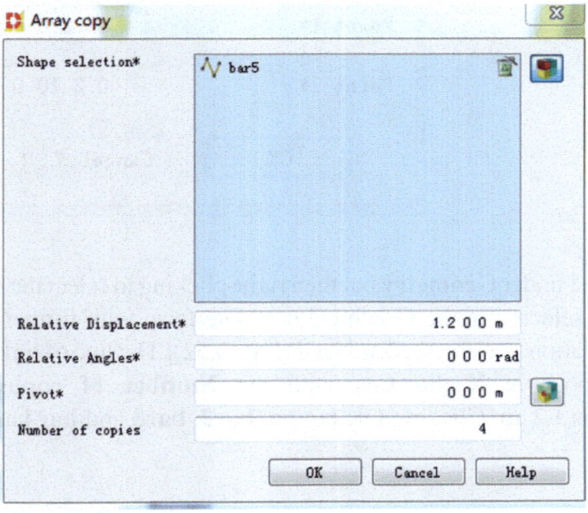

Fig. 3.53 Array copy of **bar5**

Next, **bar5** is selected and right-clicking the **select** function. It is further copied and translated in the negative X direction with the relative displacement 1.2 m and the number of copies 1. Clicking OK button to generate **bar10**, the whole geometric model of box girder with longitudinal bars is displayed (as shown in Fig. 3.54).

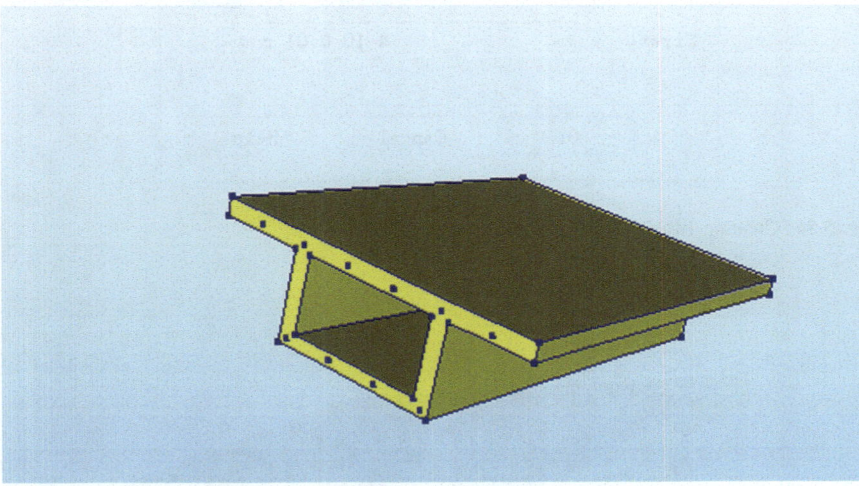

Fig. 3.54 Geometric model of box girder with longitudinal bars

In order to facilitate creating geometric model of steel plate, select the whole geometric model, then right-click to select the **Move a shape** to translate the model in positive Z direction with the relative displacement 0.01 m. Clicking shortcut icon **Adds a block solid** (see Fig. 3.55) to create a block solid with the name of **Block1**, coordinate point (0, 0, 0) is input as initial position and the size in X, Y and Z directions are specified as 4, 10 and 0.01 m, respectively, which is shown in Fig. 3.56. To generate geometric model of steel plate, as Fig. 3.57 shows, press **Ctrl** on the keyboard and rotate the mouse wheeling key, steel plate strengthening layer is enlarged and displayed.

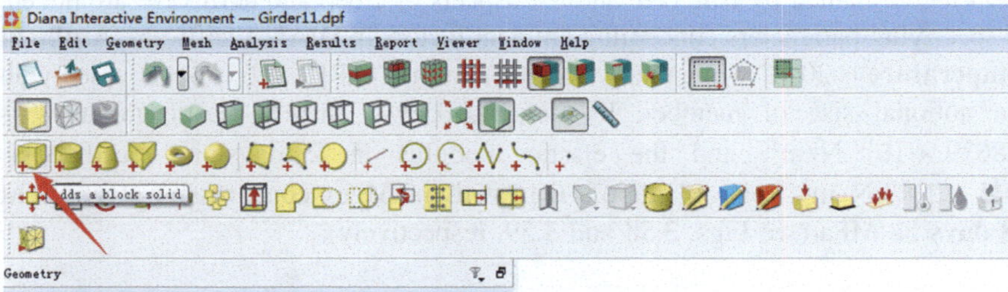

Fig. 3.55 Adds a block solid in the shortcut icon

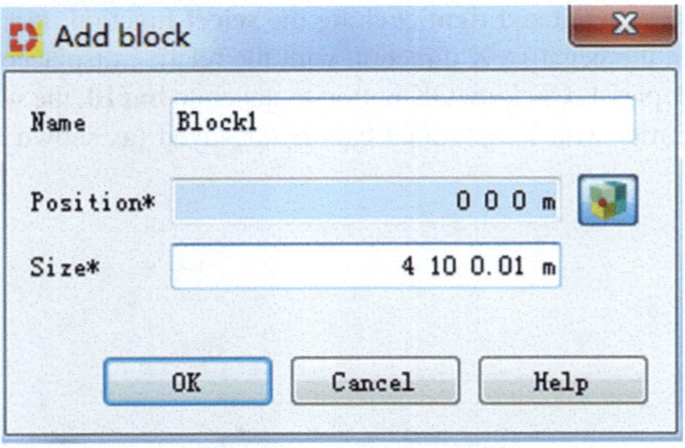

Fig. 3.56 Creating Block1

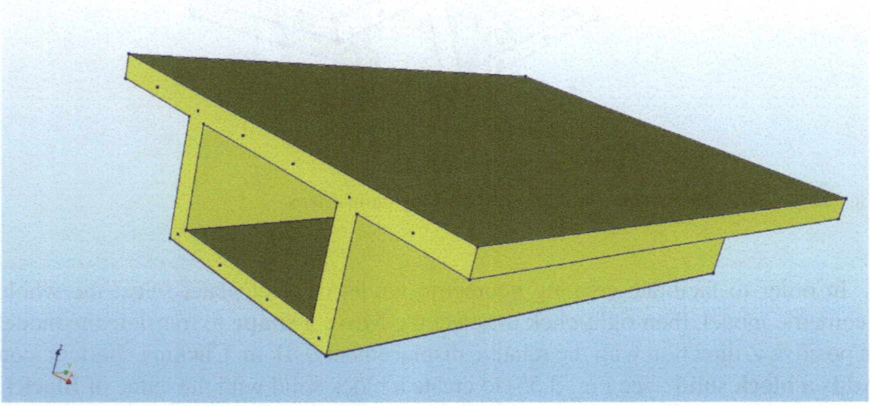

Fig. 3.57 Geometric model of box girder and steel plate strengthening layer

In order to assign material properties to 3D box girder, selecting Sheet1 and right-clicking to chose function of **Property assignments**, Structural solid is selected as element type. Time-dependent material property is European **CEB-FIP 1990** model named by **concrete** and the aspects of creep and shrinkage are ticked. In specifying parameters concerning ambient factors in DIANA software, **Ambient temperature** is 20 °C while relative ambient humidity RH in % is 60, coupled with the notional size of member, by default, 0.15. Concrete elastic modulus is 3.8629×10^{10} N/m^2, and the elastic modulus in 28 days is specified as 3.45×10^{10} N/m^2. Poisson's ratio is 0.15 with the **Mean compressive strength at 28 days** 58 MPa (see Figs. 3.58 and 3.59, respectively).

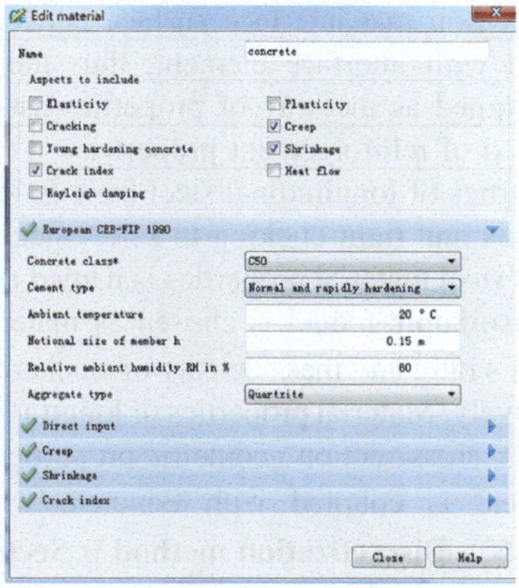

Fig. 3.58 Parameters of CEB-FIP 1990

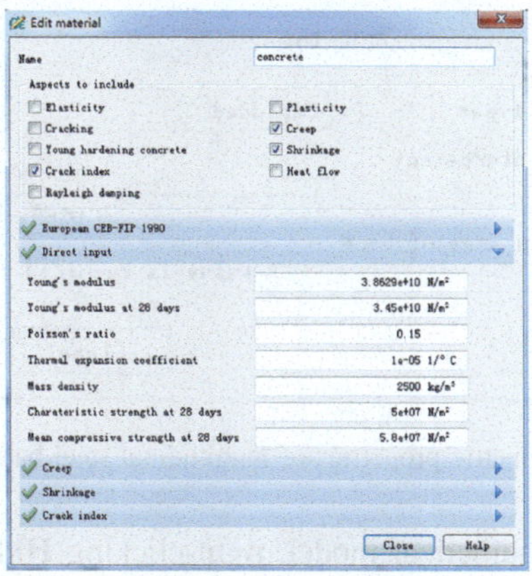

Fig. 3.59 Direct input

In the aspects of creep and shrinkage, based on the concrete aging theory, Kelvin model in which creep and shrinkage are dominant is chosen as analysis model. **Concrete age at the birth of an element** is specified as 28 days (**2,419,200 s**). **Concrete age at the end of curing period** is 1 day (**86,400 s**). Since it is unnecessary to define cross-section geometric properties in structural solid elements, click OK button to close concrete constitutive interface so that definition of material properties is accomplished. Since solid elements do not need specification of cross-section geometric properties, thus the edit of cross-section geometric properties is omitted here.

The next step is to assign steel material properties to the **Block1**. Still choosing function of **Property assignments, Steel** is selected as material **Class**, while **Linear elastic isotropic** material constitutive model is selected as material model. **Young's modulus**, **Poisson's ratio** as well as **Mass density** are 2.1×10^{11} N/m^2, 0.3 and 7800 kg/m^3, respectively. Owing to the **Block1** is also assigned with solid element, the process of defining geometric properties is also omitted too.

Note: Owing to the reason that interface surface between steel plate and box girder must be assigned with interface element, thus the definition of material properties should be assigned as the way of properties assignment like assigning concrete properties instead of reinforcement properties.

Then we define properties of longitudinal steel bars. Selecting all the geometric line representing steel bars and right-clicking to select steel material class, ejected manipulation interface of steel material properties is named as **bar**. In this numerical case, **Linear elasticity isotropic** model is chosen as material constitutive model. Elastic modulus, Poisson's ratio and mass density of steel are 2.1×10^{11} N/m^2, 0.3 and 7800 kg/m^3, respectively. Material properties of longitudinal reinforcement steel are assigned to bars. Next, cross-section geometric properties are edited. Embedded longitudinal reinforcement is coupled with concrete with the area $\frac{\pi}{4} \times 8^2 \approx 50.265$ mm^2 (see Fig. 3.60). **Discretization method** is **Section wise**.

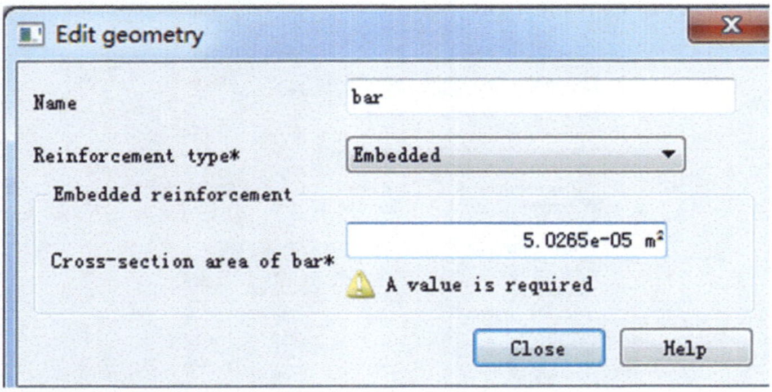

Fig. 3.60 Cross-section geometric properties of longitudinal steel bars

Selecting the whole numerical model, right-clicking **Hide** function to hide the box girder and unticking all the geometric lines of **bar** under the **Geometry** directory tree, only **Block1** is retained in the graphical user interface (see Fig. 3.61). Then we start to specify surface to surface connected interface elements.

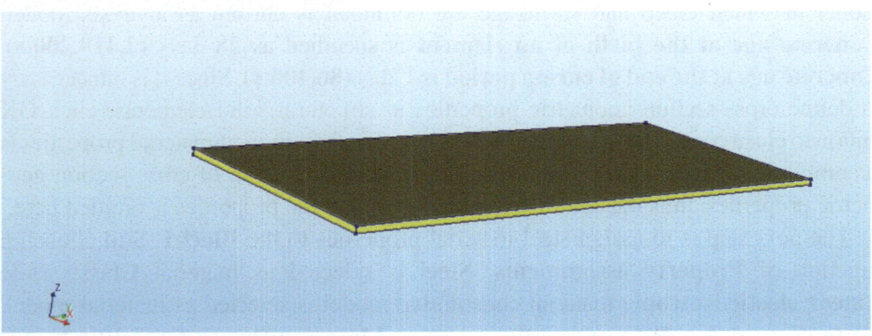

Fig. 3.61 Block1

Clicking **Edit connection property assignments** in the shortcut zone to define material constitutive properties, the ejected manipulation interface of interface element is named as **int**. **Connection type** is **Interface**; connection mode is **Auto connect**. Additionally, **Face** is chosen as selection type while **Element class** is **structural element.** Click the icon to define material properties of interface element. Since strengthening method is bonded steel strengthening, material type of contacted surface between the 3D box girder and plate is **Bondslip** (see Fig. 3.62). From the **Linear material properties** aspect, **Type** is **3D surface interface**. **Normal stiffness modulus is** 3.65×10^{16} N/m^3 and shear stiffness modulus in x and y directions is 3.65×10^{12} N/m^3 for both (see Fig. 3.63).

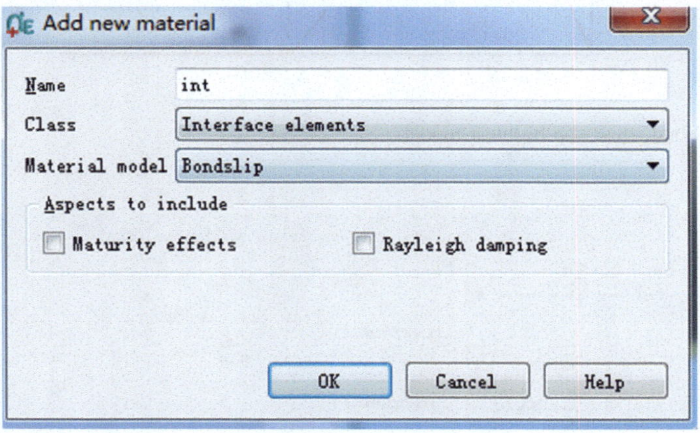

Fig. 3.62 Constitutive type of interface

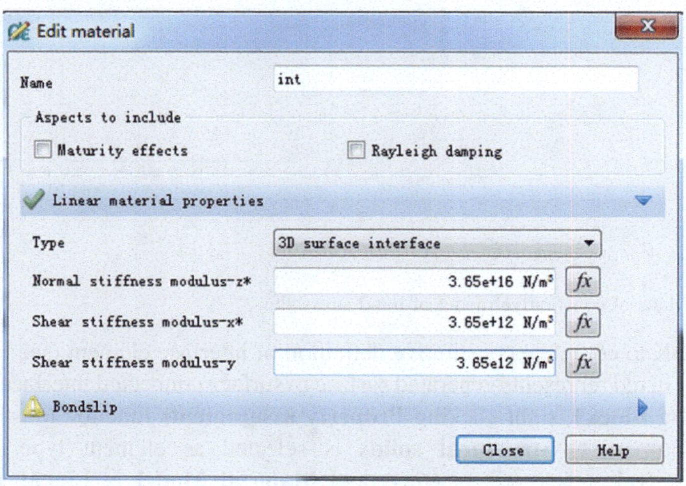

Fig. 3.63 Specification of normal and shear stiffness modulus

In the aspect of bond-slip, **Multi-Linear** is selected as bond-slip type, which is also called as stress-slip bi-linear model, where normal stiffness as well as shear stiffness can be translated (see Fig. 3.64). For the stress-slip model interface, constitutive model is displayed (Fig. 3.65).

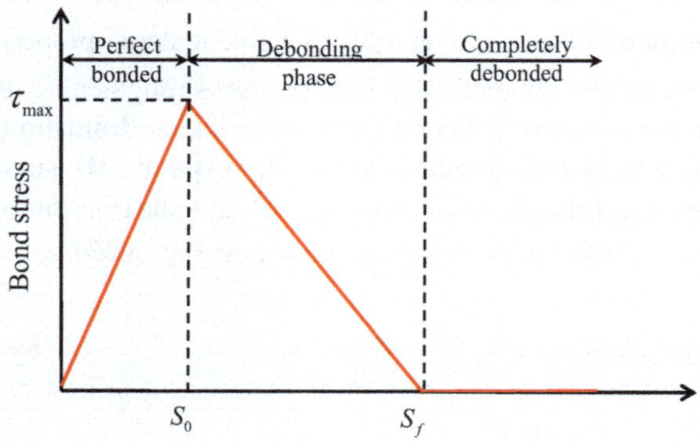

Fig. 3.64 Bond stress-slip bi-linear model

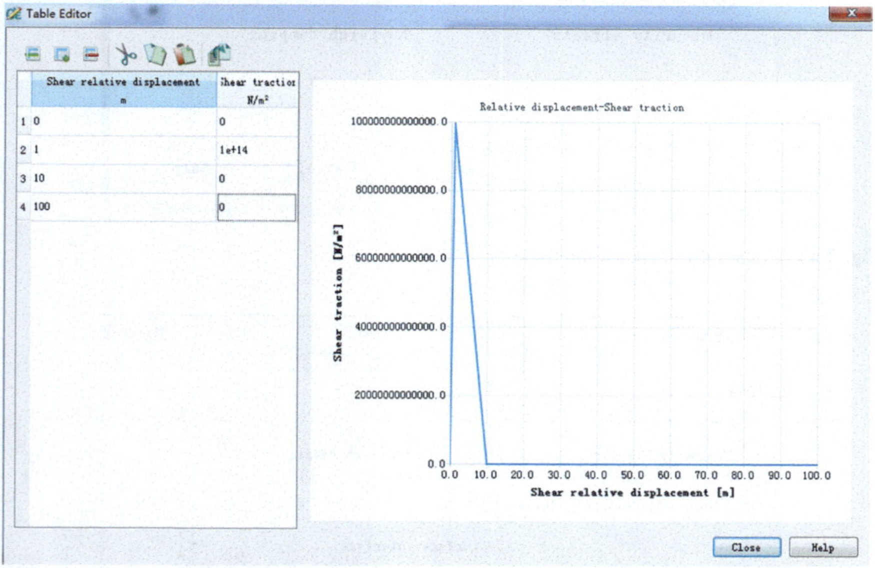

Fig. 3.65 Material constitutive model of bond stress-slip

Click OK to complete constitutive definition of interface element (see Fig. 3.66) and region in red represents generated surface to surface connected interface element.

Selecting **Block1**, right-clicking **Property assignments** function to assign steel material properties, **Structural solids** is selected as element type. Clicking **Material**, **Steel** is selected as **Class** and **Material Model** is **Linear elasticity isotropic**. Other parameters are the same as longitudinal reinforcement steel bars. Clicking OK button, assignment of steel material properties is completed.

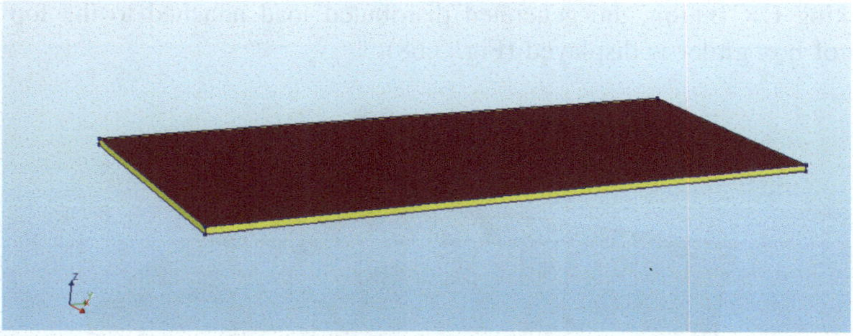

Fig. 3.66 Generation of surface to surface connected interface element

Clicking [icon] to define gravity under **Geometry** directory tree, **Load type** is **Dead weigh**. Create load case1 with the name of **load**. **Load target type** is **Face** while **Load type** is downward **Distributed force** in the negative Z direction with the value 50 kN/m^2 (see Fig. 3.67).

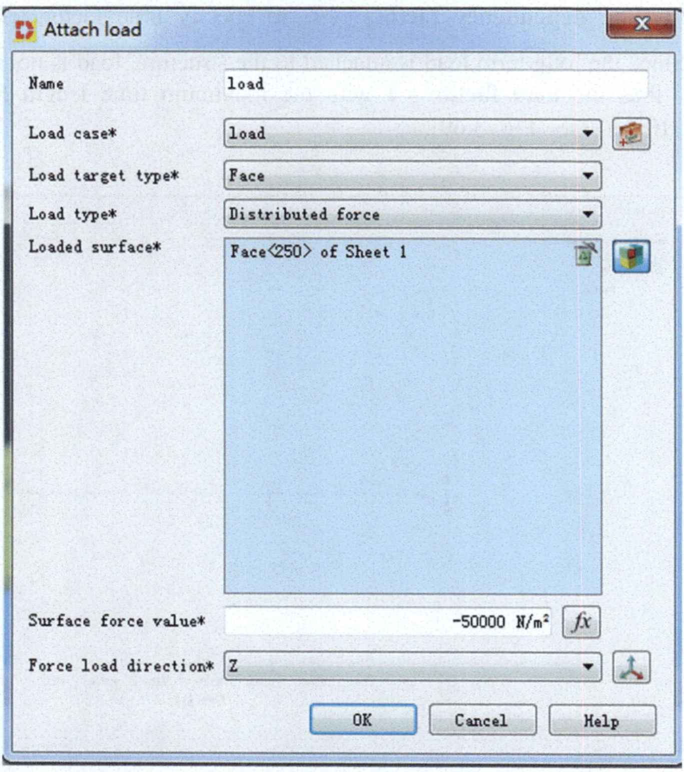

Fig. 3.67 Manipulation interface of distributed surface load

Clicking OK button, the generated distributed load attached to the top plate surface of box girder is displayed (Fig. 3.68).

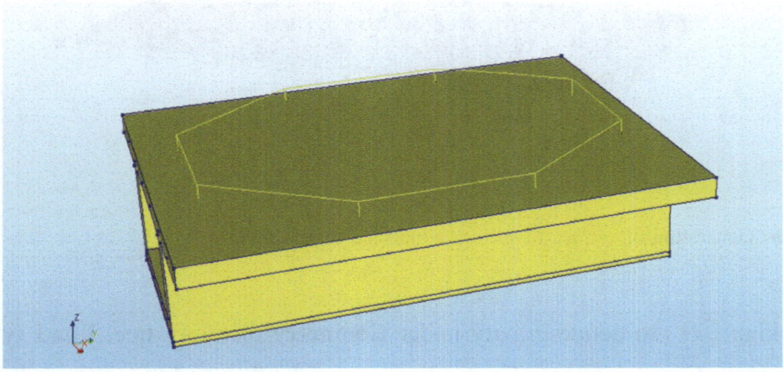

Fig. 3.68 Generation of distributed surface load

Clicking **Combination** icon, gravity and distributed load are added to the same load case with the name of loadcase1. Click the load case, and then still click icon above **Edit time dependency factors**  to specify time–factor relationship function. Since the long-term load is attached to the structure, load is not changing with time, thus the **load factor = 1** with the maximum time length **500** years (1.5768×10^{10} s) (see Fig. 3.69).

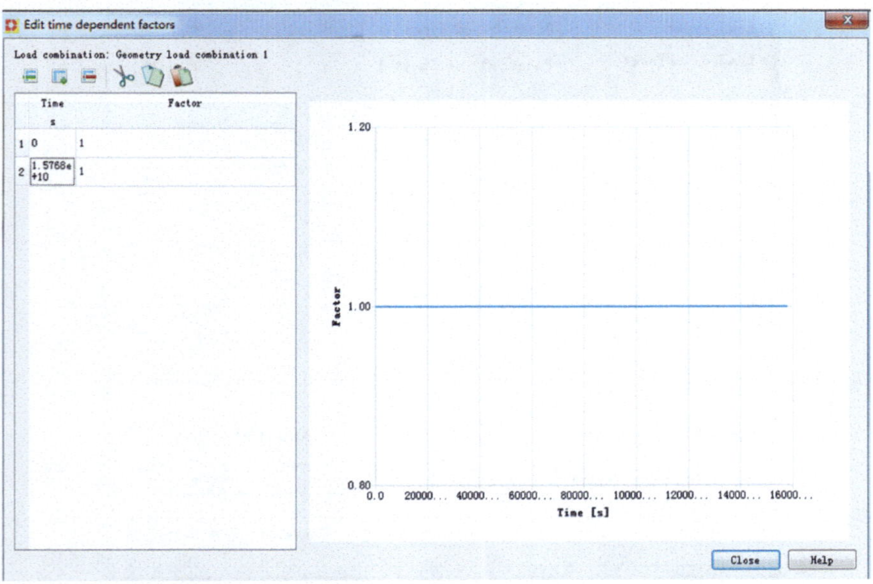

Fig. 3.69 Specification of time factor

To add the constraint set, click **Geometry—Analysis—Attach support** in the above menu bar, and translation in X, Y and Z directions is attached to the cross-section surface of box girder at the site of Y = 0 with the support set name of **co1**. **Support target type** is **Face** (see Fig. 3.70).

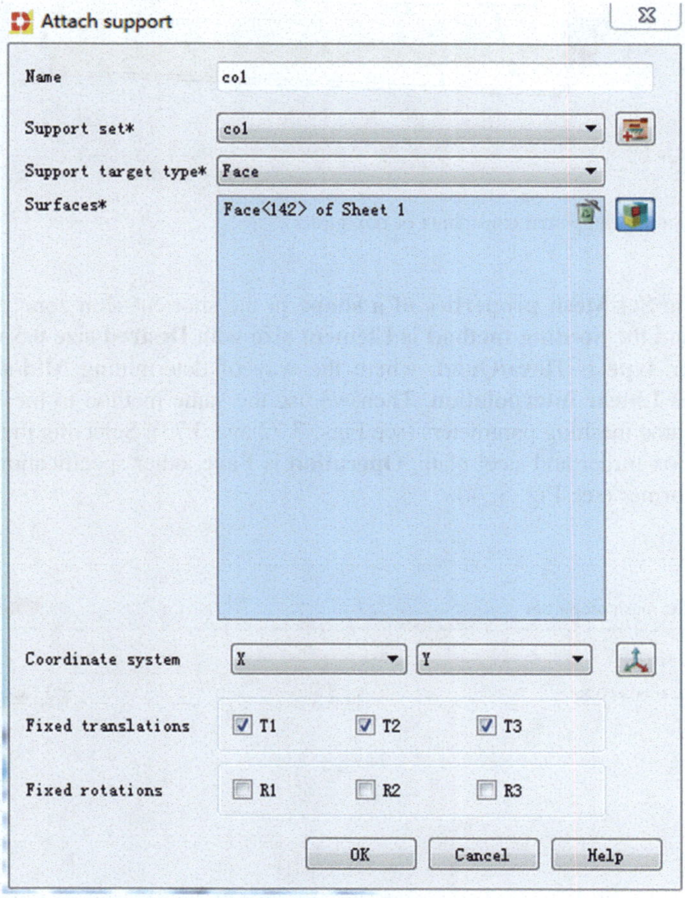

Fig. 3.70 Constraint information on the surface at Y = 0

Apply the same method to attach simply supported constraints on the cross-section surface of box girder at the site of Y = 10 with the support set name of **co2**. Constraint attachment type is the same as **co1**. Click OK button to check generated constraints on the graphical user interface (see Fig. 3.71).

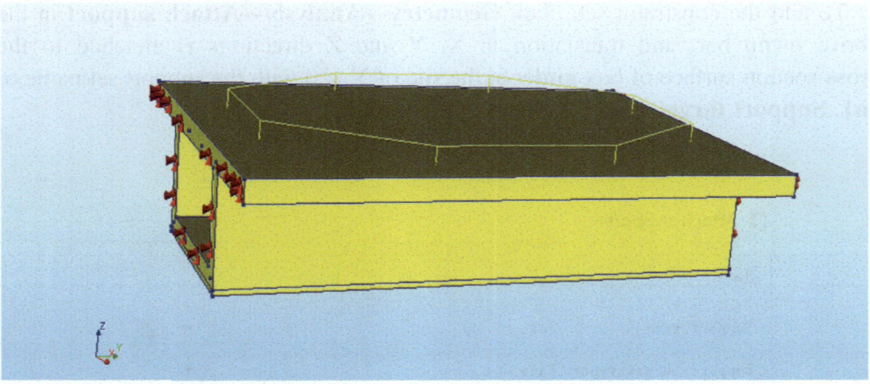

Fig. 3.71 Simply supported constraints of box girder

Clicking **Set Mesh properties of a shape** in the shortcut icon zone, **Sheet1** is selected and the **Seeding method** is **Element size** with **Desired size 0.5 m**. Shape of **Mesher type** is **Hexa/Quad,** where the way of determining **Mid-side node location** is **Linear interpolation**. Then we use the same method to mesh **Block1** with the same meshing parameters (see Figs. 3.72 and 3.73). Selecting the interface between box girder and steel plate, **Operation** is **Face**, other specifications are the same as former (see Fig. 3.74).

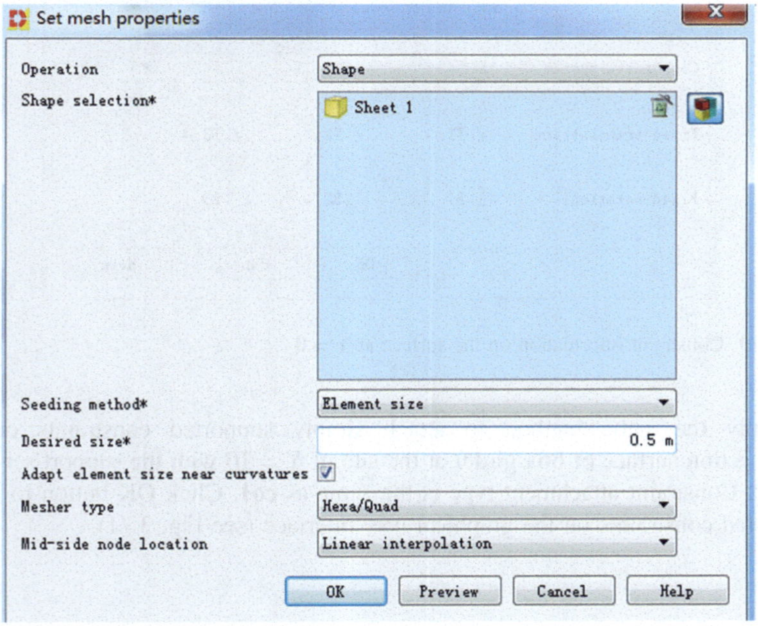

Fig. 3.72 Mesh of box girder

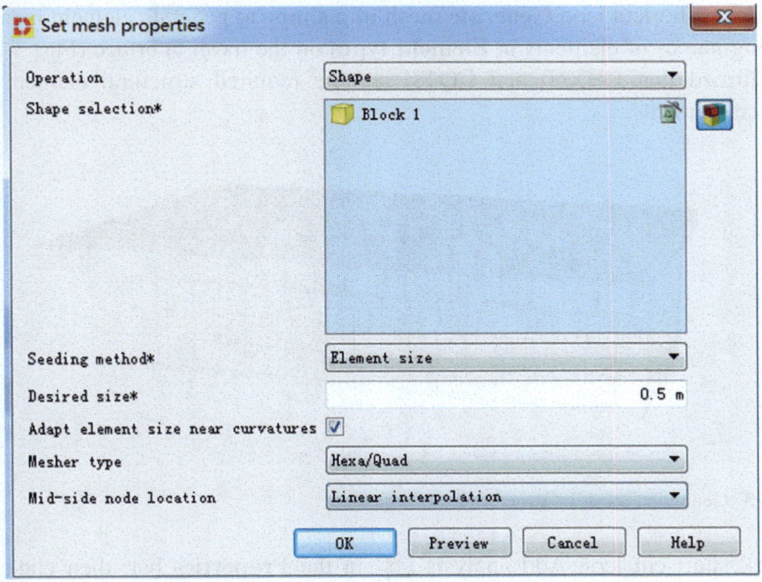

Fig. 3.73 Mesh of steel plate

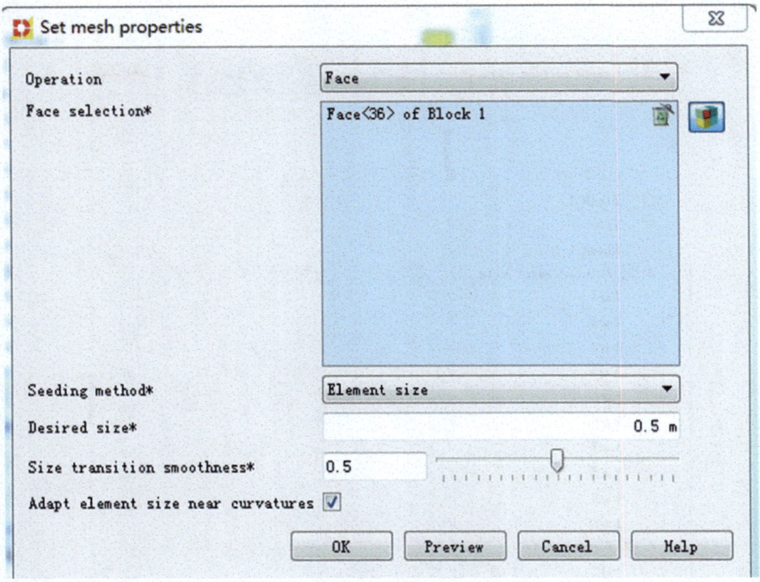

Fig. 3.74 Mesh of interface

Clicking shortcut icon **Generate mesh of a shape** to generate element mesh and checking names of elements in **Element types** on the mesh interface (Fig. 3.75), it is confirmed that CHX60 and CQ48I are the required structural elements and interface elements.

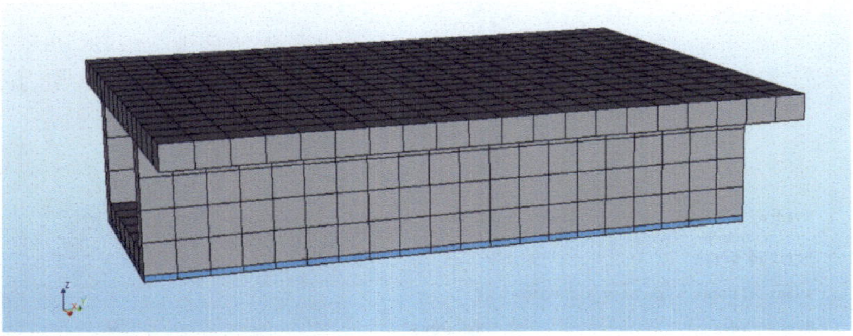

Fig. 3.75 Generated mesh of structures

Click short cut icon Add analysis  in the **Properties** bar, then click **Add analysis command** to generate module of **Analysis**. Click **Analysis** then right-click to select **Add command—Phased** to generate the first phase of phased analysis. Element **Block1** is unticked, which means that all the elements except steel plate are activated (see Fig. 3.76).

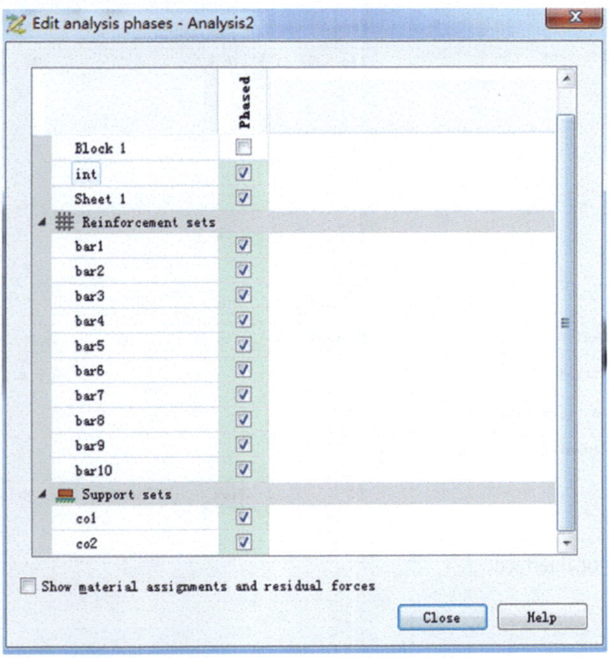

Fig. 3.76 Activated elements in the first phase

Right-clicking **Analysis** to select Structural nonlinear and still right-clicking **Add—Execute steps—load steps** create laodcase1 with the name of **load**. **Geometry load combination 1** is selected as Load set.

Selecting Structural nonlinear via clicking the **Analysis**, right-clicking **Add—Execute steps—load steps** create new load set, remaining it as **load**. **Geometry load combination 1** is chosen as load set of this block with the **User specified sizes** for **Load steps** is **1.000,** representing that both the number of load step and the total length of load factor are 1 (see Fig. 3.77).

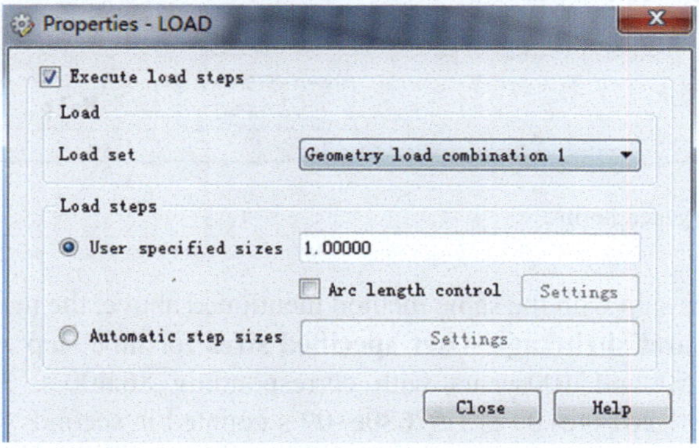

Fig. 3.77 Specification of nonlinear load block

In the modulus of equilibrium iterations, **Maximum number of iterations** is 20 while both the **Force and Displacement** are selected as **Convergence norm**, which means that the iteration calculation of current step reaches convergence when either of them reaches convergence. **Convergence tolerance** and **Abort criterion** are 0.01 and 10,000 respectively (see Figs. 3.78 and 3.79).

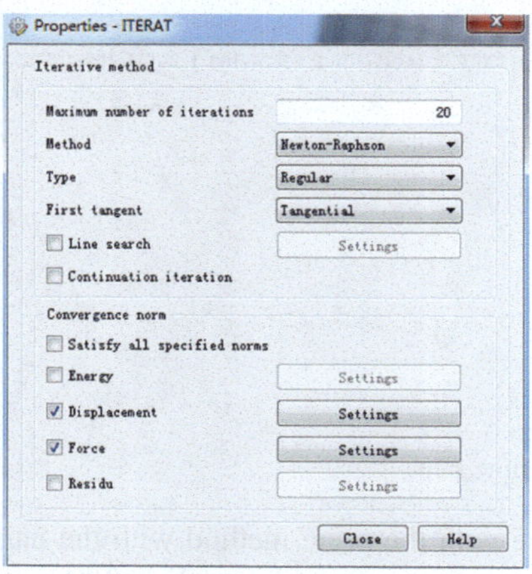

Fig. 3.78 Iterative method and convergence norm

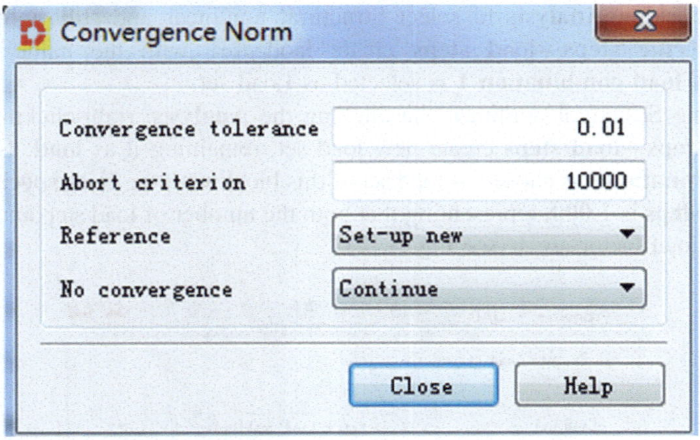

Fig. 3.79 Convergence norm

Creating time step with the same method mentioned above, the name of time step block is **creep and shrinkage**. **User specified sizes** for time step are specified as 1 day, 1, 10, 50 and 100 years with corresponding 86,400 s, 3.14496e+07 s, 2.83824e+08 s, 1.26144e+09 s, 1.57680e+09 s counted in seconds (see Fig. 3.80). Maximum number of iterations is 50, while other specifications for convergence norm and iterative parameters are the same as mentioned earlier, which is not repeated here.

Note: In DIANA software, contrary to user's initial impression, user-specified size of time step is added as time interval increment instead of the summed final total. For instance, when 86,400.0 s and 3.14496e+07 s are added, this means 1 year is added via 1 day first then 364 days rather than adding 364 days from the first to the 364th day.

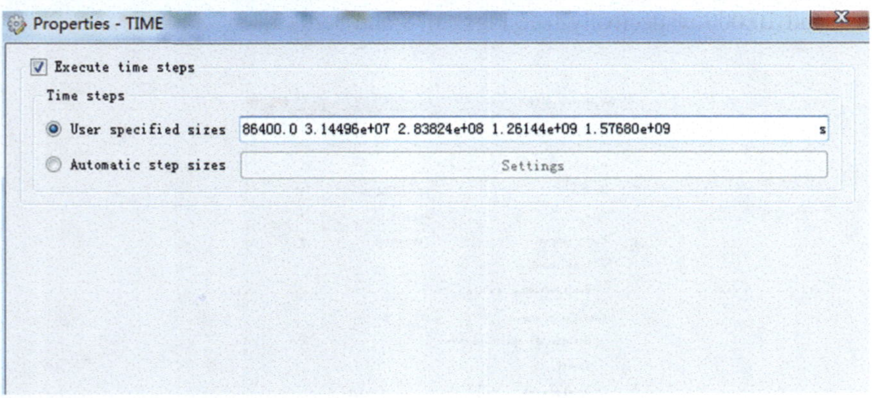

Fig. 3.80 Specification of time steps

Adding second phase with the same method with the name of **Phased1**, in this condition, all the elements are activated (see Fig. 3.81), creating structural nonlinear analysis, where former **Geometry load combination 1** is added as initial load. Specifications of iteration calculation are the same as former, where the start time of second phase is 100 years (3.1536e9s).

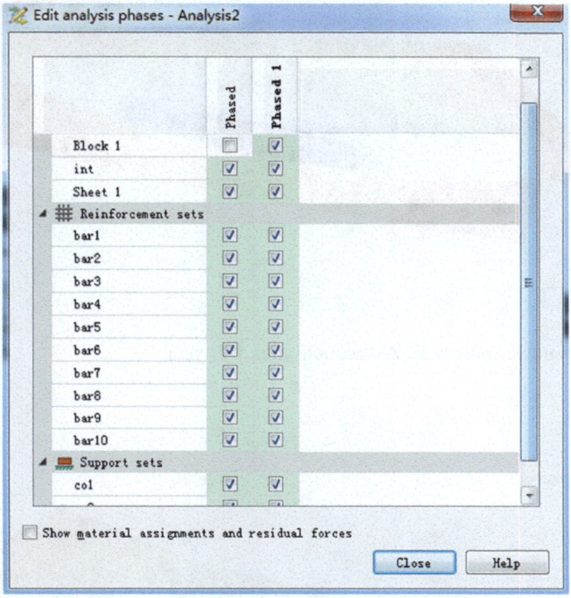

Fig. 3.81 Specification of time steps

Click **Run analysis** to start the calculation, and when the calculation is finished, displacement contour in Z direction is checked (see Figs. 3.82, 3.83 and 3.84).

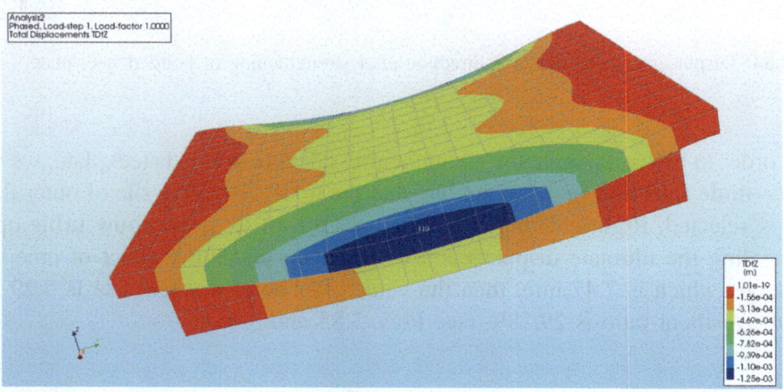

Fig. 3.82 Displacement contour in Z direction

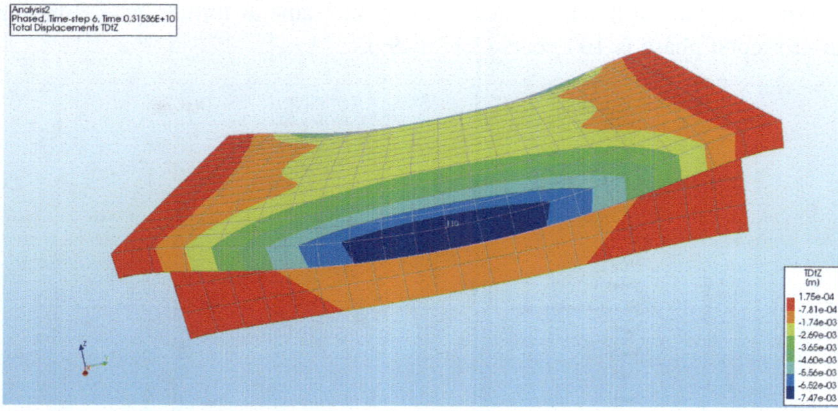

Fig. 3.83 Displacement contour in Z direction after 100 years

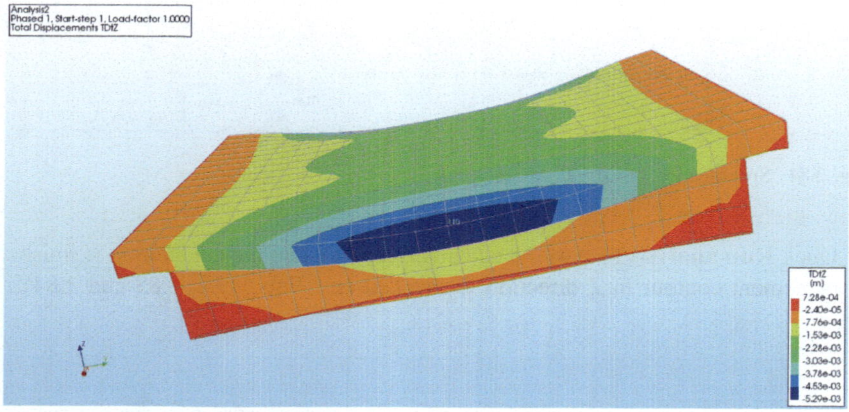

Fig. 3.84 Displacement contour in Z direction after strengthening of bonded steel plate

In order to better illustrate the strengthening effect of bonded steel plate, we click **View—node selection** in the menu bar, and the node 110 at the site of outer flange plate is selected. Right-clicking **TDtZ** below **Output** to select **show table** option and reading the ultimate displacement of the node under the effect of creep and shrinkage, which is 7.47 mm, then the value after bonded steel plate is 5.29 mm; thus the resilient ratio is 29.18% (see Figs. 3.85 and 3.86).

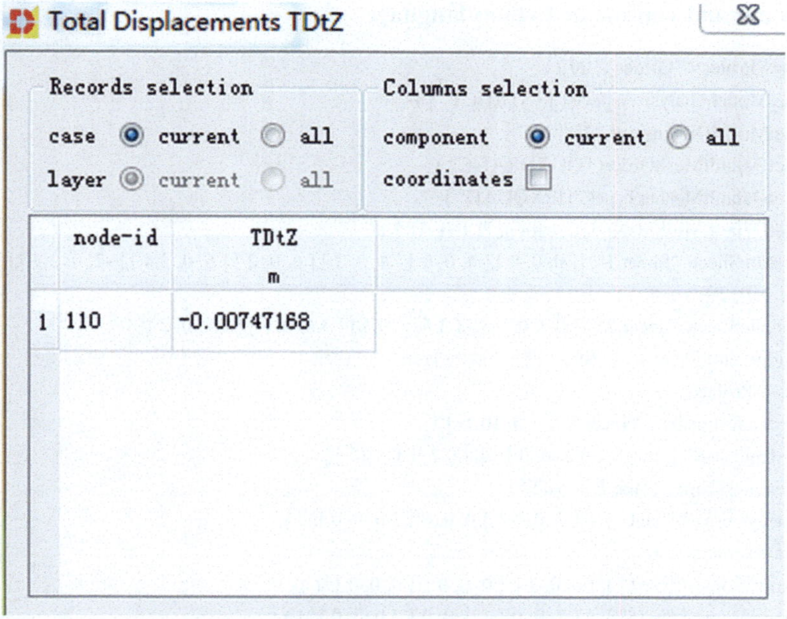

Fig. 3.85 Displacement before bonded steel strengthening

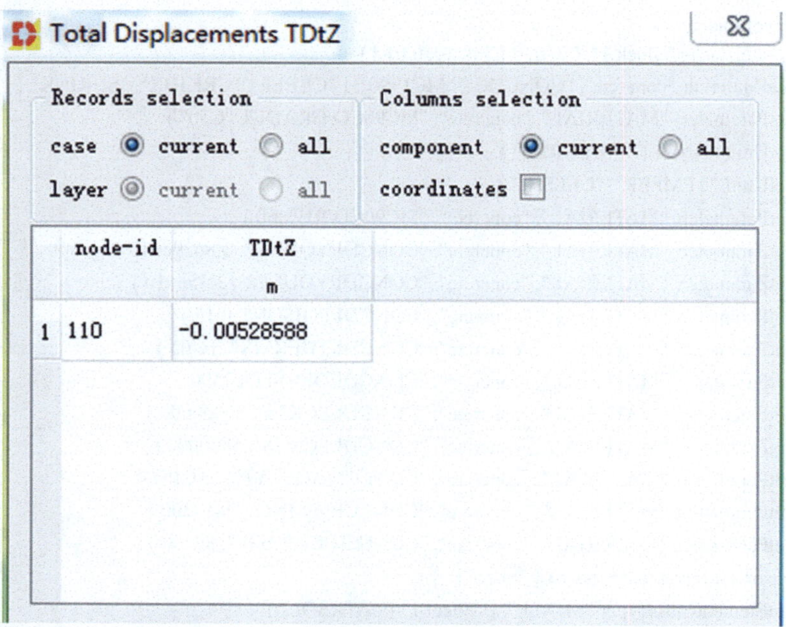

Fig. 3.86 Displacement after bonded steel strengthening

Command console in Python language

```
newProject( "Girder",100 )
setModelAnalysisAspects( [ "STRUCT" ] )
setModelDimension( "3D" )
setDefaultMeshOrder( "QUADRATIC" )
setDefaultMesherType( "HEXQUAD" )
setDefaultMidSideNodeLocation( "LINEAR" )
createSheet( "Sheet 1", [[ 0, 0, 0 ],[ 4, 0, 0 ],[ 4, 0, 2 ],[ 6, 0, 2 ],[ 6, 0, 2.4 ],[ -2, 0, 2.4 ],[ -2,
0, 2 ],[ 0, 0, 2 ]] )
createSheet( "Sheet 2", [[ 0.4, 0, 0.3 ],[ 3.6, 0, 0.3 ],[ 3.6, 0, 2 ],[ 0.4, 0, 2 ]] )
subtract( "Sheet 1", [ "Sheet 2" ], False, True )
saveProject(  )
extrudeProfile( [ "Sheet 1" ], [ 0, 10, 0 ] )
createLine( "Line 1", [ 0.2, 0, 0.15 ], [ 0.2, 10, 0.15 ] )
renameShape( "Line 1", "bar1" )
arrayCopy( [ "bar1" ], [ 1.2, 0, 0 ], [ 0, 0, 0 ], [ 0, 0, 0 ], 3 )
saveProject(  )
arrayCopy( [ "bar1" ], [ 0, 0, 2 ], [ 0, 0, 0 ], [ 0, 0, 0 ], 1 )
arrayCopy( [ "bar5" ], [ 1.2, 0, 0 ], [ 0, 0, 0 ], [ 0, 0, 0 ], 4 )
saveProject(  )
arrayCopy( [ "bar5" ], [ -1.2, 0, 0 ], [ 0, 0, 0 ], [ 0, 0, 0 ], 1 )
translate( [ "Sheet 1", "bar1", "bar2", "bar3", "bar4", "bar5", "bar6", "bar7", "bar8", "bar9",
"bar10" ], [ 0, 0, 0.01 ] )
saveProject(  )
createBlock( "Block 1", [ 0, 0, 0 ], [ 4, 10, 0.01 ] )
addMaterial( "concrete", "CONCDC", "MC1990", [ "CREEP", "CRKIDX", "SHRINK" ] )
setParameter( "MATERIAL", "concrete", "MC90CO/GRADE", "C50" )
setUnit( "ANGLE", "DEGREE" )
setUnit( "TEMPER", "CELSIU" )
setParameter( "MATERIAL", "concrete", "MC90CO/RH", 60 )
setParameter( "MATERIAL", "concrete", "CONCDI/YOUNG", 3.8629e+10 )
setParameter( "MATERIAL", "concrete", "CONCDI/YOUN28", 3.45e+10 )
setParameter( "MATERIAL", "concrete", "CONCDI/POISON", 0.15 )
setParameter( "MATERIAL", "concrete", "CONCDI/THERMX", 1e-05 )
setParameter( "MATERIAL", "concrete", "CONCDI/DENSIT", 2500 )
setParameter( "MATERIAL", "concrete", "CONCDI/FCK28", 50000000 )
setParameter( "MATERIAL", "concrete", "CONCDI/FCM28", 58000000 )
setParameter( "MATERIAL", "concrete", "CONCCP/AGETYP", "AGING" )
setParameter( "MATERIAL", "concrete", "CONCCP/AGING", 2419200 )
setParameter( "MATERIAL", "concrete", "CONCSH/CURAGE", 86400 )
clearReinforcementAspects( [ "Sheet 1" ] )
setElementClassType( "SHAPE", [ "Sheet 1" ], "STRSOL" )
```

```
assignMaterial( "concrete", "SHAPE", [ "Sheet 1" ] )
resetGeometry( "SHAPE", [ "Sheet 1" ] )
resetElementData( "SHAPE", [ "Sheet 1" ] )
hide( "SHAPE", [ "Sheet 1" ] )
hide( "SHAPE", [ "bar1" ] )
hide( "SHAPE", [ "bar2" ] )
hide( "SHAPE", [ "bar3" ] )
hide( "SHAPE", [ "bar4" ] )
hide( "SHAPE", [ "bar5" ] )
hide( "SHAPE", [ "bar6" ] )
hide( "SHAPE", [ "bar7" ] )
hide( "SHAPE", [ "bar8" ] )
hide( "SHAPE", [ "bar9" ] )
hide( "SHAPE", [ "bar10" ] )
show( "SHAPE", [ "bar1" ] )
show( "SHAPE", [ "bar2" ] )
show( "SHAPE", [ "bar3" ] )
show( "SHAPE", [ "bar4" ] )
show( "SHAPE", [ "bar5" ] )
show( "SHAPE", [ "bar6" ] )
show( "SHAPE", [ "bar7" ] )
show( "SHAPE", [ "bar8" ] )
show( "SHAPE", [ "bar9" ] )
show( "SHAPE", [ "bar10" ] )
addMaterial( "bar", "MCSTEL", "ISOTRO", [] )
setParameter( "MATERIAL", "bar", "LINEAR/ELASTI/YOUNG", 2.1e+11 )
setParameter( "MATERIAL", "bar", "LINEAR/ELASTI/POISON", 0.3 )
setParameter( "MATERIAL", "bar", "LINEAR/MASS/DENSIT", 7800 )
addGeometry( "Element geometry 1", "RELINE", "REBAR", [] )
rename( "GEOMET", "Element geometry 1", "bar" )
setParameter( "GEOMET", "bar", "REIEMB/CROSSE", 5.0265e-05 )
setReinforcementAspects( [ "bar1", "bar2", "bar3", "bar4", "bar5", "bar6", "bar7", "bar8",
"bar9", "bar10" ] )
assignMaterial( "bar", "SHAPE", [ "bar1", "bar2", "bar3", "bar4", "bar5", "bar6", "bar7",
"bar8", "bar9", "bar10" ] )
assignGeometry( "bar", "SHAPE", [ "bar1", "bar2", "bar3", "bar4", "bar5", "bar6", "bar7",
"bar8", "bar9", "bar10" ] )
resetElementData( "SHAPE", [ "bar1", "bar2", "bar3", "bar4", "bar5", "bar6", "bar7", "bar8",
"bar9", "bar10" ] )
setReinforcementDiscretization( [ "bar1", "bar2", "bar3", "bar4", "bar5", "bar6", "bar7",
"bar8", "bar9", "bar10" ], "SECTION" )
```

```
saveProject( )
hide( "SHAPE", [ "bar1" ] )
hide( "SHAPE", [ "bar2" ] )
hide( "SHAPE", [ "bar3" ] )
hide( "SHAPE", [ "bar4" ] )
hide( "SHAPE", [ "bar5" ] )
hide( "SHAPE", [ "bar6" ] )
hide( "SHAPE", [ "bar7" ] )
hide( "SHAPE", [ "bar9" ] )
hide( "SHAPE", [ "bar8" ] )
hide( "SHAPE", [ "bar10" ] )
addMaterial( "int", "INTERF", "BONDSL", [] )
setParameter( "MATERIAL", "int", "LINEAR/ELAS6/DSNZ", 3.65e+16 )
setParameter( "MATERIAL", "int", "LINEAR/ELAS6/DSSX", 3.65e+12 )
setParameter( "MATERIAL", "int", "LINEAR/ELAS6/DSSY", 3.65e+12 )
setParameter( "MATERIAL", "int", "BOSLIP/BONDSL", 3 )
setParameter( "MATERIAL", "int", "BOSLIP/BONDS3/DISTAU", [] )
setParameter( "MATERIAL", "int", "BOSLIP/BONDS3/DISTAU", [ 0, 0, 1, 1e+14, 10, 0,
100, 0 ] )
addGeometry( "Element geometry 2", "SHEET", "STRINT", [] )
remove( "GEOMET", "Element geometry 2" )
createSurfaceConnection( "int" )
setParameter( "GEOMETRYCONNECTION", "int", "CONTYP", "INTER" )
setParameter( "GEOMETRYCONNECTION", "int", "MODE", "AUTO" )
attachTo( "GEOMETRYCONNECTION", "int", "SOURCE", "Block 1", [[ 2.294292,
5.73573, 0.01 ]] )
setElementClassType( "GEOMETRYCONNECTION", "int", "STRINT" )
assignMaterial( "int", "GEOMETRYCONNECTION", "int" )
resetGeometry( "GEOMETRYCONNECTION", "int" )
resetElementData( "GEOMETRYCONNECTION", "int" )
saveProject( )
show( "SHAPE", [ "Sheet 1" ] )
show( "SHAPE", [ "bar1" ] )
show( "SHAPE", [ "bar2" ] )
show( "SHAPE", [ "bar3" ] )
show( "SHAPE", [ "bar4" ] )
show( "SHAPE", [ "bar5" ] )
hide( "SHAPE", [ "bar1" ] )
hide( "SHAPE", [ "bar2" ] )
hide( "SHAPE", [ "bar3" ] )
hide( "SHAPE", [ "bar4" ] )
```

```
hide( "SHAPE", [ "bar5" ] )
hide( "SHAPE", [ "Sheet 1" ] )
addMaterial( "steel", "MCSTEL", "ISOTRO", [] )
setParameter( "MATERIAL", "steel", "LINEAR/ELASTI/YOUNG", 2.1e+11 )
setParameter( "MATERIAL", "steel", "LINEAR/ELASTI/POISON", 0.3 )
setParameter( "MATERIAL", "steel", "LINEAR/MASS/DENSIT", 7800 )
clearReinforcementAspects( [ "Block 1" ] )
setElementClassType( "SHAPE", [ "Block 1" ], "STRSOL" )
assignMaterial( "steel", "SHAPE", [ "Block 1" ] )
resetGeometry( "SHAPE", [ "Block 1" ] )
resetElementData( "SHAPE", [ "Block 1" ] )
saveProject(  )
show( "SHAPE", [ "Sheet 1" ] )
show( "SHAPE", [ "bar1" ] )
show( "SHAPE", [ "bar2" ] )
show( "SHAPE", [ "bar4" ] )
show( "SHAPE", [ "bar3" ] )
show( "SHAPE", [ "bar5" ] )
show( "SHAPE", [ "bar6" ] )
show( "SHAPE", [ "bar7" ] )
show( "SHAPE", [ "bar8" ] )
show( "SHAPE", [ "bar9" ] )
show( "SHAPE", [ "bar10" ] )
addSet( "GEOMETRYLOADSET", "gravity" )
createModelLoad( "gravity", "gravity" )
saveProject(  )
addSet( "GEOMETRYLOADSET", "Geometry load case 2" )
rename( "GEOMETRYLOADSET", "Geometry load case 2", "load" )
createSurfaceLoad( "load", "load" )
setParameter( "GEOMETRYLOAD", "load", "FORCE/VALUE", -50000 )
setParameter( "GEOMETRYLOAD", "load", "FORCE/DIRECT", 3 )
attach( "GEOMETRYLOAD", "load", "Sheet 1", [[ 2.588584, 5.73573, 2.41 ]] )
setDefaultGeometryLoadCombinations(  )
setGeometryLoadCombinationFactor( "Geometry load combination 1", "gravity", 1 )
remove( "GEOMETRYLOADCOMBINATION", "Geometry load combination 1" )
setGeometryLoadCombinationFactor( "Geometry load combination 2", "gravity", 1 )
remove( "GEOMETRYLOADCOMBINATION", "Geometry load combination 2" )
addGeometryLoadCombination( "" )
setGeometryLoadCombinationFactor( "Geometry load combination 1", "gravity", 1 )
setGeometryLoadCombinationFactor( "Geometry load combination 1", "load", 1 )
setGeometryLoadCombinationFactor( "Geometry load combination 1", "load", 1 )
```

```
setTimeDependentLoadFactors( "GEOMETRYLOADCOMBINATION", "Geometry load combination 1", [ 0, 1.5768e+10 ], [ 1, 1 ] )
saveProject( )
setTimeDependentLoadFactors( "GEOMETRYLOADCOMBINATION", "Geometry load combination 1", [ 0, 1.5768e+10 ], [ 1, 1 ] )
addSet( "GEOMETRYSUPPORTSET", "co1" )
createSurfaceSupport( "co1", "co1" )
setParameter( "GEOMETRYSUPPORT", "co1", "AXES", [ 1, 2 ] )
setParameter( "GEOMETRYSUPPORT", "co1", "TRANSL", [ 1, 1, 1 ] )
setParameter( "GEOMETRYSUPPORT", "co1", "ROTATI", [ 0, 0, 0 ] )
attach( "GEOMETRYSUPPORT", "co1", "Sheet 1", [[ 0.1705708, -1.1202023e-16, 1.0349259 ]] )
saveProject( )
addSet( "GEOMETRYSUPPORTSET", "Geometry support set 2" )
rename( "GEOMETRYSUPPORTSET", "Geometry support set 2", "co2" )
createSurfaceSupport( "co2", "co2" )
setParameter( "GEOMETRYSUPPORT", "co2", "AXES", [ 1, 2 ] )
setParameter( "GEOMETRYSUPPORT", "co2", "TRANSL", [ 1, 1, 1 ] )
setParameter( "GEOMETRYSUPPORT", "co2", "ROTATI", [ 0, 0, 0 ] )
attach( "GEOMETRYSUPPORT", "co2", "Sheet 1", [[ 0.1705708, 10, 1.0349259 ]] )
saveProject( )
saveProject( )
setElementSize( [ "Sheet 1" ], 0.5, -1, True )
setMesherType( [ "Sheet 1" ], "HEXQUAD" )
setMidSideNodeLocation( [ "Sheet 1" ], "LINEAR" )
saveProject( )
setElementSize( [ "Block 1" ], 0.5, -1, True )
setMesherType( [ "Block 1" ], "HEXQUAD" )
setMidSideNodeLocation( [ "Block 1" ], "LINEAR" )
saveProject( )
hide( "SHAPE", [ "Sheet 1" ] )
hide( "SHAPE", [ "bar1" ] )
hide( "SHAPE", [ "bar2" ] )
hide( "SHAPE", [ "bar3" ] )
hide( "SHAPE", [ "bar4" ] )
hide( "SHAPE", [ "bar5" ] )
hide( "SHAPE", [ "bar6" ] )
hide( "SHAPE", [ "bar8" ] )
hide( "SHAPE", [ "bar7" ] )
hide( "SHAPE", [ "bar9" ] )
hide( "SHAPE", [ "bar10" ] )
```

```
setElementSize( "Block 1", 2, [[ 2.294292, 5.73573, 0.01 ]], 0.5, 0.5, True )
saveProject(  )
show( "SHAPE", [ "Sheet 1" ] )
show( "SHAPE", [ "bar1" ] )
show( "SHAPE", [ "bar2" ] )
show( "SHAPE", [ "bar3" ] )
show( "SHAPE", [ "bar4" ] )
show( "SHAPE", [ "bar5" ] )
show( "SHAPE", [ "bar6" ] )
show( "SHAPE", [ "bar7" ] )
show( "SHAPE", [ "bar8" ] )
show( "SHAPE", [ "bar9" ] )
show( "SHAPE", [ "bar10" ] )
generateMesh( [] )
hideView( "GEOM" )
showView( "MESH" )
addAnalysis( "Analysis2" )
addAnalysisCommand( "Analysis2", "PHASE", "Phased" )
renameAnalysis( "Analysis2", "Analysis2" )
setActivePhase( "Analysis2", "Phased" )
setActivePhase( "Analysis2", "Phased" )
setActiveInPhase( "Analysis2", "ELEMENTSET", [ "Block 1" ], [ "Phased" ], False )
addAnalysisCommand( "Analysis2", "NONLIN", "Structural nonlinear" )
setActivePhase( "Analysis2", "Phased" )
renameAnalysisCommandDetail( "Analysis2", "Structural nonlinear", "EXECUT(1)",
"laod" )
setAnalysisCommandDetail( "Analysis2", "Structural nonlinear",
"EXECUT(1)/ITERAT/MAXITE", 20 )
setAnalysisCommandDetail( "Analysis2", "Structural nonlinear",
"EXECUT(1)/ITERAT/MAXITE", 20 )
setAnalysisCommandDetail( "Analysis2", "Structural nonlinear",
"EXECUT(1)/ITERAT/MAXITE", 20 )
setAnalysisCommandDetail( "Analysis2", "Structural nonlinear",
"EXECUT(1)/ITERAT/CONVER/DISPLA/NOCONV", "CONTIN" )
setAnalysisCommandDetail( "Analysis2", "Structural nonlinear",
"EXECUT(1)/ITERAT/CONVER/FORCE/NOCONV", "CONTIN" )
setActivePhase( "Analysis2", "Phased" )
setAnalysisCommandDetail( "Analysis2", "Structural nonlinear", "EXECUT/EXETYP",
"TIME" )
renameAnalysisCommandDetail( "Analysis2", "Structural nonlinear", "EXECUT(2)",
"creep" )
```

renameAnalysisCommandDetail("Analysis2", "Structural nonlinear", "EXECUT(2)", "creep and shrinkage")
setAnalysisCommandDetail("Analysis2", "Structural nonlinear", "EXECUT(2)/TIME/STEPS/EXPLIC/SIZES", "86400.0 3.14496e+07 2.83824e+08 1.26144e+09 1.57680e+09")
setAnalysisCommandDetail("Analysis2", "Structural nonlinear", "EXECUT(2)/TIME/STEPS/EXPLIC/SIZES", "86400.0 3.14496e+07 2.83824e+08 1.26144e+09 1.57680e+09")
setAnalysisCommandDetail("Analysis2", "Structural nonlinear", "EXECUT(2)/TIME/STEPS/EXPLIC/SIZES", "86400.0 3.14496e+07 2.83824e+08 1.26144e+09 1.57680e+09")
setAnalysisCommandDetail("Analysis2", "Structural nonlinear", "EXECUT(2)/TIME/STEPS/EXPLIC/SIZES", "86400.0 3.14496e+07 2.83824e+08 1.26144e+09 1.57680e+09")
setAnalysisCommandDetail("Analysis2", "Structural nonlinear", "EXECUT(2)/TIME/STEPS/EXPLIC/SIZES", "86400.0 3.14496e+07 2.83824e+08 1.26144e+09 1.57680e+09")
setAnalysisCommandDetail("Analysis2", "Structural nonlinear", "EXECUT(2)/ITERAT/MAXITE", 50)
setAnalysisCommandDetail("Analysis2", "Structural nonlinear", "EXECUT(2)/ITERAT/CONVER/DISPLA/NOCONV", "CONTIN")
setAnalysisCommandDetail("Analysis2", "Structural nonlinear", "EXECUT(2)/ITERAT/CONVER/FORCE/NOCONV", "CONTIN")
addAnalysisCommand("Analysis2", "PHASE", "Phased 1")
setActivePhase("Analysis2", "Phased 1")
addAnalysisCommand("Analysis2", "NONLIN", "Structural nonlinear 1")
removeAnalysisCommandDetail("Analysis2", "Structural nonlinear 1", "EXECUT(1)")
setActivePhase("Analysis2", "Phased 1")
setAnalysisCommandDetail("Analysis2", "Structural nonlinear 1", "EXECUT/EXETYP", "START")
renameAnalysisCommand("Analysis2", "Structural nonlinear 1", "Structural nonlinear 1")
addAnalysisCommandDetail("Analysis2", "Structural nonlinear 1", "EXECUT(1)/START/INITIA/STRESS")
setAnalysisCommandDetail("Analysis2", "Structural nonlinear 1", "EXECUT(1)/START/INITIA/STRESS", True)
setAnalysisCommandDetail("Analysis2", "Structural nonlinear 1", "EXECUT(1)/START/LOAD/PREVIO", False)
saveProject( )
setAnalysisCommandDetail("Analysis2", "Structural nonlinear 1", "EXECUT(1)/ITERAT/MAXITE", 20)
setAnalysisCommandDetail("Analysis2", "Structural nonlinear 1",

```
"EXECUT(1)/ITERAT/CONVER/DISPLA/NOCONV", "CONTIN" )
    setAnalysisCommandDetail( "Analysis2", "Structural nonlinear 1",
"EXECUT(1)/ITERAT/CONVER/FORCE/NOCONV", "CONTIN" )
    addAnalysisCommandDetail( "Analysis2", "Structural nonlinear 1",
"MODEL/EVALUA/REINFO/INTERF" )
    setAnalysisCommandDetail( "Analysis2", "Structural nonlinear 1",
"MODEL/EVALUA/REINFO/INTERF", True )
    addAnalysisCommandDetail( "Analysis2", "Structural nonlinear",
"MODEL/EVALUA/REINFO/INTERF" )
    setAnalysisCommandDetail( "Analysis2", "Structural nonlinear",
"MODEL/EVALUA/REINFO/INTERF", True )
    setAnalysisCommandDetail( "Analysis2", "Structural nonlinear 1",
"EXECUT(1)/START/TIME", 3.1536e+09 )
    runSolver( "Analysis2" )
    showView( "RESULT" )
    setResultPlot( "contours", "Total Displacements/node", "TDtZ" )
    setResultCase( [ "Analysis2", "Output", "Phased, Time-step 2, Time 86400." ] )
    setResultCase( [ "Analysis2", "Output", "Phased, Time-step 3, Time 0.31536E+08" ] )
    setResultCase( [ "Analysis2", "Output", "Phased, Time-step 4, Time 0.31536E+09" ] )
    setResultCase( [ "Analysis2", "Output", "Phased, Time-step 5, Time 0.15768E+10" ] )
    setResultCase( [ "Analysis2", "Output", "Phased, Time-step 6, Time 0.31536E+10" ] )
    setResultCase( [ "Analysis2", "Output", "Phased 1, Start-step 1, Load-factor 1.0000" ] )
    setResultCase( [ "Analysis2", "Output", "Phased, Time-step 6, Time 0.31536E+10" ] )
    setResultCase( [ "Analysis2", "Output", "Phased, Time-step 5, Time 0.15768E+10" ] )
    setResultCase( [ "Analysis2", "Output", "Phased, Time-step 4, Time 0.31536E+09" ] )
    setResultCase( [ "Analysis2", "Output", "Phased, Time-step 3, Time 0.31536E+08" ] )
    setResultCase( [ "Analysis2", "Output", "Phased, Time-step 2, Time 86400." ] )
    setResultCase( [ "Analysis2", "Output", "Phased, Load-step 1, Load-factor 1.0000" ] )
    setResultCase( [ "Analysis2", "Output", "Phased, Time-step 2, Time 86400." ] )
    setResultCase( [ "Analysis2", "Output", "Phased, Load-step 1, Load-factor 1.0000" ] )
    setResultCase( [ "Analysis2", "Output", "Phased, Time-step 2, Time 86400." ] )
    setResultCase( [ "Analysis2", "Output", "Phased, Time-step 3, Time 0.31536E+08" ] )
    setResultCase( [ "Analysis2", "Output", "Phased, Time-step 4, Time 0.31536E+09" ] )
    setResultCase( [ "Analysis2", "Output", "Phased, Time-step 5, Time 0.15768E+10" ] )
    setResultCase( [ "Analysis2", "Output", "Phased, Time-step 6, Time 0.31536E+10" ] )
    setResultCase( [ "Analysis2", "Output", "Phased 1, Start-step 1, Load-factor 1.0000" ] )
    setResultCase( [ "Analysis2", "Output", "Phased, Time-step 6, Time 0.31536E+10" ] )
    setResultCase( [ "Analysis2", "Output", "Phased 1, Start-step 1, Load-factor 1.0000" ] )
    showIds( "NODE", [ 110 ] )
    setResultCase( [ "Analysis2", "Output", "Phased, Time-step 6, Time 0.31536E+10" ] )
    setResultCase( [ "Analysis2", "Output", "Phased 1, Start-step 1, Load-factor 1.0000" ] )
```

3.3 Time-Dependent Analysis of Post-tensioned Concrete Bridge

In this part, a simply supported post-tensioned concrete integral bridge with the length of 30 m in I-section shape is created, where the longitudinal and cross-section (at the middle section) sizes of the whole model are displayed in Fig. 3.87. CEB-FIP 1990 and AASHTO models are applied to compare the simulation results under 1-year time-dependent effect. Meanwhile, long-term performance of ultra-high performance concrete (UHPC) is also simulated under AASHTO time-dependent model via parametric modeling. Concrete grade is C50. Post-tensioned prestress tendon is modeled by bond-slip material constitutive model with the shape of symmetrical parabolic curve. The height of anchor end is 1.2 m, while the lowest point at the middle of the bridge is 0.3 m. Meanwhile, creating numerical model via directly editing parameters in command console in Python language is also illustrated in this numerical model. Concentrated long-term load value of 20 kN is symmetrically attached to the edge of top plate at the trisected points. Parameters of ordinary C50 concrete, UHPC concrete and post-tensioned prestress tendons are listed in Table 3.2.

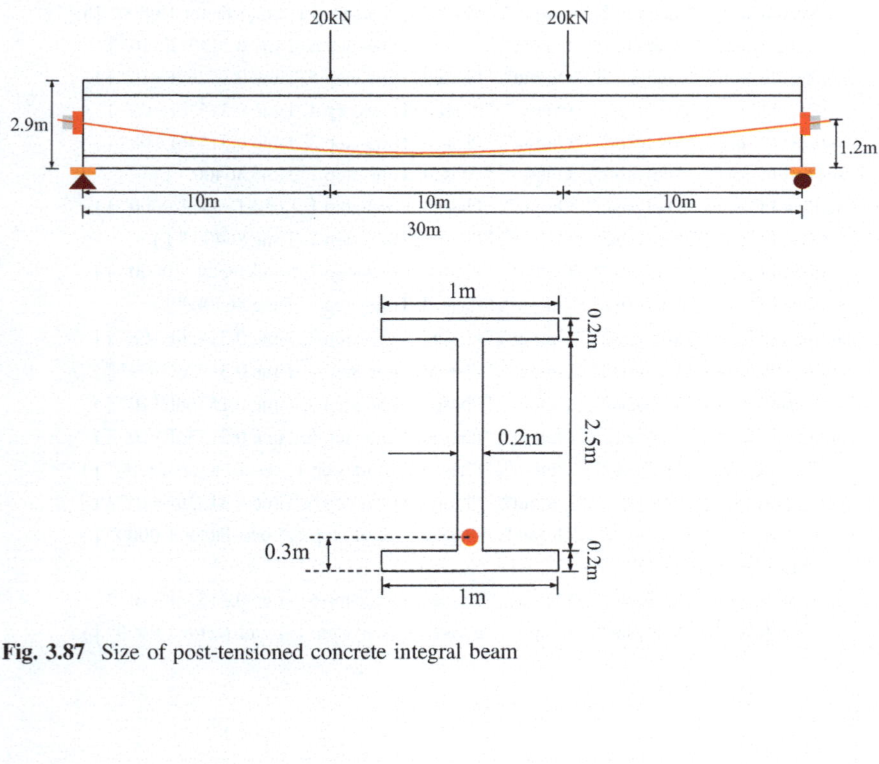

Fig. 3.87 Size of post-tensioned concrete integral beam

Table 3.2 Basic parameters of concrete, UHPC and tendon

Concrete	Elastic modulus	3.8629×10^{10} N/m^2
	Characteristic compressive strength	50 MPa
	Strength strength	2.6 MPa
	Poisson's ratio	0.15
	Mass density	2500 kg/m^3
UHPC	Elastic modulus	4.5×10^{10} N/m^2
	Tensile strength	5 MPa
	Compressive strength	120 MPa
Tendon	Elastic modulus	1.95×10^{11} N/m^2
	Yield stress	1860 MPa

Essentials of learning

(1) *Learning to modeling I-shape beam* via *regular plane stress elements*
(2) *Learning to establish curved tendon shape*
(3) *Learning to specify material constitutive and cross-section properties of bond-slip model.*
(4) *Learning to use the function of mirror shape*
(5) *Learning to specify concrete and UHPC time-dependent parameters such as notional size of member under CEB-FIP 1990 and AASHTO code.*
(6) *Mastering attachment of point imprint and projection*
(7) *Learning to master fast parametric modeling in copying, pasting, editing and modifying Python command console*
(8) *Learning to generate long-term loss chart of prestress force through Chart-view function.*

Start the graphical user interface environment; click the menu bar **"File—New"**, then create a **.dpf** document in the directory of computer G-disk area named **"I-beam"**. **Structural** analysis is selected and the number of dimensions is **three.** Maximum **Model size** is 100 m, which means that the scope of the whole graphical visualization zone ranges from –50 to 50 m in all directions of the coordinate system. **Default mesher type** is **Hexa/Quad** element so that the geometric elements shapes are all quadrilateral (2 dimensions) or hexahedron (3 dimensions) while **Default mesh order** in this case is quadratic. Determination of mid-side node location is linear interpolation. The whole manipulations mentioned above are defined and determined by clicking **OK** button; then the GUI graphical interface can be displayed immediately, as shown in Fig. 3.88.

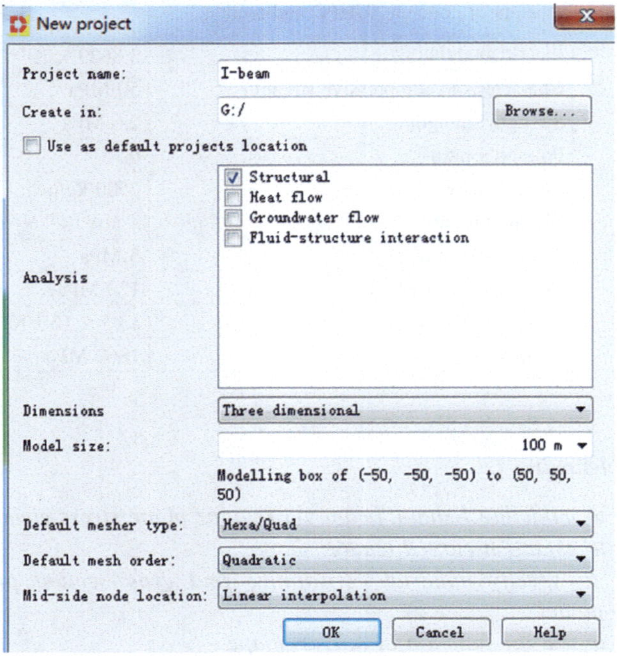

Fig. 3.88 New project interface

Inputting the coordinate values (0, 0, 0), (30, 0, 0), (30, 2.5, 0), (0, 2.5, 0) and (0, 2.5, 0), (30, 2.5, 0), (30, 2.7, 0), (0, 2.7, 0) with the default name of **Sheet1** and **Sheet2** to generate the parts of web and top plate, respectively, then selecting the **Sheet2** in the GUI zone and right-clicking to select **Array copy** function, bottom plate of **Sheet3** is created with the displacement 2.7 m in the negative Y direction and the **Number of copies** is 1 (see Fig. 3.89).

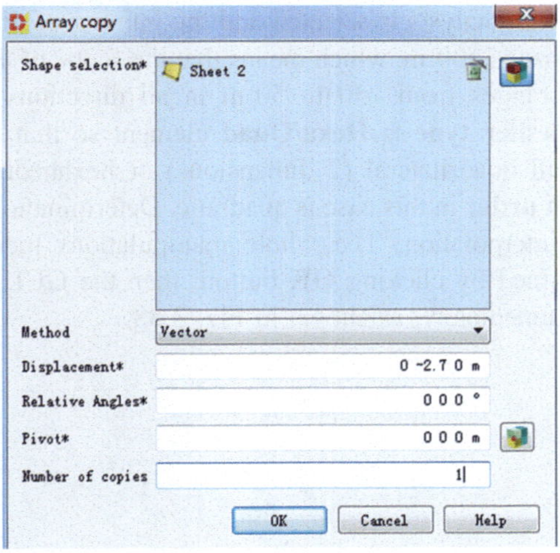

Fig. 3.89 Interface of Array copy

Clicking OK button, whole numerical geometric model is generated, which is displayed in Fig. 3.90.

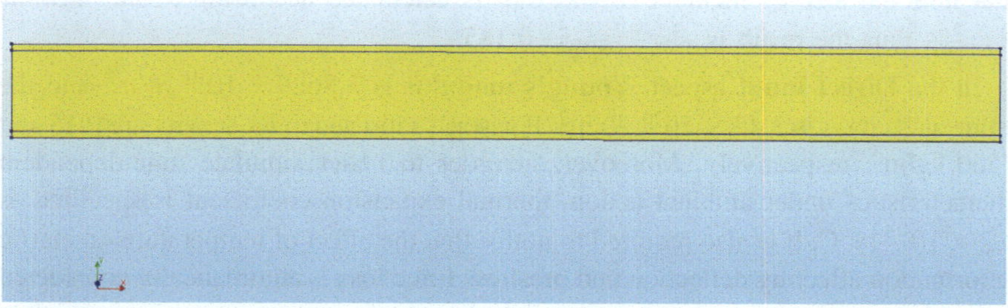

Fig. 3.90 Geometric model of beam

The following procedure is to construct symmetrical parabolic curve with the name of tendon, and the coordinate values of vertexes are shown in Fig. 3.91.

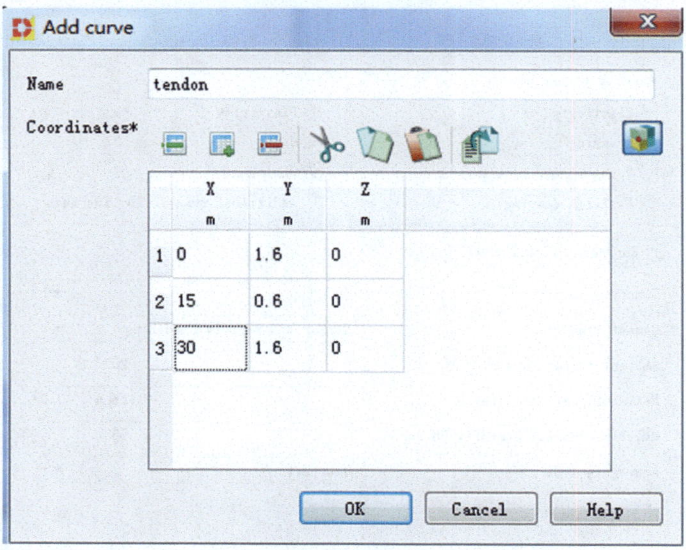

Fig. 3.91 Coordinate values of parabolic curve tendon

Modifying the units of temperature and angle from Kelvin and radian to Celsius and degree, respectively, clicking to select **Sheet1** and right-clicking it to select **Property assignments, Regular plane stress** is chosen as **Element class**, while dialog box of material specification named **block1** is opened via clicking shortcut icon . Design codes of CEB-FIP 1990 model are selected and the aspects of creep and shrinkage are ticked as **Aspects to include**. In the European CEB-FIP

1990 model, concrete class is C50 and Cement type is Normal and rapidly hardening type. **Ambient temperature** is specified as 20 °C while **Relative ambient humidity** is 55%. Aggregate type in this case is **Quartzite**. It is worth to mention that notional size of member of the web is calculated according to the formula $h = \frac{2A_c}{c}$, thus the result is $\frac{2.5 \times 0.2 \times 2}{2 \times (2.5 + 0.2)} = 0.185$.

In the **Direct input** aspect, **Young's modulus** is $3.8629 \times 10^{10}\ \mathrm{N/m^2}$ and the value at 28 days is $3.45 \times 10^{10}\ \mathrm{N/m^2}$. Poisson's ratio and mass density are 0.15 and 2500 kg/m^3, respectively. Moreover, in order to better simulate time-dependent characteristics under ambient action, thermal expansion coefficient is specified as $1.2 \times 10^{-5}\,1/°\mathrm{C}$. It is also required to notice that the effect of temperature on elastic deformation affecting deflection and prestress force loss is automatically considered in DIANA software, which means that this elastic deformation can be ignored when thermal expansion coefficient is zero. Characteristic strength at 28 days and mean compressive strength at 28 days are 50 and 58 MPa, respectively. Specifications of basic parameters in CEB-FIP 1990 are displayed in Figs. 3.92 and 3.93, respectively.

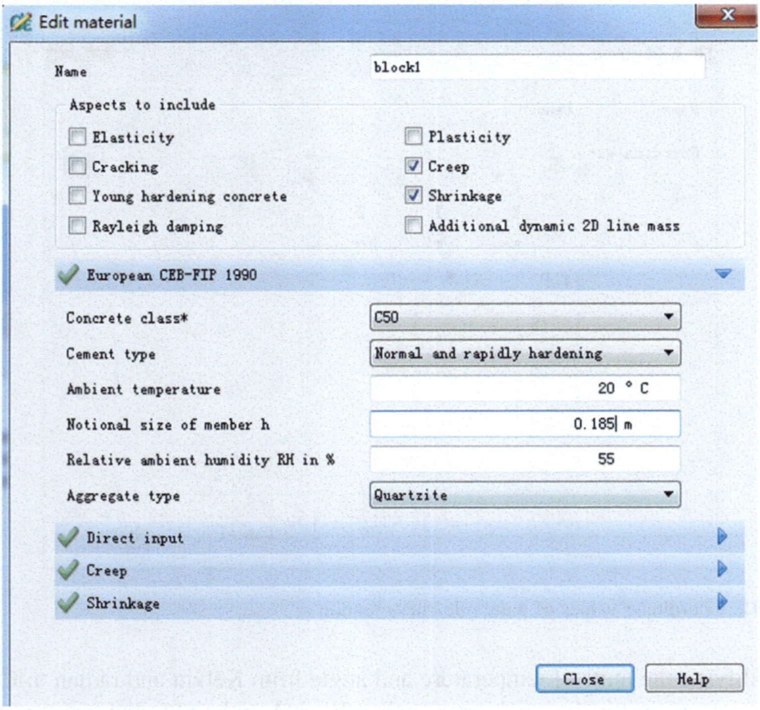

Fig. 3.92 Parameters of European CEB-FIP 1990

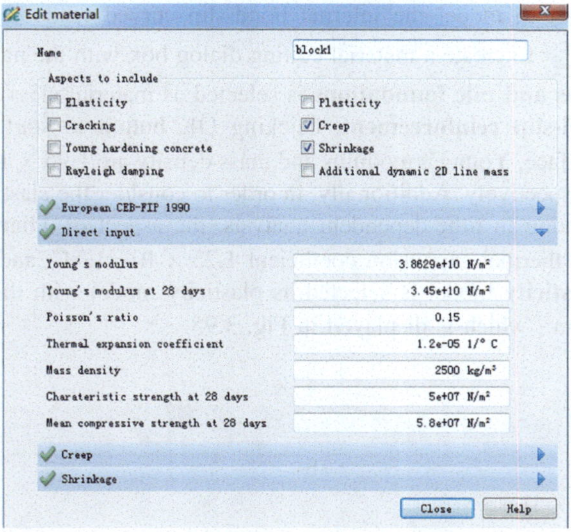

Fig. 3.93 Parameters of direct input

When material properties are specified, the cross-section geometric features of **block1** are specified. Thickness of web is defined as 0.2 m, while **Local element axis** is by default, representing that element x-axis corresponds to X direction in the global coordinate system (see Fig. 3.94).

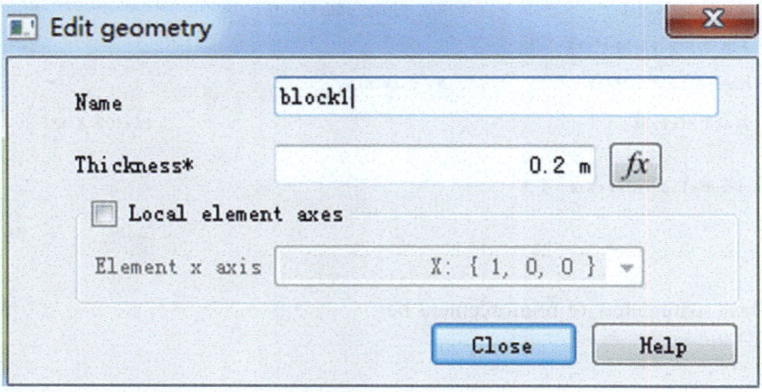

Fig. 3.94 Specification for cross-section geometry

Taking the same method to define material and geometric properties of top and bottom plates, the notional size of member is 0.167 m for both, while the thickness of both top and bottom plate is 1 m. Other parameters are the same as the ones in **block1**.

Next we start to model the internal bond-slip curved tendon. Initially, click shortcut icon to create a material editing dialog box with the name of **tendon**. **Reinforcements and pile foundations** is selected as material class while material model is **Bond-slip reinforcement**. Clicking OK button to start the bond-slip specifying interface, Young's modulus and mass density are 1.95×10^{11} N/m^2 and 7800 kg/m^3, respectively. Additionally, in order to consider the elastic deformation of prestress tendon in time-dependent analysis, the aspect of **Thermal effects** is ticked with the thermal expansion coefficient 1.25×10^{-5} 1/°C, and no hardening **von Mises plasticity** model is selected as plasticity model with the **Yield stress** 1.86×10^{9} N/m^2, which is displayed in Fig. 3.95.

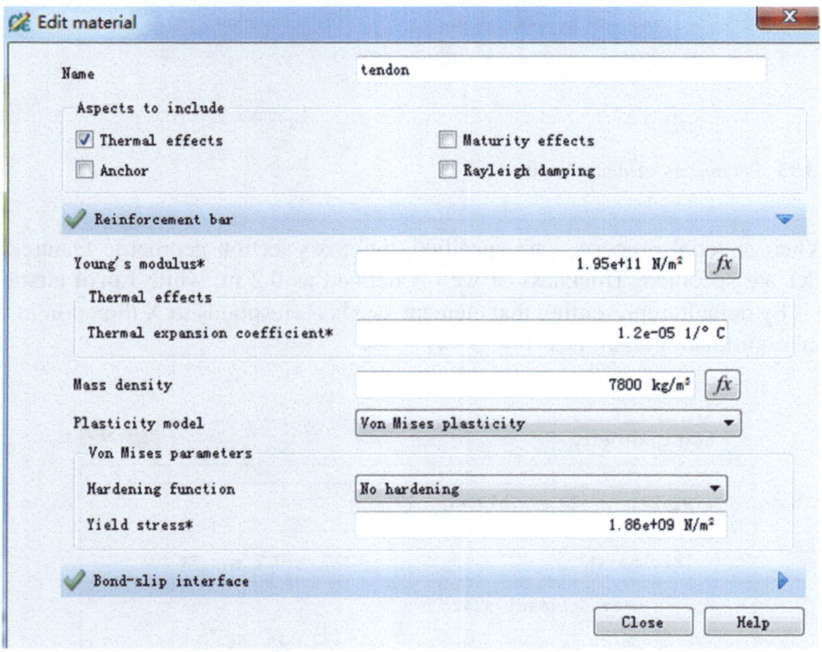

Fig. 3.95 Basic parameters of reinforcement bar

Normal and shear stiffness modulus are specified as 2×10^{12} N/m^3 and 2×10^{6} N/m^3, respectively. **Cubic bond-slip function by Doerr** is selected as **Bond-slip interface failure model**. When it comes to defining slip parameters, parameter c representing cohesion coefficient is 20 N/m^2 and the **Shear slip at start plateau** is 0.1 m (see Fig. 3.96).

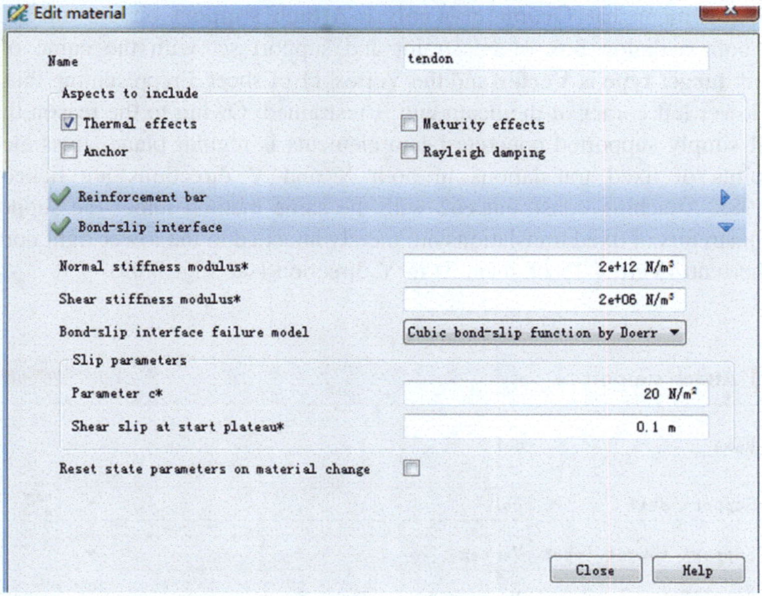

Fig. 3.96 Parameters of bond-slip model

While creating dialog box of **Edit geometry** with the name of tendon to specify cross-section geometric properties, **Truss bondslip** is selected as **Reinforcement type**. Under cross-section input, area of bar is 0.00139 m^2 with the **Contact perimeter** between tendon and bond-slip interface concrete being 0.9576 m. Bond-slip anchor of surface area is specified as 0.00278 m^2 (see Fig. 3.97).

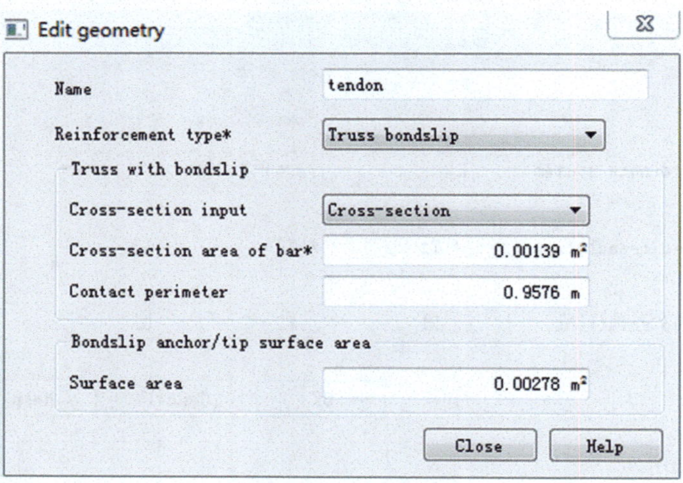

Fig. 3.97 Geometric parameters of bond-slip

Clicking menu bar **Geometry-Analysis-Attach support** on the top tool bar opens both a dialog box of constraint and support set with the name of **co1**. **Support target type** is **Vertex** and the Vertex 11 of sheet 3 representing that point at the lower left corner of the beam end is restrained. Owing to the reason that the type of simply supported concrete beam elements is regular plane stress element, constraints of fixed translations in both X and Y directions are ticked (see Fig. 3.98). Attaching constraint **co2** with the same method under the support set **co1**, constraints of fixed translations are merely attached to the lower right corner of the other end (Vertex 15 of sheet 3) in Y direction (see Fig. 3.99).

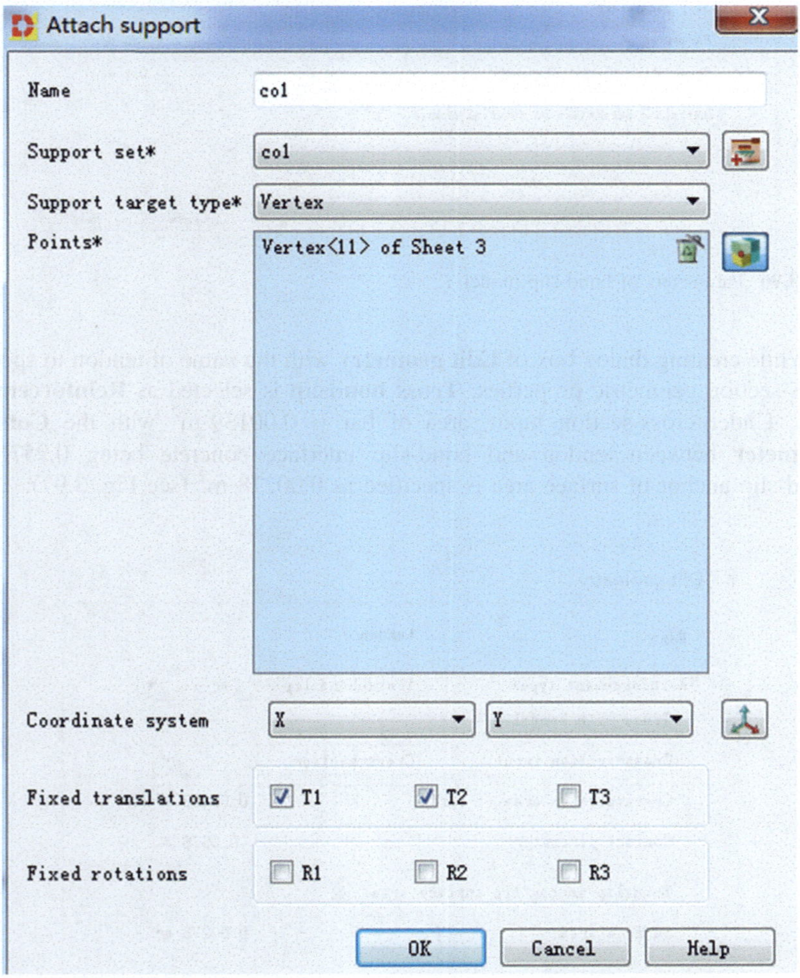

Fig. 3.98 co1

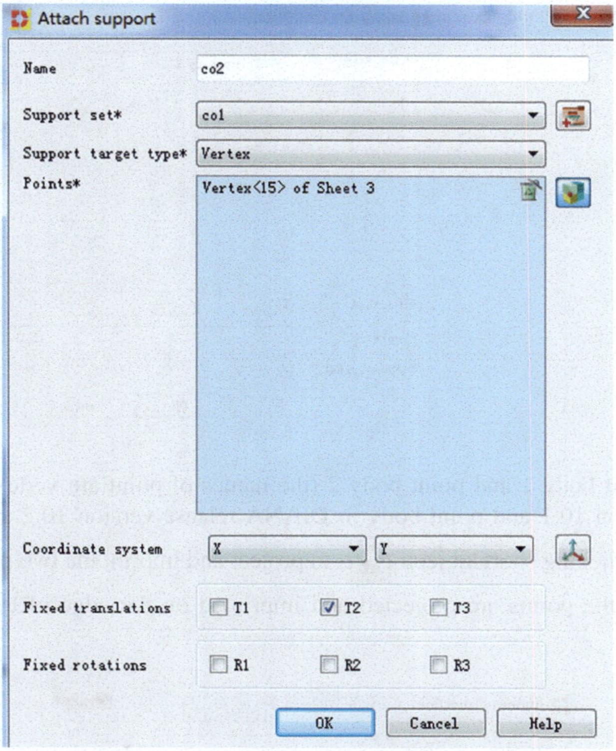

Fig. 3.99 co2

The following key step is to attach load. Before attaching symmetrical concentrated load, we begin to create loading point. Initially, clicking shortcut icon **Adds a point body,** which is also called as **Adds a vertex** in DIANA 10.1, coordinate value in Fig. 3.100 is input; then symmetrically copied in X direction via function **Mirror a shape**, where the **Pivot** is 15 m (see Fig. 3.101).

Fig. 3.100 Adding a point body

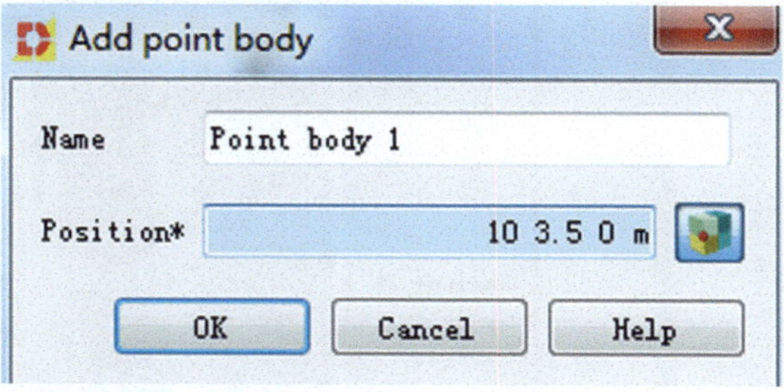

Fig. 3.101 Mirror a shape

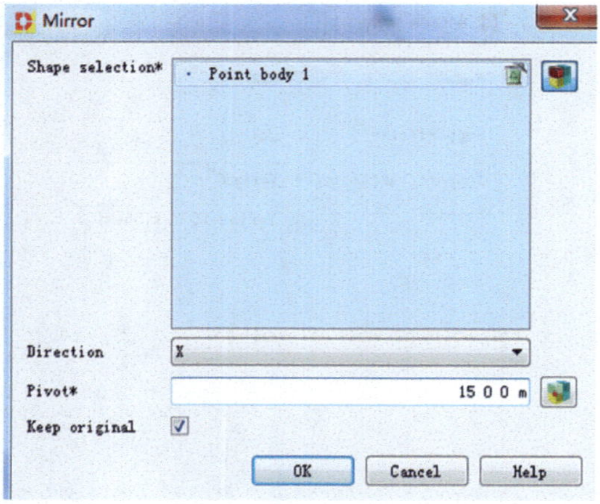

After point body 1 and point body 2 (the names of point are vertex in DIANA release version 10.1 and point body in DIANA release version 10.2, respectively) are created, clicking shortcut icon  to project and imprint the two points one by one in turn, the points are projected and imprinted on the edge of top plate (see Fig. 3.102).

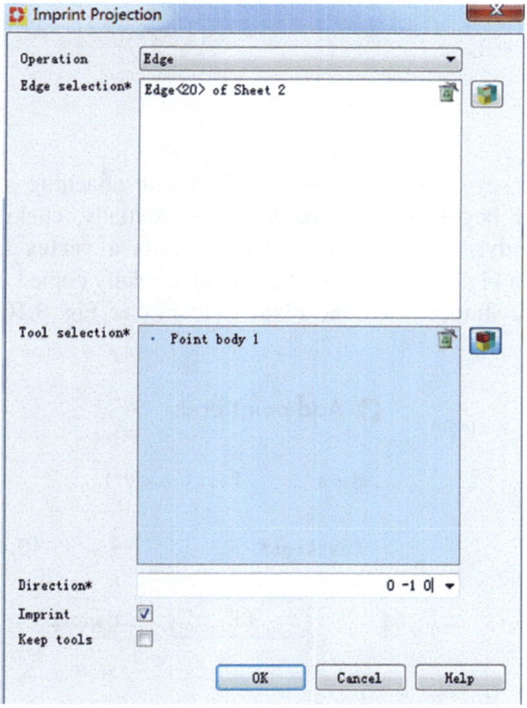

Fig. 3.102 Imprint projection of point body

Note: In the procedure of imprint projection, it is worth to mention that points should be manipulated one by one instead of simultaneous operation, which results in errors in the following meshing results.

Clicking shortcut icon **Define a global load** , load case of gravity is created while the **Load type** is **Dead weight** (see Fig. 3.103).

Fig. 3.103 Load case of **Dead weight**

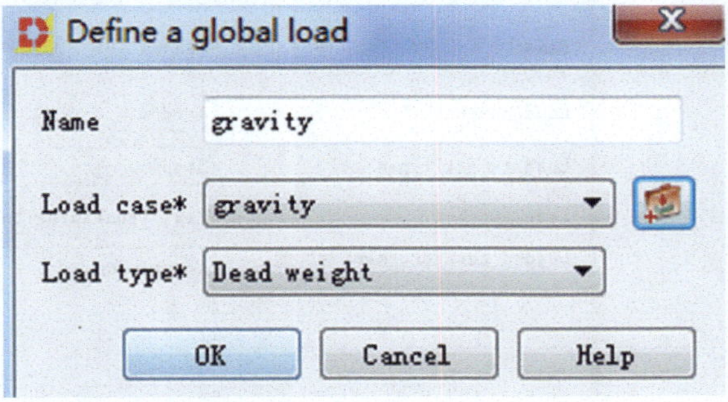

Still clicking shortcut icon **Add a new load case** , dialog box with the name of **load** is generated, where the **Load target type** is **Vertex** and the **Load type** is **Force**. Symmetrically concentrated load is applied to the former imprinted and projected points with the value of 20 kN in the negative Y direction, which is displayed in Fig. 3.104.

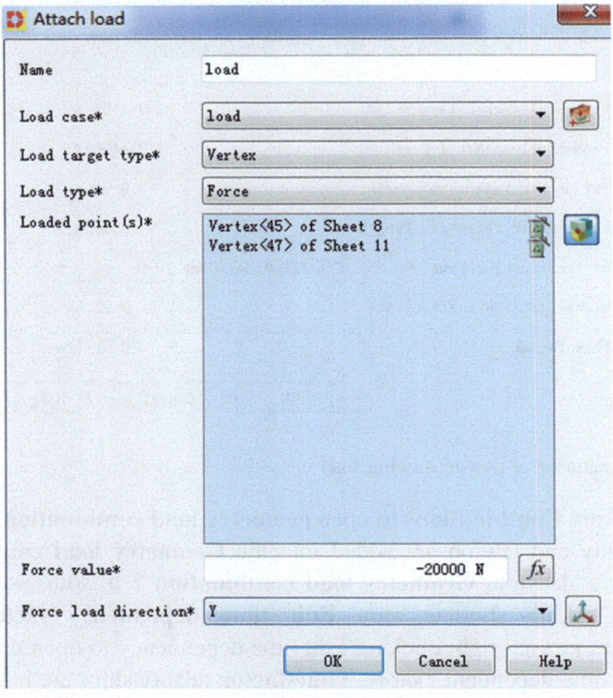

Fig. 3.104 Attachment of concentrated load

Creating post-tensioning load case with the name of **tendon** in the same way, post-tensioning load is attached to both anchor ends with the value 1940 kN, retention length of both sides is 0.01 m, with the **Coulomb friction coefficient** and **Wobble factor** are 0.22 and 0.01, respectively (see Fig. 3.105).

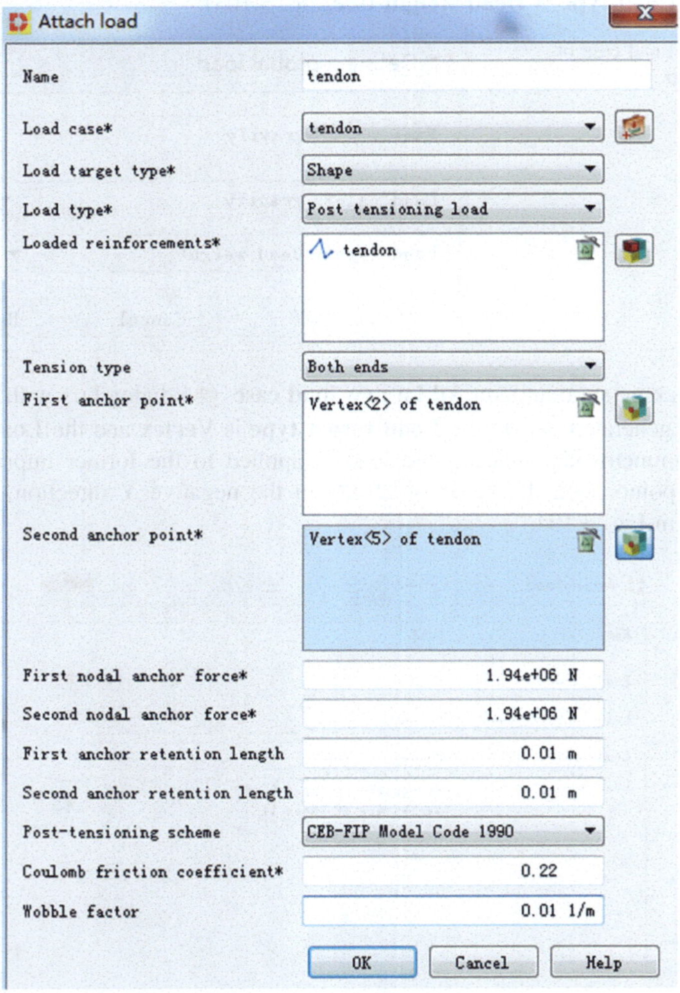

Fig. 3.105 Attachment of post-tensioning load

Right-clicking **Combinations** to open **geometry load combination tables**, load cases of gravity and tendon are added into the **Geometry load combination 1**, while **load** is added into **Geometry load combination 2** in solo, as is shown in Fig. 3.106. Clicking shortcut icon **Edit time dependency factors** to edit time-dependent factors, right-clicking Edit time dependency to open the dialog box of specifying time-dependent factors, **Time-factor** relationships are both defined as constant coefficient 1 ranging from 0 to 100 years, which is (0 s, 1) (31,536,000 s, 1) (see Fig. 3.107).

Clicking Close button, the restrained and loaded numerical model of the beam is displayed (Fig. 3.108).

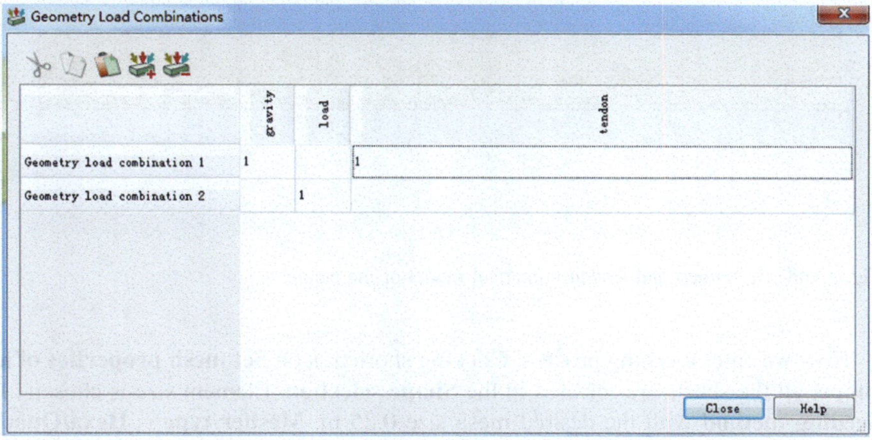

Fig. 3.106 Geometry load combination

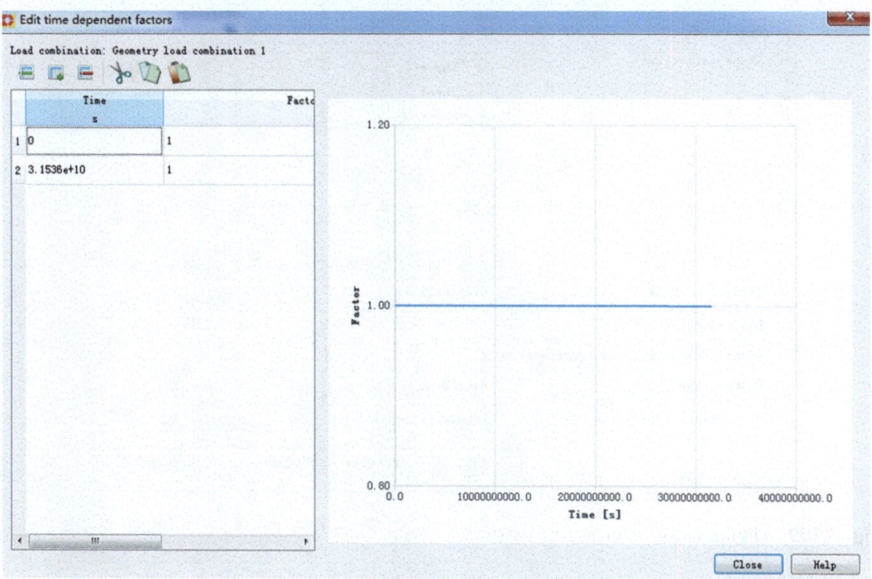

Fig. 3.107 Specification of time-dependent factors

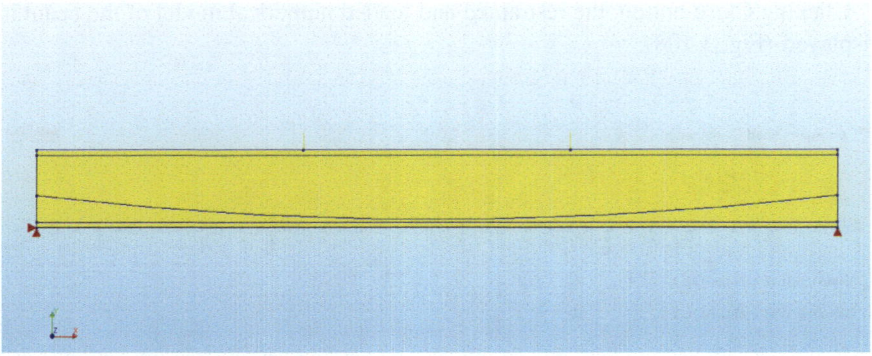

Fig. 3.108 Restrained and loaded numerical model of the beam

Now we enter meshing process. Clicking shortcut icon **Set mesh properties of a shape**, all the sheets are selected in the **Shape selection**. Element size is chosen as **Seeding method** with the desired mesh size 0.25 m. **Mesher type** is **Hexa/Quad** representing hexahedral shape in 3D and quadrilateral shape in 2D. The way of determining mid-side node location is **Linear interpolation** (see Fig. 3.109).

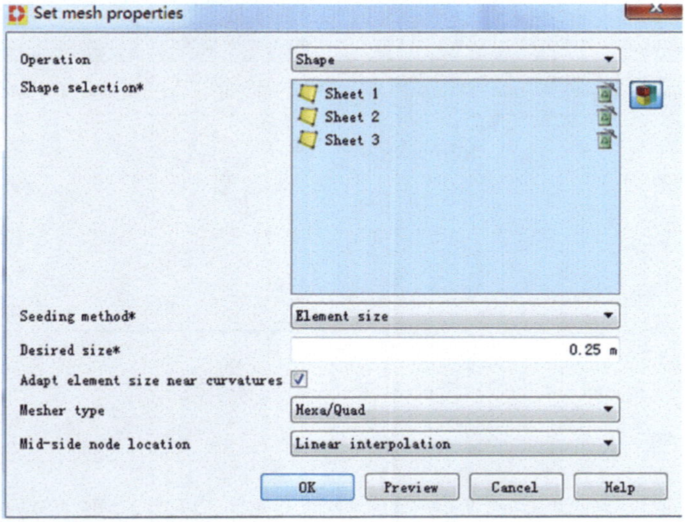

Fig. 3.109 Manipulation interface of mesh

Clicking shortcut icon **Generate mesh of a shape,** the meshed numerical model is displayed (Fig. 3.110).

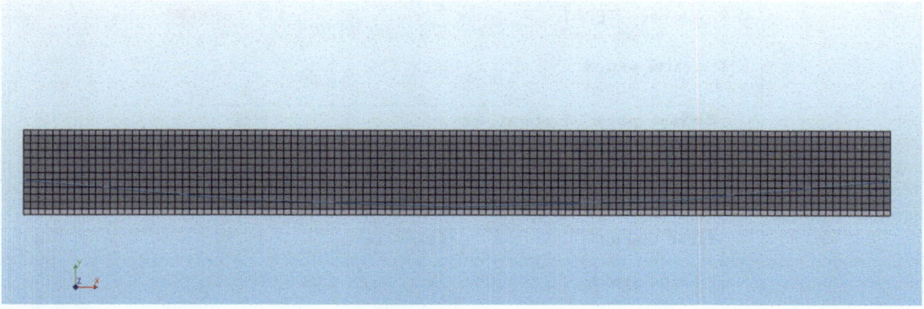

Fig. 3.110 Mesh generation

Clicking **Add an analysis** button to start a new analysis block with the name of **Analysis1**, Structural nonlinear analysis type is selected. In the **Evaluate model**, option of Evaluate reinforcements in interface elements is ticked. Deleting the original load step and creating the start step, in the aspect of start steps, **Geometry load combination 1** is added as the load set. In the **Establish equilibrium**, **User specified size** of load combination 1 is 1.00000 coupled with only one step to simulate the initial first dead load with the name of **first dead load**. Right-clicking **Equilibrium iteration** to open **Edit properties,** maximum number of iterations is 50 while iteration method is regular **Newton–Raphson**. Both **Force** and **Displacement** are selected as convergence norm with the **Convergence tolerance** and **Abort criterion** 0.001 and 10,000 respectively (see Figs. 3.111, 3.112 and 3.113).

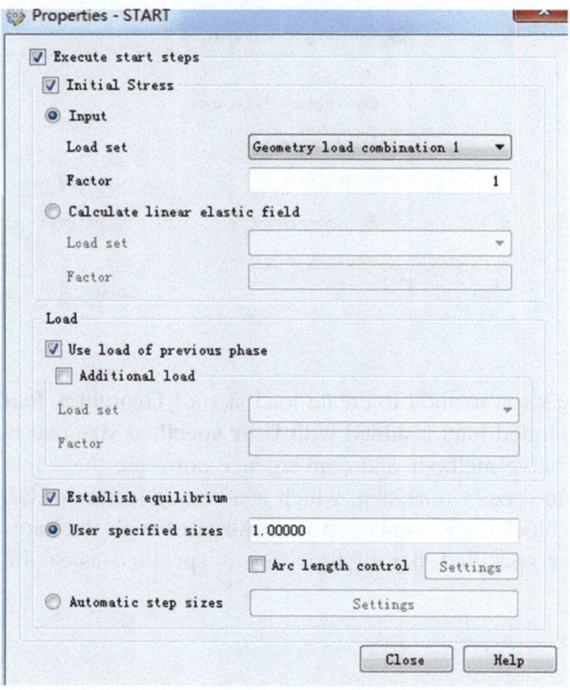

Fig. 3.111 Start step of **first dead load**

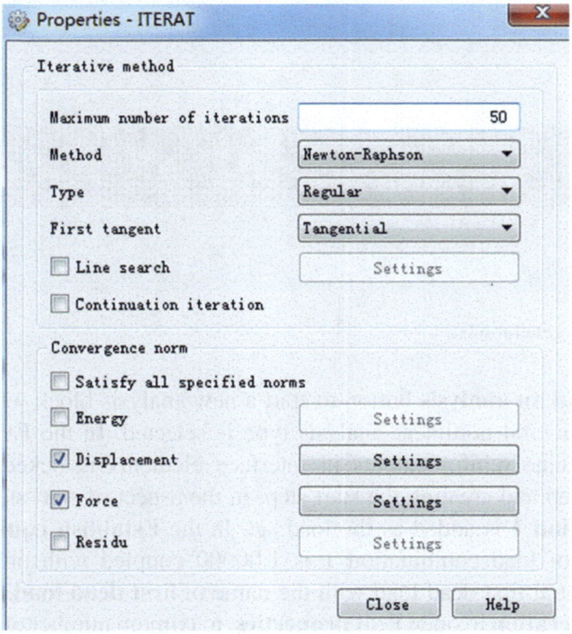

Fig. 3.112 Iteration properties

Fig. 3.113 Convergence norm

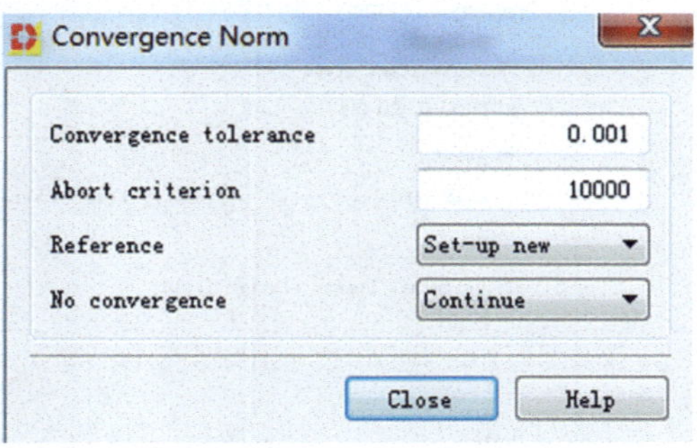

Applying the same method to create load step of **Geometry load combination 2**, where concentrated load is added with **User specified size** and **number of load step** both 1, iterative methods and convergence norm are the same as former.

Next step is to specify time step, which is a unique feature of DIANA software. A new execute block step—time step is constructed with the name of **creep and shrinkage**. **User specified size** of time-step is specified as 86,400.0 s 2.50560e

+06 s 1.31760e+07 s 1.57680e+07 s corresponding to the summed period 1 day, 30 days, half year and 1 year, respectively. Iterative methods and convergence norm are the same as former.

To run analysis click shortcut icon **Run analysis** . After calculation, initial hugging-up and ultimate 1 year time-dependent displacement contour plots in Y direction are displayed in Figs. 3.114, 3.115 and 3.116, where the displacement value in positive is upward and negative is downward. It is evident to see that after 1 year time effect, the total long-term displacement deflection except the instant loading value is $0.488 + 0.833 = 1.321$ cm.

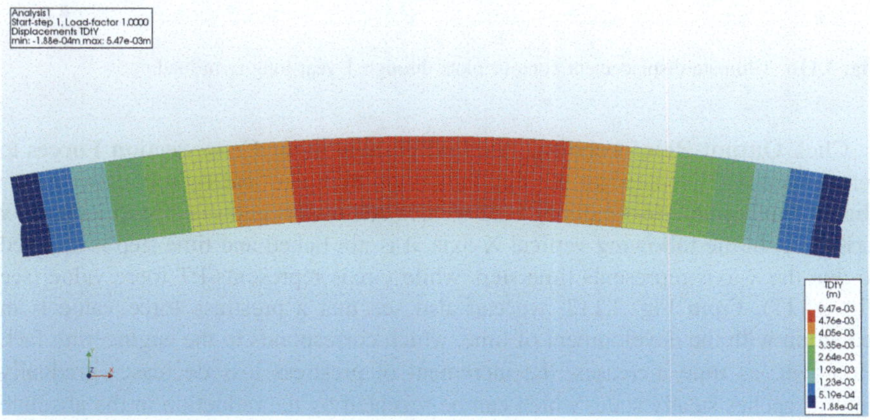

Fig. 3.114 Displacement contour of hugging-up

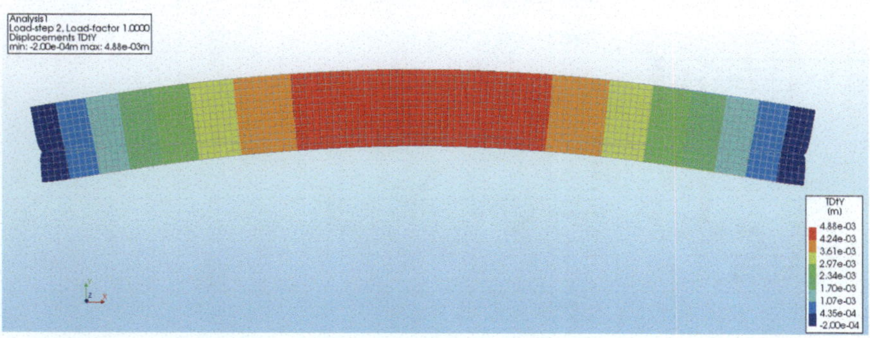

Fig. 3.115 Displacement contour after instant loading

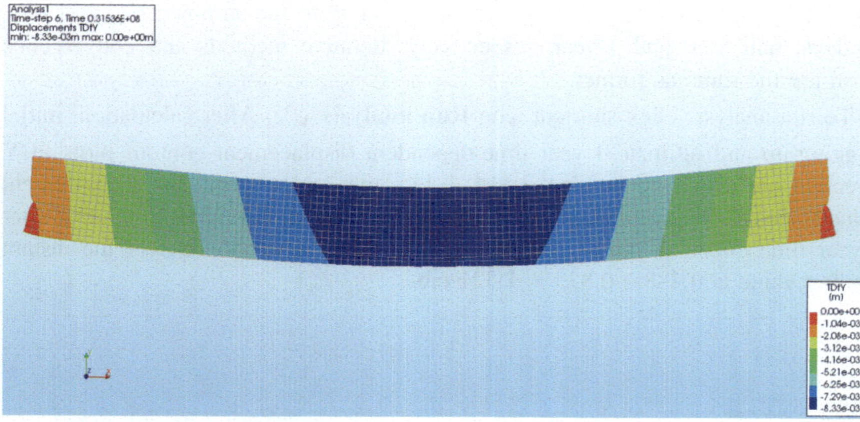

Fig. 3.116 Ultimate displacement contour plots through 1 year long-term loading

Click **Output-Reinforcement results-Reinforcement Cross-section Forces** to select node 4721 located in the middle site of the span, and right-clicking opens **show Chart** option. In the dialog box, we tick **All results** and select **step value** as **x axis date**. In the following vertical X-axis, it is not ticked and time step is selected so that the x-axis represents time-step, while y-axis represents PT force value (see Fig. 3.117). From Fig. 3.117, we can also see that a prestress force value is in reduction with the development of time, which corresponds to the engineering fact. Moreover, as time increases, the increment of prestress loss decreases gradually until it reaches steady state, which can be revealed by the reduction in the absolute slope value in the graph.

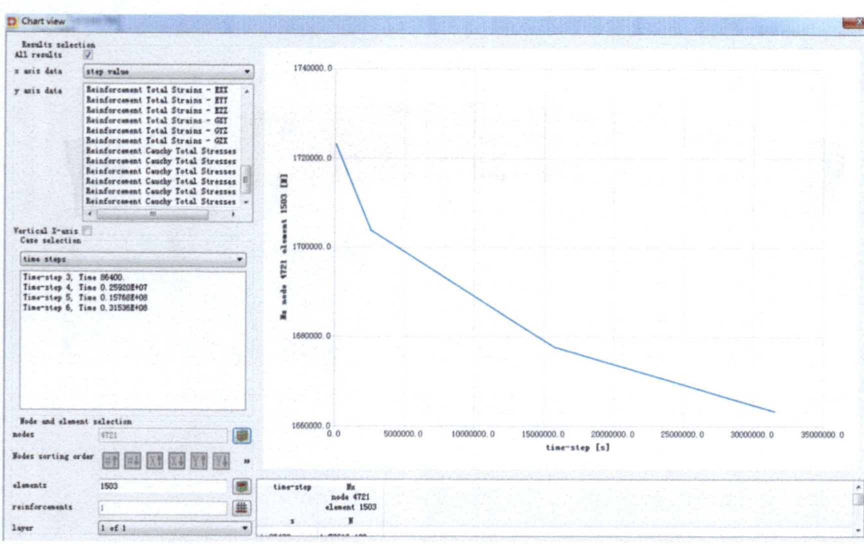

Fig. 3.117 Long-term loss curve of prestress force

Deleting dialog box of CEB-FIP 1990 and reconstructing new AASHTO editing dialog box of material and cross-section geometric properties with the name of **block1, top1** and **bot1**, respectively, the aspects of **Creep** and **Shrinkage** are selected. In AASHTO code, compressive strength at the age of 28 days is 5e7 N/m^2, while tensile strength at the age of 28 days is 2.6e6 N/m^2. Correction factor K1 for source of aggregate is 1 while Young's modulus is 3.8629×10^{10} N/m^2. Other time-dependent parameters are the same as former CEB-FIP 1990 code (see Figs. 3.118 and 3.119). It must be noted that ambient factor of temperature is not taken into consideration in AASHTO compared with CEB-FIP 1990, thus the calculation value may tend to be conservative.

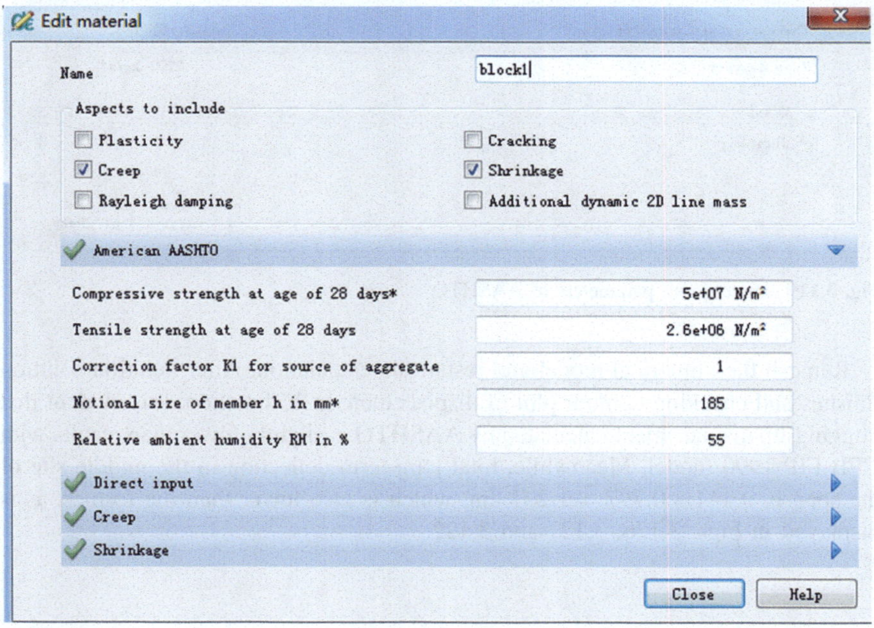

Fig. 3.118 Basic parameters of AASHTO

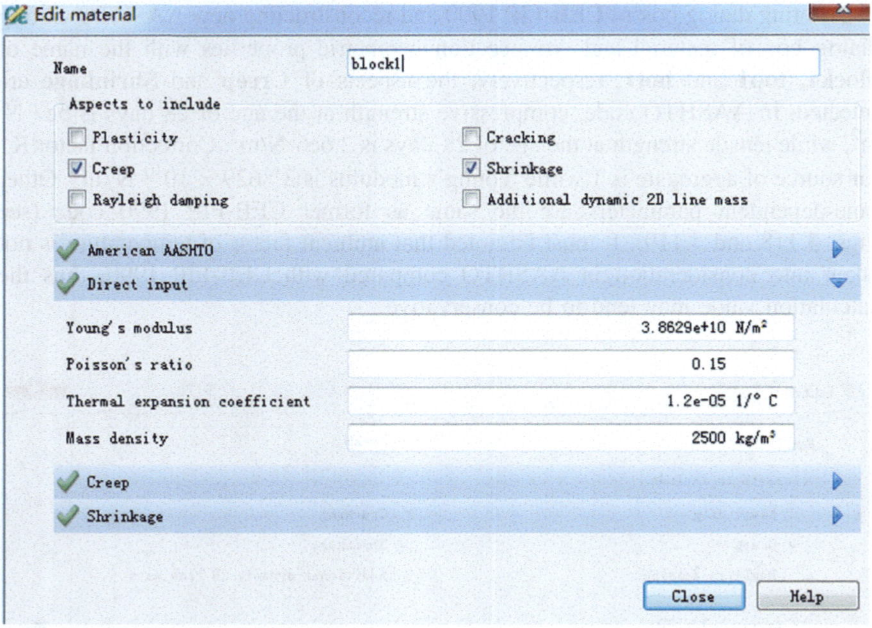

Fig. 3.119 Direct input parameters of AASHTO

Remesh the numerical model and restart the calculation. After iteration solution finishes and checking contour plot of displacement in Y direction, it is evident that hugging-up displacement calculated by AASHTO is slightly lower than results with CEB-FIP 1990 model. Meanwhile, total long-term deflection in the middle site of the span is $0.444 + 0.797 = 1.241$ cm, which is also lower than the former, as is displayed in Figs. 3.120, 3.121 and 3.122.

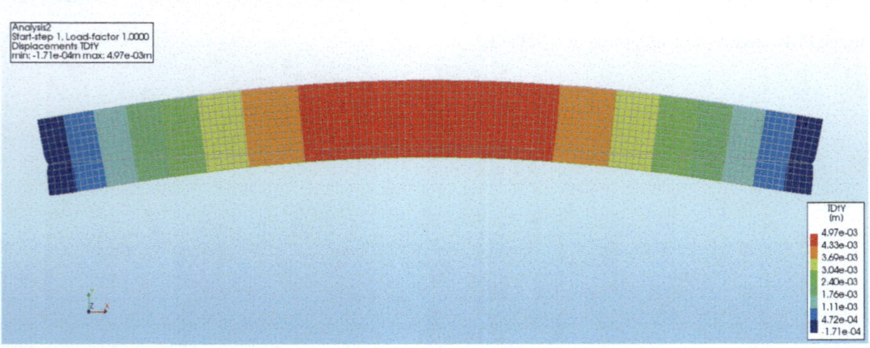

Fig. 3.120 Displacement contour of hugging-up in AASHTO

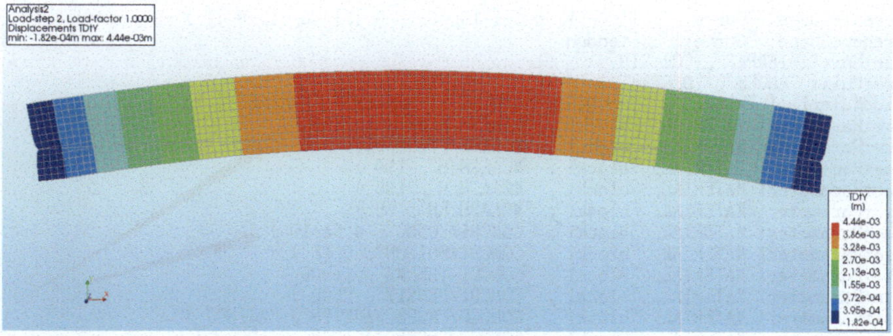

Fig. 3.121 Displacement contour after instant loading in AASHTO

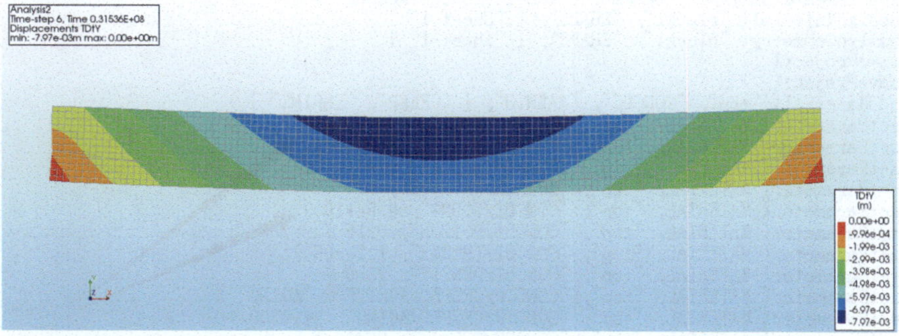

Fig. 3.122 Ultimate displacement contour plots through 1 year long-term loading in AASHTO

UHPC beam in I-shape is modeled via parameter modeling. Parametric modeling is a fast modeling method via modification of some parameters in command console then importing it into the DIANA through running a saved script without consuming time to remodel the model in graphical user interface zone. Modeling steps are generated one by one when command console in Python language is read in turn. In UHPC manuscript, Young's modulus is modified as $4.5 \times 10^{10}\ \mathrm{N/m^2}$ while compressive and tensile concrete strength at 28 days are 120 and 5 MPa, respectively (see Fig. 3.123).

```
createCurve( "Curve 1", [[ 0, 1, 0 ],[ 15, 0.1, 0 ],[ 30, 1, 0 ]] )
renameShape( "Curve 1", "tendon" )
setUnit( "TEMPER", "CELSIU" )
setUnit( "ANGLE", "DEGREE" )
addMaterial( "block1", "CONCDC", "AASHTO", [ "CREEP", "SHRINK" ] )
setParameter( MATERIAL, "block1", "MCAASH/FC28", 1.2e+08 )
setParameter( MATERIAL, "block1", "MCAASH/FT28", 5000000 )
setParameter( MATERIAL, "block1", "MCAASH/H", 150 )
setParameter( MATERIAL, "block1", "MCAASH/H", 185 )
setParameter( MATERIAL, "block1", "MCAASH/RH", 55 )
setParameter( MATERIAL, "block1", "CONCDI/YOUNG", 4.5e+10 )
setParameter( MATERIAL, "block1", "CONCDI/POISON", 0.15 )
setParameter( MATERIAL, "block1", "CONCDI/THERMX", 1e-05 )
setParameter( MATERIAL, "block1", "CONCDI/DENSIT", 2500 )
setParameter( MATERIAL, "block1", "CONCCP/CRSPEC/AGETYP", "AGING" )
setParameter( MATERIAL, "block1", "CONCCP/CRSPEC/AGING", 2419200 )
setParameter( MATERIAL, "block1", "CONCSH/CURAGE", 86400 )
addGeometry( "Element geometry 1", "SHEET", "MEMBRA", [] )
rename( GEOMET, "Element geometry 1", "block1" )
setParameter( GEOMET, "block1", "THICK", 0.2 )
setParameter( GEOMET, "block1", "LOCAXS", True )
setElementClassType( "SHAPE", [ "Sheet 1" ], "MEMBRA" )
assignMaterial( "block1", "SHAPE", [ "Sheet 1" ] )
assignGeometry( "block1", "SHAPE", [ "Sheet 1" ] )
saveProject( )
saveProject( )
addMaterial( "top", "CONCDC", "AASHTO", [ "CREEP", "SHRINK" ] )
setParameter( MATERIAL, "top", "MCAASH/FC28", 1.2e+08 )
setParameter( MATERIAL, "top", "MCAASH/FT28", 5000000 )
setParameter( MATERIAL, "top", "MCAASH/H", 167 )
setParameter( MATERIAL, "top", "MCAASH/RH", 55 )
setParameter( MATERIAL, "top", "CONCDI/YOUNG", 4.5e+10 )
setParameter( MATERIAL, "top", "CONCDI/POISON", 0.15 )
setParameter( MATERIAL, "top", "CONCDI/THERMX", 1.2e-05 )
setParameter( MATERIAL, "top", "CONCDI/DENSIT", 2500 )
setParameter( MATERIAL, "top", "CONCCP/CRSPEC/AGETYP", "AGING" )
setParameter( MATERIAL, "top", "CONCCP/CRSPEC/AGING", 2419200 )
setParameter( MATERIAL, "top", "CONCSH/CURAGE", 86400 )
addGeometry( "Element geometry 2", "SHEET", "MEMBRA", [] )
rename( GEOMET, "Element geometry 2", "top" )
setParameter( GEOMET, "top", "THICK", 1 )
setParameter( GEOMET, "top", "LOCAXS", True )
setElementClassType( "SHAPE", [ "Sheet 2" ], "MEMBRA" )
assignMaterial( "top", "SHAPE", [ "Sheet 2" ] )
assignGeometry( "top", "SHAPE", [ "Sheet 2" ] )
saveProject( )
saveProject( )
setParameter( MATERIAL, "block1", "CONCDI/THERMX", 1.2e-05 )
saveProject( )
addMaterial( "bot", "CONCDC", "AASHTO", [ "CREEP", "SHRINK" ] )
setParameter( MATERIAL, "bot", "MCAASH/FC28", 1.2e+08 )
setParameter( MATERIAL, "bot", "MCAASH/FT28", 5000000 )
setParameter( MATERIAL, "bot", "MCAASH/H", 167 )
setParameter( MATERIAL, "bot", "MCAASH/RH", 55 )
setParameter( MATERIAL, "bot", "CONCDI/YOUNG", 4.5e+10 )
setParameter( MATERIAL, "bot", "CONCDI/POISON", 0.15 )
setParameter( MATERIAL, "bot", "CONCDI/THERMX", 1.2e-05 )
setParameter( MATERIAL, "bot", "CONCDI/DENSIT", 2500 )
setParameter( MATERIAL, "bot", "CONCCP/CRSPEC/AGETYP", "AGING" )
setParameter( MATERIAL. "bot". "CONCCP/CRSPEC/AGING". 2419200 )
```

Fig. 3.123 Modification of manuscript via parametric modeling

Restarting to start solution, results of UHPC are displayed in Figs. 3.124, 3.125 to 3.126 (Table 3.3).

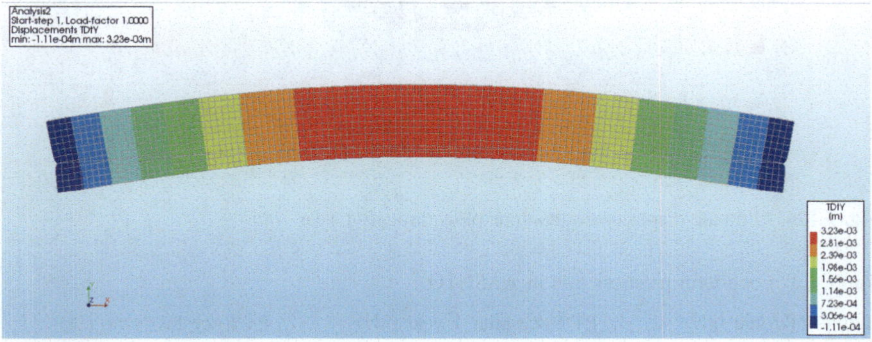

Fig. 3.124 Displacement contour of hugging-up of UHPC in AASHTO

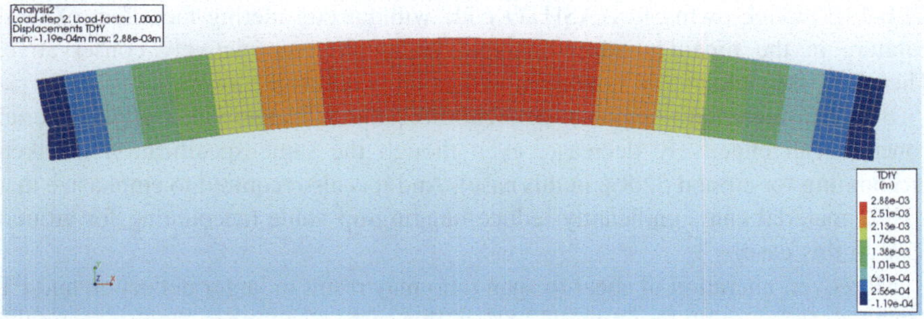

Fig. 3.125 Displacement contour after instant loading of UHPC in AASHTO

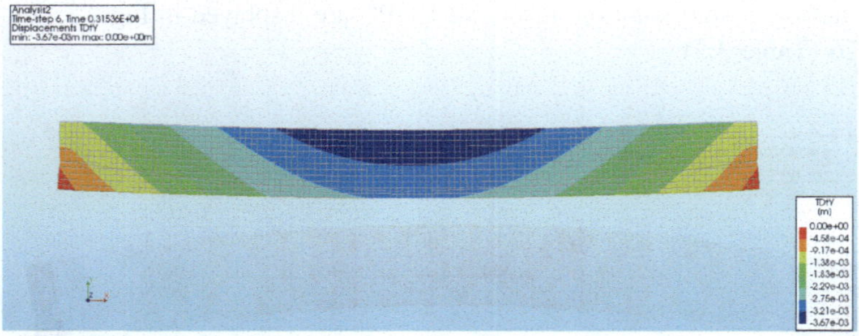

Fig. 3.126 Ultimate displacement contour plots through 1 year

Table 3.3 Long-term prestress loss in AASHTO

Initial PT force (kN)	PT force after 1 year (kN)	PT force loss ratio (%)
170.517	155.728	8.67
170.519	164.494	4.095

Above all, it is worth to mention that compared with results calculated by CEB-FIP model, owing to AASHTO code without considering the effect of temperature in the time-dependent analysis, its results are relatively conservative. Therefore, the outcome of long-term deflection is relatively lower.

Besides, maximum long-term deflection results of bridge with UHPC material concrete can effectively decrease, even though the same specification is taken (accounting for around 52.8% in this case). And it is also required to emphasize that UHPC material can significantly reduce hugging-up value (accounting for around 35% in this case).

Moreover, alteration of shear to span ratio may result in larger deflection and PT force loss (accounting for around 52.8% in this case) when other conditions are the same, which means that vehicles with longer wheels may have more deteriorated influence on long-term performance.

Finally, long-term deflection of girder increases with the increment of longitudinal length of span while the prestress force loss decreases with the increment of longitudinal length of span when other conditions are the same, which can be inter-compared and validated mutually in the following cases.

```
###########################################################################
# DianaIE 10.2 update 2018-01-19 17:07:06
# Python  3.6.1
# Session recorded at 2019-04-11 14:51:53
###########################################################################
newProject( "G:/I-beam", 100 )
setModelAnalysisAspects( [ "STRUCT" ] )
setModelDimension( "2D" )
setDefaultMeshOrder( "QUADRATIC" )
setDefaultMesherType( "HEXQUAD" )
setDefaultMidSideNodeLocation( "LINEAR" )
saveProject(  )
createSheet( "Sheet 1", [[ 0, 0, 0 ],[ 30, 0, 0 ],[ 30, 2.5, 0 ],[ 0, 2.5, 0 ]] )
createSheet( "Sheet 2", [[ 0, 2.5, 0 ],[ 30, 2.5, 0 ],[ 30, 2.7, 0 ],[ 0, 2.7, 0 ]] )
arrayCopy( [ "Sheet 2" ], [ 0, -2.7, 0 ], [ 0, 0, 0 ], [ 0, 0, 0 ], 1 )
createCurve( "Curve 1", [[ 0, 1, 0 ],[ 15, 0.1, 0 ],[ 30, 1, 0 ]] )
renameShape( "Curve 1", "tendon" )
setUnit( "TEMPER", "CELSIU" )
setUnit( "ANGLE", "DEGREE" )
addMaterial( "block1", "CONCDC", "MC1990", [ "CREEP", "SHRINK" ] )
setParameter( "MATERIAL", "block1", "MC90CO/GRADE", "C50" )
setParameter( "MATERIAL", "block1", "MC90CO/RH", 55 )
setParameter( "MATERIAL", "block1", "CONCDI/YOUNG", 3.8629e+10 )
setParameter( "MATERIAL", "block1", "CONCDI/YOUN28", 3.45e+10 )
setParameter( "MATERIAL", "block1", "CONCDI/POISON", 0.15 )
setParameter( "MATERIAL", "block1", "CONCDI/THERMX", 1.2e-05 )
setParameter( "MATERIAL", "block1", "CONCDI/DENSIT", 2500 )
setParameter( "MATERIAL", "block1", "CONCDI/FCK28", 50000000 )
setParameter( "MATERIAL", "block1", "CONCDI/FCM28", 58000000 )
setParameter( "MATERIAL", "block1", "CONCCP/CRSPEC/AGETYP", "AGING" )
setParameter( "MATERIAL", "block1", "CONCCP/CRSPEC/AGING", 2419200 )
setParameter( "MATERIAL", "block1", "CONCSH/CURAGE", 86400 )
setParameter( MATERIAL, "block1", "MC90CO/H", 0.185 )
addGeometry( "Element geometry 1", "SHEET", "MEMBRA", [] )
rename( "GEOMET", "Element geometry 1", "block1" )
setParameter( "GEOMET", "block1", "THICK", 0.2 )
setElementClassType( "SHAPE", [ "Sheet 1" ], "MEMBRA" )
assignMaterial( "block1", "SHAPE", [ "Sheet 1" ] )
assignGeometry( "block1", "SHAPE", [ "Sheet 1" ] )
saveProject(  )
addMaterial( "top1", "CONCDC", "MC1990", [ "CREEP", "SHRINK" ] )
```

```
setParameter( "MATERIAL", "top1", "MC90CO/GRADE", "C50" )
setParameter( "MATERIAL", "top1", "MC90CO/AMBTEM", 20 )
setParameter( "MATERIAL", "top1", "MC90CO/AMBTEM", 20 )
setParameter( "MATERIAL", "top1", "MC90CO/RH", 55 )
setParameter( "MATERIAL", "top1", "CONCDI/YOUNG", 3.8629e+10 )
setParameter( "MATERIAL", "top1", "CONCDI/YOUN28", 3.45e+10 )
setParameter( "MATERIAL", "top1", "CONCDI/POISON", 0.15 )
setParameter( "MATERIAL", "top1", "CONCDI/THERMX", 1.2e-05 )
setParameter( "MATERIAL", "top1", "CONCDI/DENSIT", 2500 )
setParameter( "MATERIAL", "top1", "CONCDI/FCK28", 50000000 )
setParameter( "MATERIAL", "top1", "CONCDI/FCM28", 58000000 )
setParameter( "MATERIAL", "top1", "CONCCP/CRSPEC/AGETYP", "AGING" )
setParameter( "MATERIAL", "top1", "CONCCP/CRSPEC/AGING", 86400 )
setParameter( "MATERIAL", "top1", "CONCCP/CRSPEC/AGING", 2419200 )
setParameter( "MATERIAL", "top1", "CONCSH/CURAGE", 86400 )
setParameter( MATERIAL, "top1", "MC90CO/H", 0.167 )
addGeometry( "Element geometry 2", "SHEET", "MEMBRA", [] )
rename( "GEOMET", "Element geometry 2", "top1" )
setParameter( "GEOMET", "top1", "THICK", 1 )
setElementClassType( "SHAPE", [ "Sheet 2" ], "MEMBRA" )
assignMaterial( "top1", "SHAPE", [ "Sheet 2" ] )
assignGeometry( "top1", "SHAPE", [ "Sheet 2" ] )
saveProject(   )
addMaterial( "bot1", "CONCDC", "MC1990", [ "CREEP", "SHRINK" ] )
setParameter( "MATERIAL", "bot1", "MC90CO/GRADE", "C50" )
setParameter( "MATERIAL", "bot1", "MC90CO/AMBTEM", 20 )
setParameter( "MATERIAL", "bot1", "MC90CO/AMBTEM", 20 )
setParameter( "MATERIAL", "bot1", "MC90CO/RH", 55 )
setParameter( "MATERIAL", "bot1", "CONCDI/YOUNG", 3.8629e+10 )
setParameter( "MATERIAL", "bot1", "CONCDI/YOUN28", 3.45e+10 )
setParameter( "MATERIAL", "bot1", "CONCDI/POISON", 0.15 )
setParameter( "MATERIAL", "bot1", "CONCDI/THERMX", 1.2e-05 )
setParameter( "MATERIAL", "bot1", "CONCDI/DENSIT", 2500 )
setParameter( "MATERIAL", "bot1", "CONCDI/FCK28", 50000000 )
setParameter( "MATERIAL", "bot1", "CONCDI/FCM28", 58000000 )
setParameter( "MATERIAL", "bot1", "CONCCP/CRSPEC/AGETYP", "AGING" )
setParameter( "MATERIAL", "bot1", "CONCCP/CRSPEC/AGING", 86400 )
setParameter( "MATERIAL", "bot1", "CONCCP/CRSPEC/AGING", 2419200 )
setParameter( "MATERIAL", "bot1", "CONCSH/CURAGE", 86400 )
setParameter( MATERIAL, "bot1", "MC90CO/H", 0.167 )
addGeometry( "Element geometry 3", "SHEET", "MEMBRA", [] )
rename( "GEOMET", "Element geometry 3", "bot1" )
setParameter( "GEOMET", "bot1", "THICK", 1 )
```

```
setElementClassType( "SHAPE", [ "Sheet 3" ], "MEMBRA" )
assignMaterial( "bot1", "SHAPE", [ "Sheet 3" ] )
assignGeometry( "bot1", "SHAPE", [ "Sheet 3" ] )
saveProject( )
addMaterial( "tendon", "REINFO", "REBOND", [] )
setMaterialAspects( "tendon", [ "THERMA" ] )
setParameter( MATERIAL, "tendon", "REBARS/THERMA/THERMX", 1.2e-05 )
setParameter( "MATERIAL", "tendon", "REBARS/ELASTI/YOUNG", 1.95e+11 )
setParameter( "MATERIAL", "tendon", "REBARS/MASS/DENSIT", 7800 )
setParameter( "MATERIAL", "tendon", "REBARS/PLATYP", "VMISES" )
setParameter( "MATERIAL", "tendon", "REBARS/PLASTI/YLDSTR", 1.86e+09 )
setParameter( "MATERIAL", "tendon", "RESLIP/DSNY", 2e+12 )
setParameter( "MATERIAL", "tendon", "RESLIP/DSSX", 2000000 )
setParameter( "MATERIAL", "tendon", "RESLIP/SHFTYP", "BONDS1" )
setParameter( "MATERIAL", "tendon", "RESLIP/BONDS1/SLPVAL", [ 20, 0.1 ] )
addGeometry( "Element geometry 4", "RELINE", "REBAR", [] )
setParameter( "GEOMET", "Element geometry 4", "REITYP", "REITRU" )
setParameter( "GEOMET", "Element geometry 4", "REITRU/CROSSE", 0.00139 )
setParameter( "GEOMET", "Element geometry 4", "REITRU/PERIME", 0.9576 )
setParameter( "GEOMET", "Element geometry 4", "TIPELM/SURFAC", 0.00278 )
setReinforcementAspects( [ "tendon" ] )
assignMaterial( "tendon", "SHAPE", [ "tendon" ] )
assignGeometry( "Element geometry 4", "SHAPE", [ "tendon" ] )
resetElementData( "SHAPE", [ "tendon" ] )
setReinforcementDiscretization( [ "tendon" ], "SECTION" )
saveProject( )
rename( "GEOMET", "Element geometry 4", "tendon" )
saveProject( )
createPointSupport( "co1", "co1" )
setParameter( GEOMETRYSUPPORT, "co1", "AXES", [ 1, 2 ] )
setParameter( GEOMETRYSUPPORT, "co1", "TRANSL", [ 1, 1, 0 ] )
setParameter( GEOMETRYSUPPORT, "co1", "ROTATI", [ 0, 0, 0 ] )
attach( GEOMETRYSUPPORT, "co1", "Sheet 3", [[ 0, -0.2, 0 ]] )
saveProject( )
createPointSupport( "co2", "co1" )
setParameter( GEOMETRYSUPPORT, "co2", "AXES", [ 1, 2 ] )
setParameter( GEOMETRYSUPPORT, "co2", "TRANSL", [ 0, 1, 0 ] )
setParameter( GEOMETRYSUPPORT, "co2", "ROTATI", [ 0, 0, 0 ] )
attach( GEOMETRYSUPPORT, "co2", "Sheet 3", [[ 30, -0.2, 0 ]] )
saveProject( )
createPointBody( "Point body 1", [ 10, 3.5, 0 ] )
mirror( [ "Point body 1" ], [ 15, 0, 0 ], [ True, False, False ], True )
```

```
projection( SHAPEEDGE, "Sheet 2", [[ 10, 2.7, 0 ]], [ "Point body 1" ], [ 0, -1, 0 ], True )
removeShape( [ "Point body 1" ] )
saveProject( )
projection( SHAPEEDGE, "Sheet 2", [[ 20, 2.7, 0 ]], [ "Point body 2" ], [ 0, -1, 0 ], True )
removeShape( [ "Point body 2" ] )
saveProject( )
addSet( GEOMETRYLOADSET, "gravity" )
createModelLoad( "gravity", "gravity" )
addSet( GEOMETRYLOADSET, "Geometry load case 1" )
rename( GEOMETRYLOADSET, "Geometry load case 1", "load" )
createPointLoad( "load", "load" )
setParameter( GEOMETRYLOAD, "load", "FORCE/VALUE", -20000 )
setParameter( GEOMETRYLOAD, "load", "FORCE/DIRECT", 2 )
attach( GEOMETRYLOAD, "load", "Sheet 2", [[ 10, 2.7, 0 ]] )
attach( GEOMETRYLOAD, "load", "Sheet 2", [[ 20, 2.7, 0 ]] )
addSet( GEOMETRYLOADSET, "Geometry load case 2" )
rename( GEOMETRYLOADSET, "Geometry load case 2", "tendon" )
createBodyLoad( "tendon", "tendon" )
setParameter( GEOMETRYLOAD, "tendon", "LODTYP", "POSTEN" )
setParameter( GEOMETRYLOAD, "tendon", "POSTEN/BOTHEN/FORCE1", 1940000 )
setParameter( GEOMETRYLOAD, "tendon", "POSTEN/BOTHEN/FORCE2", 1940000 )
setParameter( GEOMETRYLOAD, "tendon", "POSTEN/BOTHEN/RETLE1", 0.01 )
setParameter( GEOMETRYLOAD, "tendon", "POSTEN/BOTHEN/RETLE2", 0.01 )
setParameter( GEOMETRYLOAD, "tendon", "POSTEN/SHEAR", 0.22 )
setParameter( GEOMETRYLOAD, "tendon", "POSTEN/WOBBLE", 0.01 )
attachTo( GEOMETRYLOAD, "tendon", "POSTEN/BOTHEN/PNTS1", "tendon", [[ 0, 1.6, 0 ]] )
attachTo( GEOMETRYLOAD, "tendon", "POSTEN/BOTHEN/PNTS2", "tendon", [[ 30, 1.6, 0 ]] )
attach( GEOMETRYLOAD, "tendon", [ "tendon" ] )
saveProject( )
setDefaultGeometryLoadCombinations( )
setGeometryLoadCombinationFactor( "Geometry load combination 1", "gravity", 1 )
remove( GEOMETRYLOADCOMBINATION, "Geometry load combination 3" )
setGeometryLoadCombinationFactor( "Geometry load combination 1", "tendon", 1 )
setTimeDependentLoadFactors( "GEOMETRYLOADCOMBINATION", "Geometry load combination 1", [ 0, 3.1536e+10 ], [ 1, 1 ] )
setTimeDependentLoadFactors( "GEOMETRYLOADCOMBINATION", "Geometry load combination 2", [ 0, 3.1536e+10 ], [ 1, 1 ] )
saveProject( )
setElementSize( [ "Sheet 1", "Sheet 2", "Sheet 3" ], 0.25, -1, True )
setMesherType( [ "Sheet 1", "Sheet 2", "Sheet 3" ], "HEXQUAD" )
```

```
setMidSideNodeLocation( [ "Sheet 1", "Sheet 2", "Sheet 3" ], "LINEAR" )
generateMesh( [] )
hideView( "GEOM" )
showView( "MESH" )
addAnalysis( "Analysis1" )
renameAnalysis( "Analysis1", "Analysis1" )
addAnalysisCommand( "Analysis1", "NONLIN", "Structural nonlinear" )
addAnalysisCommandDetail( "Analysis1", "Structural nonlinear",
"MODEL/EVALUA/REINFO/INTERF" )
setAnalysisCommandDetail( "Analysis1", "Structural nonlinear",
"MODEL/EVALUA/REINFO/INTERF", True )
removeAnalysisCommandDetail( "Analysis1", "Structural nonlinear", "EXECUT(1)" )
setAnalysisCommandDetail( "Analysis1", "Structural nonlinear", "EXECUT/EXETYP",
"START" )
renameAnalysisCommand( "Analysis1", "Structural nonlinear", "Structural nonlinear" )
renameAnalysisCommandDetail( "Analysis1", "Structural nonlinear", "EXECUT(1)", "first dead
load" )
setAnalysisCommandDetail( "Analysis1", "Structural nonlinear",
"EXECUT(1)/ITERAT/MAXITE", 50 )
setAnalysisCommandDetail( "Analysis1", "Structural nonlinear",
"EXECUT(1)/ITERAT/CONVER/DISPLA/TOLCON", 0.001 )
setAnalysisCommandDetail( "Analysis1", "Structural nonlinear",
"EXECUT(1)/ITERAT/MAXITE", 50 )
setAnalysisCommandDetail( "Analysis1", "Structural nonlinear",
"EXECUT(1)/ITERAT/CONVER/FORCE/TOLCON", 0.001 )
setAnalysisCommandDetail( "Analysis1", "Structural nonlinear",
"EXECUT(1)/ITERAT/CONVER/FORCE/NOCONV", "CONTIN" )
setAnalysisCommandDetail( "Analysis1", "Structural nonlinear",
"EXECUT(1)/ITERAT/CONVER/DISPLA/NOCONV", "CONTIN" )
saveProject( )
addAnalysisCommandDetail( "Analysis1", "Structural nonlinear",
"EXECUT(1)/START/INITIA/STRESS" )
setAnalysisCommandDetail( "Analysis1", "Structural nonlinear",
"EXECUT(1)/START/INITIA/STRESS", True )
addAnalysisCommandDetail( "Analysis1", "Structural nonlinear",
"EXECUT(1)/START/INITIA/STRESS/INPUT/LOAD" )
setAnalysisCommandDetail( "Analysis1", "Structural nonlinear",
"EXECUT(1)/START/INITIA/STRESS/INPUT/LOAD", 1 )
saveProject( )
setAnalysisCommandDetail( "Analysis1", "Structural nonlinear",
"EXECUT(1)/ITERAT/MAXITE", 50 )
```

```
setAnalysisCommandDetail( "Analysis1", "Structural nonlinear", "EXECUT/EXETYP", "LOAD" )
renameAnalysisCommand( "Analysis1", "Structural nonlinear", "Structural nonlinear" )
renameAnalysisCommandDetail( "Analysis1", "Structural nonlinear", "EXECUT(2)", "load" )
addAnalysisCommandDetail( "Analysis1", "Structural nonlinear", "EXECUT(2)/LOAD/LOADNR" )
setAnalysisCommandDetail( "Analysis1", "Structural nonlinear", "EXECUT(2)/LOAD/LOADNR", 2 )
saveProject( )
setAnalysisCommandDetail( "Analysis1", "Structural nonlinear", "EXECUT(2)/ITERAT/MAXITE", 50 )
setAnalysisCommandDetail( "Analysis1", "Structural nonlinear", "EXECUT(2)/ITERAT/CONVER/DISPLA/TOLCON", 0.001 )
setAnalysisCommandDetail( "Analysis1", "Structural nonlinear", "EXECUT(2)/ITERAT/CONVER/DISPLA/NOCONV", "CONTIN" )
setAnalysisCommandDetail( "Analysis1", "Structural nonlinear", "EXECUT(2)/ITERAT/CONVER/FORCE/TOLCON", 0.001 )
setAnalysisCommandDetail( "Analysis1", "Structural nonlinear", "EXECUT(2)/ITERAT/CONVER/FORCE/NOCONV", "CONTIN" )
saveProject( )
setAnalysisCommandDetail( "Analysis1", "Structural nonlinear", "EXECUT/EXETYP", "TIME" )
renameAnalysisCommand( "Analysis1", "Structural nonlinear", "Structural nonlinear" )
renameAnalysisCommandDetail( "Analysis1", "Structural nonlinear", "EXECUT(3)", "creep and shrinkage" )
setAnalysisCommandDetail( "Analysis1", "Structural nonlinear", "EXECUT(3)/ITERAT/MAXITE", 50 )
setAnalysisCommandDetail( "Analysis1", "Structural nonlinear", "EXECUT(3)/ITERAT/CONVER/DISPLA/TOLCON", 0.001 )
setAnalysisCommandDetail( "Analysis1", "Structural nonlinear", "EXECUT(3)/ITERAT/CONVER/DISPLA/NOCONV", "CONTIN" )
setAnalysisCommandDetail( "Analysis1", "Structural nonlinear", "EXECUT(3)/ITERAT/CONVER/FORCE/TOLCON", 0.001 )
setAnalysisCommandDetail( "Analysis1", "Structural nonlinear", "EXECUT(3)/ITERAT/CONVER/FORCE/NOCONV", "CONTIN" )
saveProject( )
setAnalysisCommandDetail( "Analysis1", "Structural nonlinear", "EXECUT(3)/TIME/STEPS/EXPLIC/SIZES", "86400.0 2.50560e+06 1.3176e+07 1.57680e+07" )
runSolver( "Analysis1" )
showView( "RESULT" )
```

Command console of UHPC

```
#########################################################################################
# DianaIE 10.2 update 2018-01-19 17:07:06
# Python    3.6.1
# Session recorded at 2019-04-11 14:51:53
#########################################################################################
newProject( "G:/I-beam-UHPC", 100 )
setModelAnalysisAspects( [ "STRUCT" ] )
setModelDimension( "2D" )
setDefaultMeshOrder( "QUADRATIC" )
setDefaultMesherType( "HEXQUAD" )
setDefaultMidSideNodeLocation( "LINEAR" )
saveProject(   )
createSheet( "Sheet 1", [[ 0, 0, 0 ],[ 30, 0, 0 ],[ 30, 2.5, 0 ],[ 0, 2.5, 0 ]] )
createSheet( "Sheet 2", [[ 0, 2.5, 0 ],[ 30, 2.5, 0 ],[ 30, 2.7, 0 ],[ 0, 2.7, 0 ]] )
arrayCopy( [ "Sheet 2" ], [ 0, -2.7, 0 ], [ 0, 0, 0 ], [ 0, 0, 0 ], 1 )
createCurve( "Curve 1", [[ 0, 1, 0 ],[ 15, 0.1, 0 ],[ 30, 1, 0 ]] )
renameShape( "Curve 1", "tendon" )
setUnit( "TEMPER", "CELSIU" )
setUnit( "ANGLE", "DEGREE" )
addMaterial( "block1", "CONCDC", "AASHTO", [ "CREEP", "SHRINK" ] )
setParameter( MATERIAL, "block1", "MCAASH/FC28", 1.2e+08 )
setParameter( MATERIAL, "block1", "MCAASH/FT28", 5000000 )
setParameter( MATERIAL, "block1", "MCAASH/H", 150 )
setParameter( MATERIAL, "block1", "MCAASH/H", 185 )
setParameter( MATERIAL, "block1", "MCAASH/RH", 55 )
setParameter( MATERIAL, "block1", "CONCDI/YOUNG", 4.5e+10 )
setParameter( MATERIAL, "block1", "CONCDI/POISON", 0.15 )
setParameter( MATERIAL, "block1", "CONCDI/THERMX", 1.2e-05 )
setParameter( MATERIAL, "block1", "CONCDI/DENSIT", 2500 )
setParameter( MATERIAL, "block1", "CONCCP/CRSPEC/AGETYP", "AGING" )
setParameter( MATERIAL, "block1", "CONCCP/CRSPEC/AGING", 2419200 )
setParameter( MATERIAL, "block1", "CONCSH/CURAGE", 86400 )
addGeometry( "Element geometry 1", "SHEET", "MEMBRA", [] )
rename( GEOMET, "Element geometry 1", "block1" )
setParameter( GEOMET, "block1", "THICK", 0.2 )
setParameter( GEOMET, "block1", "LOCAXS", True )
setElementClassType( "SHAPE", [ "Sheet 1" ], "MEMBRA" )
assignMaterial( "block1", "SHAPE", [ "Sheet 1" ] )
assignGeometry( "block1", "SHAPE", [ "Sheet 1" ] )
saveProject(   )
saveProject(   )
```

```
addMaterial( "top", "CONCDC", "AASHTO", [ "CREEP", "SHRINK" ] )
setParameter( MATERIAL, "top", "MCAASH/FC28", 1.2e+08 )
setParameter( MATERIAL, "top", "MCAASH/FT28", 5000000 )
setParameter( MATERIAL, "top", "MCAASH/H", 167 )
setParameter( MATERIAL, "top", "MCAASH/RH", 55 )
setParameter( MATERIAL, "top", "CONCDI/YOUNG", 4.5e+10 )
setParameter( MATERIAL, "top", "CONCDI/POISON", 0.15 )
setParameter( MATERIAL, "top", "CONCDI/THERMX", 1.2e-05 )
setParameter( MATERIAL, "top", "CONCDI/DENSIT", 2500 )
setParameter( MATERIAL, "top", "CONCCP/CRSPEC/AGETYP", "AGING" )
setParameter( MATERIAL, "top", "CONCCP/CRSPEC/AGING", 2419200 )
setParameter( MATERIAL, "top", "CONCSH/CURAGE", 86400 )
addGeometry( "Element geometry 2", "SHEET", "MEMBRA", [] )
rename( GEOMET, "Element geometry 2", "top" )
setParameter( GEOMET, "top", "THICK", 1 )
setParameter( GEOMET, "top", "LOCAXS", True )
setElementClassType( "SHAPE", [ "Sheet 2" ], "MEMBRA" )
assignMaterial( "top", "SHAPE", [ "Sheet 2" ] )
assignGeometry( "top", "SHAPE", [ "Sheet 2" ] )
saveProject(  )
addMaterial( "bot", "CONCDC", "AASHTO", [ "CREEP", "SHRINK" ] )
setParameter( MATERIAL, "bot", "MCAASH/FC28", 1.2e+08 )
setParameter( MATERIAL, "bot", "MCAASH/FT28", 5000000 )
setParameter( MATERIAL, "bot", "MCAASH/H", 167 )
setParameter( MATERIAL, "bot", "MCAASH/RH", 55 )
setParameter( MATERIAL, "bot", "CONCDI/YOUNG", 4.5e+10 )
setParameter( MATERIAL, "bot", "CONCDI/POISON", 0.15 )
setParameter( MATERIAL, "bot", "CONCDI/THERMX", 1.2e-05 )
setParameter( MATERIAL, "bot", "CONCDI/DENSIT", 2500 )
setParameter( MATERIAL, "bot", "CONCCP/CRSPEC/AGETYP", "AGING" )
setParameter( MATERIAL, "bot", "CONCCP/CRSPEC/AGING", 2419200 )
setParameter( MATERIAL, "bot", "CONCSH/CURAGE", 86400 )
addGeometry( "Element geometry 3", "SHEET", "MEMBRA", [] )
rename( GEOMET, "Element geometry 3", "bot" )
setParameter( GEOMET, "bot", "THICK", 1 )
setParameter( GEOMET, "bot", "LOCAXS", True )
setElementClassType( "SHAPE", [ "Sheet 3" ], "MEMBRA" )
assignMaterial( "bot", "SHAPE", [ "Sheet 3" ] )
assignGeometry( "bot", "SHAPE", [ "Sheet 3" ] )
addMaterial( "tendon", "REINFO", "REBOND", [] )
setParameter( "MATERIAL", "tendon", "REBARS/ELASTI/YOUNG", 1.95e+11 )
```

```
setParameter( "MATERIAL", "tendon", "REBARS/MASS/DENSIT", 7800 )
setParameter( "MATERIAL", "tendon", "REBARS/PLATYP", "VMISES" )
setParameter( "MATERIAL", "tendon", "REBARS/PLASTI/YLDSTR", 1.86e+09 )
setParameter( "MATERIAL", "tendon", "RESLIP/DSNY", 2e+12 )
setParameter( "MATERIAL", "tendon", "RESLIP/DSSX", 2000000 )
setParameter( "MATERIAL", "tendon", "RESLIP/SHFTYP", "BONDS1" )
setParameter( "MATERIAL", "tendon", "RESLIP/BONDS1/SLPVAL", [ 20, 0.1 ] )
addGeometry( "Element geometry 4", "RELINE", "REBAR", [] )
setParameter( "GEOMET", "Element geometry 4", "REITYP", "REITRU" )
setParameter( "GEOMET", "Element geometry 4", "REITRU/CROSSE", 0.00139 )
setParameter( "GEOMET", "Element geometry 4", "REITRU/PERIME", 0.9576 )
setParameter( "GEOMET", "Element geometry 4", "TIPELM/SURFAC", 0.00278 )
setReinforcementAspects( [ "tendon" ] )
assignMaterial( "tendon", "SHAPE", [ "tendon" ] )
assignGeometry( "Element geometry 4", "SHAPE", [ "tendon" ] )
resetElementData( "SHAPE", [ "tendon" ] )
setReinforcementDiscretization( [ "tendon" ], "SECTION" )
saveProject(  )
rename( "GEOMET", "Element geometry 4", "tendon" )
saveProject(  )
createPointSupport( "co1", "co1" )
setParameter( GEOMETRYSUPPORT, "co1", "AXES", [ 1, 2 ] )
setParameter( GEOMETRYSUPPORT, "co1", "TRANSL", [ 1, 1, 0 ] )
setParameter( GEOMETRYSUPPORT, "co1", "ROTATI", [ 0, 0, 0 ] )
attach( GEOMETRYSUPPORT, "co1", "Sheet 3", [[ 0, -0.2, 0 ]] )
saveProject(  )
createPointSupport( "co2", "co1" )
setParameter( GEOMETRYSUPPORT, "co2", "AXES", [ 1, 2 ] )
setParameter( GEOMETRYSUPPORT, "co2", "TRANSL", [ 0, 1, 0 ] )
setParameter( GEOMETRYSUPPORT, "co2", "ROTATI", [ 0, 0, 0 ] )
attach( GEOMETRYSUPPORT, "co2", "Sheet 3", [[ 30, -0.2, 0 ]] )
saveProject(  )
createPointBody( "Point body 1", [ 10, 3.5, 0 ] )
mirror( [ "Point body 1" ], [ 15, 0, 0 ], [ True, False, False ], True )
projection( SHAPEEDGE, "Sheet 2", [[ 10, 2.7, 0 ]], [ "Point body 1" ], [ 0, -1, 0 ], True )
removeShape( [ "Point body 1" ] )
saveProject(  )
projection( SHAPEEDGE, "Sheet 2", [[ 20, 2.7, 0 ]], [ "Point body 2" ], [ 0, -1, 0 ], True )
removeShape( [ "Point body 2" ] )
saveProject(  )
addSet( GEOMETRYLOADSET, "gravity" )
```

```
createModelLoad( "gravity", "gravity" )
addSet( GEOMETRYLOADSET, "Geometry load case 1" )
rename( GEOMETRYLOADSET, "Geometry load case 1", "load" )
createPointLoad( "load", "load" )
setParameter( GEOMETRYLOAD, "load", "FORCE/VALUE", -20000 )
setParameter( GEOMETRYLOAD, "load", "FORCE/DIRECT", 2 )
attach( GEOMETRYLOAD, "load", "Sheet 2", [[ 10, 2.7, 0 ]] )
attach( GEOMETRYLOAD, "load", "Sheet 2", [[ 20, 2.7, 0 ]] )
addSet( GEOMETRYLOADSET, "Geometry load case 2" )
rename( GEOMETRYLOADSET, "Geometry load case 2", "tendon" )
createBodyLoad( "tendon", "tendon" )
setParameter( GEOMETRYLOAD, "tendon", "LODTYP", "POSTEN" )
setParameter( GEOMETRYLOAD, "tendon", "POSTEN/BOTHEN/FORCE1", 1940000 )
setParameter( GEOMETRYLOAD, "tendon", "POSTEN/BOTHEN/FORCE2", 1940000 )
setParameter( GEOMETRYLOAD, "tendon", "POSTEN/BOTHEN/RETLE1", 0.01 )
setParameter( GEOMETRYLOAD, "tendon", "POSTEN/BOTHEN/RETLE2", 0.01 )
setParameter( GEOMETRYLOAD, "tendon", "POSTEN/SHEAR", 0.22 )
setParameter( GEOMETRYLOAD, "tendon", "POSTEN/WOBBLE", 0.01 )
attachTo( GEOMETRYLOAD, "tendon", "POSTEN/BOTHEN/PNTS1", "tendon", [[ 0, 1.6, 0 ]] )
attachTo( GEOMETRYLOAD, "tendon", "POSTEN/BOTHEN/PNTS2", "tendon", [[ 30, 1.6,
0 ]] )
attach( GEOMETRYLOAD, "tendon", [ "tendon" ] )
saveProject(  )
setDefaultGeometryLoadCombinations(  )
setGeometryLoadCombinationFactor( "Geometry load combination 1", "gravity", 1 )
remove( GEOMETRYLOADCOMBINATION, "Geometry load combination 3" )
setGeometryLoadCombinationFactor( "Geometry load combination 1", "tendon", 1 )
setTimeDependentLoadFactors( "GEOMETRYLOADCOMBINATION", "Geometry load
combination 1", [ 0, 3.1536e+10 ], [ 1, 1 ] )
setTimeDependentLoadFactors( "GEOMETRYLOADCOMBINATION", "Geometry load
combination 2", [ 0, 3.1536e+10 ], [ 1, 1 ] )
saveProject(  )
setElementSize( [ "Sheet 1", "Sheet 2", "Sheet 3" ], 0.25, -1, True )
setMesherType( [ "Sheet 1", "Sheet 2", "Sheet 3" ], "HEXQUAD" )
setMidSideNodeLocation( [ "Sheet 1", "Sheet 2", "Sheet 3" ], "LINEAR" )
generateMesh( [] )
showView( "MESH" )
addAnalysis( "Analysis2" )
renameAnalysis( "Analysis2", "Analysis2" )
addAnalysisCommand( "Analysis2", "NONLIN", "Structural nonlinear" )
addAnalysisCommandDetail( "Analysis2", "Structural nonlinear",
```

```
"MODEL/EVALUA/REINFO/INTERF" )
setAnalysisCommandDetail( "Analysis2", "Structural nonlinear", "MODEL/EVALUA/REINFO/INTERF", True )
removeAnalysisCommandDetail( "Analysis2", "Structural nonlinear", "EXECUT(1)" )
setAnalysisCommandDetail( "Analysis2", "Structural nonlinear", "EXECUT/EXETYP", "START" )
renameAnalysisCommand( "Analysis2", "Structural nonlinear", "Structural nonlinear" )
renameAnalysisCommandDetail( "Analysis2", "Structural nonlinear", "EXECUT(1)", "first dead load" )
setAnalysisCommandDetail( "Analysis2", "Structural nonlinear", "EXECUT(1)/ITERAT/MAXITE", 50 )
setAnalysisCommandDetail( "Analysis2", "Structural nonlinear", "EXECUT(1)/ITERAT/CONVER/DISPLA/TOLCON", 0.001 )
setAnalysisCommandDetail( "Analysis2", "Structural nonlinear", "EXECUT(1)/ITERAT/MAXITE", 50 )
setAnalysisCommandDetail( "Analysis2", "Structural nonlinear", "EXECUT(1)/ITERAT/CONVER/FORCE/TOLCON", 0.001 )
setAnalysisCommandDetail( "Analysis2", "Structural nonlinear", "EXECUT(1)/ITERAT/CONVER/FORCE/NOCONV", "CONTIN" )
setAnalysisCommandDetail( "Analysis2", "Structural nonlinear", "EXECUT(1)/ITERAT/CONVER/DISPLA/NOCONV", "CONTIN" )
saveProject( )
addAnalysisCommandDetail( "Analysis2", "Structural nonlinear", "EXECUT(1)/START/INITIA/STRESS" )
setAnalysisCommandDetail( "Analysis2", "Structural nonlinear", "EXECUT(1)/START/INITIA/STRESS", True )
addAnalysisCommandDetail( "Analysis2", "Structural nonlinear", "EXECUT(1)/START/INITIA/STRESS/INPUT/LOAD" )
setAnalysisCommandDetail( "Analysis2", "Structural nonlinear", "EXECUT(1)/START/INITIA/STRESS/INPUT/LOAD", 1 )
saveProject( )
setAnalysisCommandDetail( "Analysis2", "Structural nonlinear", "EXECUT(1)/ITERAT/MAXITE", 50 )
setAnalysisCommandDetail( "Analysis2", "Structural nonlinear", "EXECUT/EXETYP", "LOAD" )
renameAnalysisCommand( "Analysis2", "Structural nonlinear", "Structural nonlinear" )
renameAnalysisCommandDetail( "Analysis2", "Structural nonlinear", "EXECUT(2)", "load" )
addAnalysisCommandDetail( "Analysis2", "Structural nonlinear", "EXECUT(2)/LOAD/LOADNR" )
setAnalysisCommandDetail( "Analysis2", "Structural nonlinear", "EXECUT(2)/LOAD/LOADNR", 2 )
```

```
saveProject( )
setAnalysisCommandDetail( "Analysis2", "Structural nonlinear",
"EXECUT(2)/ITERAT/MAXITE", 50 )
setAnalysisCommandDetail( "Analysis2", "Structural nonlinear",
"EXECUT(2)/ITERAT/CONVER/DISPLA/TOLCON", 0.001 )
setAnalysisCommandDetail( "Analysis2", "Structural nonlinear",
"EXECUT(2)/ITERAT/CONVER/DISPLA/NOCONV", "CONTIN" )
setAnalysisCommandDetail( "Analysis2", "Structural nonlinear",
"EXECUT(2)/ITERAT/CONVER/FORCE/TOLCON", 0.001 )
setAnalysisCommandDetail( "Analysis2", "Structural nonlinear",
"EXECUT(2)/ITERAT/CONVER/FORCE/NOCONV", "CONTIN" )
saveProject( )
setAnalysisCommandDetail( "Analysis2", "Structural nonlinear", "EXECUT/EXETYP",
"TIME" )
renameAnalysisCommand( "Analysis2", "Structural nonlinear", "Structural nonlinear" )
renameAnalysisCommandDetail( "Analysis2", "Structural nonlinear", "EXECUT(3)", "creep and
shrinkage" )
setAnalysisCommandDetail( "Analysis2", "Structural nonlinear",
"EXECUT(3)/ITERAT/MAXITE", 50 )
setAnalysisCommandDetail( "Analysis2", "Structural nonlinear",
"EXECUT(3)/ITERAT/CONVER/DISPLA/TOLCON", 0.001 )
setAnalysisCommandDetail( "Analysis2", "Structural nonlinear",
"EXECUT(3)/ITERAT/CONVER/DISPLA/NOCONV", "CONTIN" )
setAnalysisCommandDetail( "Analysis2", "Structural nonlinear",
"EXECUT(3)/ITERAT/CONVER/FORCE/TOLCON", 0.001 )
setAnalysisCommandDetail( "Analysis2", "Structural nonlinear",
"EXECUT(3)/ITERAT/CONVER/FORCE/NOCONV", "CONTIN" )
saveProject( )
setAnalysisCommandDetail( "Analysis2", "Structural nonlinear",
"EXECUT(3)/TIME/STEPS/EXPLIC/SIZES", "86400.0 2.50560e+06 1.3176e+07
1.57680e+07" )
runSolver( "Analysis2" )
showView( "RESULT" )
```

3.4 Cracking Analysis of Reinforced Concrete

This case originates from author's DIANA 9.4 numerical case of thesis for reliability calculation of simply supported reinforced concrete beam. Plane stress element in two dimensions is applied to simulate concrete, thus embedded reinforcement bar element is modeled for longitudinal steel bars and stirrups. Total strain-based crack model in smeared cracking is applied for crack in concrete beam.

The whole length and height of beam are 4 and 0.45 m, respectively, and the symmetrical load is applied on it. Therefore, semi-structural model shown in Fig. 3.127 is created in the numerical model, where vertical downward concentrated load is 48 kN/m in total by applying stepwise loading type on it. Spacing of stirrups is 150 mm while distance between longitudinal steel bars and edge of concrete, both in tensile and compressive zone, is 0.05 m. Specification type of longitudinal bars is 3ϕ8 while type of stirrups is ϕ6@150.

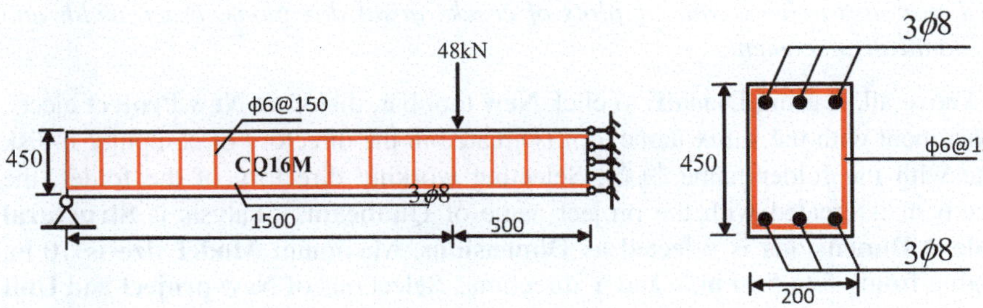

Fig. 3.127 Semi-structural model of reinforced concrete and its sectional size

Parameters of concrete and reinforcement are displayed in Table 3.4.

Table 3.4 Paramaters of this case

Items	Concrete	Reinforcement
Element types	Quadratic plane stress element in 2D	Bar element
Material parameters	Poisson's ratio 0.15 Mass density 2500 kg/m^3 Elastic modulus 3×10^{10} N/m^2 Ultimate strain $\varepsilon_{cr,ult} = 0.000311$ Shear retention $\beta = 0.01$ Tension softening model Exponential	Elastic modulus 2.1×10^{11} N/m^2 Yield stress 4.4×10^8 N/m^2
Geometric properties	2m × 0.45m × 0.2m	Equivalent cross-section area of longitudinal bars 1.51×10^{-4} m^2 Equivalent cross-section area of stirrups 5.6×10^{-5} m^2
Load	48 kN, vertical downward	

Essentials of learning

(1) *Mastering specification of plane stress elements*
(2) *Learning to specify the concrete parameters in total strain-based crack model.*
(3) *Simulating equivalent cross-section area of longitudinal bars and stirrups* via *modeling of single bar element*
(4) *Mastering attachment of semi-structural supports and constraints*
(5) *Learning to specify output results such as crack strains, summed crack strains, crack width and nodal displacement*
(6) *Learning to check contour plots of cracks in all directions, crack width and nodal displacement.*

Above all, starting DianaIE to click **New** tool bar, dialog of **New Project** ejects, a document with the suffix name .dpf is created in the directory of computer F-disk zone with the folder name 例题. Selecting working directory of the folder, the document is created with the project name of **Quabeam**. Analysis is **Structural** while **2 Dimensions** is selected as **Dimensions**. Maximum **Model size** is 10 m, ranging from –5 to 5 m in X and Y directions. Selections of **New project** and **Unit** are shown in Figs. 3.128 and 3.129, respectively.

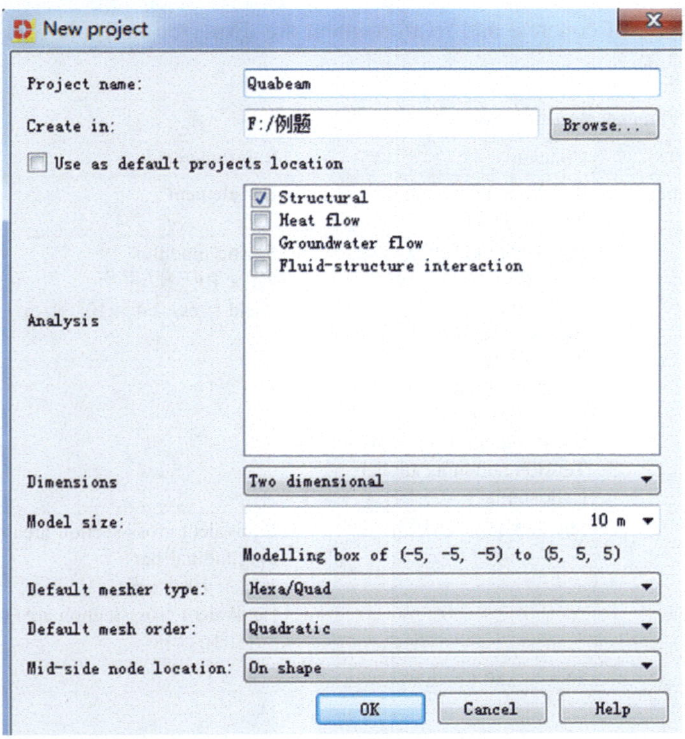

Fig. 3.128 Selection of New project

Fig. 3.129 Selection of unit

The next procedure is to create geometric model of concrete beam. Clicking shortcut icon **Adds a sheet** to create a plane, four coordinate values (0, 0, 0), (2, 0, 0), (2, 0.45, 0) and (0, 0.45, 0) are used as input. Clicking OK button, geometric model of concrete beam is displayed (Fig. 3.130).

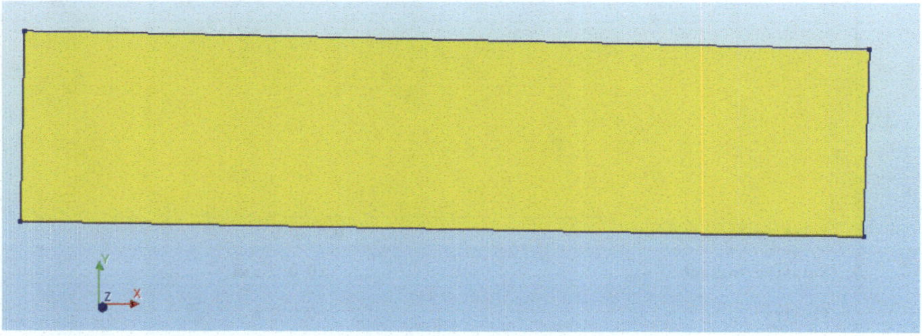

Fig. 3.130 Geometric model of concrete beam

Noting that dialog box of inputting coordinate values in DIANA software, 0 value in the third direction representing numerical value of out-of-plane direction is indispensable even though the model is 2D. In this case, owing to the fact that geometric model is 2D and created in XOY coordinate surface, values in Z direction are all set as zero. Users can make directional adjustment according to individual preferences.

Clicking shortcut icon **Adds a line** to create line with the name of **bar1**, coordinate values are shown as Fig. 3.131. Then the **bar1** is selected and the **Array copy** is right-clicked to copy and translate to the upward in Y direction with the **Relative displacement** and the **Number of copies** 0.35 m and 1, respectively, as Fig. 3.132 displays, where the name of line is **bar2**.

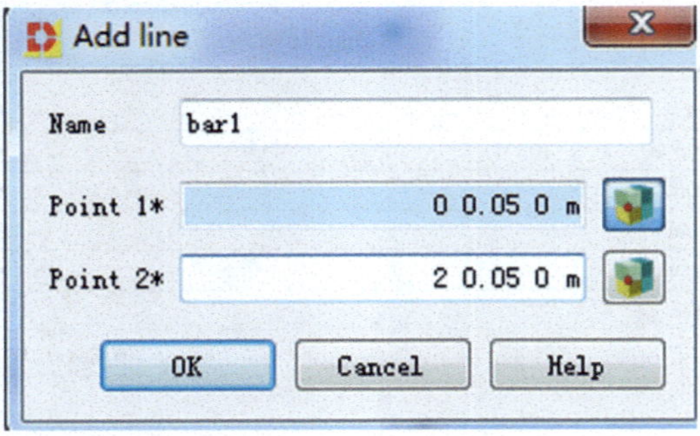

Fig. 3.131 Coordinate values of geometric model

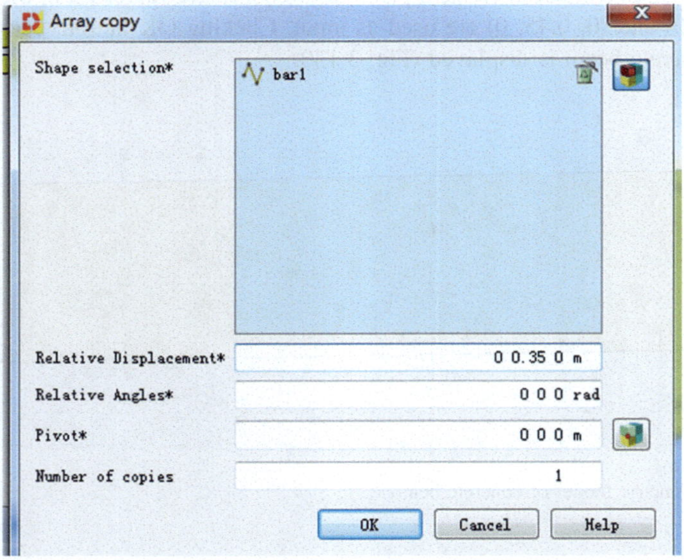

Fig. 3.132 Creating **bar2** via **Array copy**

Clicking OK button, the generated model is displayed (Fig. 3.133).

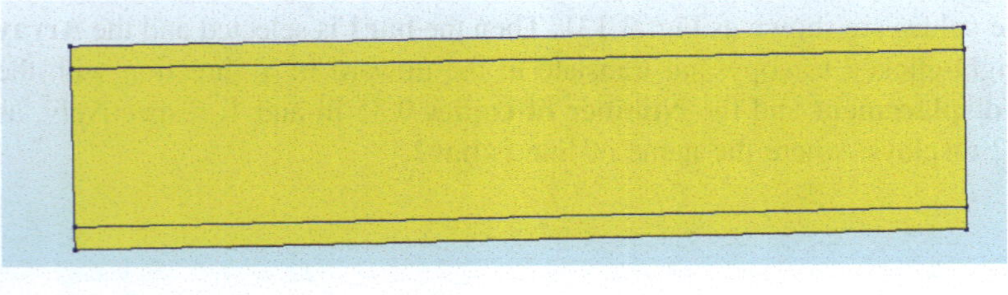

Fig. 3.133 Generation of longitudinal steel bars

Then we start to create stirrups. Above all, single line is created and then stirrups are all generated via **Array copy** with **Relative displacement** 0.15 m and the **Number of copies** 13, as Figs. 3.134 and 3.135 displays. Generated results are displayed in Fig. 3.136.

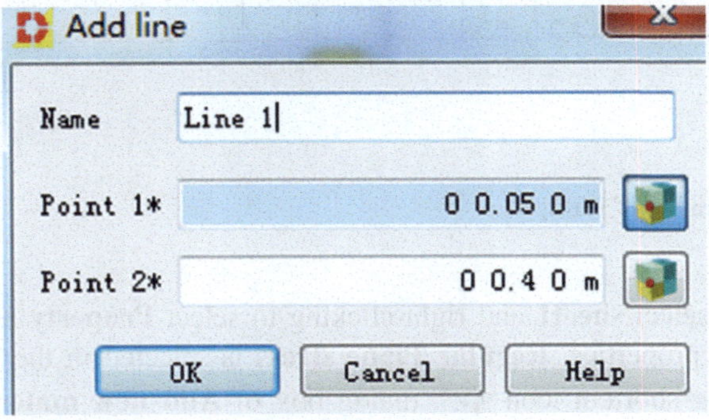

Fig. 3.134 Creation of single line

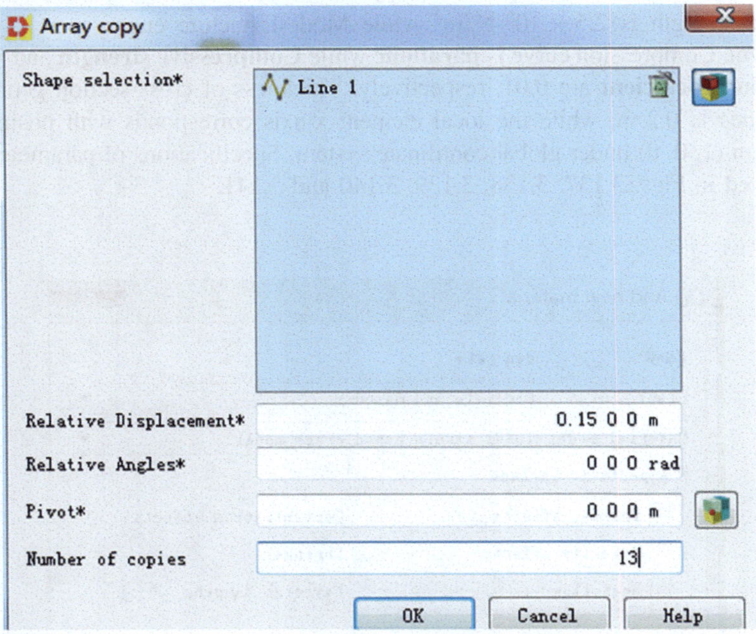

Fig. 3.135 Creation of stirrups via **Array copy**

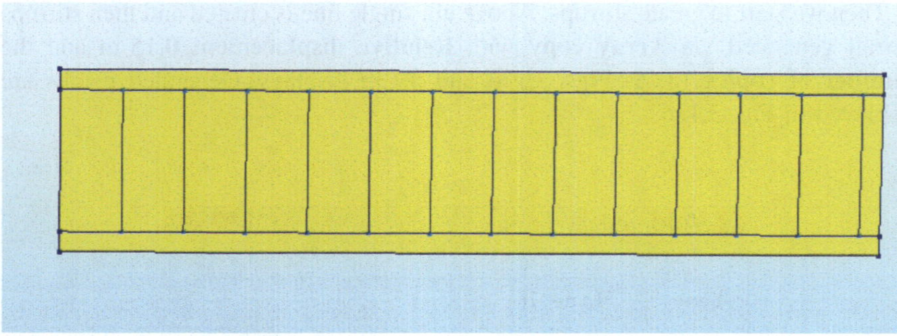

Fig. 3.136 Generation of stirrups

Clicking to select **sheet1** and right-clicking to select **Property assignments** to attach material properties, **Regular Plane stress** is selected in the **Element type** option. Clicking shortcut icon, dialog box of **Add new material** ejects, the name of which is **concrete**. **Concrete and masonry** selection is selected with the elastic modulus, Poisson's ratio and mass density 3×10^{10} N/m^2, 0.15 and 2500 kg/m^3, respectively. **Rotating to fixed Orientation** in total strain-based crack models is selected while **Exponential type** in tension softening model is chosen. Tensile strength is 2.5×10^6 N/m^2 while Mode-I fracture energy per height is 150 N/m. Compression curve is **parabolic** while **Compressive strength** and **Shear retention coefficient** are 0.01, respectively. Thickness of cross-section geometric properties is 0.2 m, while the local element x-axis corresponds with positive X direction (1, 0, 0) under global coordinate system. Specifications of parameters are displayed in Figs. 3.137, 3.138, 3.139, 3.140 and 3.141.

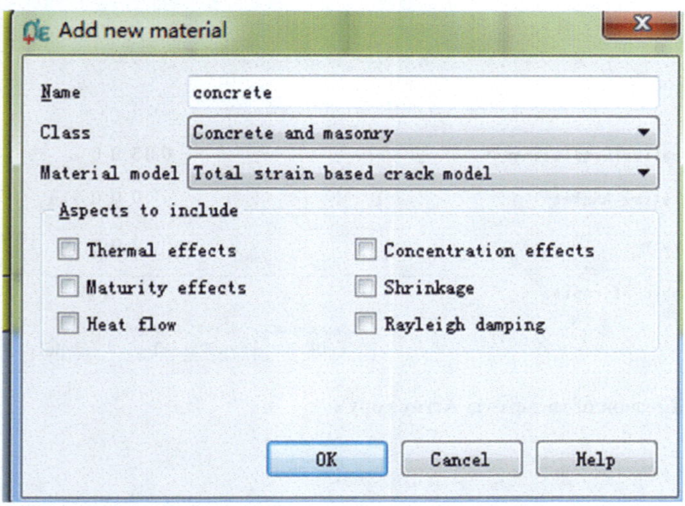

Fig. 3.137 Adding material properties of concrete

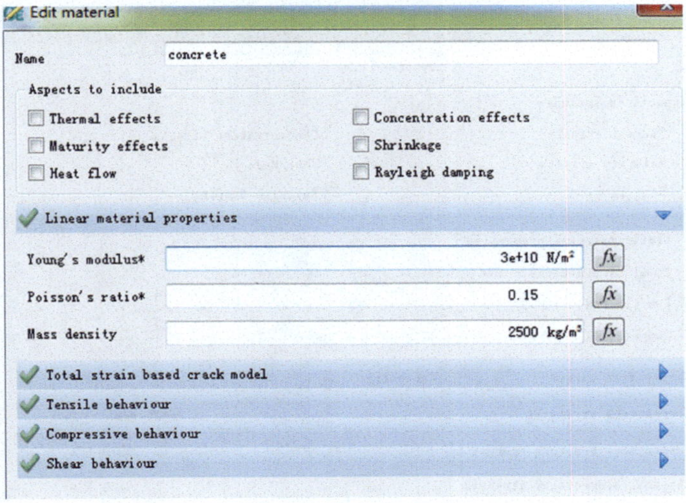

Fig. 3.138 Spefification of material properties

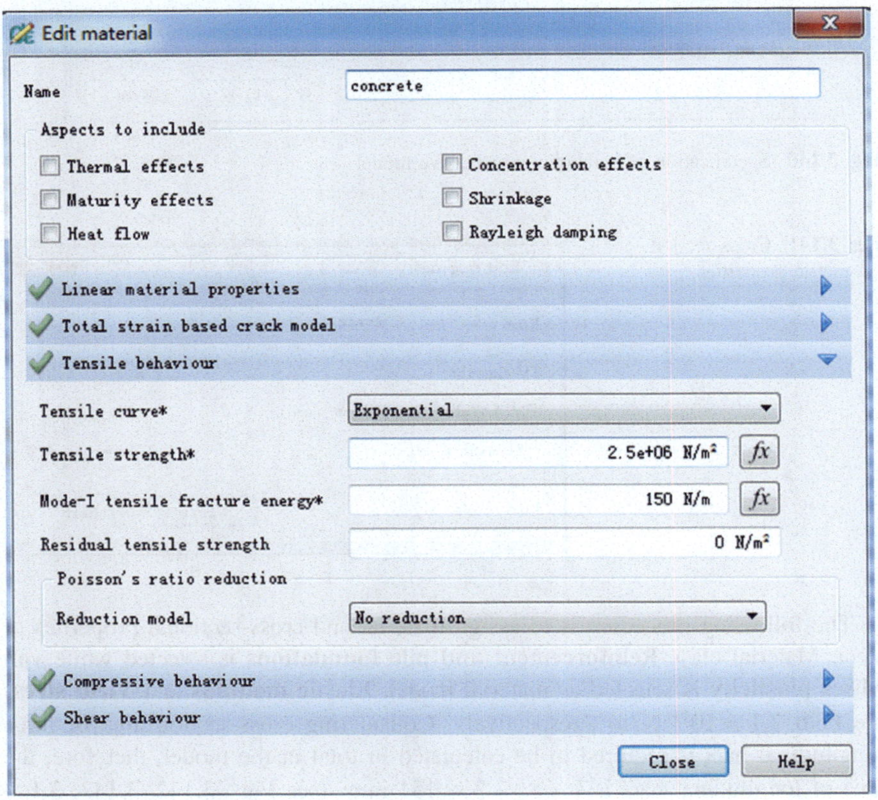

Fig. 3.139 Exponential tension softening model

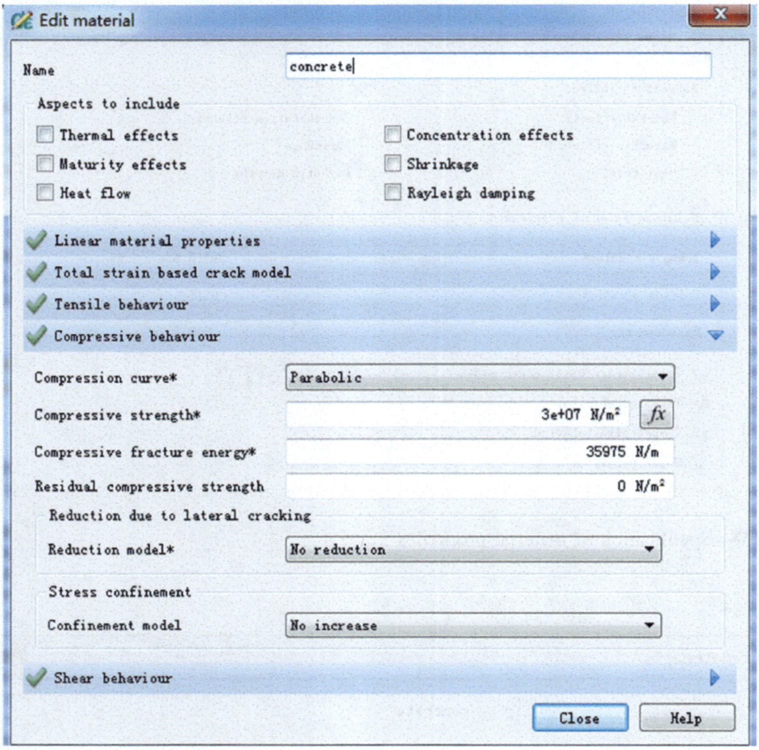

Fig. 3.140 Specification of parabolic compressive model

Fig. 3.141 Cross-section geometric properties

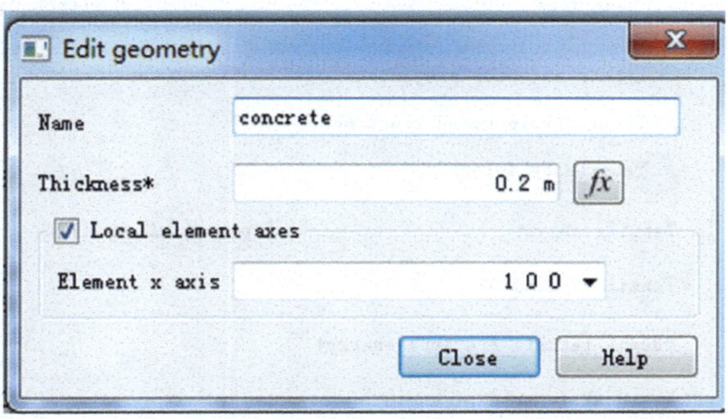

The following procedure is to assign material and cross-sectional properties of steel. Material class **Reinforcement and pile foundations** is selected while **von Mises plasticity** is selected as material model. **Elastic modulus** and **Yield stress** are both 2.1×10^{11} $\mathrm{N/m^2}$, respectively. Considering cross-section area of three longitudinal bars is required to be calculated in total in the model, therefore, the area of longitudinal bars is $\frac{\pi}{4} \times 8^2 \times 3 \approx 151$ $\mathrm{mm^2}$ (see Figs. 3.142, 3.143, 3.144

and 3.145). Defining properties of stirrups in the same way with the name of **Gird**, area of which is calculated as the superposition of two layers, the value is calculated as $\frac{\pi}{4} \times 6^2 \times 2 \approx 56.5487$ mm^2. As the material constitutive model and parameter specification of stirrup are the same as former, it is not repeated here.

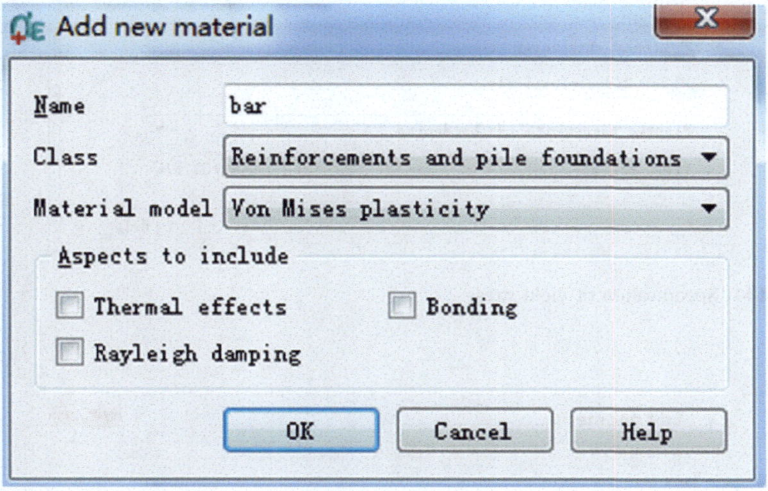

Fig. 3.142 Reinforcement material class

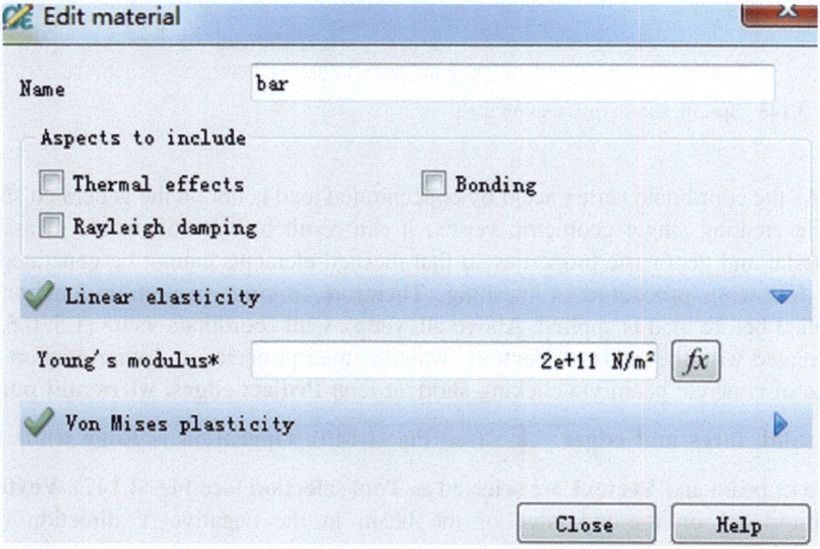

Fig. 3.143 Specification of elastic modulus

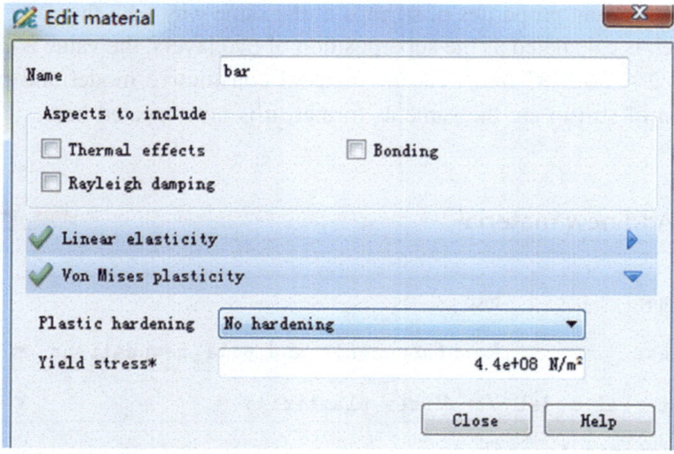

Fig. 3.144 Specification of yield stress

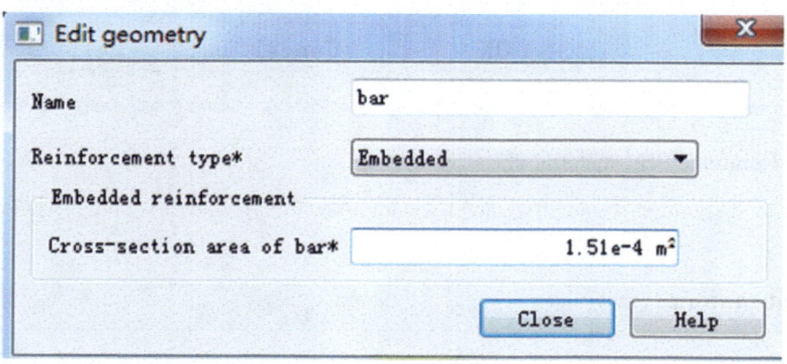

Fig. 3.145 Specification cross-section area

As the coordinate vertex acted by concentrated load is not on the generated sheet while creating single geometric vertex, it can result in error of unable to assign material and geometric properties so that meshed elements cannot be generated in the following procedure of meshing. Therefore, imprint projection function is applied before load is applied. Above all, vertex with coordinate value (1.5, 0.5, 0) is created with the name of **Vertex1**, which is then projected and imprinted on the edge of concrete beam via clicking shortcut icon **Project edges, wires and points on solid, faces and edges**  (see Fig. 3.146). **Operation** is **Edge** while top edge of beam and **Vertex1** are selected as **Tool selection** (see Fig. 3.147). **Vextex1** is imprinted on the top edge of the beam in the negative Y direction (see Fig. 3.148).

Fig. 3.146 Shortcut icon **Project edges, wires and points on solid, faces and edges**

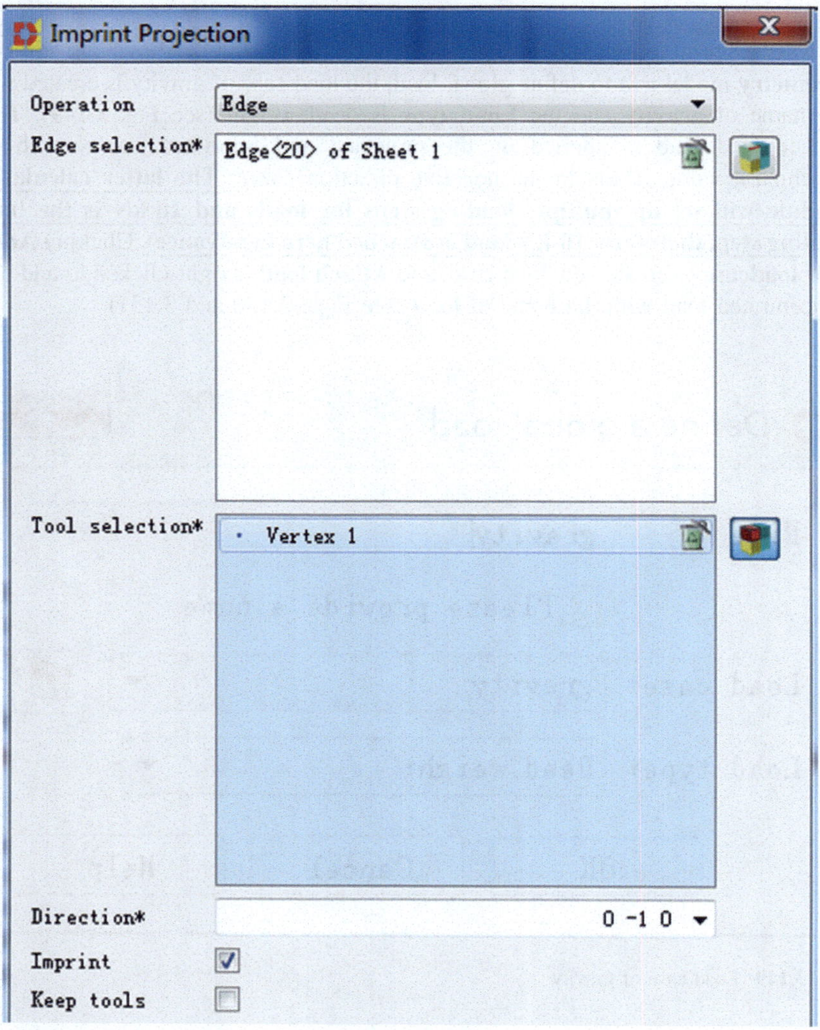

Fig. 3.147 Manipulation interface of imprint projection

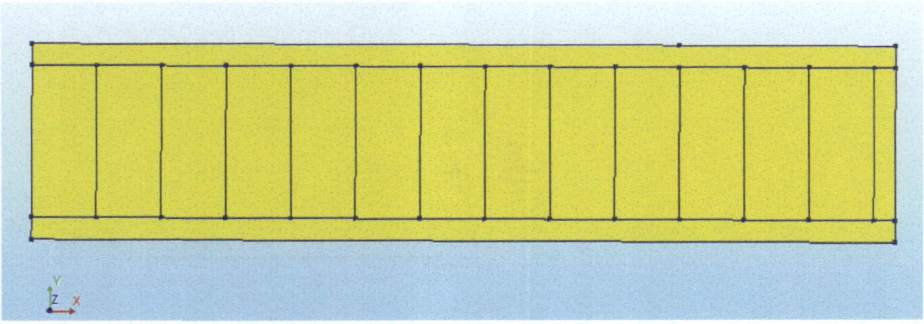

Fig. 3.148 Numerical model of beam after imprint projection

After imprint projection is completed, load is attached. Above all, clicking **Geometry** model tree to define global load, the load case of gravity is created with the name of **gravity** and the **Load type** is dead weight (see Fig. 3.149). Then concentrated load is applied on the projected and imprinted vertex with the benchmark value 10 kN in the negative direction (*Note*: **The latter calculation module will set up multiple loading steps for loads and 10 kN is the basic loading step, therefore, 10 kN load is attached here in advance**). Clicking **Add a new loadcase** to create new load case, and **Attach load** is right-clicked to add new concentrated load with the name of **load** (see Figs. 3.150 and 3.151).

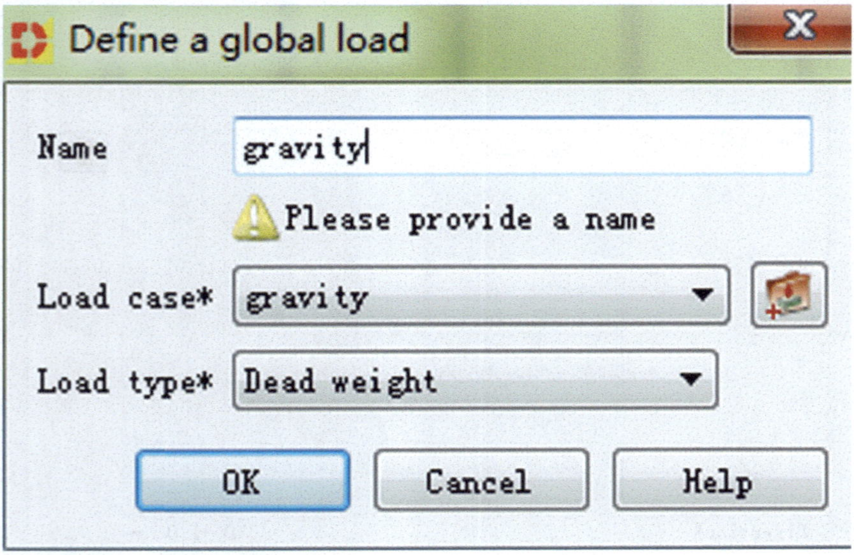

Fig. 3.149 Load case of gravity

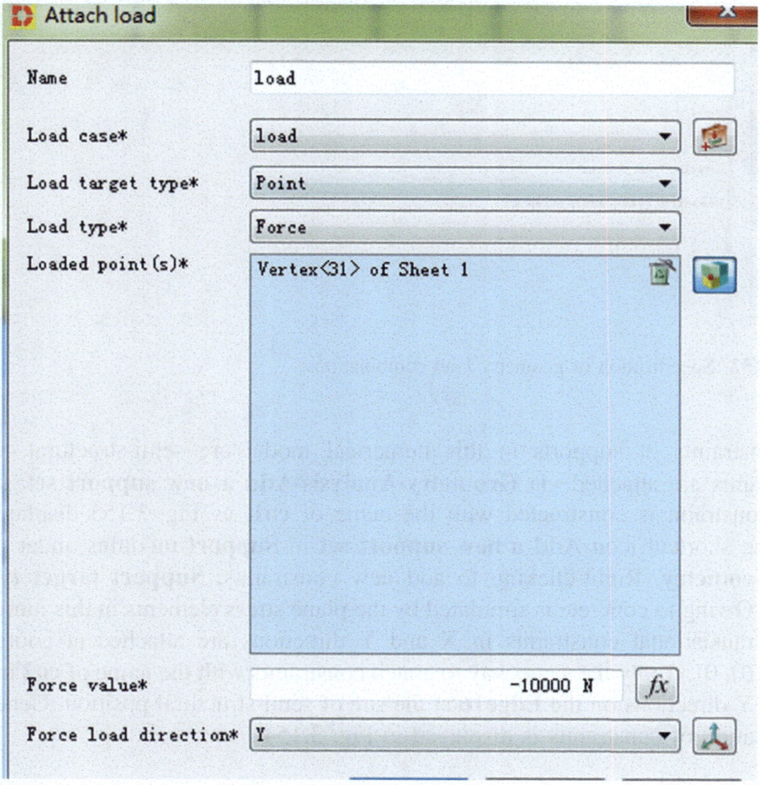

Fig. 3.150 Attaching concentrated load case

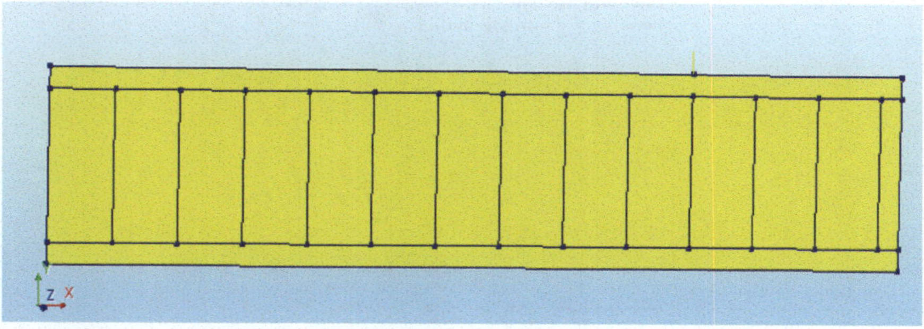

Fig. 3.151 Geometric model after attaching load

When it comes to adding geometry load combinations, load case of gravity is specified as **Geometry load combination 1** and the load case of **load** is classified as **Geometry load combination 2** (see Fig. 3.152).

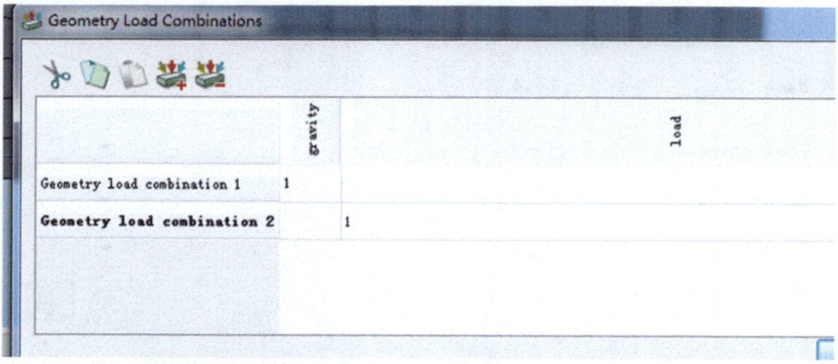

Fig. 3.152 Specification of geometry load combinations

Constraints of supports in this numerical model are semi-structural, where constraints are attached via **Geometry-Analysis-Add a new support set**, then a new constraint is constructed with the name of **co1,** as Fig. 3.153 displays, via clicking shortcut icon **Add a new support set** in **Support** modulus under model tree **Geometry**. Right-clicking to add new constraints, **Support target type** is **point**. Owing to concrete is simulated by the plane stress elements in this numerical case, translational constraints in X and Y directions are attached at coordinate vertex (0, 0). Apply the same way to attach constraints with the name of **co2** in both X and Y directions on the **Edge16** at the site of semi-structural position. Generated information of constraints is displayed in Fig. 3.154.

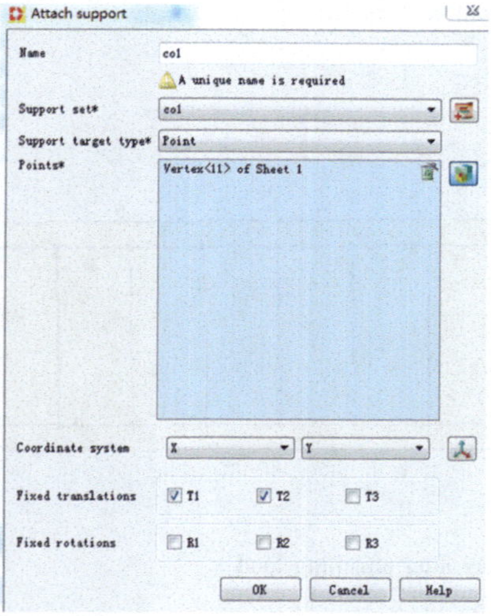

Fig. 3.153 Constraint attachment of **co1**

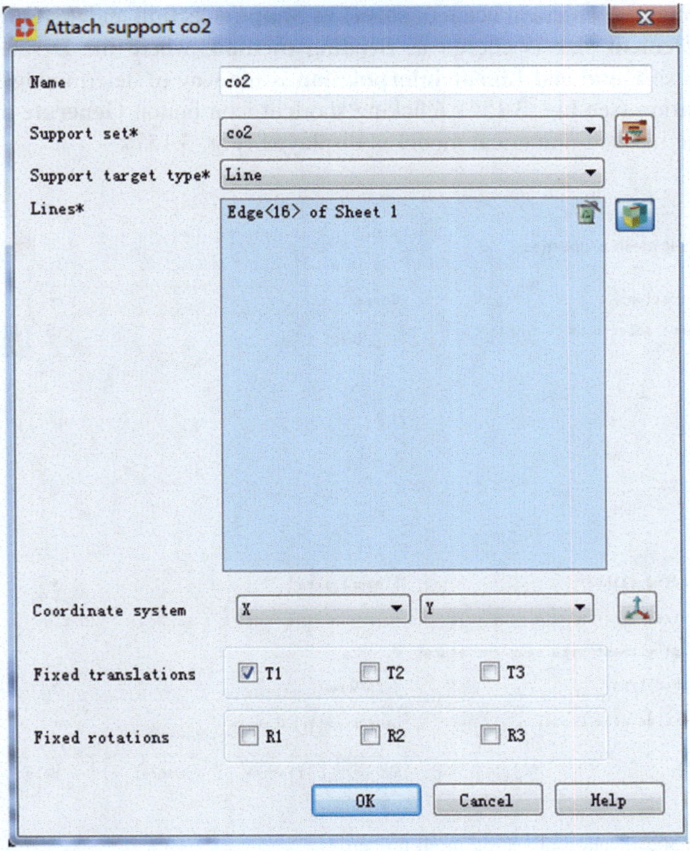

Fig. 3.154 Constraint attachment of **co2**

Click OK button to complete procedure of constraint attachment, and check the constraints in positive symmetric semi-structure. The results of constraint attachment are shown in Fig. 3.155.

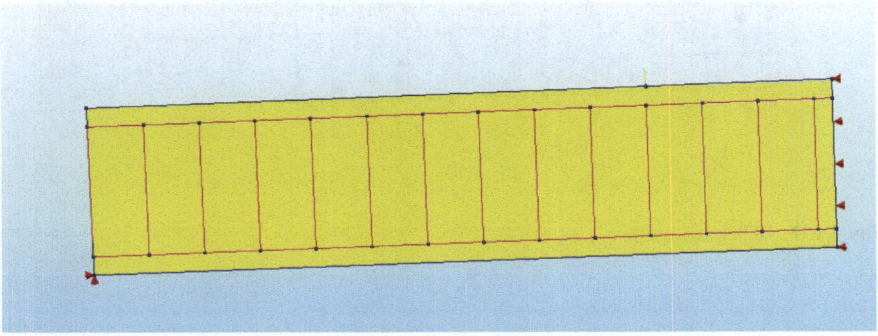

Fig. 3.155 Constraint attachment of positive symmetric semi-structure

Selecting the numerical beam of sheet1 as **Shape selection** and the **Operation** is **Shape**, **Element size** is chosen as **Seeding method**, where the **Desired size** is 0.1 m. **Hexa/Quad** and **Linear interpolation** is the way of determining **Mid-side node location** (see Fig. 3.156). Clicking shortcut icon button **Generate mesh of a shape,** the meshed numerical model is displayed (Fig. 3.157).

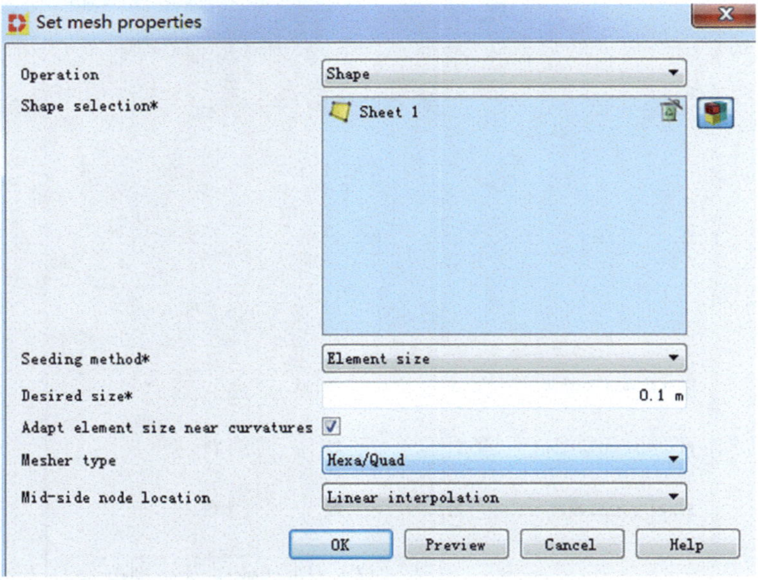

Fig. 3.156 Specification of setting mesh properties

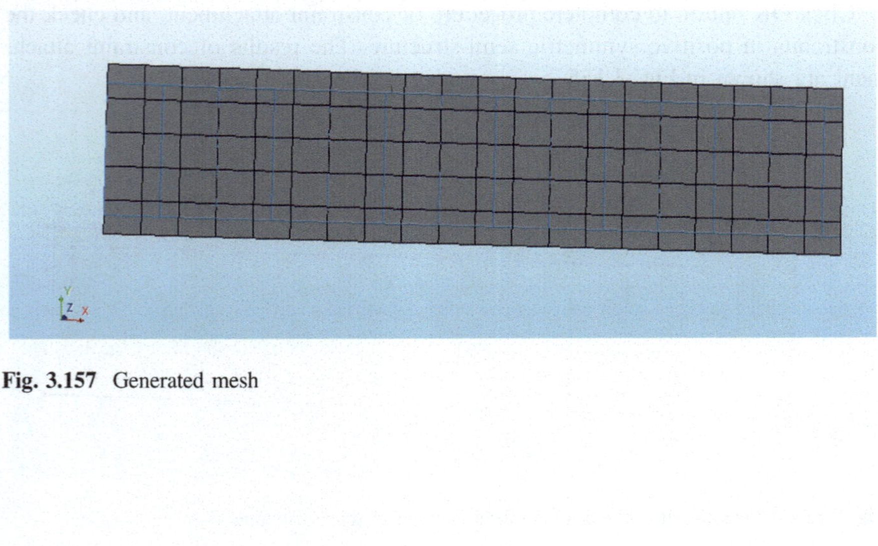

Fig. 3.157 Generated mesh

After mesh is completed, check the mesh type CQ16M, which is confirmed as desired plane stress element, as Fig. 3.158 displays.

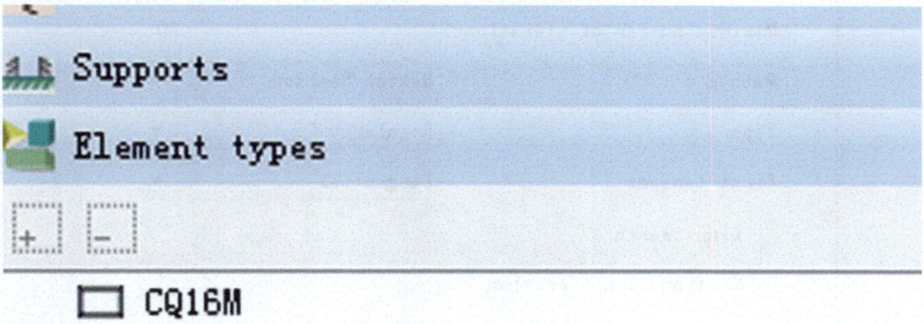

Fig. 3.158 Plane stress element CQ16M

Add analysis module **Add an analysis** to generate **Analysis1**. Right-click to select **Add command** to add calculation controlling instructions. **Structural nonlinear** is selected as analysis type in calculation controlling instructions, then load step is selected to be added in Structural nonlinear analysis. Constructing user-specified sizes of gravity and concentrated load, where load combination 1 is selected for gravity case, number of load step is 1 and the total step length of load is 1. Maximum number of iterations is 20 while **Displacement** and **Force** are both selected as convergence norm (see Figs. 3.159 and 3.160).

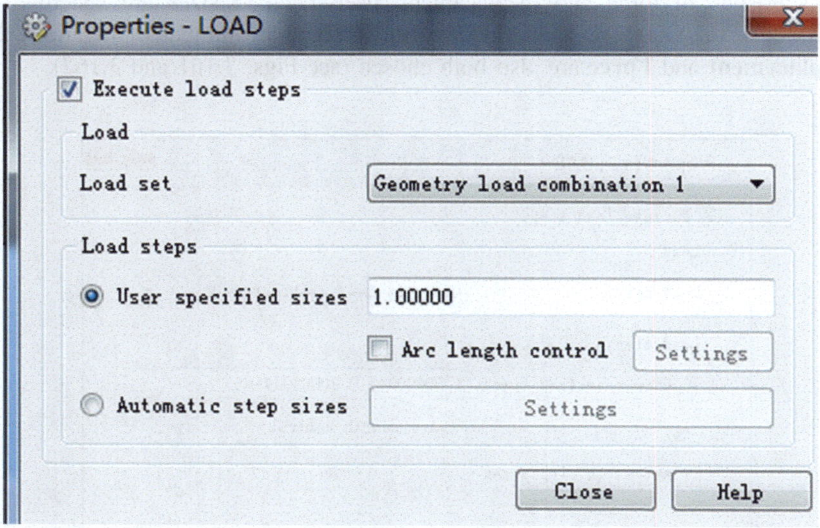

Fig. 3.159 Specification of gravity case

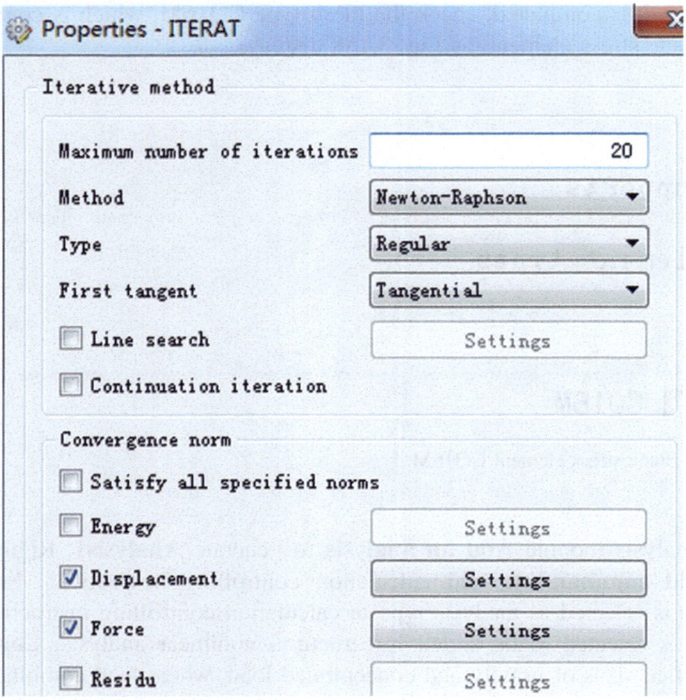

Fig. 3.160 Calculation module of gravity case

Geometry load combination 2 corresponds with **load.** Considering nonlinear characteristics of force and displacement, user-specified sizes are specified as 1.00000(4), 0.200000(4). Then still inputting maximum number of iterations 20, **Displacement** and **Force** are also both chosen (see Figs. 3.161 and 3.162).

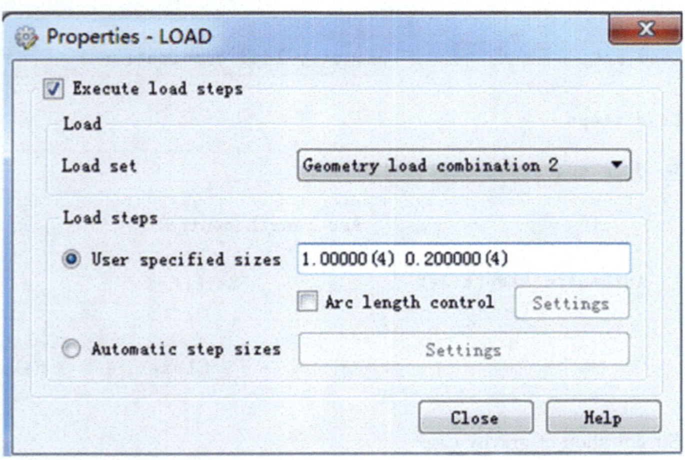

Fig. 3.161 Specification of **load** case

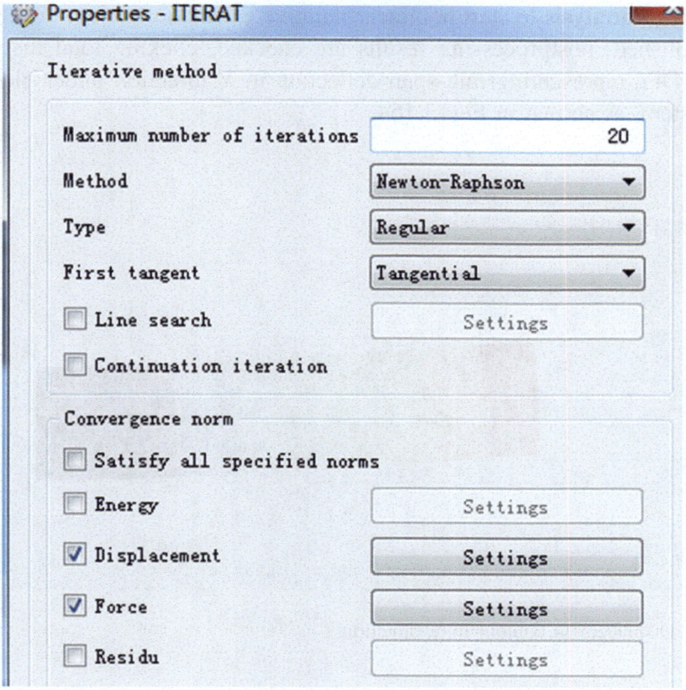

Fig. 3.162 Calculation module of concentrated load

Specifying output properties, instead of default choice **All primaries**, the output of calculation is selected as **User selection**, and the selected output items are listed in Fig. 3.163.

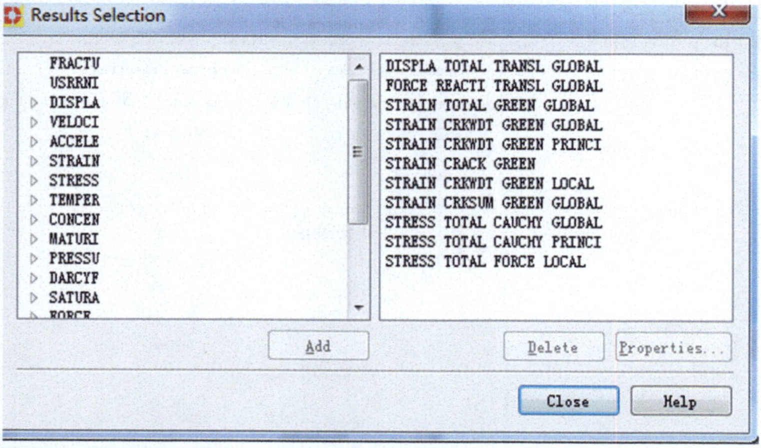

Fig. 3.163 Output items

Click **Run analysis** to start nonlinear iterative calculation; when the calculation is accomplished, postprocessing results are checked, clicking total displacement contour TDtY representing mid-span deflection in Y direction under global coordinate system, as shown in Fig. 3.164.

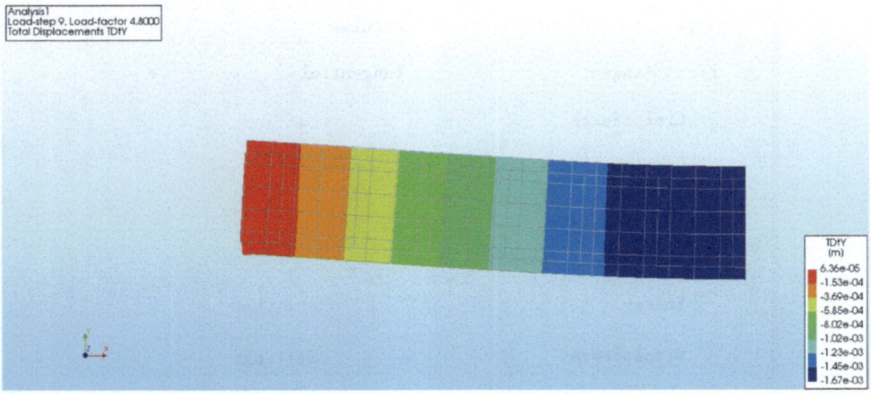

Fig. 3.164 Displacement contour in Y direction

In order to better check displacement in the middle site, nodal displacement in the middle site is selected to check. Clicking shortcut menu bar **View**—selecting **node selection**. After selecting the node then right-click **Show ids** to display node ids; selecting TDty to choose displacement in Y direction; right-clicking to select **Show table** to check displacement of this node. It is found that the node id is 3 and the displacement is 1.66 mm downward in the negative Y direction, as Fig. 3.165 displays.

Fig. 3.165 Displacement of node id 3 in Y direction

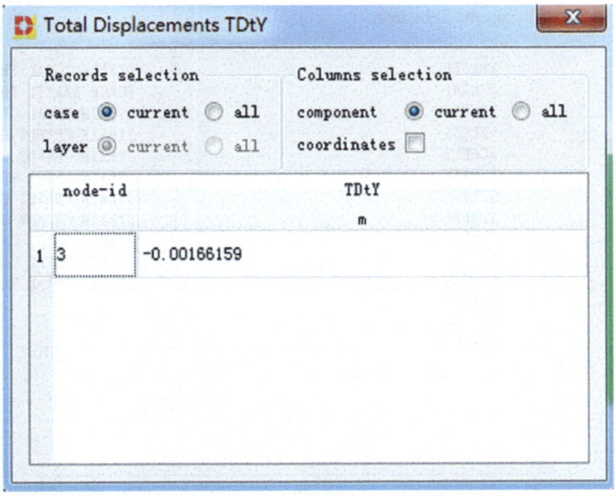

Checking contours of crack width and crack strains, click **Element results** of **Output.** When the load step reaches fifth step, three postprocessing new options: Crack-widths, Crack Strains and Summed Crack Strains related with crack appear. Above all, EcwXX under Crack-widths is clicked to check crack width in X direction under global coordinate system, as Fig. 3.166 displays.

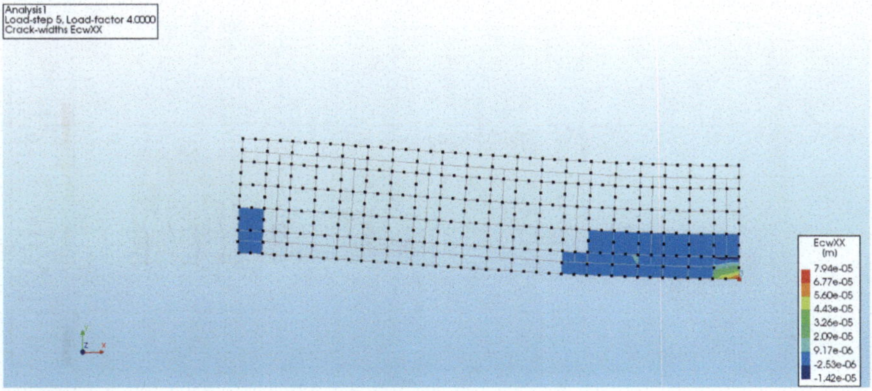

Fig. 3.166 Crack-width contour EcwXX under global coordinate system for load step 5

Clicking the loadstep options until ultimate load step—load step 9. To check contour in X direction under local coordinate system, it can be vividly depicted that cracks are propagating and new cracks are created on the top of initial cracking site. It is also evident to see that ultimate maximum crack width lies in the tensile zone of middle site of the beam (see Fig. 3.167).

Fig. 3.167 Crack-width contour EcwXX under global coordinate system for load step 9

Back to load step 5, clicking Eknn under Crack Strain to check contour of summed crack strains when initial crack is created, it reveals that cracks appear at the bearing support and middle site of the beam in lower tensile zone (see Fig. 3.168).

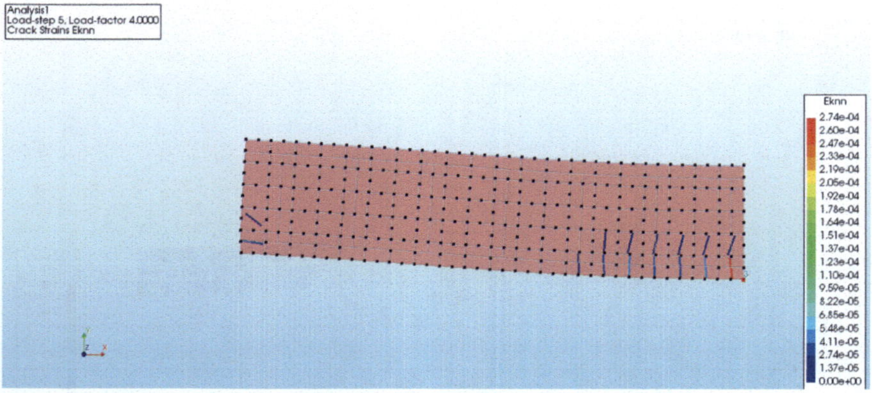

Fig. 3.168 Crack strain contour Eknn for load step 5

Selecting load step 7, the current user-specified load factor is 4.4, and the contour of crack strain is shown in Fig. 3.169.

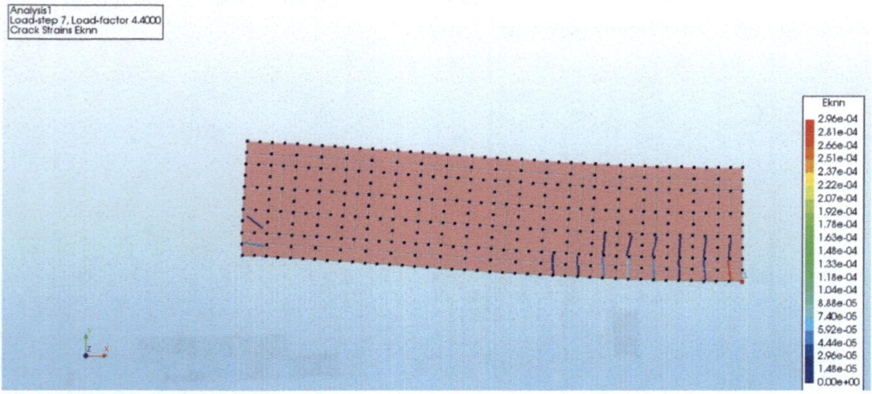

Fig. 3.169 Crack strain contour Eknn for load step 7

Checking crack strain contour under ultimate state—load step 9, it can be seen that the maximum crack strain still lies in the lower site of the beam in tensile zone, as Fig. 3.170 displays.

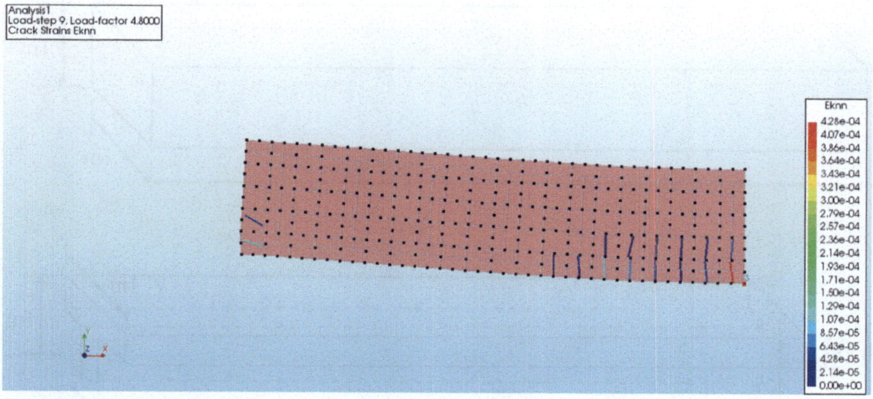

Fig. 3.170 Crack strain contour Eknn for load step 9

Users having interest can add loading value via number of steps or user-specified load factor in DIANA software to further investigate cracking features and properties for simply supported beam under ultimate bearing state.

3.5 Comparisons of Ultimate Bearing Capacity for Concrete and UHPC Integral-Cast Box Girder

This case is focused on a rectangular beam with ultra-high performance concrete (UHPC) as well as C50 grade concrete, which is modeled by 3D solid elements. Symmetric concentrated load is applied on the surface of beam. 3D sizes of length, width as well as height are 10, 1 and 1 m, respectively (see Fig. 3.171a). Compressive strength of UHPC and C50 concrete are 120 and 32.4 MPa. Stepwise loading is attached to the beam and the results such as initial cracking load, ultimate bending capacity as well as displacement are investigated and compared. Meanwhile, in order to better validate priority of UHCP, UHPC beam in T shape shown in Fig. 3.171b with less reinforcement ratio is also modeled in this case and the initial cracking load as well as ultimate bearing capacity are further investigated and compared.

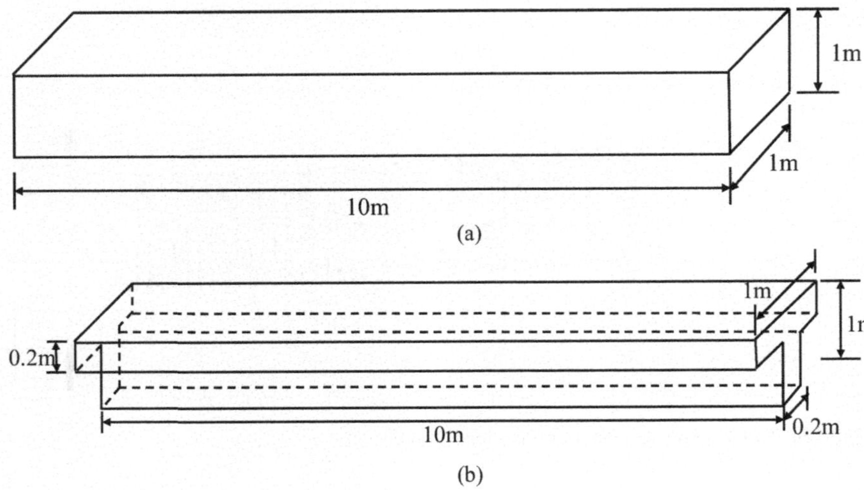

Fig. 3.171 Size of rectangular beam and T beam

Essentials of learning

(1) *Learning Boolean addition and subtraction for constructing hollow model plane in geometric modeling.*
(2) *Learning to construct reinforcement bars directly reapplications of the manipulation* ***Mirror a shape****.*
(3) *Learning to specify tensile and compressive model to simulate mechanic behaviors of UHPC concrete with steel fibers according to the international UHPC code.*
(4) *Learning to create concentration point load on the solid elements*
(5) *Learning to create T shape beam with subtract Boolean logic manipulation in solid element*
(6) *Learning to edit, duplicate, paste and run command console manuscripts in Python language.*

Above all, creating a new project with the name of UHPCbeam in the working directory of F disk in computer with the name of UHPCbeam, **Structural** analysis is selected as **Analysis** type. Number of dimensions is three with the maximum scope of model sizes 100 m, meaning that the graphical user interface zone ranges from –50 to 50 m in the X, Y and Z directions. Default mesher type is **Hexa/Quad**. **Default mesh order** is **quadratic** and **Mid-side node location** is **linear interpolation** (see Fig. 3.172).

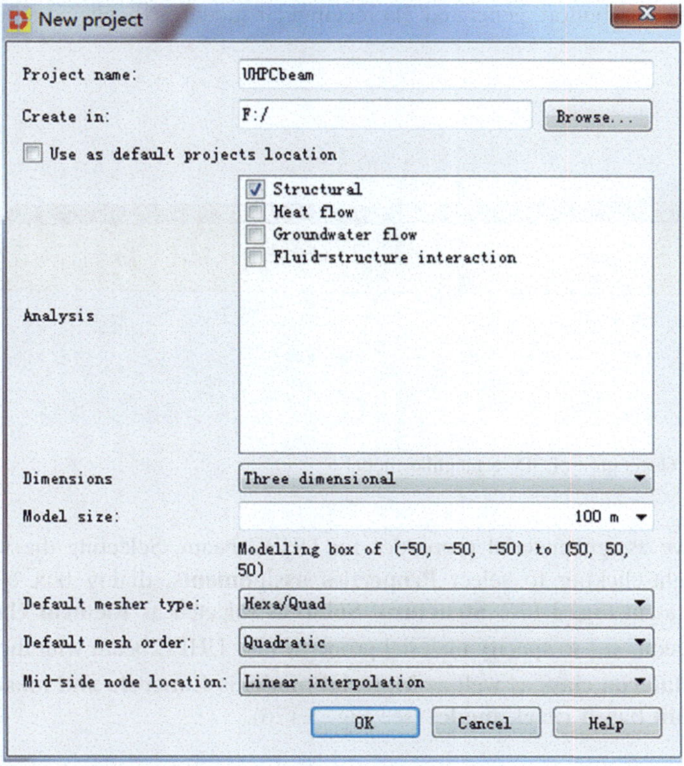

Fig. 3.172 New project of UHPCbeam

Clicking shortcut icon **Adds a block solid**  with the name of UHPC beam, **Position** representing starting point is (0, 0, 0) and dimensional sizes in three dimensions are 1, 10 and 1 m, respectively (see Fig. 3.173).

Fig. 3.173 Adding shortcut icon

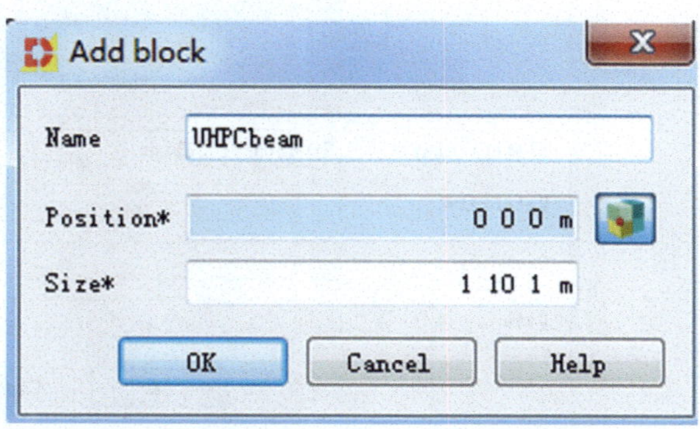

Clicking OK button, generated 3D rectangular model is displayed (Fig. 3.174).

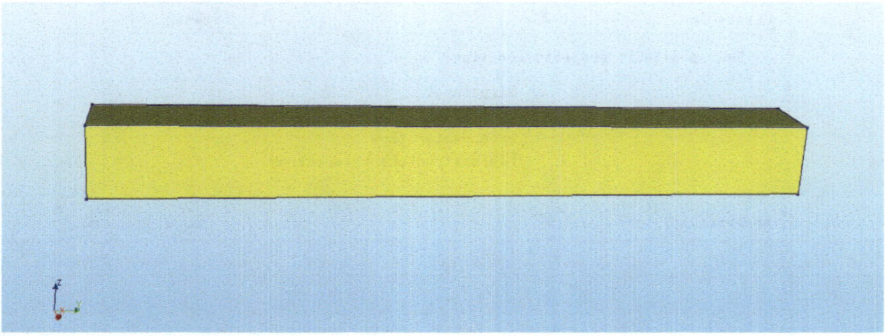

Fig. 3.174 Generation of 3D rectangular model

Then we assign material properties for UHPC beam. Selecting the whole 3D model, right-clicking to select **Properties assignments,** dialog box ejecting is displayed as in Fig. 3.175. **Structural Solids** is selected as **Element class**. Then still click icon to specify material properties for UHPC beam with the name of **UHPC**. **Material class** as well as **Material model** is **Concrete and masonry** and **Total strain based crack model** (see Fig. 3.176).

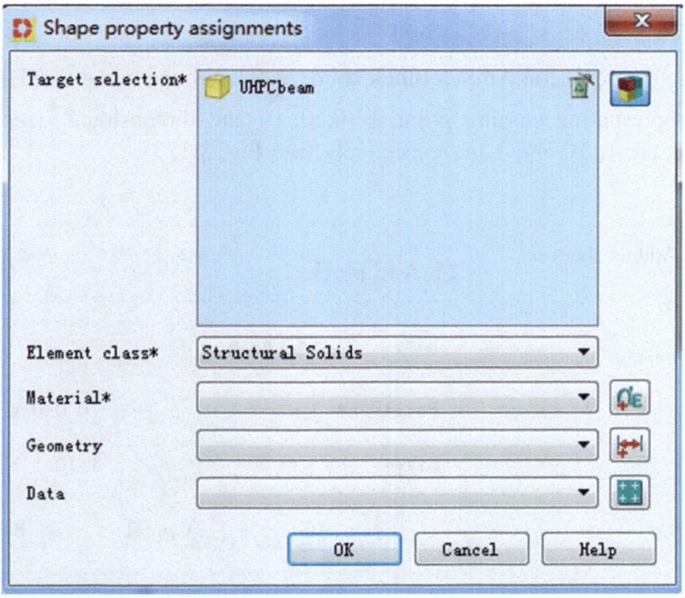

Fig. 3.175 Properties assignments for 3D rectangular model

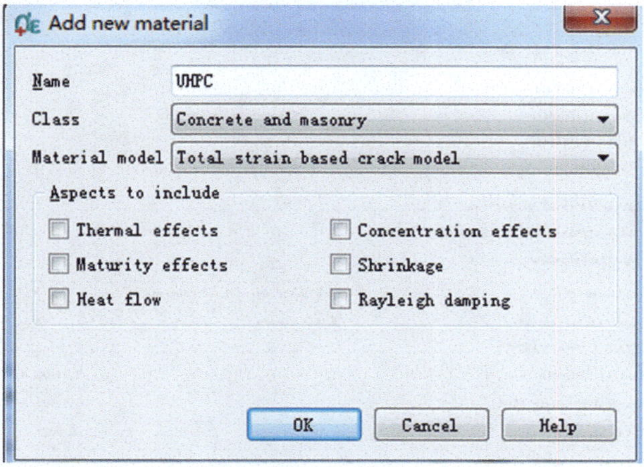

Fig. 3.176 Material model and class

Clicking OK button to enter material dialog box, Young modulus is 4.35×10^{10} N/m^2, **Poisson's ratio** as well as **Mass density** are 0.15 and 2500 kg/m^3 (see Fig. 3.177). **Rotating** is selected as **Crack orientation** option in **Total strain based crack model** under smearing cracks. It is worth to mention that **fib fiber reinforced concrete** is specifically selected as **Tensile curve** and **Total strain** is selected as **CMOD or strain curve** to simulate mechanic behaviors of UHPC with steel fibers. According to the reference of fiber-reinforced concrete model in DIANA manual, **Tensile strength fL** is 5.5 Mpa and **Residual strength fRi** is 5.5 MPa with corresponding **Total strain at fRi** 0.0024. **Residual strength fRj** is 5.5 MPa with **Total strain at fRj** and **Ultimate total strain** 0.0038 and 0.004, respectively (see Fig. 3.178).

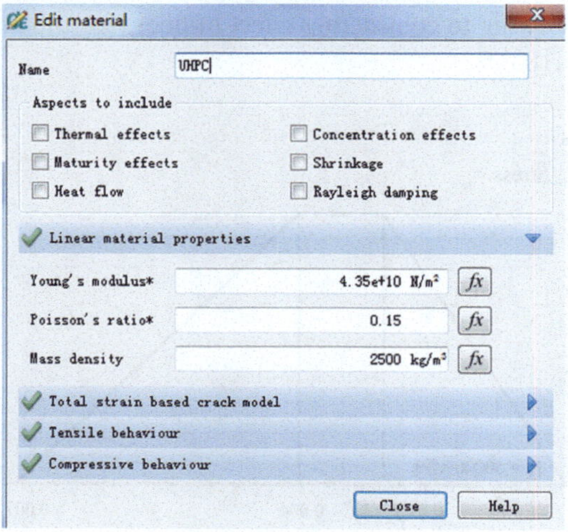

Fig. 3.177 Basic parameters in linear material properties

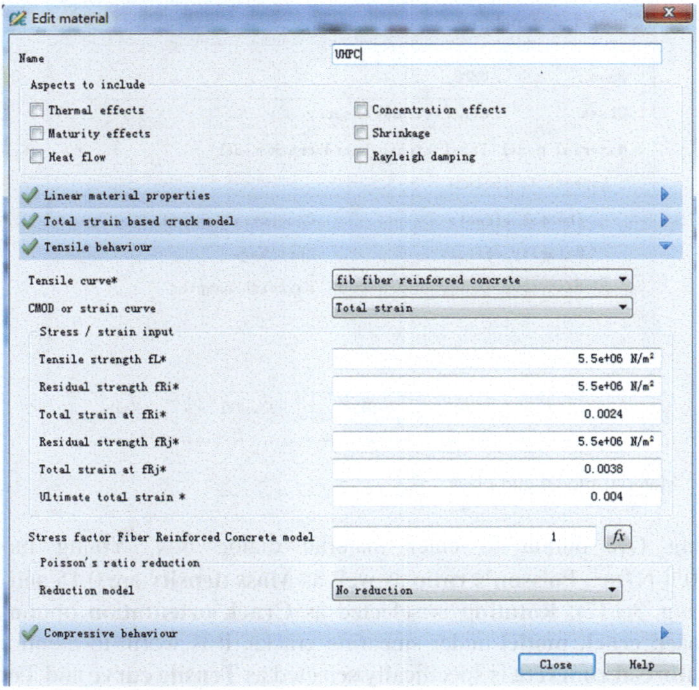

Fig. 3.178 Tensile behavior of UHPC

From the aspect of compressive behavior, **EN1992-1-2** is selected as **Compression curve** with compressive strength 120 MPa. Strain at maximum stress and ultimate stress are 0.004 and 0.007 according to the Australia UHPC code [1] (Design Guidelines for Ductal Prestressed Concrete Beams), where the linear elastic state is around $0.85f_c/E_c = 0.00237$ after calculation, compared with design stress–strain relationship owing to considering effect of fiber redistribution in UHPC (see Figs. 3.179 and 3.180).

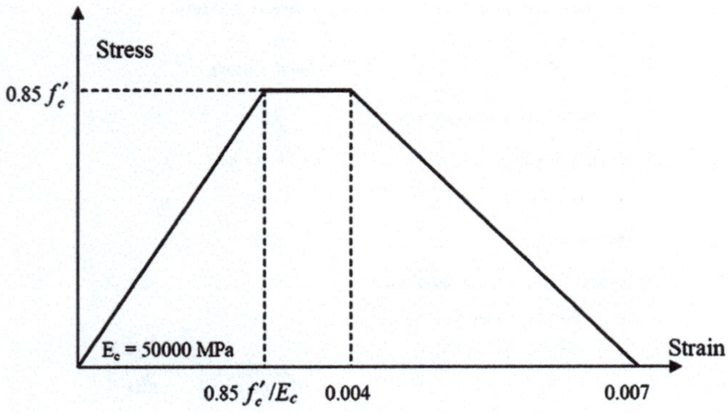

Fig. 3.179 Stress–strain relationship in Australia UHPC code [1]

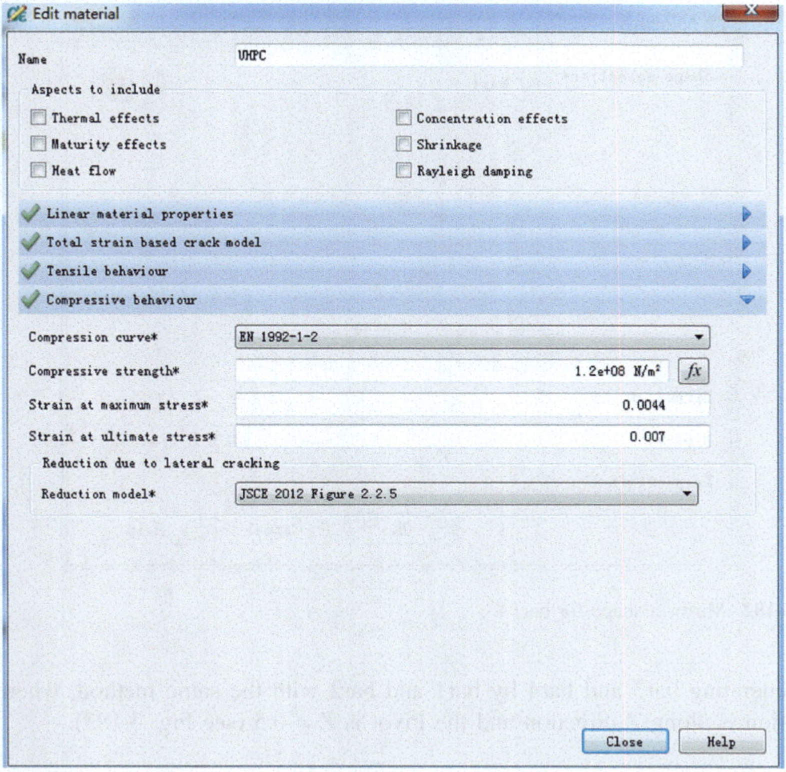

Fig. 3.180 Compressive specification for UHPC beam

After that, longitudinal steel bars are created via shortcut icon **Adds a line**. Coordinate values of first longitudinal bar are displayed in Fig. 3.181. Then we select bar1 to generate bar2 via the function of **Mirror a shape**, where the direction is in X-axis under global coordinate system with the **Pivot** representing mirror symmetric axis as X = 0.5 m, which is (0.5, 0, 0) m (see Fig. 3.182).

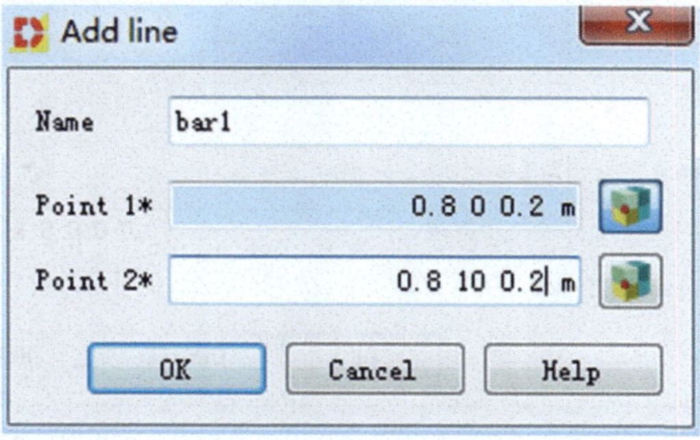

Fig. 3.181 Coordinate values of first longitudinal values

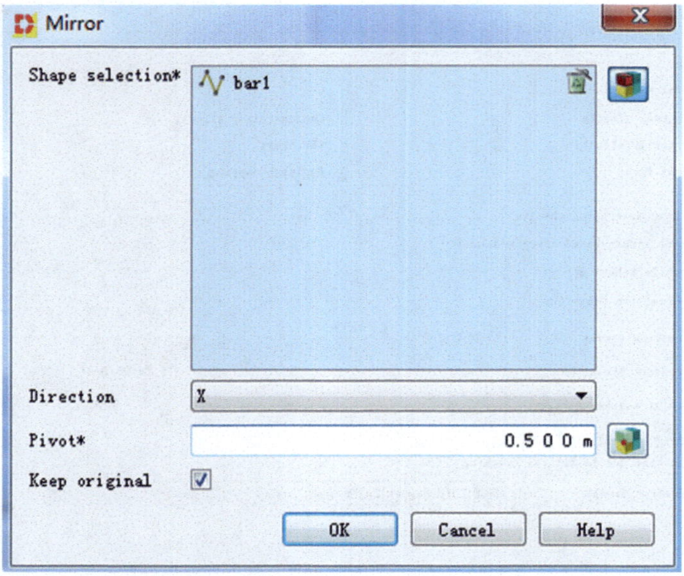

Fig. 3.182 Mirror a shape for bar1

Generating bar3 and bar4 by bar1 and bar2 with the same method, where the direction is along Z direction and the Pivot is Z = 0.5 (see Fig. 3.183).

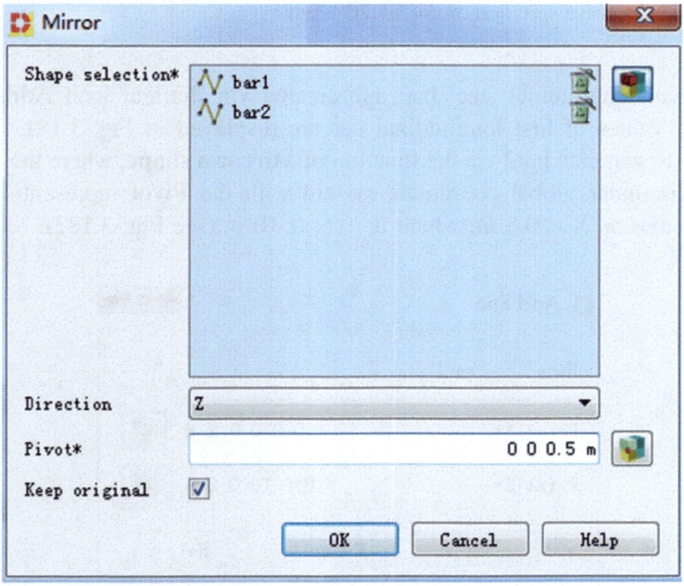

Fig. 3.183 Generation of bar3 and bar4

Material properties of steel bars are assigned with **Reinforcement and pile foundations** material class as well as **von Mises plasticity** material model (see Fig. 3.184). Elastic modulus and yielding stress are 2.1×10^{11} N/m^2 and 440 MPa. **Cross-section area of bar** is 1.1397e-4 m^2 according to the reason in UHPC code that diameters of reinforcement in HPPC concrete is no less than 12 mm (see Fig. 3.185).

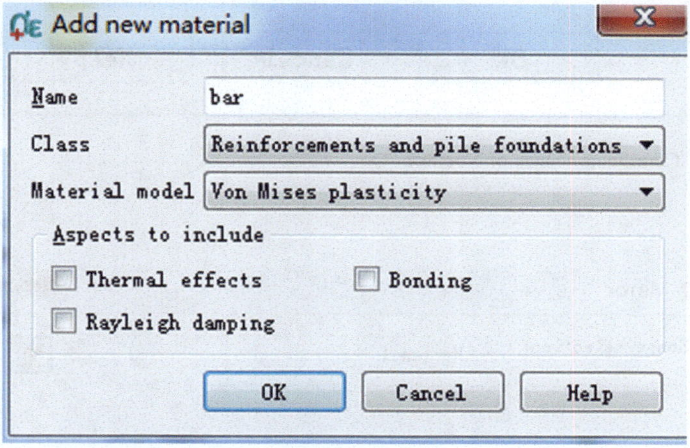

Fig. 3.184 Material class and model for longitudinal steel bars

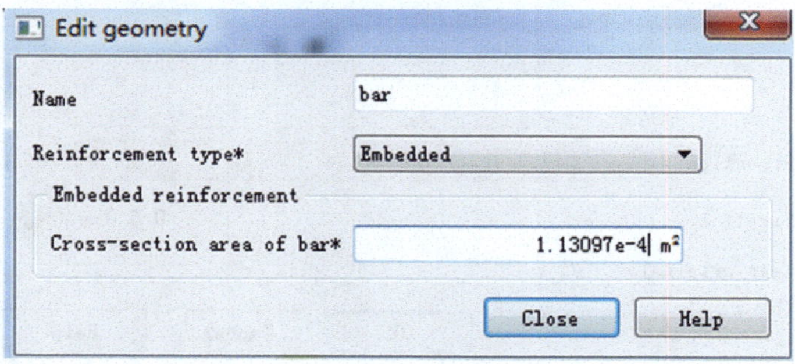

Fig. 3.185 Cross-section geometric properties for longitudinal steel bar

In order to create attaching point of concentrated point load, **Vertex1** is created as Fig. 3.186 shows, which is then selected to create **Vertex2** still with the manipulation of **Mirror a shape**, where direction is along Y axis and **Pivot** is at Y = 5 m (see Fig. 3.187).

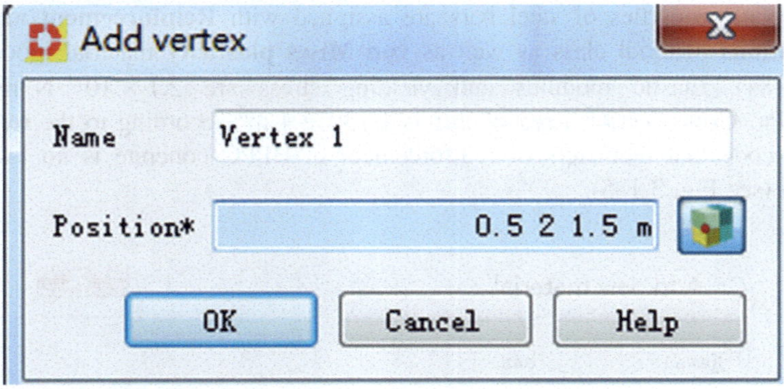

Fig. 3.186 Coordinate values of **Vertex1**

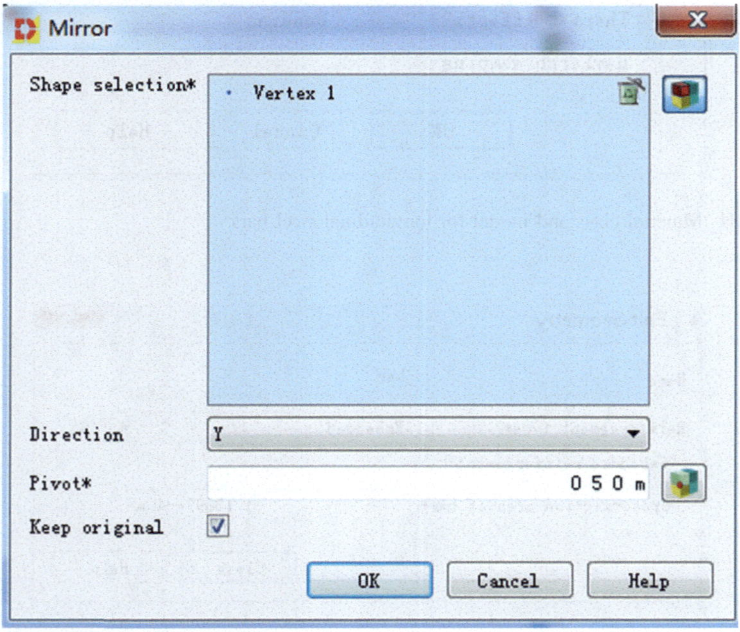

Fig. 3.187 Generation of **Vertex2** via manipulation of **Mirror a shape** for **Vertex1**

After the manipulation mentioned above is completed, advisable readers can sense that the following step is to imprint and project these two points. **Operation** is **Face**, top plane of UHPC beam is selected as **Face selection** while **Vertex1** and **Vertex2** are chosen as **Tool selection** with the direction in the negative Z direction. Manipulation interface is shown in Fig. 3.188.

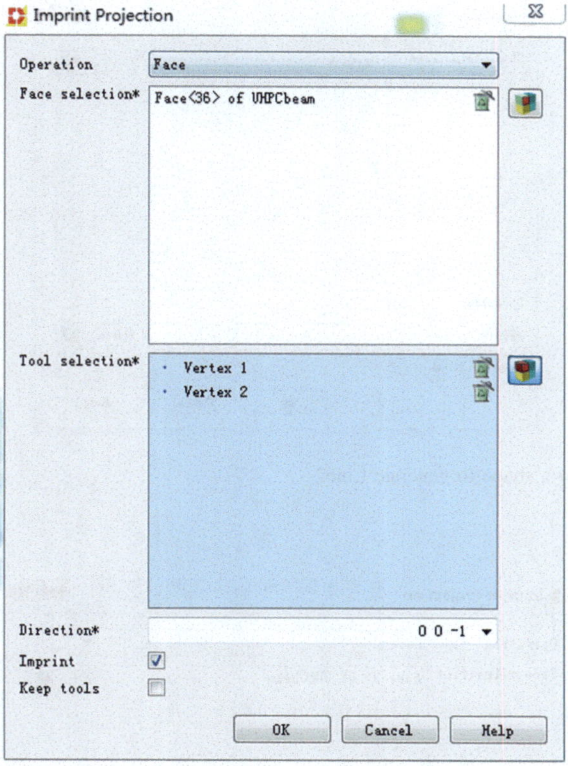

Fig. 3.188 Manipulation interface of imprint and projection

Create **Line1** with the coordinate values displayed in Fig. 3.189, which are further selected to choose the manipulation of **Mirror a shape** again to generate Line2. After the generation of these two lines, they are both selected **Tool selection** to be imprinted and projected on the bottom plane still acting as **Face selection** (see Fig. 3.190). It is noted that owing to the coordinate values in Z direction of these two lines are lower than the bottom plane of UHPC rectangular beam, therefore, the imprint and projection is in the positive Z direction (see Fig. 3.191).

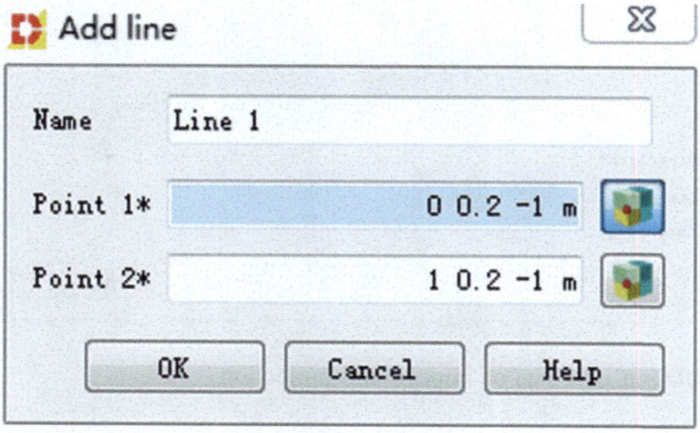

Fig. 3.189 Coordinate values of Line1

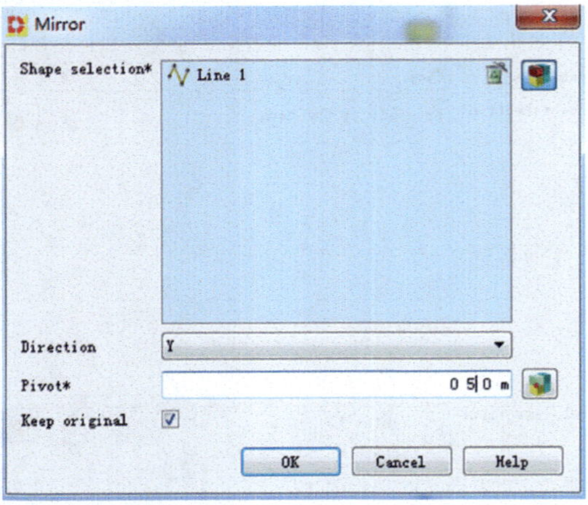

Fig. 3.190 Mirror a shape to generate Line2

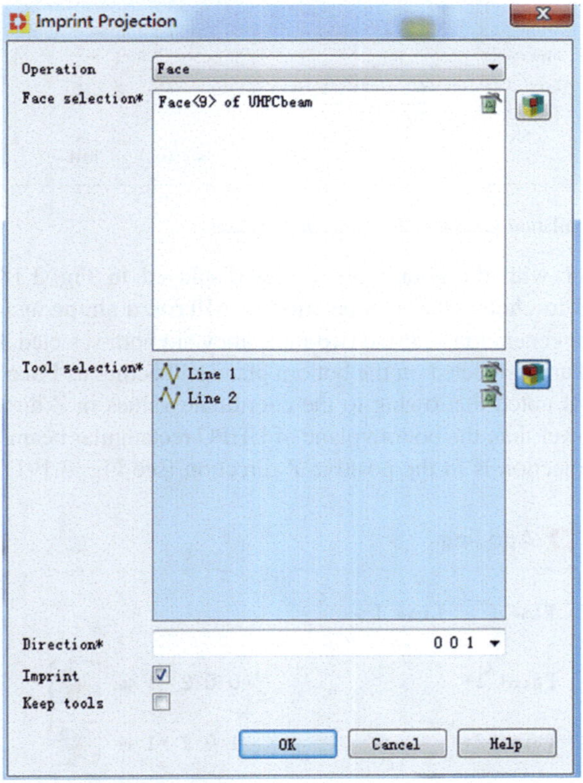

Fig. 3.191 Manipulation interface of imprint and projection

Click **Geometry-Analysis-Attach support** under menu bar to create constraints. **Line1** is selected to attach translational constraints in X, Y and Z directions while translational constraints in X and Z directions are attached on the **Line2** with the name of **co1** and **co2** (see Figs. 3.192 and 3.193). Clicking OK button generated simply-supported constraints which are displayed in Fig. 3.194.

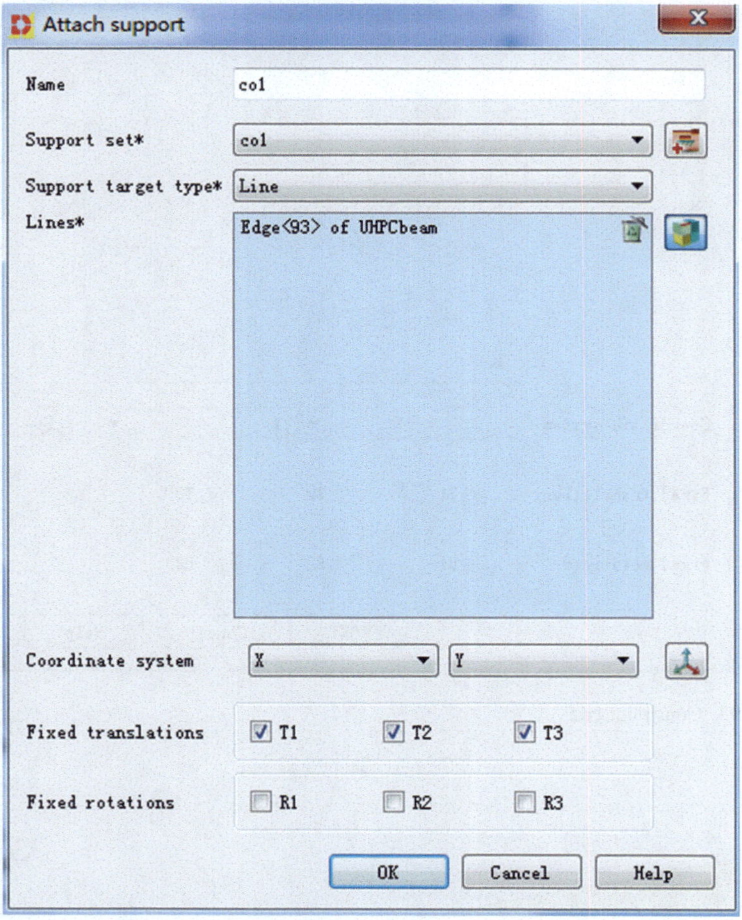

Fig. 3.192 Constraint **co1**

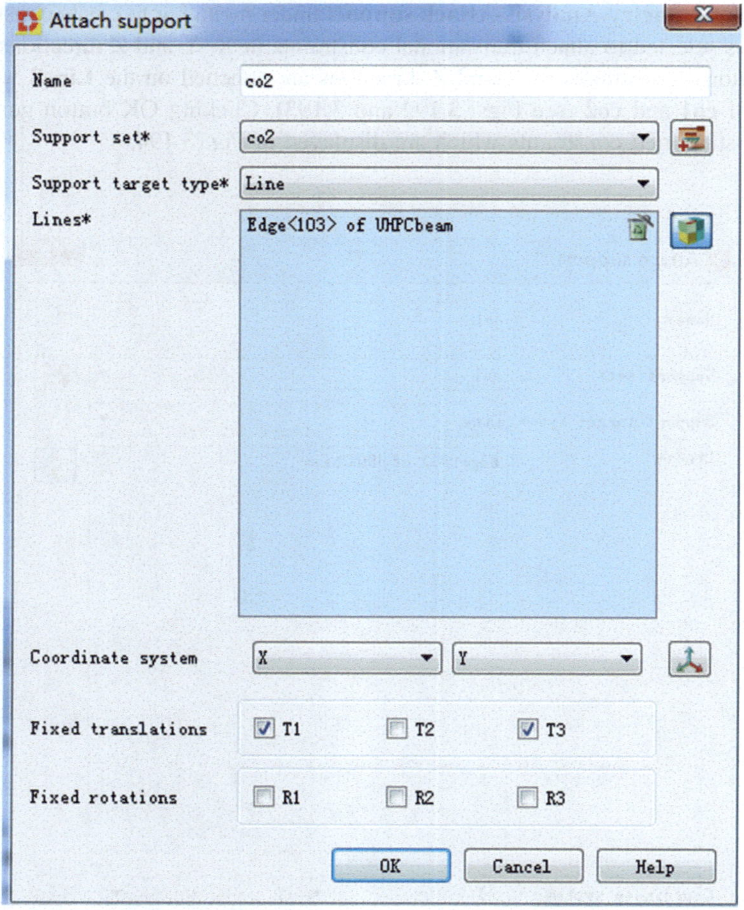

Fig. 3.193 Constraint **co2**

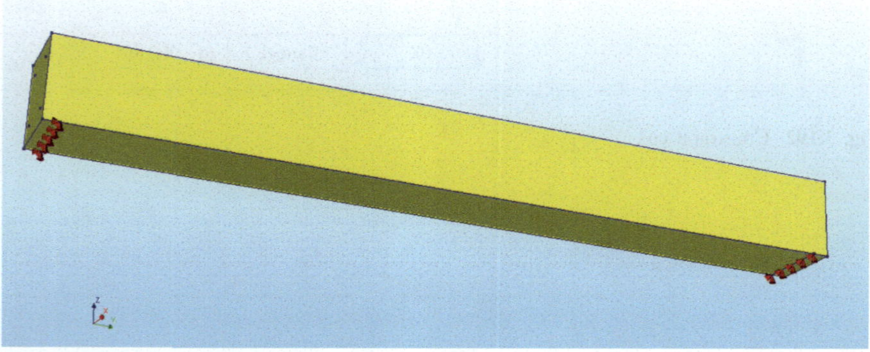

Fig. 3.194 Generation of simply supported constraints

The following step is to attach load. Attachment of gravity is the same as former part and it is not repeated here. Creating point concentrated load with the name of load case **load**, **Load target type** is **Point** and **Load type** is **Force.** Imprinted and projected vertexes mentioned above are selected as attaching point with the value of 20 kN in the negative Z direction (see Fig. 3.195). Gravity is added into **Geometry load combination 1** while concentrated point load is added into **Geometry load combination 2**.

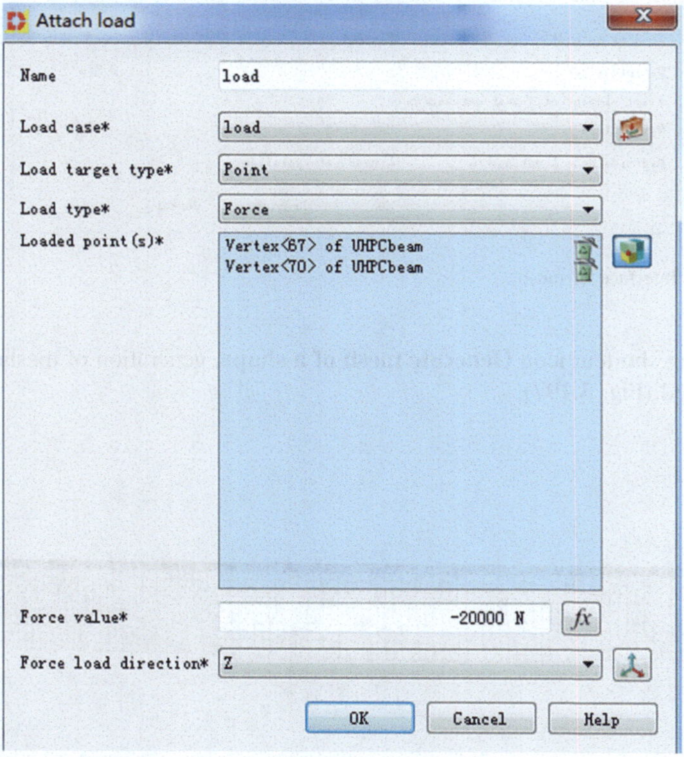

Fig. 3.195 Manipulation interface of attaching point load

When the manipulations mentioned above are completed, we start to mesh the numerical model. Selecting the 3D model UHPCbeam, **Operation** is Shape and **Element size** is selected as **Seeding method** with **Desired size** 0.05 m in order to be meshed as 20, 200 and 20 elements in length, width and height, which is proportional to corresponding geometric sizes. **Mesher type** is **Hexa/Quad** and **Linear interpolation** is chosen as **Mid-side node location** (see Fig. 3.196).

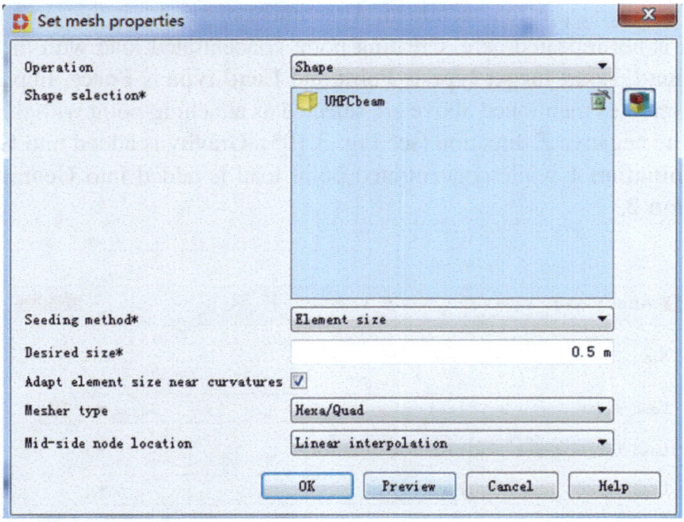

Fig. 3.196 Interface of mesh

Clicking shortcut icon **Generate mesh of a shape**, generation of meshed model is displayed (Fig. 3.197).

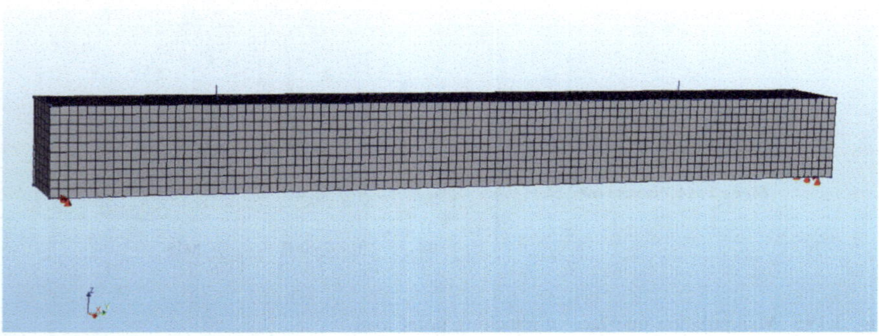

Fig. 3.197 Generated meshed shape

Adding a new analysis for 3D rectangular model with the name of Analysis5, structural nonlinear analysis is created. Deleting the former default load step block, we can construct starting steps with **Geometry load combination 1** added into load set. Apply the same method to add **Geometry load combination 1** into a new

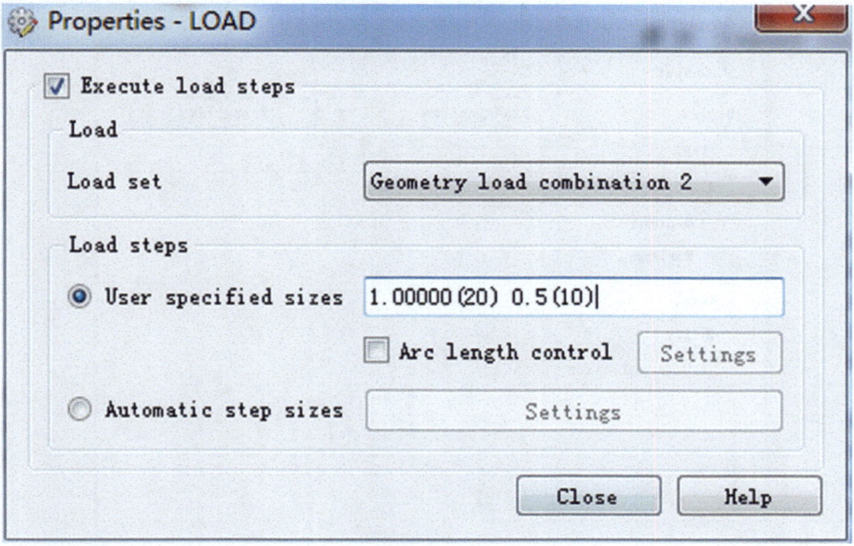

Fig. 3.198 User-specified sizes

created ordinary load step, with user-specified sizes 1.00000(20) 0.5(10) (see Fig. 3.198). In the **Equilibrium iteration** module of both execute step blocks, **maximum** number of iterations is set as 50 and method of iteration is **Newton–Raphson** with the **Type** and **First tangent** selecting **Regular** and **Tangential,** respectively. **Force** and **Displacement** are both selected as convergence norm. Convergence tolerance is still set at default value 0.01, while Abort criterion is kept as 10,000.

Translational displacement in all directions under global coordinate system (DISPLA TOTAL TRANSL GLOBAL), cracking strains in all directions (STRAIN CRACK GREEN), summed crack strains under local coordinate system (STRAIN CRKSUM GREEN LOCAL), summed crack strains under global coordinate system (STRAIN CRKSUM GREEN GLOBAL), crack width in all directions under global and local coordinate system (STRAIN CRKWDT GREEN GLOBAL/LOCAL) are selected as the outcomes of output, shown in Fig. 3.199.

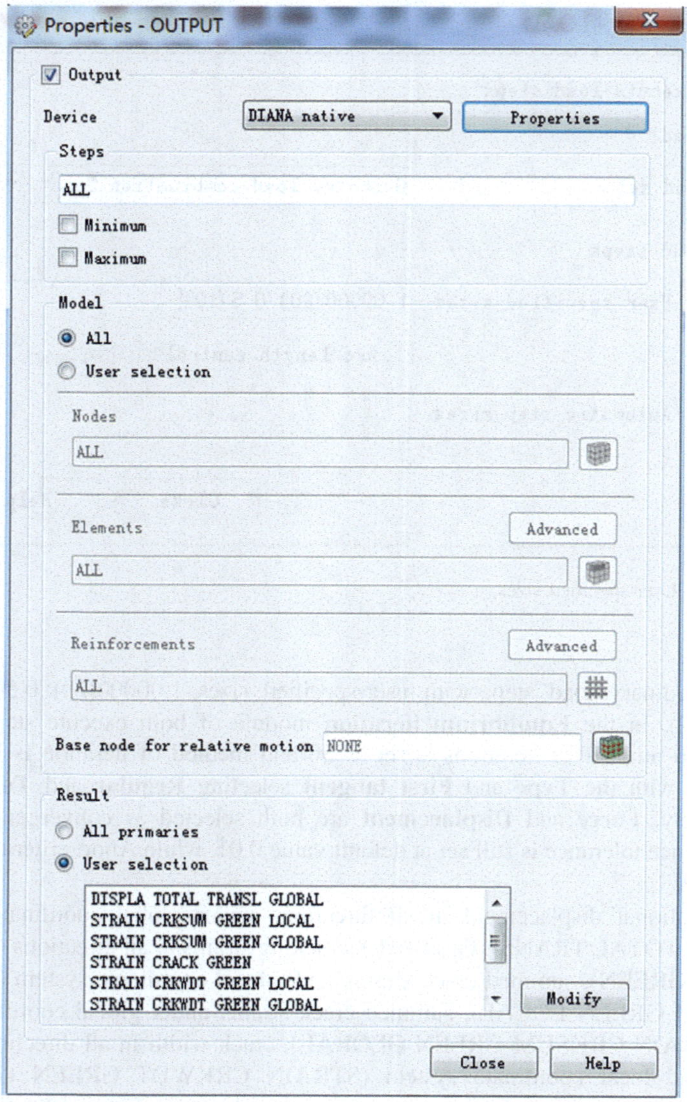

Fig. 3.199 Items of output

Click run an analysis; when divergence occurs in iteration calculation, it means the beam reaches ultimate bearing capacity. Contour plot results after calculation are displayed in Figs. 3.200, 3.201, 3.202, 3.203, 3.204, 3.205 and 3.206.

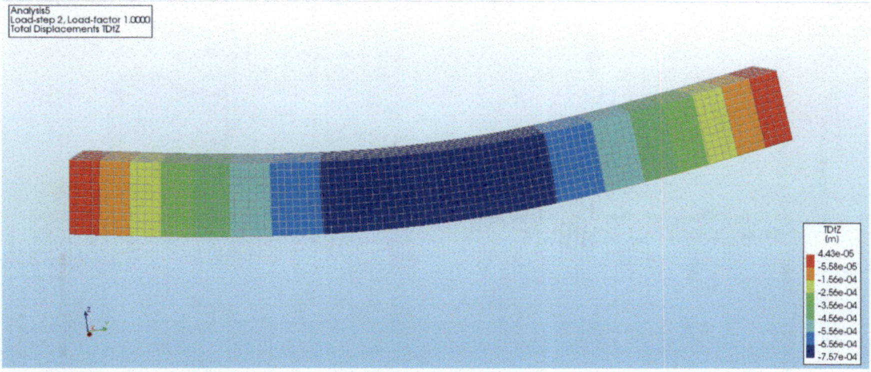

Fig. 3.200 Contour plot of displacement in Z directions after 20 kN

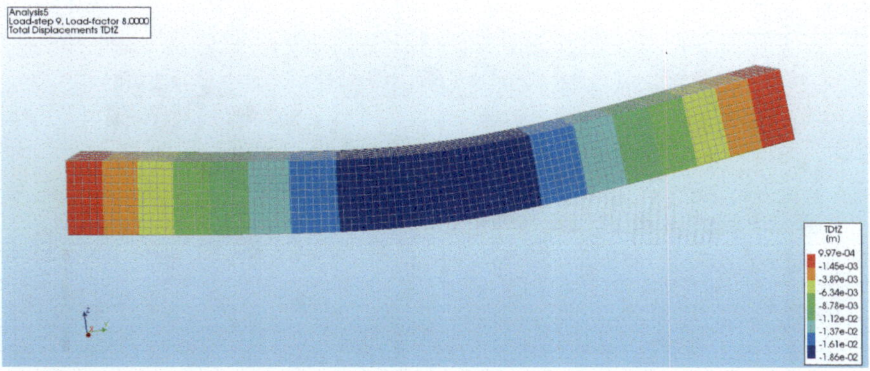

Fig. 3.201 Contour plot of displacement in Z directions after 160 kN before crushing

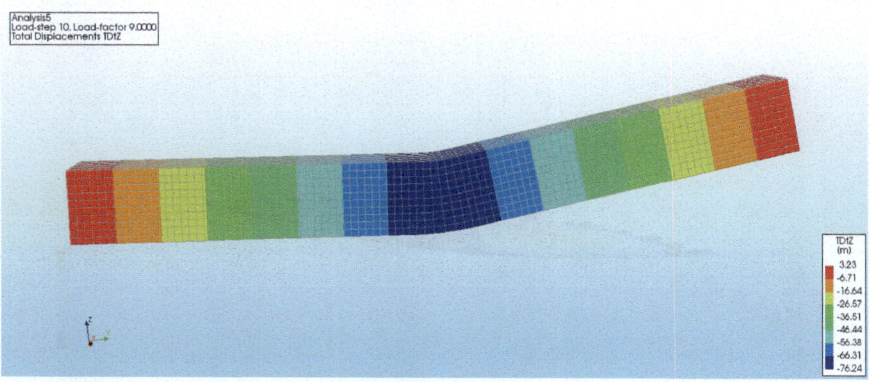

Fig. 3.202 Contour plot of displacement in Z directions after 160 kN when crushing

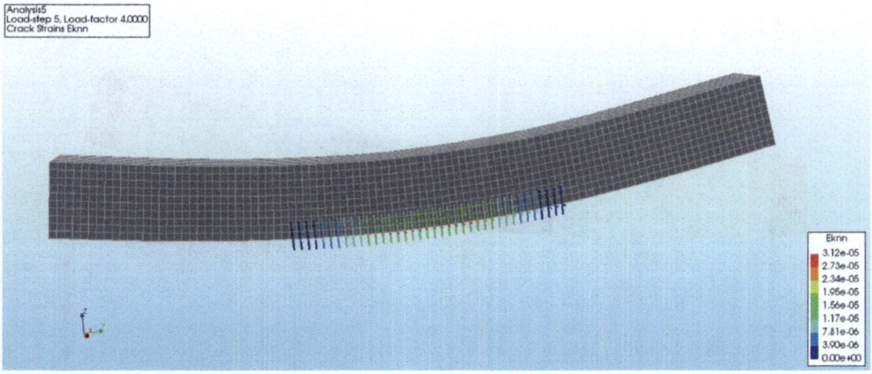

Fig. 3.203 Contour plot of strains in normal direction after 80 kN

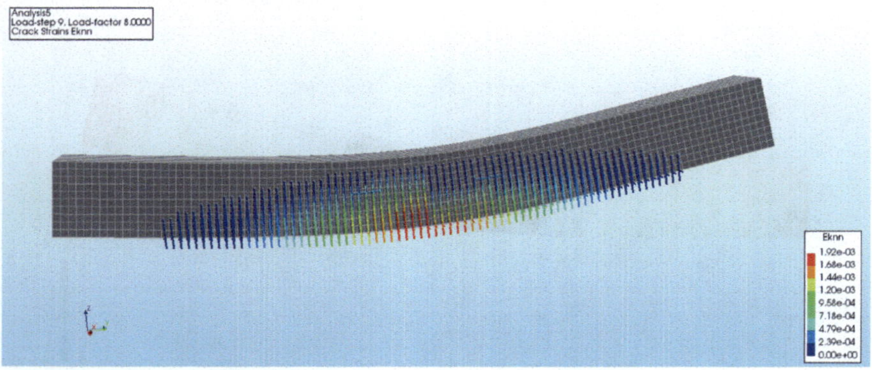

Fig. 3.204 Contour plot of strains in normal direction after 160 kN

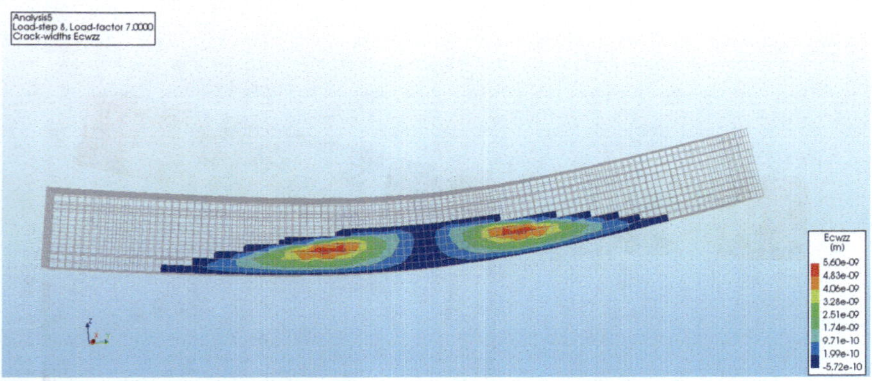

Fig. 3.205 Contour plot of crack widths in local z direction (Ecwzz) after 140 kN

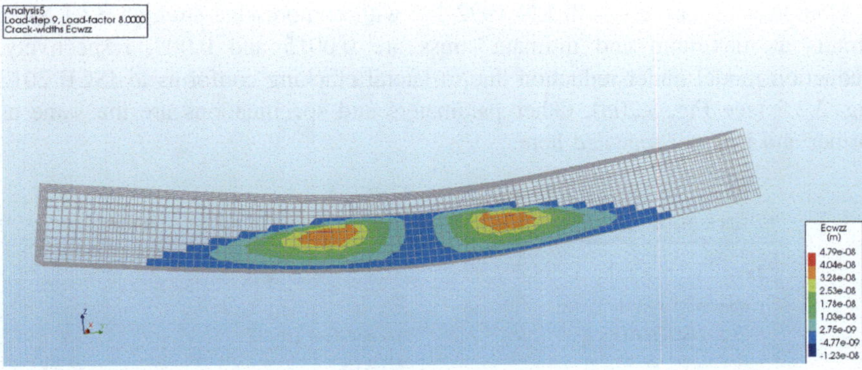

Fig. 3.206 Contour plot of crack widths in local z direction (Ecwzz) after 160 kN before crushing

Altering the tension softening constitutive properties into Hordijk model, UHPC concrete is replaced by C50 concrete, where tensile strength is 2.6 MPa with the Mode-I tensile fracture energy 150 N/m. Poisson's ratio reduction shows no reduction (see Fig. 3.207).

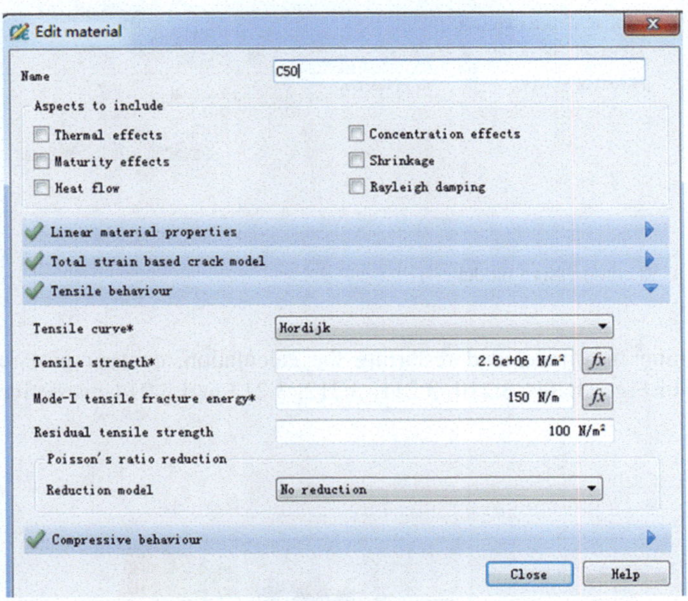

Fig. 3.207 Tensile behavior of C50 beam

Compression curve is still EN 1992-1-2 with compressive strength 32.4 MPa. Strains at maximum and ultimate stress are 0.0015 and 0.003, respectively. Reduction model under reduction due to lateral cracking conforms to JSCE 2012 Fig. 2.2.5 (see Fig. 3.208). Other parameters and specifications are the same as former and it is not repeated here.

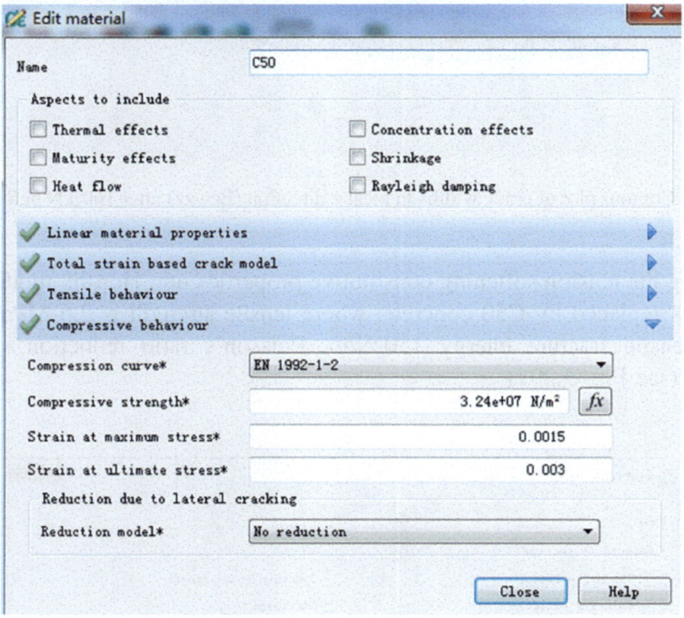

Fig. 3.208 C50 compressive model

Remeshing the model and restarting the calculation, contour plot results are displayed in Figs. 3.209, 3.210, 3.211, 3.212, 3.213 and 3.214 respectively.

Fig. 3.209 Contour plot of displacement in Z directions after 20 kN

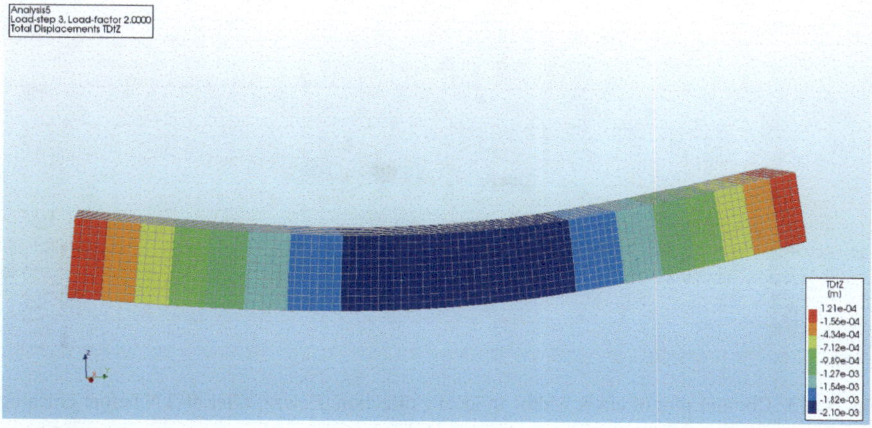

Fig. 3.210 Contour plot of displacement in Z directions after 40 kN before crushing

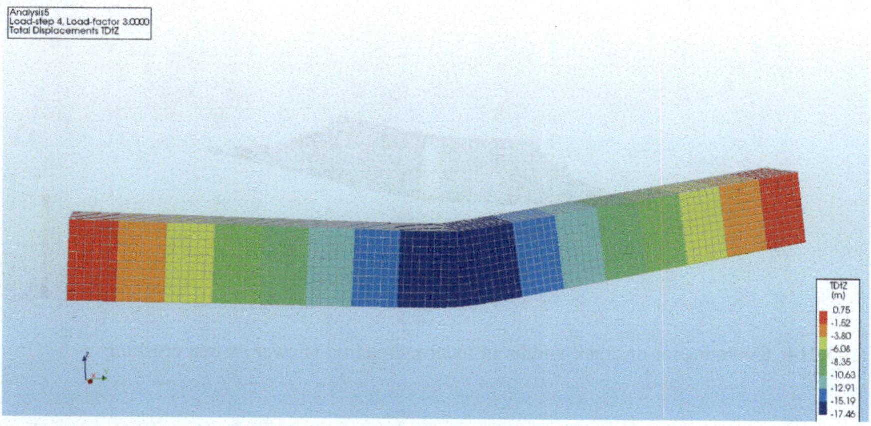

Fig. 3.211 Contour plot of displacement in Z directions when crushing

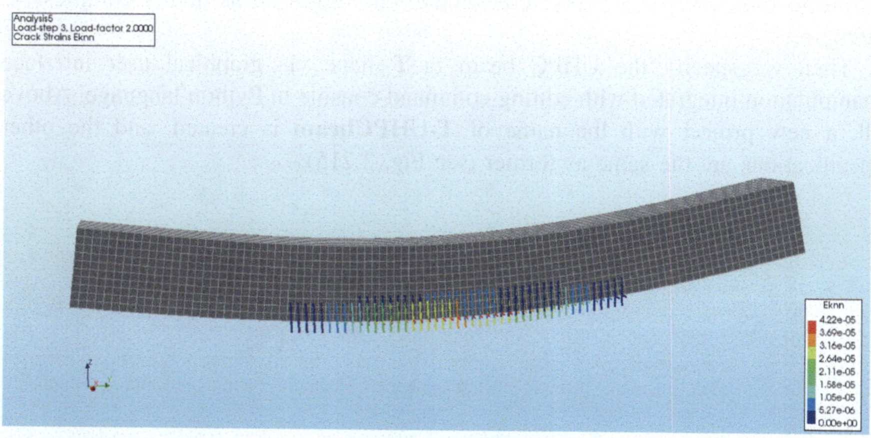

Fig. 3.212 Contour plot of strains in normal direction after 40 kN before crushing

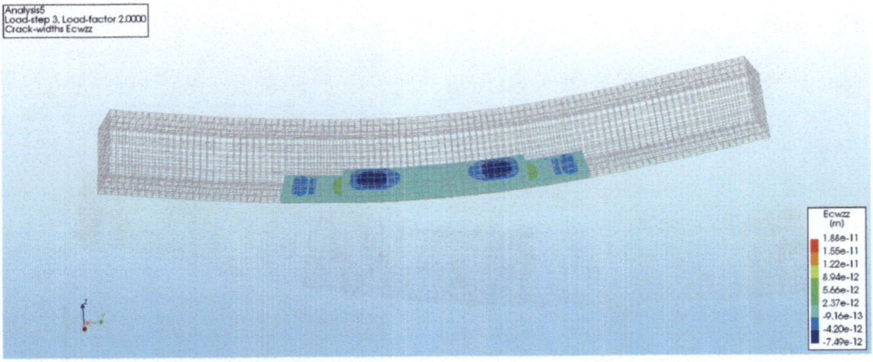

Fig. 3.213 Contour plot of crack widths in local z direction (Ecwzz) after 40 kN before crushing

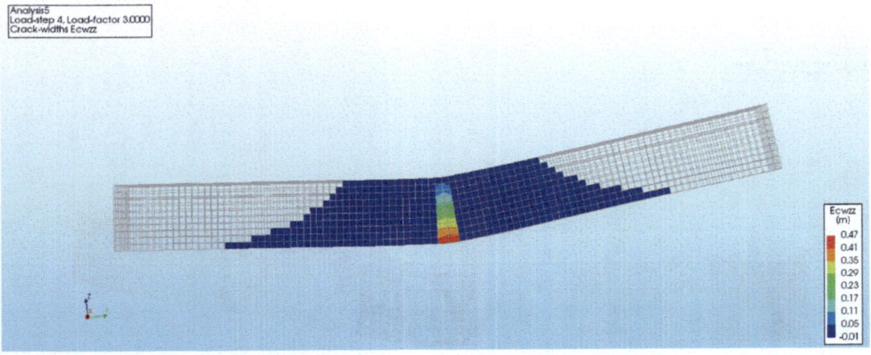

Fig. 3.214 Contour plot of crack widths in local z direction (Ecwzz) when crushing

Initial cracking load of C50 concrete occurs at 20 kN while the ultimate loading value is 80 kN. Meanwhile, ductility of UHPC is much better than the C50 concrete owing to more excellent tensile constitutive curve as well as higher compressive strength.

Then we specify the UHPC beam in T-shape via graphical user interface manipulation integrated with editing command console in Python language. Above all, a new project with the name of **T-UHPCbeam** is created, and the other specifications are the same as former (see Fig. 3.215).

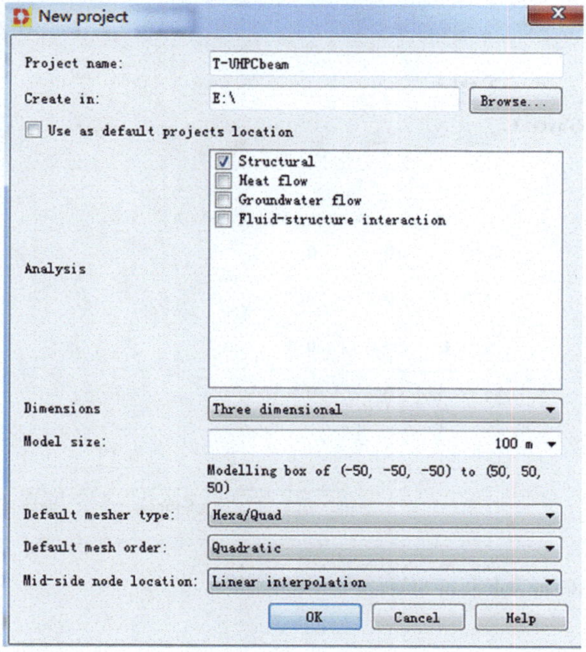

Fig. 3.215 New project of T-UHPC beam

Sheet 1 is created and the coordinate is displayed (Fig. 3.216). Then Sheet 2 is also created and the coordinate is displayed (Fig. 3.217).

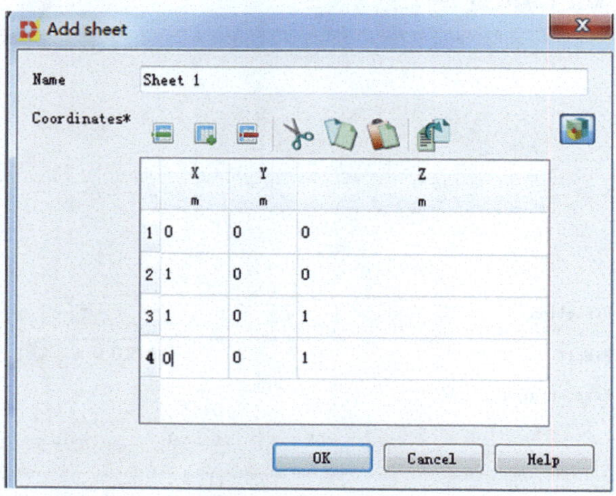

Fig. 3.216 Coordinate values of **Sheet 1**

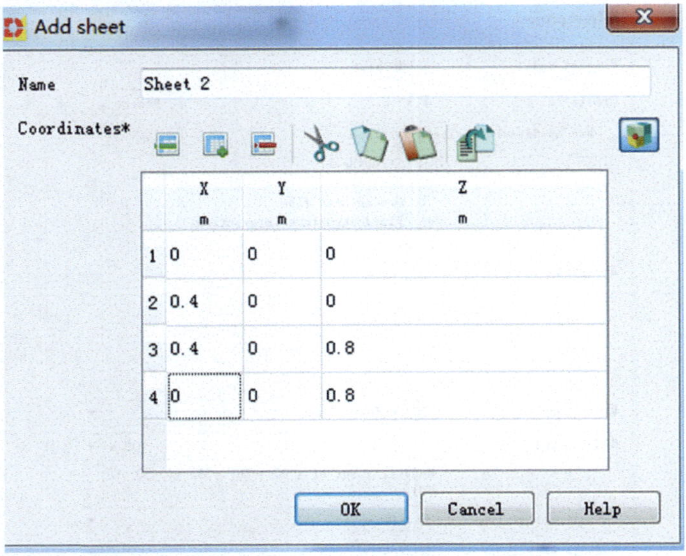

Fig. 3.217 Coordinate values of **Sheet 2**

Then we create **Sheet 3** which is mirror symmetric to **Sheet 2** via the manipulation of **Mirror a shape**. Pivot of **Sheet 2** is X = 0.5 m in the X direction (see Fig. 3.218).

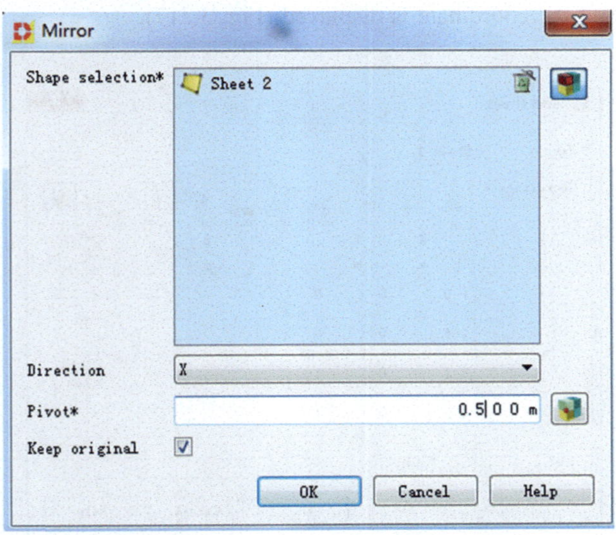

Fig. 3.218 Generation of **Sheet 3** via manipulation **Mirror a shape**

Sheet 1 is selected as **Target selection**, while **Sheet 2** and **Sheet 3** are selected as **Tool selection**. Operation is **Subtract** in Boolean logic operation (see Fig. 3.219). Clicking OK button to finish the subtraction, it is observed that **Sheet 1** after subtraction is only in T shape, and the generated cross-section in T shape is displayed in Fig. 3.220.

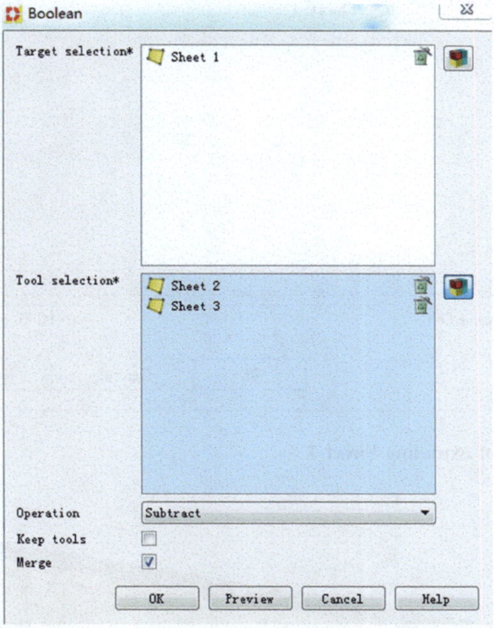

Fig. 3.219 Interface of subtraction in Boolean logic manipulation

Fig. 3.220 Generation cross-section in T shape

Selecting the **Sheet 1** in T shape, right-clicking to select the function **Extrude a shape**, it is extruded in longitudinal Y direction with the displacement 10 m (see Fig. 3.221). Also clicking OK button, T beam is displayed as in Fig. 3.222.

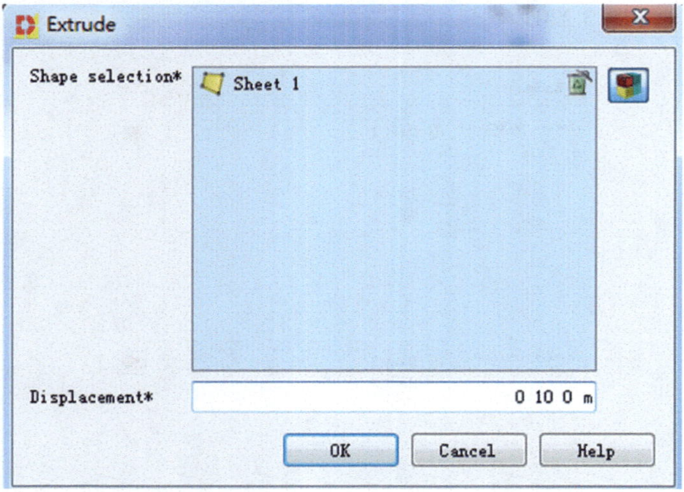

Fig. 3.221 Interface of extruding **Sheet 1**

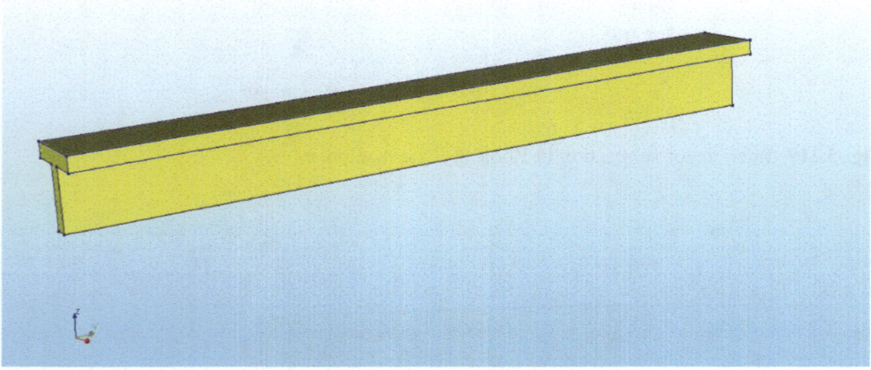

Fig. 3.222 Generation of I-shape beam

Then we duplicate command console of assigning material properties by pasting this section into command console region to generate corresponding assignment manipulation (see Fig. 3.223). After that, considering the variation in cross-section, longitudinal steel bar is only single one with the coordinate values [0.5, 0, 0.2], [0.5, 10, 0.2] created by the syntax "**createLine("bar1", [0.5, 0, 0.2], [0.5, 10, 0.2])**"; also see Fig. 3.223. After that, remaining command console in Python language is directly duplicated and pasted into command console zone to complete other manipulations, where the supported lines imprint and projection are displayed in Fig. 3.224. Figure 3.225 also demonstrates numerical model after vertex imprint and projection and loading attachment. Figure 3.226 displays numerical model after attachment of constraints, while Fig. 3.227 shows generation of meshed elements.

```
createSheet( "Sheet 1", [[ 0, 0, 0 ],[ 1, 0, 0 ],[ 1, 0, 1 ],[ 0, 0, 1 ]] )
saveProject( )
createSheet( "Sheet 2", [[ 0, 0, 0 ],[ 0.4, 0, 0 ],[ 0.4, 0, 0.8 ],[ 0, 0, 0.8 ]] )
saveProject( )
mirror( [ "Sheet 2" ], [ 0.5, 0, 0 ], [ True, False, False ], True )
saveProject( )
subtract( "Sheet 1", [ "Sheet 2", "Sheet 3" ], False, True )
saveProject( )
extrudeProfile( [ "Sheet 1" ], [ 0, 10, 0 ] )
renameShape( "Sheet 1", "UHPCbeam" )
addMaterial( "UHPC", "CONCR", "TSCR", [] )
setParameter( MATERIAL, "UHPC", "LINEAR/ELASTI/YOUNG", 4.35e+10 )
setParameter( MATERIAL, "UHPC", "LINEAR/ELASTI/POISON", 0.15 )
setParameter( MATERIAL, "UHPC", "LINEAR/MASS/DENSIT", 2500 )
setParameter( MATERIAL, "UHPC", "MODTYP/TOTCRK", "ROTATE" )
setParameter( MATERIAL, "UHPC", "TENSIL/TENCRV", "FRCCON" )
setParameter( MATERIAL, "UHPC", "TENSIL/FRCTYP", "STRAIN" )
setParameter( MATERIAL, "UHPC", "TENSIL/FRCEPS", [ 5500000, 5500000, 0.0024, 5500000, 0.0038, 0.004 ] )
setParameter( MATERIAL, "UHPC", "COMPRS/COMCRV", "EN1992" )
setParameter( MATERIAL, "UHPC", "COMPRS/COMSTR", 1.2e+08 )
setParameter( MATERIAL, "UHPC", "COMPRS/EPSC1", 0.004 )
setParameter( MATERIAL, "UHPC", "COMPRS/EPSCU", 0.007 )
setParameter( MATERIAL, "UHPC", "COMPRS/REDUCT/REDCRV", "JSCE12" )
clearReinforcementAspects( [ "UHPCbeam" ] )
setElementClassType( SHAPE, [ "UHPCbeam" ], "STRSOL" )
assignMaterial( "UHPC", SHAPE, [ "UHPCbeam" ] )
resetGeometry( SHAPE, [ "UHPCbeam" ] )
resetElementData( SHAPE, [ "UHPCbeam" ] )
saveProject( )
createLine( "bar1", [ 0.5, 0, 0.2 ], [ 0.5, 10, 0.2 ] )
addMaterial( "bar", "REINFO", "VMISES", [] )
setParameter( MATERIAL, "bar", "LINEAR/ELASTI/YOUNG", 2.1e+11 )
setParameter( MATERIAL, "bar", "PLASTI/HARDI1/YLDSTR", 4.4e+08 )
addGeometry( "Element geometry 1", "RELINE", "REBAR", [] )
rename( GEOMET, "Element geometry 1", "bar" )
setParameter( GEOMET, "bar", "REIEMB/CROSSE", 0.000113097 )
setParameter( GEOMET, "bar", "REIEMB/CROSSE", 0.000113097 )
setParameter( GEOMET, "bar", "REIEMB/CROSSE", 0.000113097 )
setReinforcementAspects( [ "bar1"] )
assignMaterial( "bar", SHAPE, [ "bar1" ] )
assignGeometry( "bar", SHAPE, [ "bar1" ] )
resetElementData( SHAPE, [ "bar1" ] )
```

Fig. 3.223 Command console required to be duplicated and pasted

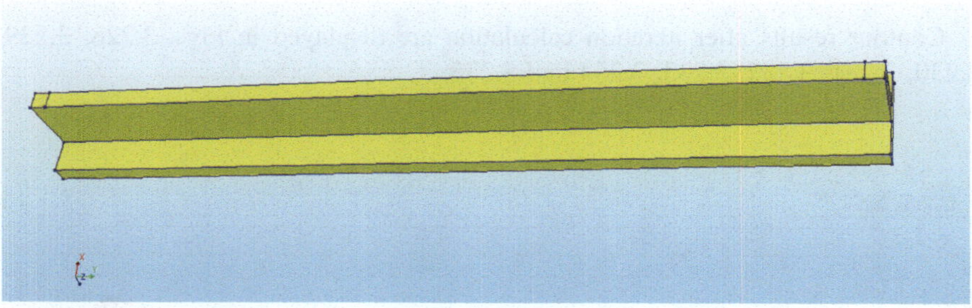

Fig. 3.224 Imprinted and projected lines

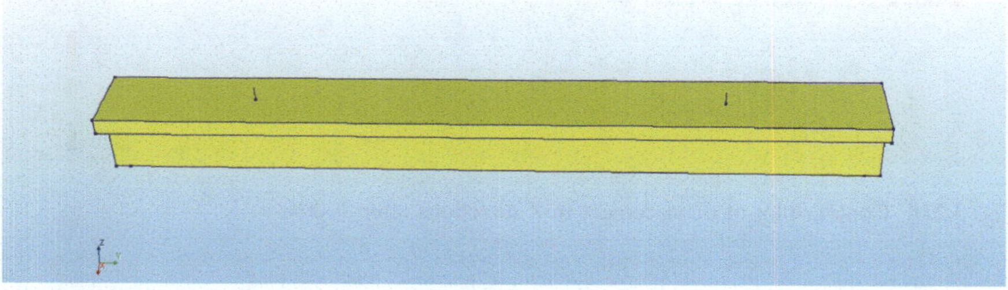

Fig. 3.225 Numerical model after vertex imprint and projection and loading attachment

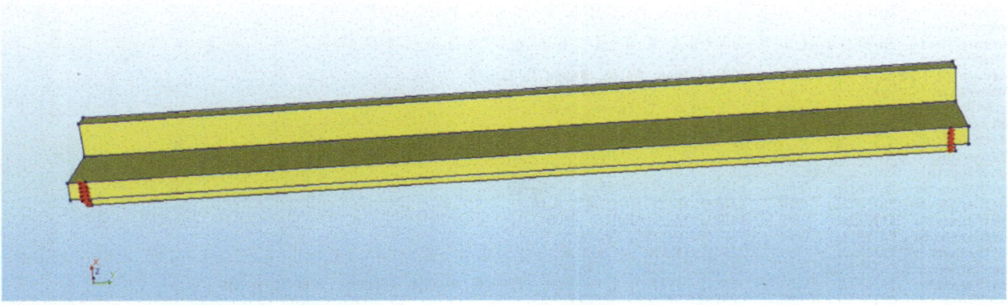

Fig. 3.226 Numerical model after attachment of constraints

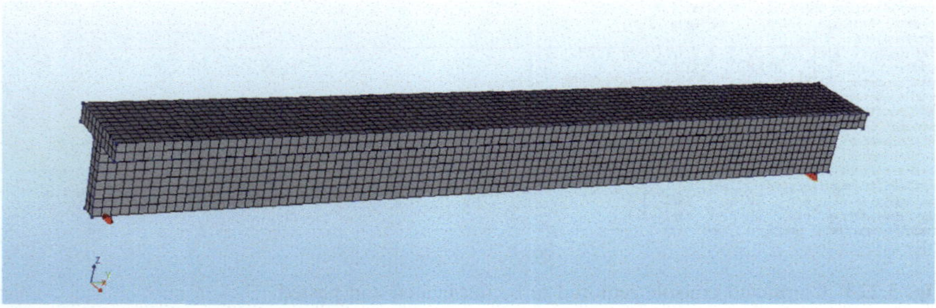

Fig. 3.227 Generation of meshed elements

Contour results after iteration calculation are displayed in Figs. 3.228, 3.229, 3.230, 3.231, 3.232, 3.233, 3.234 and 3.235.

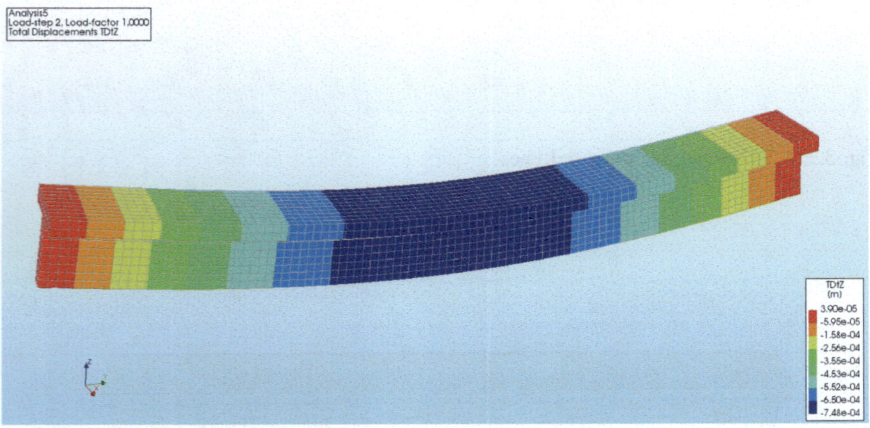

Fig. 3.228 Contour plot of displacement in Z directions after 20 kN

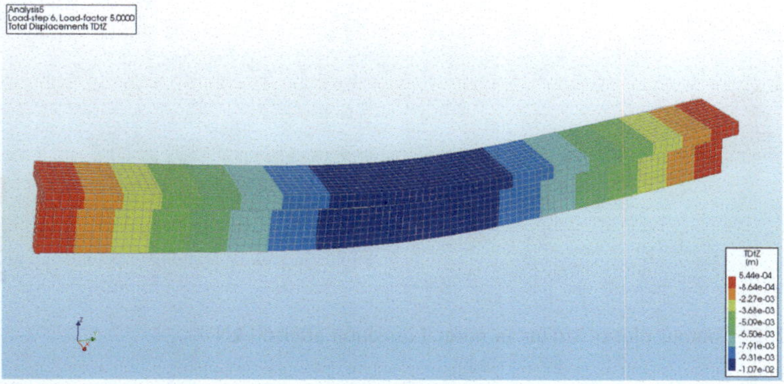

Fig. 3.229 Contour plot of displacement in Z directions after 100 kN before crushing

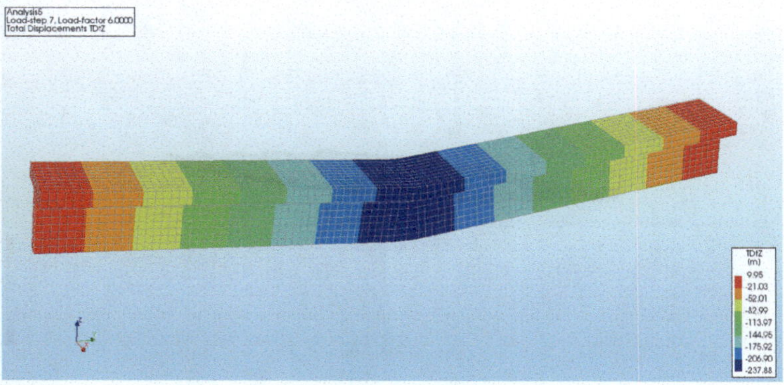

Fig. 3.230 Contour plot of displacement in Z directions after 100 kN when crushing

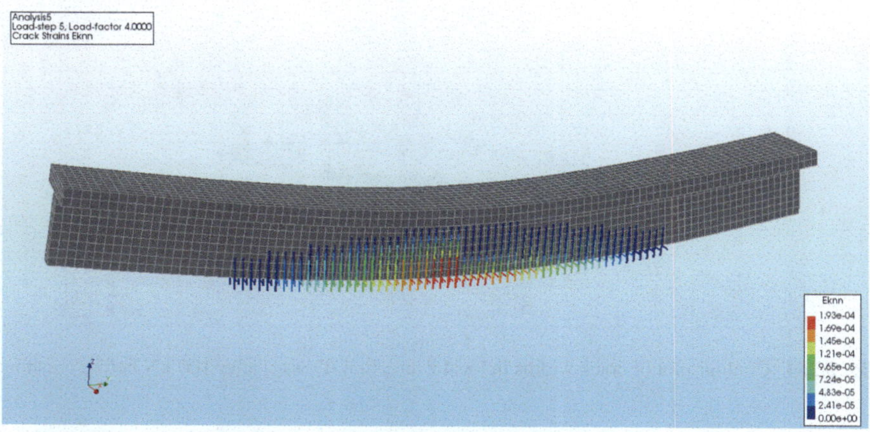

Fig. 3.231 Contour plot of displacement in Z directions after 100 kN when crushing

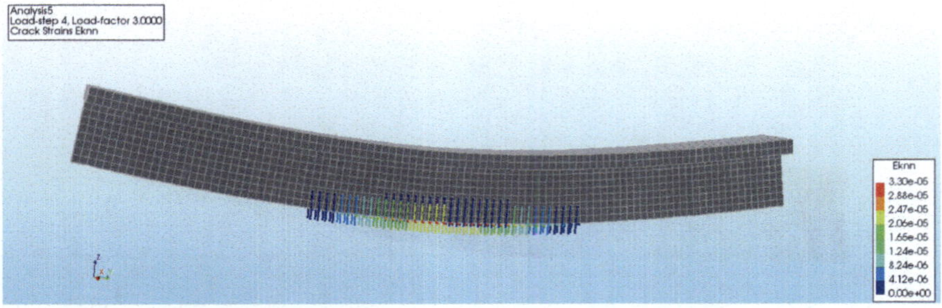

Fig. 3.232 Contour plot of strains in normal direction after 60 kN

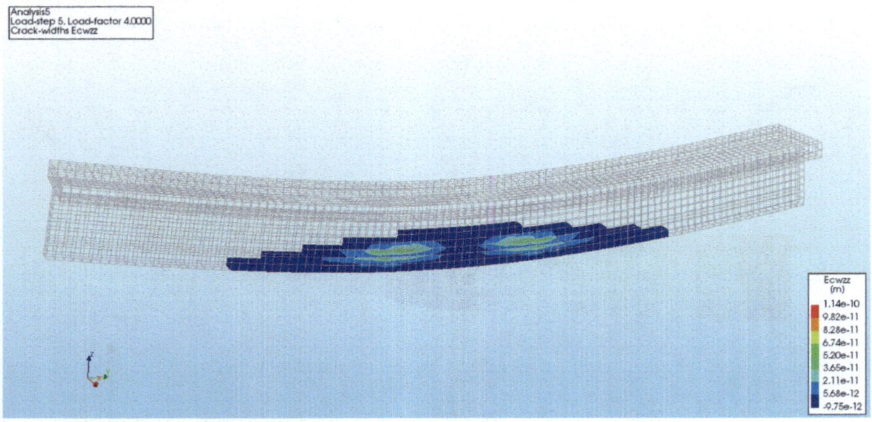

Fig. 3.233 Contour plot of crack widths in local z direction (Ecwzz) after 80 kN

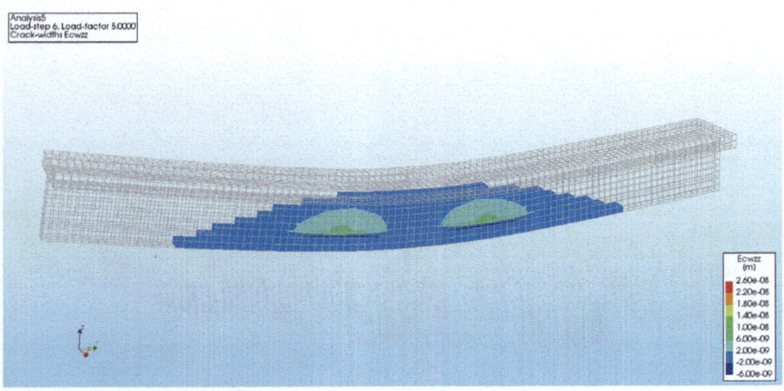

Fig. 3.234 Contour plot of crack widths in local z direction (Ecwzz) after 100 kN before crushing

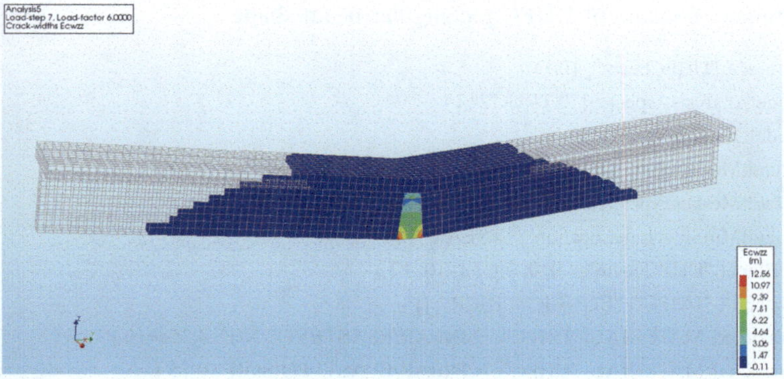

Fig. 3.235 Contour plot of crack widths in local z direction (Ecwzz) when crushing

In order to better compare and validate the priority of UHPC beam, initial cracking load and ultimate bearing capacity of UHPC rectangular beam, C50 rectangular beam, as well as UHPC beam in T shape are listed in Table 3.5.

Table 3.5 Comparison of three beams

	Initial cracking load (kN)	Ultimate bearing capacity (kN)
UHPC rectangular beam	80	180
C50 rectangular beam	20	80
UHPC T beam	60	120

It is observed that both initial cracking load and ultimate bearing capacity of UHPC beam are far higher than the C50 reinforced concrete beam when other conditions are the same, exceeding around 4 and 2.25 times, respectively. Furthermore, initial cracking load of UHPC beam in T shape is 60 kN, while ultimate load is 100 kN, which is still higher than C50 concrete rectangular beam, although reinforcement ratio and size of cross-section area are in vast decrease, which indicates that the dimensional sizes of beam and reinforcement ratio can be decreased in applying UHPC material and renders a reference for construction in reality.

Command console of UHPC rectangular beam shape

```
newProject( "UHPCbeam", 100 )
setModelAnalysisAspects( [ "STRUCT" ] )
setModelDimension( "3D" )
setDefaultMeshOrder( "QUADRATIC" )
setDefaultMesherType( "HEXQUAD" )
setDefaultMidSideNodeLocation( "LINEAR" )
createBlock( "UHPCbeam", [ 0, 0, 0 ], [ 1, 10, 1 ] )
addMaterial( "UHPC", "CONCR", "TSCR", [] )
setParameter( MATERIAL, "UHPC", "LINEAR/ELASTI/YOUNG", 4.35e+10 )
setParameter( MATERIAL, "UHPC", "LINEAR/ELASTI/POISON", 0.15 )
setParameter( MATERIAL, "UHPC", "LINEAR/MASS/DENSIT", 2500 )
setParameter( MATERIAL, "UHPC", "MODTYP/TOTCRK", "ROTATE" )
setParameter( MATERIAL, "UHPC", "TENSIL/TENCRV", "FRCCON" )
setParameter( MATERIAL, "UHPC", "TENSIL/FRCTYP", "STRAIN" )
setParameter( MATERIAL, "UHPC", "TENSIL/FRCEPS", [ 5500000, 5500000, 0.0024,
5500000, 0.0038, 0.004 ] )
setParameter( MATERIAL, "UHPC", "COMPRS/COMCRV", "EN1992" )
setParameter( MATERIAL, "UHPC", "COMPRS/COMSTR", 1.2e+08 )
setParameter( MATERIAL, "UHPC", "COMPRS/EPSC1", 0.004 )
setParameter( MATERIAL, "UHPC", "COMPRS/EPSCU", 0.007 )
setParameter( MATERIAL, "UHPC", "COMPRS/REDUCT/REDCRV", "JSCE12" )
clearReinforcementAspects( [ "UHPCbeam" ] )
setElementClassType( SHAPE, [ "UHPCbeam" ], "STRSOL" )
assignMaterial( "UHPC", SHAPE, [ "UHPCbeam" ] )
resetGeometry( SHAPE, [ "UHPCbeam" ] )
resetElementData( SHAPE, [ "UHPCbeam" ] )
saveProject(  )
createLine( "bar1", [ 0.8, 0, 0.2 ], [ 0.8, 10, 0.2 ] )
mirror( [ "bar1" ], [ 0.5, 0, 0 ], [ True, False, False ], True )
saveProject(  )
mirror( [ "bar1", "bar2" ], [ 0, 0, 0.5 ], [ False, False, True ], True )
saveProject(  )
saveProject(  )
addMaterial( "bar", "REINFO", "VMISES", [] )
setParameter( MATERIAL, "bar", "LINEAR/ELASTI/YOUNG", 2.1e+11 )
setParameter( MATERIAL, "bar", "PLASTI/HARDI1/YLDSTR", 4.4e+08 )
addGeometry( "Element geometry 1", "RELINE", "REBAR", [] )
rename( GEOMET, "Element geometry 1", "bar" )
setParameter( GEOMET, "bar", "REIEMB/CROSSE", 0.000113097 )
setParameter( GEOMET, "bar", "REIEMB/CROSSE", 0.000113097 )
setParameter( GEOMET, "bar", "REIEMB/CROSSE", 0.000113097 )
```

```
setReinforcementAspects( [ "bar1", "bar2", "bar3", "bar4" ] )
assignMaterial( "bar", SHAPE, [ "bar1", "bar2", "bar3", "bar4" ] )
assignGeometry( "bar", SHAPE, [ "bar1", "bar2", "bar3", "bar4" ] )
resetElementData( SHAPE, [ "bar1", "bar2", "bar3", "bar4" ] )
setReinforcementDiscretization( [ "bar1", "bar2", "bar3", "bar4" ], "SECTION" )
saveProject(  )
createVertex( "Vertex 1", [ 0.5, 2, 1.5 ] )
mirror( [ "Vertex 1" ], [ 0, 5, 0 ], [ False, True, False ], True )
saveProject(  )
projection( SHAPEFACE, "UHPCbeam", [[ 0.573573, 5.73573, 1 ]], [ "Vertex 1", "Vertex 2" ],
[ 0, 0, -1 ], True )
removeShape( [ "Vertex 1", "Vertex 2" ] )
saveProject(  )
createLine( "co1", [ 0, 1, -1 ], [ 1, 1, -1 ] )
undo( 1 )
createLine( "Line 1", [ 0, 0.2, -1 ], [ 1, 0.2, -1 ] )
mirror( [ "Line 1" ], [ 0, 5, 0 ], [ False, True, False ], True )
projection( SHAPEFACE, "UHPCbeam", [[ 0.426427, 5.73573, 0 ]], [ "Line 1", "Line 2" ], [ 0,
0, 1 ], True )
removeShape( [ "Line 1", "Line 2" ] )
addSet( GEOMETRYLOADSET, "Geometry load case 1" )
rename( GEOMETRYLOADSET, "Geometry load case 1", "load" )
createPointLoad( "load", "load" )
setParameter( GEOMETRYLOAD, "load", "FORCE/VALUE", -20000 )
setParameter( GEOMETRYLOAD, "load", "FORCE/DIRECT", 3 )
attach( GEOMETRYLOAD, "load", "UHPCbeam", [[ 0.5, 2, 1 ],[ 0.5, 8, 1 ]] )
saveProject(  )
addSet( GEOMETRYLOADSET, "gravity" )
createModelLoad( "gravity", "gravity" )
setDefaultGeometryLoadCombinations(  )
setGeometryLoadCombinationFactor( "Geometry load combination 1", "load", 1 )
addSet( GEOMETRYSUPPORTSET, "co1" )
createLineSupport( "co1", "co1" )
setParameter( GEOMETRYSUPPORT, "co1", "AXES", [ 1, 2 ] )
setParameter( GEOMETRYSUPPORT, "co1", "TRANSL", [ 1, 1, 1 ] )
setParameter( GEOMETRYSUPPORT, "co1", "ROTATI", [ 0, 0, 0 ] )
attach( GEOMETRYSUPPORT, "co1", "UHPCbeam", [[ 0.5, 0.2, 0 ]] )
saveProject(  )
addSet( GEOMETRYSUPPORTSET, "Geometry support set 2" )
rename( GEOMETRYSUPPORTSET, "Geometry support set 2", "co2" )
createLineSupport( "co2", "co2" )
```

```
setParameter( GEOMETRYSUPPORT, "co2", "AXES", [ 1, 2 ] )
setParameter( GEOMETRYSUPPORT, "co2", "TRANSL", [ 1, 0, 1 ] )
setParameter( GEOMETRYSUPPORT, "co2", "ROTATI", [ 0, 0, 0 ] )
attach( GEOMETRYSUPPORT, "co2", "UHPCbeam", [[ 0.5, 9.8, 0 ]] )
setElementSize( [ "UHPCbeam" ], 0.1, -1, True )
setMesherType( [ "UHPCbeam" ], "HEXQUAD" )
setMidSideNodeLocation( [ "UHPCbeam" ], "LINEAR" )
saveProject( )
generateMesh( [] )
addAnalysis( "Analysis5" )
addAnalysisCommand( "Analysis5", "NONLIN", "Structural nonlinear" )
renameAnalysis( "Analysis5", "Analysis5" )
removeAnalysisCommandDetail( "Analysis5", "Structural nonlinear", "EXECUT(1)" )
setAnalysisCommandDetail( "Analysis5", "Structural nonlinear", "EXECUT/EXETYP",
"START" )
renameAnalysisCommand( "Analysis5", "Structural nonlinear", "Structural nonlinear" )
renameAnalysisCommandDetail( "Analysis5", "Structural nonlinear", "EXECUT(1)",
"gravity" )
addAnalysisCommandDetail( "Analysis5", "Structural nonlinear",
"EXECUT(1)/START/INITIA/STRESS" )
setAnalysisCommandDetail( "Analysis5", "Structural nonlinear",
"EXECUT(1)/START/INITIA/STRESS", True )
saveProject( )
setAnalysisCommandDetail( "Analysis5", "Structural nonlinear",
"EXECUT(1)/ITERAT/MAXITE", 50 )
setAnalysisCommandDetail( "Analysis5", "Structural nonlinear",
"EXECUT(1)/ITERAT/CONVER/DISPLA/NOCONV", "CONTIN" )
setAnalysisCommandDetail( "Analysis5", "Structural nonlinear",
"EXECUT(1)/ITERAT/CONVER/FORCE/NOCONV", "CONTIN" )
saveProject( )
renameAnalysisCommandDetail( "Analysis5", "Structural nonlinear", "EXECUT(1)",
"gravity" )
setAnalysisCommandDetail( "Analysis5", "Structural nonlinear", "EXECUT/EXETYP",
"LOAD" )
renameAnalysisCommandDetail( "Analysis5", "Structural nonlinear", "EXECUT(2)", "load" )
setAnalysisCommandDetail( "Analysis5", "Structural nonlinear",
"EXECUT(2)/LOAD/LOADNR", 2 )
setAnalysisCommandDetail( "Analysis5", "Structural nonlinear",
"EXECUT(2)/LOAD/STEPS/EXPLIC/SIZES", "1.00000(20) 0.5(10)" )
setAnalysisCommandDetail( "Analysis5", "Structural nonlinear",
"EXECUT(2)/LOAD/STEPS/EXPLIC/SIZES", "1.00000(20) 0.500000(10)" )
```

```
setAnalysisCommandDetail( "Analysis5", "Structural nonlinear", "EXECUT(2)/LOAD/STEPS/EXPLIC/SIZES", "1.00000(20) 0.500000(10)" )
setAnalysisCommandDetail( "Analysis5", "Structural nonlinear", "EXECUT(2)/ITERAT/MAXITE", 50 )
setAnalysisCommandDetail( "Analysis5", "Structural nonlinear", "OUTPUT(1)/SELTYP", "USER" )
addAnalysisCommandDetail( "Analysis5", "Structural nonlinear", "OUTPUT(1)/USER" )
addAnalysisCommandDetail( "Analysis5", "Structural nonlinear", "OUTPUT(1)/USER/DISPLA(1)/TOTAL/TRANSL/GLOBAL" )
addAnalysisCommandDetail( "Analysis5", "Structural nonlinear", "OUTPUT(1)/USER/STRAIN(1)/CRKSUM/GREEN/LOCAL" )
addAnalysisCommandDetail( "Analysis5", "Structural nonlinear", "OUTPUT(1)/USER/STRAIN(2)/CRKSUM/GREEN/GLOBAL" )
addAnalysisCommandDetail( "Analysis5", "Structural nonlinear", "OUTPUT(1)/USER/STRAIN(3)/CRACK/GREEN" )
addAnalysisCommandDetail( "Analysis5", "Structural nonlinear", "OUTPUT(1)/USER/STRAIN(4)/CRKWDT/GREEN/LOCAL" )
addAnalysisCommandDetail( "Analysis5", "Structural nonlinear", "OUTPUT(1)/USER/STRAIN(5)/CRKWDT/GREEN/GLOBAL" )
addAnalysisCommandDetail( "Analysis5", "Structural nonlinear", "OUTPUT(1)/USER/STRESS(1)/TOTAL/FORCE/LOCAL" )
saveProject( )
runSolver( "Analysis5" )
Command sonsole of UHPC T beam shape
newProject( "UHPCbeam", 100 )
setModelAnalysisAspects( [ "STRUCT" ] )
setModelDimension( "3D" )
setDefaultMeshOrder( "QUADRATIC" )
setDefaultMesherType( "HEXQUAD" )
setDefaultMidSideNodeLocation( "LINEAR" )
setModelAnalysisAspects( [ "STRUCT" ] )
setModelDimension( "3D" )
setDefaultMeshOrder( "QUADRATIC" )
setDefaultMesherType( "HEXQUAD" )
setDefaultMidSideNodeLocation( "LINEAR" )
saveProject( )
createSheet( "Sheet 1", [[ 0, 0, 0 ],[ 1, 0, 0 ],[ 1, 0, 1 ],[ 0, 0, 1 ]] )
saveProject( )
createSheet( "Sheet 2", [[ 0, 0, 0 ],[ 0.4, 0, 0 ],[ 0.4, 0, 0.8 ],[ 0, 0, 0.8 ]] )
saveProject( )
mirror( [ "Sheet 2" ], [ 0.5, 0, 0 ], [ True, False, False ], True )
```

```
saveProject( )
subtract( "Sheet 1", [ "Sheet 2", "Sheet 3" ], False, True )
saveProject( )
extrudeProfile( [ "Sheet 1" ], [ 0, 10, 0 ] )
renameShape( "Sheet 1", "UHPCbeam" )
addMaterial( "UHPC", "CONCR", "TSCR", [] )
setParameter( MATERIAL, "UHPC", "LINEAR/ELASTI/YOUNG", 4.35e+10 )
setParameter( MATERIAL, "UHPC", "LINEAR/ELASTI/POISON", 0.15 )
setParameter( MATERIAL, "UHPC", "LINEAR/MASS/DENSIT", 2500 )
setParameter( MATERIAL, "UHPC", "MODTYP/TOTCRK", "ROTATE" )
setParameter( MATERIAL, "UHPC", "TENSIL/TENCRV", "FRCCON" )
setParameter( MATERIAL, "UHPC", "TENSIL/FRCTYP", "STRAIN" )
setParameter( MATERIAL, "UHPC", "TENSIL/FRCEPS", [ 5500000, 2800000, 0.0024, 
4200000, 0.0038, 0.004 ] )
setParameter( MATERIAL, "UHPC", "COMPRS/COMCRV", "EN1992" )
setParameter( MATERIAL, "UHPC", "COMPRS/COMSTR", 1.2e+08 )
setParameter( MATERIAL, "UHPC", "COMPRS/EPSC1", 0.0044 )
setParameter( MATERIAL, "UHPC", "COMPRS/EPSCU", 0.007 )
setParameter( MATERIAL, "UHPC", "COMPRS/REDUCT/REDCRV", "JSCE12" )
clearReinforcementAspects( [ "UHPCbeam" ] )
setElementClassType( SHAPE, [ "UHPCbeam" ], "STRSOL" )
assignMaterial( "UHPC", SHAPE, [ "UHPCbeam" ] )
resetGeometry( SHAPE, [ "UHPCbeam" ] )
resetElementData( SHAPE, [ "UHPCbeam" ] )
saveProject( )
createLine( "bar1", [ 0.5, 0, 0.2 ], [ 0.5, 10, 0.2 ] )
addMaterial( "bar", "REINFO", "VMISES", [] )
setParameter( MATERIAL, "bar", "LINEAR/ELASTI/YOUNG", 2.1e+11 )
setParameter( MATERIAL, "bar", "PLASTI/HARDI1/YLDSTR", 4.4e+08 )
addGeometry( "Element geometry 1", "RELINE", "REBAR", [] )
rename( GEOMET, "Element geometry 1", "bar" )
setParameter( GEOMET, "bar", "REIEMB/CROSSE", 0.000113097 )
setParameter( GEOMET, "bar", "REIEMB/CROSSE", 0.000113097 )
setParameter( GEOMET, "bar", "REIEMB/CROSSE", 0.000113097 )
setReinforcementAspects( [ "bar1"] )
assignMaterial( "bar", SHAPE, [ "bar1" ] )
assignGeometry( "bar", SHAPE, [ "bar1" ] )
resetElementData( SHAPE, [ "bar1" ] )
setReinforcementDiscretization( [ "bar1" ], "SECTION" )
saveProject( )
createVertex( "Vertex 1", [ 0.5, 2, 1.5 ] )
```

```
mirror( [ "Vertex 1" ], [ 0, 5, 0 ], [ False, True, False ], True )
saveProject(  )
projection( SHAPEFACE, "UHPCbeam", [[ 0.573573, 5.73573, 1 ]], [ "Vertex 1", "Vertex 2" ],
[ 0, 0, -1 ], True )
removeShape( [ "Vertex 1", "Vertex 2" ] )
saveProject(  )
createLine( "Line 1", [ 0, 0.2, -1 ], [ 1, 0.2, -1 ] )
mirror( [ "Line 1" ], [ 0, 5, 0 ], [ False, True, False ], True )
projection( SHAPEFACE, "UHPCbeam", [[ 0.426427, 5.73573, 0 ]], [ "Line 1", "Line 2" ], [ 0,
0, 1 ], True )
removeShape( [ "Line 1", "Line 2" ] )
addSet( GEOMETRYLOADSET, "Geometry load case 1" )
rename( GEOMETRYLOADSET, "Geometry load case 1", "load" )
createPointLoad( "load", "load" )
setParameter( GEOMETRYLOAD, "load", "FORCE/VALUE", -20000 )
setParameter( GEOMETRYLOAD, "load", "FORCE/DIRECT", 3 )
attach( GEOMETRYLOAD, "load", "UHPCbeam", [[ 0.5, 2, 1 ],[ 0.5, 8, 1 ]] )
saveProject(  )
addSet( GEOMETRYLOADSET, "gravity" )
createModelLoad( "gravity", "gravity" )
setDefaultGeometryLoadCombinations(  )
setGeometryLoadCombinationFactor( "Geometry load combination 1", "load", 1 )
addSet( GEOMETRYSUPPORTSET, "co1" )
createLineSupport( "co1", "co1" )
setParameter( GEOMETRYSUPPORT, "co1", "AXES", [ 1, 2 ] )
setParameter( GEOMETRYSUPPORT, "co1", "TRANSL", [ 1, 1, 1 ] )
setParameter( GEOMETRYSUPPORT, "co1", "ROTATI", [ 0, 0, 0 ] )
attach( GEOMETRYSUPPORT, "co1", "UHPCbeam", [[ 0.5, 0.2, 0 ]] )
saveProject(  )
addSet( GEOMETRYSUPPORTSET, "Geometry support set 2" )
rename( GEOMETRYSUPPORTSET, "Geometry support set 2", "co2" )
createLineSupport( "co2", "co2" )
setParameter( GEOMETRYSUPPORT, "co2", "AXES", [ 1, 2 ] )
setParameter( GEOMETRYSUPPORT, "co2", "TRANSL", [ 1, 0, 1 ] )
setParameter( GEOMETRYSUPPORT, "co2", "ROTATI", [ 0, 0, 0 ] )
attach( GEOMETRYSUPPORT, "co2", "UHPCbeam", [[ 0.5, 9.8, 0 ]] )
setElementSize( [ "UHPCbeam" ], 0.1, -1, True )
setMesherType( [ "UHPCbeam" ], "HEXQUAD" )
setMidSideNodeLocation( [ "UHPCbeam" ], "LINEAR" )
saveProject(  )
generateMesh( [] )
addAnalysis( "Analysis5" )
```

```
addAnalysisCommand( "Analysis5", "NONLIN", "Structural nonlinear" )
renameAnalysis( "Analysis5", "Analysis5" )
removeAnalysisCommandDetail( "Analysis5", "Structural nonlinear", "EXECUT(1)" )
setAnalysisCommandDetail( "Analysis5", "Structural nonlinear", "EXECUT/EXETYP", "START" )
renameAnalysisCommand( "Analysis5", "Structural nonlinear", "Structural nonlinear" )
renameAnalysisCommandDetail( "Analysis5", "Structural nonlinear", "EXECUT(1)", "gravity" )
addAnalysisCommandDetail( "Analysis5", "Structural nonlinear", "EXECUT(1)/START/INITIA/STRESS" )
setAnalysisCommandDetail( "Analysis5", "Structural nonlinear", "EXECUT(1)/START/INITIA/STRESS", True )
saveProject( )
setAnalysisCommandDetail( "Analysis5", "Structural nonlinear", "EXECUT(1)/ITERAT/MAXITE", 50 )
setAnalysisCommandDetail( "Analysis5", "Structural nonlinear", "EXECUT(1)/ITERAT/CONVER/DISPLA/NOCONV", "CONTIN" )
setAnalysisCommandDetail( "Analysis5", "Structural nonlinear", "EXECUT(1)/ITERAT/CONVER/FORCE/NOCONV", "CONTIN" )
saveProject( )
renameAnalysisCommandDetail( "Analysis5", "Structural nonlinear", "EXECUT(1)", "gravity" )
setAnalysisCommandDetail( "Analysis5", "Structural nonlinear", "EXECUT/EXETYP", "LOAD" )
renameAnalysisCommandDetail( "Analysis5", "Structural nonlinear", "EXECUT(2)", "load" )
setAnalysisCommandDetail( "Analysis5", "Structural nonlinear", "EXECUT(2)/LOAD/LOADNR", 2 )
setAnalysisCommandDetail( "Analysis5", "Structural nonlinear", "EXECUT(2)/LOAD/STEPS/EXPLIC/SIZES", "1.00000(20) 0.5(10)" )
setAnalysisCommandDetail( "Analysis5", "Structural nonlinear", "EXECUT(2)/LOAD/STEPS/EXPLIC/SIZES", "1.00000(20) 0.500000(10)" )
setAnalysisCommandDetail( "Analysis5", "Structural nonlinear", "EXECUT(2)/LOAD/STEPS/EXPLIC/SIZES", "1.00000(20) 0.500000(10)" )
setAnalysisCommandDetail( "Analysis5", "Structural nonlinear", "EXECUT(2)/ITERAT/MAXITE", 50 )
setAnalysisCommandDetail( "Analysis5", "Structural nonlinear", "OUTPUT(1)/SELTYP", "USER" )
addAnalysisCommandDetail( "Analysis5", "Structural nonlinear", "OUTPUT(1)/USER" )
addAnalysisCommandDetail( "Analysis5", "Structural nonlinear", "OUTPUT(1)/USER/DISPLA(1)/TOTAL/TRANSL/GLOBAL" )
addAnalysisCommandDetail( "Analysis5", "Structural nonlinear",
```

```
"OUTPUT(1)/USER/STRAIN(1)/CRKSUM/GREEN/LOCAL" )
  addAnalysisCommandDetail(   "Analysis5",   "Structural   nonlinear",
"OUTPUT(1)/USER/STRAIN(2)/CRKSUM/GREEN/GLOBAL" )
  addAnalysisCommandDetail(   "Analysis5",   "Structural   nonlinear",
"OUTPUT(1)/USER/STRAIN(3)/CRACK/GREEN" )
  addAnalysisCommandDetail(   "Analysis5",   "Structural   nonlinear",
"OUTPUT(1)/USER/STRAIN(4)/CRKWDT/GREEN/LOCAL" )
  addAnalysisCommandDetail(   "Analysis5",   "Structural   nonlinear",
"OUTPUT(1)/USER/STRAIN(5)/CRKWDT/GREEN/GLOBAL" )
  addAnalysisCommandDetail(   "Analysis5",   "Structural   nonlinear",
"OUTPUT(1)/USER/STRESS(1)/TOTAL/FORCE/LOCAL" )
  saveProject( )
  runSolver( "Analysis5" )
```

3.6 Hysteresis Analysis of Shear Wall

This numerical case is a hysteresis analysis of shear wall, which can be classified as beam at the top, shear wall in the middle site and bearing beam at the bottom site. The length, height and thickness values of loading beam are 2, 0.4 and 0.6 m, respectively, with an internal FRP tendon and vertical distributed load 30 kN/m on it, while the length, height and thickness values of bearing beam are 2.6, 0.4 and 0.6 m, respectively. The total length, height and thickness of shear wall are 1.8, 2 m and 0.4 m, respectively, where the 2D size is displayed in Fig. 3.236. There are reinforcement gird and stirrups in the model and the von Mises plasticity model is applied for the reinforcement. Concrete plane stress elements are applied for the numerical model. Hordijk tension softening model is applied for tensile behaviors while Maekawa–Fukuura model is applied for compressive mechanic behavior. Cross-section area of single steel bar is 157 mm^2 and cross-section area of stirrups in two layers is 100.53 mm^2. Modeling parameters in every section is displayed in Table 3.6.

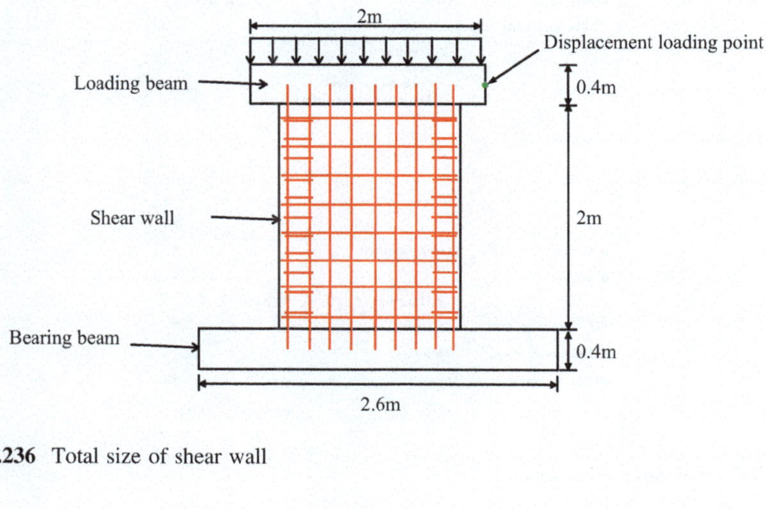

Fig. 3.236 Total size of shear wall

Table 3.6 Parameters of every section

Shear wall	Concrete elastic modulus	3.7×10^{10} N/m^2
	Concrete tensile strength	2.8×10^{6} N/m^2
	Concrete compressive strength	4.5×10^{7} N/m^2
	Shear retention coefficient	0.1
Loading and bearing beam	Concrete elastic modulus	3.25×10^{10} N/m^2
FRP tendon	Concrete elastic modulus	1.36×10^{11} N/m^2
	Concrete tensile strength	1.2×10^{9} N/m^2
Reinforcement steel	Concrete elastic modulus	2.1×10^{11} N/m^2
	Yield stress	4.4×10^{8} N/m^2

Essentials of learning

(1) *Learning to specify the parameters in Hordijk tension softening model.*
(2) *Learning to specify the compressive parameters in Maekawa-Fukuura model.*
(3) *Specification of constitutive parameters in FRP*
(4) *Attaching method of indirect constraints on point displacement and prescribed deformation*
(5) *Attaching method of cyclic displacement loading step.*

Above all, start DIANA 10.1 manipulation interface, and the graphical user interface shown in Fig. 3.237 ejects. A document with the suffix name .dpf and the project name shear wall is generated in the directory of computer G-disk zone. 2D model is selected and the maximum size of model file is 10 m, representing the scope of the whole graphical visualization zone ranging from –5 to 5 m in all directions of the coordinate system. **Default mesher type** is chosen as **Hexa/Quad** element while the **Default mesh order** in this case is **Quadratic**. Determination of mid-side node location is linear interpolation.

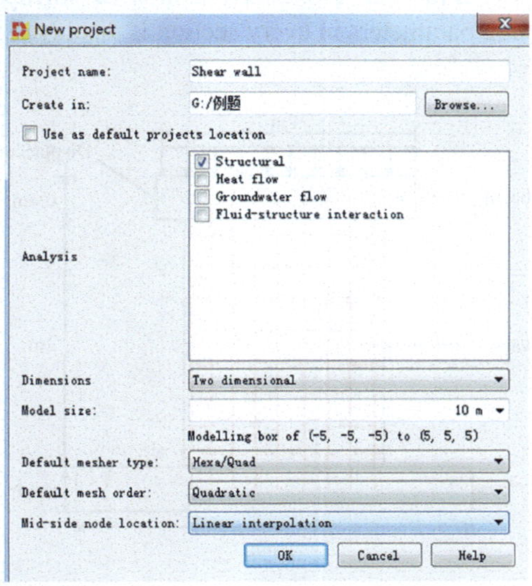

Fig. 3.237 New project interface

Click shortcut icon **Adds a sheet** and coordinate values of every point are the input to create geometric model of bearing beam at the bottom with the name of **support**; coordinate values of every point are displayed in Fig. 3.238; then click OK button to generate the geometric model of bearing beam.

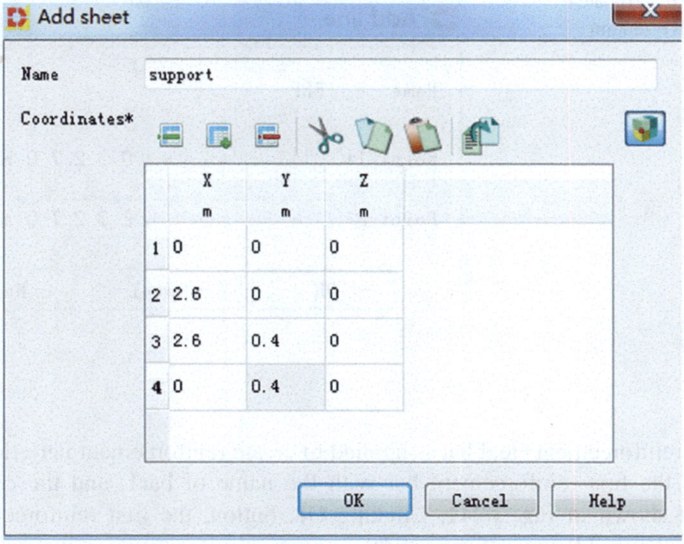

Fig. 3.238 Coordinate values of bearing support beam

Input coordinate values (0.4, 0.4), (2.2, 0.4), (2.2, 2.6) and (0.4, 2.6) to generate the geometric model of shear wall, then again input coordinate values (0.3, 2.6), (2.3, 2.6), (2.3, 3) and (0.3, 3), respectively; click OK button, the geometric model of loading beam at the top is generated, and the whole geometric model is displayed in Fig. 3.239.

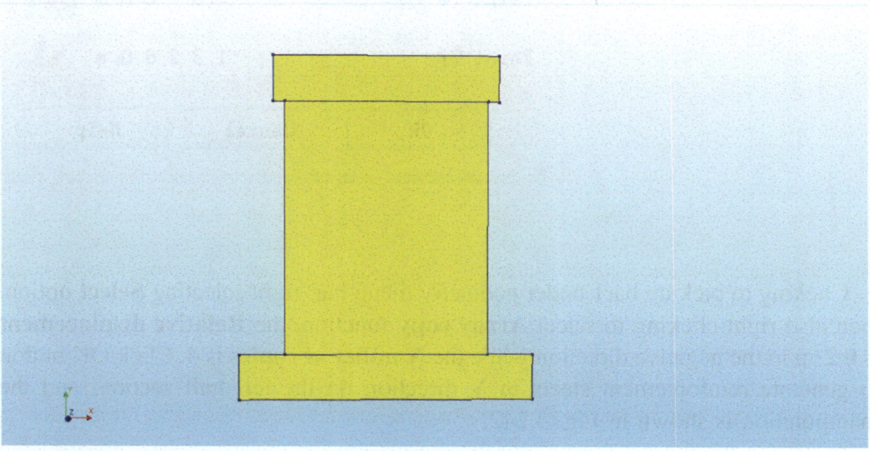

Fig. 3.239 Geometric model of shear wall loading integral

Add shortcut icon **Adds a line**  to create geometric coordinate values of FRP tendon, and the manipulation interface is shown in Fig. 3.240.

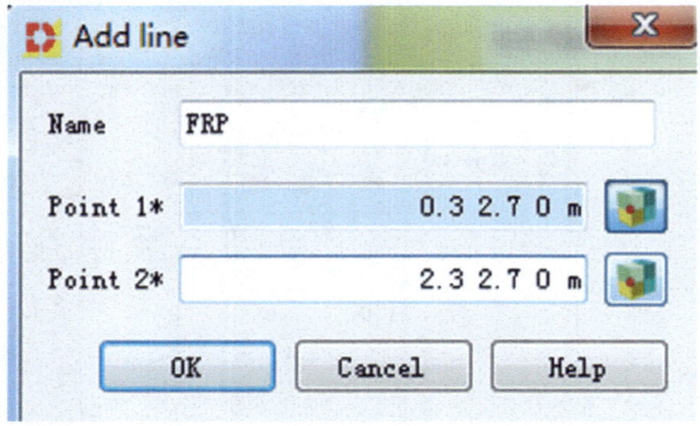

Fig. 3.240 Coordinate values of FRP tendon

Single reinforcement steel bar is applied to create reinforcement net grid. Above all, create the first reinforcement bar with the name of bar1, and the coordinate values are shown in Fig. 3.241. Clicking OK button, the first reinforcement bar appears in the middle site of the model.

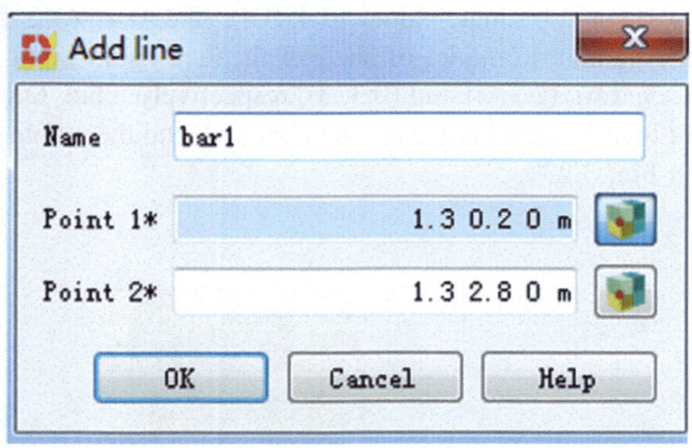

Fig. 3.241 Geometric coordinate value of bar1

Clicking to pick up bar1 under geometry menu bar, right selecting **Select** option, then also right-clicking to select **Array copy** function, the **Relative displacement** is 0.2 m in the negative direction while the **Number of copies** is 4. Click OK button to generate reinforcement steels in Y direction on the left half section, and the manipulation is shown in Fig. 3.242.

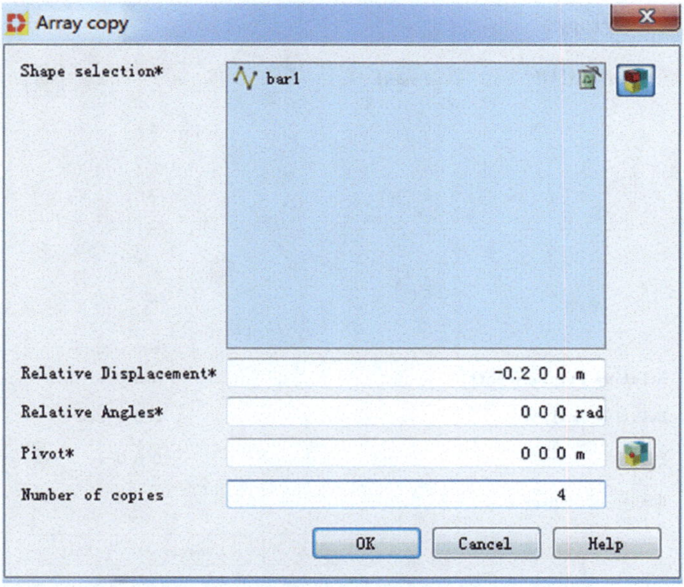

Fig. 3.242 Reinforcement steels in Y direction in the left half section

Following the same way and relative displacement also with the number of copies 4 to generate reinforcement steels in Y direction on the right half section, we can create the first transverse steel bar along the global X direction with the name of barx1, and the coordinate values are shown in Fig. 3.243.

Fig. 3.243 Geometric coordinate value of barx1

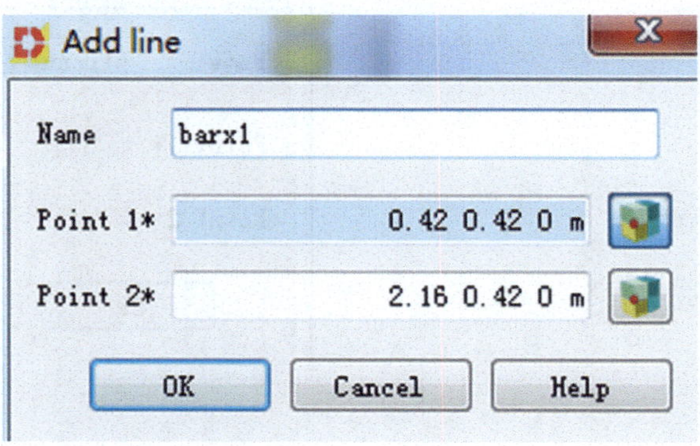

Applying the **Array copy** function again to generate distributed steel bars along the X direction with the **Relative displacement** 0.3 m in the positive Y direction, the **Number of copies** is 7 (see Fig. 3.244).

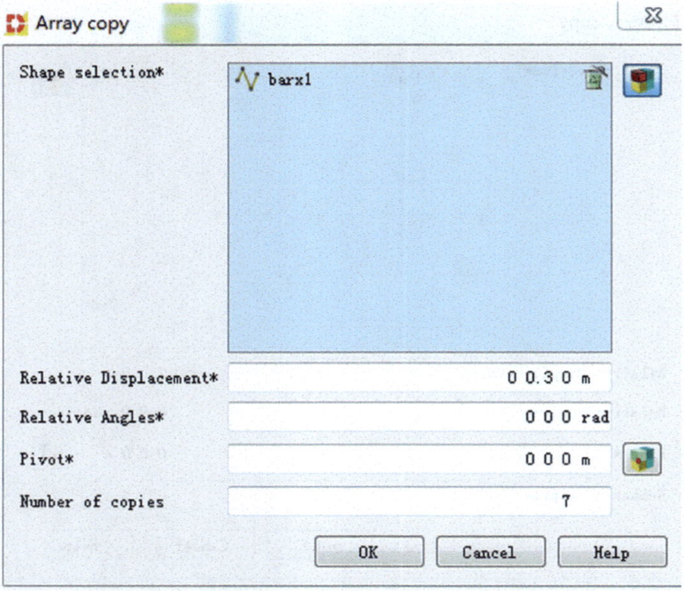

Fig. 3.244 Manipulation of **Array copy**

Create geometric model of stirrups and input coordinate values of first stirrups as in Fig. 3.245 to generate the first stirrup, and the name is **stirrup**.

Fig. 3.245 Coordinate values of the first stirrup

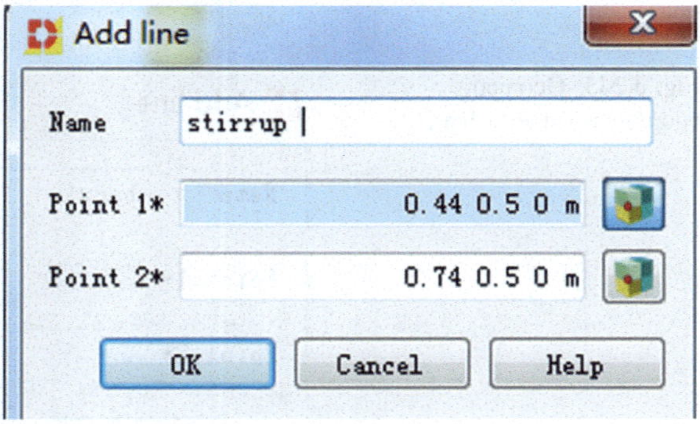

As Fig. 3.246 shows, applying **Array copy** function again generates stirrups on the left side. **Relative displacement** is 0.2 m in the positive Y direction with the **Number of copies** 10. Click OK button to generate geometric model of stirrups on the left side.

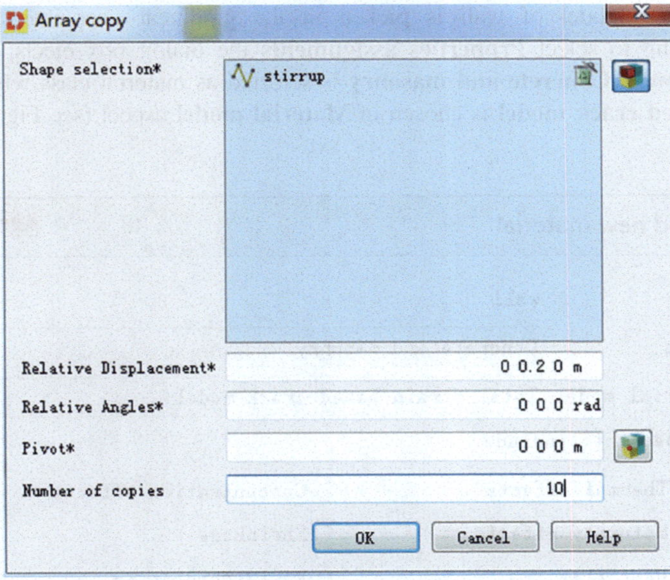

Fig. 3.246 Manipulation interface of Array copy function for stirrups on the left side

Applying the same method to generate stirrups on the right side in the positive X direction, the **Relative displacement** and **Number of copies** are 1.4 m and 1, respectively (see Fig. 3.247).

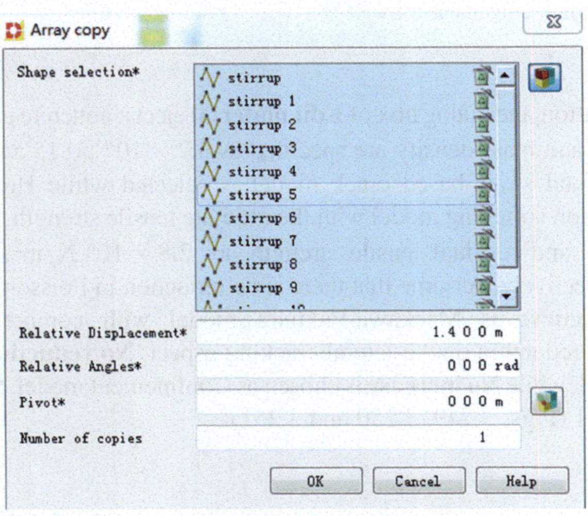

Fig. 3.247 Array copy function for stirrups on the right side

Geometric model of wall is picked in the graphical user interface, then right-clicking to select **Properties assignments** the dialog box ejects, which is named as **wall**. **Concrete and masonry** is selected as material class while **Total strain based crack model** is chosen in **Material model** aspect (see Fig. 3.248).

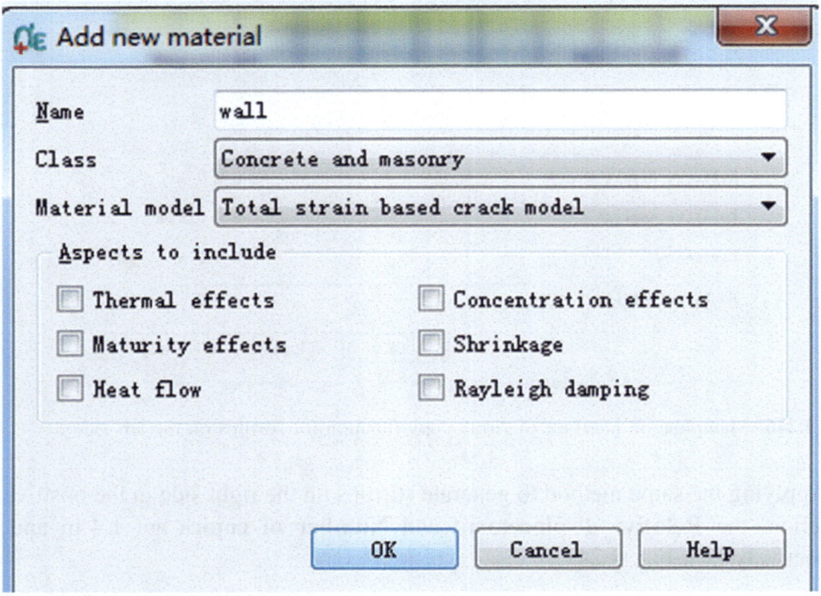

Fig. 3.248 Properties assignments of wall

Click OK button, the dialog box of **Edit material** ejects; concrete elastic modulus, Poisson's ratio and mass density are specified as 3.7×10^{10}, 0.15 and 2500 kg/m^3, respectively. Total strain-based crack model is selected while Hordijk model is selected as tension softening model with the ultimate tensile strength. Mode-I tensile fracture energy and residual tensile strength are 2.8×10^6 N/m^2, 200 N/m and 100 N/m^2, respectively. Assume that there is no reduction in Poisson's ratio and the **Compression curve** is Maekawa-Fukuura model with compressive strength 45 MPa. In the reduction due to lateral cracking aspect, **No reduction** is chosen as reduction model, while **No increase** is chosen as Confinement model. Shear retention coefficient is 0.1 (Figs. 3.249, 3.250 and 3.251).

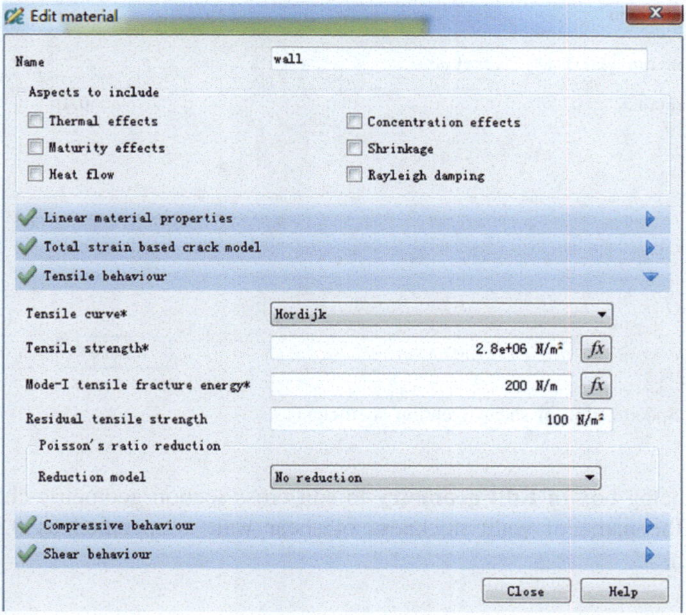

Fig. 3.249 Specification of Hordijk tension softening model

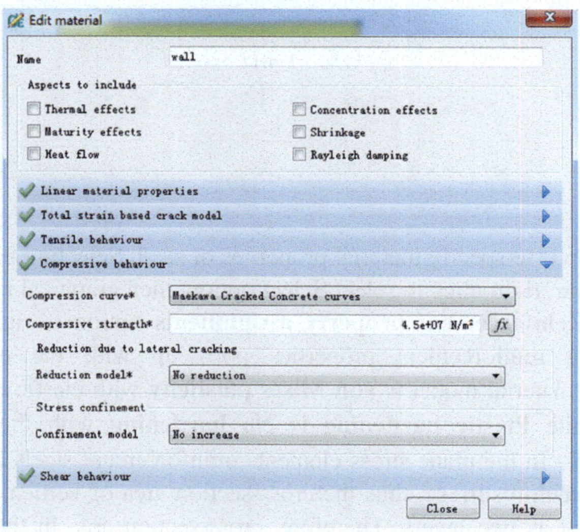

Fig. 3.250 Specification of compressive Maekawa–Fukuura model

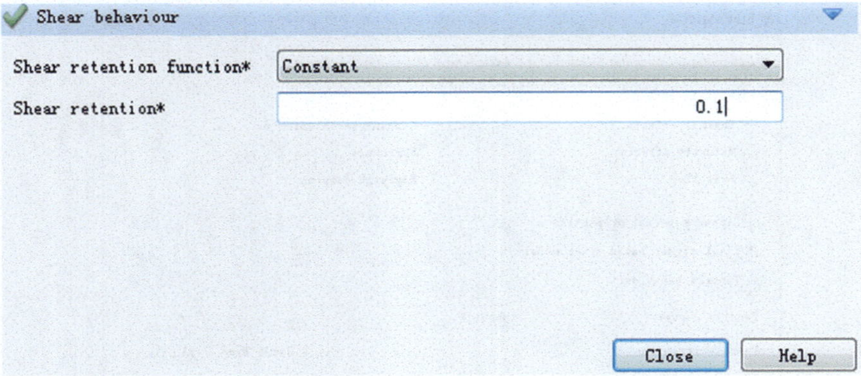

Fig. 3.251 Specification of shear retention coefficient

Open dialog box of **Edit geometry** to edit cross-section geometric characteristics with the name of wall; thickness of shear wall is specified as 0.4 m (see Fig. 3.252).

Fig. 3.252 Specification of cross-section geometric characteristics

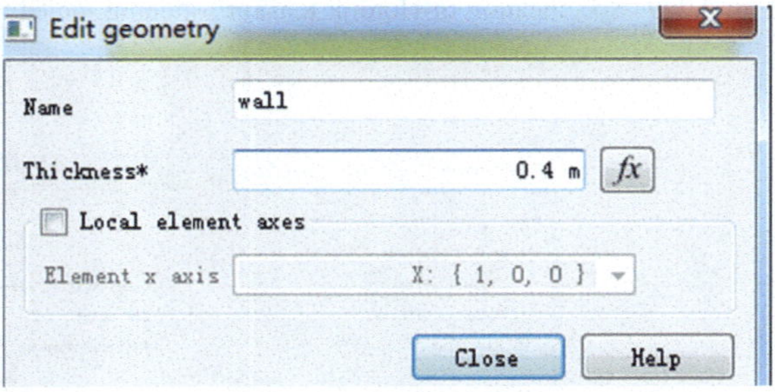

The next step is to specify material properties of steel. Selecting bar1-bar9 under the Geometry bar, right-click to select **Select** option, then graphical user interface is right-clicked. **Reinforcement property assignments** option is right-clicked and editing box of reinforcement properties pops up with the name of **bar**. Reinforcement material model is **von Mises plasticity** with elastic modulus 2.1×10^{11} N/m^2 while **Plastic hardening** is **No hardening** with the **Yield stress** 4.4×10^8 N/m^2. In the plane stress elements, reinforcement steels are modeled at the geometric neutral surface, thus the cross-section area of vertical reinforcement bar is calculated as two layers. Therefore, cross-section area in the cross-section geometric properties editing box is defined as 157 mm^2 according to the calculation $2 \times \frac{1}{4}\pi \times 10^2 \approx 157$ mm^2. Manipulation is shown as following figures: Figs. 3.253, 3.254 and 3.255.

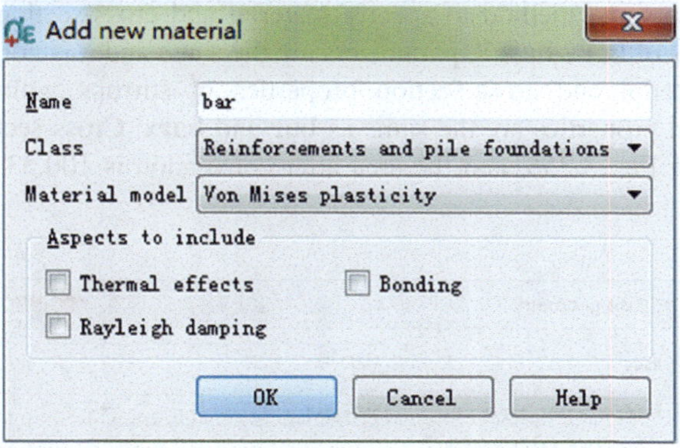

Fig. 3.253 Specification of von Mises plasticity model

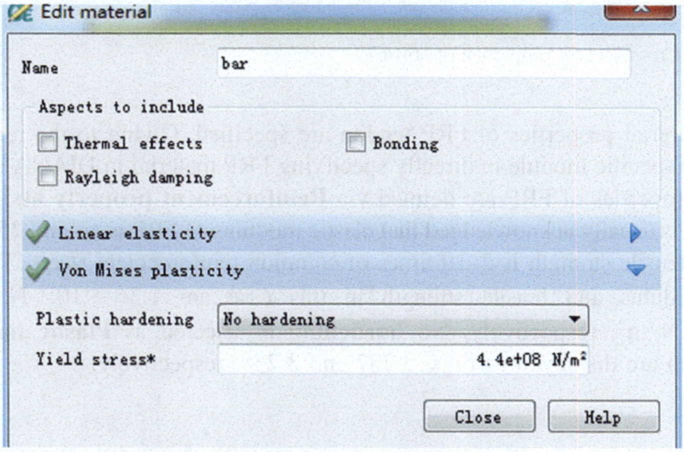

Fig. 3.254 Constitutive parameters and selections of von Mises plasticity model

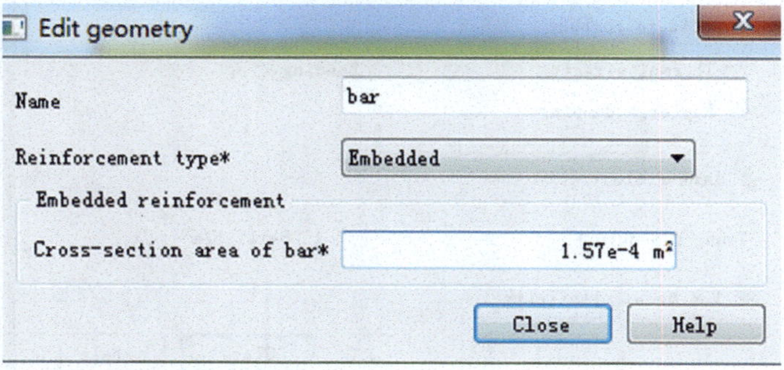

Fig. 3.255 Cross-section geometric properties of bar

Applying the same method to specify material and cross-section properties of transverse steels of **barx-barx8**, parameters are the same and it is not repeated here.

Specify material and cross-section properties of stirrups with the name of stirrup. Material properties are the same as **bar** and **barx**. Cross-section properties are displayed in Fig. 3.256, and the area after conversion is 100.53 mm^2.

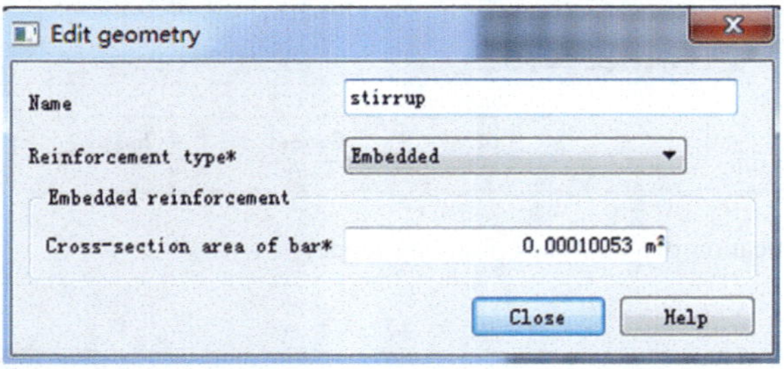

Fig. 3.256 Cross-section properties of stirrups

The material properties of FRP tendon are specified. Owing to the reason that there is no specific module in directly specifying FRP material in DIANA software, material properties of FRP are defined via **Reinforcement property assignments** module. It is usually acknowledged that elastic modulus of FRP is around 25%–75%, while the tensile strength is 2–10 times of common reinforcement steels. Therefore, elastic modulus and tensile strength in this case are $1.36 \times 10^{11}\ \mathrm{N/m^2}$ and $1.2 \times 10^{9}\ \mathrm{N/m^2}$, respectively. **No hardening** is selected as **Plastic hardening** type, which are displayed as Figs. 3.257 and 3.258, respectively.

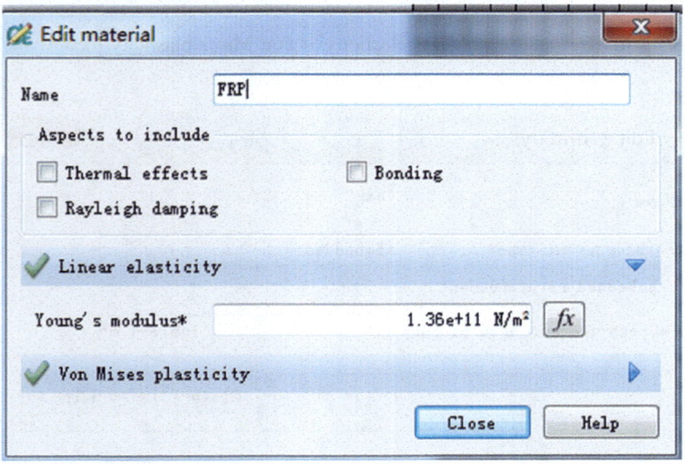

Fig. 3.257 Elastic modulus specification of FRP tendon

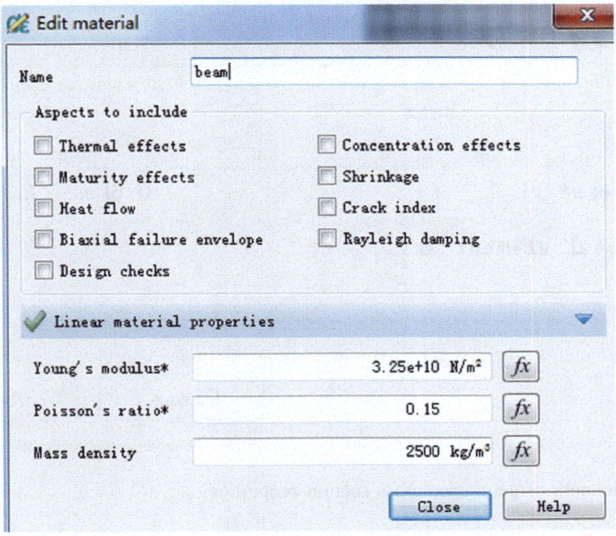

Fig. 3.258 Yield stress of FRP tendon

Specifying constitutive material model of upper loading beam and bearing beam at the bottom, both material models are elastic without considering the impacts of cracking or plastic behaviors in order to avoid accumulated cracking damage under cyclic back-to-back displacement. Elastic modulus, Poisson's ratio and mass density of both beams are 3.25×10^{10} N/m^2, 0.15 and 2500 kg/m^3, respectively, with the thickness value 0.6 m. The concrete constitutive parameters are displayed in Figs. 3.259 and 3.260.

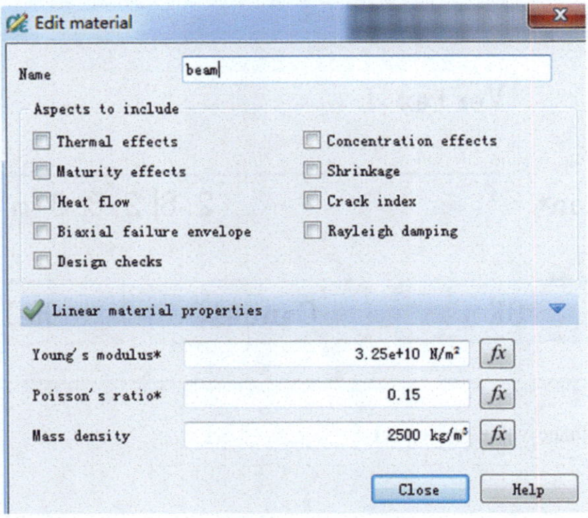

Fig. 3.259 Constitutive specification of upper loading beam

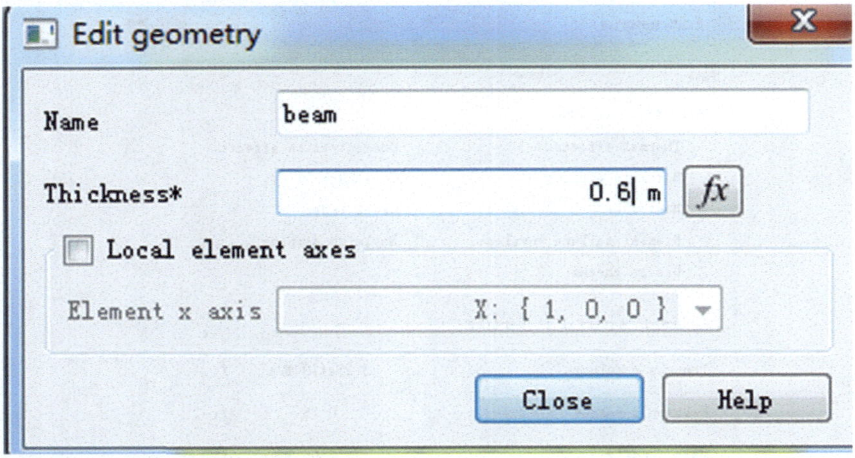

Fig. 3.260 Thickness of geometric cross-section properties

The following procedure is to create displacement acting point. The initial work is to project and imprint an acting point so that horizontal displacement can be acted at this point. Similar to previous cases, a point named Vertex 1 is created through adding coordinate values as input. The direction of imprint projection is in the negative X direction under global coordinate system, and specific manipulation is displayed in Figs. 3.261 and 3.262, respectively.

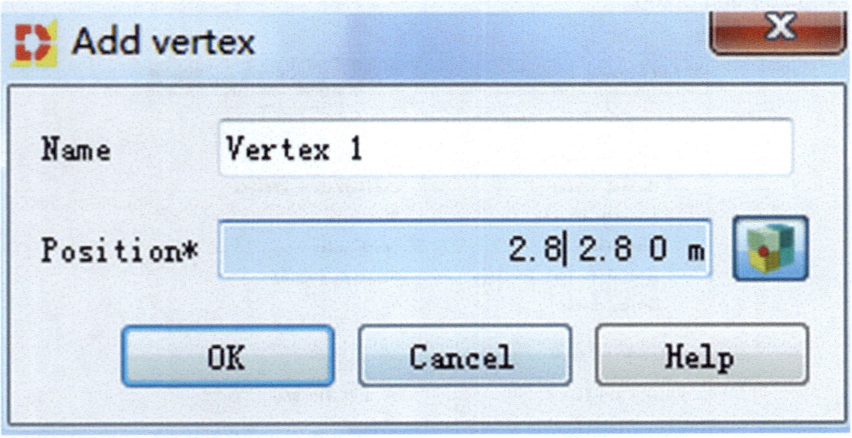

Fig. 3.261 Coordinate value of Vertex1

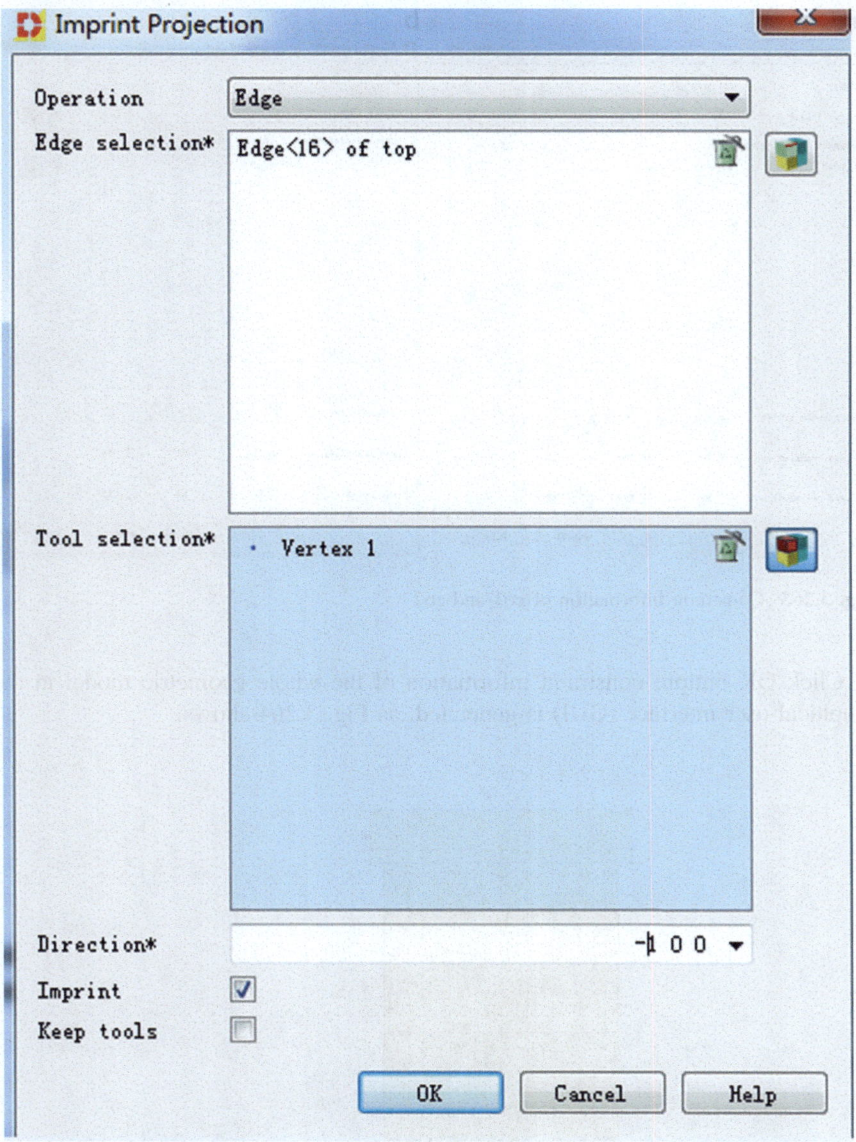

Fig. 3.262 Manipulation interface of imprint projection

Attaching translation constraints in both X and Y direction on the bottom edge of bearing beam with the name of co1, owing to displacement action must applied on the constraint point, point constraint co2 in X direction is also applied on the displacement loading point immediately when constraint co1 is attached. Information of constraint attachment is shown as Fig. 3.263a, b respectively.

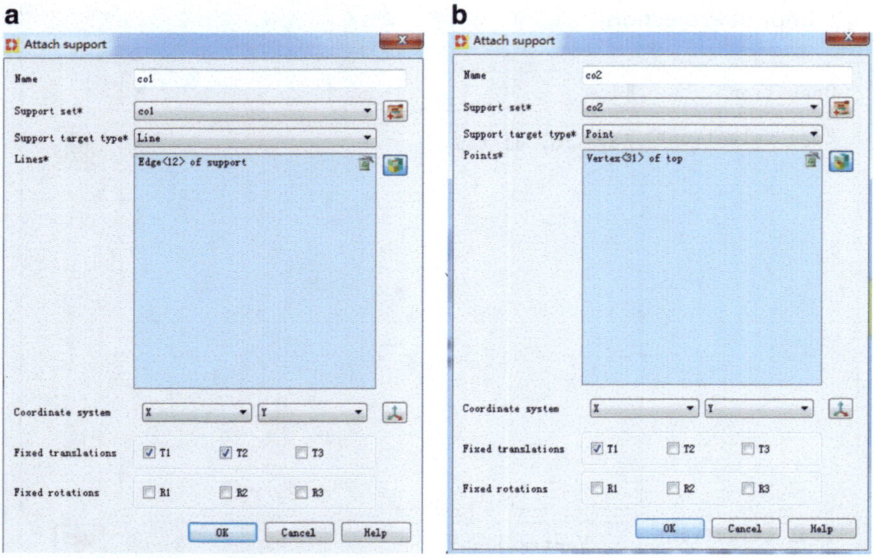

Fig. 3.263 Constraint information of co1 and co2

Click OK button; constraint information of the whole geometric model in the graphical user interface (GUI) is generated, as Fig. 3.264 shows.

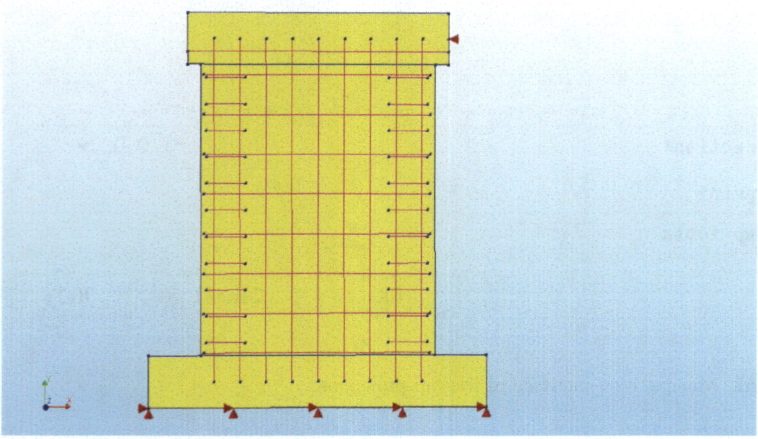

Fig. 3.264 Constraint information of the whole geometric model

Click icon **Define a global load** in the **Load** bar under the model tree **Geometry** to attach gravity, then **Load target type** of line load is attached at the top edge of upper loading beam with the distributed value 30 kN/m in the vertical negative Y direction, where the name is **load** (see Fig. 3.265).

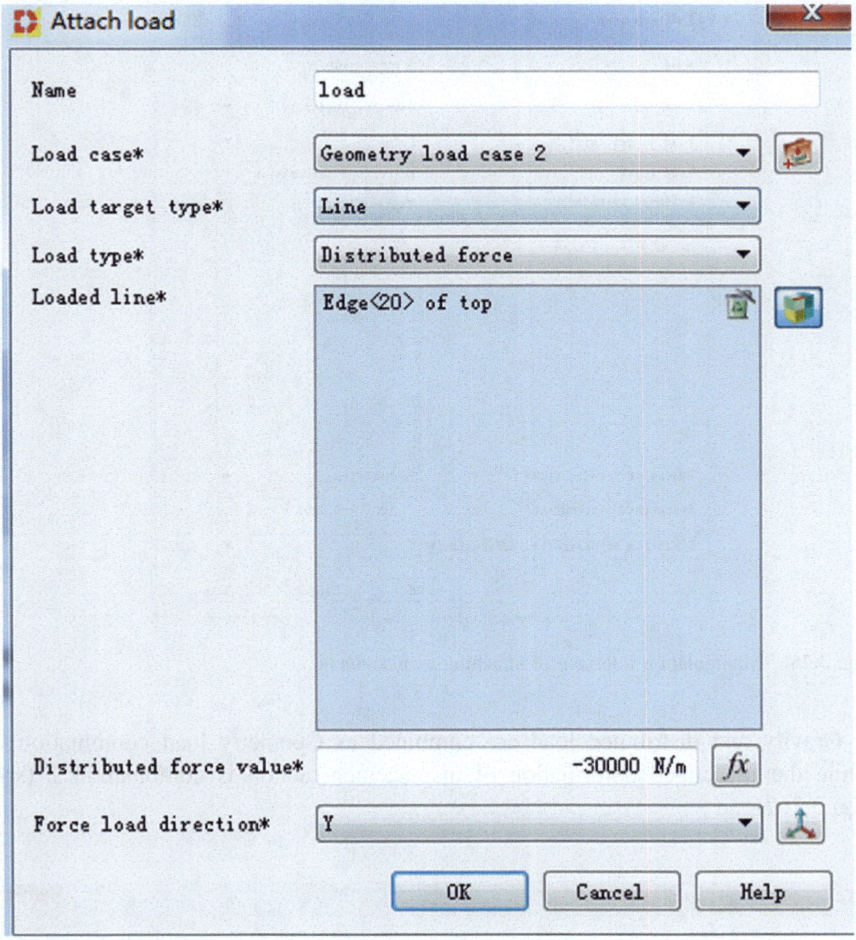

Fig. 3.265 Attaching line load

The next key step is to define displacement load. Point displacement is attached to the predefined horizontal point constraint co2 with the name of load case **displacement**. Point is chosen as **Load target typ**e while Load type is **Prescribed deformation**. The value of prescribed deformation is 0.5 mm in the negative X direction (see Fig. 3.266).

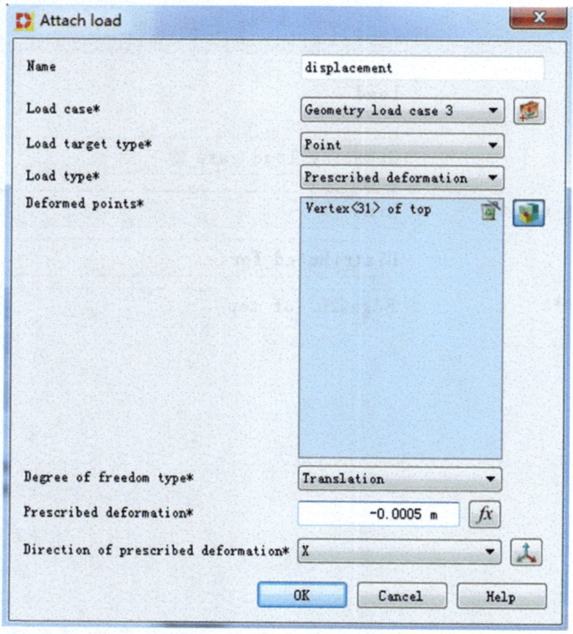

Fig. 3.266 Manipulation interface of attaching displacement

Gravity and distributed load are combined as Geometry load combination 1 while the prescribed deformation of displacement action is combination 2 (see Fig. 3.267).

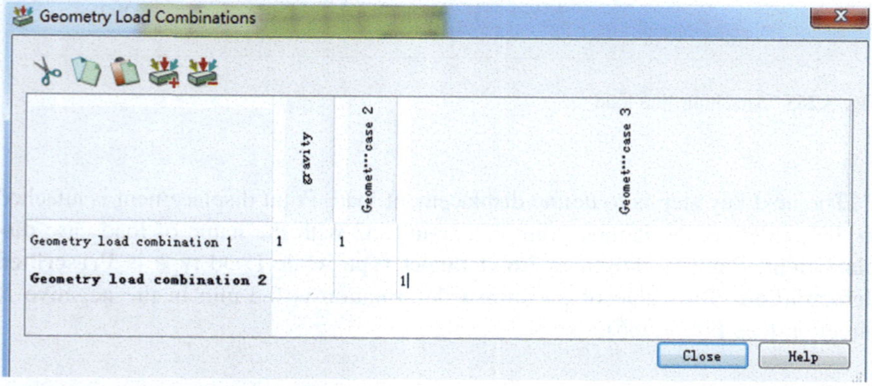

Fig. 3.267 Geometry load combination

After adding geometry load combinations, selecting the whole model and clicking shortcut icon **Set mesh properties of a shape**, dialog box of mesh interface, as Fig. 3.268 shows, ejects. **Operation** is **Shape** while **Element size** is chosen as **Seeding method** with **Desired size** 0.1 m. **Mesher type** is quadrilateral or hexahedron (**Hexa/Quad**). **Linear interpolation** is a way of determining mid-side node location.

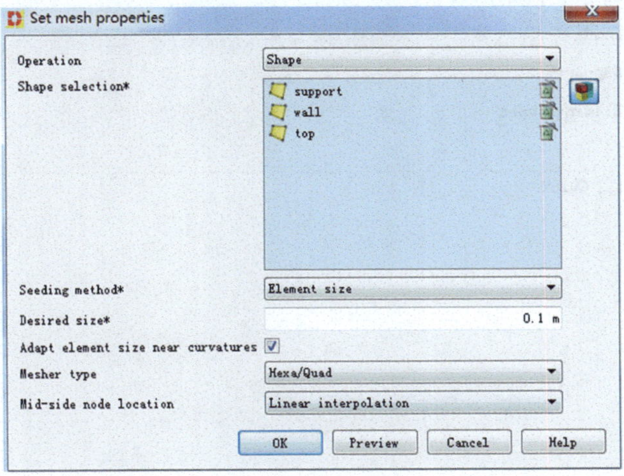

Fig. 3.268 Mesh setting interface

Click shortcut icon **Generate mesh of a shape**; the meshed numerical model is shown in Fig. 3.269.

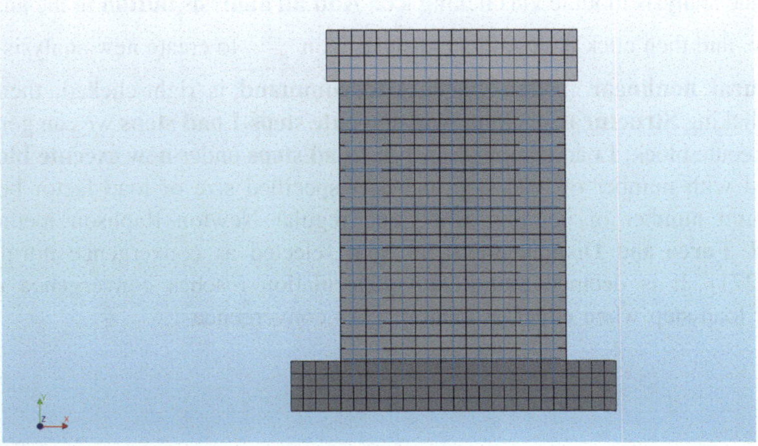

Fig. 3.269 Meshed numerical model

After meshing procedure is completed, click **Element types** bar under the mesh module. It can be found that CQ16M is the required structural element type (see Fig. 3.270).

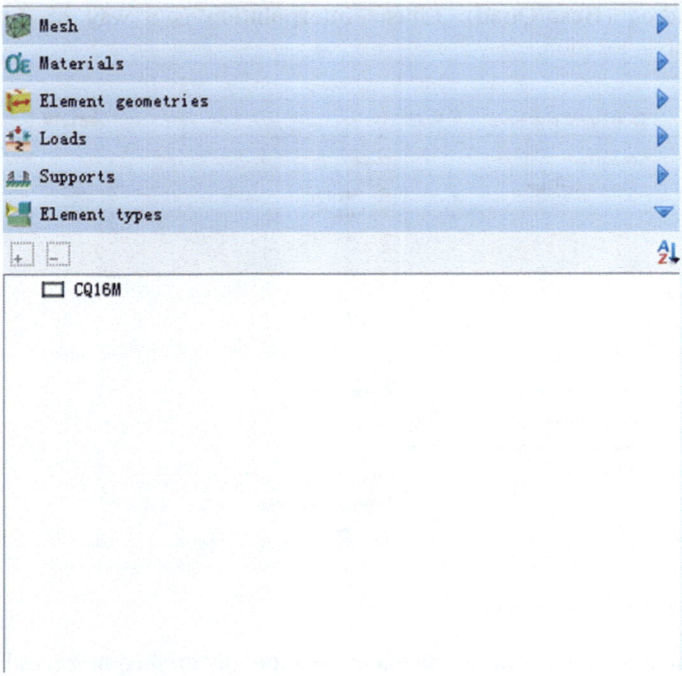

Fig. 3.270 Element type CQ16M

Create analysis module via clicking icon **Add an analysis button** in the analysis module, and then click icon Add an analysis icon , to create new analysis case. **Structural nonlinear** module under **Add command** is right-clicked, then still right-clicking **Structural**, clicking **Add-Execute steps-Load steps** we can generate new execute block; Load combination 1 in **Load steps** under **new execute block** is selected with number of load step and user specified size of load factor both 1. Maximum number of iteration is 50 and regular Newton–Raphson method is applied. **Force** and **Displacement** are both selected as convergence norm (see Fig. 3.271). It is deemed that iterative calculation reaches convergence under current load step when either of them reaches convergence

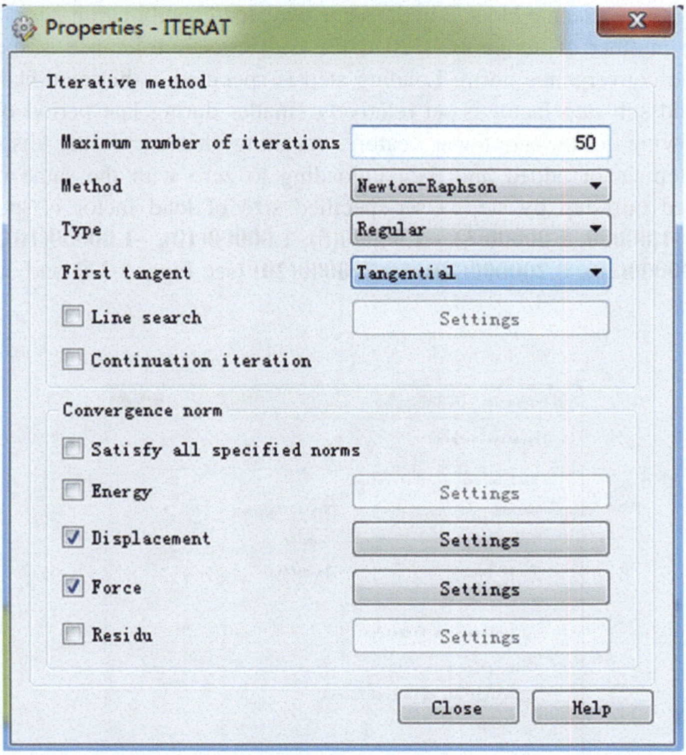

Fig. 3.271 Specification of iteration

On clicking Setting button of both **Displacement** and **Force**, dialog box of manipulation interface, as shown in Fig. 3.272, ejects. **Convergence tolerance** is, by default, **0.01;** while **Abort criterion** is also kept as **10,000** unchanged, meaning that calculation continues until 10,000 times if iterative result does not reach convergence.

Fig. 3.272 Convergence norm of load combination 1

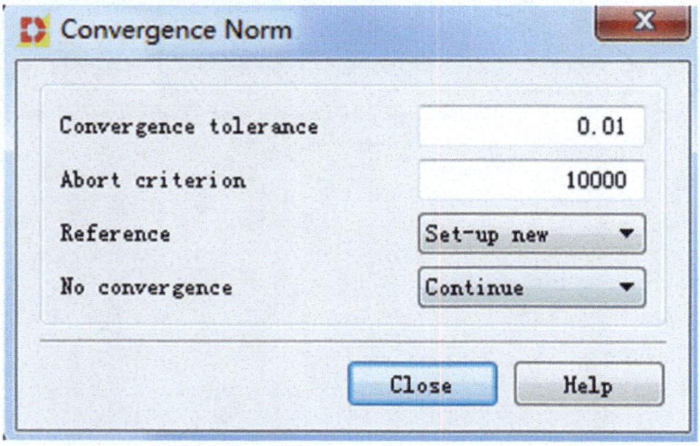

Creating new execute block 2 in the same way, load combination 2 is chosen and maximum number of iterations is 20 while convergence tolerance is specified as 0.05 in the convergence norm. Loading step is specified as bidirectional periodic loading and sub-step factor is set relatively smaller during late period of loading process owing to the nonlinear feature, meaning that increasing displacement loading step factor added and then unloading to zero with the same increment was carried out step by step. User-specified size of load factor is specified as **1.00000, –1.00000, 1.00000(5), –1.00000(5), 1.00000(10), –1.00000(10), 1.00000 (20), –1.00000(20), 0.200000(20), –0.200000(20)** (see Figs. 3.273 and 3.274).

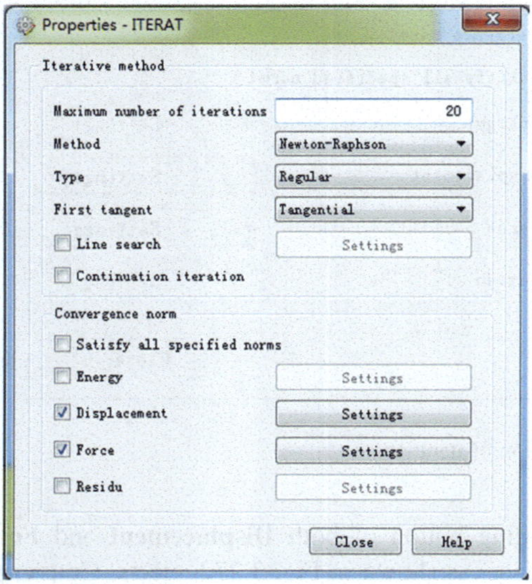

Fig. 3.273 Properties of iteration

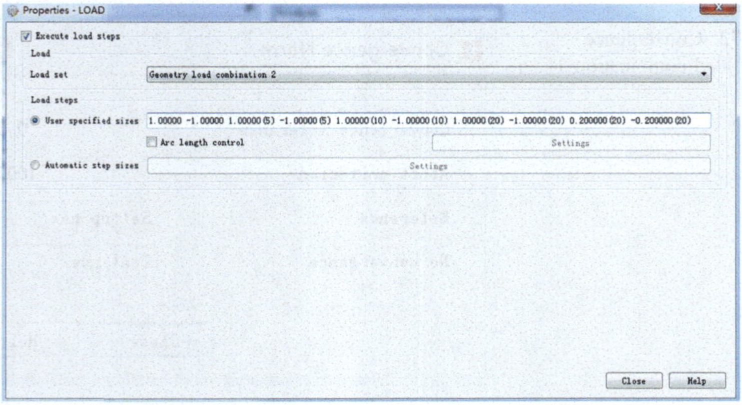

Fig. 3.274 Specification of displacement loading step factor

The next step is to specify the output of nonlinear iteration results before running analysis. Instead of directly selecting all primaries, translational displacement in all directions under global coordinate system (DISOLA TOTAL TRANSL GLOBAL), cracking strains in all directions (STRAIN CRACK GREEN), summed cracking strains under global coordinate system (STRAIN CRKSUM GREEN GLOBAL), summed cracking strains in all principal stress directions (STRAIN CRKSUM GREEN PRINCI), crack width in all directions under global and local coordinate system (STRAIN CRKWDT GREEN GLOBAL/LOCAL) are all selected (see Fig. 3.275).

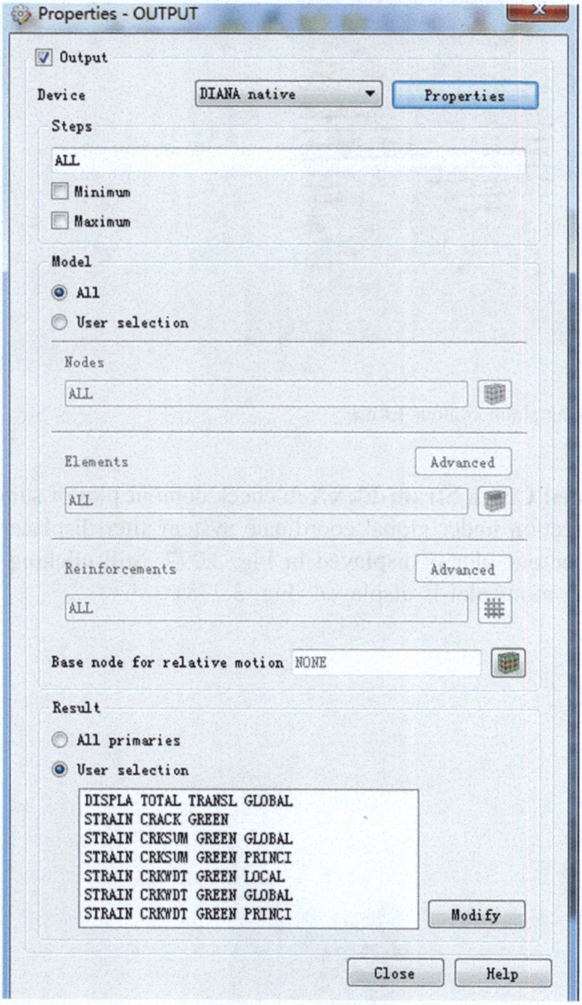

Fig. 3.275 Specification of output

Clicking run an analysis button, the contour output is displayed after iterative calculation finished. Selecting the ultimate load step and clicking **Results-Element results-Crack Strain-Eknn** option, contour plot of crack strain **Eknn** is displayed (Fig. 3.276).

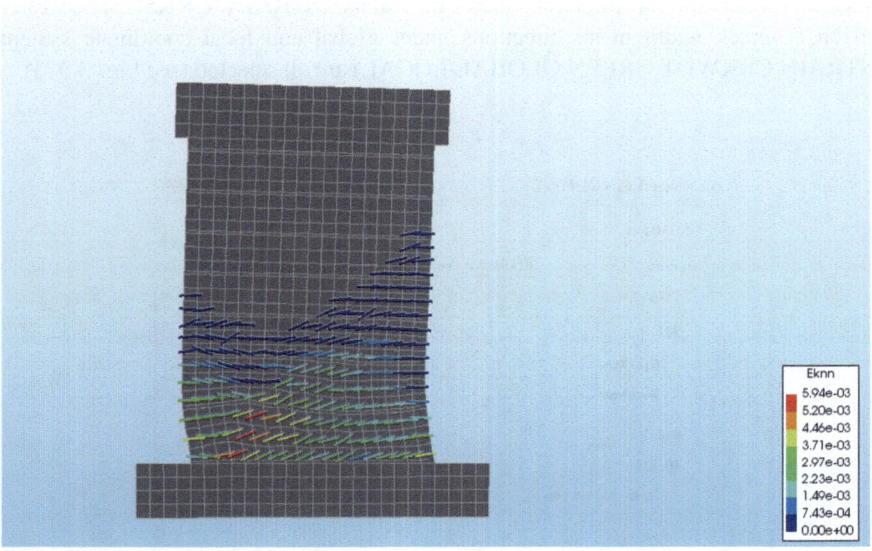

Fig. 3.276 Cracking strain contour **Eknn**

Click **Summed Crack Strain-EkXX** to check contour plot of summed cracking strains in X direction under global coordinate system after displacement action is finished. The contour plot is displayed in Fig. 3.277. Still clicking below EkYY option, EkYY contour plot is displayed (Fig. 3.278).

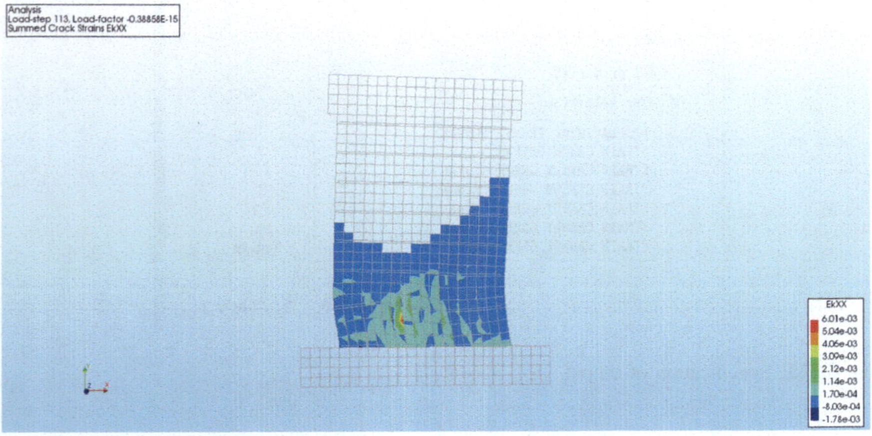

Fig. 3.277 Contour of summed cracking strains **EkXX** after displacement action finished

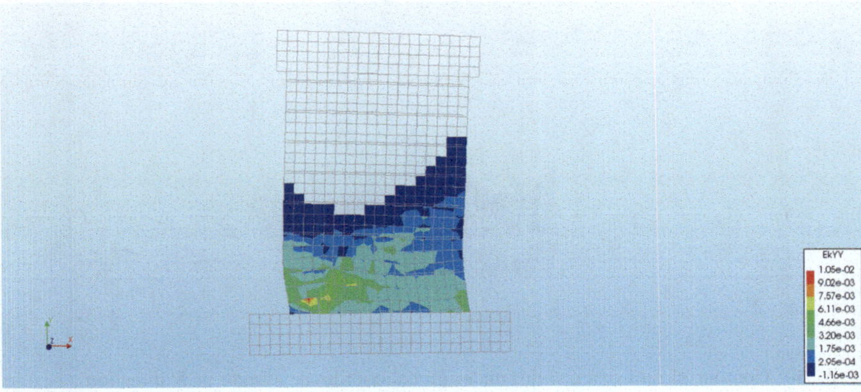

Fig. 3.278 Contour of summed cracking strains **EkYY** after displacement action finished

Click **Element results-Crack widths-Ecw1** to check distribution contour plot of crack width in the first principal stress direction, displayed as Fig. 3.279. Contour plot of crack width under global coordinate system in X direction, EcwXX, is also checked (Fig. 3.280).

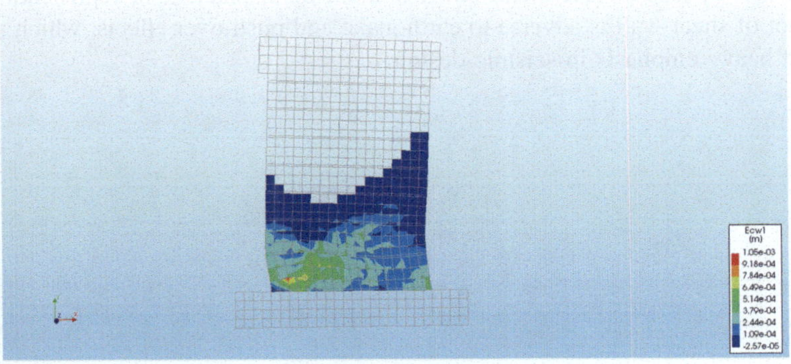

Fig. 3.279 Contour plot of crack width Ecw1 in the first principal stress direction

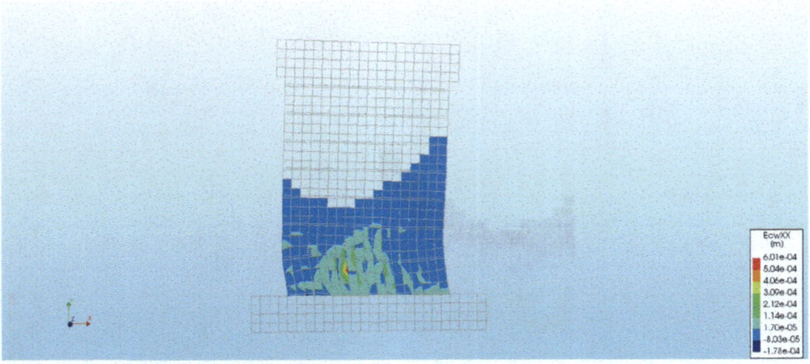

Fig. 3.280 Contour plot of crack width under global coordinate system in X direction

Judging from the contours mentioned above in this case, seriously accumulated damage under nonlinear hysteretic analysis is mainly concentrated in the middle root of shear wall and the tendency of cracking propagation is developing from the root to the upper section as the displacement action continues even though on the conditions of reinforcement and small amount of displacement, representing that the root of shear wall is adverse to earthquake and push-over effects, which should be put heavy emphasis in seismic design.

Python console:

```
##############################################################################
# DianaIE 10.1 update 2017-04-25 13:38:53
# Python   3.3.4
# Session recorded at 2018-04-24 18:30:38
##############################################################################
newProject( "Shear wall", 10 )
setModelAnalysisAspects( [ "STRUCT" ] )
setModelDimension( "2D" )
setDefaultMeshOrder( "QUADRATIC" )
setDefaultMesherType( "HEXQUAD" )
setDefaultMidSideNodeLocation( "LINEAR" )
createSheet( "support", [[ 0, 0, 0 ],[ 2.6, 0, 0 ],[ 2.6, 0.4, 0 ],[ 0, 0.4, 0 ]] )
createSheet( "wall", [[ 0.4, 0.4, 0 ],[ 2.2, 0.4, 0 ],[ 2.2, 2.6, 0 ],[ 0.4, 2.6, 0 ]] )
createSheet( "top", [[ 0.3, 2.6, 0 ],[ 2.3, 2.6, 0 ],[ 2.3, 3, 0 ],[ 0.3, 3, 0 ]] )
createLine( "FRP", [ 0.3, 2.7, 0 ], [ 2.3, 2.7, 0 ] )
createLine( "bar1", [ 1.3, 0.2, 0 ], [ 1.3, 2.8, 0 ] )
arrayCopy( [ "bar1" ], [ -0.2, 0, 0 ], [ 0, 0, 0 ], [ 0, 0, 0 ], 4 )
arrayCopy( [ "bar1" ], [ 0.2, 0, 0 ], [ 0, 0, 0 ], [ 0, 0, 0 ], 4 )
createLine( "stirrup", [ 0.44, 0.5, 0 ], [ 0.74, 0.5, 0 ] )
arrayCopy( [ "stirrup" ], [ 0, 0.2, 0 ], [ 0, 0, 0 ], [ 0, 0, 0 ], 10 )
arrayCopy( [ "stirrup", "stirrup 1", "stirrup 2", "stirrup 3", "stirrup 4", "stirrup 5", "stirrup 6",
"stirrup 7", "stirrup 8", "stirrup 9", "stirrup 10" ], [ 1.4, 0, 0 ], [ 0, 0, 0 ], [ 0, 0, 0 ], 1 )
createLine( "barx1", [ 0.42, 0.42, 0 ], [ 2.16, 0.42, 0 ] )
arrayCopy( [ "barx1" ], [ 0, 0.3, 0 ], [ 0, 0, 0 ], [ 0, 0, 0 ], 7 )
addMaterial( "wall", "CONCR", "TSCR", [] )
setParameter( "MATERIAL", "wall", "LINEAR/ELASTI/YOUNG", 3e+10 )
setParameter( "MATERIAL", "wall", "LINEAR/ELASTI/YOUNG", 3.7e+10 )
setParameter( "MATERIAL", "wall", "LINEAR/ELASTI/POISON", 0.15 )
setParameter( "MATERIAL", "wall", "LINEAR/MASS/DENSIT", 2500 )
setParameter( "MATERIAL", "wall", "TENSIL/TENCRV", "HORDYK" )
setParameter( "MATERIAL", "wall", "TENSIL/TENSTR", 3 )
setParameter( "MATERIAL", "wall", "TENSIL/TENSTR", 2.8 )
setParameter( "MATERIAL", "wall", "TENSIL/TENSTR", 2800000 )
setParameter( "MATERIAL", "wall", "TENSIL/GF1", 200 )
setParameter( "MATERIAL", "wall", "TENSIL/RESTST", 100 )
setParameter( "MATERIAL", "wall", "COMPRS/COMCRV", "MAEKCC" )
setParameter( "MATERIAL", "wall", "COMPRS/COMSTR", 45000000 )
setParameter( "MATERIAL", "wall", "SHEAR/BETA", 0.1 )
addGeometry( "Element geometry 1", "SHEET", "MEMBRA", [] )
rename( "GEOMET", "Element geometry 1", "wall" )
setParameter( "GEOMET", "wall", "THICK", 0.4 )
```

```
clearReinforcementAspects( [ "wall" ] )
setElementClassType( "SHAPE", [ "wall" ], "MEMBRA" )
assignMaterial( "wall", "SHAPE", [ "wall" ] )
assignGeometry( "wall", "SHAPE", [ "wall" ] )
resetElementData( "SHAPE", [ "wall" ] )
addMaterial( "beam", "CONCR", "LEI", [] )
setParameter( "MATERIAL", "beam", "LINEAR/ELASTI/YOUNG", 3.25e+10 )
setParameter( "MATERIAL", "beam", "LINEAR/ELASTI/POISON", 0.15 )
setParameter( "MATERIAL", "beam", "LINEAR/MASS/DENSIT", 2500 )
addGeometry( "Element geometry 2", "SHEET", "MEMBRA", [] )
rename( "GEOMET", "Element geometry 2", "beam" )
setParameter( "GEOMET", "beam", "THICK", 0.6 )
clearReinforcementAspects( [ "support", "top" ] )
setElementClassType( "SHAPE", [ "support", "top" ], "MEMBRA" )
assignMaterial( "beam", "SHAPE", [ "support", "top" ] )
assignGeometry( "beam", "SHAPE", [ "support", "top" ] )
resetElementData( "SHAPE", [ "support", "top" ] )
setParameter( "GEOMET", "beam", "THICK", 0.6 )
addMaterial( "bar", "REINFO", "VMISES", [] )
setParameter( "MATERIAL", "bar", "LINEAR/ELASTI/YOUNG", 2.1e+11 )
setParameter( "MATERIAL", "bar", "PLASTI/HARDI1/YLDSTR", 4.4e+08 )
addGeometry( "Element geometry 3", "RELINE", "REBAR", [] )
rename( "GEOMET", "Element geometry 3", "bar" )
setParameter( "GEOMET", "bar", "REIEMB/CROSSE", 0.000157 )
setReinforcementAspects( [ "bar1", "bar2", "bar3", "bar4", "bar5", "bar6", "bar7", "bar8",
"bar9" ] )
assignMaterial( "bar", "SHAPE", [ "bar1", "bar2", "bar3", "bar4", "bar5", "bar6", "bar7",
"bar8", "bar9" ] )
assignGeometry( "bar", "SHAPE", [ "bar1", "bar2", "bar3", "bar4", "bar5", "bar6", "bar7",
"bar8", "bar9" ] )
resetElementData( "SHAPE", [ "bar1", "bar2", "bar3", "bar4", "bar5", "bar6", "bar7", "bar8",
"bar9" ] )
setReinforcementDiscretization( [ "bar1", "bar2", "bar3", "bar4", "bar5", "bar6", "bar7",
"bar8", "bar9" ], "SECTION" )
addMaterial( "barx", "REINFO", "VMISES", [] )
setParameter( "MATERIAL", "barx", "LINEAR/ELASTI/YOUNG", 2.1e+11 )
setParameter( "MATERIAL", "barx", "PLASTI/HARDI1/YLDSTR", 4.4e+08 )
addGeometry( "Element geometry 4", "RELINE", "REBAR", [] )
rename( "GEOMET", "Element geometry 4", "barx" )
setParameter( "GEOMET", "barx", "REIEMB/CROSSE", 0.000157 )
setReinforcementAspects( [ "barx1", "barx2", "barx3", "barx4", "barx5", "barx6", "barx7",
"barx8" ] )
```

assignMaterial("barx", "SHAPE", ["barx1", "barx2", "barx3", "barx4", "barx5", "barx6", "barx7", "barx8"])

assignGeometry("barx", "SHAPE", ["barx1", "barx2", "barx3", "barx4", "barx5", "barx6", "barx7", "barx8"])

resetElementData("SHAPE", ["barx1", "barx2", "barx3", "barx4", "barx5", "barx6", "barx7", "barx8"])

setReinforcementDiscretization(["barx1", "barx2", "barx3", "barx4", "barx5", "barx6", "barx7", "barx8"], "SECTION")

saveProject( )

saveProject( )

addMaterial("stirrup", "REINFO", "VMISES", [])

setParameter("MATERIAL", "stirrup", "LINEAR/ELASTI/YOUNG", 2.1e+11)

setParameter("MATERIAL", "stirrup", "PLASTI/HARDI1/YLDSTR", 4.4e+08)

addGeometry("Element geometry 5", "RELINE", "REBAR", [])

rename("GEOMET", "Element geometry 5", "stirrup")

setParameter("GEOMET", "stirrup", "REIEMB/CROSSE", 1.0053e-04)

setReinforcementAspects(["stirrup", "stirrup 1", "stirrup 2", "stirrup 3", "stirrup 4", "stirrup 5", "stirrup 6", "stirrup 7", "stirrup 8", "stirrup 9", "stirrup 10", "stirrup 11", "stirrup 12", "stirrup 13", "stirrup 14", "stirrup 15", "stirrup 16", "stirrup 17", "stirrup 18", "stirrup 19", "stirrup 20", "stirrup 21"])

assignMaterial("stirrup", "SHAPE", ["stirrup", "stirrup 1", "stirrup 2", "stirrup 3", "stirrup 4", "stirrup 5", "stirrup 6", "stirrup 7", "stirrup 8", "stirrup 9", "stirrup 10", "stirrup 11", "stirrup 12", "stirrup 13", "stirrup 14", "stirrup 15", "stirrup 16", "stirrup 17", "stirrup 18", "stirrup 19", "stirrup 20", "stirrup 21"])

assignGeometry("stirrup", "SHAPE", ["stirrup", "stirrup 1", "stirrup 2", "stirrup 3", "stirrup 4", "stirrup 5", "stirrup 6", "stirrup 7", "stirrup 8", "stirrup 9", "stirrup 10", "stirrup 11", "stirrup 12", "stirrup 13", "stirrup 14", "stirrup 15", "stirrup 16", "stirrup 17", "stirrup 18", "stirrup 19", "stirrup 20", "stirrup 21"])

resetElementData("SHAPE", ["stirrup", "stirrup 1", "stirrup 2", "stirrup 3", "stirrup 4", "stirrup 5", "stirrup 6", "stirrup 7", "stirrup 8", "stirrup 9", "stirrup 10", "stirrup 11", "stirrup 12", "stirrup 13", "stirrup 14", "stirrup 15", "stirrup 16", "stirrup 17", "stirrup 18", "stirrup 19", "stirrup 20", "stirrup 21"])

setReinforcementDiscretization(["stirrup", "stirrup 1", "stirrup 2", "stirrup 3", "stirrup 4", "stirrup 5", "stirrup 6", "stirrup 7", "stirrup 8", "stirrup 9", "stirrup 10", "stirrup 11", "stirrup 12", "stirrup 13", "stirrup 14", "stirrup 15", "stirrup 16", "stirrup 17", "stirrup 18", "stirrup 19", "stirrup 20", "stirrup 21"], "SECTION")

saveProject( )

addMaterial("FRP", "REINFO", "VMISES", [])

setParameter("MATERIAL", "FRP", "LINEAR/ELASTI/YOUNG", 1.36e+11)

setParameter("MATERIAL", "FRP", "PLASTI/HARDI1/YLDSTR", 1.2e+09)

addGeometry("Element geometry 6", "RELINE", "REBAR", [])

rename("GEOMET", "Element geometry 6", "FRP")

```
setParameter( "GEOMET", "FRP", "REIEMB/CROSSE", 0.000139 )
setReinforcementAspects( [ "FRP" ] )
assignMaterial( "FRP", "SHAPE", [ "FRP" ] )
assignGeometry( "FRP", "SHAPE", [ "FRP" ] )
resetElementData( "SHAPE", [ "FRP" ] )
setReinforcementDiscretization( [ "FRP" ], "ELEMENT" )
saveProject(  )
addSet( "GEOMETRYSUPPORTSET", "Geometry support set 1" )
rename( "GEOMETRYSUPPORTSET", "Geometry support set 1", "co1" )
createLineSupport( "co1", "co1" )
setParameter( "GEOMETRYSUPPORT", "co1", "AXES", [ 1, 2 ] )
setParameter( "GEOMETRYSUPPORT", "co1", "TRANSL", [ 1, 1, 0 ] )
setParameter( "GEOMETRYSUPPORT", "co1", "ROTATI", [ 0, 0, 0 ] )
attach( "GEOMETRYSUPPORT", "co1", "support", [[ 1.3, 7.4981504e-34, 1.2245938e-17 ]] )
saveProject(  )
createVertex( "Vertex 1", [ 2.8, 2.8, 0 ] )
projection( "SHAPEEDGE", "top", [[ 2.3, 2.8, -5.6944687e-19 ]], [ "Vertex 1" ], [ -1, 0, 0 ], True )
removeShape( [ "Vertex 1" ] )
addSet( "GEOMETRYSUPPORTSET", "Geometry support set 2" )
rename( "GEOMETRYSUPPORTSET", "Geometry support set 2", "co2" )
createPointSupport( "co2", "co2" )
setParameter( "GEOMETRYSUPPORT", "co2", "AXES", [ 1, 2 ] )
setParameter( "GEOMETRYSUPPORT", "co2", "TRANSL", [ 1, 0, 0 ] )
setParameter( "GEOMETRYSUPPORT", "co2", "ROTATI", [ 0, 0, 0 ] )
attach( "GEOMETRYSUPPORT", "co2", "top", [[ 2.3, 2.8, -5.6944687e-19 ]] )
saveProject(  )
addSet( "GEOMETRYLOADSET", "gravity" )
createModelLoad( "gravity", "gravity" )
saveProject(  )
addSet( "GEOMETRYLOADSET", "Geometry load case 2" )
createLineLoad( "load", "Geometry load case 2" )
setParameter( "GEOMETRYLOAD", "load", "FORCE/VALUE", -30000 )
setParameter( "GEOMETRYLOAD", "load", "FORCE/DIRECT", 2 )
attach( "GEOMETRYLOAD", "load", "top", [[ 1.3, 3, -1.2245938e-17 ]] )
saveProject(  )
addSet( "GEOMETRYLOADSET", "Geometry load case 3" )
createPointLoad( "displacement", "Geometry load case 3" )
setParameter( "GEOMETRYLOAD", "displacement", "LODTYP", "DEFORM" )
setParameter( "GEOMETRYLOAD", "displacement", "DEFORM/TR/VALUE", -0.0005 )
setParameter( "GEOMETRYLOAD", "displacement", "DEFORM/TR/DIRECT",1 )
```

```
attach( "GEOMETRYLOAD", "displacement", "top", [[ 2.3, 2.8, -5.6944687e-19 ]] )
saveProject( )
setDefaultGeometryLoadCombinations( )
setGeometryLoadCombinationFactor( "Geometry load combination 1", "gravity", 1 )
remove( "GEOMETRYLOADCOMBINATION", "Geometry load combination 1" )
remove( "GEOMETRYLOADCOMBINATION", "Geometry load combination 2" )
remove( "GEOMETRYLOADCOMBINATION", "Geometry load combination 3" )
addGeometryLoadCombination( "" )
setGeometryLoadCombinationFactor( "Geometry load combination 1", "gravity", 1 )
setGeometryLoadCombinationFactor( "Geometry load combination 1", "Geometry load case 2", 1 )
addGeometryLoadCombination( "" )
setGeometryLoadCombinationFactor( "Geometry load combination 2", "Geometry load case 3", 1 )
setGeometryLoadCombinationFactor( "Geometry load combination 1", "gravity", 1 )
setElementSize( [ "support", "wall", "top" ], 0.1, -1, True )
setMesherType( [ "support", "wall", "top" ], "HEXQUAD" )
setMidSideNodeLocation( [ "support", "wall", "top" ], "LINEAR" )
generateMesh( [] )
hideView( "GEOM" )
showView( "MESH" )
addAnalysis( "Analysis7" )
renameAnalysis( "Analysis7", "Analysis" )
addAnalysisCommand( "Analysis", "NONLIN", "Structural nonlinear" )
renameAnalysis( "Analysis", "Analysis" )
setAnalysisCommandDetail( "Analysis", "Structural nonlinear", "EXECUT(1)/ITERAT/MAXITE", 50 )
setAnalysisCommandDetail( "Analysis", "Structural nonlinear", "EXECUT(1)/ITERAT/CONVER/DISPLA/NOCONV", "CONTIN" )
setAnalysisCommandDetail( "Analysis", "Structural nonlinear", "EXECUT(1)/ITERAT/CONVER/FORCE/NOCONV", "CONTIN" )
setAnalysisCommandDetail( "Analysis", "Structural nonlinear", "EXECUT/EXETYP", "LOAD" )
setAnalysisCommandDetail( "Analysis", "Structural nonlinear", "EXECUT(2)/LOAD/LOADNR", 2 )
setAnalysisCommandDetail( "Analysis", "Structural nonlinear", "EXECUT(2)/LOAD/STEPS/EXPLIC/SIZES", "1 -1 1(5) -1(5) 1(10) -1(10) 1(20) -1(20) 0.2(20) -0.2(20)" )
setAnalysisCommandDetail( "Analysis", "Structural nonlinear", "EXECUT(2)/ITERAT/MAXITE", 20 )
setAnalysisCommandDetail( "Analysis", "Structural nonlinear",
```

```
"EXECUT(2)/ITERAT/CONVER/DISPLA/NOCONV", "CONTIN" )
    setAnalysisCommandDetail( "Analysis", "Structural nonlinear",
"EXECUT(2)/ITERAT/CONVER/FORCE/NOCONV", "CONTIN" )
    setAnalysisCommandDetail( "Analysis", "Structural nonlinear",
"EXECUT(2)/ITERAT/CONVER/DISPLA/TOLCON", 0.05 )
    setAnalysisCommandDetail( "Analysis", "Structural nonlinear",
"EXECUT(2)/ITERAT/CONVER/FORCE/TOLCON", 0.05 )
    setAnalysisCommandDetail( "Analysis", "Structural nonlinear", "OUTPUT(1)/SELTYP",
"USER" )
    addAnalysisCommandDetail( "Analysis", "Structural nonlinear", "OUTPUT(1)/USER" )
    addAnalysisCommandDetail( "Analysis", "Structural nonlinear",
"OUTPUT(1)/USER/DISPLA(1)/TOTAL/TRANSL/GLOBAL" )
    addAnalysisCommandDetail( "Analysis", "Structural nonlinear",
"OUTPUT(1)/USER/STRAIN(1)/CRACK/GREEN" )
    addAnalysisCommandDetail( "Analysis", "Structural nonlinear",
"OUTPUT(1)/USER/STRAIN(2)/CRKSUM/GREEN/GLOBAL" )
    addAnalysisCommandDetail( "Analysis", "Structural nonlinear",
"OUTPUT(1)/USER/STRAIN(3)/CRKSUM/GREEN/PRINCI" )
    addAnalysisCommandDetail( "Analysis", "Structural nonlinear",
"OUTPUT(1)/USER/STRAIN(4)/CRKWDT/GREEN/LOCAL" )
    addAnalysisCommandDetail( "Analysis", "Structural nonlinear",
"OUTPUT(1)/USER/STRAIN(5)/CRKWDT/GREEN/GLOBAL" )
    addAnalysisCommandDetail( "Analysis", "Structural nonlinear",
"OUTPUT(1)/USER/STRAIN(6)/CRKWDT/GREEN/PRINCI" )
    runSolver( "Analysis" )
    showView( "RESULT" )
    setResultCase( [ "Analysis", "Output", "Load-step 113, Load-factor -0.38858E-15" ] )
    setResultPlot( "contours", "Crack-widths/node", "Ecwxx" )
    setResultPlot( "cracks", "Crack Strains/mappedcrack", "Eknn" )
    setResultPlot( "contours", "Crack-widths/node", "Ecw3" )
    setResultPlot( "contours", "Crack-widths/node", "Ecw2" )
    setResultPlot( "contours", "Crack-widths/node", "Ecw1" )
    setResultPlot( "contours", "Crack-widths/node", "EcwYY" )
    setResultPlot( "contours", "Crack-widths/node", "EcwXX" )
    setResultPlot( "contours", "Crack-widths/node", "Ecw1" )
    setResultPlot( "contours", "Crack-widths/node", "Ecw2" )
    setResultPlot( "contours", "Crack-widths/node", "Ecwxx" )
    setResultPlot( "contours", "Summed Crack Strains/node", "EkXX" )
    setResultPlot( "contours", "Summed Crack Strains/node", "EkYY" )
    setResultPlot( "contours", "Crack-widths/node", "Ecwxx" )
    setResultPlot( "contours", "Crack-widths/node", "EcwXX" )
    setResultPlot( "contours", "Crack-widths/node", "EcwYY" )
```

```
setResultPlot( "contours", "Total Displacements/node", "TDtX" )
setResultPlot( "contours", "Total Displacements/node", "TDtY" )
setResultPlot( "contours", "Summed Crack Strains/node", "Ek1" )
setResultPlot( "contours", "Summed Crack Strains/node", "EkXX" )
setResultPlot( "contours", "Summed Crack Strains/node", "EkYY" )
setResultPlot( "contours", "Crack-widths/node", "EcwYY" )
setResultPlot( "contours", "Crack-widths/node", "Ecw1" )
setResultPlot( "contours", "Crack-widths/node", "Ecw3" )
setResultPlot( "contours", "Crack-widths/node", "EcwXX" )
```

3.7 Time-History Dynamic Analysis of Pier

This numerical case is a concrete pier with vertical distributed load value 50 kN/m^2. Pier is constituted by two cuboids and a cylinder, and 3D size is displayed in Fig. 3.281. Diameter of cylinder pier column is 3 m. The whole modeling procedure is based on the DIANA 10.2 platform. Six longitudinal reinforcement steel bars are embedded into the concrete column with user-specified plasticity model. Structural solid elements are applied to simulate the pier, and base excitation load **seismic\H24 H24_T1-II-1_2003_TOKACHI-Coast_EW.dat** is adopted to conduct structural dynamic analysis.

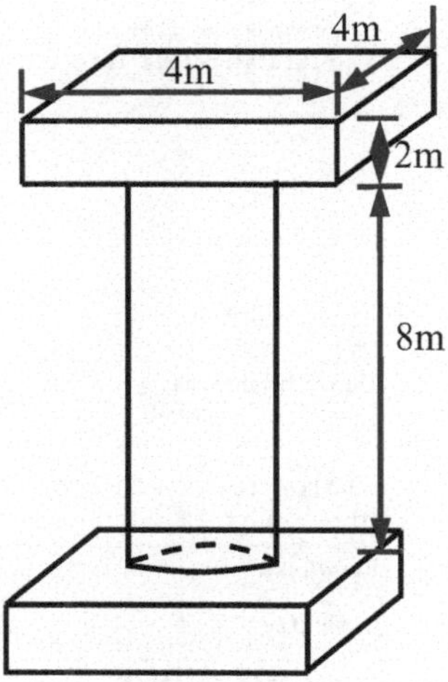

Fig. 3.281 3D sizes of piers

Essentials of learning

(1) *Learning to concrete geometry cylinder shape*
(2) *Learning to use remanipulation of* ***Mirror a shape*** *to construct a pier model*

(3) *Learning to steel hardening stress–strain constitutive curve* via *user-specified model*
(4) *Learning to specify time load of base excitation.*
(5) *Relearning to master specify parameters of and frequency and eigenvalue analysis.*

Above all, opening DianaIE and a new project name with the name of **Bridge Pier** is constructed in a document with the Chinese name of 10.2例题 with working directory of G-disk in computer. **Structural** is selected as analysis type while option of **Dimensions** is **Three**. Maximum model size of numerical model is 100 m, indicating that the scope of graphical user interface ranges from –50 to 50 m in the X, Y and Z directions under global coordinate system. **Default mesher type** is **Hexa/Quad** and **Quadratic** is selected as **Default mesh order**. Similarly, the way of determining mid-side node location is **Linear interpolation** (see Fig. 3.282).

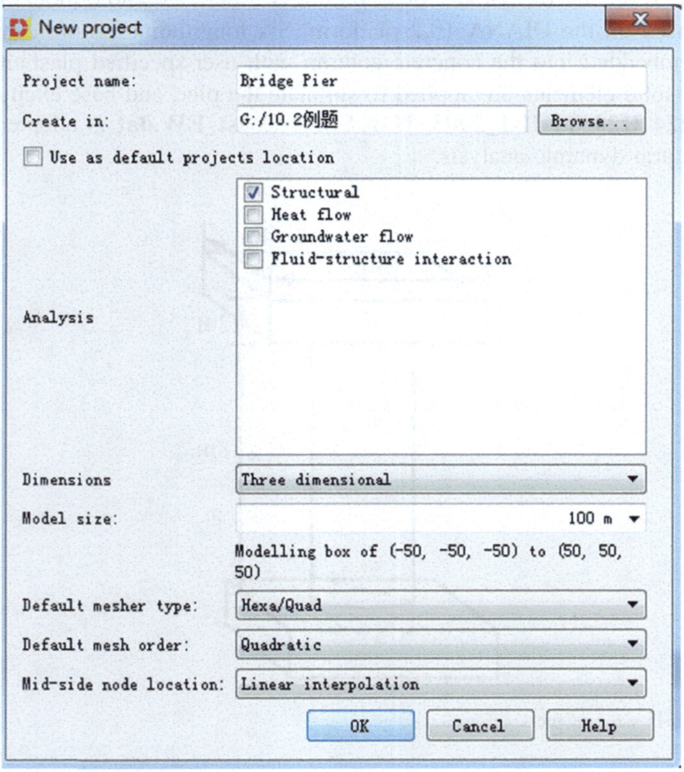

Fig. 3.282 Interface of new project

Clicking shortcut icon **Adds a block solid** to construct a new volume with the name of top, coordinate value of **Position** representing starting point is (0, 0, 0) and the length, width as well as height in **Size** is 4, 4 and 2 m, which is displayed in Fig. 3.283.

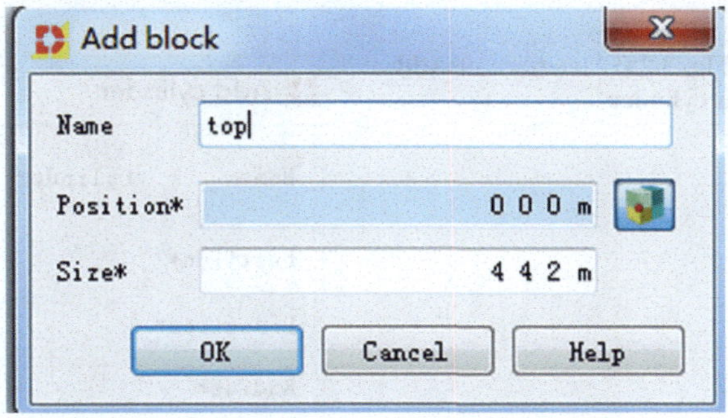

Fig. 3.283 Coordinate values of **Adds a block**

In order to render a convenient position for following cylinder, geometric model of **top** is selected, then right-click to select manipulation of **Move a shape** to translate it 8 m in the positive Z direction (see Fig. 3.284).

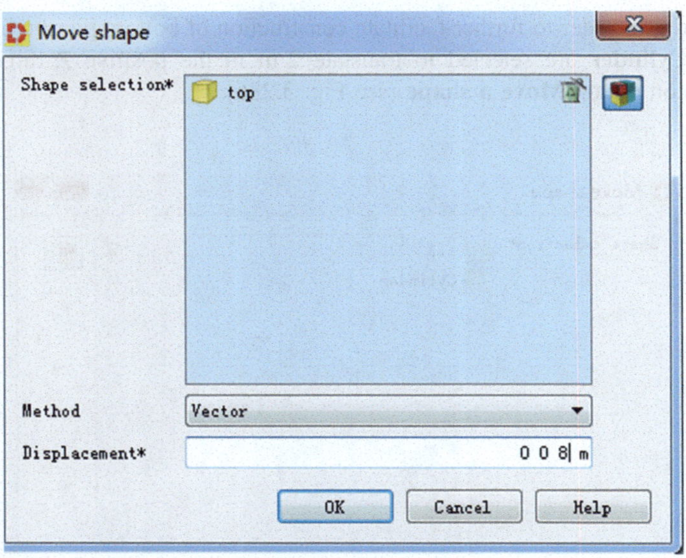

Fig. 3.284 Interface of **Move a shape**

The next step is to create cylinder column. Clicking shortcut icon **Adds a cylinder**, dialog ejects with the name of **Cylinder**. **Position** representing modeling starting point is (2, 2, 0) and the **Direction** is in the Z direction under global coordinate system (0, 0, 1). **Radius** and **Height** of this cylinder are 1.5 and 8 m, respectively (see Fig. 3.285).

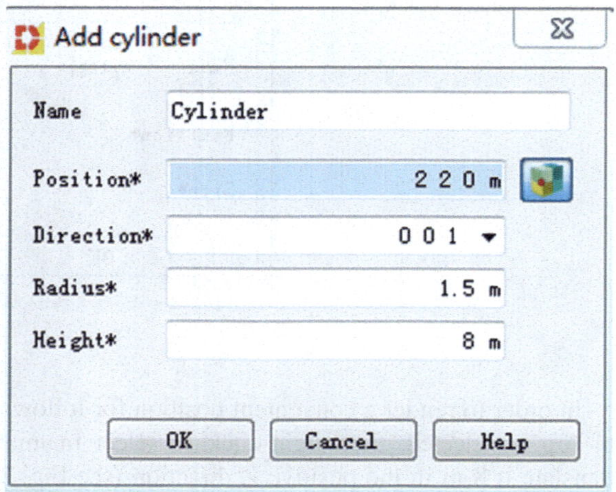

Fig. 3.285 Interface of **Adds a cylinder**

Similarly, in order to further facilitate construction of bottom pier buttress, both **top** and **Cylinder** are selected to translate 2 m in the positive Z direction via shortcut icon button **Move a shape** (see Fig. 3.286).

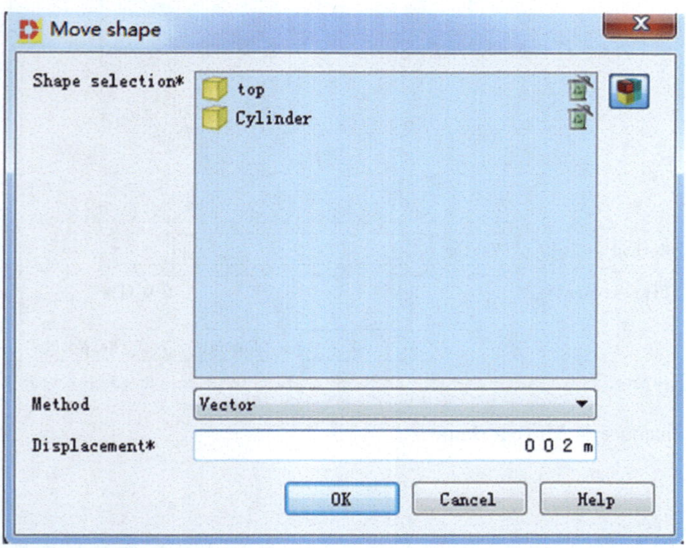

Fig. 3.286 Interface of **Move a shape**

Then we create bottom pier buttress with the same method again; **Position** and **Size** are displayed in Fig. 3.287.

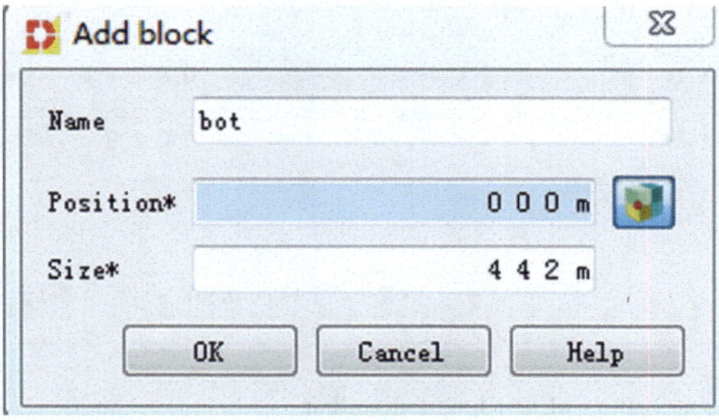

Fig. 3.287 Interface of constructing bottom pier

Click OK button, generated pier is displayed as in Fig. 3.288.

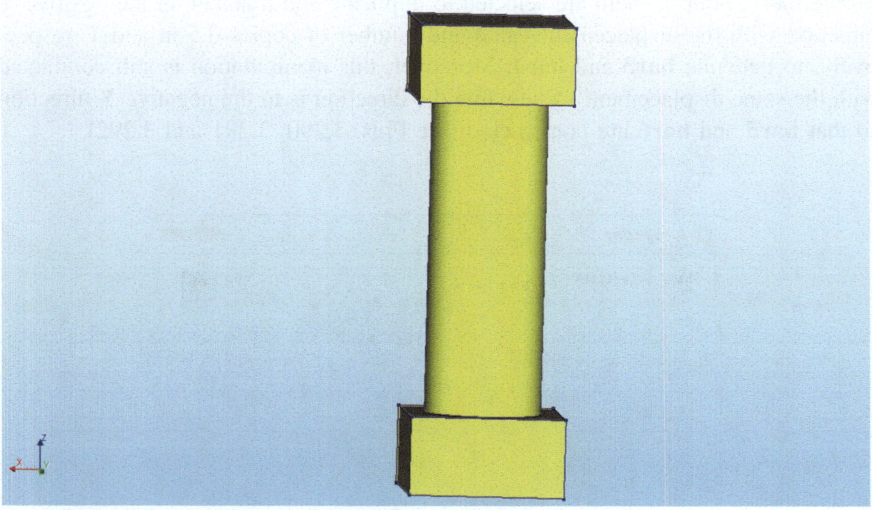

Fig. 3.288 Generation of pier

Next, we create longitudinal reinforcement steel bars. Click **Adds a line** shortcut icon button to construct the first longitudinal reinforcement steel bar with the name of **bar1**; **Method** is **Absolute** and coordinate values of **point 1** and **point 2** are displayed in Fig. 3.289.

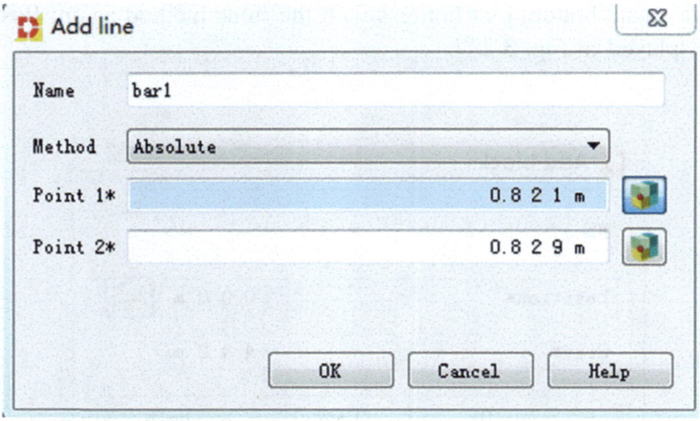

Fig. 3.289 Coordinate values of first longitudinal bar

Then other longitudinal steel bars are created via repeatedly applying the manipulation of **Array copy**. Initially, single **bar1** is selected to duplicate and translate in the positive X direction with the **Displacement** 2.4 m while the number of copies is specified as 1 to generate a new steel bar with the name of **bar2**. Furthermore, both of them are selected to duplicate and translate in the positive Y direction with the displacement value and number of copies 0.5 m and 1, respectively, to generate **bar3** and **bar4**. Moreover, this manipulation is still conducted with the same displacement value while the direction is in the negative Y direction so that **bar5** and **bar6** are constructed (see Figs. 3.290, 3.291 and 3.292).

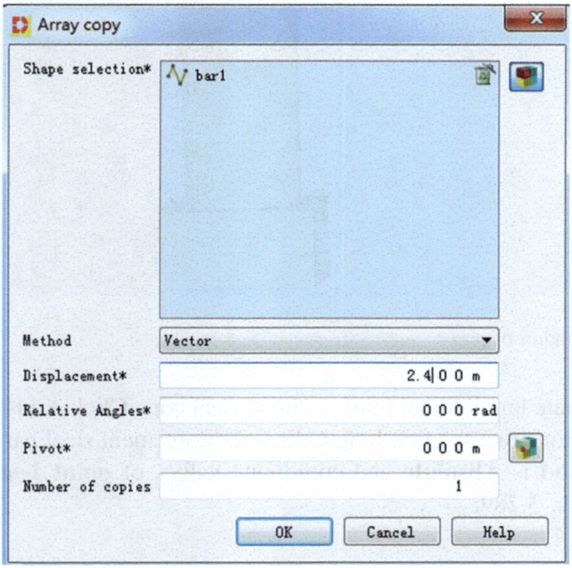

Fig. 3.290 Generation of **bar2** via **Array copy**

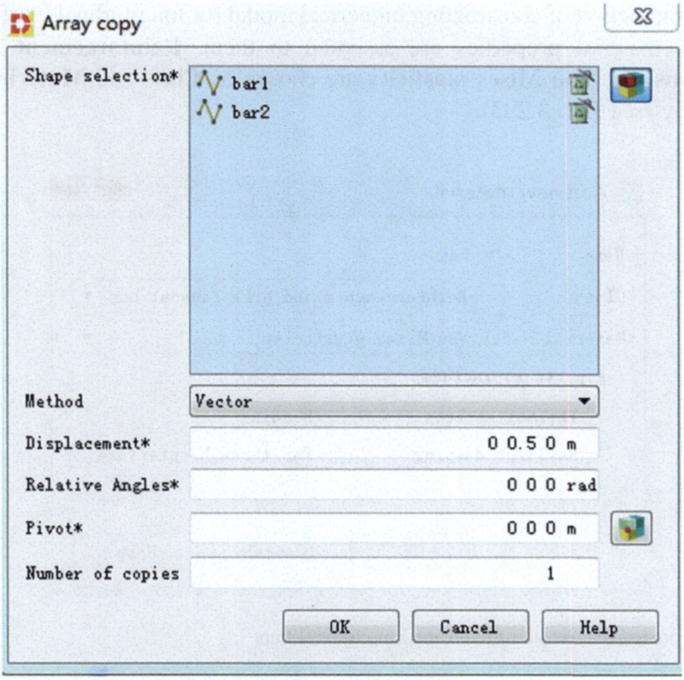

Fig. 3.291 Generation of **bar3** and **bar4** via **Array copy**

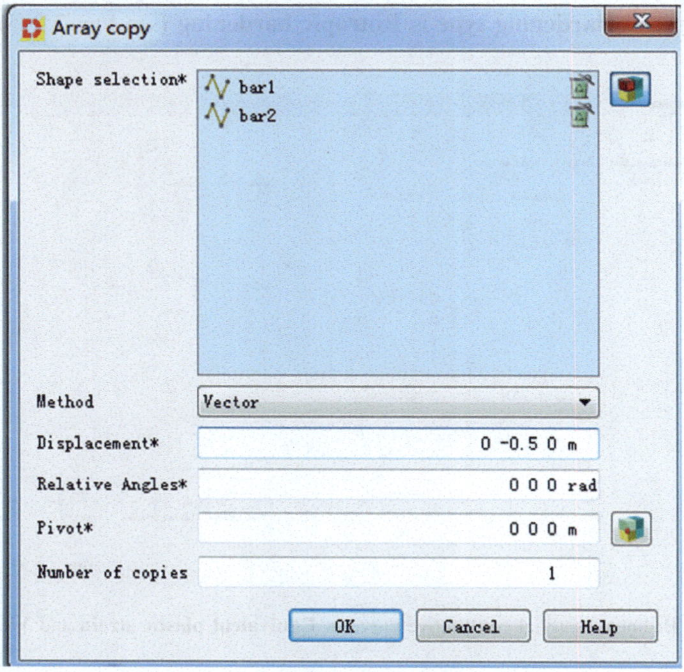

Fig. 3.292 Generation of **bar5** and **bar6** via **Array copy**

After completion of constructing numerical model for longitudinal reinforcement steel bars, material properties are assigned to them. **Reinforcement and pile foundations** and **Von Mises plasticity** are chosen as **Class** and **Material model,** respectively (see Fig. 3.293).

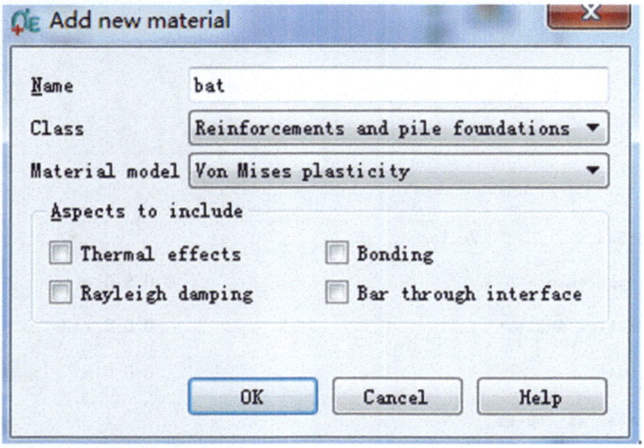

Fig. 3.293 Material class and model for longitudinal bars

Elastic modulus of steel is $2.1 \times 10^{11}\ \mathrm{N/m^2}$ and **Poisson's ratio** is 0.3. From the aspect of von Mises plasticity, **Plastic strain-yield stress** in the **Plastic hardening** option is chosen, where bi-linear material constitutive curve of **Equivalent plastic strain** and **Yield stress** is specified (Fig. 3.294). **Hardening hypothesis** is **Strain hardening** and **Hardening type** is **Isotropic hardening** (see Fig. 3.295).

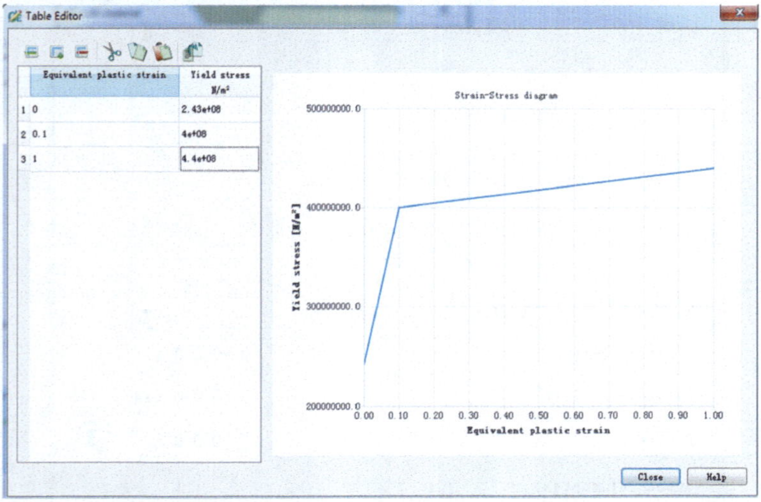

Fig. 3.294 Bi-linear material constitutive curve of **Equivalent plastic strain** and **Yield stress**

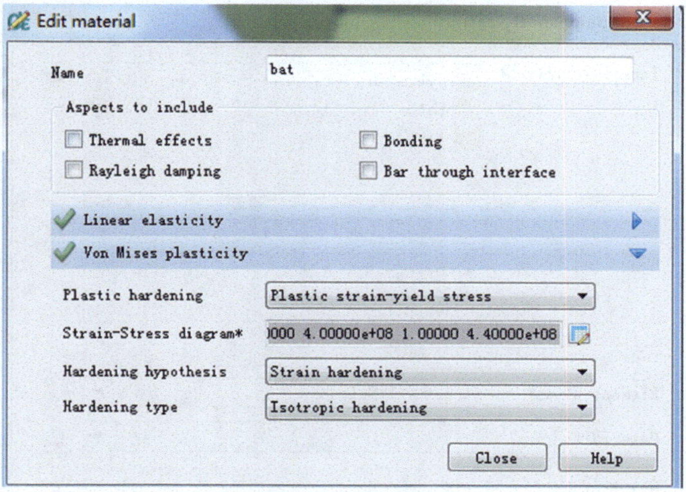

Fig. 3.295 User-specified selections under **von Mises plasticity**

After completion of material properties assignments, cross-section geometric properties are also specified, where **Reinforcement type** is **Embedded** and **Cross section area of bar** is 1000 mm^2, which is displayed in Fig. 3.296.

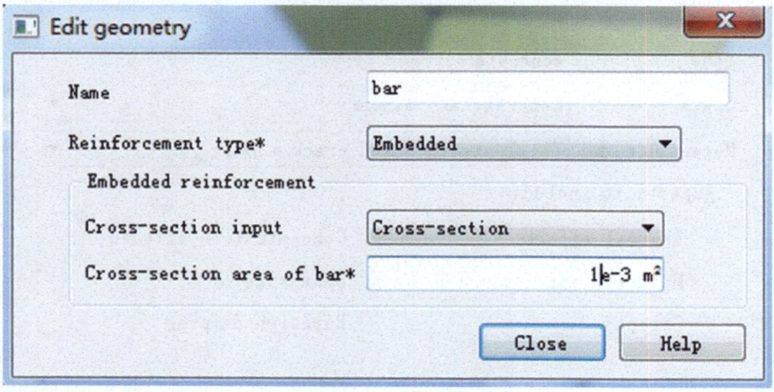

Fig. 3.296 Cross-section geometric properties assignments for bars

Then we start to define concrete properties for the pier. Solid element is selected as element type (see Fig. 3.297). Clicking icon , a new material dialog box with the name of **concrete** is created. **Concrete and masonry** as well as **Total strain based crack model** are selected as **Class** and **Material model**, respectively (see Fig. 3.298).

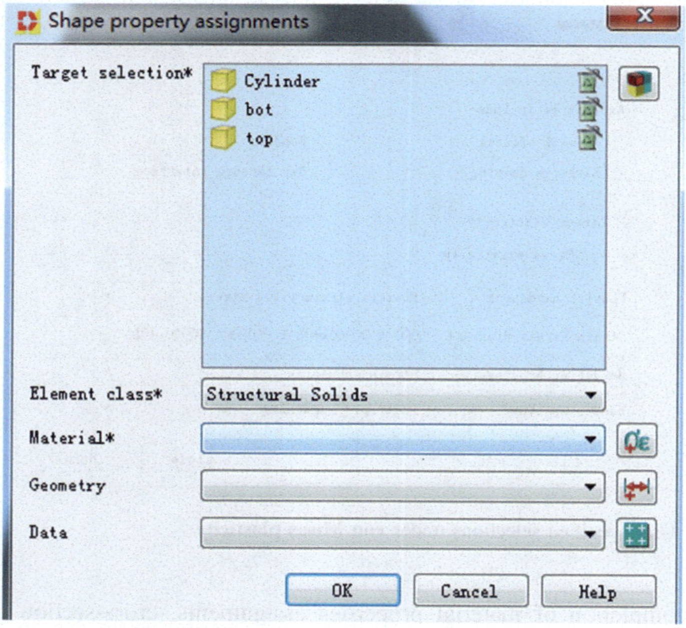

Fig. 3.297 Structural elements

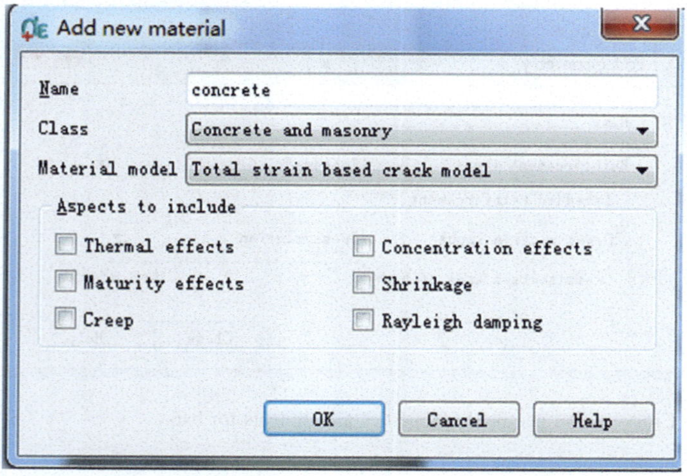

Fig. 3.298 Material class and model for concrete

Clicking OK button to enter the dialog of material specifications, in the **Linear material properties** aspect, elastic modulus is 3.45×10^{10} N/m^2 and Poisson's ratio is 0.15. Mass density is 2500 kg/m^3, which is displayed in Fig. 3.299. For **Tensile behavior model**, **CEB-FIP Model Code 1990** is selected as tensile curve option with tensile strength as well as Mode-I tensile fracture energy 2.6e6 and 150 N/m, respectively. **Compression curve** is **Maekawa Cracked Concrete curves** with compressive strength 3.24e7, respectively (see Figs. 3.300 and 3.301).

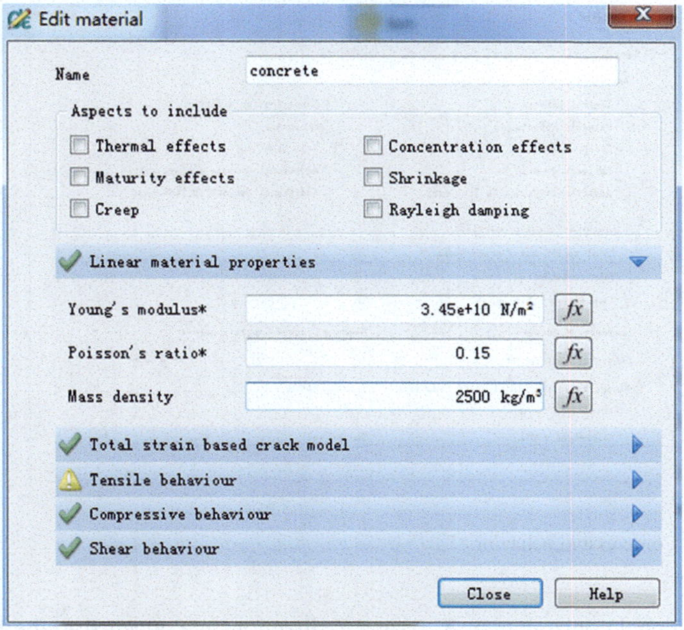

Fig. 3.299 Parameters in linear material properties

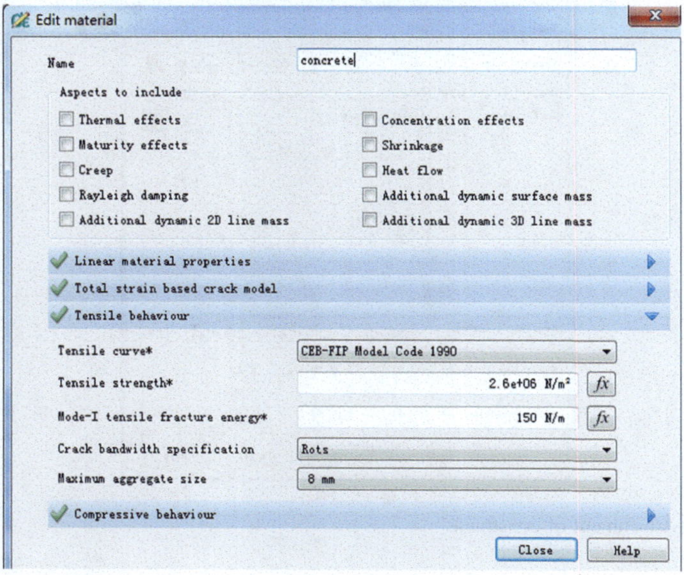

Fig. 3.300 Parameters and specifications for tensile behavior

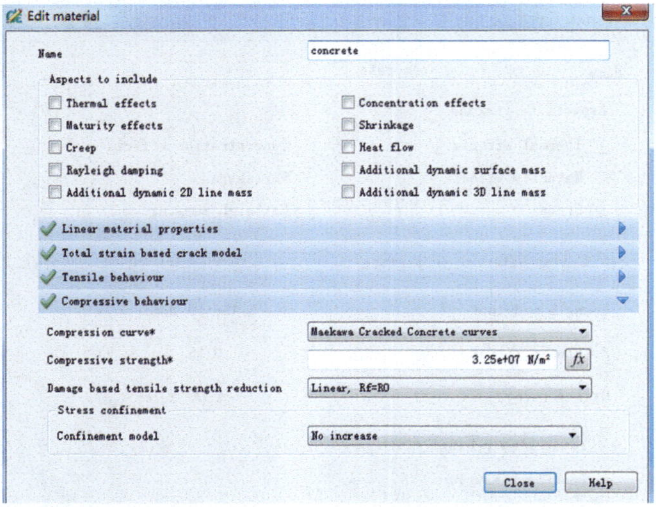

Fig. 3.301 Compressive behavior

Translational constraints in X, Y and Z directions are attached to the boundary surface to simulate the rigid consolidated contact surface with ground, displayed in Fig. 3.302. Clicking OK button, generated boundary interface constraint is displayed (Fig. 3.303).

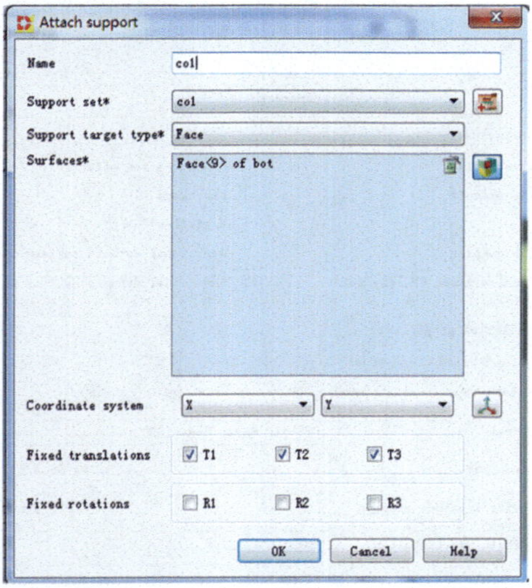

Fig. 3.302 Attachment of constraint

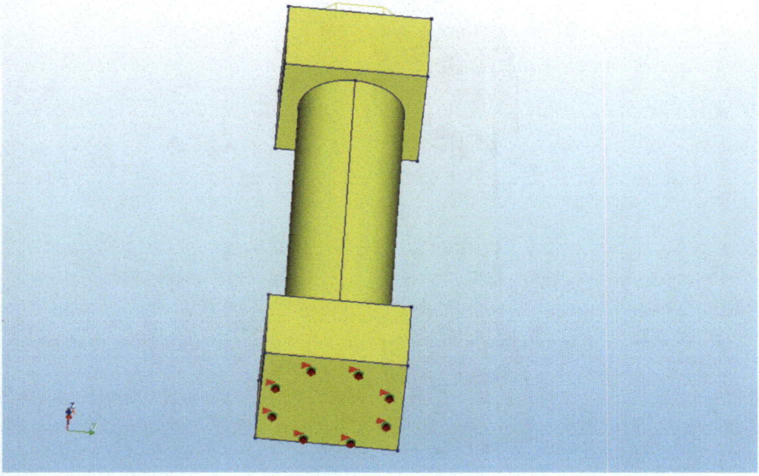

Fig. 3.303 Generation of boundary constraints

The following step is to construct load case. A new load case with the name of **lo1** is created where the **Load target type** and **Load type** are **Face** and **Distributed force,** respectively. **Surface force value** is 50 kN/m^2 in the vertical negative Z direction (see Fig. 3.304). Clicking OK button, attachment of distributed force is displayed (Fig. 3.305).

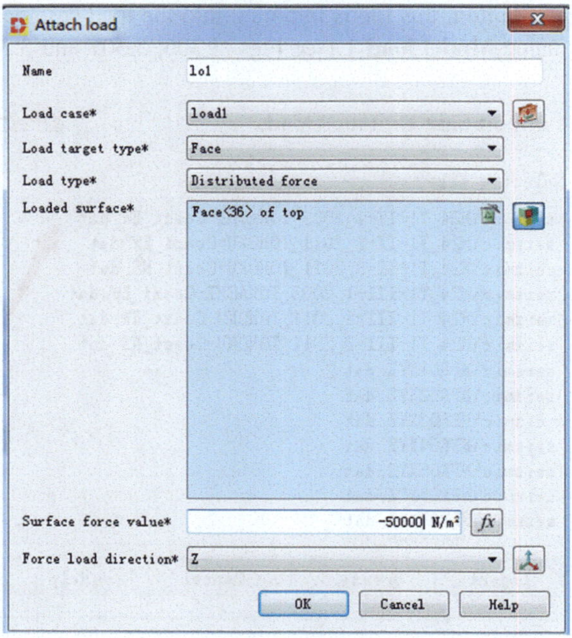

Fig. 3.304 Definition of vertical surface distributed force

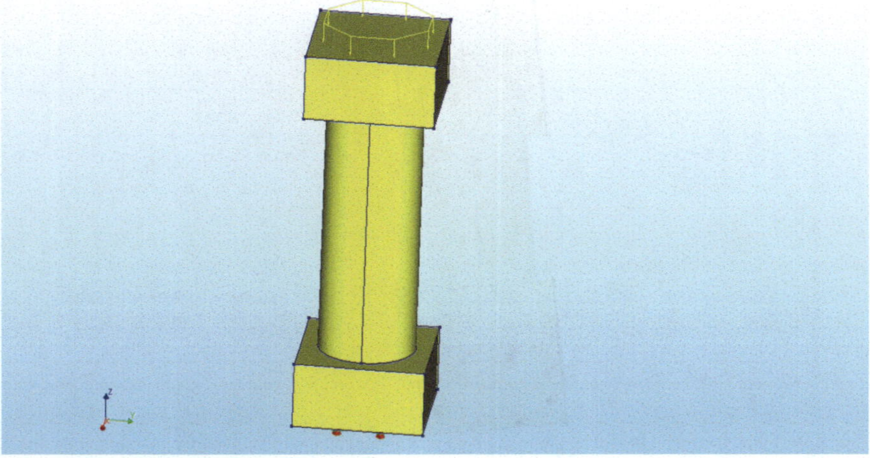

Fig. 3.305 Attachment of distributed force

Then gravity is created, where the manipulation is the same as former and it is not repeated here. After that, a kind of special action called base excitation load is applied. We click icon **Import time loads** under the **Load** bar. When the dialog box ejects, the first line of **.dat** file named "**seismic\H24 H24_T1-II-1_2003_TOKACHI-Coast_EW.dat**" stored in the following working directory of program is selected as base excitation.

("C:/Program Files/Diana https://doi.org/10.2/share/lib/seismic/h24_t1-ii-1_2003_tokachi-coast_ew.dat)

Clicking Import button, this file is imported into the load case named **BASE_X** with the action name **Model load 1** (see Figs. 3.306, 3.307 and 3.308).

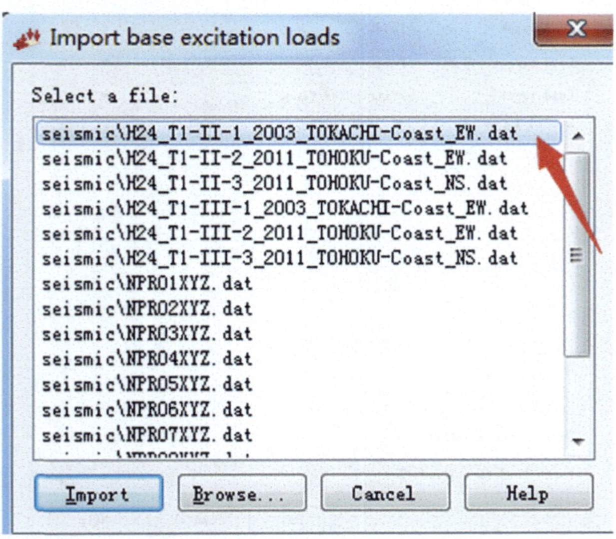

Fig. 3.306 seismic\H24 H24_T1-II-1_2003_TOKACHI-Coast_EW.dat

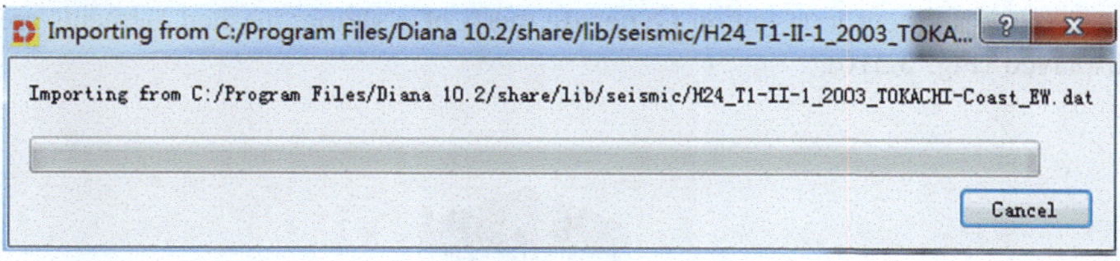

Fig. 3.307 Importing file from storage working directory

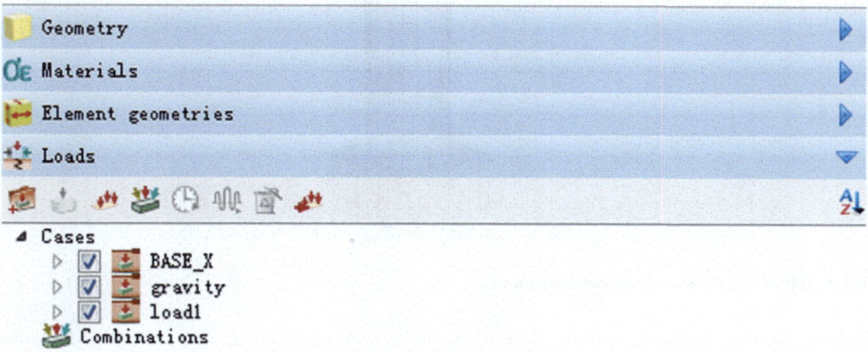

Fig. 3.308 Generated **BASE_X** and **Mode load 1**

Selecting the whole model of concrete pier, **seeding method** is **Element size** with **Desired size** 0.5 m. **Mesher type** is **Hexa/Quad**, and determination of **Mid-side node location** is **Linear interpolation** (see Fig. 3.309).

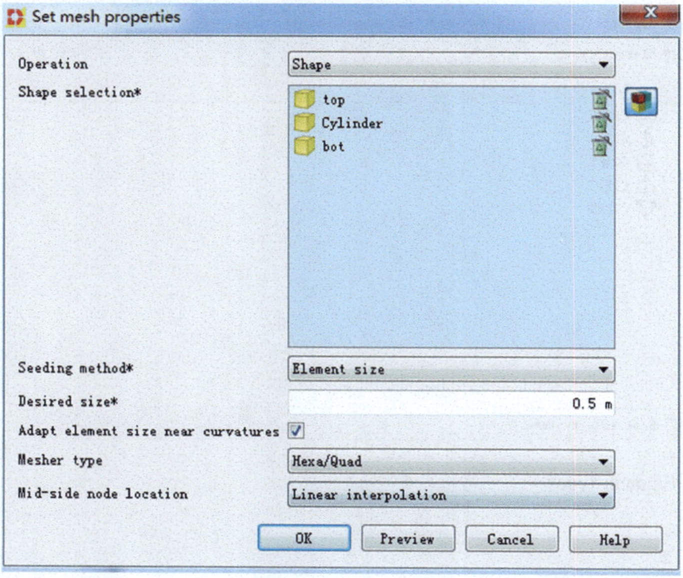

Fig. 3.309 Meshing interface

Clicking shortcut icon button **Generate mesh of a shape**, meshed elements are displayed (Fig. 3.310).

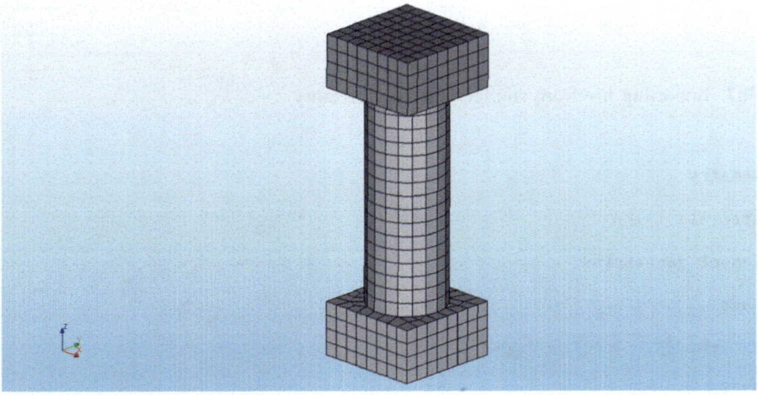

Fig. 3.310 Generation of meshed elements

Element types can be checked as in Fig. 3.311.

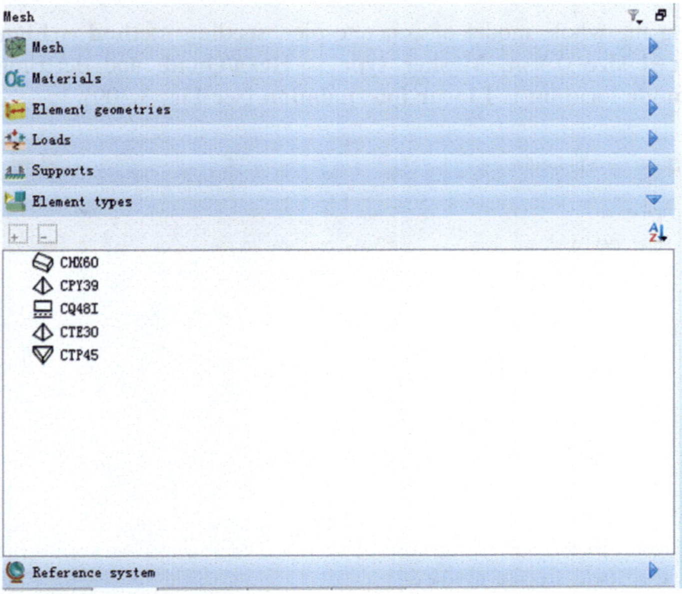

Fig. 3.311 Element types

Click **Add an analysis** button to construct a new analysis case named Analysis3. Structural nonlinear is initially constructed and load case **gravity** is added into load set 1 with user-specified size 1.00000 (see Fig. 3.312). In the aspect of **Equilibrium iteration**, maximum number of iterations is 20 and **Newton–Raphson** iteration method is applied. Both **Force** and **Displacement** are chosen as convergence norm with convergence tolerance and abort criterion 0.01 and 10,000 respectively (see Figs. 3.313 and 3.314).

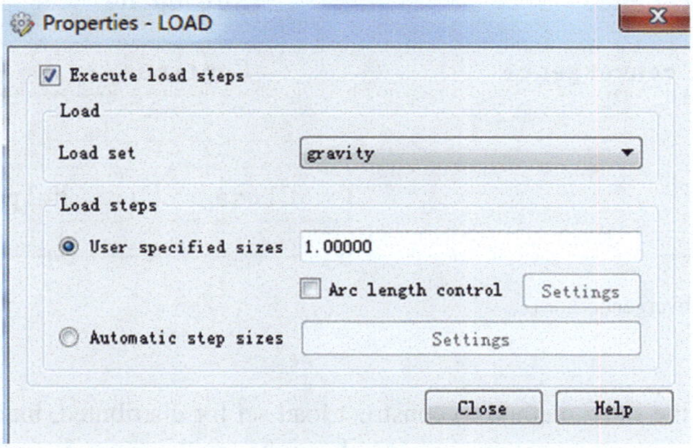

Fig. 3.312 Specification for **gravity load set**

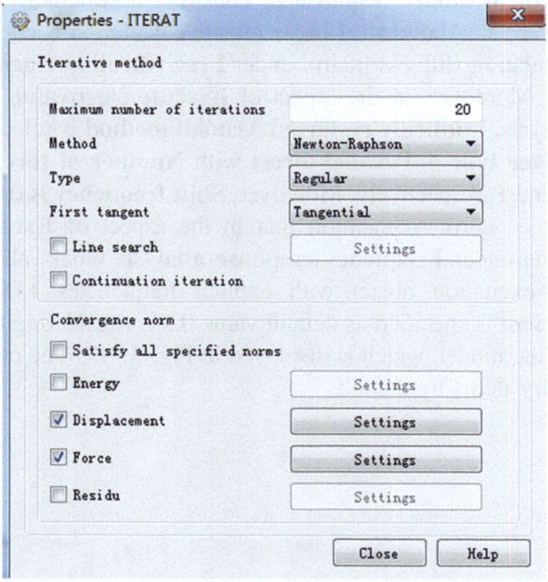

Fig. 3.313 Specification for iteration

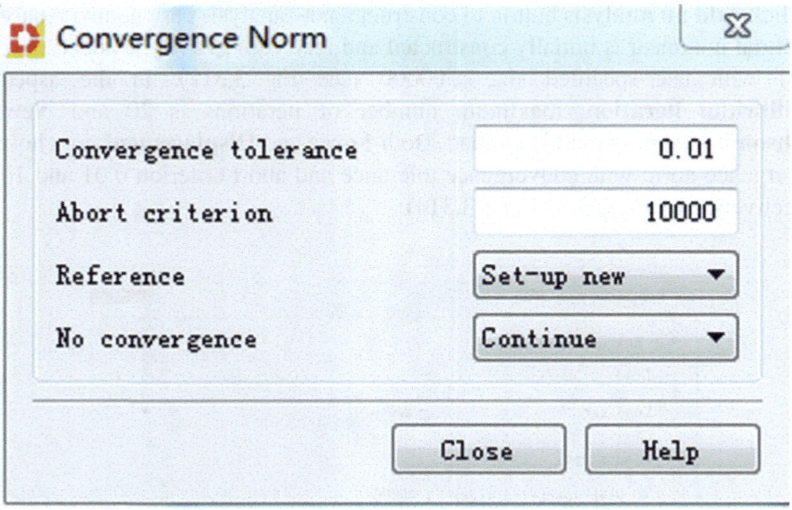

Fig. 3.314 Convergence norm

Applying the same method to construct load set for distributed, **load1** is chosen as load set 2. Since parameters and specifications are the same as former, it is not repeated here.

After completion of constructing structural nonlinear analysis, dynamic analysis is constructed in the **Analysis 3** block. Right-click **Analysis 3-Add command-Structural model response** to construct a structural dynamic model response for **BASE_X**. Above all, **Linear elastic** calculation is chosen as solution property for calculating stiffness matrix under **Free vibration eigenvalue analysis** (see Fig. 3.315). Moreover, in the aspect of **Execute eigenvalue analysis** under **Eigenvalue analysis, Implicitly restarted Arnoldi method** is selected as **Solution method** and **Solver type** is **Parallel direct** with **Number of threads** and **eigen-frenquencies** 3 and 1, respectively. Moreover, **Shift frequency** is 0.1 Hz, as shown in Fig. 3.316. It is worth to mention that in the aspect of **Execute frequency response analysis** under **Frequency response analysis** block, all the modes are selected as the excitation object with explicit frequencies 1.0000–2(0.01)Hz. **Damping coefficient** is specified as default value 0.01, considering the influence of air resistance on the model, which is displayed in Fig. 3.317. The output results are kept as all primary items by default.

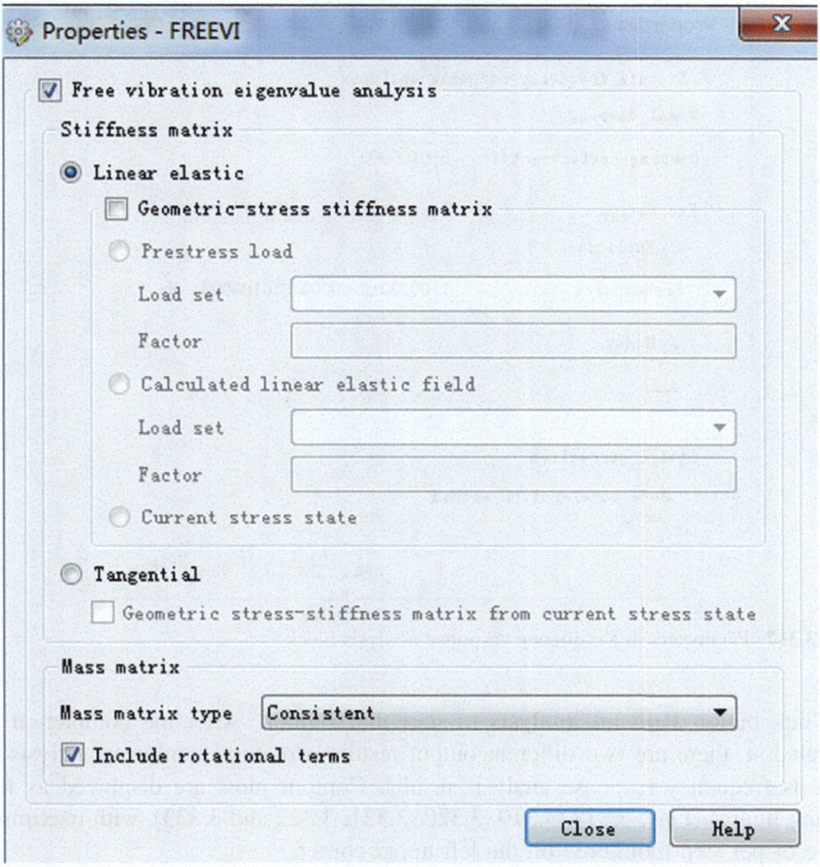

Fig. 3.315 **Linear elastic** calculation for stiffness matrix

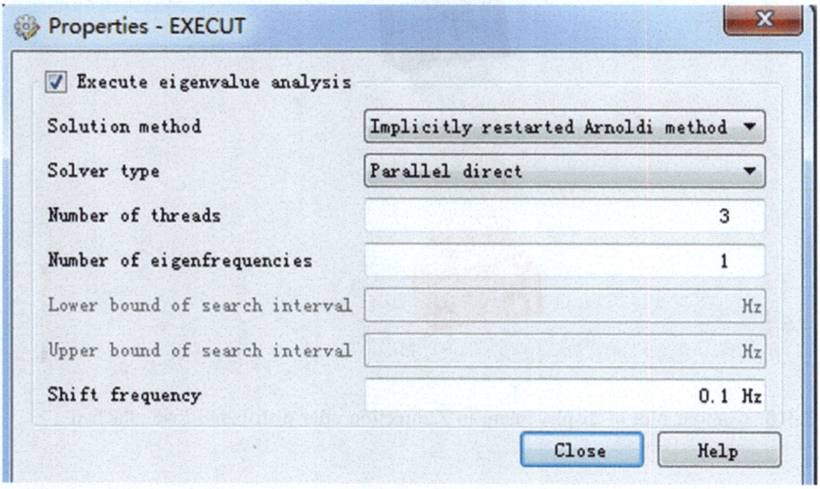

Fig. 3.316 Specification for **Execute eigenvalue analysis**

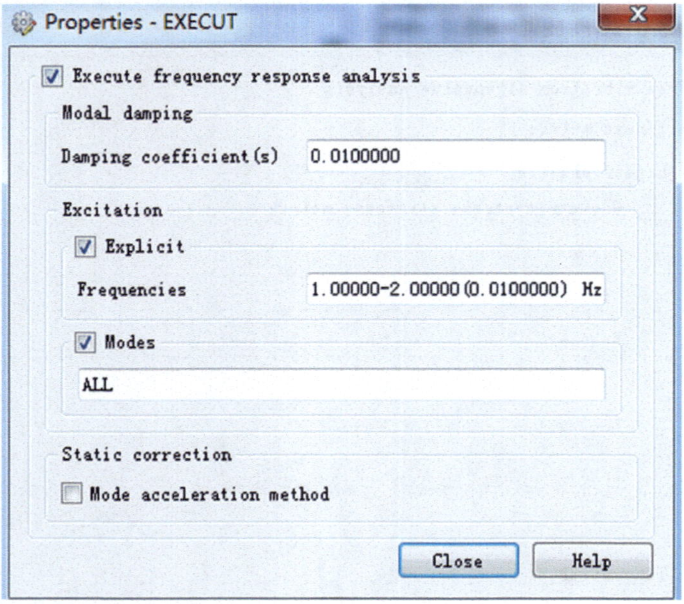

Fig. 3.317 Parameters in **Frequency response analysis** block

Click button **Run an analysis** to start the solution. After the completion of calculation, there are two different output results: structural nonlinear analysis as well as frequency response analysis results. Contour plots are displayed as following figures (Figs. 3.318, 3.319, 3.320, 3.321, 3.322 and 3.323), with maximum value of per step monitored on the left upper corner.

Fig. 3.318 Contour plot of displacement in Z direction after distributed load attached

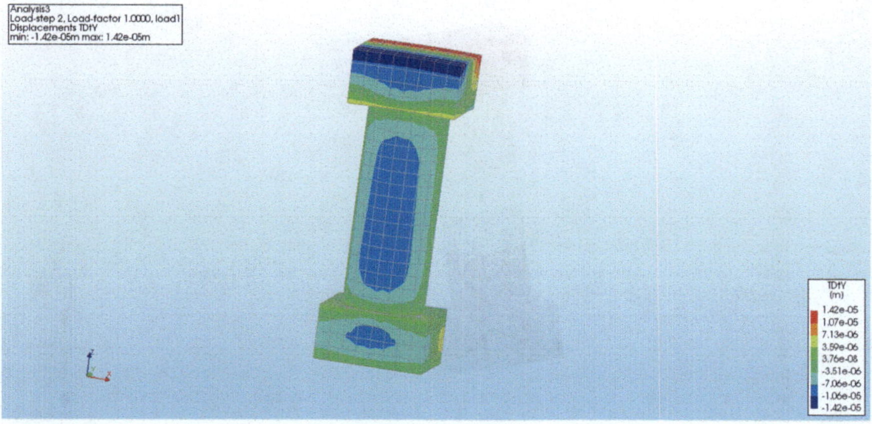

Fig. 3.319 Contour plot of displacement in Y direction after distributed load attached

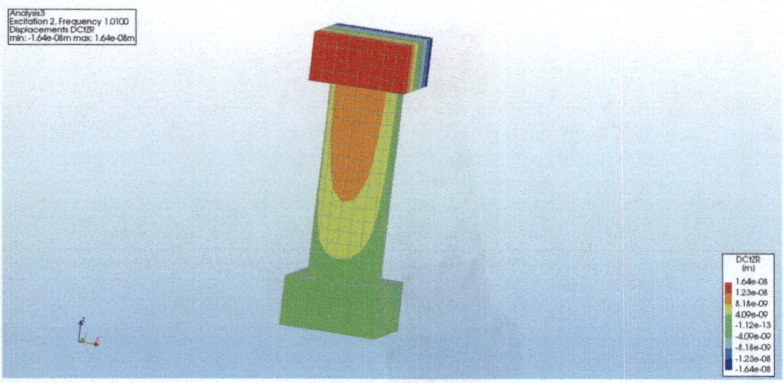

Fig. 3.320 Contour plot of displacement in Z direction when frequency is 1.01

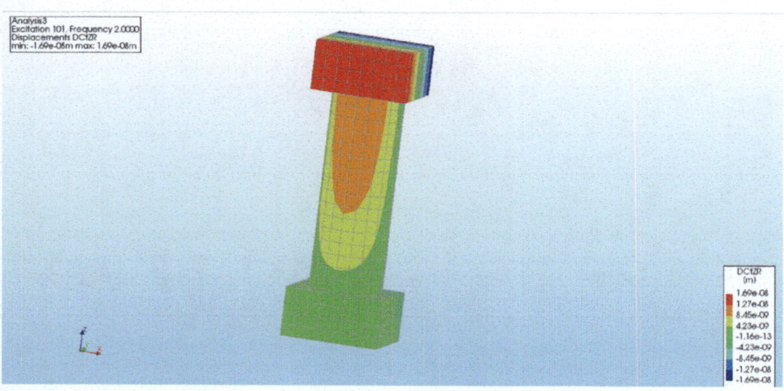

Fig. 3.321 Contour plot of displacement in Z direction when frequency is 2

Fig. 3.322 Contour plot of displacement in Y direction when frequency is 1.01

Fig. 3.323 Contour plot of displacement in Y direction when frequency is 2

Command console in Python language is shown as follows:

```
newProject( "Bridge Pier.dpf", 100 )
setModelAnalysisAspects( [ "STRUCT" ] )
setModelDimension( "3D" )
setDefaultMeshOrder( "QUADRATIC" )
setDefaultMesherType( "HEXQUAD" )
setDefaultMidSideNodeLocation( "LINEAR" )
createBlock( "top", [ 0, 0, 0 ], [ 4, 4, 2 ] )
translate( [ "top" ], [ 0, 0, 8 ] )
saveProject(   )
createCylinder( "Cylinder", [ 2, 2, 0 ], [ 0, 0, 1 ], 1.5, 8 )
saveProject(   )
translate( [ "top", "Cylinder" ], [ 0, 0, 2 ] )
saveProject(   )
createBlock( "bot", [ 0, 0, 0 ], [ 4, 4, 2 ] )
saveProject(   )
fitAll(   )
saveProject(   )
createLine( "bar1", [ 0.8, 2, 1 ], [ 0.8, 2, 9 ] )
arrayCopy( [ "bar1" ], [ 2.4, 0, 0 ], [ 0, 0, 0 ], [ 0, 0, 0 ], 1 )
saveProject(   )
hide( "SHAPE", [ "bot" ] )
hide( "SHAPE", [ "Cylinder" ] )
hide( "SHAPE", [ "top" ] )
show( "SHAPE", [ "bot" ] )
show( "SHAPE", [ "Cylinder" ] )
show( "SHAPE", [ "top" ] )
hide( "SHAPE", [ "Cylinder" ] )
hide( "SHAPE", [ "top" ] )
arrayCopy( [ "bar1", "bar2" ], [ 0, 0.5, 0 ], [ 0, 0, 0 ], [ 0, 0, 0 ], 1 )
saveProject(   )
show( "SHAPE", [ "top" ] )
show( "SHAPE", [ "Cylinder" ] )
arrayCopy( [ "bar1", "bar2" ], [ 0, -0.5, 0 ], [ 0, 0, 0 ], [ 0, 0, 0 ], 1 )
saveProject(   )
addMaterial( "bat", "REINFO", "VMISES", [] )
setParameter( "MATERIAL", "bat", "LINEAR/ELASTI/YOUNG", 2.1e+11 )
setParameter( "MATERIAL", "bat", "PLASTI/YLDTYP", "KAPSIG" )
setParameter( "MATERIAL", "bat", "PLASTI/YLDTYP", "NONE" )
setParameter( "MATERIAL", "bat", "PLASTI/YLDTYP", "KAPSIG" )
setParameter( "MATERIAL", "bat", "PLASTI/HARDI2/KAPSIG", [] )
setParameter( "MATERIAL", "bat", "PLASTI/HARDI2/KAPSIG", [ 0, 2.43e+08, 0.1, 4e+08, 1,
4.4e+08 ] )
```

```
addGeometry( "Element geometry 1", "RELINE", "REBAR", [] )
rename( "GEOMET", "Element geometry 1", "bar" )
setParameter( "GEOMET", "bar", "REIEMB/CROSSE", 0.001 )
setParameter( "GEOMET", "bar", "REIEMB/CROSSE", 0.001 )
setParameter( "GEOMET", "bar", "REIEMB/CROSSE", 0.001 )
setReinforcementAspects( [ "bar1", "bar2", "bar3", "bar4", "bar5", "bar6" ] )
assignMaterial( "bat", "SHAPE", [ "bar1", "bar2", "bar3", "bar4", "bar5", "bar6" ] )
assignGeometry( "bar", "SHAPE", [ "bar1", "bar2", "bar3", "bar4", "bar5", "bar6" ] )
resetElementData( "SHAPE", [ "bar1", "bar2", "bar3", "bar4", "bar5", "bar6" ] )
setReinforcementDiscretization( [ "bar1", "bar2", "bar3", "bar4", "bar5", "bar6" ], "SECTION" )
saveProject(  )
addMaterial( "concrete", "CONCR", "TSCR", [] )
setParameter( "MATERIAL", "concrete", "LINEAR/ELASTI/YOUNG", 3.45e+10 )
setParameter( "MATERIAL", "concrete", "LINEAR/ELASTI/POISON", 0.15 )
setParameter( "MATERIAL", "concrete", "LINEAR/MASS/DENSIT", 2500 )
setParameter( "MATERIAL", "concrete", "LINEAR/MASS/DENSIT", 2500 )
setParameter( "MATERIAL", "concrete", "LINEAR/MASS/DENSIT", 2500 )
setParameter( "MATERIAL", "concrete", "MODTYP/TOTCRK", "ROTATE" )
setParameter( MATERIAL, "concrete", "TENSIL/TENCRV", "MC1990" )
setParameter( MATERIAL, "concrete", "TENSIL/TENSTR", 2600000 )
setParameter( MATERIAL, "concrete", "TENSIL/GF1", 150 )
setParameter( "MATERIAL", "concrete", "TENSIL/DMAX", 8 )
setParameter( "MATERIAL", "concrete", "COMPRS/COMCRV", "MAEKCC" )
setParameter( "MATERIAL", "concrete", "COMPRS/COMSTR", 32500000 )
setParameter( "MATERIAL", "concrete", "COMPRS/REDTEN", "LINEAR" )
setElementClassType( "SHAPE", [ "Cylinder", "bot", "top" ], "STRSOL" )
assignMaterial( "concrete", "SHAPE", [ "Cylinder", "bot", "top" ] )
addSet( "GEOMETRYSUPPORTSET", "co1" )
createSurfaceSupport( "co1", "co1" )
setParameter( "GEOMETRYSUPPORT", "co1", "AXES", [ 1, 2 ] )
setParameter( "GEOMETRYSUPPORT", "co1", "TRANSL", [ 1, 1, 1 ] )
setParameter( "GEOMETRYSUPPORT", "co1", "ROTATI", [ 0, 0, 0 ] )
attach( "GEOMETRYSUPPORT", "co1", "bot", [[ 1.705708, 2.294292, 0 ]] )
addSet( "GEOMETRYLOADSET", "Geometry load case 1" )
rename( "GEOMETRYLOADSET", "Geometry load case 1", "load1" )
createSurfaceLoad( "lo1", "load1" )
setParameter( "GEOMETRYLOAD", "lo1", "FORCE/VALUE", -500000 )
setParameter( "GEOMETRYLOAD", "lo1", "FORCE/DIRECT", 3 )
attach( "GEOMETRYLOAD", "lo1", "top", [[ 2.294292, 2.294292, 12 ]] )
addSet( "GEOMETRYLOADSET", "gravity" )
createModelLoad( "gravity", "gravity" )
saveProject(  )
```

```
setAnalysisCommandDetail( "Analysis3", "Structural modal response",
"RESPON/EXECUT/DAMPIN", "0.0100000" )
addAnalysisCommandDetail( "Analysis3", "Structural modal response",
"RESPON/EXECUT/MODES" )
setAnalysisCommandDetail( "Analysis3", "Structural modal response",
"RESPON/EXECUT/MODES", True )
setAnalysisCommandDetail( "Analysis3", "Structural modal response",
"RESPON/EXECUT/EXPLIC/FREQUE", "1.00000-2(0.01)" )
runSolver( "Analysis3" )
showView( "RESULT" )
```

3.8 Nonlinear Dynamic Analysis for Reinforced Concrete

A semi-structural portal frame model [2] is created as the object of dynamic time-history analysis in this case (see Fig. 3.324). Newmark-β and Wilson-θ method are applied as implicit integral algorithms in nonlinear dynamic analysis. Cracking model is total strain-based crack model, where **Maekawa Cracked Concrete curve** in DIANA is chosen as compression model and tension softening curve is **fib Model Code for Concrete Structures 2010** with elastic modulus and compressive strength $3.45 \times 10^{10}\ \mathrm{N/m^2}$ and 32.5 MPa, respectively. The first 5 s of 72,448 earthquake wave is selected and the interval is specified as 0.1 s as a substep length for time–load curve. von Mises plasticity model is applied for reinforcement, and the damping coefficients for mass matrix as well as stiffness matrix are 1.1042 and 0.00165, respectively [3] (see Dynamics of Structures, R. Claugh and J. Penzien 2006) to investigate cracking issues under seismic effect. Meanwhile, results of two methods under seismic effect are compared.

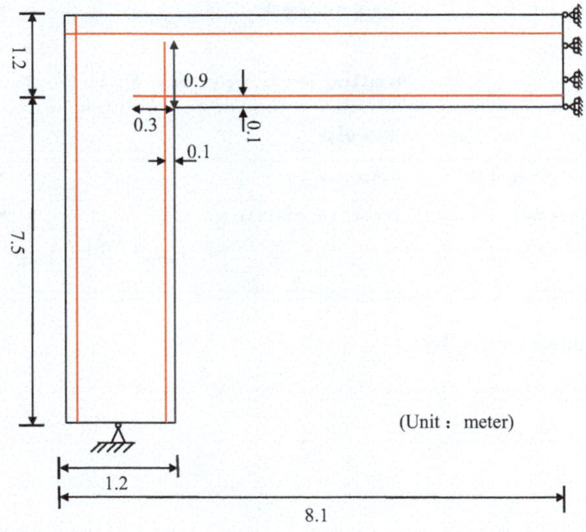

Fig. 3.324 Semi-structural size

Essentials of learning

(1) *Mastering specification of tension softening curve in **fib Model Code for Concrete Structures 2010** as well as **Maekawa Cracked Concrete curve** in compressive model.*
(2) *Learning to specify time–load curve under time-dependent model.*
(3) *Specifications of nonlinear parameters in* Newmark-β and Wilson-θ *methods.*

Starting DianaIE and clicking menu bar **File—New, New project** dialog box appears, which is named as **ChaiShun-Newmark**. **Analysis** type is **Structural** and **Two dimensional** is selected as **Dimensions** with maximum **Model size** 100 m, ranging from –50 to 50 m in X and Y directions.

Default mesher type is **Hexa/Quad** while **Quadratic** is selected as **Default mesh order** and the determination of mid-side node location is the same as former parts—**Linear interpolation** (see Fig. 3.325).

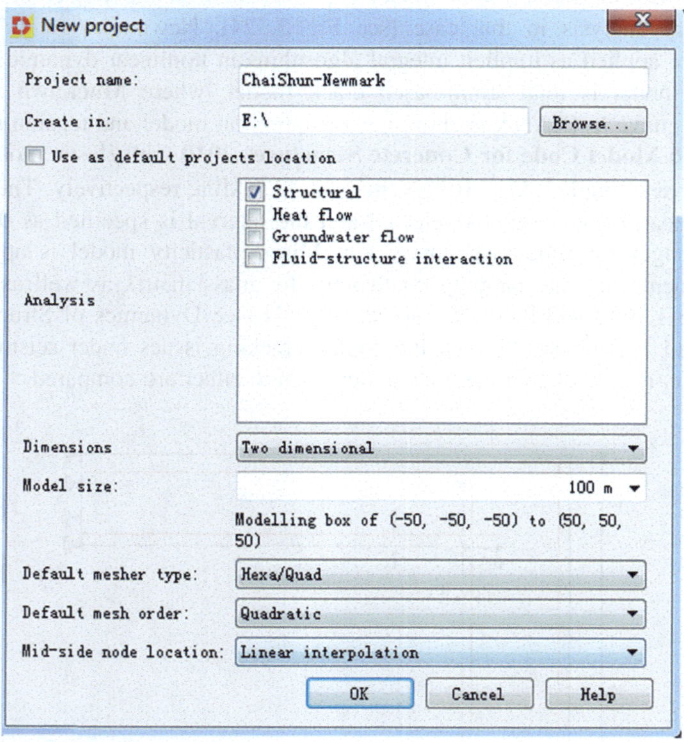

Fig. 3.325 New project dialog box

Click shortcut icon button **Adds a sheet**  and input coordinate values [0, 0, 0], [0.6, 0, 0], [1.2, 0, 0], [1.2, 7.5, 0], [8.1, 7.5, 0], [8.1, 8.7, 0], [0, 8.7, 0] in turn, respectively, to create Sheet1. Clicking OK button, the sheet 1 is generated. Then click shortcut icon button **Adds a line**  to create geometric model of reinforcement; the coordinate values, displayed in Table 3.7, is given as input in turn and the reinforcement geometric models are generated (see Fig. 3.326).

Table 3.7 Geometric coordinate values of reinforcement bars

bar1	[0.1, 0, 0], [0.1, 8.7, 0]
bar2	[1.1, 0, 0], [1.1, 8.4, 0]
bar3	[0.9, 7.6, 0], [8.1, 7.6, 0]
bar4	[0, 8.6, 0], [8.1, 8.6, 0]

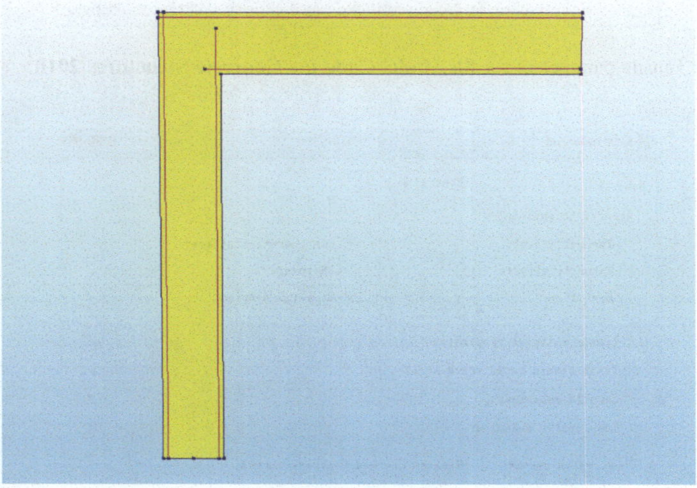

Fig. 3.326 Geometric model of concrete and reinforcement bars

For concrete portal frame, plane stress element is selected as **Element class**. The next step is to specify concrete parameters, where concrete elastic modulus is 3.45×10^{10} N/m^2 with Poisson's ratio 0.15. Since Rayleigh damping parameters should be taken into account in the nonlinear dynamic analysis, **Rayleigh damping** aspect is ticked. The factor for mass matrix and stiffness matrix in Rayleigh damping parameters is 1.10412 and 0.00165 [3], respectively. Total strain-based crack model is selected as smeared cracking model in this case, and tension softening curve is **fib Model Code for Concrete Structures 2010** with tensile strength and Mode-I tensile fracture energy 2.6 MPa and 500 N/m, respectively. **Maekawa Cracked Concrete curve** in DIANA is chosen as compression model, which is applicable for low-period cyclic loading and dynamic nonlinear analysis, and can be perfectly combined with Rayleigh damping module (see Figs. 3.327 and 3.328). Thickness value in cross-section geometric properties is 0.7 m.

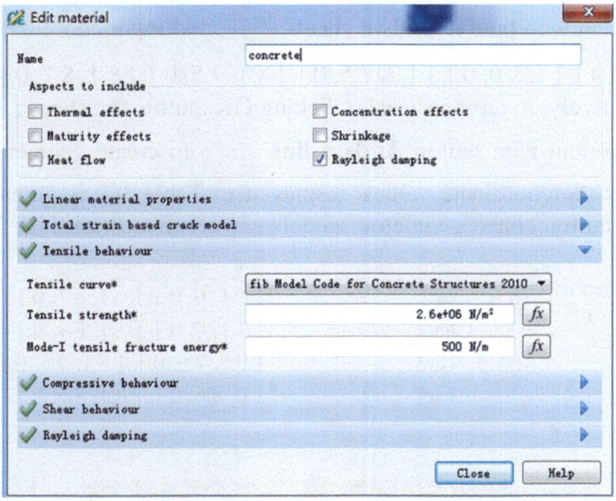

Fig. 3.327 Tensile parameters for **fib Model Code for Concrete Structures 2010**

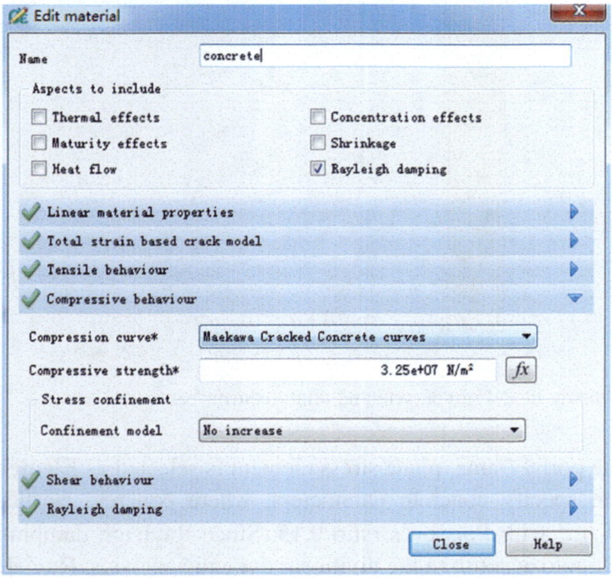

Fig. 3.328 Parameter specifications for **Maekawa Cracked Concrete curve**

Defining material properties of reinforcement, Rayleigh damping module is ticked and **von Mises plasticity** is selected as material model with Young's modulus 2.1×10^{11} N/m^2 and yield stress 4.5×10^{8} N/m^2. Node at the middle site of bottom edge is selected as constraint point, which is attached with translational constraints both in X and Y direction. Meanwhile, translational constraint in horizontal X direction is attached to the upper right-side edge (see Fig. 3.329).

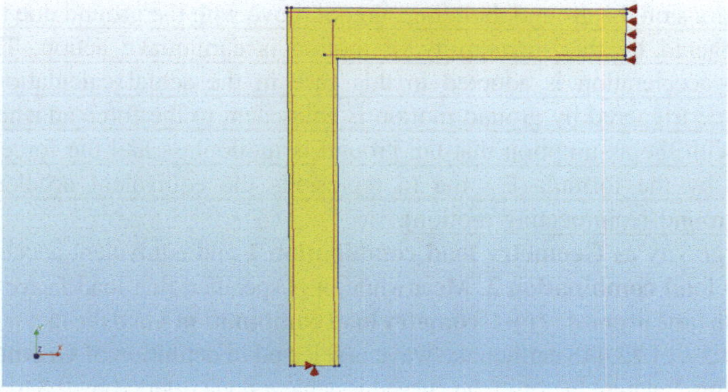

Fig. 3.329 Attached constraints on the model

Manipulation of defining gravity is the same as former, which is not repeated here. Attaching in the type of **Equivalent acceleration** with the name of **lo2** and **Load target type** is **Solid. Loaded body** is the whole **Sheet 1.** Since the earthquake wave required to be input to the following time–load factor curve is taking g as a unit, thus the value of equivalent acceleration is 9.8 m/s^2 in negative X direction (see Fig. 3.330).

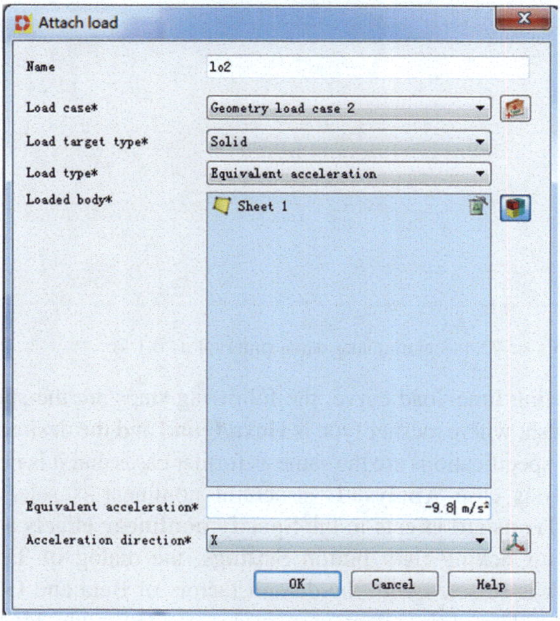

Fig. 3.330 Specification interface of equivalent acceleration

Note: According to the First Law of Newton Mechanics, when there are horizontal motions on the ground, buildings do not move with the ground due to inertia, action triggered by this incongruity of motion is earthquake action. Therefore, equivalent acceleration is adopted in this case in the actual calculation, where inertial force triggered by ground motion is equivalent to the force attached on the structure with the assumption that the ground is motionless and the force value is calculated by the formula F = ma (a represents the equivalent acceleration of constant ground reciprocating motion).

Define gravity as **Geometry load combination 1** and equivalent acceleration is **Geometry load combination 2.** Meanwhile, it is specified that load factor does not change with time in one day for **Geometry load combination 1** and the factor is always 1. The first 5 s of 72,448 earthquake wave is selected in definition of **Geometry load combination 2**, where factor in the curved is achieved via divided by 9.8 to represent the numerical relationship between equivalent acceleration and gravity. Data interval is specified as 0.1 s, as a substep length, for time–load curve (see Fig. 3.331).

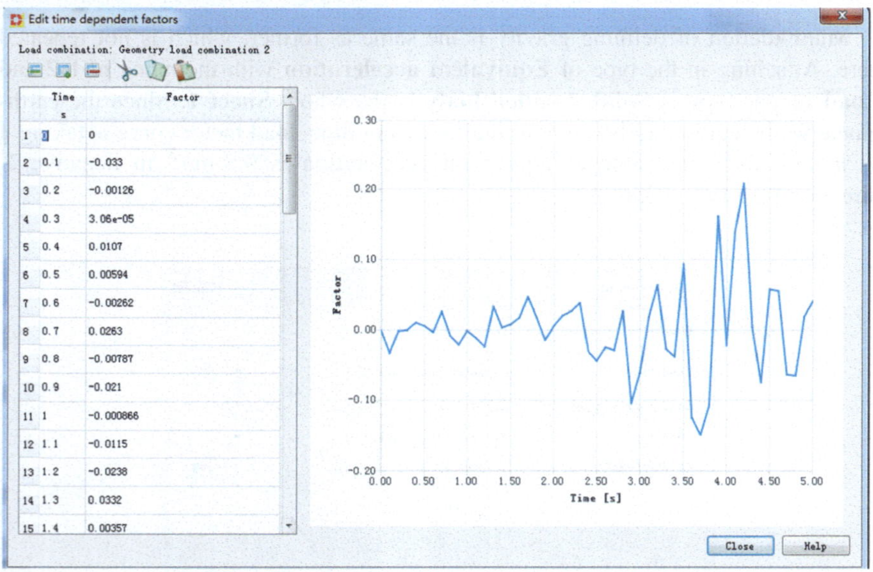

Fig. 3.331 First 5 s of 72,448 earthquake wave (interval as 0.1 s)

After completing time–load curve, the following steps are the specifications and generation of mesh, where mesher type is **Hexa/Quad** and the desired element size is 0.1 m. Meshing specifications are the same as former cases and it is not repeated here.

Adding analysis case **Analysis1**, structural nonlinear is selected. **Physically nonlinear** and **Transient effects** in the **Specify nonlinear effects** aspect are ticked (see Fig. 3.332). Clicking right button **Settings**, the dialog of **Transient effects** ejects; Newmark is selected with the default factors of Beta and Gamma 0.25 and 0.5, respectively. **Dynamic effects** aspect is selected with **Mass matrix and Damping matrix** both **Consistent.** Meanwhile, **Time derivative effects option** is also selected (see Fig. 3.333).

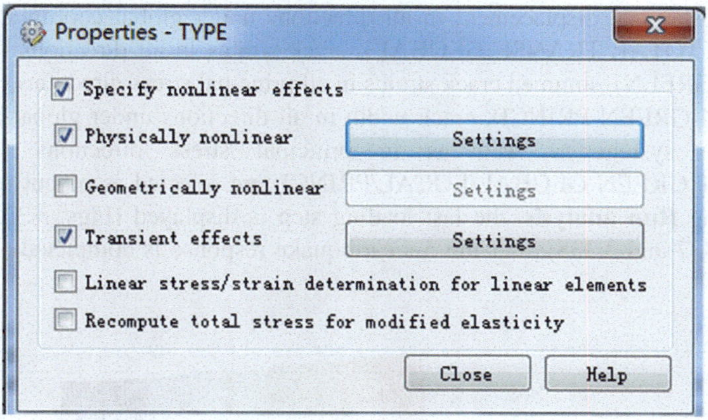

Fig. 3.332 Specifying nonlinear effects

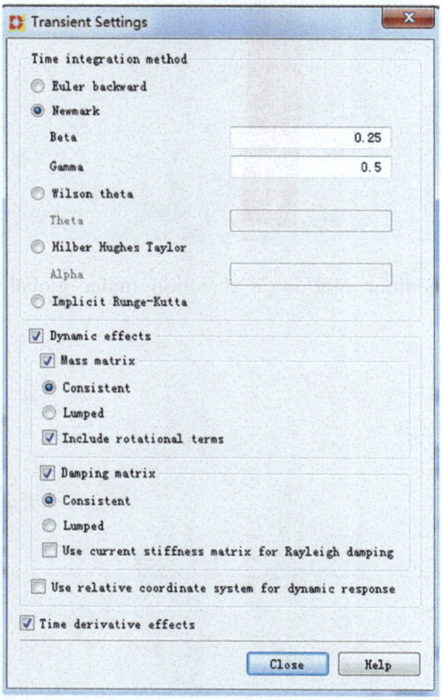

Fig. 3.333 Specifications for Newmark-β

Geometry load combination 1, where the gravity is located, is specified as new execute block. Iterative method is **Newton–Raphson** with the maximum number of iterations 50. **Force** and **Displacement** are ticked as **Convergence norm** with the convergence tolerance 0.01. Specifying **Geometry load combination 2** and time step with the same method, the number of time steps and substep length is 0.100000

(50). Translational displacement in all directions under global coordinate system (DISPLA TOTAL TRANSL GLOBAL), crack strains in all directions (STRAIN CRACK GREEN), summed crack strains in all principal stress directions (STRAIN CRKSUM GREEN PRINCI), crack width in all directions under global and local coordinate system as well as in principal stress directions (STRAIN CRKWDT GREEN GLOBAL/LOCAL/PRINCI) are selected as output results.

Clicking **Run analysis**, the last loading step is displayed (Figs. 3.334, 3.335, 3.336, 3.337 and 3.338) after the 5 s earthquake response is completed.

Fig. 3.334 Displacement contour plot in Y direction under global coordinate system with Newmark-β

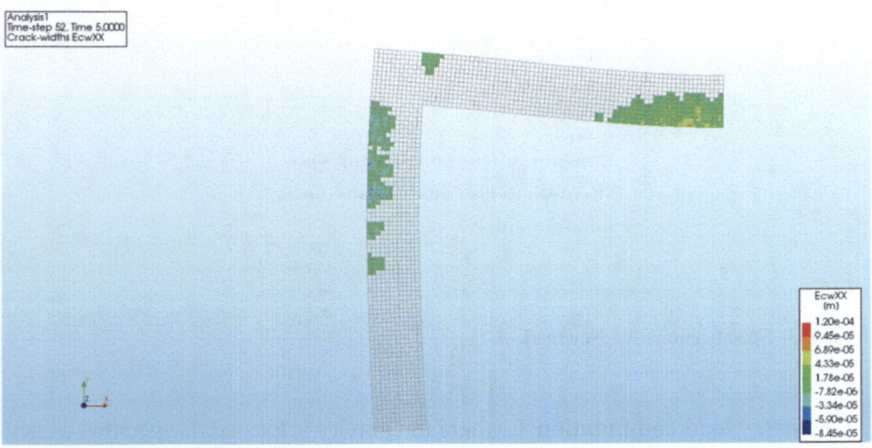

Fig. 3.335 Crack width contour plot in X direction under global coordinate system (EcwXX) with Newmark-β

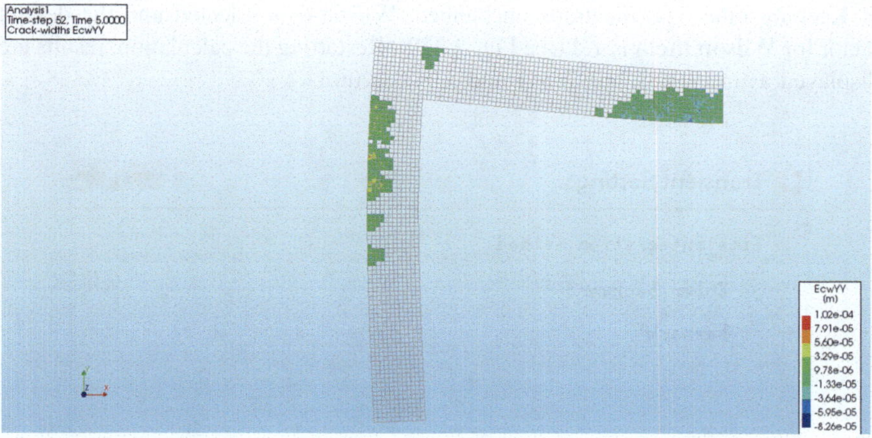

Fig. 3.336 Crack width contour plot in Y direction under global coordinate system (EcwYY) with Newmark-β

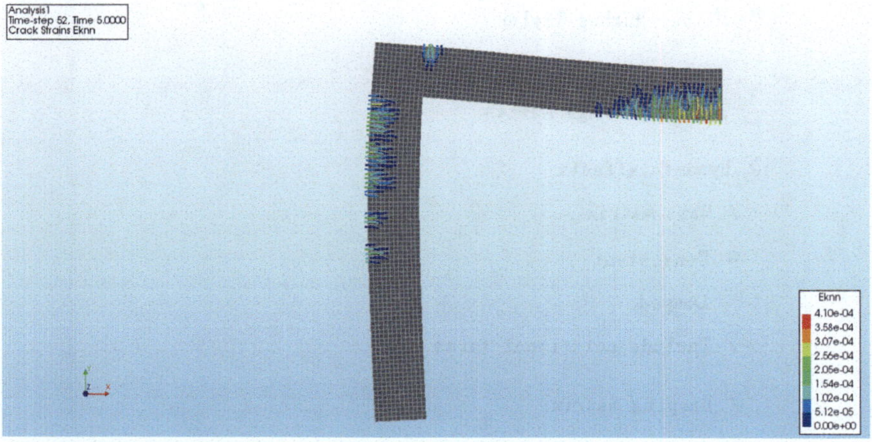

Fig. 3.337 Crack strain contour plot in normal direction (Eknn) with Newmark-β with Newmark-β

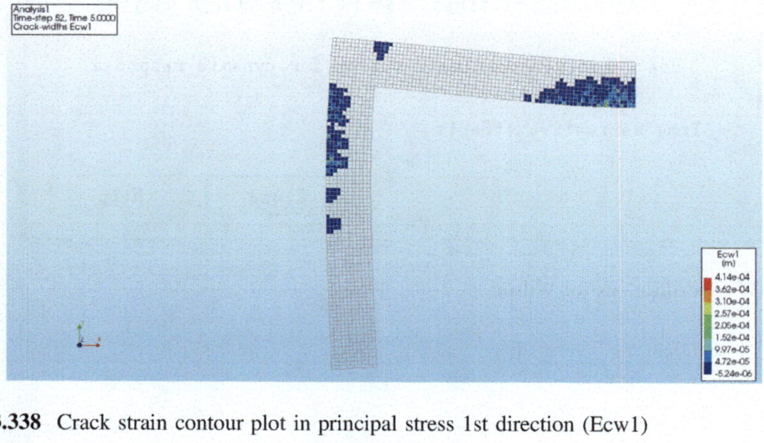

Fig. 3.338 Crack strain contour plot in principal stress 1st direction (Ecw1)

Keeping other specifications unchanged, Wilson-θ is selected and the default factor for **Wilson theta** is 1.4 (see Fig. 3.339). Restarting the calculation, results are displayed as in Figs. 3.340, 3.341 and 3.342 in turn.

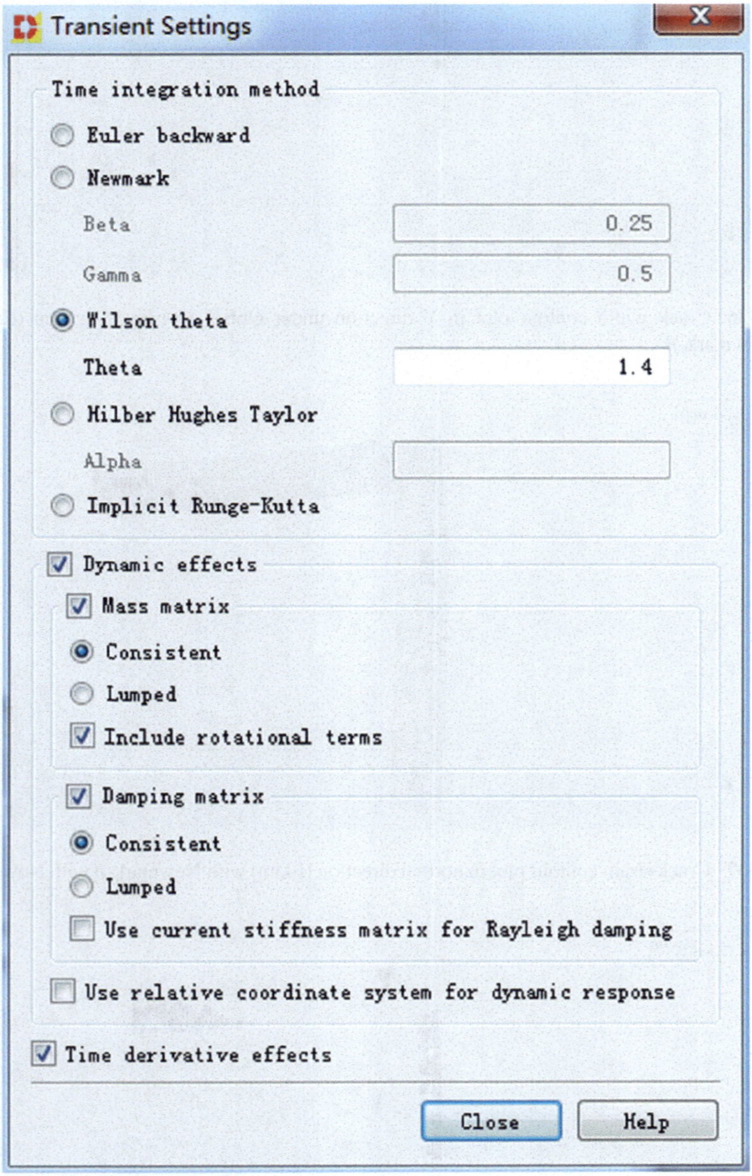

Fig. 3.339 Specifications for Wilson-θ

Fig. 3.340 Displacement contour plot in Y direction under global coordinate system with Wilson-θ

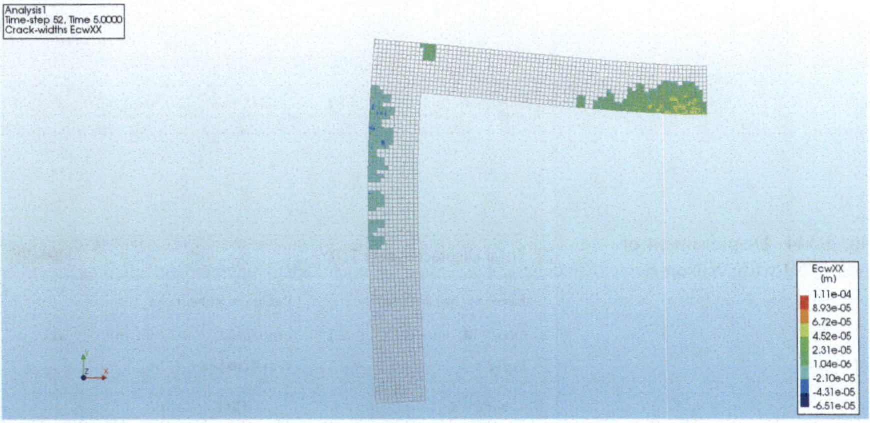

Fig. 3.341 Crack width contour plot in X direction under global coordinate system (EcwXX) with Wilson-θ

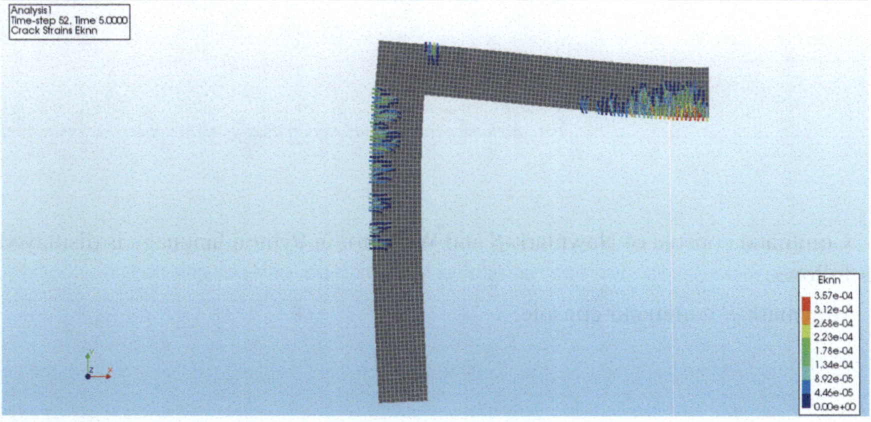

Fig. 3.342 Crack strain contour plot in normal direction (Eknn) with Wilson-θ

Selecting node id 2289, check displacement values in Y direction after 5 s earthquake wave under two methods, which are 7.65648 and 7.49014 mm, respectively (see Figs. 3.343 and 3.344). Comparison validates that simulation results in DIANA are similar in dynamic nonlinear analysis with these two methods.

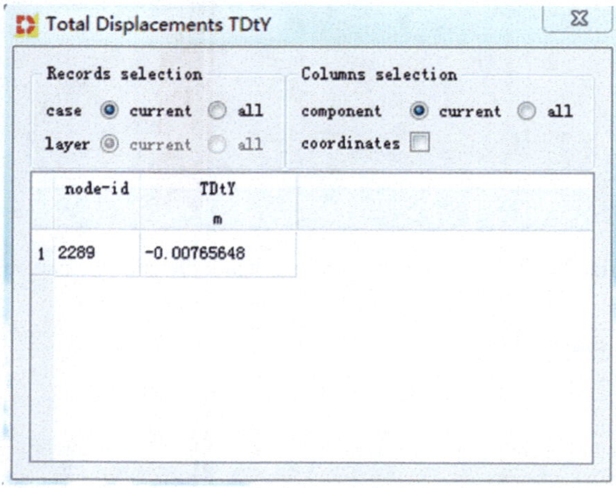

Fig. 3.343 Displacement of node 2889 with Newmark-β

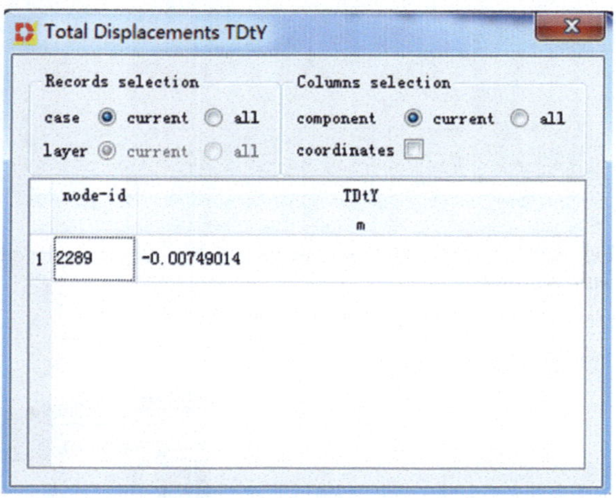

Fig. 3.344 Displacement of node 2889 with Wilson-θ

Command console of Newmark-β and Wilson-θ in Python language is displayed as follows:

Newmark-β command console:

```
newProject( "ChaiShun-Newmark", 100 )
setModelAnalysisAspects( [ "STRUCT" ] )
setModelDimension( "2D" )
setDefaultMeshOrder( "QUADRATIC" )
setDefaultMesherType( "HEXQUAD" )
setDefaultMidSideNodeLocation( "ONSHAP" )
setUnit( "TEMPER", "CELSIU" )
setUnit( "ANGLE", "DEGREE" )
saveProject( )
createSheet( "Sheet 1", [[ 0, 0, 0 ],[ 0.6, 0, 0 ],[ 1.2, 0, 0 ],[ 1.2, 7.5, 0 ],[ 8.1, 7.5, 0 ],[ 8.1, 8.7,
0 ],[ 0, 8.7, 0 ]] )
createLine( "bar1", [ 0.1, 0, 0 ], [ 0.1, 8.7, 0 ] )
saveProject( )
createLine( "bar2", [ 1.1, 0, 0 ], [ 1.1, 8.4, 0 ] )
saveProject( )
createLine( "bar3", [ 0.9, 7.6, 0 ], [ 0.9, 8.1, 0 ] )
removeShape( [ "bar3" ] )
createLine( "bar3", [ 0.9, 7.6, 0 ], [ 8.1, 7.6, 0 ] )
saveProject( )
createLine( "bar4", [ 0, 8.6, 0 ], [ 8.1, 8.6, 0 ] )
saveProject( )
addMaterial( "concrete", "CONCR", "TSCR", [ "RAYDAM" ] )
setParameter( MATERIAL, "concrete", "LINEAR/ELASTI/YOUNG", 3.45e+10 )
setParameter( MATERIAL, "concrete", "LINEAR/ELASTI/POISON", 0.15 )
setParameter( MATERIAL, "concrete", "LINEAR/MASS/DENSIT", 2500 )
setParameter( MATERIAL, "concrete", "TENSIL/TENCRV", "MC2010" )
setParameter( MATERIAL, "concrete", "TENSIL/TENSTR", 2600000 )
setParameter( MATERIAL, "concrete", "TENSIL/GF1", 500 )
setParameter( MATERIAL, "concrete", "COMPRS/COMCRV", "MAEKCC" )
setParameter( MATERIAL, "concrete", "COMPRS/COMSTR", 32500000 )
setParameter( MATERIAL, "concrete", "RAYDAM/RAYLEI", [ 1.1042, 0.00165 ] )
addGeometry( "Element geometry 1", "SHEET", "MEMBRA", [] )
rename( GEOMET, "Element geometry 1", "concrete" )
setParameter( GEOMET, "concrete", "THICK", 0.7 )
clearReinforcementAspects( [ "Sheet 1" ] )
setElementClassType( SHAPE, [ "Sheet 1" ], "MEMBRA" )
assignMaterial( "concrete", SHAPE, [ "Sheet 1" ] )
assignGeometry( "concrete", SHAPE, [ "Sheet 1" ] )
resetElementData( SHAPE, [ "Sheet 1" ] )
addMaterial( "bar", "REINFO", "VMISES", [ "RAYDAM" ] )
setParameter( MATERIAL, "bar", "LINEAR/ELASTI/YOUNG", 2.1e+11 )
setParameter( MATERIAL, "bar", "PLASTI/HARDI1/YLDSTR", 4.5e+08 )
```

```
setParameter( MATERIAL, "bar", "RAYDAM/RAYLEI", [ 1.1042, 0.00165 ] )
addGeometry( "Element geometry 2", "RELINE", "REBAR", [] )
rename( GEOMET, "Element geometry 2", "bar" )
setParameter( GEOMET, "bar", "REITYP", "REITRU" )
setParameter( GEOMET, "bar", "REITRU/CROSSE", 0.000157 )
setParameter( GEOMET, "bar", "REITRU/PERIME", 0.012 )
setReinforcementAspects( [ "bar1", "bar2", "bar3", "bar4" ] )
assignMaterial( "bar", SHAPE, [ "bar1", "bar2", "bar3", "bar4" ] )
assignGeometry( "bar", SHAPE, [ "bar1", "bar2", "bar3", "bar4" ] )
resetElementData( SHAPE, [ "bar1", "bar2", "bar3", "bar4" ] )
setReinforcementDiscretization( [ "bar1", "bar2", "bar3", "bar4" ], "ELEMENT" )
saveProject( )
addSet( "GEOMETRYSUPPORTSET", "co1" )
createPointSupport( "co1", "co1" )
setParameter( "GEOMETRYSUPPORT", "co1", "AXES", [ 1, 2 ] )
setParameter( "GEOMETRYSUPPORT", "co1", "TRANSL", [ 1, 1, 0 ] )
setParameter( "GEOMETRYSUPPORT", "co1", "ROTATI", [ 0, 0, 0 ] )
attach( "GEOMETRYSUPPORT", "co1", "Sheet 1", [[ 0.6, 1.8378494e-34, -5.2871687e-18 ]] )
addSet( "GEOMETRYSUPPORTSET", "Geometry support set 2" )
rename( "GEOMETRYSUPPORTSET", "Geometry support set 2", "co2" )
createLineSupport( "co2", "co2" )
setParameter( "GEOMETRYSUPPORT", "co2", "AXES", [ 1, 2 ] )
setParameter( "GEOMETRYSUPPORT", "co2", "TRANSL", [ 1, 0, 0 ] )
setParameter( "GEOMETRYSUPPORT", "co2", "ROTATI", [ 0, 0, 0 ] )
attach( "GEOMETRYSUPPORT", "co2", "Sheet 1", [[ 8.1, 8.1, -1.4938131e-17 ]] )
saveProject( )
addSet( "GEOMETRYLOADSET", "gravity" )
createModelLoad( "gravity", "gravity" )
createBodyLoad( "lo2", "Geometry load case 2" )
setParameter( GEOMETRYLOAD, "lo2", "LODTYP", "EQUIAC" )
setParameter( GEOMETRYLOAD, "lo2", "EQUIAC/ACCELE", -9.8 )
setParameter( GEOMETRYLOAD, "lo2", "EQUIAC/DIRECT", 1 )
attach( GEOMETRYLOAD, "lo2", [ "Sheet 1" ] )
setDefaultGeometryLoadCombinations( )
setGeometryLoadCombinationFactor( "Geometry load combination 1", "gravity", 1 )
remove( GEOMETRYLOADCOMBINATION, "Geometry load combination 2" )
setGeometryLoadCombinationFactor( "Geometry load combination 1", "gravity", 1 )
remove( GEOMETRYLOADCOMBINATION, "Geometry load combination 1" )
addGeometryLoadCombination( "" )
setGeometryLoadCombinationFactor( "Geometry load combination 1", "gravity", 1 )
addGeometryLoadCombination( "" )
```

```
setGeometryLoadCombinationFactor( "Geometry load combination 2", "Geometry load case 2", 1 )
setTimeDependentLoadFactors( GEOMETRYLOADCOMBINATION, "Geometry load combination 1", [ 0, 86400 ], [ 1, 1 ] )
setTimeDependentLoadFactors( GEOMETRYLOADCOMBINATION, "Geometry load combination 2", [ 0, 0.1, 0.2, 0.3, 0.4, 0.5, 0.6, 0.7, 0.8, 0.9, 1, 1.1, 1.2, 1.3, 1.4, 1.5, 1.6, 1.7, 1.8, 1.9, 2, 2.1, 2.2, 2.3, 2.4, 2.5, 2.6, 2.7, 2.8, 2.9, 3, 3.1, 3.2, 3.3, 3.4, 3.5, 3.6, 3.7, 3.8, 3.9, 4, 4.1, 4.2, 4.3, 4.4, 4.5, 4.6, 4.7, 4.8, 4.9, 5 ], [ 0, -0.033, -0.00126, 3.06e-05, 0.0107, 0.00594, -0.00262, 0.0263, -0.00787, -0.021, -0.000866, -0.0115, -0.0238, 0.0332, 0.00357, 0.00819, 0.0176, 0.0468, 0.018, -0.0139, 0.0062, 0.0206, 0.0271, 0.0385, -0.0307, -0.0442, -0.0246, -0.0288, 0.0267, -0.104, -0.0613, 0.0187, 0.0637, -0.0269, -0.0381, 0.0935, -0.124, -0.148, -0.107, 0.162, -0.0218, 0.141, 0.208, 0.0046, -0.0751, 0.0576, 0.0553, -0.0639, -0.0653, 0.0194, 0.041 ] )
setElementSize( [ "Sheet 1" ], 0.1, -1, True )
setMesherType( [ "Sheet 1" ], "HEXQUAD" )
setMidSideNodeLocation( [ "Sheet 1" ], "LINEAR" )
generateMesh( [] )
hideView( "GEOM" )
showView( "MESH" )
addAnalysis( "Analysis1" )
addAnalysisCommand( "Analysis1", "NONLIN", "Structural nonlinear" )
renameAnalysis( "Analysis1", "Analysis1" )
setAnalysisCommandDetail( "Analysis1", "Structural nonlinear", "EXECUT(1)/ITERAT/MAXITE", 50 )
setAnalysisCommandDetail( "Analysis1", "Structural nonlinear", "EXECUT(1)/ITERAT/CONVER/DISPLA/NOCONV", "CONTIN" )
setAnalysisCommandDetail( "Analysis1", "Structural nonlinear", "EXECUT(1)/ITERAT/CONVER/FORCE/NOCONV", "CONTIN" )
setAnalysisCommandDetail( "Analysis1", "Structural nonlinear", "EXECUT(1)/ITERAT/MAXITE", 50 )
addAnalysisCommandDetail( "Analysis1", "Structural nonlinear", "TYPE/TRANSI" )
setAnalysisCommandDetail( "Analysis1", "Structural nonlinear", "TYPE/TRANSI", True )
addAnalysisCommandDetail( "Analysis1", "Structural nonlinear", "TYPE/TRANSI/DYNAMI" )
setAnalysisCommandDetail( "Analysis1", "Structural nonlinear", "TYPE/TRANSI/DYNAMI", True )
setAnalysisCommandDetail( "Analysis1", "Structural nonlinear", "EXECUT(1)/ITERAT/MAXITE", 50 )
setAnalysisCommandDetail( "Analysis1", "Structural nonlinear", "EXECUT/EXETYP", "LOAD" )
renameAnalysisCommand( "Analysis1", "Structural nonlinear", "Structural nonlinear" )
setAnalysisCommandDetail( "Analysis1", "Structural nonlinear", "EXECUT(2)/ITERAT/MAXITE", 10 )
```

```
setAnalysisCommandDetail( "Analysis1", "Structural nonlinear", "EXECUT(2)/ITERAT/MAXITE", 50 )
setAnalysisCommandDetail( "Analysis1", "Structural nonlinear", "EXECUT(2)/ITERAT/CONVER/DISPLA/NOCONV", "CONTIN" )
setAnalysisCommandDetail( "Analysis1", "Structural nonlinear", "EXECUT(2)/ITERAT/CONVER/FORCE/NOCONV", "CONTIN" )
setAnalysisCommandDetail( "Analysis1", "Structural nonlinear", "EXECUT(2)/ITERAT/MAXITE", 50 )
setAnalysisCommandDetail( "Analysis1", "Structural nonlinear", "EXECUT(2)/ITERAT/MAXITE", 50 )
setAnalysisCommandDetail( "Analysis1", "Structural nonlinear", "EXECUT(1)/ITERAT/MAXITE", 50 )
addAnalysisCommandDetail( "Analysis1", "Structural nonlinear", "TYPE/TRANSI/DYNAMI/DAMPIN" )
setAnalysisCommandDetail( "Analysis1", "Structural nonlinear", "TYPE/TRANSI/DYNAMI/DAMPIN", True )
setAnalysisCommandDetail( "Analysis1", "Structural nonlinear", "EXECUT(2)/LOAD/LOADNR", 2 )
setAnalysisCommandDetail( "Analysis1", "Structural nonlinear", "EXECUT/EXETYP", "TIME" )
renameAnalysisCommandDetail( "Analysis1", "Structural nonlinear", "EXECUT(3)", "new execute block 3" )
setAnalysisCommandDetail( "Analysis1", "Structural nonlinear", "EXECUT(3)/TIME/STEPS/EXPLIC/SIZES", "0.1(50)" )
setAnalysisCommandDetail( "Analysis1", "Structural nonlinear", "EXECUT(3)/ITERAT/MAXITE", 50 )
setAnalysisCommandDetail( "Analysis1", "Structural nonlinear", "EXECUT(3)/ITERAT/CONVER/DISPLA/NOCONV", "CONTIN" )
setAnalysisCommandDetail( "Analysis1", "Structural nonlinear", "EXECUT(3)/ITERAT/CONVER/FORCE/NOCONV", "CONTIN" )
setAnalysisCommandDetail( "Analysis1", "Structural nonlinear", "OUTPUT(1)/SELTYP", "USER" )
addAnalysisCommandDetail( "Analysis1", "Structural nonlinear", "OUTPUT(1)/USER" )
addAnalysisCommandDetail( "Analysis1", "Structural nonlinear", "OUTPUT(1)/USER/DISPLA(1)/TOTAL/TRANSL/GLOBAL" )
addAnalysisCommandDetail( "Analysis1", "Structural nonlinear", "OUTPUT(1)/USER/STRAIN(1)/CRKSUM/GREEN/GLOBAL" )
addAnalysisCommandDetail( "Analysis1", "Structural nonlinear", "OUTPUT(1)/USER/STRAIN(2)/CRKSUM/GREEN/PRINCI" )
addAnalysisCommandDetail( "Analysis1", "Structural nonlinear", "OUTPUT(1)/USER/STRAIN(3)/CRACK/GREEN" )
```

```
addAnalysisCommandDetail( "Analysis1", "Structural nonlinear", "OUTPUT(1)/USER/STRAIN(4)/CRKWDT/GREEN/GLOBAL" )
addAnalysisCommandDetail( "Analysis1", "Structural nonlinear", "OUTPUT(1)/USER/STRAIN(5)/CRKWDT/GREEN/LOCAL" )
addAnalysisCommandDetail( "Analysis1", "Structural nonlinear", "OUTPUT(1)/USER/STRAIN(6)/CRKWDT/GREEN/PRINCI" )
runSolver( "Analysis1" )
```

Wilson-θ command console:

```
newProject( "ChaiShun-Wilson", 100 )
setModelAnalysisAspects( [ "STRUCT" ] )
setModelDimension( "2D" )
setDefaultMeshOrder( "QUADRATIC" )
setDefaultMesherType( "HEXQUAD" )
setDefaultMidSideNodeLocation( "ONSHAP" )
setUnit( "TEMPER", "CELSIU" )
setUnit( "ANGLE", "DEGREE" )
saveProject( )
createSheet( "Sheet 1", [[ 0, 0, 0 ],[ 0.6, 0, 0 ],[ 1.2, 0, 0 ],[ 1.2, 7.5, 0 ],[ 8.1, 7.5, 0 ],[ 8.1, 8.7, 0 ],[ 0, 8.7, 0 ]] )
createLine( "bar1", [ 0.1, 0, 0 ], [ 0.1, 8.7, 0 ] )
saveProject( )
createLine( "bar2", [ 1.1, 0, 0 ], [ 1.1, 8.4, 0 ] )
saveProject( )
createLine( "bar3", [ 0.9, 7.6, 0 ], [ 0.9, 8.1, 0 ] )
removeShape( [ "bar3" ] )
createLine( "bar3", [ 0.9, 7.6, 0 ], [ 8.1, 7.6, 0 ] )
saveProject( )
createLine( "bar4", [ 0, 8.6, 0 ], [ 8.1, 8.6, 0 ] )
saveProject( )
addMaterial( "concrete", "CONCR", "TSCR", [ "RAYDAM" ] )
setParameter( MATERIAL, "concrete", "LINEAR/ELASTI/YOUNG", 3.45e+10 )
setParameter( MATERIAL, "concrete", "LINEAR/ELASTI/POISON", 0.15 )
setParameter( MATERIAL, "concrete", "LINEAR/MASS/DENSIT", 2500 )
setParameter( MATERIAL, "concrete", "TENSIL/TENCRV", "MC2010" )
setParameter( MATERIAL, "concrete", "TENSIL/TENSTR", 2600000 )
setParameter( MATERIAL, "concrete", "TENSIL/GF1", 500 )
setParameter( MATERIAL, "concrete", "COMPRS/COMCRV", "MAEKCC" )
setParameter( MATERIAL, "concrete", "COMPRS/COMSTR", 32500000 )
setParameter( MATERIAL, "concrete", "RAYDAM/RAYLEI", [ 1.1042, 0.00165 ] )
addGeometry( "Element geometry 1", "SHEET", "MEMBRA", [] )
rename( GEOMET, "Element geometry 1", "concrete" )
setParameter( GEOMET, "concrete", "THICK", 0.7 )
```

```
clearReinforcementAspects( [ "Sheet 1" ] )
setElementClassType( SHAPE, [ "Sheet 1" ], "MEMBRA" )
assignMaterial( "concrete", SHAPE, [ "Sheet 1" ] )
assignGeometry( "concrete", SHAPE, [ "Sheet 1" ] )
resetElementData( SHAPE, [ "Sheet 1" ] )
addMaterial( "bar", "REINFO", "VMISES", [ "RAYDAM" ] )
setParameter( MATERIAL, "bar", "LINEAR/ELASTI/YOUNG", 2.1e+11 )
setParameter( MATERIAL, "bar", "PLASTI/HARDI1/YLDSTR", 4.5e+08 )
setParameter( MATERIAL, "bar", "RAYDAM/RAYLEI", [ 1.1042, 0.00165 ] )
addGeometry( "Element geometry 2", "RELINE", "REBAR", [] )
rename( GEOMET, "Element geometry 2", "bar" )
setParameter( GEOMET, "bar", "REITYP", "REITRU" )
setParameter( GEOMET, "bar", "REITRU/CROSSE", 0.000157 )
setParameter( GEOMET, "bar", "REITRU/PERIME", 0.012 )
setReinforcementAspects( [ "bar1", "bar2", "bar3", "bar4" ] )
assignMaterial( "bar", SHAPE, [ "bar1", "bar2", "bar3", "bar4" ] )
assignGeometry( "bar", SHAPE, [ "bar1", "bar2", "bar3", "bar4" ] )
resetElementData( SHAPE, [ "bar1", "bar2", "bar3", "bar4" ] )
setReinforcementDiscretization( [ "bar1", "bar2", "bar3", "bar4" ], "ELEMENT" )
saveProject(  )
addSet( "GEOMETRYSUPPORTSET", "co1" )
createPointSupport( "co1", "co1" )
setParameter( "GEOMETRYSUPPORT", "co1", "AXES", [ 1, 2 ] )
setParameter( "GEOMETRYSUPPORT", "co1", "TRANSL", [ 1, 1, 0 ] )
setParameter( "GEOMETRYSUPPORT", "co1", "ROTATI", [ 0, 0, 0 ] )
attach( "GEOMETRYSUPPORT", "co1", "Sheet 1", [[ 0.6, 1.8378494e-34,
-5.2871687e-18 ]] )
addSet( "GEOMETRYSUPPORTSET", "Geometry support set 2" )
rename( "GEOMETRYSUPPORTSET", "Geometry support set 2", "co2" )
createLineSupport( "co2", "co2" )
setParameter( "GEOMETRYSUPPORT", "co2", "AXES", [ 1, 2 ] )
setParameter( "GEOMETRYSUPPORT", "co2", "TRANSL", [ 1, 0, 0 ] )
setParameter( "GEOMETRYSUPPORT", "co2", "ROTATI", [ 0, 0, 0 ] )
attach( "GEOMETRYSUPPORT", "co2", "Sheet 1", [[ 8.1, 8.1, -1.4938131e-17 ]] )
saveProject(  )
addSet( "GEOMETRYLOADSET", "gravity" )
createModelLoad( "gravity", "gravity" )
createBodyLoad( "lo2", "Geometry load case 2" )
setParameter( GEOMETRYLOAD, "lo2", "LODTYP", "EQUIAC" )
setParameter( GEOMETRYLOAD, "lo2", "EQUIAC/ACCELE", -9.8 )
setParameter( GEOMETRYLOAD, "lo2", "EQUIAC/DIRECT", 1 )
attach( GEOMETRYLOAD, "lo2", [ "Sheet 1" ] )
```

setDefaultGeometryLoadCombinations( )
setGeometryLoadCombinationFactor("Geometry load combination 1", "gravity", 1)
remove(GEOMETRYLOADCOMBINATION, "Geometry load combination 2")
setGeometryLoadCombinationFactor("Geometry load combination 1", "gravity", 1)
remove(GEOMETRYLOADCOMBINATION, "Geometry load combination 1")
addGeometryLoadCombination("")
setGeometryLoadCombinationFactor("Geometry load combination 1", "gravity", 1)
addGeometryLoadCombination("")
setGeometryLoadCombinationFactor("Geometry load combination 2", "Geometry load case 2", 1)
setTimeDependentLoadFactors(GEOMETRYLOADCOMBINATION, "Geometry load combination 1", [0, 86400], [1, 1])
setTimeDependentLoadFactors(GEOMETRYLOADCOMBINATION, "Geometry load combination 2", [0, 0.1, 0.2, 0.3, 0.4, 0.5, 0.6, 0.7, 0.8, 0.9, 1, 1.1, 1.2, 1.3, 1.4, 1.5, 1.6, 1.7, 1.8, 1.9, 2, 2.1, 2.2, 2.3, 2.4, 2.5, 2.6, 2.7, 2.8, 2.9, 3, 3.1, 3.2, 3.3, 3.4, 3.5, 3.6, 3.7, 3.8, 3.9, 4, 4.1, 4.2, 4.3, 4.4, 4.5, 4.6, 4.7, 4.8, 4.9, 5], [0, -0.033, -0.00126, 3.06e-05, 0.0107, 0.00594, -0.00262, 0.0263, -0.00787, -0.021, -0.000866, -0.0115, -0.0238, 0.0332, 0.00357, 0.00819, 0.0176, 0.0468, 0.018, -0.0139, 0.0062, 0.0206, 0.0271, 0.0385, -0.0307, -0.0442, -0.0246, -0.0288, 0.0267, -0.104, -0.0613, 0.0187, 0.0637, -0.0269, -0.0381, 0.0935, -0.124, -0.148, -0.107, 0.162, -0.0218, 0.141, 0.208, 0.0046, -0.0751, 0.0576, 0.0553, -0.0639, -0.0653, 0.0194, 0.041])
setElementSize(["Sheet 1"], 0.1, -1, True)
setMesherType(["Sheet 1"], "HEXQUAD")
setMidSideNodeLocation(["Sheet 1"], "LINEAR")
generateMesh([])
hideView("GEOM")
showView("MESH")
addAnalysis("Analysis1")
addAnalysisCommand("Analysis1", "NONLIN", "Structural nonlinear")
renameAnalysis("Analysis1", "Analysis1")
setAnalysisCommandDetail("Analysis1", "Structural nonlinear", "TYPE/TRANSI/METHOD/INTTYP", "WILSON")
setAnalysisCommandDetail("Analysis1", "Structural nonlinear", "EXECUT(1)/ITERAT/MAXITE", 50)
setAnalysisCommandDetail("Analysis1", "Structural nonlinear", "EXECUT(1)/ITERAT/CONVER/DISPLA/NOCONV", "CONTIN")
setAnalysisCommandDetail("Analysis1", "Structural nonlinear", "EXECUT(1)/ITERAT/CONVER/FORCE/NOCONV", "CONTIN")
setAnalysisCommandDetail("Analysis1", "Structural nonlinear", "EXECUT(1)/ITERAT/MAXITE", 50)
addAnalysisCommandDetail("Analysis1", "Structural nonlinear", "TYPE/TRANSI")
setAnalysisCommandDetail("Analysis1", "Structural nonlinear", "TYPE/TRANSI", True)
addAnalysisCommandDetail("Analysis1", "Structural nonlinear",

```
"TYPE/TRANSI/DYNAMI" )
    setAnalysisCommandDetail( "Analysis1", "Structural nonlinear", "TYPE/TRANSI/DYNAMI", True )
    setAnalysisCommandDetail( "Analysis1", "Structural nonlinear", "EXECUT(1)/ITERAT/MAXITE", 50 )
    setAnalysisCommandDetail( "Analysis1", "Structural nonlinear", "EXECUT/EXETYP", "LOAD" )
    renameAnalysisCommand( "Analysis1", "Structural nonlinear", "Structural nonlinear" )
    setAnalysisCommandDetail( "Analysis1", "Structural nonlinear", "EXECUT(2)/ITERAT/MAXITE", 10 )
    setAnalysisCommandDetail( "Analysis1", "Structural nonlinear", "EXECUT(2)/ITERAT/MAXITE", 50 )
    setAnalysisCommandDetail( "Analysis1", "Structural nonlinear", "EXECUT(2)/ITERAT/CONVER/DISPLA/NOCONV", "CONTIN" )
    setAnalysisCommandDetail( "Analysis1", "Structural nonlinear", "EXECUT(2)/ITERAT/CONVER/FORCE/NOCONV", "CONTIN" )
    setAnalysisCommandDetail( "Analysis1", "Structural nonlinear", "EXECUT(2)/ITERAT/MAXITE", 50 )
    setAnalysisCommandDetail( "Analysis1", "Structural nonlinear", "EXECUT(2)/ITERAT/MAXITE", 50 )
    setAnalysisCommandDetail( "Analysis1", "Structural nonlinear", "EXECUT(1)/ITERAT/MAXITE", 50 )
    addAnalysisCommandDetail( "Analysis1", "Structural nonlinear", "TYPE/TRANSI/DYNAMI/DAMPIN" )
    setAnalysisCommandDetail( "Analysis1", "Structural nonlinear", "TYPE/TRANSI/DYNAMI/DAMPIN", True )
    setAnalysisCommandDetail( "Analysis1", "Structural nonlinear", "EXECUT(2)/LOAD/LOADNR", 2 )
    setAnalysisCommandDetail( "Analysis1", "Structural nonlinear", "EXECUT/EXETYP", "TIME" )
    renameAnalysisCommandDetail( "Analysis1", "Structural nonlinear", "EXECUT(3)", "new execute block 3" )
    setAnalysisCommandDetail( "Analysis1", "Structural nonlinear", "EXECUT(3)/TIME/STEPS/EXPLIC/SIZES", "0.1(50)" )
    setAnalysisCommandDetail( "Analysis1", "Structural nonlinear", "EXECUT(3)/ITERAT/MAXITE", 50 )
    setAnalysisCommandDetail( "Analysis1", "Structural nonlinear", "EXECUT(3)/ITERAT/CONVER/DISPLA/NOCONV", "CONTIN" )
    setAnalysisCommandDetail( "Analysis1", "Structural nonlinear", "EXECUT(3)/ITERAT/CONVER/FORCE/NOCONV", "CONTIN" )
    setAnalysisCommandDetail( "Analysis1", "Structural nonlinear", "OUTPUT(1)/SELTYP", "USER" )
```

```
addAnalysisCommandDetail( "Analysis1", "Structural nonlinear", "OUTPUT(1)/USER" )
addAnalysisCommandDetail( "Analysis1", "Structural nonlinear", "OUTPUT(1)/USER/DISPLA(1)/TOTAL/TRANSL/GLOBAL" )
addAnalysisCommandDetail( "Analysis1", "Structural nonlinear", "OUTPUT(1)/USER/STRAIN(1)/CRKSUM/GREEN/GLOBAL" )
addAnalysisCommandDetail( "Analysis1", "Structural nonlinear", "OUTPUT(1)/USER/STRAIN(2)/CRKSUM/GREEN/PRINCI" )
addAnalysisCommandDetail( "Analysis1", "Structural nonlinear", "OUTPUT(1)/USER/STRAIN(3)/CRACK/GREEN" )
addAnalysisCommandDetail( "Analysis1", "Structural nonlinear", "OUTPUT(1)/USER/STRAIN(4)/CRKWDT/GREEN/GLOBAL" )
addAnalysisCommandDetail( "Analysis1", "Structural nonlinear", "OUTPUT(1)/USER/STRAIN(5)/CRKWDT/GREEN/LOCAL" )
addAnalysisCommandDetail( "Analysis1", "Structural nonlinear", "OUTPUT(1)/USER/STRAIN(6)/CRKWDT/GREEN/PRINCI" )
runSolver( "Analysis1" )
```

3.9 Discrete Cracking Analysis of Plain Concrete Beam

This case illustrates a segment of plain concrete beam, the length of which is 16 m, the height is 2 m while the thickness is 1 m, and the geometry of the beam is shown in Fig. 3.345. Concrete is simulated by the plane quadratic stress elements while the cracking in the middle site of the beam is simulated by the line to line connected interface elements. Discrete cracking model is applied to simulate the whole process of cracking and the material constitutive model of line to line connected interface element is applied to mechanic behavior of discrete cracking [2]. The whole model sustains distributed force with the loading concentration value 20 kN/m. Simulating processing procedure is based on the platform of DIANA release 10.2.

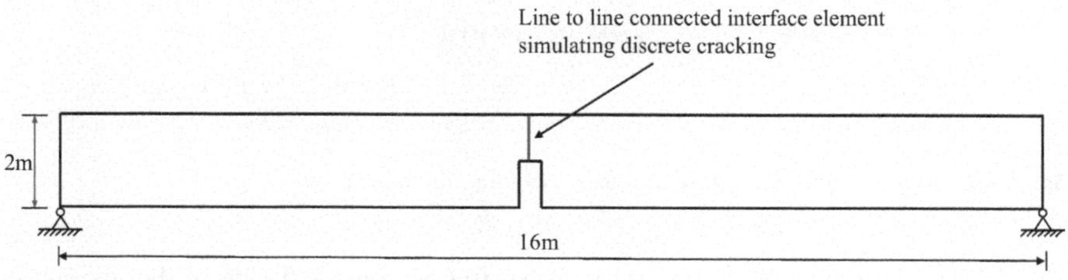

Fig. 3.345 Geometric model of beam

Essentials of learning

(1) *Definition of line to line connected interface element*
(2) *Material discrete cracking constitutive model for line to line connected interface element*
(3) *Familiar with Boolean addition and subtraction logic operation in DIANA 10.2.*

Above all, opening DianaIE operational interface, clicking **File—New**, the dialog box **New project** ejects. Then a document with the suffix name .dpf and the project name Discrete cracking is created in the directory of computer G-disk area with the document name 10.2例题. The type of **Analysis** is structural and modeling dimension of this document is **Two dimensional**. Maximum model size is 100 m, meaning that the range of coordinate values in the coordinate system is from (–50, –50) to (50, 50). **Default mesher type** is **Hexa/Quad** while **Default mesh order** is **Quadratic,** as Fig. 3.346 displays.

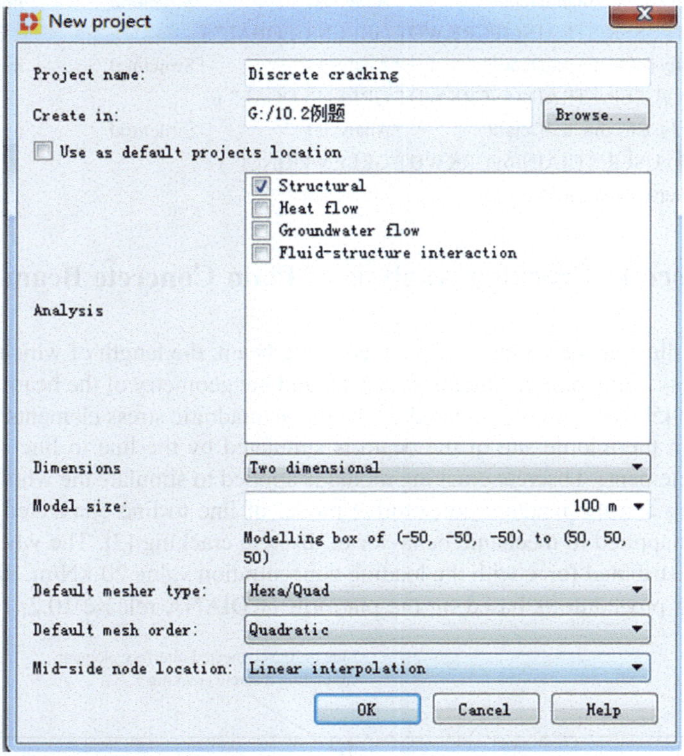

Fig. 3.346 New project dialog box to create modeling document

Click shortcut icon  under menu bar to create the left side of the geometric plane with the name of **left**; then click **OK** button to ensure and finish the process. **Sheet 2** is still created in the same way and the coordinate values of geometric points are shown in Figs. 3.347 and 3.348, respectively.

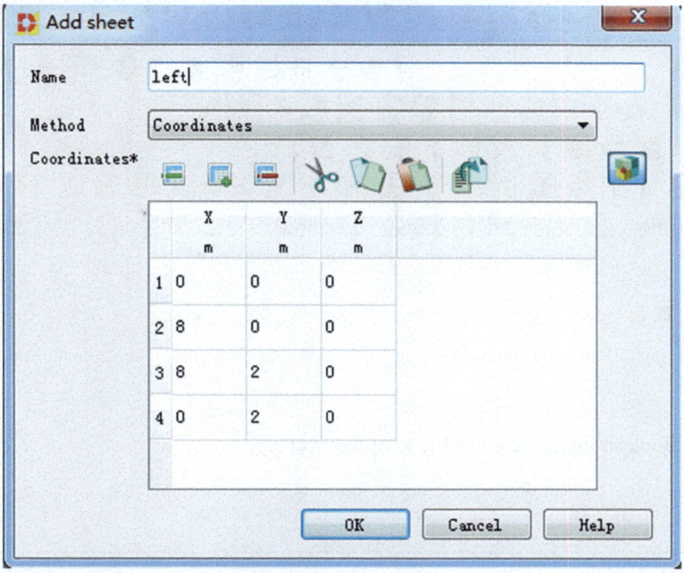

Fig. 3.347 Coordinate values on the left side

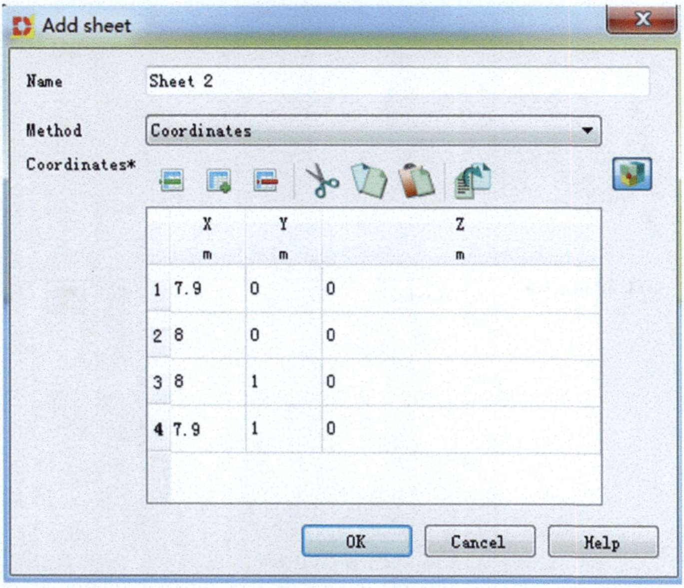

Fig. 3.348 Coordinate values on the sheet 2

Adopt the subtraction function of Boolean logic operation mentioned above to subtract the Sheet2 from the geometric model of **left**. The specific icons in the DianaIE and manipulating interface are shown in Figs. 3.349 and 3.350, respectively.

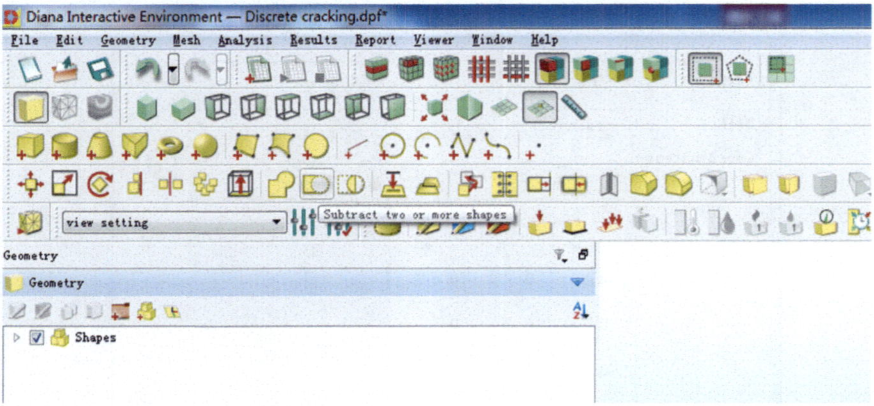

Fig. 3.349 Boolean logic operational icon of Subtract

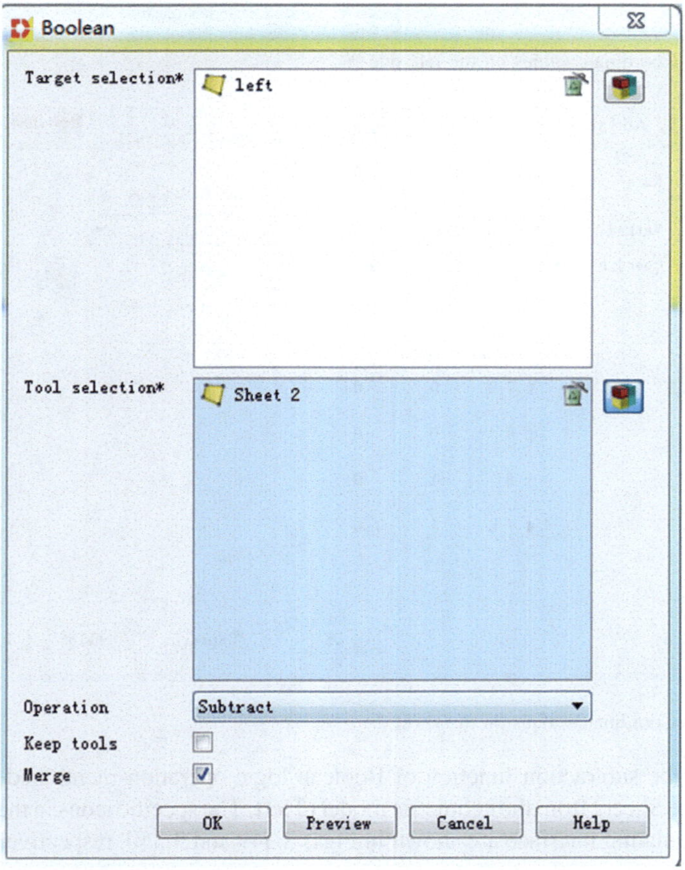

Fig. 3.350 Manipulating interface of Boolean logic operation-**Subtract**

Clicking OK button, the incomplete plane, as Fig. 3.351 shows, is created, where the section that omitting Sheet 2 is used to simulate the concrete cracking features in the lower half part.

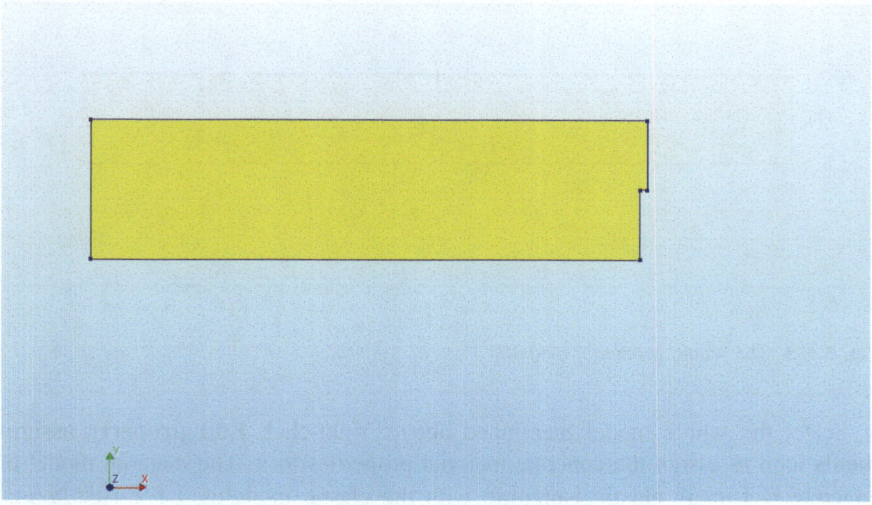

Fig. 3.351 Left part of beam after Boolean logic subtract function

The right part of the beam is started to establish, then directly input coordinate values of every point: [8, 2, 0], [16, 2, 0], [16, 0, 0], [8.1, 0, 0], [8.1, 1, 0], [8, 1, 0] to create right part of geometric model, as Fig. 3.352 displays.

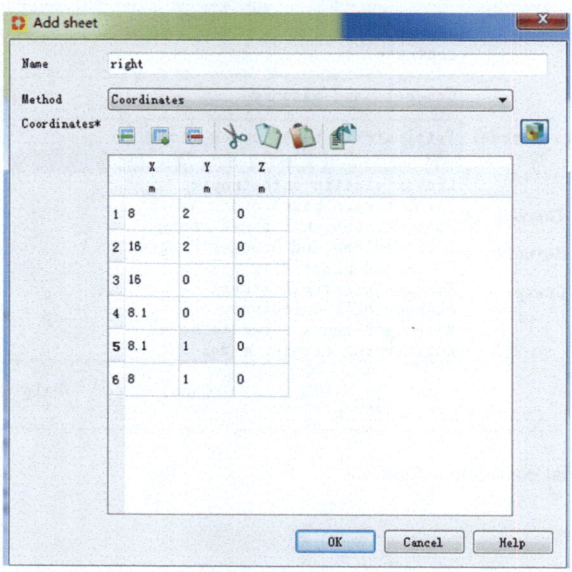

Fig. 3.352 Coordinate values of right part

Then OK button is clicked to generate the whole geometric model of beam (Fig. 3.353).

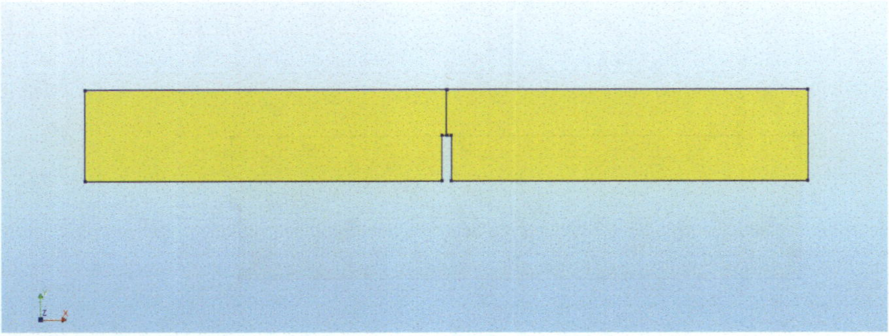

Fig. 3.353 The whole geometric model

Select the whole model mentioned above; right-click **Edit property assignments** icon to assign the concrete material properties to it. The material model of concrete is **Linear elastic isotropic** with the elastic modulus 3.1×10^{10} N/m^2, Poisson's ratio 0.15 and density 2500 kg/m^3 respectively, as shown in Figs. 3.354, 3.355 and 3.356. Clicking icon to edit geometric sectional characteristics with the thickness 1 m, local element x-axis corresponds to the positive X-axis direction under global coordinate system, displayed as Fig. 3.357.

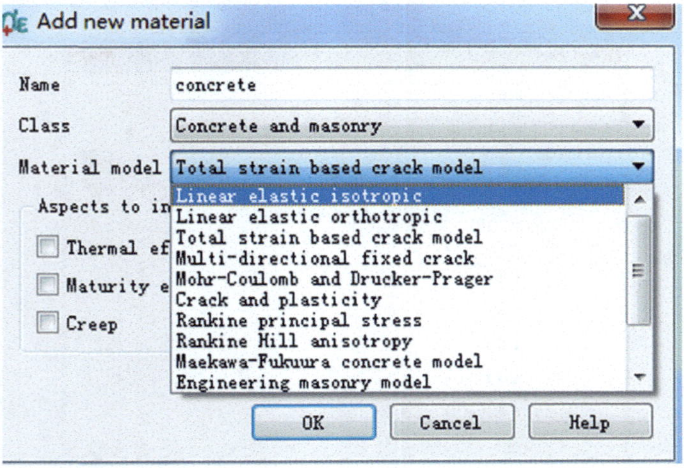

Fig. 3.354 Material constitutive definition

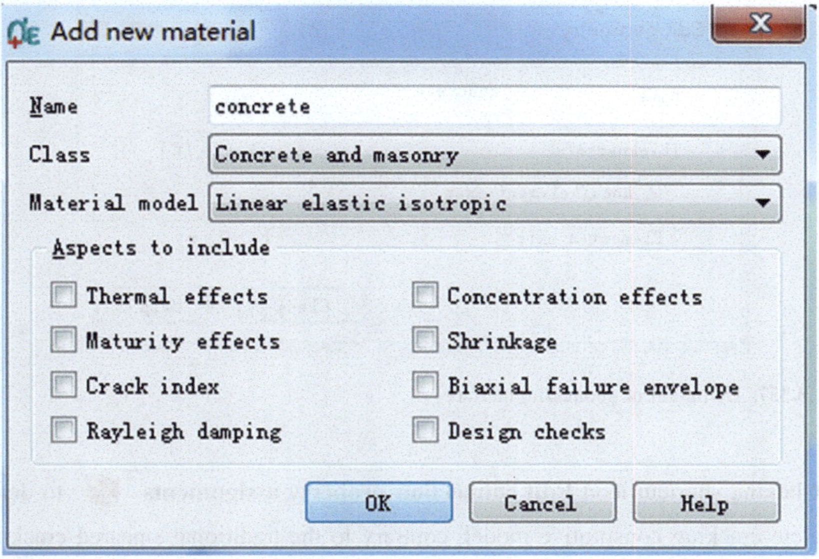

Fig. 3.355 Linear elastic isotropic material model

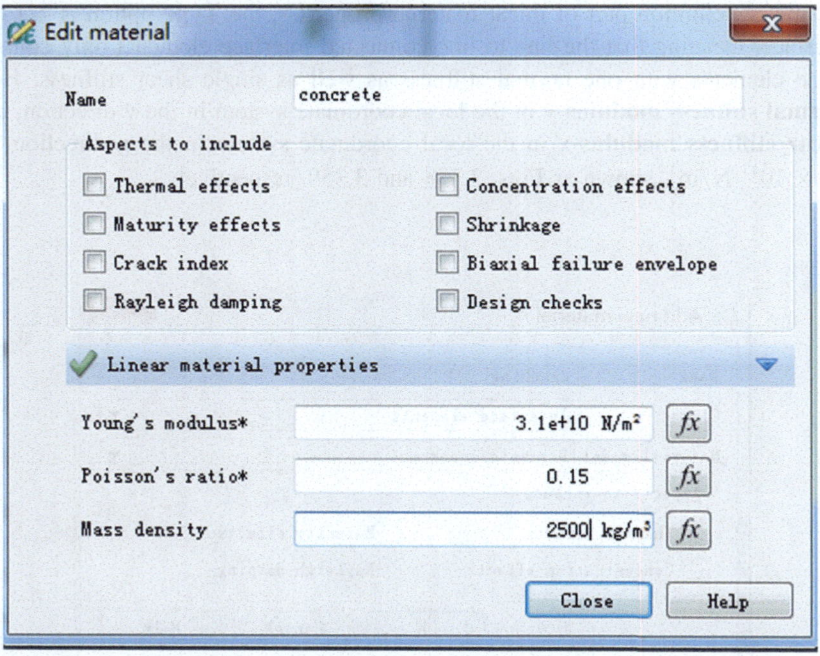

Fig. 3.356 Concrete material parameters

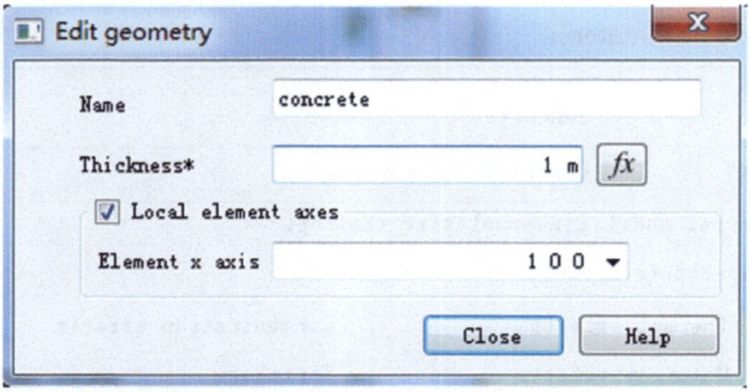

Fig. 3.357 Definition of geometric interface

Clicking shortcut icon **Edit connection property assignments** to define discrete cracking constitutive model, contrary to the traditional smeared cracking, cracks are simulated by the line to line connected interface elements. Defining the discrete cracking constitutive model by clicking shortcut icon, the dialog box of editing material model pops up and the name of this material dialog box was named as cracking. In the **Class** aspect, **Interface elements** is selected while the **Discrete cracking** is chosen as the material model of the interface elements. In the parameter definition part of linear material properties, the **Type** option is 2D line interface, meaning that the line to line connected interface elements only connect plane elements with one normal stiffness as well as single shear stiffness. Both **Normal stiffness modulus-y** in the local coordinate system in the y direction, and **Shear stiffness modulus-x** in the local coordinate system in the x direction are $1.1 \times 10^{17}\ \mathrm{N/m^3}$, shown as Figs. 3.358 and 3.359, respectively.

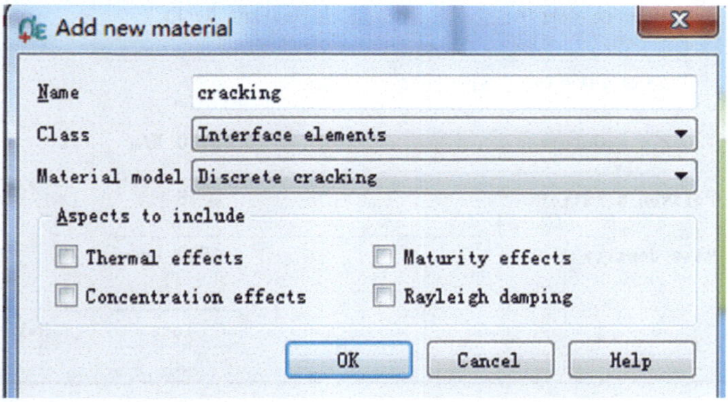

Fig. 3.358 Definition of discrete cracking

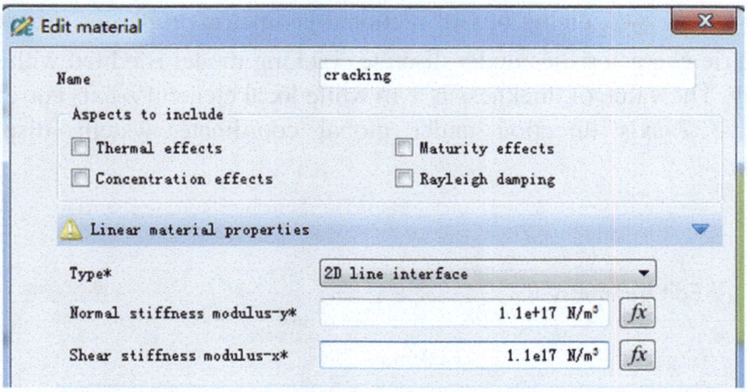

Fig. 3.359 Parameter definition of 2D line interface element

The following procedure is to input the parameters of discrete cracking material model. Tensile strength of concrete is 2.4 MPa while **Mode-I tension softening criterion** selects **JSCE tensioning** model with the fracture energy per unit width 80 N/m [2]. **Mode-I unloading/reloading model** is **Secant** while **Mode-II shear criterion for crack development** selects **Constant shear modulus**. In order to simulate characteristics of rapid descending tensile strength of plain concrete after cracking more precisely, **Shear modulus after cracking** is 1100 N/m^3 [2], as Fig. 3.360 shows.

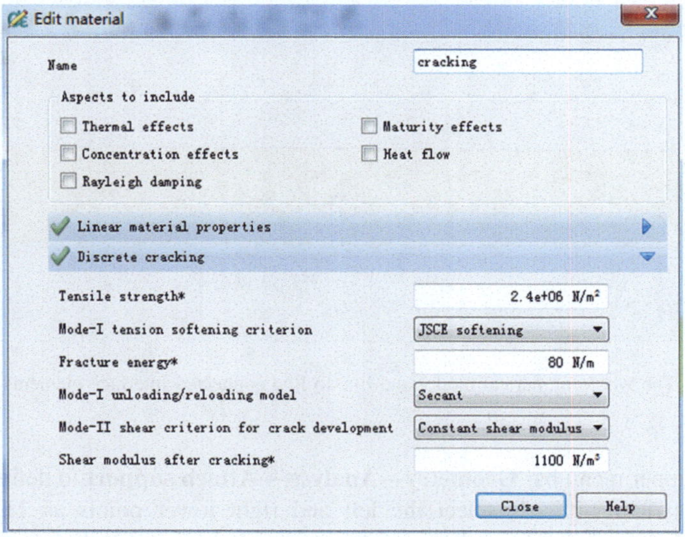

Fig. 3.360 Parameters of discrete cracking module

Clicking icon [icon], dialog box of sectional geometric properties shows up, then the geometric characteristics under discrete cracking model is edited with the name of cracking. The value of thickness is 1 m while local element z-axis corresponds to the positive Z-axis direction under global coordinate system, displayed as Fig. 3.361.

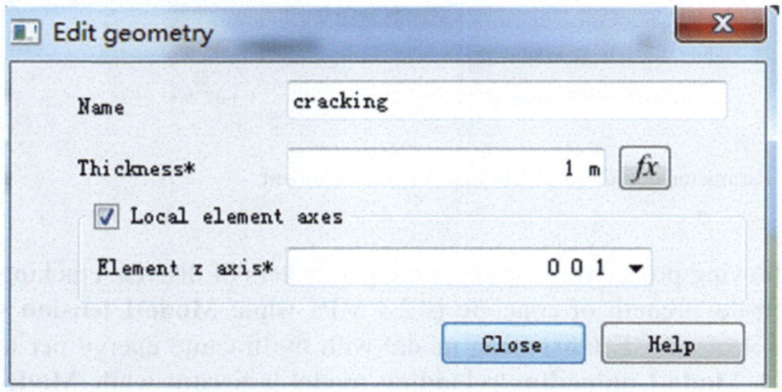

Fig. 3.361 Specification of sectional geometric properties of interface element

Clicking OK button to accomplish the whole process of defining interface element, the whole numerical model is generated in the GUI interface zone. Line to line connected interface element is identified in red shown in Fig. 3.362.

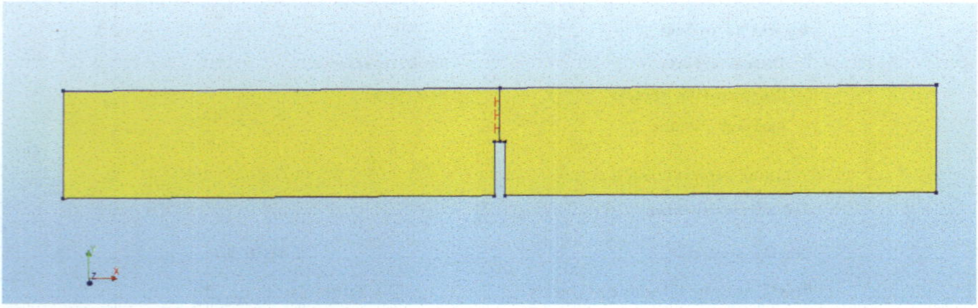

Fig. 3.362 The whole numerical model and line to line connected interface elements

Click upper menu bar **Geometry—Analysis—Attach support** to define type of constraints; name it co1; select the left and right lower points as 11 and 19, respectively; attach fixed translation point constraints in X and Y direction—T1, T2 (see Fig. 3.363).

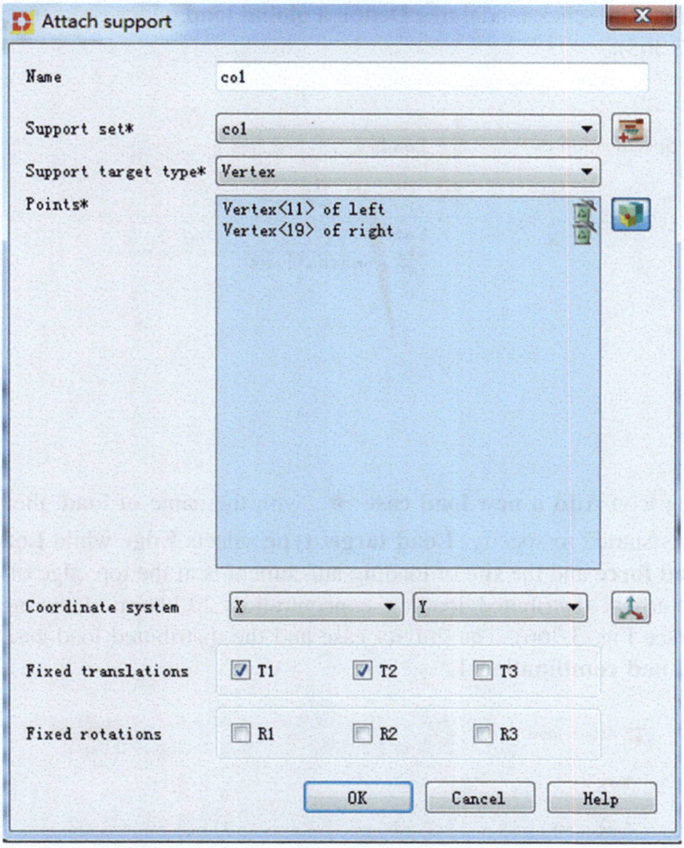

Fig. 3.363 Defining type of supports

Clicking OK button, the finite-element numerical model with constraints is displayed (Fig. 3.364).

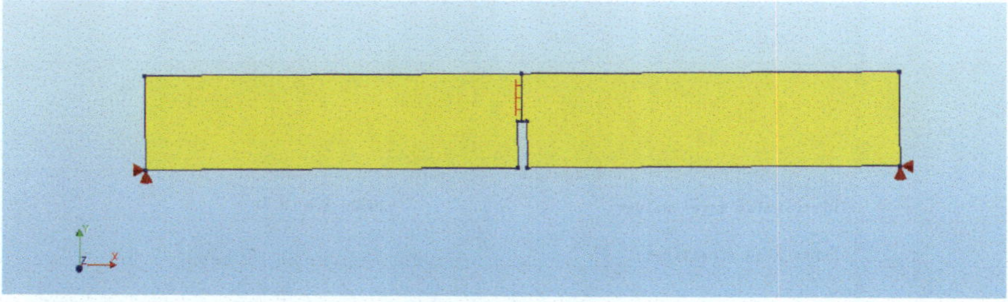

Fig. 3.364 Finite-element numerical model with constraints

Attaching load, click model tree **Define a global load** under geometry bar **Load** (see Fig. 3.365).

Fig. 3.365 Specification of gravity load

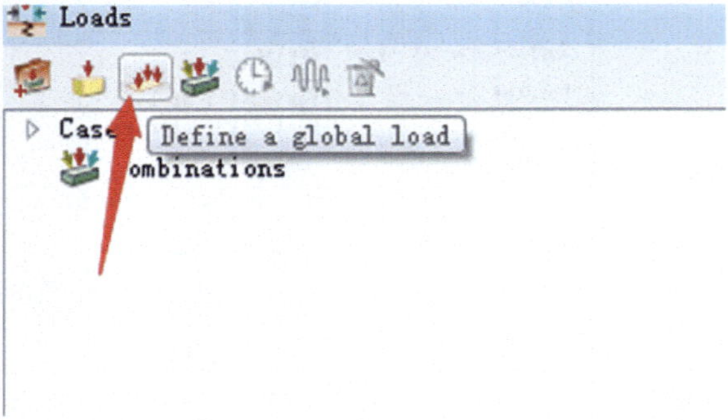

Clicking icon **Add a new load case** , with the name of **load**, then the load case **load** is started to specify. **Load target type** selects **Edge** while **Load type** is **Distributed force** and the site of loading attachment is at the top edge of the beam with the vertical distributed loading concentration 20 kN/m in the negative Y direction (see Fig. 3.366). The gravity case and the distributed load case are both added as **Load combination 1.**

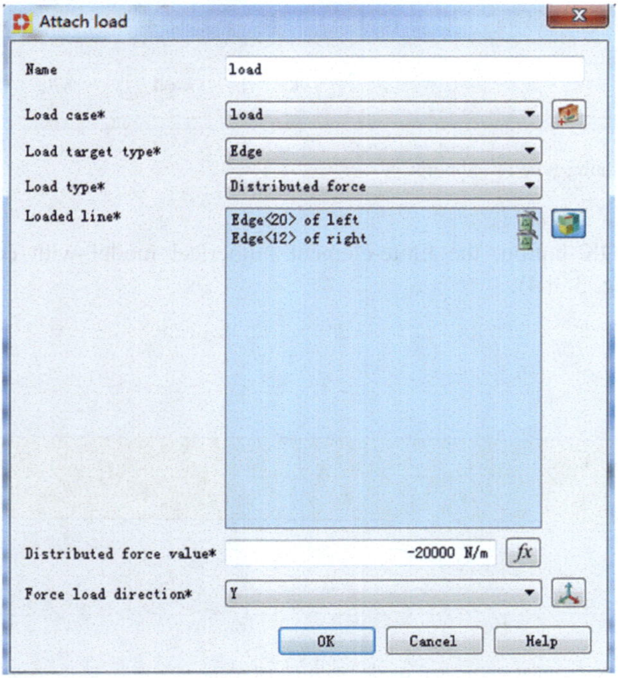

Fig. 3.366 Specification of distributed load case

Click shortcut tool button **Set mesh properties of a shape** to mesh the geometric model. **Operation** is **Shape** while **Seeding method** selects **Element size** with the **Desired size** 0.5 m. **Mesher type** representing the shape of the meshed elements is **Hexa/Quad** and **Linear interpolation** is the way of determining **Mid-side node location** at the same time (see Fig. 3.367).

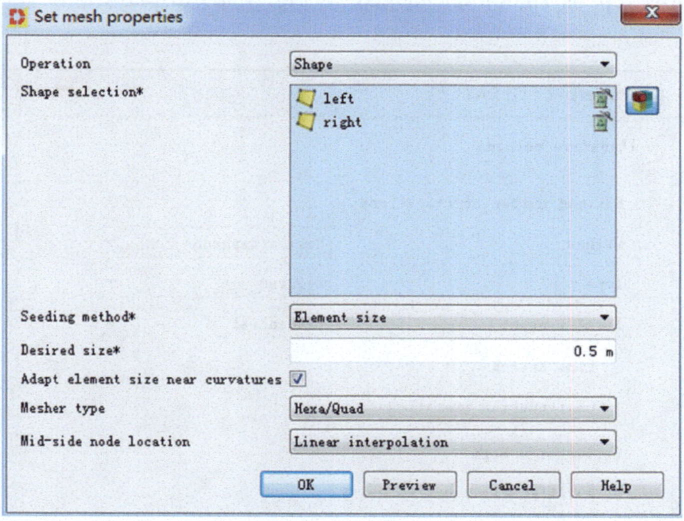

Fig. 3.367 Specifying parameters of mesh

Adopting the same meshing method to specify line to line connected interface elements, the desired size and the specification of meshing are the same as mentioned above, which will not be repeated here. Click shortcut icon **Generate mesh of a shape,** meshed plane stress elements as Fig. 3.368 shows are generated.

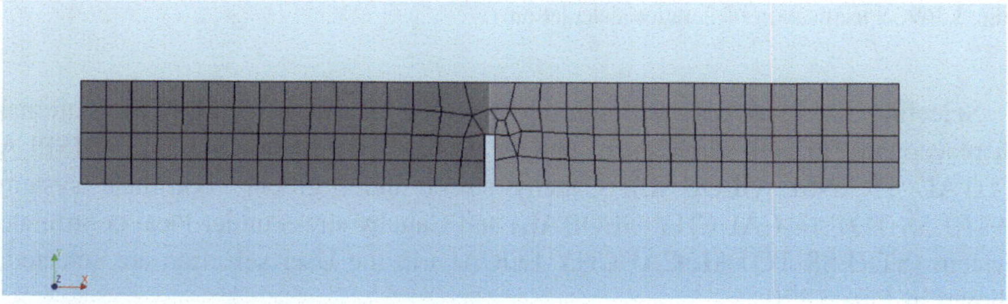

Fig. 3.368 Adopted mesh presentation

Click icon **Add an analysis** button in the Analysis module to create new nonlinear analysis. Meanwhile, kick off the original default load set setting. Right-clicking **Structural nonlinear** option the combination 1 is included in the **load steps** under **new execute block** and **User specified size** of load factor is 1. Under the load case **load**, **Maximum number of iterations** is also set as 20 in the **Equilibrium iterations**. **Force** and **Displacement** are both selected as convergence norm. **Convergence tolerance** is still set at default value **0.01**, while **Abort criterion** is also kept as **10,000** unchanged, as Fig. 3.369 displays.

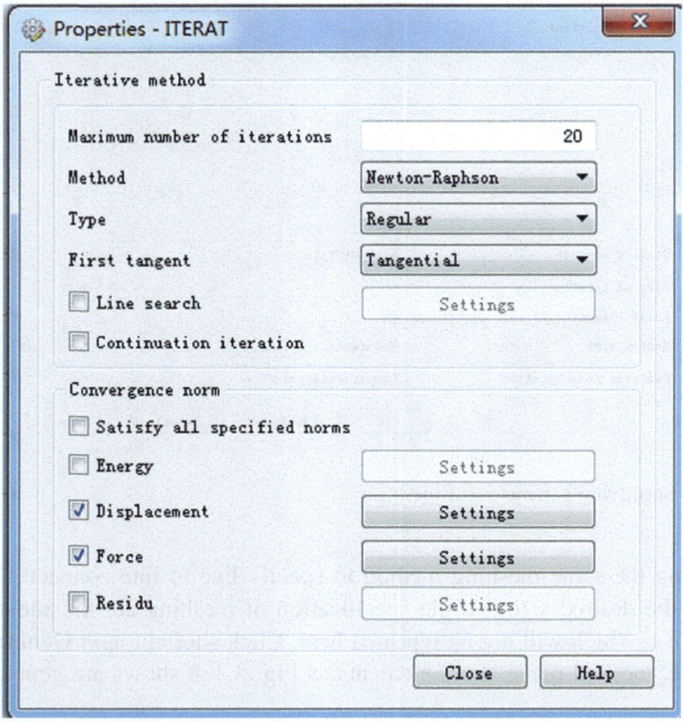

Fig. 3.369 Specification of iteration calculation

Selecting native option representing the type of output is .dpf, translational displacement in all directions under global coordinate system (DISPLA TOTAL TRANSL GLOBAL), Cauchy stress under global coordinate system (STRESS TOTAL CAUCHY GLOBAL) and Cauchy stress under local coordinate system (STRESS TOTAL CAUCHY LOCAL) in the User selection are selected, shown in Fig. 3.370.

Fig. 3.370 Specification of output selection

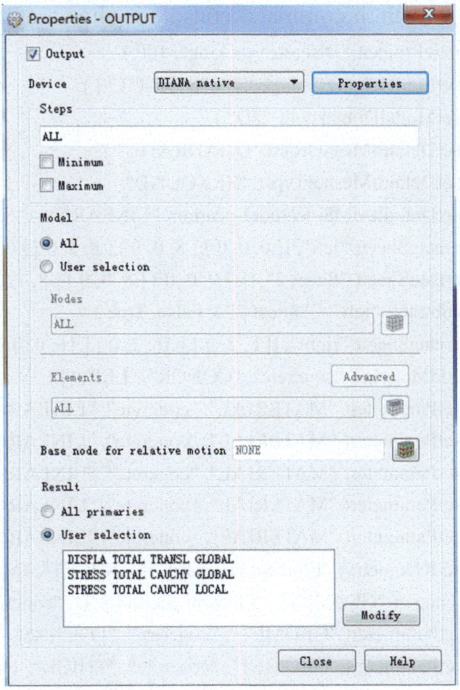

Click **Run analysis** to start nonlinear calculation. After calculation is finished, ultimate translation displacement contour in the Y direction is displayed (Fig. 3.371). According to the monitor tab of Analysis, it is easy to see that the maximum and minimum displacements are 1.8 and 0.0256 mm, respectively, which means that the deflection both at the discrete cracking site and middle site of the beam is the maximum value 1.8 mm; the negative sign represents the displacement is downward deflection while the positive sign represents the upward displacement. Besides, generally speaking, displacement in the Y direction is enlarging from the end to the middle.

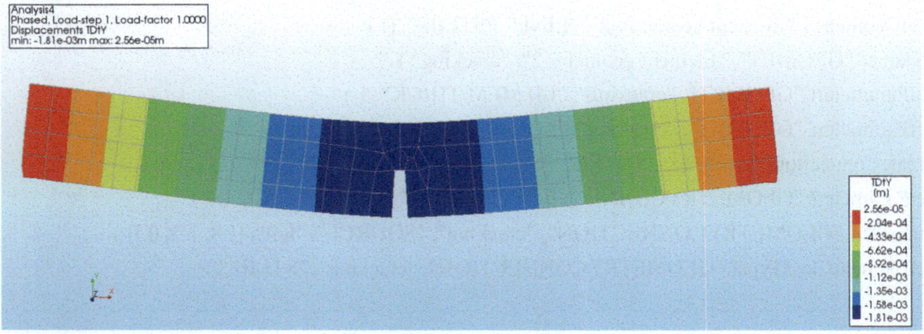

Fig. 3.371 Traditional displacement

Python command console is shown as follows:

```
newProject( "Discrete cracking", 100 )
setModelAnalysisAspects( [ "STRUCT" ] )
setModelDimension( "2D" )
setDefaultMeshOrder( "QUADRATIC" )
setDefaultMesherType( "HEXQUAD" )
setDefaultMidSideNodeLocation( "LINEAR" )
createSheet( "left", [[ 0, 0, 0 ],[ 8, 0, 0 ],[ 8, 2, 0 ],[ 0, 2, 0 ]] )
createSheet( "Sheet 2", [[ 7.9, 0, 0 ],[ 8, 0, 0 ],[ 8, 1, 0 ],[ 7.9, 1, 0 ]] )
subtract( "left", [ "Sheet 2" ], False, True )
createSheet( "right", [[ 8, 2, 0 ],[ 16, 2, 0 ],[ 16, 0, 0 ],[ 8.1, 0, 0 ],[ 8.1, 1, 0 ],[ 8, 1, 0 ]] )
addMaterial( "concrete", "CONCR", "LEI", [] )
setParameter( "MATERIAL", "concrete", "LINEAR/ELASTI/YOUNG", 3.1e+10 )
setParameter( "MATERIAL", "concrete", "LINEAR/ELASTI/POISON", 0.15 )
setParameter( "MATERIAL", "concrete", "LINEAR/MASS/DENSIT", 2500 )
setParameter( "MATERIAL", "concrete", "LINEAR/MASS/DENSIT", 2500 )
setParameter( "MATERIAL", "concrete", "LINEAR/MASS/DENSIT", 2500 )
addGeometry( "Element geometry 1", "SHEET", "MEMBRA", [] )
rename( "GEOMET", "Element geometry 1", "concrete" )
setParameter( "GEOMET", "concrete", "LOCAXS", True )
setParameter( "GEOMET", "concrete", "THICK", 1 )
setParameter( "GEOMET", "concrete", "THICK", 1 )
setParameter( "GEOMET", "concrete", "THICK", 1 )
setElementClassType( "SHAPE", [ "right", "left" ], "MEMBRA" )
assignMaterial( "concrete", "SHAPE", [ "right", "left" ] )
assignGeometry( "concrete", "SHAPE", [ "right", "left" ] )
addMaterial( "cracking", "INTERF", "DISCRA", [] )
setParameter( "MATERIAL", "cracking", "LINEAR/IFTYP", "LIN2D" )
setParameter( "MATERIAL", "cracking", "LINEAR/ELAS2/DSNY", 1.1e+17 )
setParameter( "MATERIAL", "cracking", "LINEAR/ELAS2/DSSX", 1.1e+17 )
setParameter( "MATERIAL", "cracking", "DCRACK/DCRVAL", 2400000 )
setParameter( "MATERIAL", "cracking", "DCRACK/MODE2", 1 )
setParameter( "MATERIAL", "cracking", "DCRACK/MO2VAL", 0.0011 )
setParameter( "MATERIAL", "cracking", "DCRACK/MODE11/MO1VAL", 80 )
setParameter( "MATERIAL", "cracking", "DCRACK/MODE1", 4 )
addGeometry( "Element geometry 2", "LINE", "STLIIF", [] )
rename( "GEOMET", "Element geometry 2", "cracking" )
setParameter( "GEOMET", "cracking", "LIFMEM/THICK", 1 )
setParameter( "GEOMET", "cracking", "LOCAXS", True )
createConnection( "cracking", "INTER", "SHAPEEDGE" )
setParameter( "GEOMETRYCONNECTION", "cracking", "MODE", "AUTO" )
attachTo( "GEOMETRYCONNECTION", "cracking", "SOURCE", "left", [[ 8, 1.5, 0 ]] )
setElementClassType( "GEOMETRYCONNECTION", "cracking", "STLIIF" )
```

```
assignMaterial( "cracking", "GEOMETRYCONNECTION", "cracking" )
assignGeometry( "cracking", "GEOMETRYCONNECTION", "cracking" )
setParameter( "GEOMETRYCONNECTION", "cracking", "FLIP", False )
resetElementData( "GEOMETRYCONNECTION", "cracking" )
addSet( "GEOMETRYSUPPORTSET", "co1" )
createPointSupport( "co1", "co1" )
setParameter( "GEOMETRYSUPPORT", "co1", "AXES", [ 1, 2 ] )
setParameter( "GEOMETRYSUPPORT", "co1", "TRANSL", [ 1, 1, 0 ] )
setParameter( "GEOMETRYSUPPORT", "co1", "ROTATI", [ 0, 0, 0 ] )
attach( "GEOMETRYSUPPORT", "co1", "left", [[ 0, 0, 0 ]] )
attach( "GEOMETRYSUPPORT", "co1", "right", [[ 16, 0, 0 ]] )
saveProject( )
addSet( "GEOMETRYLOADSET", "gravity" )
createModelLoad( "gravity", "gravity" )
addSet( "GEOMETRYLOADSET", "Geometry load case 1" )
rename( "GEOMETRYLOADSET", "Geometry load case 1", "load" )
createLineLoad( "load", "load" )
setParameter( "GEOMETRYLOAD", "load", "FORCE/VALUE", -20000 )
setParameter( "GEOMETRYLOAD", "load", "FORCE/DIRECT", 2 )
attach( "GEOMETRYLOAD", "load", "left", [[ 4, 2, 0 ]] )
attach( "GEOMETRYLOAD", "load", "right", [[ 12, 2, 0 ]] )
saveProject( )
setDefaultGeometryLoadCombinations( )
setGeometryLoadCombinationFactor( "Geometry load combination 1", "gravity", 1 )
remove( GEOMETRYLOADCOMBINATION, "Geometry load combination 2" )
setGeometryLoadCombinationFactor( "Geometry load combination 1", "load", 1 )
setElementSize( [ "left", "right" ], 0.5, -1, True )
setMesherType( [ "left", "right" ], "HEXQUAD" )
setMidSideNodeLocation( [ "left", "right" ], "LINEAR" )
generateMesh( [] )
hideView( "GEOM" )
showView( "MESH" )
addAnalysis( "Analysis4" )
addAnalysisCommand( "Analysis4", "NONLIN", "Structural nonlinear" )
renameAnalysis( "Analysis4", "Analysis4" )
addAnalysisCommandDetail( "Analysis4", "Structural nonlinear",
"EXECUT(1)/LOAD/LOADNR" )
setAnalysisCommandDetail( "Analysis4", "Structural nonlinear",
"EXECUT(1)/LOAD/LOADNR", 1 )
setAnalysisCommandDetail( "Analysis4", "Structural nonlinear",
"EXECUT(1)/ITERAT/MAXITE", 20 )
```

```
setAnalysisCommandDetail( "Analysis4", "Structural nonlinear", "EXECUT(1)/ITERAT/CONVER/DISPLA/NOCONV", "CONTIN" )
setAnalysisCommandDetail( "Analysis4", "Structural nonlinear", "EXECUT(1)/ITERAT/CONVER/FORCE/NOCONV", "CONTIN" )
setAnalysisCommandDetail( "Analysis4", "Structural nonlinear", "OUTPUT(1)/SELTYP", "USER" )
addAnalysisCommandDetail( "Analysis4", "Structural nonlinear", "OUTPUT(1)/USER" )
addAnalysisCommandDetail( "Analysis4", "Structural nonlinear", "OUTPUT(1)/USER/DISPLA(1)/TOTAL/TRANSL/GLOBAL" )
addAnalysisCommandDetail( "Analysis4", "Structural nonlinear", "OUTPUT(1)/USER/STRESS(1)/TOTAL/CAUCHY/GLOBAL" )
addAnalysisCommandDetail( "Analysis4", "Structural nonlinear", "OUTPUT(1)/USER/STRESS(2)/TOTAL/CAUCHY/LOCAL" )
runSolver( "Analysis4" )
showView( "RESULT" )
```

3.10 Strengthening Case of Twin Box with Single-Chamber Girder Bridge

This is an active and passive strengthening case on twin box girder bridge with single chamber. The size is shown in Fig. 3.372a–c, respectively. The total longitudinal length is 20 m while the total height of the section is 2.4 m, including the thickness of top plate 0.2 m. Widths of the top plate and bottom plate are 6 and 4 m, respectively. Internal bonded prestress tendons are in harp shape with the elastic modulus and nominal stress $1.95 \times 10^{11}\ \text{N/m}^2$ and 1860 MPa, respectively, while the external tendons with the same material parameter are in the shape of straight line. Active strengthening case (external and internal prestress strengthening methods) and passive strengthening case (bonded steel strengthening method) are illustrated in this case, and the corresponding long-term deflection-controlled effects after strengthening are compared. Solid elements are applied in the DIANA release 10.2 to establish finite-element numerical model of box girder bridge. External strengthening target and maximum deflection control index are aimed at 1/1000 span of the bridge, which is 2 cm around at the time point of 100 years. After external strengthening, further deflection under time-dependent effect is investigated in this numerical case.

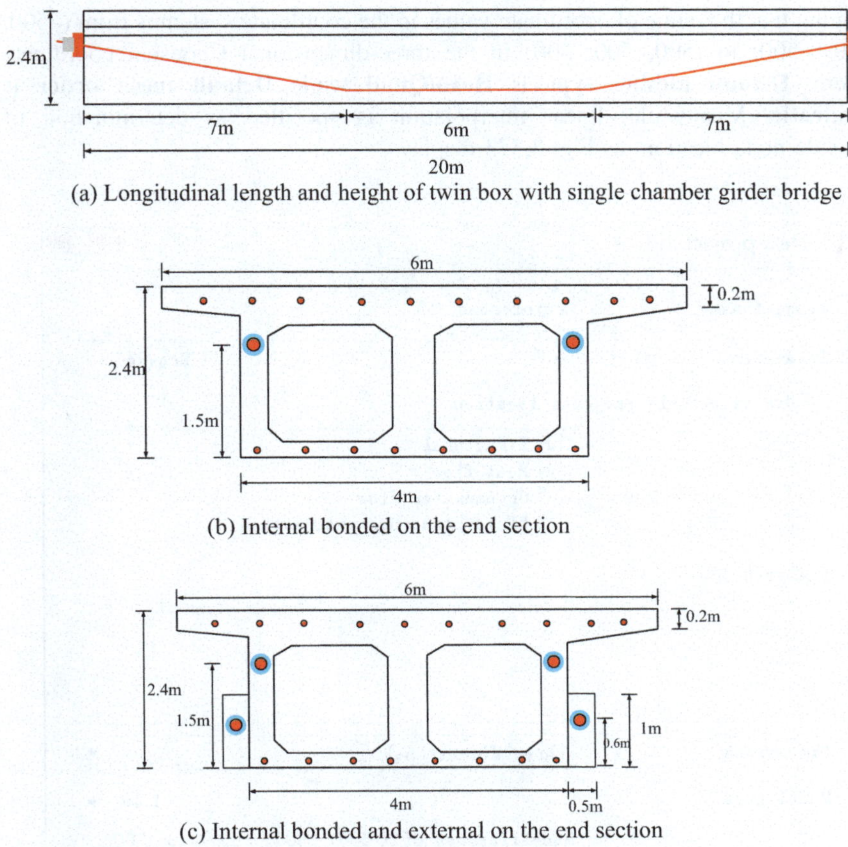

Fig. 3.372 Size of twin box with single-chamber girder bridge at the end section

Essentials of learning

(1) *Learning to create geometric model of twin box with single-chamber box girder* via *structural solid element of DIANA release 10.2.*
(2) *Learning to use the function of subtraction to generate hollow section.*
(3) *Learning to use the functions of mirror shape and extrusion to generate 3D box girder structures.*
(4) *Learning to use the phase analysis and settings of activated elements in the strengthening cases.*
(5) *Mastering active and passive strengthening cases in the DIANA release 10.2.*

Opening DianaIE operational interface, clicking **File—New** menu bar to start the preprocessing working window, the dialog box **New project** ejects. Then a document with the suffix name .dpf and the project name reinforced is generated in the directory of computer G-disk zone. The type of **Analysis** is structural and modeling dimension of this document is **Two dimensional**. Maximum model size is 1 m, meaning that the range of coordinate values in the coordinate system is from (–500, –500, –500) to (500, 500, 500) in the three-dimensional Cartesian coordinate system. **Default mesher type** is **Hexa/Quad** while **Default mesh order** is **Quadratic.** Meanwhile, linear interpolation is specified as determination of mid-side node location, as Fig. 3.373 displays.

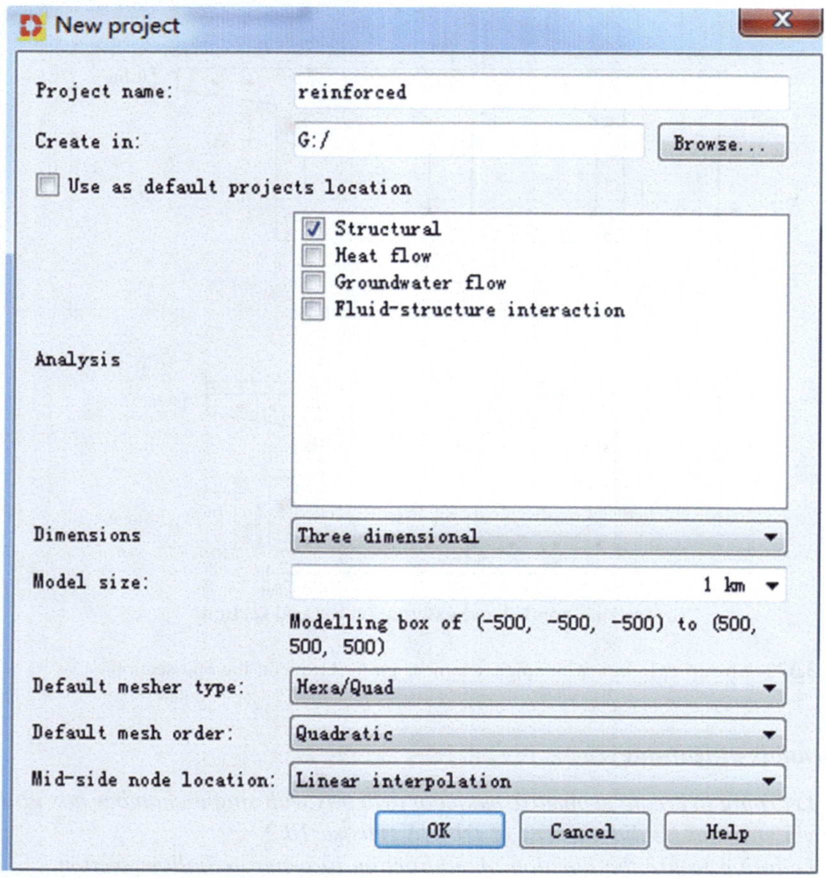

Fig. 3.373 New project operational interface

Click shortcut icon **Add a sheet** , and coordinate values as input, as Fig. 3.374 shows, to generate the initial external outline of box girder section, which is shown in Fig. 3.375.

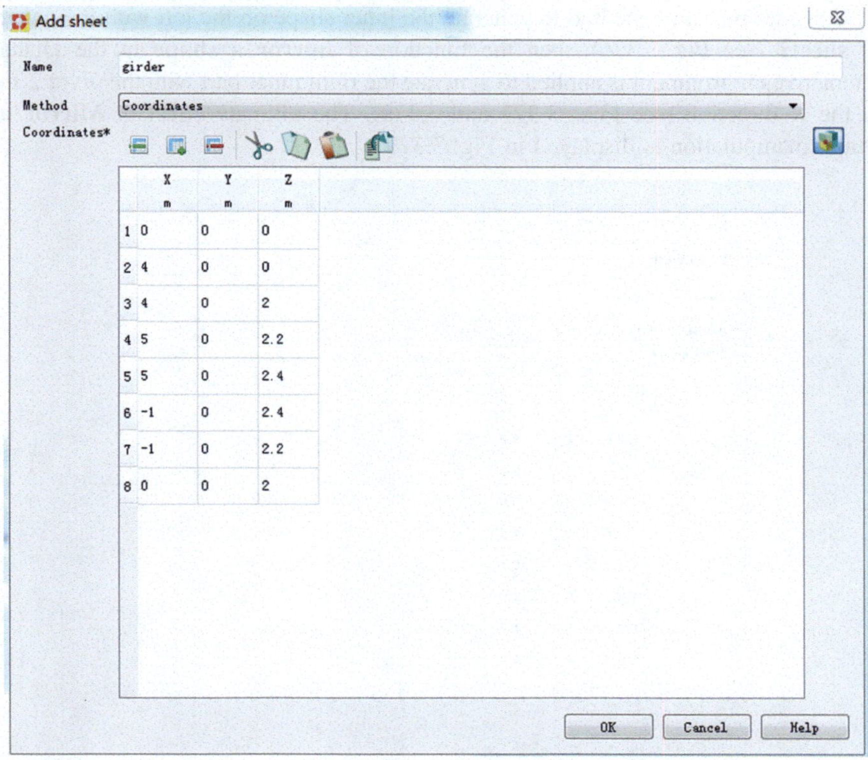

Fig. 3.374 Coordinate values of external section outline

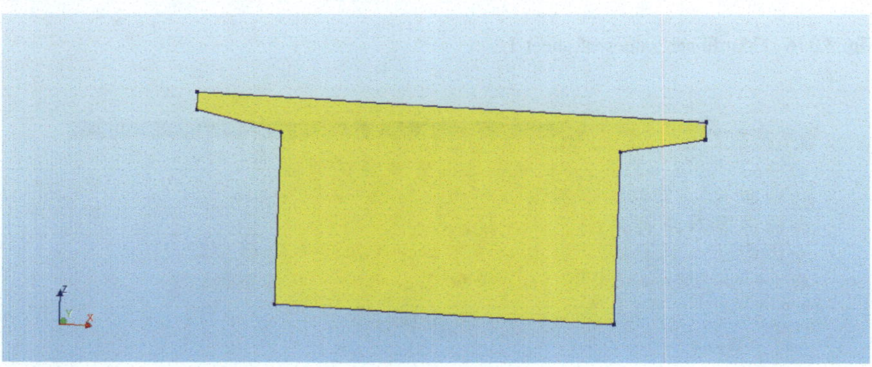

Fig. 3.375 Initial external outline of box girder section

Consider the same method to generate the inner shape on the left with the name of **sheet1** (see Fig. 3.376), then the function of **mirror a shape** in the Diana interactive environment is applied to generate the right inner part with the pivot 2 m in the X direction (see Figs. 3.377 and 3.378). The ultimate effect of **Mirror a shape** manipulation is displayed in Fig. 3.379.

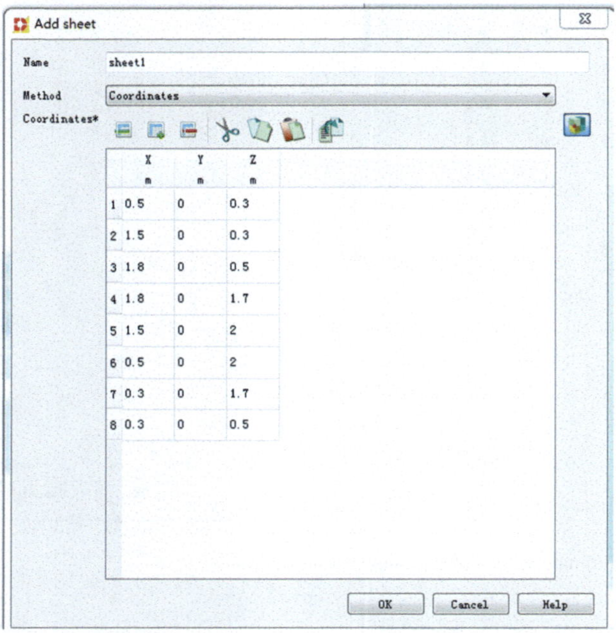

Fig. 3.376 Coordinate values of sheet 1

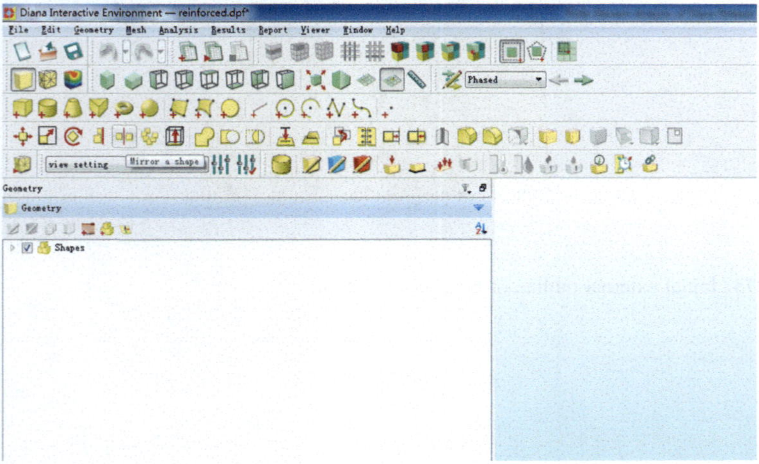

Fig. 3.377 Shortcut icon of mirror a shape

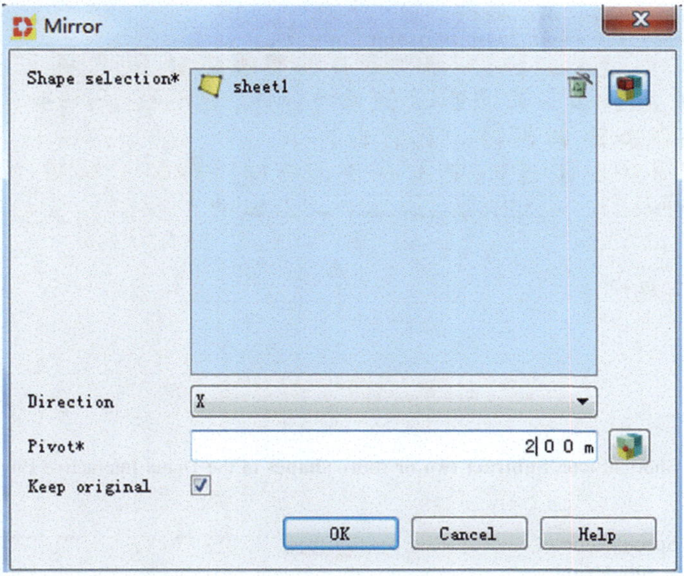

Fig. 3.378 Specification of mirror a shape

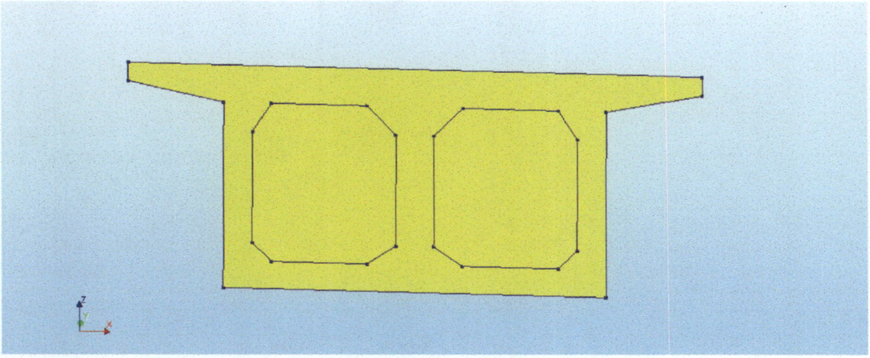

Fig. 3.379 Ultimate effect of **Mirror a shape** manipulation

Note: The principle of Mirror a shape in the DIANA means making a shape mirror symmetric to the other side like looking at a mirror via this operation while the pivot in the dialog box represents the site of symmetrical axis, which is expressed by direction and the coordinates of the symmetric axis corresponding to the direction of mirror symmetry, which is functional in the DIANA modeling manipulation in practice.

Click shortcut icon **Subtract two or more shapes** in the Diana interactive environment to enter the Boolean aspect (see Fig. 3.380); the whole external outline-**girder** is selected as **Target selection** while both **sheet1** and **sheet2** are chosen as **Tool selection**. Meanwhile, **Operation** option is **Subtract,** as Fig. 3.381 displays.

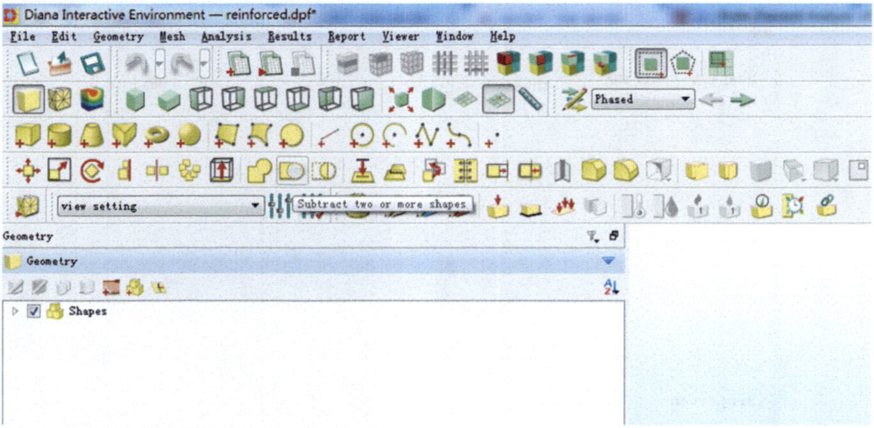

Fig. 3.380 Shortcut icon **Subtract two or more shapes** in the Diana Interactive Environment

Fig. 3.381 Specification of Boolean logic subtraction

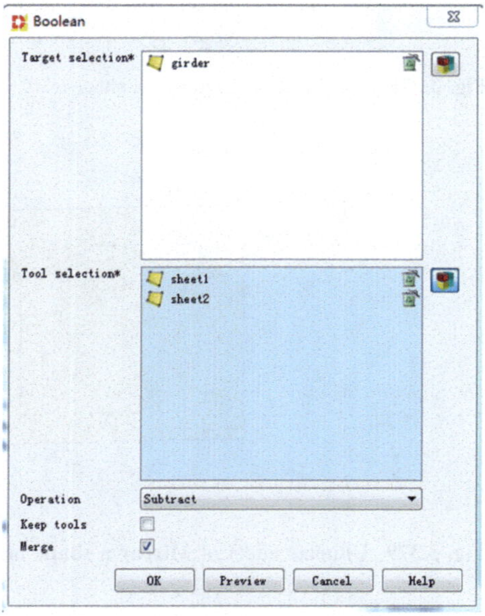

Clicking **OK** button, the section of twin box with single-chamber girder bridge is shown (Fig. 3.382).

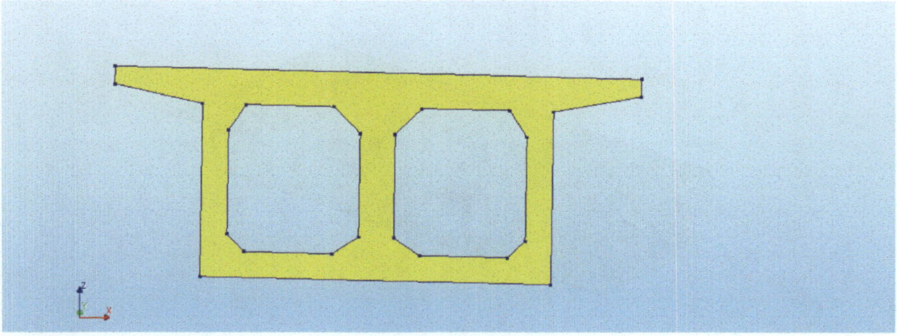

Fig. 3.382 Twin box with single-chamber girder bridge

Click shortcut icon **Extrude a shape** to extrude the two-dimensional section of box girder into a three-dimensional girder bridge. The extrude displacement is specified as 20 m in the longitudinal Y direction under the coordinate system (see Fig. 3.383), and the ultimate geometric model of twin box with single-chamber girder bridge is displayed (Fig. 3.384).

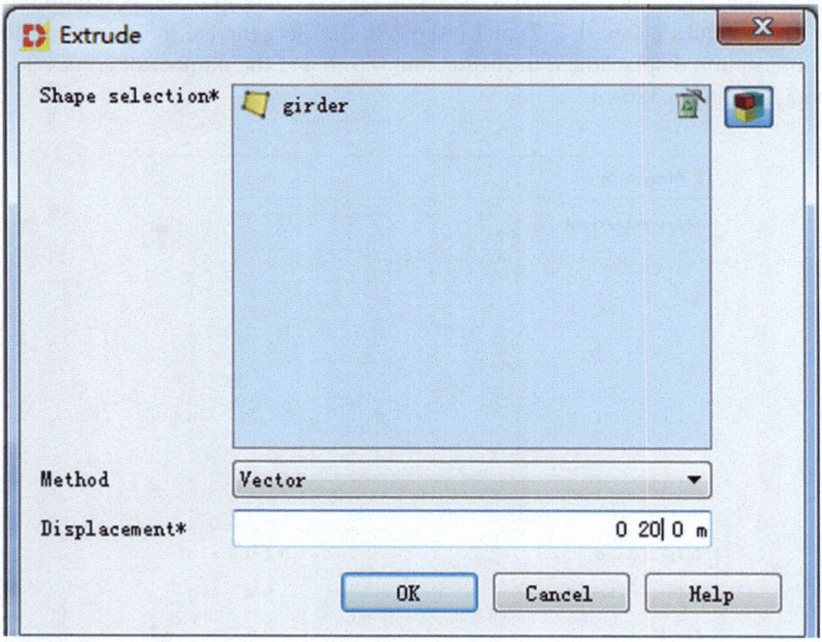

Fig. 3.383 Specification of extrusion displacement

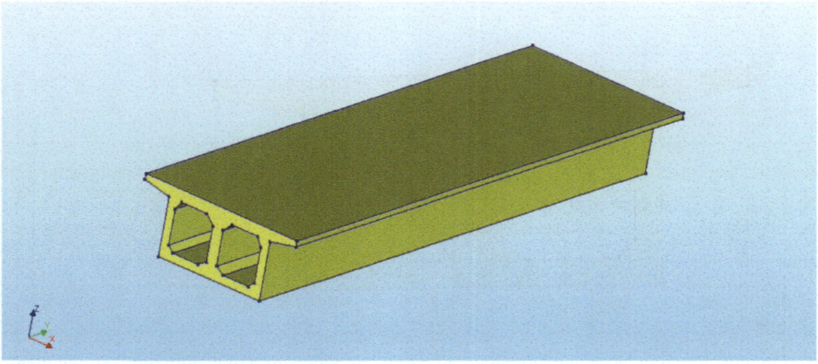

Fig. 3.384 The ultimate geometric model of twin box with single-chamber girder bridge

The following step is to create geometric model of longitudinal steel bars. Click shortcut icon **Adds a line** to establish the first geometric line; input coordinate values (0.4, 0, 0.2) and (0.4, 20, 0.2) to create the first geometric line of steel bars with the name of **bar,** which is then copied and translated according to the manipulation **Array copy**; displacement is 0.4 m in the positive X direction while the number of copies is 8 in order to generate all the longitudinal bars in the bottom plate, then use the same method to generate bar9 on the top plate by inputting the coordinate values (–0.6, 0, 2.2) and (–0.6, 20, 2.2) to generate the geometric line, with translation displacement in X direction 0.6 m and the number of copies 9 (see Figs. 3.385 and 3.386).

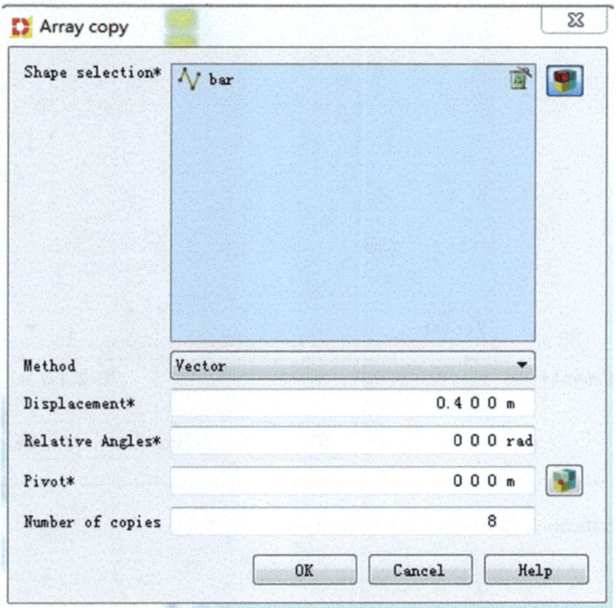

Fig. 3.385 Array copy of **bar**

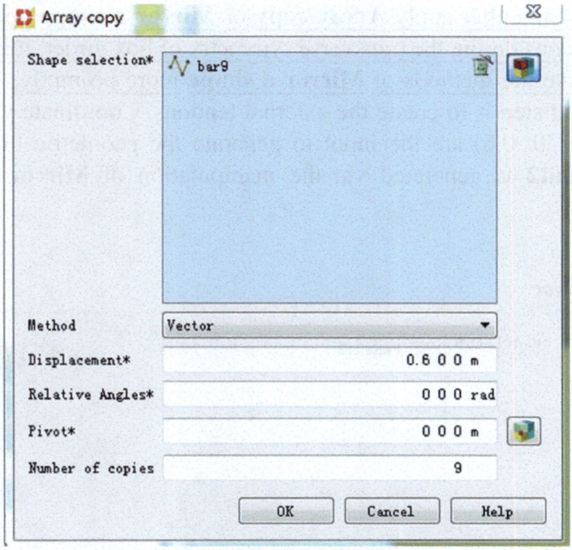

Fig. 3.386 Array copy of **bar9**

Then we start to create the geometric harp line of internal bonded prestress tendons. The shortcut icon **Adds a polyline** is clicked with the coordinate values (0.15, 0, 1.5), (0.15, 7, 0.5), (0.15, 13, 0.5) and (0.15, 20, 1.5), respectively, as input in the name of **tenin1**. Select polyline of **tenin1** and right-click to further select the option of **Mirror a shape** to generate the polyline of harp prestress tendon on the other side of the web with the name of **tenin2**. Pivot is specified as 2 m in the X direction in the **Mirror** aspect (see Fig. 3.387).

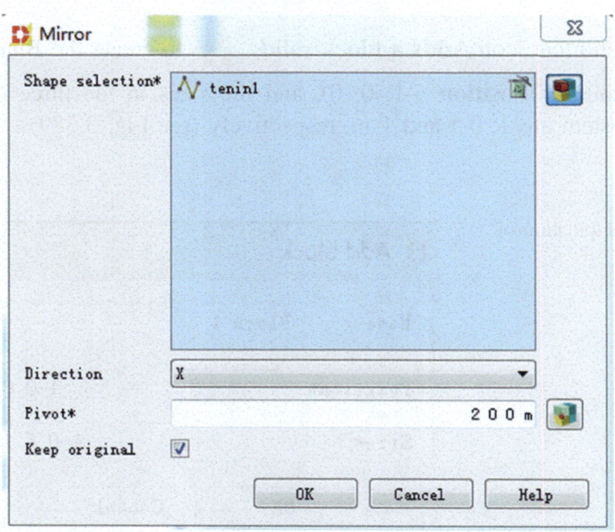

Fig. 3.387 Specification of Mirror a shape

Note: Users can either apply **Array copy** or **Mirror a shape** operation on this step; however, considering the transverse symmetry of box girder, it is suggested by the author to take the methods of **Mirror a shape** more promptly.

The following step is to create the external tendons. Coordinate values (–0.5, 0, 0.6) and (–0.5, 20, 0.6) are the input to generate the geometric line of **tenout1**, where the **tenout2** is generated via the manipulation of **Mirror a shape** (see Fig. 3.388).

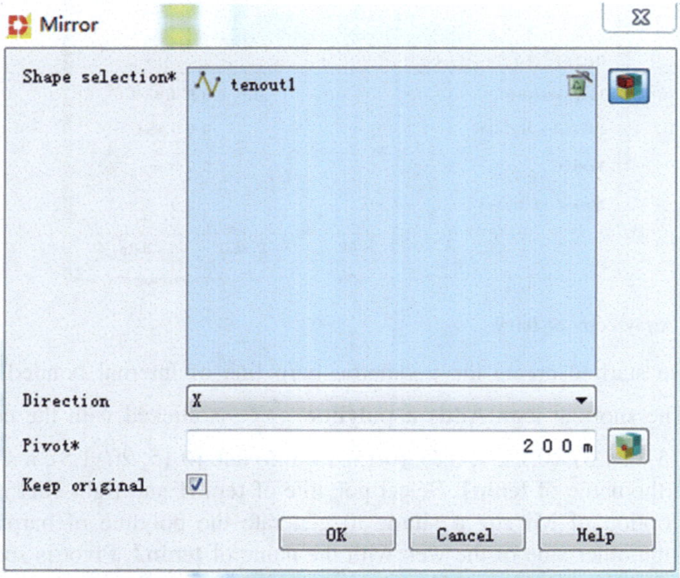

Fig. 3.388 Interface of Mirror a shape

Click the shortcut icon **Adds a block solid** to create the **Block1** with the coordinate value of **Position** (–1, 0, 0), and the sizes in the three directions of coordinate system are 1, 0.5 and 1 m, respectively (see Fig. 3.389).

Fig. 3.389 Manipulation of **Adds a block**

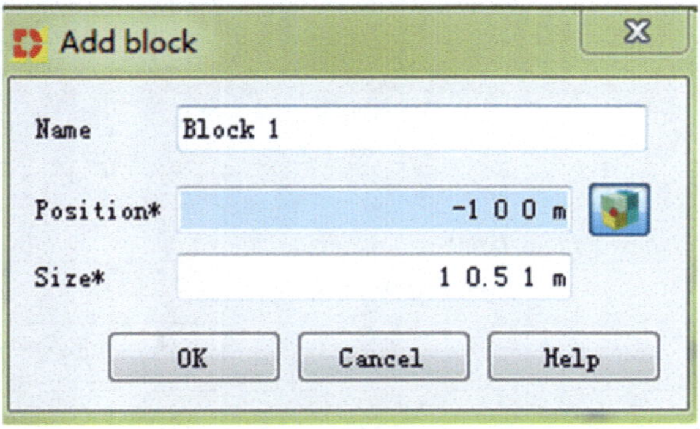

Note: In the aspect of **Adds a block solid** operation, hexahedrons, cubes and blocks are determined by the parameters of initial start point and the dimensional sizes in the three directions, which means that **Position** represents the three directional coordinate values of start point in creating the hexahedron, cube or block while the **Size** determines the 3D sizes such as length, width and height.

Still using the **Mirror a shape** operation to generate the blocks in the four outer corners of box girder in turn to simulate the concrete block of external prestressing tendon, which are activated elements in the following second phase analysis, the directions are in the X and Y directions, respectively, with corresponding pivots 2 and 10 m, respectively, as Figs. 3.390 and 3.391 displays.

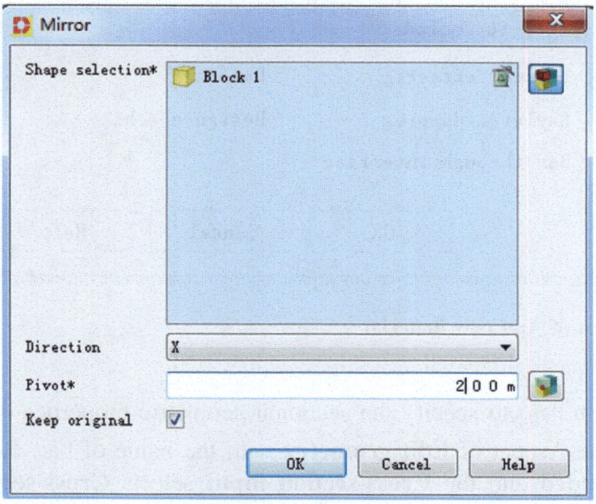

Fig. 3.390 Specification of Mirroring **Block1** in X direction

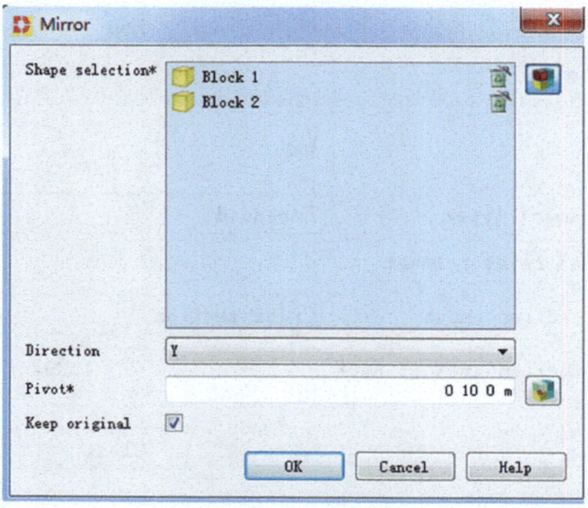

Fig. 3.391 Specification of Mirroring **Block1** and **Block2** in Y direction

Defining the material properties of bar, in the dialog box aspect of **Add new material** with the name **bar, Reinforcements and pile foundations** are chosen as **Class** option while **Material model** is Linear elasticity. Clicking OK button, **Young's modulus** is specified as 2.1×10^{11} $\mathbf{N/m^2}$ (Fig. 3.392).

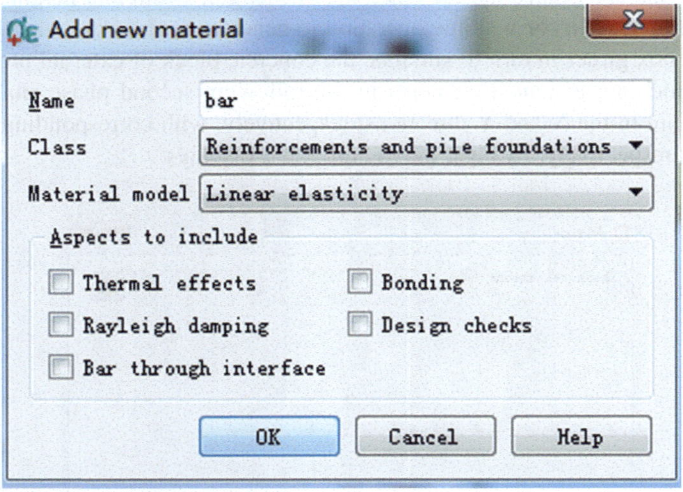

Fig. 3.392 Aspect of **Add new material**

Clicking icon to specify the sectional geometric properties of longitudinal steel bars in the Aspect of **Edit geometry** with the name of bar, **Reinforcement type** is **Embedded** and the **Cross-section input** selects **Cross-section** with the **Cross-section area of bar** 1.39×10^{-4} m^2 (see Fig. 3.393).

Section wise is chosen as discretization method; clicking OK button to complete the definition of longitudinal steel bars (see Fig. 3.394).

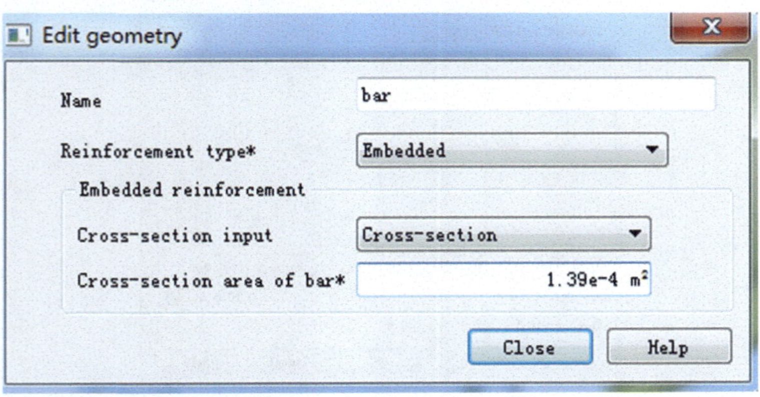

Fig. 3.393 Specification of sectional geometric properties of longitudinal steel bars

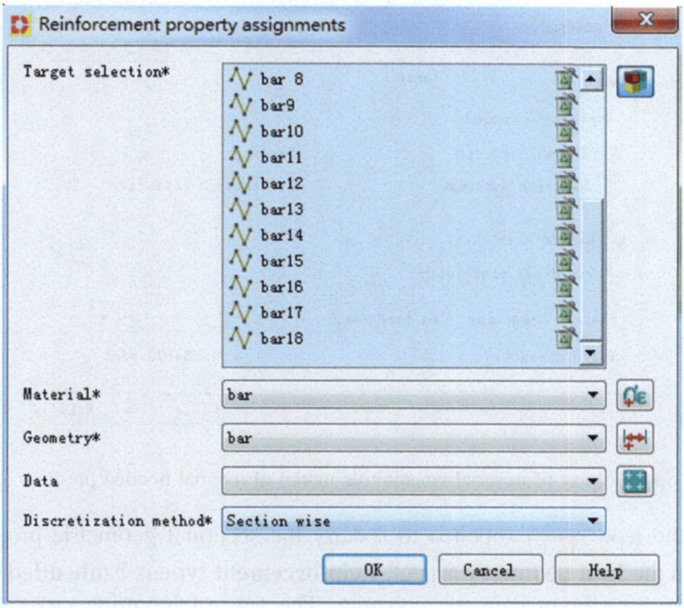

Fig. 3.394 Material specification of longitudinal steel bars

Define the material properties of the internal bonded prestress tendons, establish **Add new material** aspect with the name of **tenin;** reinforcement and pile foundations is specified as material class while the **von Mises** plasticity is specified as material constitutive model. Clicking OK button to enter the **Edit material** aspect, Young's modulus under the tab of Linear elasticity is set as 1.95×10^{11} N/m^2 while **No hardening** option is selected as **Plastic hardening** type with Yield stress 1.86×10^9 N/m^2, as Figs. 3.395 and 3.396 display. It is also worth noticing that **Bonding** option should not be ticked in the **Aspects to include** if you want to keep the tendons in bonded state.

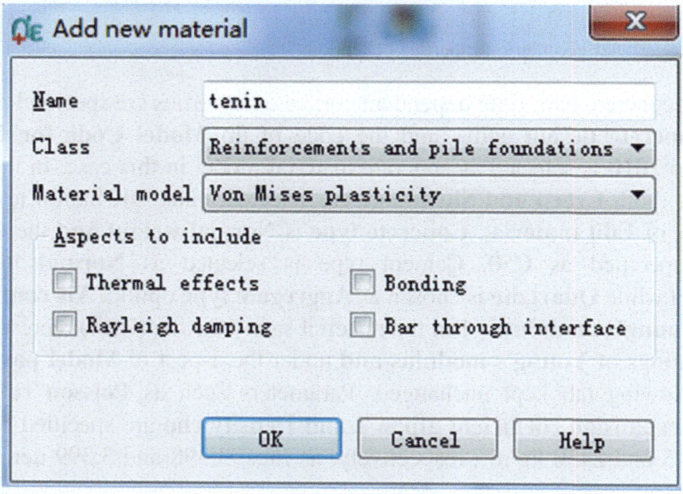

Fig. 3.395 Aspect of **tenin1** in defining material class

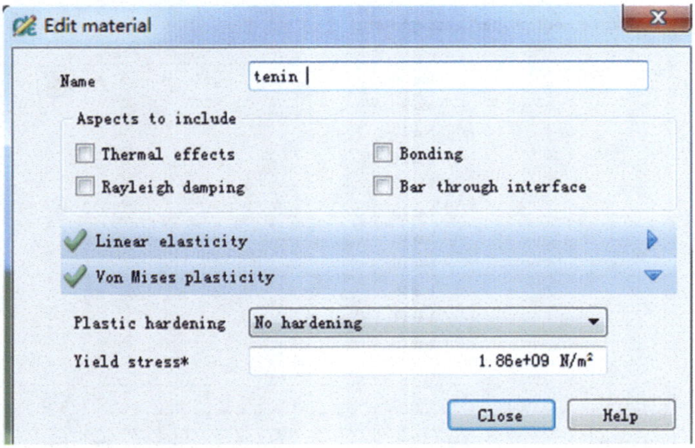

Fig. 3.396 Specification of material constitutive model of internal bonded prestress tendons

Then, the icon  is opened to specify the sectional geometric properties of tendons. In the **Edit geometry** aspect**, Reinforcement type** is **Embedded** to ensure the tendons fully bonded in the concrete. The way of inputting cross section is Cross-section and the area of internal tendons is 0.002886 m^2 (see Fig. 3.397).

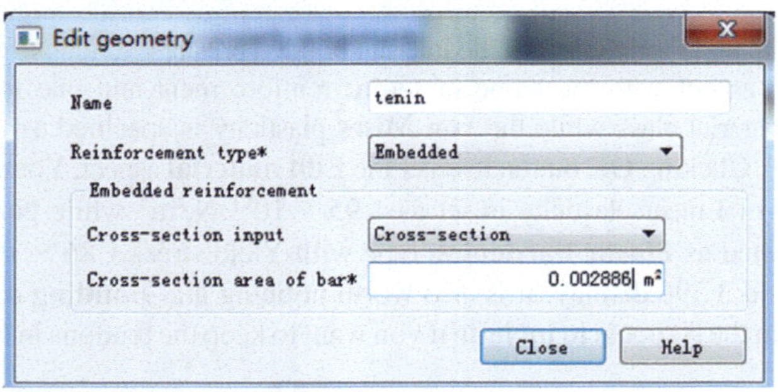

Fig. 3.397 Sectional geometric properties of tendons

For the concrete part, time-dependent concrete properties are specified according to the **Concrete design codes** and the code of **fib Model Code for Concrete Structures 2010** is selected as concrete material model in this case. In the **Aspect to include**, both **Creep and Shrinkage** are selected; click OK button to open the dialog box of **Edit material. Concrete type** is **Normal weight and** the **Concrete class** is specified as C50. Cement type is selected as **Normal hardening CE.52.5 N** while **Quartzite** is chosen as **Aggregate type** option. **Air content in %** is 2 and **Young's modulus** is 1 in the material safety factors (ULS) representing the default settings of **Young's modulus and** under the aspect of **Model parameters,** in the following tab kept unchanged. Parameters such as **Poisson ratio nu_c, Thermal expansion coefficient alpha_t and Density rho** are specified by default as 0.2, 1e-5 and 2450 kg/m^3, respectively, as Figs. 3.398 and 3.399 demonstrate.

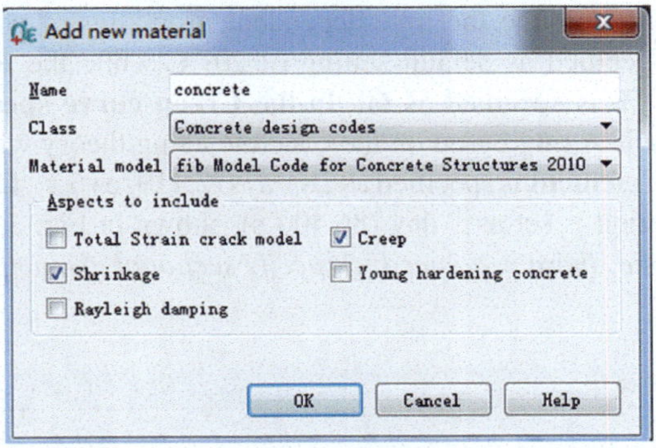

Fig. 3.398 Material class of fib model code for concrete structures 2010

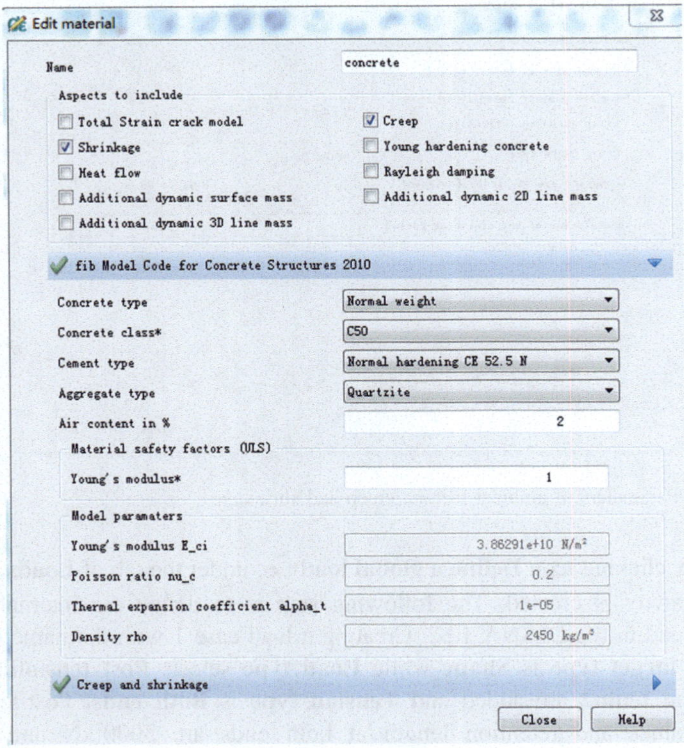

Fig. 3.399 Parameters of fib model code for concrete structures 2010

The key step is to define the time-dependent properties of concrete. **Ambient temperature** is specified as default setting of 20 °C while the **Relative ambient humidity RN in %** is specified as 60. In the **Creep curve specification** aspect, **Creep curve type** is **Aging** based on the concrete aging theory while the **Concrete age at the birth of element** is specified as 28 days (2.4192e+6 s). Concrete age at the end of curving period is set as 1 day (86,400 s), shown in Fig. 3.400. *As concrete element type is solid, there is no need to specify sectional geometric properties.*

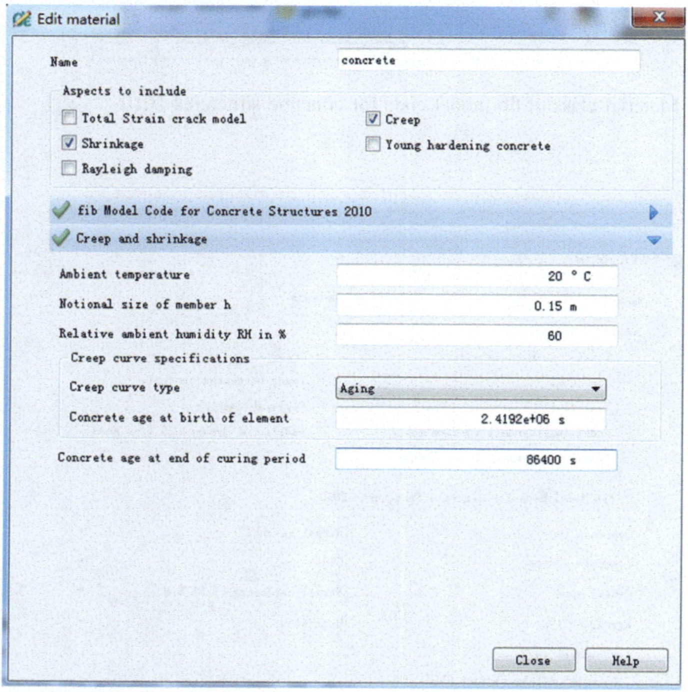

Fig. 3.400 Parameters of ambient factors, creep and shrinkage

Directly clicking icon **Define a global load** under the tab of **Loads,** the load case of gravity is created. The following step is to define the internal bonded prestress load in the DIANA 10.2. Creating a load case 1 with the name of **tenin,** the **Load target type** is **Shape** while **Load type** selects **Post tensioning load**. **Tenin1** and **tenin2** are added and **Tension type** is **Both ends.** Post-tensioning loading values and retention length at both ends are 3000 kN and 0.01 m, respectively. **Post-tensioning scheme** is **CEB-FIP Model Code 1990** with **Coulomb friction coefficient** and **Wobble factor** 0.22 and 0.001/m, respectively (see Fig. 3.401). Both gravity and post-tensioning load are added as the same load case as the initial attaching first dead load in the following nonlinear calculation with the name of **postte**.

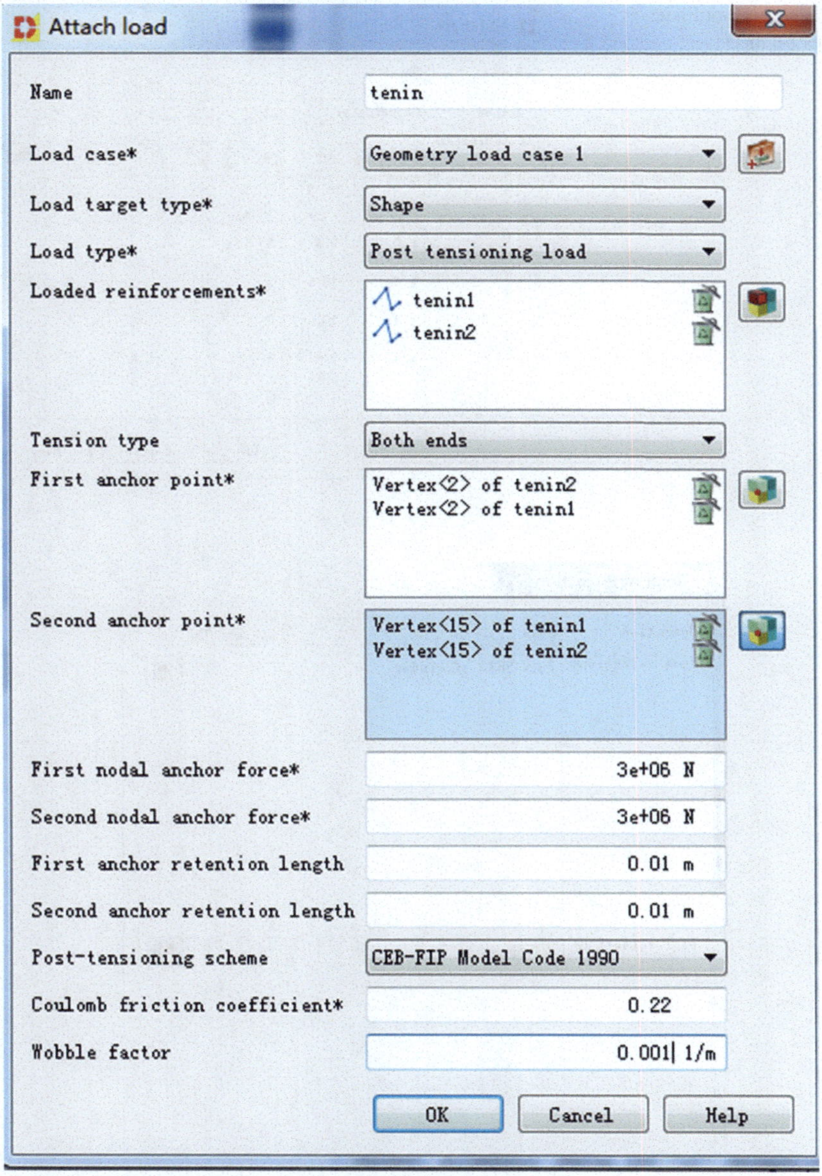

Fig. 3.401 Parameters of internal post-tensioning load

Create a sheet above the girder bridge waiting to be imprinted with the name of sheet1, and the coordinate values are shown in Fig. 3.402. Clicking shortcut icon **Project edges, wires and points on solid, faces and edges** , **Operation** is **Face** and plane of top plate is selected as **Face selection**. Tool selection is the Sheet1 above the girder bridge waiting to be imprinted in the negative Z direction, represented as (0, 0, –1) (see Fig. 3.403).

Fig. 3.402 Coordinate values of sheet1

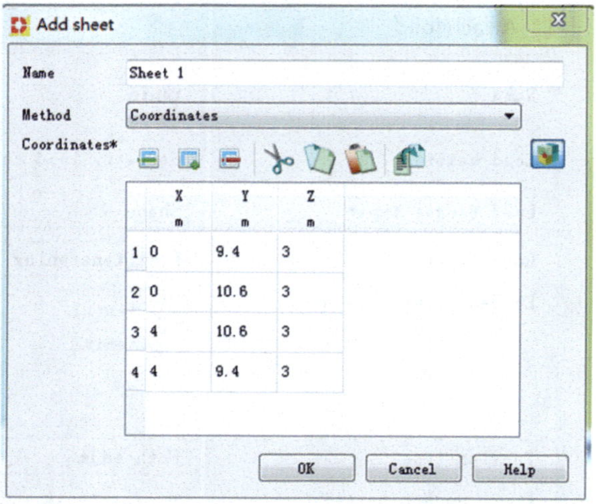

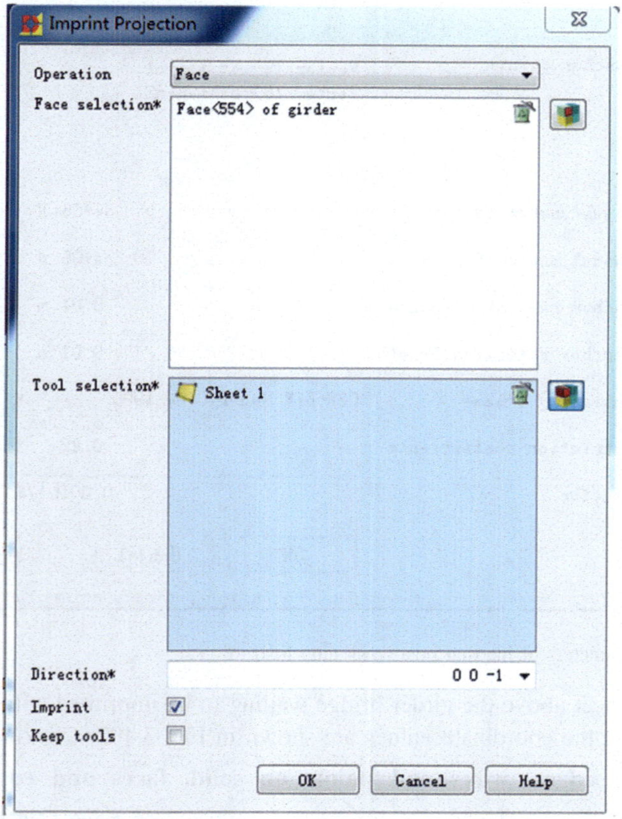

Fig. 3.403 Imprint projection

Attaching distributed load on the imprinted zone with the name of **load, load case** is **Geometry load case 2** while the **Load target type** is **Face. Load type** selects Distributed force with the **Surface force value 50 kN/m^2** in the negative Z direction (Fig. 3.404).

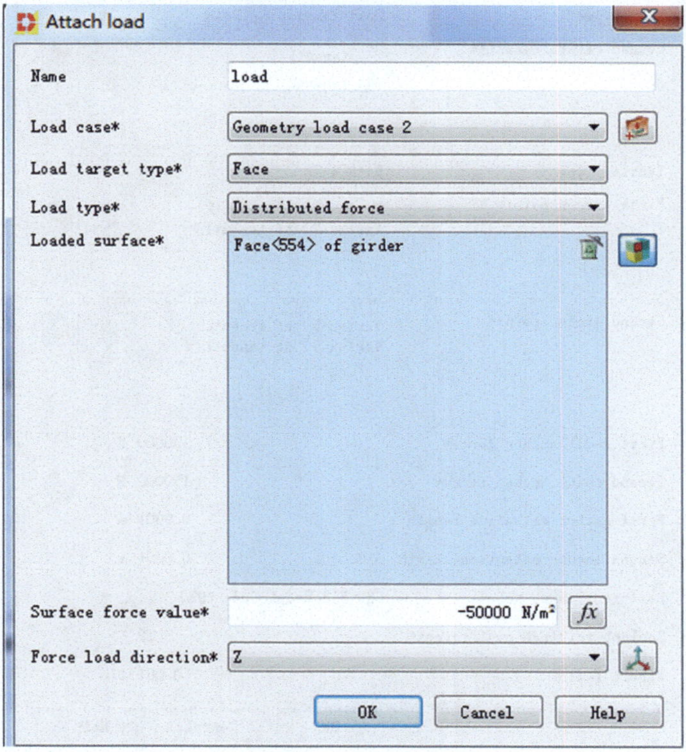

Fig. 3.404 Specification of distributed load

Attaching the external post-tensioning load, **Tenout1** and **tenout2** are added and **Tension type** is **Both ends.** Considering **tenout** as the role of eternal strengthening tendons with only one strand on both sides, post-tensioning loading values and retention length at both ends are 150 kN and 0.0001 m, respectively, and the other parameters are the same as **tenin** (see Fig. 3.405).

Note: Considering the material bonding function of external tendons represents no prestress loss under time-dependent effects, which is not corresponding to the actual condition, and the external tendons are bonded with concrete blocks at both ends, the retention length on both sides is set near to zero but should not be neglected as none, so the value is set as 0.0001, but the actual retention length is according to the actual condition.

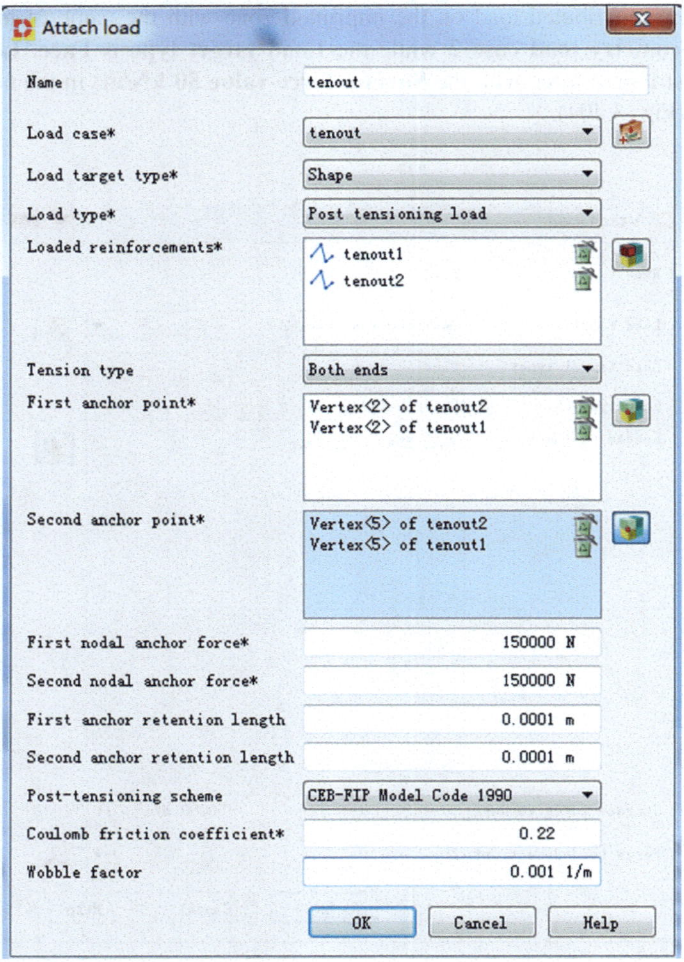

Fig. 3.405 Parameters of external post-tensioning load

The following step is to define the geometry load combinations, which is a key point in this case. **Postte** load case is set as **Geometry load combination 1** while **load** case and **tenout** case are specified as **Geometry load combination 2** and **Geometry load combination 3**, respectively. It is deserved to notice that all the load cases are added in the **Geometry load combination 4** because these load cases are added as a whole execute block in the following nonlinear analysis phase1 to further simulate the time-dependent effect after external strengthening. All the factors of the geometry load combination are specified as 1 (see Fig. 3.406).

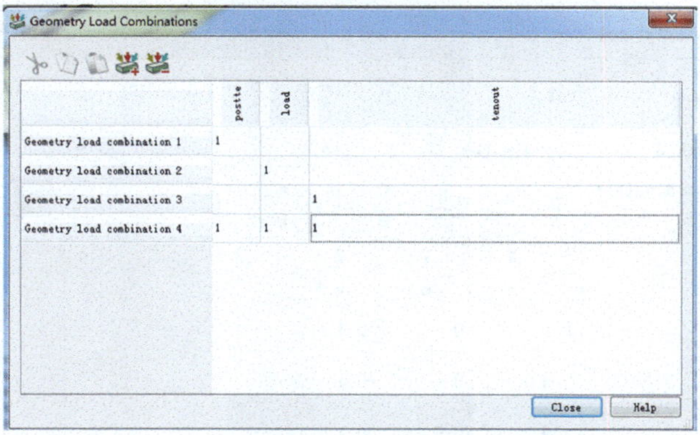

Fig. 3.406 Specifications of geometry load combination

Then we start to edit time-dependent factors. Clicking the short cut icon **Edit time dependency factors**, then right-clicking **Edit time dependency**, the dialog box pops up. All the time-dependency factors of the four geometry load combinations are specified as constant and invariable load according to (86,400 s, 1) and (3.1536e10 s, 1), as Fig. 3.407 shows.

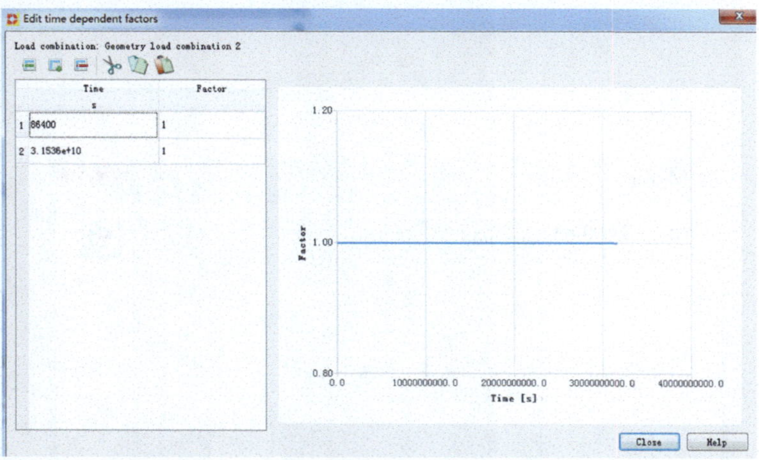

Fig. 3.407 Specification of time-dependency factors

Clicking menu bar **Geometry-Analysis-Attach support** to create support set, in order to correspond with the bottom plane size of four blocks so as to attach supports uniformly, a sheet naming **co1** is created via inputting the coordinate values according to the Fig. 3.408. Then the co1 sheet is mirrored to the other end of the girder with the name of **co2** and the pivot is 10 m in the Y direction (see Fig. 3.409. Ultimately, co1 and co2 are both imprinted on the surface of girder in the positive Z direction with the function of **Imprint projection,** as Fig. 3.410 displays.

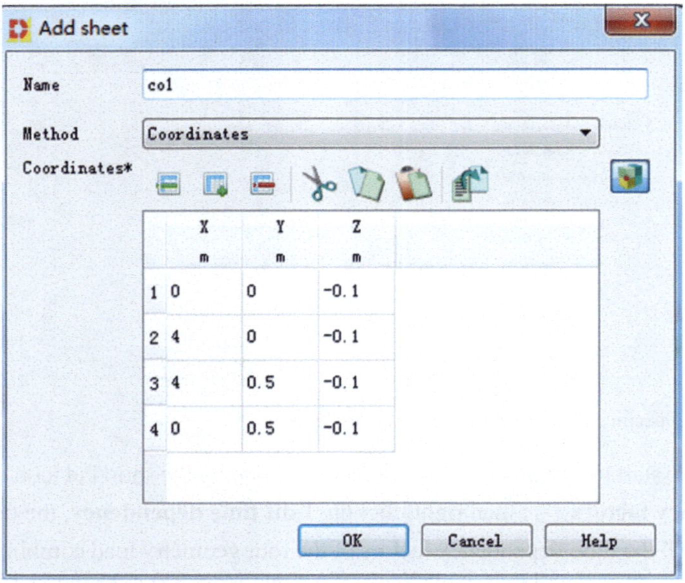

Fig. 3.408 Coordinate values of **col**

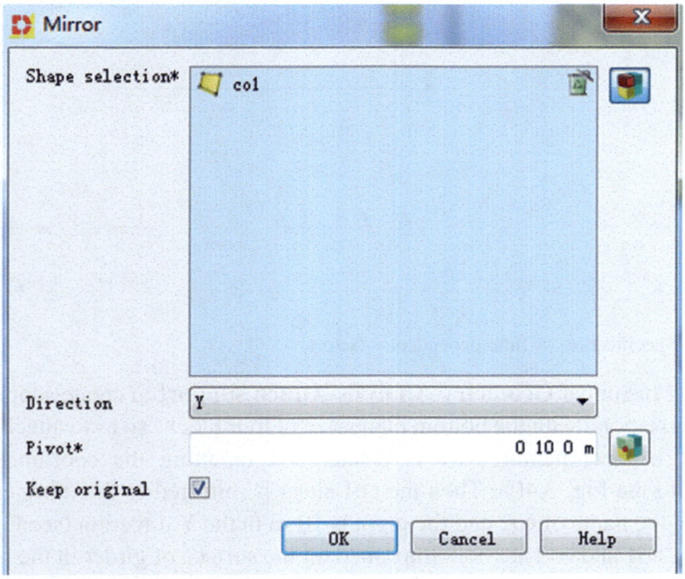

Fig. 3.409 Specification of col **Mirror shape**

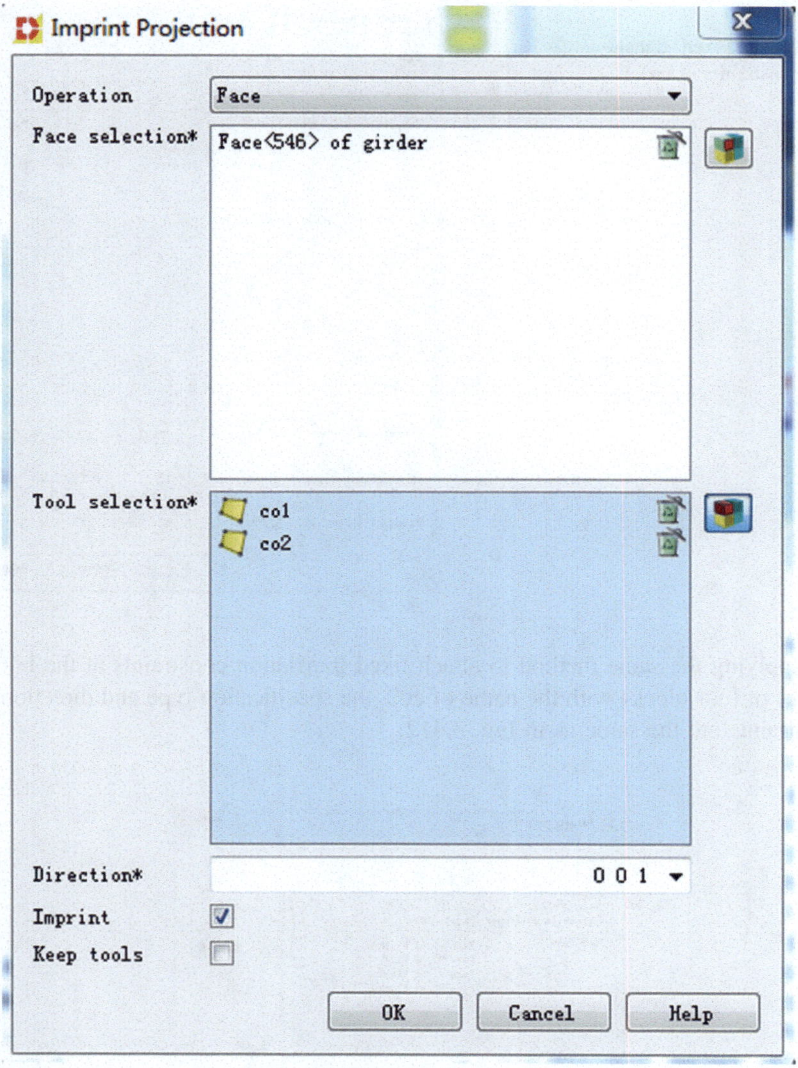

Fig. 3.410 Imprint projection of **co1** and **co2**

Then we start to attach supports. The imprinted sheets of co1 and co2 are selected to attach supports with the name of co1 in the translation X, Y and Z directions to fix the translation displacement in the three directions, the name of which are T1, T2 and T3, respectively (see Fig. 3.411).

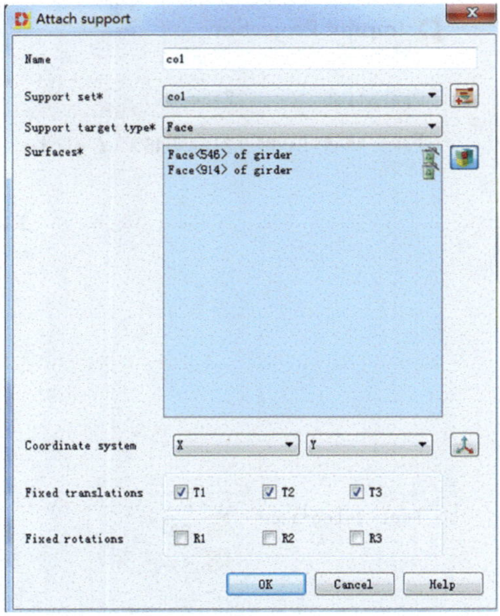

Fig. 3.411 Fixed translation constraints of co1

Applying the same method to attach fixed translation constraints at the bottom planes of four blocks with the name of co2, the specification type and directions of constraints are the same as in Fig. 3.412.

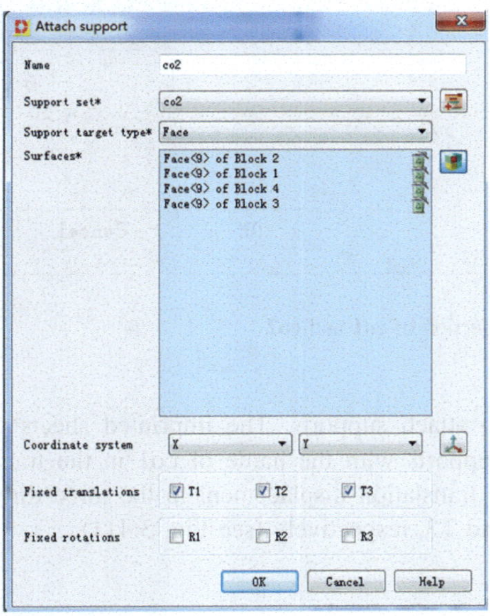

Fig. 3.412 Fixed translation constraints of co2

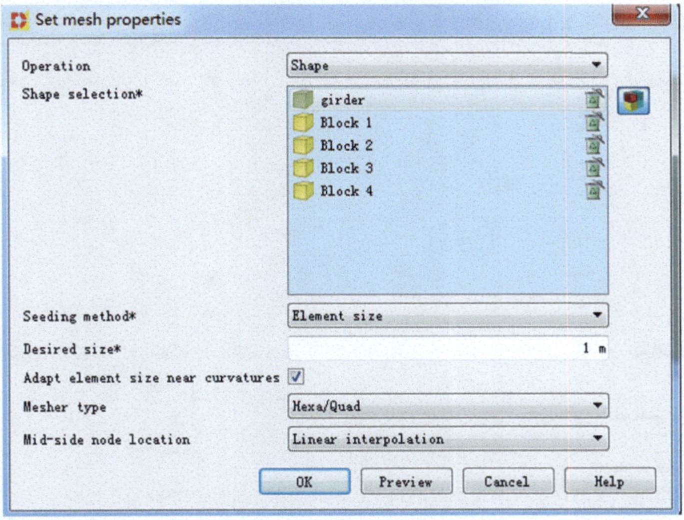

Fig. 3.413 Meshing specification of girder and blocks

All the four blocks and girder are meshed with the seeding method 1 m, while the **Mesher type** and **Mid-side node location** are selected as Hexa/Quad and Linear interpolation, respectively (see Fig. 3.413). Clicking the shortcut icon **Generate mesh of a shape** to generate the meshed elements, the meshed structure is shown (Fig. 3.414).

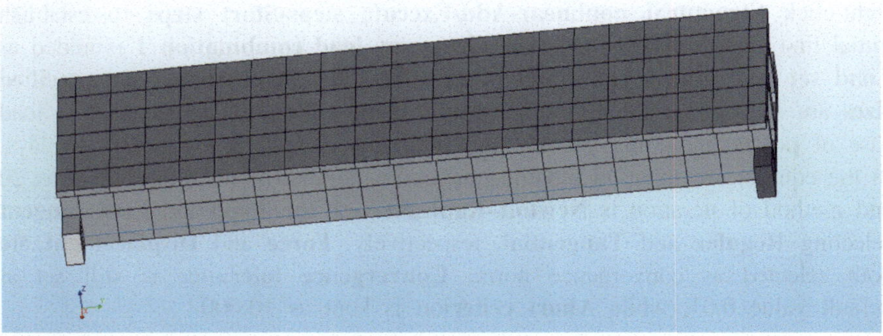

Fig. 3.414 Generation of meshed structure

The following step is to set the nonlinear analysis module. Click **Add an analysis** button  to establish a new nonlinear analysis block with the name of

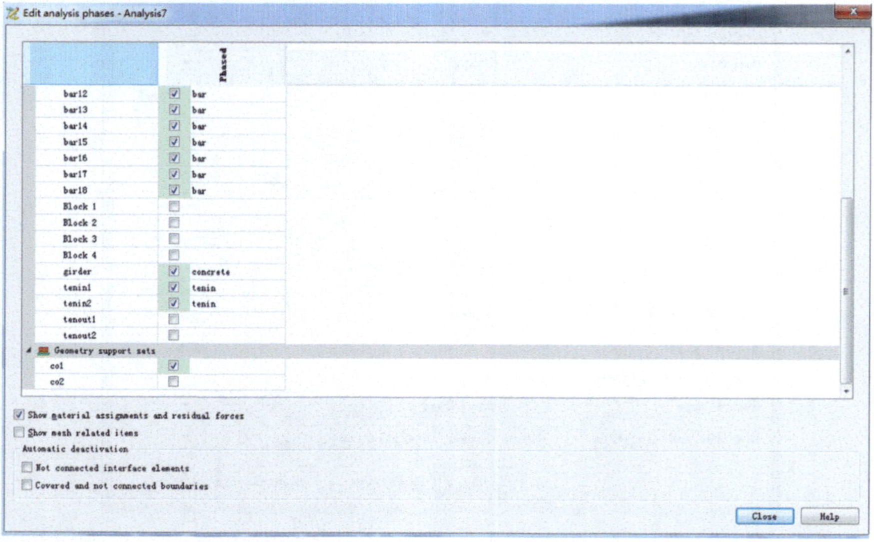

Fig. 3.415 Activated element and support sets in **Phased**

Analysis7, then right-click **Analysis7-Add command-Phased** to establish phased analysis. In the **Phased** aspect, elements block1 to block4, **tenout** external tendon set and **co2** support set are unticked as inactivated element and support sets for further strengthening phase (see Fig. 3.415).

Right-click **Analysis7-Add command-Structural nonlinear** to create nonlinear analysis of first phase, deleting the default execute block, and also right-click **Structural nonlinear-Add-Execute steps-Start steps** to establish initial first stage of dead load. The **Geometry load combination 1** is added as **Load set** under the **input** aspect of **execute start steps** while **User specified sizes** are set as 1 in the **Establish equilibrium** aspect, which means that load case of **postte**, including **gravity** and **tenin** are added, as Fig. 3.416 displays. In the equilibrium iteration module, maximum number of iterations is set as 20 and method of iteration is **Newton–Raphson** with the **Type** and **First tangent** selecting **Regular** and **Tangential,** respectively. **Force** and **Displacement** are both selected as convergence norm. **Convergence tolerance** is still set at default value 0.01, while **Abort criterion** is kept as 10,000.

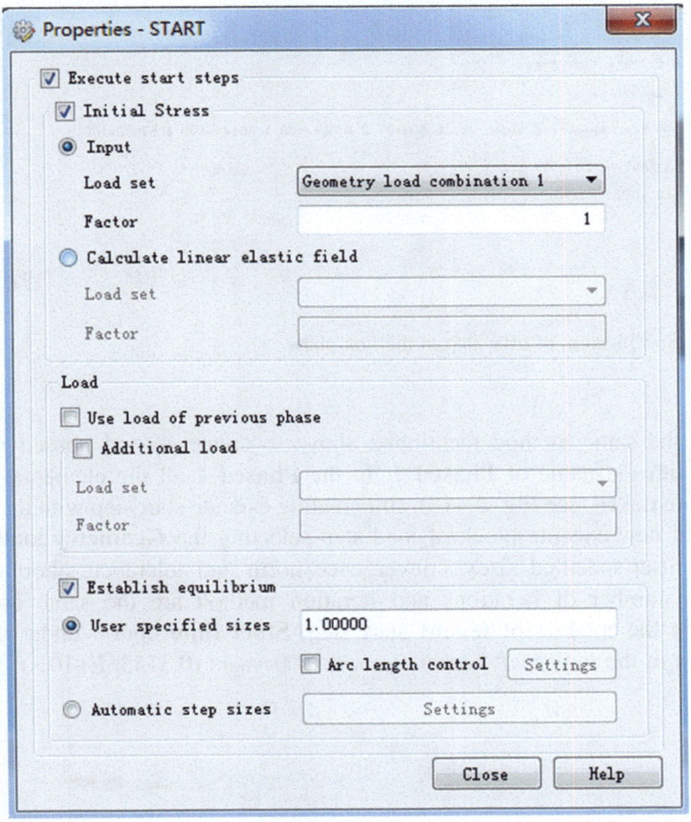

Fig. 3.416 Specification of start step

Apply the same method to create new execute block of ordinary load step and specify the **Geometry load combination 1** of load case with **User specified sizes** also 1. Convergence norm and tolerance, abort criterion, maximum number of iterations and iteration method are all kept the same as former.

Apply the same method to create new execute block of time step with the name of creep and shrinkage, in the **time step** module. **User specified sizes** are specified as 1, 10, 50 and 100 years while the corresponding adding time steps required to be specified are 3.15360e+07 s, 2.83824e+08 s, 1.26144e+09 s and 1.57680e+09 s, respectively (see Fig. 3.417).

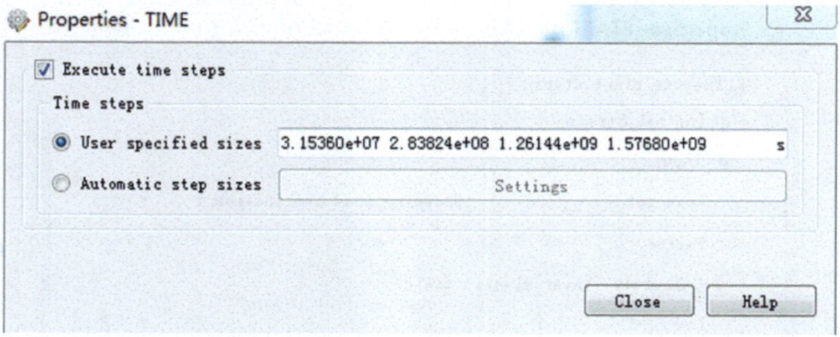

Fig. 3.417 Specification of time step in the first phase

Apply the same method mentioned above to create second phased nonlinear analysis with the name of **Phased 1**. In the **Phased 1** all the elements, sets and supports are ticked (see Fig. 3.418), still creating execute start step with the name of **tenout** and new execute block of load step selecting the **Geometry load combination 4**, user-specified sizes, convergence norm and tolerance, abort criterion, maximum number of iterations and iteration method are the same as former. Besides, in the module of **tenout** start step, **Start time** option under aspect of **Properties** in the lower left corner starts as 100 years (0.31536E+10 s).

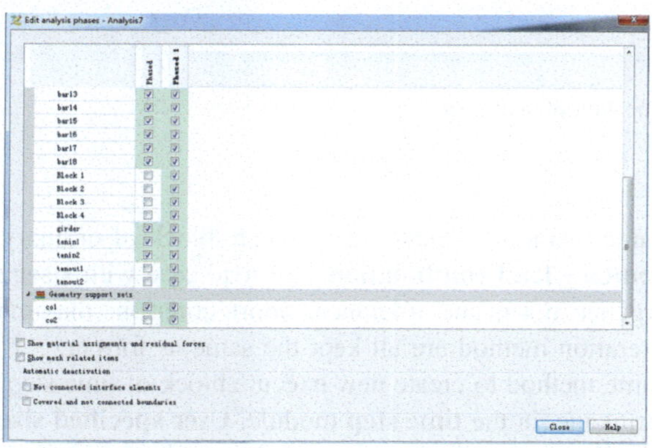

Fig. 3.418 Activated element and support sets in **Phased 1**

Continuously adding new execute blocks of time step with the name of **creep and shrinkage1**, user-specified size of time step is specified as 3.15360e+08(10) s, meaning the interval of time loading step is 10 years per step. Other specifications are the same as former.

Click **Run analysis** button to calculate the time-dependent nonlinear analysis, Displacement contours of initial hugging-up state (phased, start-step 1, load-factor 1.0000), the state after 100 years (phased, time steps 6, time 0.31536E+10) belonging to the first phase in the Z translation direction and ultimate state after strengthening (phased 1, time steps 12, Time 0.63072E+10) belonging to the second phase in the Z translation direction are shown in Figs. 3.419, 3.420 and 3.421, respectively.

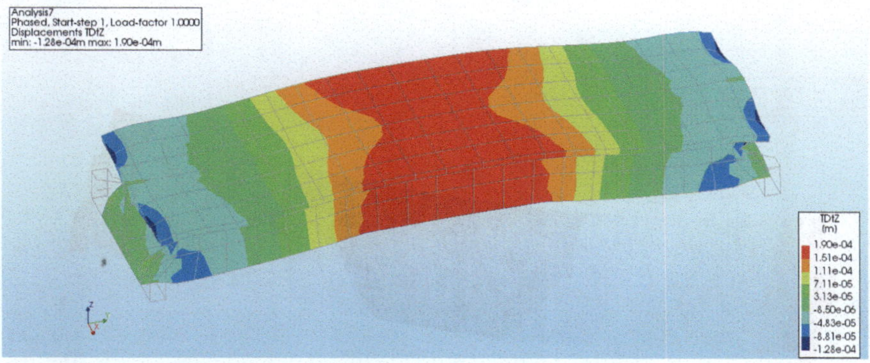

Fig. 3.419 Displacement contour of initial hugging-up state belonging to the first phase

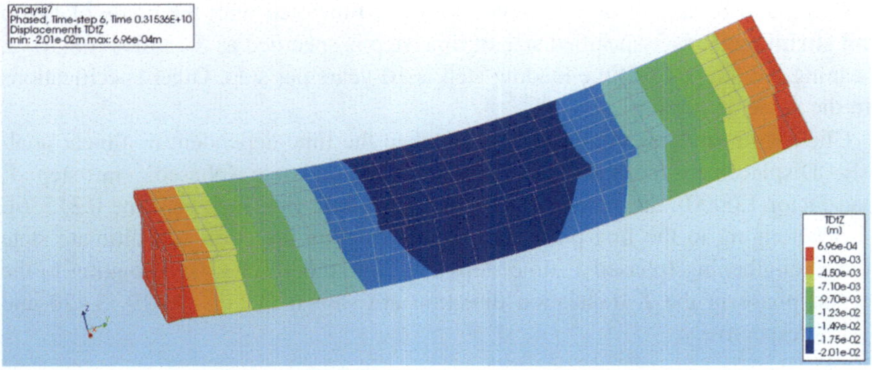

Fig. 3.420 Displacement contour of 100 years belonging to the first phase

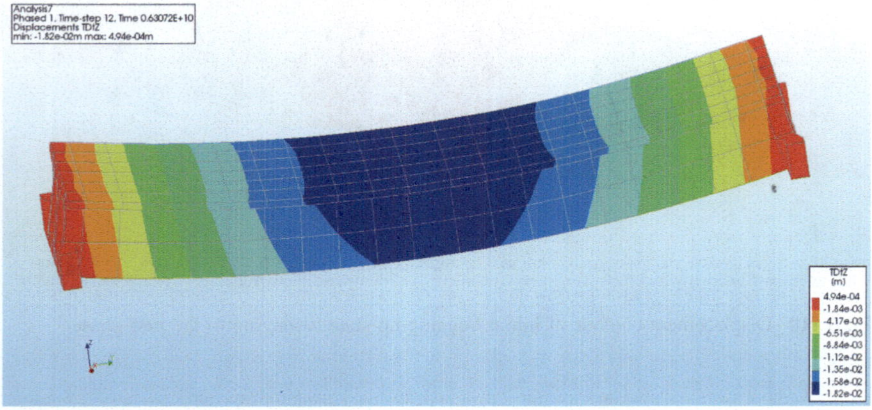

Fig. 3.421 Displacement contour after strengthening belonging to the second phase

According to the monitor of Analyais7, the negative value represents downward displacement while the positive one represents the hugging-up displacement. Thus it is obvious to judge that the maximum deflection after 100 years is located at the middle site of the span with the long-term deflection 2.01 cm while the maximum deflection after strengthening is 1.82 cm, thus the resilient effect of external tendons is $\frac{2.01-1.82}{2.01} = 9.45\%$.

Command console of **"reinforced"** in Python language is as follows

```
newProject( "reinforced", 1000 )
setModelAnalysisAspects( [ "STRUCT" ] )
setModelDimension( "3D" )
setDefaultMeshOrder( "QUADRATIC" )
setDefaultMesherType( "HEXQUAD" )
setDefaultMidSideNodeLocation( "LINEAR" )
createSheet( "girder", [[ 0, 0, 0 ],[ 4, 0, 0 ],[ 4, 0, 2 ],[ 5, 0, 2.2 ],[ 5, 0, 2.4 ],[ -1, 0, 2.4 ],[ -1,
0, 2.2 ],[ 0, 0, 2 ]] )
createSheet( "sheet1", [[ 0.5, 0, 0.3 ],[ 1.5, 0, 0.3 ],[ 1.8, 0, 0.5 ],[ 1.8, 0, 1.7 ],[ 1.5, 0,
2 ],[ 0.5, 0, 2 ],[ 0.3, 0, 1.7 ],[ 0.3, 0, 0.5 ]] )
mirror( [ "sheet1" ], [ 2, 0, 0 ], [ True, False, False ], True )
subtract( "girder", [ "sheet1", "sheet2" ], False, True )
extrudeProfile( [ "girder" ], [ 0, 20, 0 ] )
saveProject( )
createLine( "bar", [ 0.4, 0, 0.2 ], [ 0.4, 20, 0.2 ] )
show( SHAPE, [ "bar" ] )
arrayCopy( [ "bar" ], [ 0.4, 0, 0 ], [ 0, 0, 0 ], [ 0, 0, 0 ], 8 )
saveProject( )
createLine( "bar9", [ -0.6, 0, 2.2 ], [ -0.6, 20, 2.2 ] )
arrayCopy( [ "bar9" ], [ 0.6, 0, 0 ], [ 0, 0, 0 ], [ 0, 0, 0 ], 9 )
createPolyline( "tenin1", [[ 0.15, 0, 1.5 ],[ 0.15, 7, 0.5 ],[ 0.15, 13, 0.5 ],[ 0.15, 20, 1.5 ]],
False )
hide( SHAPE, [ "bar 1" ] )
mirror( [ "tenin1" ], [ 2, 0, 0 ], [ True, False, False ], True )
createLine( "tenout1", [ -0.5, 0, 0.6 ], [ -0.5, 20, 0.6 ] )
mirror( [ "tenout1" ], [ 2, 0, 0 ], [ True, False, False ], True )
createBlock( "Block 1", [ -1, 0, 0 ], [ 1, 0.5, 1 ] )
mirror( [ "Block 1" ], [ 2, 0, 0 ], [ True, False, False ], True )
saveProject( )
mirror( [ "Block 1", "Block 2" ], [ 0, 10, 0 ], [ False, True, False ], True )
saveProject( )
addMaterial( "bar", "REINFO", "LINEAR", [] )
setParameter( MATERIAL, "bar", "LINEAR/ELASTI/YOUNG", 2.1e+11 )
addGeometry( "Element geometry 1", "RELINE", "REBAR", [] )
rename( GEOMET, "Element geometry 1", "bar" )
setParameter( GEOMET, "bar", "REIEMB/CROSSE", 0.000139 )
setParameter( GEOMET, "bar", "REIEMB/CROSSE", 0.000139 )
setParameter( GEOMET, "bar", "REIEMB/CROSSE", 0.000139 )
setReinforcementAspects( [ "bar", "bar 1", "bar 2", "bar 3", "bar 4", "bar 5", "bar 6", "bar 7",
"bar 8", "bar9", "bar10", "bar11", "bar12", "bar13", "bar14", "bar15", "bar16", "bar17", "bar18" ] )
assignMaterial( "bar", "SHAPE", [ "bar", "bar 1", "bar 2", "bar 3", "bar 4", "bar 5", "bar 6",
"bar 7", "bar 8", "bar9", "bar10", "bar11", "bar12", "bar13", "bar14", "bar15", "bar16", "bar17",
```

```
"bar18" ] )
    assignGeometry( "bar", "SHAPE", [ "bar", "bar 1", "bar 2", "bar 3", "bar 4", "bar 5", "bar 6",
"bar 7", "bar 8", "bar9", "bar10", "bar11", "bar12", "bar13", "bar14", "bar15", "bar16", "bar17",
"bar18" ] )
    resetElementData( "SHAPE", [ "bar", "bar 1", "bar 2", "bar 3", "bar 4", "bar 5", "bar 6", "bar
7", "bar 8", "bar9", "bar10", "bar11", "bar12", "bar13", "bar14", "bar15", "bar16", "bar17",
"bar18" ] )
    setReinforcementDiscretization( [ "bar", "bar 1", "bar 2", "bar 3", "bar 4", "bar 5", "bar 6",
"bar 7", "bar 8", "bar9", "bar10", "bar11", "bar12", "bar13", "bar14", "bar15", "bar16", "bar17",
"bar18" ], "SECTION" )
    saveProject(  )
    addMaterial( "tenin", "REINFO", "VMISES", [] )
    setParameter( MATERIAL, "tenin", "LINEAR/ELASTI/YOUNG", 1.95e+11 )
    setParameter( MATERIAL, "tenin", "PLASTI/HARDI1/YLDSTR", 1.86e+09 )
    setParameter( MATERIAL, "tenin", "PLASTI/HARDI1/YLDSTR", 1.86e+09 )
    setParameter( MATERIAL, "tenin", "PLASTI/HARDI1/YLDSTR", 1.86e+09 )
    addGeometry( "Element geometry 2", "RELINE", "REBAR", [] )
    rename( GEOMET, "Element geometry 2", "tenin" )
    setParameter( GEOMET, "tenin", "REIEMB/CROSSE", 0.002886 )
    setParameter( GEOMET, "tenin", "REIEMB/CROSSE", 0.002886 )
    setParameter( GEOMET, "tenin", "REIEMB/CROSSE", 0.002886 )
    setParameter( GEOMET, "tenin", "REIEMB/CROSSE", 0.002886 )
    setReinforcementAspects( [ "tenin1", "tenin2" ] )
    assignMaterial( "tenin", SHAPE, [ "tenin1", "tenin2" ] )
    assignGeometry( "tenin", SHAPE, [ "tenin1", "tenin2" ] )
    resetElementData( SHAPE, [ "tenin1", "tenin2" ] )
    setReinforcementDiscretization( [ "tenin1", "tenin2" ], "SECTION" )
    saveProject(  )
    addMaterial( "tenout", "REINFO", "VMISES", [] )
    setParameter( MATERIAL, "tenout", "LINEAR/ELASTI/YOUNG", 1.95e+11 )
    setParameter( MATERIAL, "tenout", "PLASTI/HARDI1/YLDSTR", 1.86e+09 )
    addGeometry( "Element geometry 3", "RELINE", "REBAR", [] )
    rename( GEOMET, "Element geometry 3", "tenout" )
    setParameter( GEOMET, "tenout", "REIEMB/CROSSE", 0.000139 )
    setParameter( GEOMET, "tenout", "REIEMB/CROSSE", 0.000139 )
    setParameter( GEOMET, "tenout", "REIEMB/CROSSE", 0.000139 )
    setReinforcementAspects( [ "tenin1", "tenin2", "tenout1", "tenout2" ] )
    assignMaterial( "tenout", SHAPE, ["tenout1", "tenout2" ] )
    assignGeometry( "tenout", SHAPE, ["tenout1", "tenout2" ] )
    resetElementData( SHAPE, ["tenout1", "tenout2" ] )
    setReinforcementDiscretization( ["tenout1", "tenout2" ], "SECTION" )
    saveProject(  )
```

```
addMaterial( "concrete", "CONCDC", "MC1990", [ "CRACKI", "CREEP", "SHRINK" ] )
setUnit( "ANGLE", "DEGREE" )
setUnit( "TEMPER", "CELSIU" )
setParameter( MATERIAL, "concrete", "MC90CO/RH", 60 )
remove( MATERIAL, "concrete" )
addMaterial( "concrete", "CONCDC", "MC2010", [ "CREEP", "SHRINK" ] )
setParameter( MATERIAL, "concrete", "MC10CO/NORMAL/GRADE", "C50" )
setParameter( MATERIAL, "concrete", "CONCCP/RH", 60 )
setParameter( MATERIAL, "concrete", "CONCCP/CREEP/CRSPEC/AGETYP",
"AGING" )
setParameter( MATERIAL, "concrete", "CONCCP/CREEP/CRSPEC/AGING", 2419200 )
setParameter( MATERIAL, "concrete", "CONCCP/SHRINK/CURAGE", 86400 )
setParameter( MATERIAL, "concrete", "CONCCP/SHRINK/CURAGE", 86400 )
setParameter( MATERIAL, "concrete", "CONCCP/SHRINK/CURAGE", 86400 )
setParameter( MATERIAL, "concrete", "MC10CO/NORMAL/CEMTYP", "CE52N" )
setElementClassType( SHAPE, [ "girder" ], "STRSOL" )
assignMaterial( "concrete", SHAPE, [ "girder" ] )
saveProject(  )
addMaterial( "concrete1", "CONCDC", "MC2010", [ "CREEP", "SHRINK" ] )
setParameter( MATERIAL, "concrete1", "MC10CO/NORMAL/GRADE", "C50" )
setParameter( MATERIAL, "concrete1", "CONCCP/RH", 60 )
setParameter( MATERIAL, "concrete1", "CONCCP/CREEP/CRSPEC/AGETYP",
"AGING" )
setParameter( MATERIAL, "concrete1", "CONCCP/CREEP/CRSPEC/AGING", 2419200 )
setParameter( MATERIAL, "concrete1", "CONCCP/SHRINK/CURAGE", 86400 )
setParameter( MATERIAL, "concrete1", "MC10CO/NORMAL/CEMTYP", "CE52N" )
setElementClassType( SHAPE, [ "Block 1", "Block 2", "Block 4", "Block 3" ], "STRSOL" )
assignMaterial( "concrete1", SHAPE, [ "Block 1", "Block 2", "Block 4", "Block 3" ] )
saveProject(  )
saveProject(  )
addSet( GEOMETRYLOADSET, "Geometry load case 1" )
createModelLoad( "gravity", "Geometry load case 1" )
saveProject(  )
createBodyLoad( "tenin", "Geometry load case 1" )
setParameter( GEOMETRYLOAD, "tenin", "LODTYP", "POSTEN" )
setParameter( GEOMETRYLOAD, "tenin", "POSTEN/BOTHEN/FORCE1", 3000000 )
setParameter( GEOMETRYLOAD, "tenin", "POSTEN/BOTHEN/FORCE2", 3000000 )
setParameter( GEOMETRYLOAD, "tenin", "POSTEN/BOTHEN/RETLE1", 0.01 )
setParameter( GEOMETRYLOAD, "tenin", "POSTEN/BOTHEN/RETLE2", 0.01 )
setParameter( GEOMETRYLOAD, "tenin", "POSTEN/SHEAR", 0.22 )
setParameter( GEOMETRYLOAD, "tenin", "POSTEN/WOBBLE", 0.001 )
attachTo( GEOMETRYLOAD, "tenin", "POSTEN/BOTHEN/PNTS1", "tenin1", [[ 0.15, 0,
1.5 ]] )
```

```
attachTo( GEOMETRYLOAD, "tenin", "POSTEN/BOTHEN/PNTS1", "tenin2", [[ 3.85, 0, 1.5 ]] )
attachTo( GEOMETRYLOAD, "tenin", "POSTEN/BOTHEN/PNTS2", "tenin1", [[ 0.15, 20, 1.5 ]] )
attachTo( GEOMETRYLOAD, "tenin", "POSTEN/BOTHEN/PNTS2", "tenin2", [[ 3.85, 20, 1.5 ]] )
attach( GEOMETRYLOAD, "tenin", [ "tenin1", "tenin2" ] )
saveProject(  )
createSheet( "Sheet 1", [[ 0, 9.4, 3 ],[ 0, 10.6, 3 ],[ 4, 10.6, 3 ],[ 4, 9.4, 3 ]] )
projection( SHAPEFACE, "girder", [[ 2.441438, 11.47146, 2.4 ]], [ "Sheet 1" ], [ 0, 0, -1 ], True )
removeShape( [ "Sheet 1" ] )
addSet( GEOMETRYLOADSET, "Geometry load case 2" )
createSurfaceLoad( "load", "Geometry load case 2" )
setParameter( GEOMETRYLOAD, "load", "FORCE/VALUE", -50000 )
setParameter( GEOMETRYLOAD, "load", "FORCE/DIRECT", 3 )
attach( GEOMETRYLOAD, "load", "girder", [[ 2.294292, 10.088288, 2.4 ]] )
addSet( GEOMETRYLOADSET, "Geometry load case 3" )
rename( GEOMETRYLOADSET, "Geometry load case 1", "postte" )
rename( GEOMETRYLOADSET, "Geometry load case 3", "tenout" )
rename( GEOMETRYLOADSET, "Geometry load case 2", "load" )
setDefaultGeometryLoadCombinations(  )
setGeometryLoadCombinationFactor( "Geometry load combination 1", "postte", 1 )
remove( GEOMETRYLOADCOMBINATION, "Geometry load combination 3" )
remove( GEOMETRYLOADCOMBINATION, "Geometry load combination 2" )
setGeometryLoadCombinationFactor( "Geometry load combination 1", "postte", 1 )
addGeometryLoadCombination( "" )
setGeometryLoadCombinationFactor( "Geometry load combination 2", "load", 1 )
addGeometryLoadCombination( "" )
setGeometryLoadCombinationFactor( "Geometry load combination 3", "tenout", 1 )
addGeometryLoadCombination( "" )
setGeometryLoadCombinationFactor( "Geometry load combination 4", "postte", 1 )
setGeometryLoadCombinationFactor( "Geometry load combination 4", "load", 1 )
setGeometryLoadCombinationFactor( "Geometry load combination 4", "tenout", 1 )
createSheet( "co1", [[ 0, 0, -0.1 ],[ 4, 0, -0.1 ],[ 4, 0.5, -0.1 ],[ 0, 0.5, -0.1 ]] )
mirror( [ "co1" ], [ 0, 10, 0 ], [ False, True, False ], True )
projection( SHAPEFACE, "girder", [[ 1.705708, 11.47146, 0 ]], [ "co1", "co2" ], [ 0, 0, 1 ], True )
removeShape( [ "co1", "co2" ] )
setTimeDependentLoadFactors( GEOMETRYLOADCOMBINATION, "Geometry load combination 1", [ 86400, 3.1536e+10 ], [ 1, 1 ] )
setTimeDependentLoadFactors( GEOMETRYLOADCOMBINATION, "Geometry load combination 2", [ 86400, 3.1536e+10 ], [ 1, 1 ] )
```

setTimeDependentLoadFactors(GEOMETRYLOADCOMBINATION, "Geometry load combination 3", [86400, 3.1536e+10], [1, 1])
setTimeDependentLoadFactors(GEOMETRYLOADCOMBINATION, "Geometry load combination 4", [86400, 3.1536e+10], [1, 1])
saveProject( )
createBodyLoad("tenout", "tenout")
setParameter(GEOMETRYLOAD, "tenout", "LODTYP", "POSTEN")
setParameter(GEOMETRYLOAD, "tenout", "POSTEN/BOTHEN/FORCE1", 150000)
setParameter(GEOMETRYLOAD, "tenout", "POSTEN/BOTHEN/FORCE2", 150000)
setParameter(GEOMETRYLOAD, "tenout", "POSTEN/BOTHEN/RETLE1", 0.0001)
setParameter(GEOMETRYLOAD, "tenout", "POSTEN/BOTHEN/RETLE2", 0.0001)
setParameter(GEOMETRYLOAD, "tenout", "POSTEN/SHEAR", 0.22)
setParameter(GEOMETRYLOAD, "tenout", "POSTEN/WOBBLE", 0.001)
attachTo(GEOMETRYLOAD, "tenout", "POSTEN/BOTHEN/PNTS1", "tenout1", [[-0.5, 0, 0.6]])
attachTo(GEOMETRYLOAD, "tenout", "POSTEN/BOTHEN/PNTS1", "tenout2", [[4.5, 0, 0.6]])
attachTo(GEOMETRYLOAD, "tenout", "POSTEN/BOTHEN/PNTS2", "tenout1", [[-0.5, 20, 0.6]])
attachTo(GEOMETRYLOAD, "tenout", "POSTEN/BOTHEN/PNTS2", "tenout2", [[4.5, 20, 0.6]])
attach(GEOMETRYLOAD, "tenout", ["tenout1", "tenout2"])
addSet(GEOMETRYSUPPORTSET, "co1")
createSurfaceSupport("co1", "co1")
setParameter(GEOMETRYSUPPORT, "co1", "AXES", [1, 2])
setParameter(GEOMETRYSUPPORT, "co1", "TRANSL", [1, 1, 1])
setParameter(GEOMETRYSUPPORT, "co1", "ROTATI", [0, 0, 0])
attach(GEOMETRYSUPPORT, "co1", "girder", [[1.705708, 0.2867865, 0],[1.705708, 19.786787, 0]])
addSet(GEOMETRYSUPPORTSET, "Geometry support set 1")
rename(GEOMETRYSUPPORTSET, "Geometry support set 1", "co2")
createSurfaceSupport("co2", "co2")
setParameter(GEOMETRYSUPPORT, "co2", "AXES", [1, 2])
setParameter(GEOMETRYSUPPORT, "co2", "TRANSL", [1, 1, 1])
setParameter(GEOMETRYSUPPORT, "co2", "ROTATI", [0, 0, 0])
attach(GEOMETRYSUPPORT, "co2", "Block 1", [[-0.573573, 0.2867865, 0]])
attach(GEOMETRYSUPPORT, "co2", "Block 2", [[4.573573, 0.2132135, 0]])
attach(GEOMETRYSUPPORT, "co2", "Block 3", [[-0.573573, 19.786787, 0]])
attach(GEOMETRYSUPPORT, "co2", "Block 4", [[4.573573, 19.713213, 0]])
saveProject( )
setElementSize(["girder", "Block 1", "Block 2", "Block 3", "Block 4"], 1, -1, True)
setMesherType(["girder", "Block 1", "Block 2", "Block 3", "Block 4"], "HEXQUAD")
setMidSideNodeLocation(["girder", "Block 1", "Block 2", "Block 3", "Block 4"], "LINEAR")

```
saveProject( )
setGeometryLoadCombinationFactor( "Geometry load combination 1", "postte", 1 )
generateMesh( [] )
hideView( "GEOM" )
showView( "MESH" )
addAnalysis( "Analysis7" )
addAnalysisCommand( "Analysis7", "PHASE", "Phased" )
renameAnalysis( "Analysis7", "Analysis7" )
setActivePhase( "Analysis7", "Phased" )
setActivePhase( "Analysis7", "Phased" )
setActiveInPhase( "Analysis7", SHAPE, [ "tenout2" ], [ "Phased" ], False )
setActiveInPhase( "Analysis7", SHAPE, [ "tenout1" ], [ "Phased" ], False )
setActiveInPhase( "Analysis7", GEOMETRYSUPPORTSET, [ "co2" ], [ "Phased" ], False )
setActiveInPhase( "Analysis7", SHAPE, [ "Block 4" ], [ "Phased" ], False )
setActiveInPhase( "Analysis7", SHAPE, [ "Block 3" ], [ "Phased" ], False )
setActiveInPhase( "Analysis7", SHAPE, [ "Block 2" ], [ "Phased" ], False )
setActiveInPhase( "Analysis7", SHAPE, [ "Block 1" ], [ "Phased" ], False )
renameAnalysisCommand( "Analysis7", "Phased", "Phased" )
addAnalysisCommand( "Analysis7", "NONLIN", "Structural nonlinear" )
removeAnalysisCommandDetail( "Analysis7", "Structural nonlinear", "EXECUT(1)" )
setActivePhase( "Analysis7", "Phased" )
setAnalysisCommandDetail( "Analysis7", "Structural nonlinear", "EXECUT/EXETYP", "START" )
renameAnalysisCommand( "Analysis7", "Structural nonlinear", "Structural nonlinear" )
addAnalysisCommandDetail( "Analysis7", "Structural nonlinear", "EXECUT(1)/START/INITIA/STRESS" )
setAnalysisCommandDetail( "Analysis7", "Structural nonlinear", "EXECUT(1)/START/INITIA/STRESS", True )
setAnalysisCommandDetail( "Analysis7", "Structural nonlinear", "EXECUT(1)/START/LOAD/PREVIO", False )
addAnalysisCommandDetail( "Analysis7", "Structural nonlinear", "EXECUT(1)/START/INITIA/STRESS/INPUT/LOAD" )
setAnalysisCommandDetail( "Analysis7", "Structural nonlinear", "EXECUT(1)/START/INITIA/STRESS/INPUT/LOAD", 1 )
setAnalysisCommandDetail( "Analysis7", "Structural nonlinear", "EXECUT(1)/ITERAT/MAXITE", 20 )
setAnalysisCommandDetail( "Analysis7", "Structural nonlinear", "EXECUT(1)/ITERAT/CONVER/DISPLA/NOCONV", "CONTIN" )
setAnalysisCommandDetail( "Analysis7", "Structural nonlinear", "EXECUT(1)/ITERAT/CONVER/FORCE/NOCONV", "CONTIN" )
addAnalysisCommandDetail( "Analysis7", "Structural nonlinear", "EXECUT(1)/PHYSIC" )
setAnalysisCommandDetail( "Analysis7", "Structural nonlinear", "EXECUT(1)/PHYSIC/LIQUEF", False )
```

setAnalysisCommandDetail("Analysis7", "Structural nonlinear", "EXECUT(1)/PHYSIC/BOND", True)

saveProject( )

setActivePhase("Analysis7", "Phased")

setAnalysisCommandDetail("Analysis7", "Structural nonlinear", "EXECUT/EXETYP", "LOAD")

renameAnalysisCommand("Analysis7", "Structural nonlinear", "Structural nonlinear")

setAnalysisCommandDetail("Analysis7", "Structural nonlinear", "EXECUT(2)/ITERAT/MAXITE", 20)

setAnalysisCommandDetail("Analysis7", "Structural nonlinear", "EXECUT(2)/ITERAT/CONVER/DISPLA/NOCONV", "CONTIN")

setAnalysisCommandDetail("Analysis7", "Structural nonlinear", "EXECUT(2)/ITERAT/CONVER/FORCE/NOCONV", "CONTIN")

addAnalysisCommandDetail("Analysis7", "Structural nonlinear", "EXECUT(2)/LOAD/LOADNR")

setAnalysisCommandDetail("Analysis7", "Structural nonlinear", "EXECUT(2)/LOAD/LOADNR", 2)

setActivePhase("Analysis7", "Phased")

setAnalysisCommandDetail("Analysis7", "Structural nonlinear", "EXECUT/EXETYP", "TIME")

renameAnalysisCommandDetail("Analysis7", "Structural nonlinear", "EXECUT(3)", "creep and shrinkage")

setAnalysisCommandDetail("Analysis7", "Structural nonlinear", "EXECUT(3)/TIME/STEPS/EXPLIC/SIZES", "3.15360e+07 2.83824e+08 1.26144e+09 1.57680e+09")

setAnalysisCommandDetail("Analysis7", "Structural nonlinear", "EXECUT(3)/ITERAT/MAXITE", 20)

setAnalysisCommandDetail("Analysis7", "Structural nonlinear", "EXECUT(3)/ITERAT/CONVER/DISPLA/NOCONV", "CONTIN")

setAnalysisCommandDetail("Analysis7", "Structural nonlinear", "EXECUT(3)/ITERAT/CONVER/FORCE/NOCONV", "CONTIN")

addAnalysisCommand("Analysis7", "PHASE", "Phased 1")

setActivePhase("Analysis7", "Phased 1")

renameAnalysisCommand("Analysis7", "Phased 1", "Phased 1")

addAnalysisCommand("Analysis7", "NONLIN", "Structural nonlinear 1")

renameAnalysis("Analysis7", "Analysis7")

removeAnalysisCommandDetail("Analysis7", "Structural nonlinear 1", "EXECUT(1)")

setActivePhase("Analysis7", "Phased 1")

setAnalysisCommandDetail("Analysis7", "Structural nonlinear 1", "EXECUT/EXETYP", "START")

renameAnalysisCommand("Analysis7", "Structural nonlinear 1", "Structural nonlinear 1")

renameAnalysisCommandDetail("Analysis7", "Structural nonlinear 1", "EXECUT(1)", "tenout")

addAnalysisCommandDetail("Analysis7", "Structural nonlinear 1", "EXECUT(1)/START/INITIA/STRESS")

```
setAnalysisCommandDetail( "Analysis7", "Structural nonlinear 1", "EXECUT(1)/START/INITIA/STRESS", True )
addAnalysisCommandDetail( "Analysis7", "Structural nonlinear 1", "EXECUT(1)/START/INITIA/STRESS/INPUT/LOAD" )
setAnalysisCommandDetail( "Analysis7", "Structural nonlinear 1", "EXECUT(1)/START/INITIA/STRESS/INPUT/LOAD", 3 )
setAnalysisCommandDetail( "Analysis7", "Structural nonlinear 1", "EXECUT(1)/START/LOAD/PREVIO", False )
setAnalysisCommandDetail( "Analysis7", "Structural nonlinear 1", "EXECUT(1)/ITERAT/MAXITE", 20 )
setAnalysisCommandDetail( "Analysis7", "Structural nonlinear 1", "EXECUT(1)/ITERAT/CONVER/DISPLA/NOCONV", "CONTIN" )
setAnalysisCommandDetail( "Analysis7", "Structural nonlinear 1", "EXECUT(1)/ITERAT/CONVER/FORCE/NOCONV", "CONTIN" )
setActivePhase( "Analysis7", "Phased 1" )
setAnalysisCommandDetail( "Analysis7", "Structural nonlinear 1", "EXECUT/EXETYP", "LOAD" )
addAnalysisCommandDetail( "Analysis7", "Structural nonlinear 1", "EXECUT(2)/LOAD/LOADNR" )
setAnalysisCommandDetail( "Analysis7", "Structural nonlinear 1", "EXECUT(2)/LOAD/LOADNR", 4 )
setAnalysisCommandDetail( "Analysis7", "Structural nonlinear 1", "EXECUT(2)/ITERAT/MAXITE", 20 )
setAnalysisCommandDetail( "Analysis7", "Structural nonlinear 1", "EXECUT(2)/ITERAT/CONVER/DISPLA/NOCONV", "CONTIN" )
setAnalysisCommandDetail( "Analysis7", "Structural nonlinear 1", "EXECUT(2)/ITERAT/CONVER/FORCE/NOCONV", "CONTIN" )
setActivePhase( "Analysis7", "Phased 1" )
setAnalysisCommandDetail( "Analysis7", "Structural nonlinear 1", "EXECUT/EXETYP", "TIME" )
renameAnalysisCommandDetail( "Analysis7", "Structural nonlinear 1", "EXECUT(3)", "creep and shrinkage 1" )
setAnalysisCommandDetail( "Analysis7", "Structural nonlinear 1", "EXECUT(3)/TIME/STEPS/EXPLIC/SIZES", "315360000(10)" )
setAnalysisCommandDetail( "Analysis7", "Structural nonlinear 1", "EXECUT(3)/ITERAT/MAXITE", 20 )
setAnalysisCommandDetail( "Analysis7", "Structural nonlinear 1", "EXECUT(3)/ITERAT/CONVER/DISPLA/NOCONV", "CONTIN" )
setAnalysisCommandDetail( "Analysis7", "Structural nonlinear 1", "EXECUT(3)/ITERAT/CONVER/FORCE/NOCONV", "CONTIN" )
setAnalysisCommandDetail( "Analysis7", "Structural nonlinear 1", "EXECUT(1)/START/TIME", 3.1536e+09 )
saveProject( )
runSolver( "Analysis7" )
```

Delete the four external blocks in the corner of the box girder bridge, and now, we start to apply the passive strengthening method via the platform of DIANA release 10.2. Considering the steel plate is paste on the surface of the bottom plane of the box girder bridge, thus it is essential to move the whole geometric model, including the reinforcements 0.01 m, upward in the positive Z direction via the manipulation of **Move shape** (see Fig. 3.422).

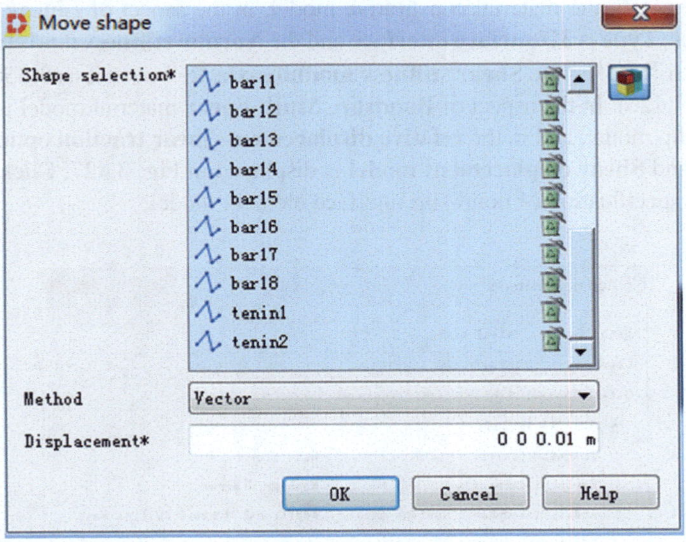

Fig. 3.422 Move shape

Click shortcut icon **Adds a block solid**  and create a new solid block with the name of steel plate. The coordinate value of **Position** is (0, 2, 0) while size in the X, Y and Z directions are 4, 16 and 0.01 m, respectively, as Fig. 3.423 displays.

Fig. 3.423 Specification of **steel plate** block

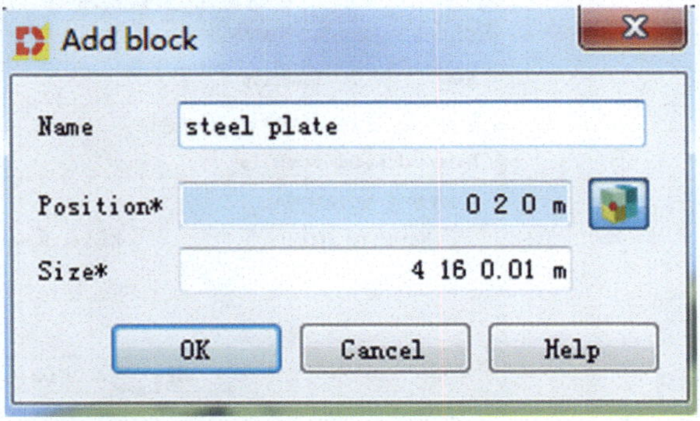

The next step is to define material properties of pasted steel plate. Creating a dialog of material properties with the name of s**teel plate** while the **Class** selecting **Steel** to be assigned with steel constitutive model, **Linear elastic isotropic** is chosen as material model. Clicking OK button to enter the **Edit material** aspect, Young's modulus, Poisson's ratio and mass density are 2.1×10^{11} N/m^2, 0.3 and 7800 kg/m^3, respectively, as Fig. 3.424 displays. Selecting the interface between the bottom plane of the box girder bridge and top surface steel plate as the interface element with the name of **bondslip** (as Fig. 3.425 shows), applying the same method mentioned above to specify the material constitutive model, in the aspect of **Linear material properties**, **Type** is **3D surface interface** and the **Normal stiffness modulus-y** is set as 3.65e16 N/m^3 while **Shear stiffness modulus-x/z** are both set as 3.65e6 N/m^3 (see Fig. 3.426). In the aspect of **Bondslip**, **Multi-linear** material model is selected as bond-slip model, and in the **relative displacement-Shear traction** option, **Shear traction** and **Shear displacement model** is displayed in Fig. 3.427. Click Close to finish the specification of bond-slip interface element model.

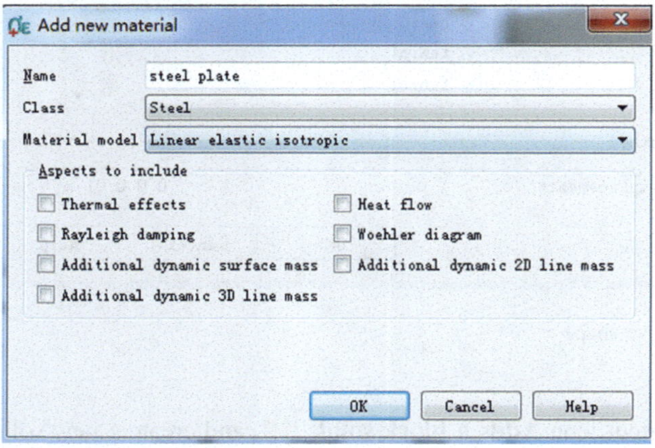

Fig. 3.424 Steel material properties

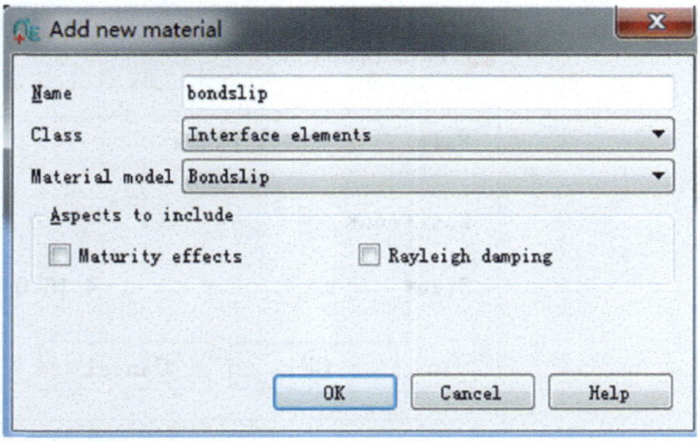

Fig. 3.425 Interface element material properties

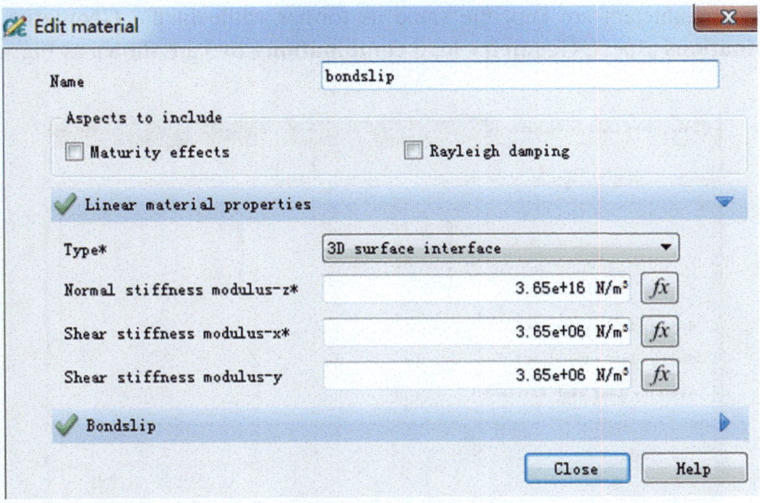

Fig. 3.426 Specification of bond-slip interface material model

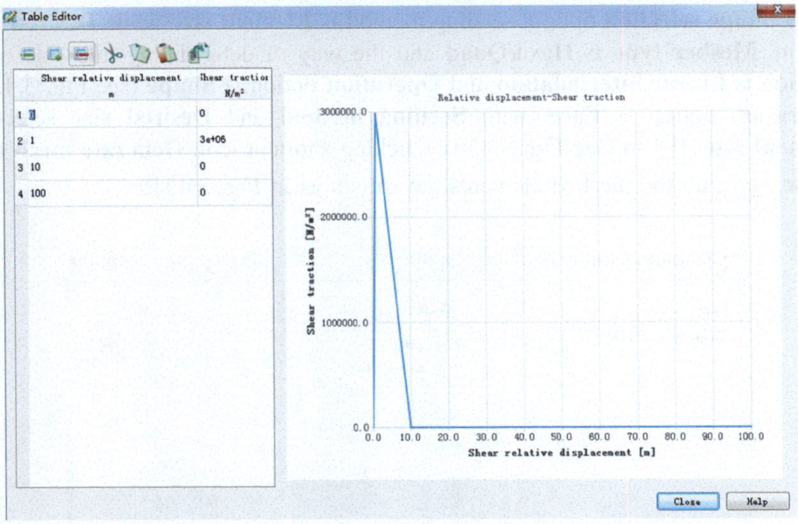

Fig. 3.427 Specification of shear traction–relative displacement relationship

Other parameters are kept the same as former while in the **Geometry Load Combinations** aspect, **Geometry load combination1 to 3** are shown as Fig. 3.428.

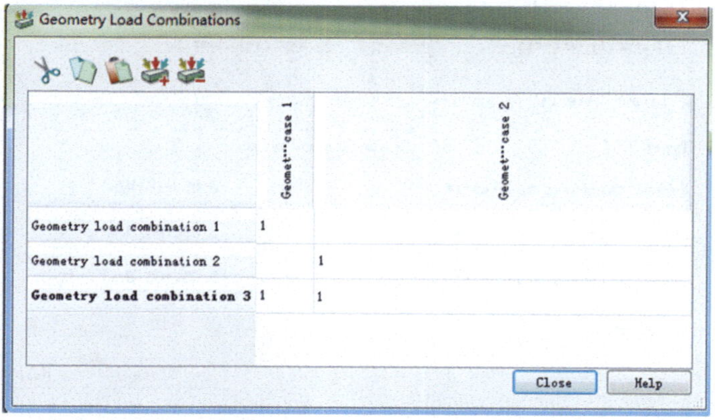

Fig. 3.428 Specification of geometry load combinations

During the procedure of meshing settings, both girder and steel plate are selected in the **Shape selection** option; seeding method is **Element size** while **Desired size** is 1 m. **Mesher type** is **Hexa/Quad** and the way of determining mid-side node location is **Linear interpolation** and **Operation** option is **Shape** (see Fig. 3.429). **Operation** option is **Face**, with **Seeding method** and **Desired size** selecting **Element sizes** 0.5 m (see Fig. 3.430). Clicking shortcut icon **Generate mesh of a shape** , all the meshed elements are shown as in Fig. 3.431.

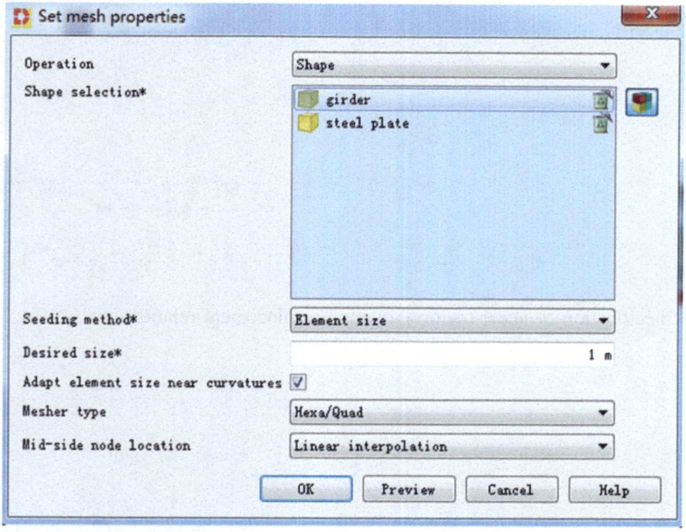

Fig. 3.429 Specification of girder and steel plate

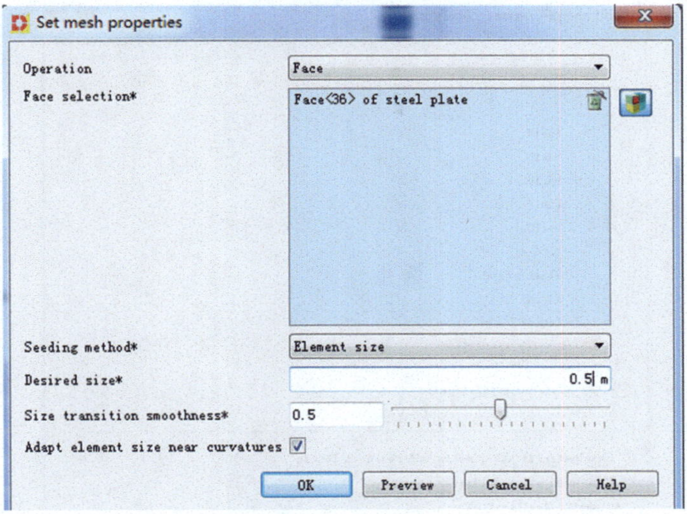

Fig. 3.430 Specification of meshed interface surface elements

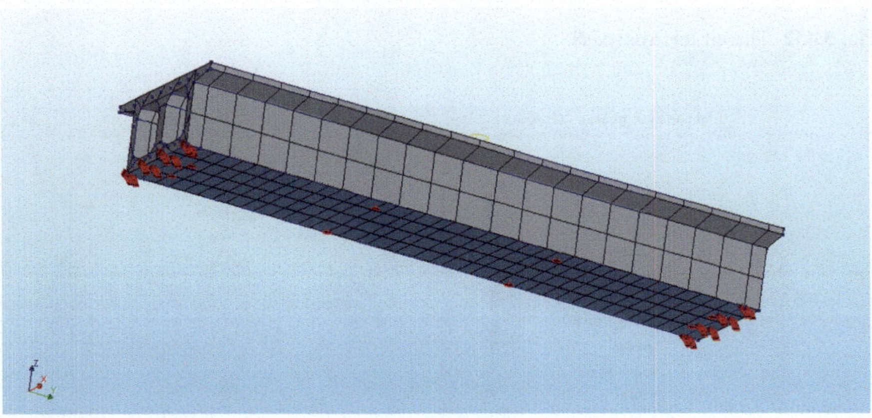

Fig. 3.431 Generated meshed structural elements

Create **Analysis8** to create nonlinear analysis of first phase with the name of **Phased**. In the **Phased** aspect, elements steel plate set is unticked as inactivated element for further second strengthening phase—**Phased 1** (see Figs. 3.432 and 3.433 respectively).

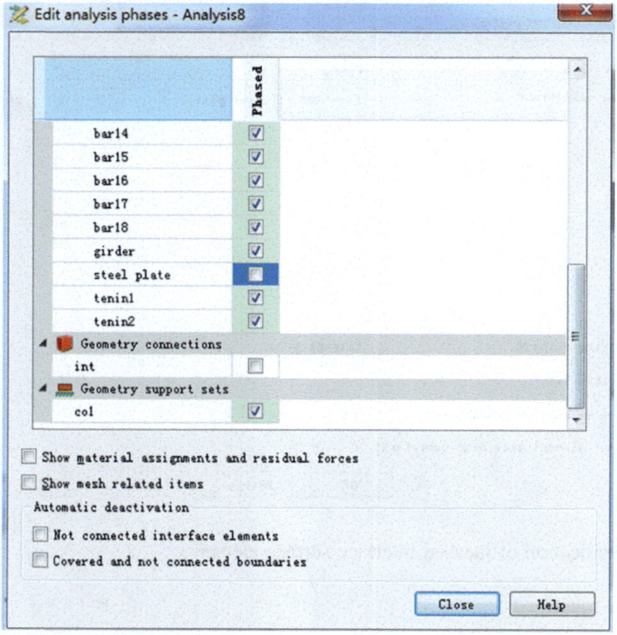

Fig. 3.432 Phased for **Analysis8**

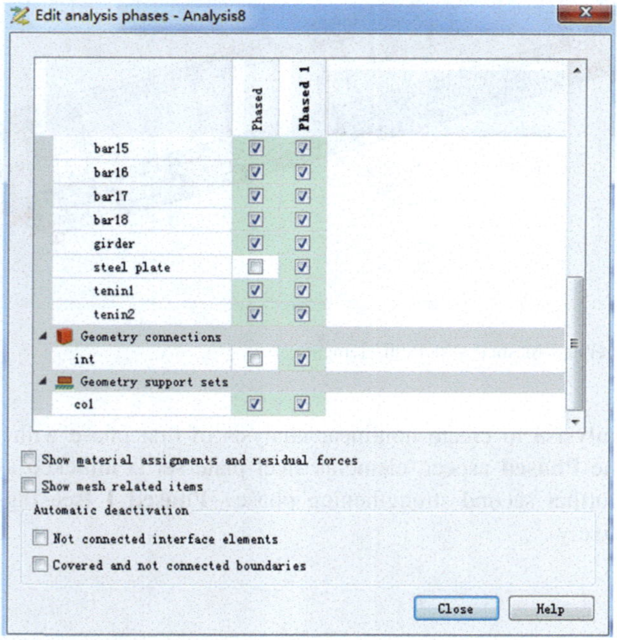

Fig. 3.433 Phased 1 for **Analysis8**

Other parameters and specifications of **Analysis8** are the same except start step **in the Phased 1.** The **Load set** selects **Geometry load combination 3** with the **factor 1** under the aspect of **Input** while **User specified size** is **1.0000** in the **Establish equilibrium**, as Fig. 3.434 demonstrates. All the time steps are specified as former.

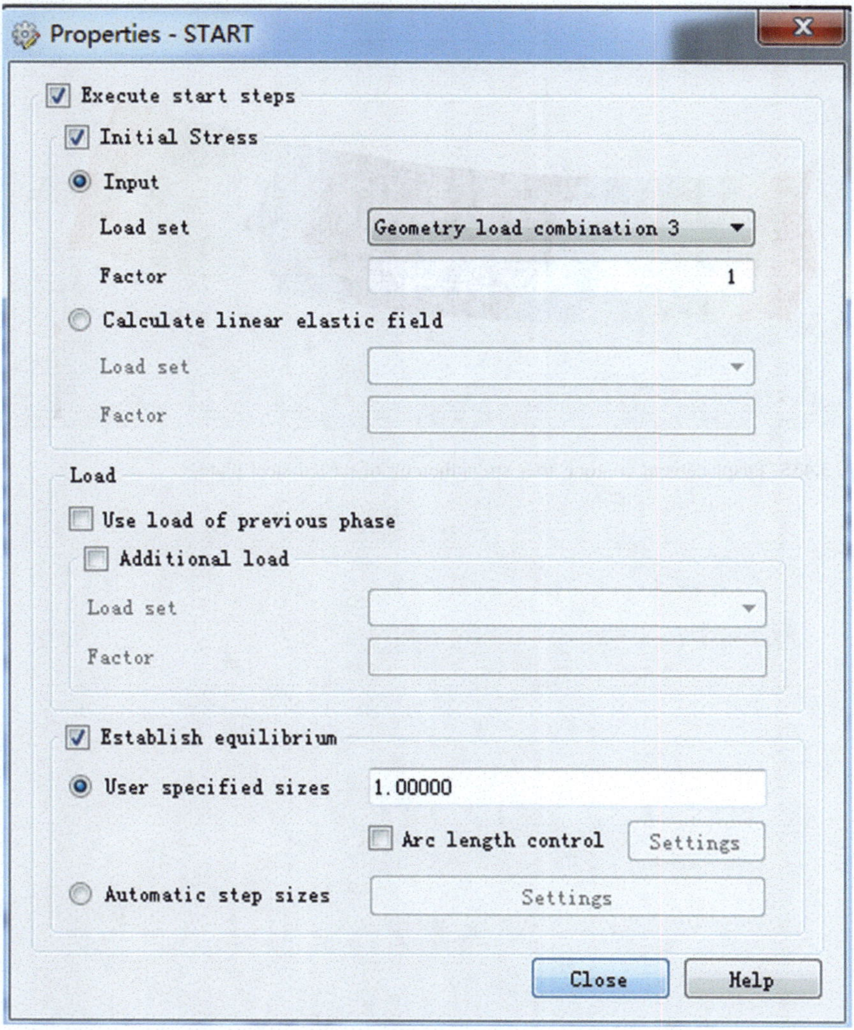

Fig. 3.434 Start step of **Phased 1**

Clicking **Run analysis** button to calculate the time-dependent nonlinear analysis. Displacement contour after strengthening of pasted steel plate (Phased 1, Time-steps 12, Time 0.63072E+10 s) belonging to the second phase in the Z translation direction is shown in Fig. 3.435. It is not hard to see that the maximum displacement downward is 1.88 cm and the resilient effect is $\frac{2.01-1.88}{2.01}=6.46\%$.

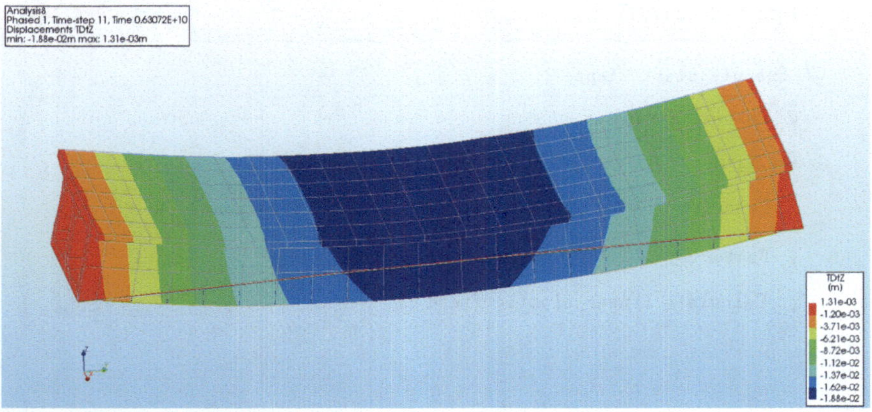

Fig. 3.435 Displacement contour after strengthening of pasted steel plate

Command console of pasted steel plate in Python language

```
newProject( "reinforced-steel", 1000 )
setModelAnalysisAspects( [ "STRUCT" ] )
setModelDimension( "3D" )
setDefaultMeshOrder( "QUADRATIC" )
setDefaultMesherType( "HEXQUAD" )
setDefaultMidSideNodeLocation( "LINEAR" )
createSheet( "girder", [[ 0, 0, 0 ],[ 4, 0, 0 ],[ 4, 0, 2 ],[ 5, 0, 2.2 ],[ 5, 0, 2.4 ],[ -1, 0, 2.4 ],[ -1, 0, 2.2 ],[ 0, 0, 2 ]] )
createSheet( "sheet1", [[ 0.5, 0, 0.3 ],[ 1.5, 0, 0.3 ],[ 1.8, 0, 0.5 ],[ 1.8, 0, 1.7 ],[ 1.5, 0, 2 ],[ 0.5, 0, 2 ],[ 0.3, 0, 1.7 ],[ 0.3, 0, 0.5 ]] )
mirror( [ "sheet1" ], [ 2, 0, 0 ], [ True, False, False ], True )
subtract( "girder", [ "sheet1", "sheet2" ], False, True )
extrudeProfile( [ "girder" ], [ 0, 20, 0 ] )
saveProject(   )
createLine( "bar", [ 0.4, 0, 0.2 ], [ 0.4, 20, 0.2 ] )
show( SHAPE, [ "bar" ] )
arrayCopy( [ "bar" ], [ 0.4, 0, 0 ], [ 0, 0, 0 ], [ 0, 0, 0 ], 8 )
saveProject(   )
createLine( "bar9", [ -0.6, 0, 2.2 ], [ -0.6, 20, 2.2 ] )
arrayCopy( [ "bar9" ], [ 0.6, 0, 0 ], [ 0, 0, 0 ], [ 0, 0, 0 ], 9 )
createPolyline( "tenin1", [[ 0.15, 0, 1.5 ],[ 0.15, 7, 0.5 ],[ 0.15, 13, 0.5 ],[ 0.15, 20, 1.5 ]], False )
hide( SHAPE, [ "bar 1" ] )
mirror( [ "tenin1" ], [ 2, 0, 0 ], [ True, False, False ], True )
saveProject(   )
addMaterial( "bar", "REINFO", "LINEAR", [] )
setParameter( MATERIAL, "bar", "LINEAR/ELASTI/YOUNG", 2.1e+11 )
addGeometry( "Element geometry 1", "RELINE", "REBAR", [] )
rename( GEOMET, "Element geometry 1", "bar" )
setParameter( GEOMET, "bar", "REIEMB/CROSSE", 0.000139 )
setParameter( GEOMET, "bar", "REIEMB/CROSSE", 0.000139 )
setParameter( GEOMET, "bar", "REIEMB/CROSSE", 0.000139 )
setReinforcementAspects( [ "bar", "bar 1", "bar 2", "bar 3", "bar 4", "bar 5", "bar 6", "bar 7", "bar 8", "bar9", "bar10", "bar11", "bar12", "bar13", "bar14", "bar15", "bar16", "bar17", "bar18" ] )
assignMaterial( "bar", "SHAPE", [ "bar", "bar 1", "bar 2", "bar 3", "bar 4", "bar 5", "bar 6", "bar 7", "bar 8", "bar9", "bar10", "bar11", "bar12", "bar13", "bar14", "bar15", "bar16", "bar17", "bar18" ] )
assignGeometry( "bar", "SHAPE", [ "bar", "bar 1", "bar 2", "bar 3", "bar 4", "bar 5", "bar 6", "bar 7", "bar 8", "bar9", "bar10", "bar11", "bar12", "bar13", "bar14", "bar15", "bar16", "bar17", "bar18" ] )
resetElementData( "SHAPE", [ "bar", "bar 1", "bar 2", "bar 3", "bar 4", "bar 5", "bar 6", "bar 7", "bar 8", "bar9", "bar10", "bar11", "bar12", "bar13", "bar14", "bar15", "bar16", "bar17", "bar18" ] )
```

```
setReinforcementDiscretization( [ "bar", "bar 1", "bar 2", "bar 3", "bar 4", "bar 5", "bar 6", "bar 7", "bar 8", "bar9", "bar10", "bar11", "bar12", "bar13", "bar14", "bar15", "bar16", "bar17", "bar18" ], "SECTION" )
saveProject( )
addMaterial( "tenin", "REINFO", "VMISES", [] )
setParameter( MATERIAL, "tenin", "LINEAR/ELASTI/YOUNG", 1.95e+11 )
setParameter( MATERIAL, "tenin", "PLASTI/HARDI1/YLDSTR", 1.86e+09 )
setParameter( MATERIAL, "tenin", "PLASTI/HARDI1/YLDSTR", 1.86e+09 )
setParameter( MATERIAL, "tenin", "PLASTI/HARDI1/YLDSTR", 1.86e+09 )
addGeometry( "Element geometry 2", "RELINE", "REBAR", [] )
rename( GEOMET, "Element geometry 2", "tenin" )
setParameter( GEOMET, "tenin", "REIEMB/CROSSE", 0.002886 )
setParameter( GEOMET, "tenin", "REIEMB/CROSSE", 0.002886 )
setParameter( GEOMET, "tenin", "REIEMB/CROSSE", 0.002886 )
setParameter( GEOMET, "tenin", "REIEMB/CROSSE", 0.002886 )
setReinforcementAspects( [ "tenin1", "tenin2" ] )
assignMaterial( "tenin", SHAPE, [ "tenin1", "tenin2" ] )
assignGeometry( "tenin", SHAPE, [ "tenin1", "tenin2" ] )
resetElementData( SHAPE, [ "tenin1", "tenin2" ] )
setReinforcementDiscretization( [ "tenin1", "tenin2" ], "SECTION" )
saveProject( )
addMaterial( "concrete", "CONCDC", "MC1990", [ "CRACKI", "CREEP", "SHRINK" ] )
setUnit( "ANGLE", "DEGREE" )
setUnit( "TEMPER", "CELSIU" )
setParameter( MATERIAL, "concrete", "MC90CO/RH", 60 )
remove( MATERIAL, "concrete" )
addMaterial( "concrete", "CONCDC", "MC2010", [ "CREEP", "SHRINK" ] )
setParameter( MATERIAL, "concrete", "MC10CO/NORMAL/GRADE", "C50" )
setParameter( MATERIAL, "concrete", "CONCCP/RH", 60 )
setParameter( MATERIAL, "concrete", "MC10CO/NORMAL/CEMTYP", "CE52N" )
setParameter( MATERIAL, "concrete", "CONCCP/CREEP/CRSPEC/AGETYP", "AGING" )
setParameter( MATERIAL, "concrete", "CONCCP/CREEP/CRSPEC/AGING", 2419200 )
setParameter( MATERIAL, "concrete", "CONCCP/SHRINK/CURAGE", 86400 )
setParameter( MATERIAL, "concrete", "CONCCP/SHRINK/CURAGE", 86400 )
setParameter( MATERIAL, "concrete", "CONCCP/SHRINK/CURAGE", 86400 )
setElementClassType( SHAPE, [ "girder" ], "STRSOL" )
assignMaterial( "concrete", SHAPE, [ "girder" ] )
saveProject( )
translate( [ "girder", "bar", "bar 1", "bar 2", "bar 3", "bar 4", "bar 5", "bar 6", "bar 7", "bar 8", "bar9", "bar10", "bar11", "bar12", "bar13", "bar14", "bar15", "bar16", "bar17", "bar18", "tenin1", "tenin2" ], [ 0, 0, 0.01 ] )
saveProject( )
```

```
createSheet( "co1", [[ 0, 0, -0.1 ],[ 4, 0, -0.1 ],[ 4, 0.5, -0.1 ],[ 0, 0.5, -0.1 ]] )
mirror( [ "co1" ], [ 0, 10, 0 ], [ False, True, False ], True )
projection( SHAPEFACE, "girder", [[ 1.705708, 11.47146, 0 ]], [ "co1", "co2" ], [ 0, 0, 1 ],
True )
removeShape( [ "co1", "co2" ] )
addSet( GEOMETRYSUPPORTSET, "co1" )
createSurfaceSupport( "co1", "co1" )
setParameter( GEOMETRYSUPPORT, "co1", "AXES", [ 1, 2 ] )
setParameter( GEOMETRYSUPPORT, "co1", "TRANSL", [ 1, 1, 1 ] )
setParameter( GEOMETRYSUPPORT, "co1", "ROTATI", [ 0, 0, 0 ] )
attach( GEOMETRYSUPPORT, "co1", "girder", [[ 1.705708, 0.2867865, 0 ],[ 1.705708,
19.786787, 0 ]] )
createBlock( "steel plate", [ 0, 0, 0 ], [ 4, 20, 0.01 ] )
addMaterial( "steel", "MCSTEL", "TRESCA", [] )
setParameter( MATERIAL, "steel", "LINEAR/ELASTI/YOUNG", 2.1e+11 )
setParameter( MATERIAL, "steel", "LINEAR/ELASTI/POISON", 0.3 )
setParameter( MATERIAL, "steel", "LINEAR/MASS/DENSIT", 7800 )
setParameter( MATERIAL, "steel", "TREPLA/YLDSTR", 4.4e+08 )
setElementClassType( SHAPE, [ "steel plate" ], "STRSOL" )
assignMaterial( "steel", SHAPE, [ "steel plate" ] )
addMaterial( "bondslip", "INTERF", "BONDSL", [] )
setParameter( MATERIAL, "bondslip", "LINEAR/ELAS6/DSNZ", 3.65e+12 )
setParameter( MATERIAL, "bondslip", "LINEAR/ELAS6/DSSX", 3.65e+08 )
setParameter( MATERIAL, "bondslip", "LINEAR/ELAS6/DSSY", 3.65e+08 )
setParameter( MATERIAL, "bondslip", "BOSLIP/BONDSL", 3 )
setParameter( MATERIAL, "bondslip", "BOSLIP/BONDS3/DISTAU", [] )
setParameter( MATERIAL, "bondslip", "BOSLIP/BONDS3/DISTAU", [ 0, 0, 1, 3e+6, 10, 0,
100, 0 ] )
createConnection( "int", "INTER", SHAPEFACE )
setParameter( GEOMETRYCONNECTION, "int", "MODE", "AUTO" )
attachTo( GEOMETRYCONNECTION, "int", "SOURCE", "girder", [[ 1.705708, 11.47146,
0.01 ]] )
setElementClassType( GEOMETRYCONNECTION, "int", "STPLIF" )
assignMaterial( "bondslip", GEOMETRYCONNECTION, "int" )
resetGeometry( GEOMETRYCONNECTION, "int" )
setParameter( GEOMETRYCONNECTION, "int", "FLIP", False )
resetElementData( GEOMETRYCONNECTION, "int" )
addSet( GEOMETRYLOADSET, "Geometry load case 1" )
createModelLoad( "gravity", "Geometry load case 1" )
createBodyLoad( "tenin", "Geometry load case 1" )
setParameter( GEOMETRYLOAD, "tenin", "LODTYP", "POSTEN" )
setParameter( GEOMETRYLOAD, "tenin", "POSTEN/BOTHEN/FORCE1", 3000000 )
setParameter( GEOMETRYLOAD, "tenin", "POSTEN/BOTHEN/FORCE2", 3000000 )
```

```
setParameter( GEOMETRYLOAD, "tenin", "POSTEN/BOTHEN/RETLE1", 0.01 )
setParameter( GEOMETRYLOAD, "tenin", "POSTEN/BOTHEN/RETLE2", 0.01 )
setParameter( GEOMETRYLOAD, "tenin", "POSTEN/SHEAR", 0.22 )
setParameter( GEOMETRYLOAD, "tenin", "POSTEN/WOBBLE", 0.001 )
attachTo( GEOMETRYLOAD, "tenin", "POSTEN/BOTHEN/PNTS1", "tenin1", [[ 0.15, 0, 1.51 ]] )
attachTo( GEOMETRYLOAD, "tenin", "POSTEN/BOTHEN/PNTS1", "tenin2", [[ 3.85, 0, 1.51 ]] )
attachTo( GEOMETRYLOAD, "tenin", "POSTEN/BOTHEN/PNTS2", "tenin1", [[ 0.15, 20, 1.51 ]] )
attachTo( GEOMETRYLOAD, "tenin", "POSTEN/BOTHEN/PNTS2", "tenin2", [[ 3.85, 20, 1.51 ]] )
attach( GEOMETRYLOAD, "tenin", [ "tenin1", "tenin2" ] )
saveProject(  )
createSheet( "Sheet 1", [[ 0, 9.4, 3 ],[ 0, 10.6, 3 ],[ 4, 10.6, 3 ],[ 4, 9.4, 3 ]] )
projection( SHAPEFACE, "girder", [[ 2.441438, 11.47146, 2.41 ]], [ "Sheet 1" ], [ 0, 0, -1 ], True )
removeShape( [ "Sheet 1" ] )
addSet( GEOMETRYLOADSET, "Geometry load case 2" )
createSurfaceLoad( "load", "Geometry load case 2" )
setParameter( GEOMETRYLOAD, "load", "FORCE/VALUE", -50000 )
setParameter( GEOMETRYLOAD, "load", "FORCE/DIRECT", 3 )
attach( GEOMETRYLOAD, "load", "girder", [[ 2.294292, 10.088288, 2.41 ]] )
setDefaultGeometryLoadCombinations(  )
setGeometryLoadCombinationFactor( "Geometry load combination 1", "Geometry load case 1", 1 )
remove( GEOMETRYLOADCOMBINATION, "Geometry load combination 1" )
remove( GEOMETRYLOADCOMBINATION, "Geometry load combination 2" )
addGeometryLoadCombination( "" )
setGeometryLoadCombinationFactor( "Geometry load combination 1", "Geometry load case 1", 1 )
addGeometryLoadCombination( "" )
setGeometryLoadCombinationFactor( "Geometry load combination 2", "Geometry load case 2", 1 )
addGeometryLoadCombination( "" )
setGeometryLoadCombinationFactor( "Geometry load combination 3", "Geometry load case 1", 1 )
setGeometryLoadCombinationFactor( "Geometry load combination 3", "Geometry load case 2", 1 )
setTimeDependentLoadFactors( GEOMETRYLOADCOMBINATION, "Geometry load combination 1", [ 86400, 3.1536e+10 ], [ 1, 1 ] )
setTimeDependentLoadFactors( GEOMETRYLOADCOMBINATION, "Geometry load combination 2", [ 86400, 3.1536e+10 ], [ 1, 1 ] )
```

```
setTimeDependentLoadFactors( GEOMETRYLOADCOMBINATION, "Geometry load combination 3", [ 86400, 3.1536e+10 ], [ 1, 1 ] )
saveProject( )
setElementSize( [ "girder", "steel plate" ], 1, -1, True )
setMesherType( [ "girder", "steel plate" ], "HEXQUAD" )
setMidSideNodeLocation( [ "girder", "steel plate" ], "LINEAR" )
saveProject( )
setElementSize( "girder", 2, [[ 1.705708, 11.47146, 0.01 ]], 1, 0.5, True )
generateMesh( [] )
hideView( "GEOM" )
showView( "MESH" )
addAnalysis( "Analysis8" )
addAnalysisCommand( "Analysis8", "PHASE", "Phased" )
renameAnalysis( "Analysis8", "Analysis8" )
setActivePhase( "Analysis8", "Phased" )
setActivePhase( "Analysis8", "Phased" )
renameAnalysisCommand( "Analysis8", "Phased", "Phased" )
setActiveInPhase( "Analysis8", GEOMETRYCONNECTION, [ "int" ], [ "Phased" ], False )
setActiveInPhase( "Analysis8", SHAPE, [ "steel plate" ], [ "Phased" ], False )
addAnalysisCommand( "Analysis8", "NONLIN", "Structural nonlinear" )
renameAnalysis( "Analysis8", "Analysis8" )
removeAnalysisCommandDetail( "Analysis8", "Structural nonlinear", "EXECUT(1)" )
setActivePhase( "Analysis8", "Phased" )
setAnalysisCommandDetail( "Analysis8", "Structural nonlinear", "EXECUT/EXETYP", "START" )
renameAnalysisCommandDetail( "Analysis8", "Structural nonlinear", "EXECUT(1)", "tenin" )
addAnalysisCommandDetail( "Analysis8", "Structural nonlinear", "EXECUT(1)/START/INITIA/STRESS" )
setAnalysisCommandDetail( "Analysis8", "Structural nonlinear", "EXECUT(1)/START/INITIA/STRESS", True )
addAnalysisCommandDetail( "Analysis8", "Structural nonlinear", "EXECUT(1)/START/INITIA/STRESS/INPUT/LOAD" )
setAnalysisCommandDetail( "Analysis8", "Structural nonlinear", "EXECUT(1)/START/INITIA/STRESS/INPUT/LOAD", 1 )
setAnalysisCommandDetail( "Analysis8", "Structural nonlinear", "EXECUT(1)/START/LOAD/PREVIO", False )
setAnalysisCommandDetail( "Analysis8", "Structural nonlinear", "EXECUT(1)/ITERAT/MAXITE", 20 )
setAnalysisCommandDetail( "Analysis8", "Structural nonlinear", "EXECUT(1)/ITERAT/CONVER/DISPLA/NOCONV", "CONTIN" )
setAnalysisCommandDetail( "Analysis8", "Structural nonlinear", "EXECUT(1)/ITERAT/CONVER/FORCE/NOCONV", "CONTIN" )
```

```
setActivePhase( "Analysis8", "Phased" )
setAnalysisCommandDetail( "Analysis8", "Structural nonlinear", "EXECUT/EXETYP", "LOAD" )
renameAnalysisCommand( "Analysis8", "Structural nonlinear", "Structural nonlinear" )
renameAnalysisCommandDetail( "Analysis8", "Structural nonlinear", "EXECUT(2)", "load" )
addAnalysisCommandDetail( "Analysis8", "Structural nonlinear", "EXECUT(2)/LOAD/LOADNR" )
setAnalysisCommandDetail( "Analysis8", "Structural nonlinear", "EXECUT(2)/LOAD/LOADNR", 2 )
setAnalysisCommandDetail( "Analysis8", "Structural nonlinear", "EXECUT(2)/ITERAT/MAXITE", 20 )
setAnalysisCommandDetail( "Analysis8", "Structural nonlinear", "EXECUT(2)/ITERAT/CONVER/DISPLA/NOCONV", "CONTIN" )
setAnalysisCommandDetail( "Analysis8", "Structural nonlinear", "EXECUT(2)/ITERAT/CONVER/FORCE/NOCONV", "CONTIN" )
saveProject(  )
setActivePhase( "Analysis8", "Phased" )
setAnalysisCommandDetail( "Analysis8", "Structural nonlinear", "EXECUT/EXETYP", "TIME" )
renameAnalysisCommand( "Analysis8", "Structural nonlinear", "Structural nonlinear" )
renameAnalysisCommandDetail( "Analysis8", "Structural nonlinear", "EXECUT(3)", "creep and shrinkage" )
setAnalysisCommandDetail( "Analysis8", "Structural nonlinear", "EXECUT(3)/ITERAT/MAXITE", 20 )
setAnalysisCommandDetail( "Analysis8", "Structural nonlinear", "EXECUT(3)/ITERAT/CONVER/DISPLA/NOCONV", "CONTIN" )
setAnalysisCommandDetail( "Analysis8", "Structural nonlinear", "EXECUT(3)/ITERAT/CONVER/FORCE/NOCONV", "CONTIN" )
setAnalysisCommandDetail( "Analysis8", "Structural nonlinear", "EXECUT(3)/TIME/STEPS/EXPLIC/SIZES", "3.15360e+07 2.83824e+08 1.26144e+09 1.57680e+09" )
addAnalysisCommand( "Analysis8", "PHASE", "Phased 1" )
setActivePhase( "Analysis8", "Phased 1" )
setActivePhase( "Analysis8", "Phased 1" )
renameAnalysisCommand( "Analysis8", "Phased 1", "Phased 1" )
addAnalysisCommand( "Analysis8", "NONLIN", "Structural nonlinear 1" )
renameAnalysis( "Analysis8", "Analysis8" )
removeAnalysisCommandDetail( "Analysis8", "Structural nonlinear 1", "EXECUT(1)" )
setActivePhase( "Analysis8", "Phased" )
setActivePhase( "Analysis8", "Phased 1" )
renameAnalysisCommand( "Analysis8", "Structural nonlinear 1", "Structural nonlinear 1" )
setActivePhase( "Analysis8", "Phased 1" )
setAnalysisCommandDetail( "Analysis8", "Structural nonlinear 1", "EXECUT/EXETYP", "START" )
```

```
renameAnalysisCommand( "Analysis8", "Structural nonlinear 1", "Structural nonlinear 1" )
renameAnalysisCommandDetail( "Analysis8", "Structural nonlinear 1", "EXECUT(1)", "new execute block 2" )
addAnalysisCommandDetail( "Analysis8", "Structural nonlinear 1", "EXECUT(1)/START/INITIA/STRESS" )
setAnalysisCommandDetail( "Analysis8", "Structural nonlinear 1", "EXECUT(1)/START/INITIA/STRESS", True )
addAnalysisCommandDetail( "Analysis8", "Structural nonlinear 1", "EXECUT(1)/START/INITIA/STRESS/INPUT/LOAD" )
setAnalysisCommandDetail( "Analysis8", "Structural nonlinear 1", "EXECUT(1)/START/INITIA/STRESS/INPUT/LOAD", 3 )
setAnalysisCommandDetail( "Analysis8", "Structural nonlinear 1", "EXECUT(1)/START/LOAD/PREVIO", False )
setAnalysisCommandDetail( "Analysis8", "Structural nonlinear 1", "EXECUT(1)/ITERAT/MAXITE", 20 )
setAnalysisCommandDetail( "Analysis8", "Structural nonlinear 1", "EXECUT(1)/ITERAT/CONVER/DISPLA/NOCONV", "CONTIN" )
setAnalysisCommandDetail( "Analysis8", "Structural nonlinear 1", "EXECUT(1)/ITERAT/CONVER/FORCE/NOCONV", "CONTIN" )
saveProject( )
setActivePhase( "Analysis8", "Phased 1" )
setAnalysisCommandDetail( "Analysis8", "Structural nonlinear 1", "EXECUT/EXETYP", "TIME" )
renameAnalysisCommand( "Analysis8", "Structural nonlinear 1", "Structural nonlinear 1" )
renameAnalysisCommandDetail( "Analysis8", "Structural nonlinear 1", "EXECUT(2)", "creep and shrinkage" )
setAnalysisCommandDetail( "Analysis8", "Structural nonlinear 1", "EXECUT(2)/ITERAT/MAXITE", 20 )
setAnalysisCommandDetail( "Analysis8", "Structural nonlinear 1", "EXECUT(2)/ITERAT/CONVER/DISPLA/TOLCON", 0.01 )
setAnalysisCommandDetail( "Analysis8", "Structural nonlinear 1", "EXECUT(2)/ITERAT/CONVER/FORCE/TOLCON", 0.01 )
setAnalysisCommandDetail( "Analysis8", "Structural nonlinear 1", "EXECUT(2)/ITERAT/CONVER/DISPLA/NOCONV", "CONTIN" )
setAnalysisCommandDetail( "Analysis8", "Structural nonlinear 1", "EXECUT(2)/ITERAT/CONVER/FORCE/NOCONV", "CONTIN" )
setAnalysisCommandDetail( "Analysis8", "Structural nonlinear 1", "EXECUT(2)/TIME/STEPS/EXPLIC/SIZES", "315360000(10)" )
setAnalysisCommandDetail( "Analysis8", "Structural nonlinear 1", "EXECUT(1)/START/TIME", 3.1536e+09 )
saveProject( )
setAnalysisCommandDetail( "Analysis8", "Structural nonlinear 1", "EXECUT(2)/ITERAT/MAXITE", 20 )
runSolver( "Analysis8" )
```

The third is internal strengthening method. Create a new model with the name of **reinforced-tenin**; strengthening tendons are straight line internally bonded with the coordinate values (0.15, 0, 0.6), (0.15, 20, 0.6) on the one side, and the internal bonded strengthening tendon on the other side is generated via mirror shape with pivot 2 m in the X direction, both of which are specified as inactivated elements in the second phase. All the parameters, manipulations and specifications are the same. The whole numerical is shown in Fig. 3.436.

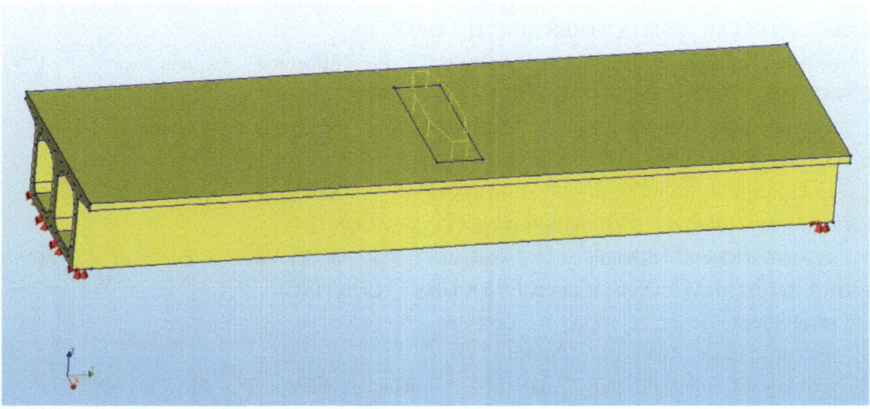

Fig. 3.436 The whole numerical model of internal strengthening method

Clicking **Run analysis** button, the ultimate results and resilient effects in the second phase (Phased 1, Time-steps 12, Time 0.63072E+10 s) are the same as external strengthening ones (see Fig. 3.437).

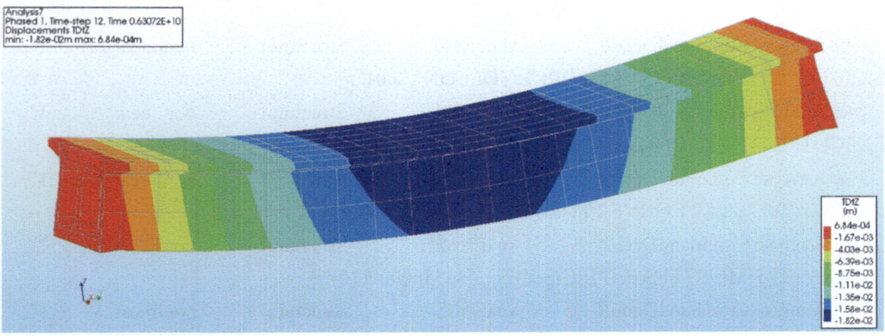

Fig. 3.437 Displacement contour after internal strengthening belonging to the second phase

Command console of internal strengthening in Python language

```
newProject( "reinforced-tenin", 1000 )
setModelAnalysisAspects( [ "STRUCT" ] )
setModelDimension( "3D" )
setDefaultMeshOrder( "QUADRATIC" )
setDefaultMesherType( "HEXQUAD" )
setDefaultMidSideNodeLocation( "LINEAR" )
createSheet( "girder", [[ 0, 0, 0 ],[ 4, 0, 0 ],[ 4, 0, 2 ],[ 5, 0, 2.2 ],[ 5, 0, 2.4 ],[ -1, 0, 2.4 ],[ -1, 0, 2.2 ],[ 0, 0, 2 ]] )
createSheet( "sheet1", [[ 0.5, 0, 0.3 ],[ 1.5, 0, 0.3 ],[ 1.8, 0, 0.5 ],[ 1.8, 0, 1.7 ],[ 1.5, 0, 2 ],[ 0.5, 0, 2 ],[ 0.3, 0, 1.7 ],[ 0.3, 0, 0.5 ]] )
mirror( [ "sheet1" ], [ 2, 0, 0 ], [ True, False, False ], True )
subtract( "girder", [ "sheet1", "sheet2" ], False, True )
extrudeProfile( [ "girder" ], [ 0, 20, 0 ] )
saveProject(  )
createLine( "bar", [ 0.4, 0, 0.2 ], [ 0.4, 20, 0.2 ] )
show( SHAPE, [ "bar" ] )
arrayCopy( [ "bar" ], [ 0.4, 0, 0 ], [ 0, 0, 0 ], [ 0, 0, 0 ], 8 )
saveProject(  )
createLine( "bar9", [ -0.6, 0, 2.2 ], [ -0.6, 20, 2.2 ] )
arrayCopy( [ "bar9" ], [ 0.6, 0, 0 ], [ 0, 0, 0 ], [ 0, 0, 0 ], 9 )
createPolyline( "tenin1", [[ 0.15, 0, 1.5 ],[ 0.15, 7, 0.5 ],[ 0.15, 13, 0.5 ],[ 0.15, 20, 1.5 ]], False )
hide( SHAPE, [ "bar 1" ] )
mirror( [ "tenin1" ], [ 2, 0, 0 ], [ True, False, False ], True )
createLine( "tenin-backup1", [ 0.15, 0, 0.6 ], [ 0.15, 20, 0.6 ] )
mirror( [ "tenin-backup1" ], [ 2, 0, 0 ], [ True, False, False ], True )
addMaterial( "bar", "REINFO", "LINEAR", [] )
setParameter( MATERIAL, "bar", "LINEAR/ELASTI/YOUNG", 2.1e+11 )
addGeometry( "Element geometry 1", "RELINE", "REBAR", [] )
rename( GEOMET, "Element geometry 1", "bar" )
setParameter( GEOMET, "bar", "REIEMB/CROSSE", 0.000139 )
setParameter( GEOMET, "bar", "REIEMB/CROSSE", 0.000139 )
setParameter( GEOMET, "bar", "REIEMB/CROSSE", 0.000139 )
setReinforcementAspects( [ "bar", "bar 1", "bar 2", "bar 3", "bar 4", "bar 5", "bar 6", "bar 7", "bar 8", "bar9", "bar10", "bar11", "bar12", "bar13", "bar14", "bar15", "bar16", "bar17", "bar18" ] )
assignMaterial( "bar", "SHAPE", [ "bar", "bar 1", "bar 2", "bar 3", "bar 4", "bar 5", "bar 6", "bar 7", "bar 8", "bar9", "bar10", "bar11", "bar12", "bar13", "bar14", "bar15", "bar16", "bar17", "bar18" ] )
assignGeometry( "bar", "SHAPE", [ "bar", "bar 1", "bar 2", "bar 3", "bar 4", "bar 5", "bar 6", "bar 7", "bar 8", "bar9", "bar10", "bar11", "bar12", "bar13", "bar14", "bar15", "bar16", "bar17", "bar18" ] )
resetElementData( "SHAPE", [ "bar", "bar 1", "bar 2", "bar 3", "bar 4", "bar 5", "bar 6", "bar 7", "bar 8", "bar9", "bar10", "bar11", "bar12", "bar13", "bar14", "bar15", "bar16", "bar17", "bar18" ] )
```

```
setParameter( MATERIAL, "concrete", "CONCCP/CREEP/CRSPEC/AGING", 2419200 )
setParameter( MATERIAL, "concrete", "CONCCP/SHRINK/CURAGE", 86400 )
setParameter( MATERIAL, "concrete", "CONCCP/SHRINK/CURAGE", 86400 )
setParameter( MATERIAL, "concrete", "CONCCP/SHRINK/CURAGE", 86400 )
setParameter( MATERIAL, "concrete", "MC10CO/NORMAL/CEMTYP", "CE52N" )
setElementClassType( SHAPE, [ "girder" ], "STRSOL" )
assignMaterial( "concrete", SHAPE, [ "girder" ] )
saveProject(  )
addSet( GEOMETRYLOADSET, "Geometry load case 1" )
createModelLoad( "gravity", "Geometry load case 1" )
saveProject(  )
createBodyLoad( "tenin", "Geometry load case 1" )
setParameter( GEOMETRYLOAD, "tenin", "LODTYP", "POSTEN" )
setParameter( GEOMETRYLOAD, "tenin", "POSTEN/BOTHEN/FORCE1", 3000000 )
setParameter( GEOMETRYLOAD, "tenin", "POSTEN/BOTHEN/FORCE2", 3000000 )
setParameter( GEOMETRYLOAD, "tenin", "POSTEN/BOTHEN/RETLE1", 0.01 )
setParameter( GEOMETRYLOAD, "tenin", "POSTEN/BOTHEN/RETLE2", 0.01 )
setParameter( GEOMETRYLOAD, "tenin", "POSTEN/SHEAR", 0.22 )
setParameter( GEOMETRYLOAD, "tenin", "POSTEN/WOBBLE", 0.001 )
attachTo( GEOMETRYLOAD, "tenin", "POSTEN/BOTHEN/PNTS1", "tenin1", [[ 0.15, 0,
1.5 ]] )
attachTo( GEOMETRYLOAD, "tenin", "POSTEN/BOTHEN/PNTS1", "tenin2", [[ 3.85, 0,
1.5 ]] )
attachTo( GEOMETRYLOAD, "tenin", "POSTEN/BOTHEN/PNTS2", "tenin1", [[ 0.15, 20,
1.5 ]] )
attachTo( GEOMETRYLOAD, "tenin", "POSTEN/BOTHEN/PNTS2", "tenin2", [[ 3.85, 20,
1.5 ]] )
attach( GEOMETRYLOAD, "tenin", [ "tenin1", "tenin2" ] )
saveProject(  )
createSheet( "Sheet 1", [[ 0, 9.4, 3 ],[ 0, 10.6, 3 ],[ 4, 10.6, 3 ],[ 4, 9.4, 3 ]] )
projection( SHAPEFACE, "girder", [[ 2.441438, 11.47146, 2.4 ]], [ "Sheet 1" ], [ 0, 0, -1 ],
True )
removeShape( [ "Sheet 1" ] )
addSet( GEOMETRYLOADSET, "Geometry load case 2" )
createSurfaceLoad( "load", "Geometry load case 2" )
setParameter( GEOMETRYLOAD, "load", "FORCE/VALUE", -50000 )
setParameter( GEOMETRYLOAD, "load", "FORCE/DIRECT", 3 )
attach( GEOMETRYLOAD, "load", "girder", [[ 2.294292, 10.088288, 2.4 ]] )
addSet( GEOMETRYLOADSET, "Geometry load case 3" )
rename( GEOMETRYLOADSET, "Geometry load case 1", "postte" )
rename( GEOMETRYLOADSET, "Geometry load case 3", "tenin-backup" )
rename( GEOMETRYLOADSET, "Geometry load case 2", "load" )
setDefaultGeometryLoadCombinations(  )
setGeometryLoadCombinationFactor( "Geometry load combination 1", "postte", 1 )
```

```
setReinforcementDiscretization( [ "bar", "bar 1", "bar 2", "bar 3", "bar 4", "bar 5", "bar 6", "bar 7", "bar 8", "bar9", "bar10", "bar11", "bar12", "bar13", "bar14", "bar15", "bar16", "bar17", "bar18" ], "SECTION" )
saveProject( )
addMaterial( "tenin", "REINFO", "VMISES", [] )
setParameter( MATERIAL, "tenin", "LINEAR/ELASTI/YOUNG", 1.95e+11 )
setParameter( MATERIAL, "tenin", "PLASTI/HARDI1/YLDSTR", 1.86e+09 )
setParameter( MATERIAL, "tenin", "PLASTI/HARDI1/YLDSTR", 1.86e+09 )
setParameter( MATERIAL, "tenin", "PLASTI/HARDI1/YLDSTR", 1.86e+09 )
addGeometry( "Element geometry 2", "RELINE", "REBAR", [] )
rename( GEOMET, "Element geometry 2", "tenin" )
setParameter( GEOMET, "tenin", "REIEMB/CROSSE", 0.002886 )
setParameter( GEOMET, "tenin", "REIEMB/CROSSE", 0.002886 )
setParameter( GEOMET, "tenin", "REIEMB/CROSSE", 0.002886 )
setParameter( GEOMET, "tenin", "REIEMB/CROSSE", 0.002886 )
setReinforcementAspects( [ "tenin1", "tenin2" ] )
assignMaterial( "tenin", SHAPE, [ "tenin1", "tenin2" ] )
assignGeometry( "tenin", SHAPE, [ "tenin1", "tenin2" ] )
resetElementData( SHAPE, [ "tenin1", "tenin2" ] )
setReinforcementDiscretization( [ "tenin1", "tenin2" ], "SECTION" )
saveProject( )
addMaterial( "tenin-backup", "REINFO", "VMISES", [] )
setParameter( MATERIAL, "tenin-backup", "LINEAR/ELASTI/YOUNG", 1.95e+11 )
setParameter( MATERIAL, "tenin-backup", "PLASTI/HARDI1/YLDSTR", 1.86e+09 )
addGeometry( "Element geometry 3", "RELINE", "REBAR", [] )
rename( GEOMET, "Element geometry 3", "tenin-backup" )
setParameter( GEOMET, "tenin-backup", "REIEMB/CROSSE", 0.000139 )
setParameter( GEOMET, "tenin-backup", "REIEMB/CROSSE", 0.000139 )
setParameter( GEOMET, "tenin-backup", "REIEMB/CROSSE", 0.000139 )
setReinforcementAspects( [ "tenin-backup1", "tenin-backup2" ] )
assignMaterial( "tenin-backup", SHAPE, ["tenin-backup1", "tenin-backup2" ] )
assignGeometry( "tenin-backup", SHAPE, ["tenin-backup1", "tenin-backup2" ] )
resetElementData( SHAPE, ["tenin-backup1", "tenin-backup2" ] )
setReinforcementDiscretization( ["tenin-backup1", "tenin-backup2" ], "SECTION" )
saveProject( )
addMaterial( "concrete", "CONCDC", "MC1990", [ "CRACKI", "CREEP", "SHRINK" ] )
setUnit( "ANGLE", "DEGREE" )
setUnit( "TEMPER", "CELSIU" )
setParameter( MATERIAL, "concrete", "MC90CO/RH", 60 )
remove( MATERIAL, "concrete" )
addMaterial( "concrete", "CONCDC", "MC2010", [ "CREEP", "SHRINK" ] )
setParameter( MATERIAL, "concrete", "MC10CO/NORMAL/GRADE", "C50" )
setParameter( MATERIAL, "concrete", "CONCCP/RH", 60 )
setParameter( MATERIAL, "concrete", "CONCCP/CREEP/CRSPEC/AGETYP", "AGING" )
```

```
remove( GEOMETRYLOADCOMBINATION, "Geometry load combination 3" )
remove( GEOMETRYLOADCOMBINATION, "Geometry load combination 2" )
setGeometryLoadCombinationFactor( "Geometry load combination 1", "postte", 1 )
addGeometryLoadCombination( "" )
setGeometryLoadCombinationFactor( "Geometry load combination 2", "load", 1 )
addGeometryLoadCombination( "" )
setGeometryLoadCombinationFactor( "Geometry load combination 3", "tenin-backup", 1 )
addGeometryLoadCombination( "" )
setGeometryLoadCombinationFactor( "Geometry load combination 4", "postte", 1 )
setGeometryLoadCombinationFactor( "Geometry load combination 4", "load", 1 )
setGeometryLoadCombinationFactor( "Geometry load combination 4", "tenin-backup", 1 )
setTimeDependentLoadFactors( GEOMETRYLOADCOMBINATION, "Geometry load combination 1", [ 86400, 3.1536e+10 ], [ 1, 1 ] )
setTimeDependentLoadFactors( GEOMETRYLOADCOMBINATION, "Geometry load combination 2", [ 86400, 3.1536e+10 ], [ 1, 1 ] )
setTimeDependentLoadFactors( GEOMETRYLOADCOMBINATION, "Geometry load combination 3", [ 86400, 3.1536e+10 ], [ 1, 1 ] )
setTimeDependentLoadFactors( GEOMETRYLOADCOMBINATION, "Geometry load combination 4", [ 86400, 3.1536e+10 ], [ 1, 1 ] )
saveProject( )
createBodyLoad( "tenin-backup", "tenin-backup" )
setParameter( GEOMETRYLOAD, "tenin-backup", "LODTYP", "POSTEN" )
setParameter( GEOMETRYLOAD, "tenin-backup", "POSTEN/BOTHEN/FORCE1", 150000 )
setParameter( GEOMETRYLOAD, "tenin-backup", "POSTEN/BOTHEN/FORCE2", 150000 )
setParameter( GEOMETRYLOAD, "tenin-backup", "POSTEN/BOTHEN/RETLE1", 0.0001 )
setParameter( GEOMETRYLOAD, "tenin-backup", "POSTEN/BOTHEN/RETLE2", 0.0001 )
setParameter( GEOMETRYLOAD, "tenin-backup", "POSTEN/SHEAR", 0.22 )
setParameter( GEOMETRYLOAD, "tenin-backup", "POSTEN/WOBBLE", 0.001 )
attachTo( GEOMETRYLOAD, "tenin-backup", "POSTEN/BOTHEN/PNTS1", "tenin-backup1", [[ 0.15, 0, 0.6 ]] )
attachTo( GEOMETRYLOAD, "tenin-backup", "POSTEN/BOTHEN/PNTS1", "tenin-backup2", [[ 3.85, 0, 0.6 ]] )
attachTo( GEOMETRYLOAD, "tenin-backup", "POSTEN/BOTHEN/PNTS2", "tenin-backup1", [[ 0.15, 20, 0.6 ]] )
attachTo( GEOMETRYLOAD, "tenin-backup", "POSTEN/BOTHEN/PNTS2", "tenin-backup2", [[ 3.85, 20, 0.6 ]] )
attach( GEOMETRYLOAD, "tenin-backup", [ "tenin-backup1", "tenin-backup2" ] )
createSheet( "co1", [[ 0, 0, -0.1 ],[ 4, 0, -0.1 ],[ 4, 0.5, -0.1 ],[ 0, 0.5, -0.1 ]] )
mirror( [ "co1" ], [ 0, 10, 0 ], [ False, True, False ], True )
projection( SHAPEFACE, "girder", [[ 1.705708, 11.47146, 0 ]], [ "co1", "co2" ], [ 0, 0, 1 ], True )
```

```
removeShape( [ "co1", "co2" ] )
addSet( GEOMETRYSUPPORTSET, "co1" )
createSurfaceSupport( "co1", "co1" )
setParameter( GEOMETRYSUPPORT, "co1", "AXES", [ 1, 2 ] )
setParameter( GEOMETRYSUPPORT, "co1", "TRANSL", [ 1, 1, 1 ] )
setParameter( GEOMETRYSUPPORT, "co1", "ROTATI", [ 0, 0, 0 ] )
attach( GEOMETRYSUPPORT, "co1", "girder", [[ 1.705708, 0.2867865, 0 ],[ 1.705708,
19.786787, 0 ]] )
saveProject( )
setElementSize( [ "girder" ], 1, -1, True )
setMesherType( [ "girder" ], "HEXQUAD" )
setMidSideNodeLocation( [ "girder" ], "LINEAR" )
saveProject( )
setGeometryLoadCombinationFactor( "Geometry load combination 1", "postte", 1 )
generateMesh( [] )
hideView( "GEOM" )
showView( "MESH" )
addAnalysis( "Analysis7" )
addAnalysisCommand( "Analysis7", "PHASE", "Phased" )
renameAnalysis( "Analysis7", "Analysis7" )
setActivePhase( "Analysis7", "Phased" )
setActivePhase( "Analysis7", "Phased" )
setActiveInPhase( "Analysis7", SHAPE, [ "tenin-backup2" ], [ "Phased" ], False )
setActiveInPhase( "Analysis7", SHAPE, [ "tenin-backup1" ], [ "Phased" ], False )
renameAnalysisCommand( "Analysis7", "Phased", "Phased" )
addAnalysisCommand( "Analysis7", "NONLIN", "Structural nonlinear" )
removeAnalysisCommandDetail( "Analysis7", "Structural nonlinear", "EXECUT(1)" )
setActivePhase( "Analysis7", "Phased" )
setAnalysisCommandDetail( "Analysis7", "Structural nonlinear", "EXECUT/EXETYP",
"START" )
renameAnalysisCommand( "Analysis7", "Structural nonlinear", "Structural nonlinear" )
addAnalysisCommandDetail( "Analysis7", "Structural nonlinear",
"EXECUT(1)/START/INITIA/STRESS" )
setAnalysisCommandDetail( "Analysis7", "Structural nonlinear",
"EXECUT(1)/START/INITIA/STRESS", True )
setAnalysisCommandDetail( "Analysis7", "Structural nonlinear",
"EXECUT(1)/START/LOAD/PREVIO", False )
addAnalysisCommandDetail( "Analysis7", "Structural nonlinear",
"EXECUT(1)/START/INITIA/STRESS/INPUT/LOAD" )
setAnalysisCommandDetail( "Analysis7", "Structural nonlinear",
"EXECUT(1)/START/INITIA/STRESS/INPUT/LOAD", 1 )
setAnalysisCommandDetail( "Analysis7", "Structural nonlinear",
"EXECUT(1)/ITERAT/MAXITE", 20 )
setAnalysisCommandDetail( "Analysis7", "Structural nonlinear",
```

```
"EXECUT(1)/ITERAT/CONVER/DISPLA/NOCONV", "CONTIN" )
    setAnalysisCommandDetail( "Analysis7", "Structural nonlinear", "EXECUT(1)/ITERAT/CONVER/FORCE/NOCONV", "CONTIN" )
    saveProject( )
    setActivePhase( "Analysis7", "Phased" )
    setAnalysisCommandDetail( "Analysis7", "Structural nonlinear", "EXECUT/EXETYP", "LOAD" )
    renameAnalysisCommand( "Analysis7", "Structural nonlinear", "Structural nonlinear" )
    setAnalysisCommandDetail( "Analysis7", "Structural nonlinear", "EXECUT(2)/ITERAT/MAXITE", 20 )
    setAnalysisCommandDetail( "Analysis7", "Structural nonlinear", "EXECUT(2)/ITERAT/CONVER/DISPLA/NOCONV", "CONTIN" )
    setAnalysisCommandDetail( "Analysis7", "Structural nonlinear", "EXECUT(2)/ITERAT/CONVER/FORCE/NOCONV", "CONTIN" )
    addAnalysisCommandDetail( "Analysis7", "Structural nonlinear", "EXECUT(2)/LOAD/LOADNR" )
    setAnalysisCommandDetail( "Analysis7", "Structural nonlinear", "EXECUT(2)/LOAD/LOADNR", 2 )
    setActivePhase( "Analysis7", "Phased" )
    setAnalysisCommandDetail( "Analysis7", "Structural nonlinear", "EXECUT/EXETYP", "TIME" )
    renameAnalysisCommandDetail( "Analysis7", "Structural nonlinear", "EXECUT(3)", "creep and shrinkage" )
    setAnalysisCommandDetail( "Analysis7", "Structural nonlinear", "EXECUT(3)/TIME/STEPS/EXPLIC/SIZES", "3.15360e+07 2.83824e+08 1.26144e+09 1.57680e+09" )
    setAnalysisCommandDetail( "Analysis7", "Structural nonlinear", "EXECUT(3)/ITERAT/MAXITE", 20 )
    setAnalysisCommandDetail( "Analysis7", "Structural nonlinear", "EXECUT(3)/ITERAT/CONVER/DISPLA/NOCONV", "CONTIN" )
    setAnalysisCommandDetail( "Analysis7", "Structural nonlinear", "EXECUT(3)/ITERAT/CONVER/FORCE/NOCONV", "CONTIN" )
    addAnalysisCommand( "Analysis7", "PHASE", "Phased 1" )
    setActivePhase( "Analysis7", "Phased 1" )
    renameAnalysisCommand( "Analysis7", "Phased 1", "Phased 1" )
    addAnalysisCommand( "Analysis7", "NONLIN", "Structural nonlinear 1" )
    renameAnalysis( "Analysis7", "Analysis7" )
    removeAnalysisCommandDetail( "Analysis7", "Structural nonlinear 1", "EXECUT(1)" )
    setActivePhase( "Analysis7", "Phased 1" )
    setAnalysisCommandDetail( "Analysis7", "Structural nonlinear 1", "EXECUT/EXETYP", "START" )
    renameAnalysisCommand( "Analysis7", "Structural nonlinear 1", "Structural nonlinear 1" )
    renameAnalysisCommandDetail( "Analysis7", "Structural nonlinear 1", "EXECUT(1)", "tenin-backup" )
    addAnalysisCommandDetail( "Analysis7", "Structural nonlinear 1",
```

```
"EXECUT(1)/START/INITIA/STRESS" )
    setAnalysisCommandDetail( "Analysis7", "Structural nonlinear 1", "EXECUT(1)/START/INITIA/STRESS", True )
    addAnalysisCommandDetail( "Analysis7", "Structural nonlinear 1", "EXECUT(1)/START/INITIA/STRESS/INPUT/LOAD" )
    setAnalysisCommandDetail( "Analysis7", "Structural nonlinear 1", "EXECUT(1)/START/INITIA/STRESS/INPUT/LOAD", 3 )
    setAnalysisCommandDetail( "Analysis7", "Structural nonlinear 1", "EXECUT(1)/START/LOAD/PREVIO", False )
    setAnalysisCommandDetail( "Analysis7", "Structural nonlinear 1", "EXECUT(1)/ITERAT/MAXITE", 20 )
    setAnalysisCommandDetail( "Analysis7", "Structural nonlinear 1", "EXECUT(1)/ITERAT/CONVER/DISPLA/NOCONV", "CONTIN" )
    setAnalysisCommandDetail( "Analysis7", "Structural nonlinear 1", "EXECUT(1)/ITERAT/CONVER/FORCE/NOCONV", "CONTIN" )
    setActivePhase( "Analysis7", "Phased 1" )
    setAnalysisCommandDetail( "Analysis7", "Structural nonlinear 1", "EXECUT/EXETYP", "LOAD" )
    addAnalysisCommandDetail( "Analysis7", "Structural nonlinear 1", "EXECUT(2)/LOAD/LOADNR" )
    setAnalysisCommandDetail( "Analysis7", "Structural nonlinear 1", "EXECUT(2)/LOAD/LOADNR", 4 )
    setAnalysisCommandDetail( "Analysis7", "Structural nonlinear 1", "EXECUT(2)/ITERAT/MAXITE", 20 )
    setAnalysisCommandDetail( "Analysis7", "Structural nonlinear 1", "EXECUT(2)/ITERAT/CONVER/DISPLA/NOCONV", "CONTIN" )
    setAnalysisCommandDetail( "Analysis7", "Structural nonlinear 1", "EXECUT(2)/ITERAT/CONVER/FORCE/NOCONV", "CONTIN" )
    setActivePhase( "Analysis7", "Phased 1" )
    setAnalysisCommandDetail( "Analysis7", "Structural nonlinear 1", "EXECUT/EXETYP", "TIME" )
    renameAnalysisCommandDetail( "Analysis7", "Structural nonlinear 1", "EXECUT(3)", "creep and shrinkage 1" )
    setAnalysisCommandDetail( "Analysis7", "Structural nonlinear 1", "EXECUT(3)/TIME/STEPS/EXPLIC/SIZES", "315360000(10)" )
    setAnalysisCommandDetail( "Analysis7", "Structural nonlinear 1", "EXECUT(3)/ITERAT/MAXITE", 20 )
    setAnalysisCommandDetail( "Analysis7", "Structural nonlinear 1", "EXECUT(3)/ITERAT/CONVER/DISPLA/NOCONV", "CONTIN" )
    setAnalysisCommandDetail( "Analysis7", "Structural nonlinear 1", "EXECUT(3)/ITERAT/CONVER/FORCE/NOCONV", "CONTIN" )
    setAnalysisCommandDetail( "Analysis7", "Structural nonlinear 1", "EXECUT(1)/START/TIME", 3.1536e+09 )
    saveProject(  )
    runSolver( "Analysis7" )
```

References

1. Gowripalan N, Gilbert RI (2000) Design guidelines for ductal prestressed concrete beams. Design guide. Civil & Environmental Engineering School, University of NSW, Sydney, Australia
2. DIANA concrete modeling and analysis tutorials and experiences
3. Claugh R, Peng J (2006) Jie gou dong li xue (结构动力学, Structural dynamics) (translated by Wang Guangyuan's). Higher Education Press, Beijing

Chapter 4
Hydration Analysis for Mass Concrete in DIANA

Abstract Hydration reactions in mass concrete during the forming stage are often required to be taken into account owing to its high heat release. During this stage, the heat released by hydration reaction at this stage has a great influence on the performance of the concrete structures. This chapter will focus on the key feature in DIANA numerical simulation of "business card"—hydration heat simulation. Based on the two numerical cases—concrete pipe gallery segments as well as square pile blocks in large volume—heat flow module and international common specifications are used via DIANA to study the influence of hydration reaction on the structure.

4.1 Transient Hydration Analysis for Mass Segment of Pipe Gallery

This model is a simplified hollow pipe gallery without tensioning prestressing tendons. The length, width and height are 5.7, 3.5 and 1.5 m, respectively. The pipe gallery is hollow with reinforcement stirrups in it, which is simulated by reinforcement grid elements in DIANA. Distance between grids is 2.45 m. Solid elements are applied to simulate the hydration effect on mass concrete. Concrete and reinforcement parameters are displayed in Table 4.1 and three-dimensional picture of pipe gallery as well as elevation drawing of reinforcement are shown in Figs. 4.1 and 4.2, respectively.

S. Chai, *Finite Element Analysis for Civil Engineering with DIANA Software*,
https://doi.org/10.1007/978-981-15-2945-0_4

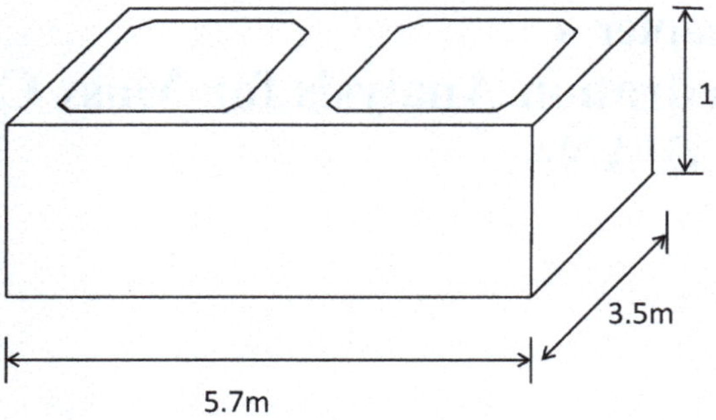

Fig. 4.1 Three-dimensional picture of pipe gallery

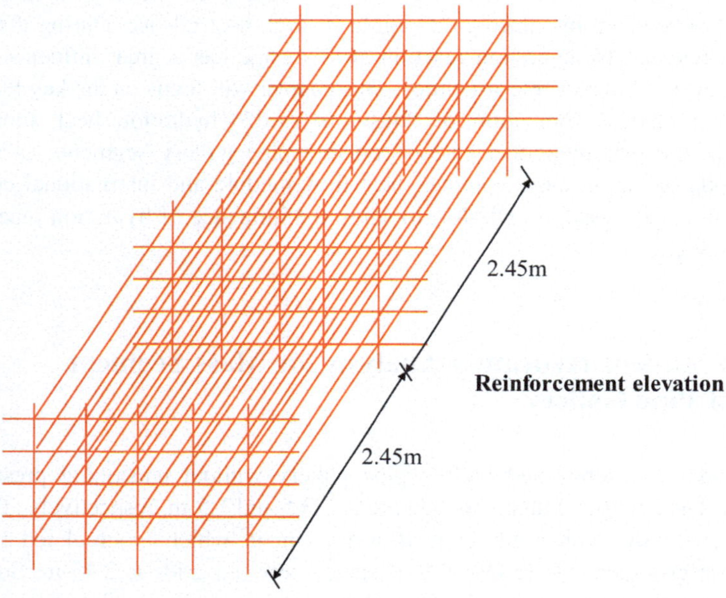

Fig. 4.2 Evaluation drawing of reinforcement

Table 4.1 Parameters for concrete and reinforcement

Concrete	Parameter	Unit
Elastic modulus	3.45×10^{10}	N/m^2
Poisson's ratio	0.15	–
Thermal expansion	1×10^{-5}	1/°C
Heat conductivity	1.73×10^{15}	kg m/day^3 °C
Heat capacity	2.0×10^{16}	kg^2/mday^2 °C
Reinforcement	Parameter	
Elastic modulus	2.1×10^{11}	kg/mday^2
Mass density	7800	kg/m^3
Yield stress	4.4×10^8	kg/mday^2

Note: For the convenience of operation, issue of pre-embedded cooling pipes in the construction of large concrete structures is not taken into account for the time being. Meanwhile, as an analysis case in DIANA, analysis model and conditions in this case are all based on assumed values, thus the accuracy of analytical results should be regarded as a different matter.

Essentials of learning

(1) *Learning to specify material constitutive properties to heat flow material*
(2) *Learning to add boundary conditions of heat flow*
(3) *Learning to add initial temperature*
(4) *Learning to construct reinforcement grid elements in DIANA*
(5) *Learning to master convergence norm and time step for hydration thermal reaction time under load cases in temperature nonlinear analysis.*

Above all, start and open DianaIE interface via clicking **File—New** to create new model file with suffix name **.dpf** in computer F-disk. The name of the file is **hydration analysis for pipe gallery** in Chinese. Since hydration effect on structures should be taken into account in this case, **Structural** and **Heat flow** modules are both ticked as analysis type. **Three dimensional** is selected as **Dimensions** owing to the solid elements for following concrete modeling. Model size is 100 m with **Default mesher type Hexa/Quad**. Additionally, **Default mesh order** is **Linear** (see Fig. 4.3).

Fig. 4.3 New project interface

First of all, an external large plane is created as bottom plane of the pipe gallery segment. Click shortcut icon button **Adds a sheet** to create a plane, coordinate values (0, 0, 0), (0, 5.7, 0) (0, 5.7, 3.5) (0, 0, 3.5) are input one by one, then OK button is clicked to generate such a plane.

Creating left outline of the inner edge with the name of **Sheet2**, coordinate size values for inner shape are displayed (Fig. 4.4).

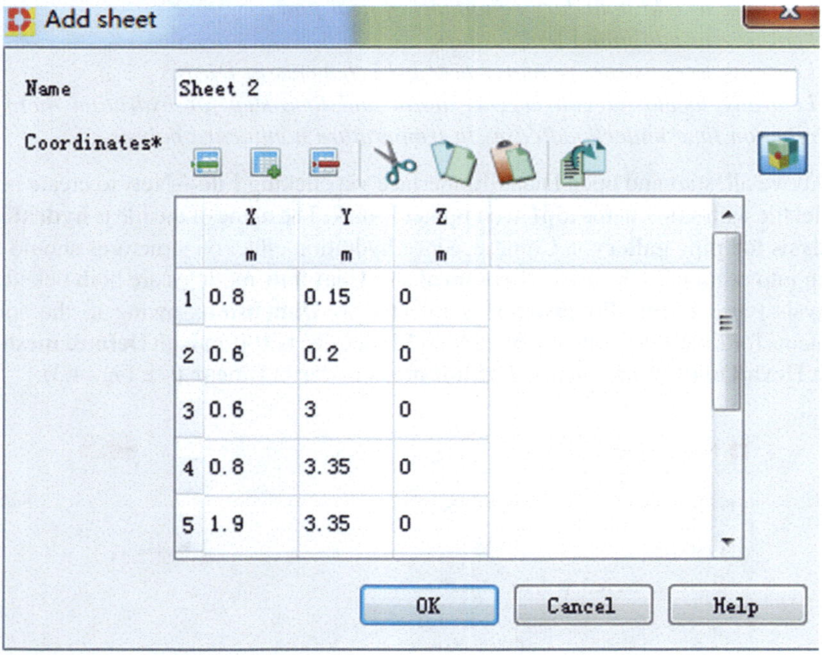

Fig. 4.4 Coordinate size values for inner shape

Select **Sheet2** in the **Geometry tree directory** and right-click to select **Array Copy** to copy and translate the generated left outline of the inner edge to generate the right outline of the inner edge. **Relative displacement** of **Array copy** is 3 m in the positive X direction with the name of **Sheet3**. Number of copies is 1 (see Fig. 4.5).

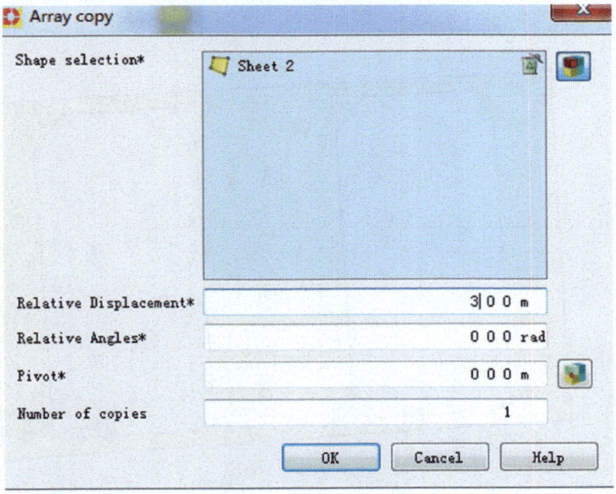

Fig. 4.5 Manipulation of **Array copy**

The next step is to conduct Boolean logic operation. External plane **Sheet1** is selected as **Target selection** in Boolean logic operation. **Sheet2** and **Sheet3** are selected as **Tool selection** and the operation is **Subtract**. Meanwhile, outline after subtraction is merged (see Fig. 4.6). The outline generated by deducting **Sheet2** and **Sheet3** from the whole plane is displayed in Fig. 4.7.

Note: Targets are viewed as subtracted objects in Boolean operations, while tools are subtracting objects in graphics as a whole

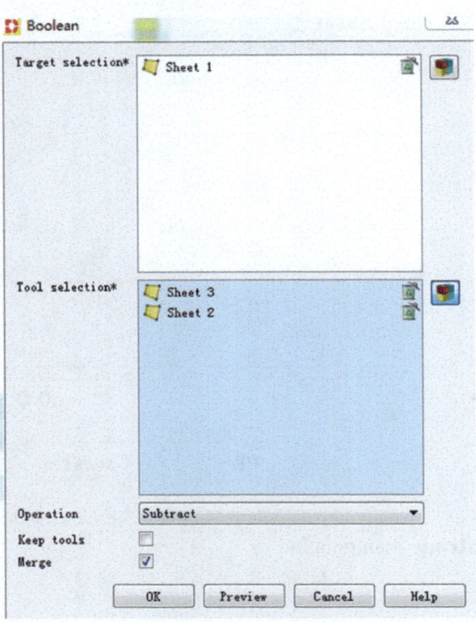

Fig. 4.6 Boolean logic operation interface

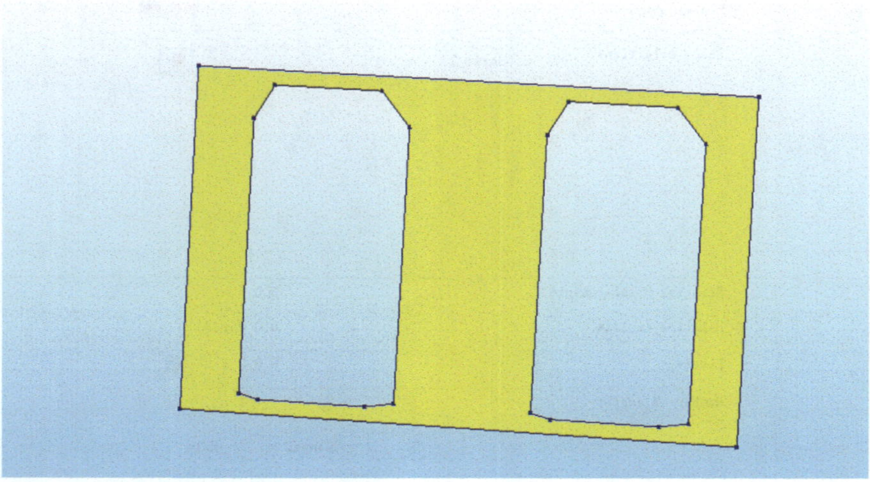

Fig. 4.7 The whole plane after subtraction

Clicking shortcut icon **Extrude**, interface as shown in Fig. 4.8 ejects; select the height of the whole volume unit by extruding **Sheet1** along Z direction with the displacement 1.5 m. The volume after extrusion is displayed as Fig. 4.9.

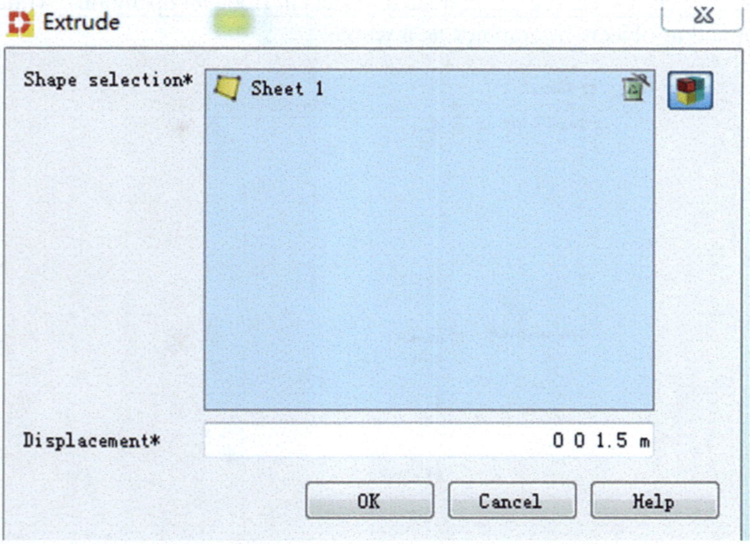

Fig. 4.8 Interface of **Extrude** manipulation

Fig. 4.9 Volume after extrusion

Select **Sheet1**, and right-click to select **property assignments** in order to assign material properties. **Concrete and masonry** is selected with the name of new dialog box **concrete,** while **Element class** is **Structural Solid**. **Total strain based crack model** is selected as smeared cracking model and **Thermal** and **Heat flow** aspects are both ticked at the same time with concrete elastic modulus and Poisson's ratio $3.45 \times 10^{10}\,\text{N/m}^2$ and 0.15, respectively. Hordijk curve type is selected as tension softening curve with tensile strength and Mode-I tensile fracture energy 2.6 MPa and 150 N/m, respectively. All the thermal expansion coefficients are $1 \times 10^{-5}\,1/^{\circ}\text{C}$. Compressive constitutive model for concrete conforms with European CEB-FIP 1990 code for compressive strength $5.8 \times 10^{7}\,\text{N/m}^2$ (see Figs. 4.10, 4.11, 4.12 and 4.13).

Note: Parameter specification mentioned above has little influence on hydration and it has high relationship with parameters in heat flow aspect!

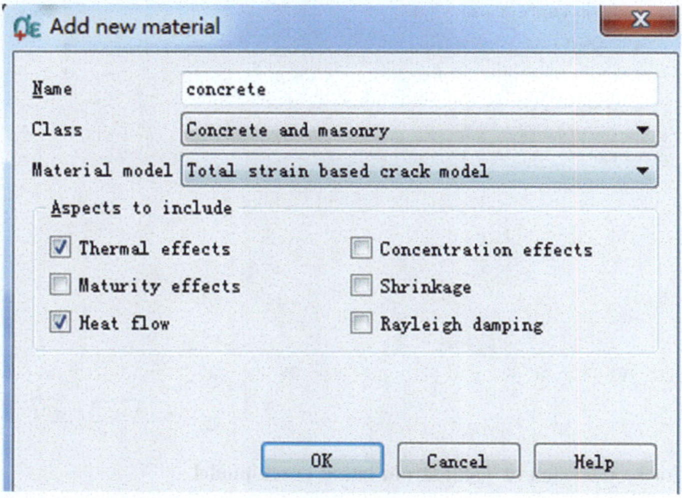

Fig. 4.10 Selection of concrete material model

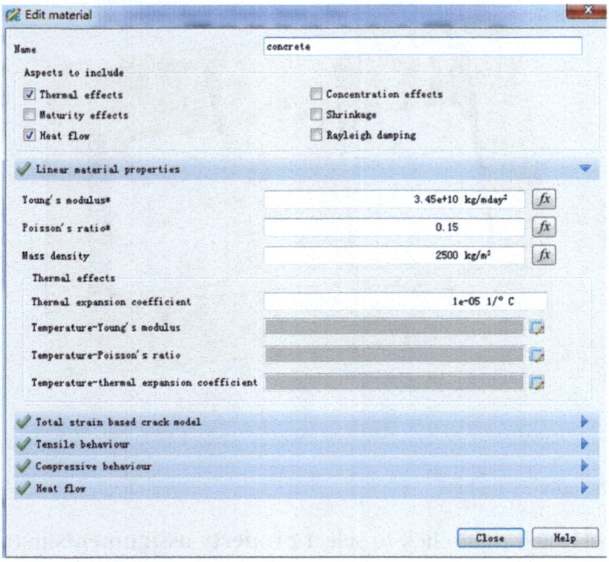

Fig. 4.11 Linear material properties in **Total strain based crack model**

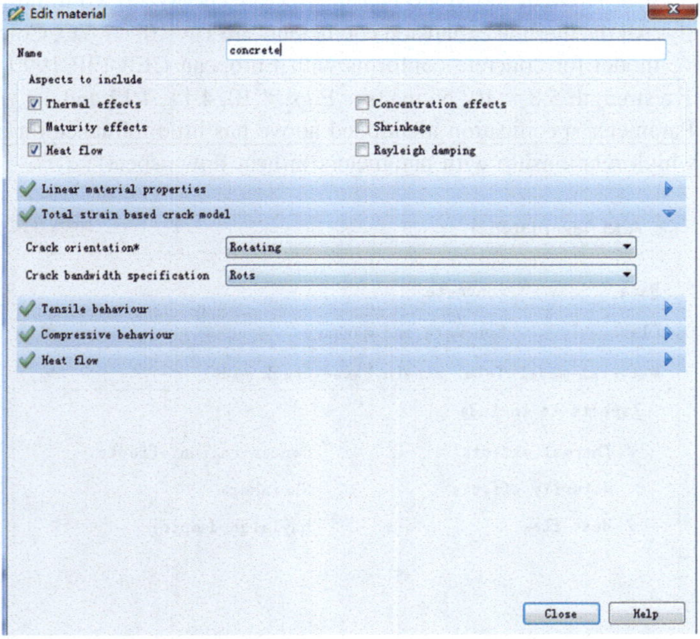

Fig. 4.12 Crack orientation of **Total strain based crack model**

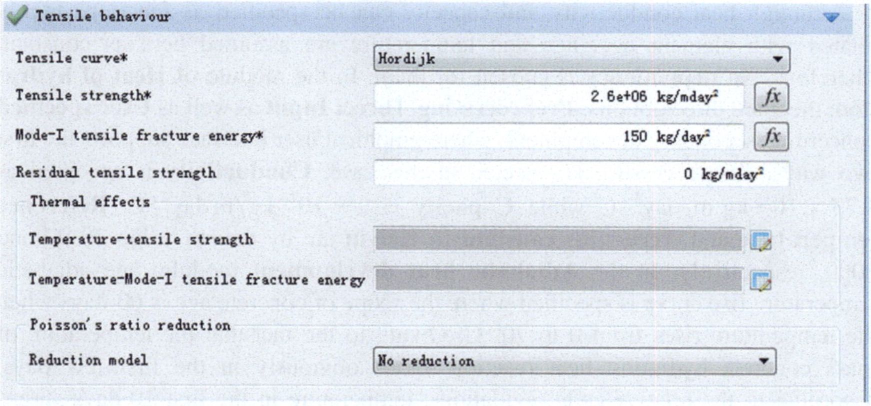

Fig. 4.13 Tensile behaviors of **Total strain based crack model**

For convenience parameter specification in heat flow aspect, units of temperature and time are set as **Celsius** and **day**, while force unit is automatically altered into kgm/day^2 (see Fig. 4.14).

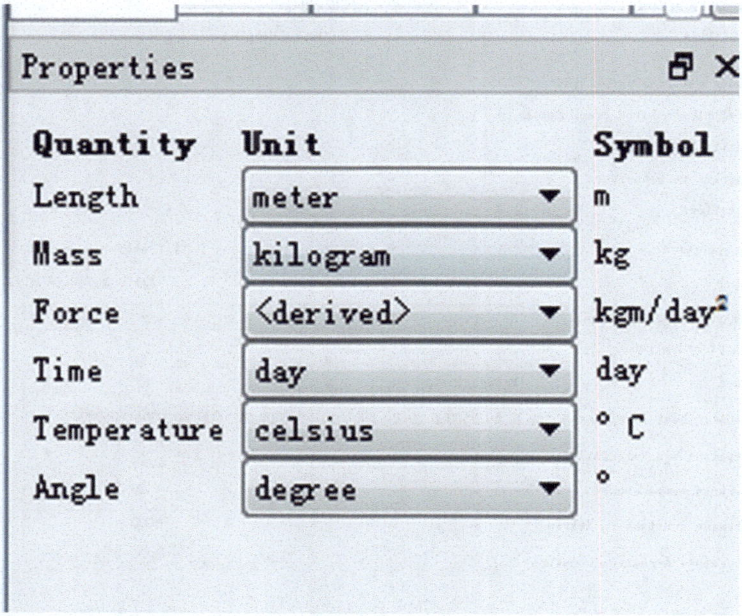

Fig. 4.14 Unit alteration

Although heat conductivity and capacity can be specified as functions highly related with element age, age and temperature are assumed here as constant. Therefore, **No dependency** is chosen for them. In the module of **Heat of hydration**, there are three options: **Preprocessing**, **Direct Input** as well as **User specified** concerning secondary development, where graphical user interface supports the first two ways. **Preprocessing** is selected in this case. **Conductivity** is specified as 1.73×10^{15} kg m/day^3 °C while **Capacity** is 2×10^{16} kg/m day^2 °C. **Reference temperature** and **Arrhenius constant in Kelvin** are by default value 20 °C and 6000, respectively. In the **Adiabatic heat development** module, age–adiabatic temperature rise curve is specified where the scope of concrete age is 60 days when the temperature rises from 0 to 70 °C. Owing to the fact that the temperature of mass concrete hydration heat reaction varies obviously in the first few days, according to the relative code regulations, temperature in the first 10 days surges fast, reaching 70 °C. In concrete age ranging from 10 to 60 days, temperature is stable at 70 C. Specifications of **Heat flow** aspect and age–temperature curve are displayed in Figs. 4.15 and 4.16, respectively.

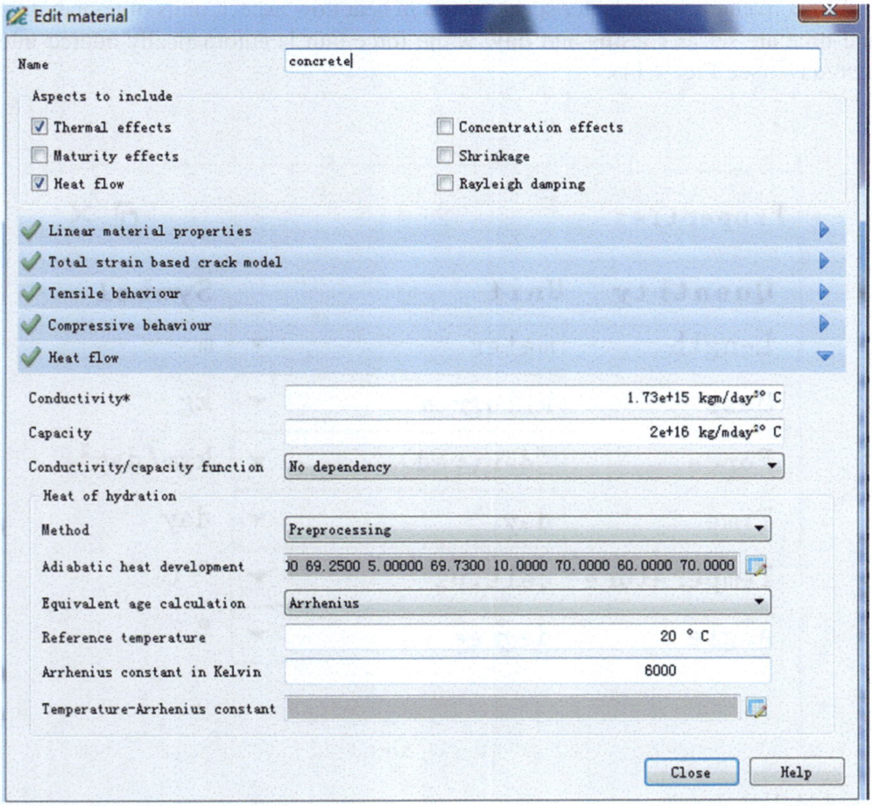

Fig. 4.15 Parameter specifications for **Heat flow** aspect

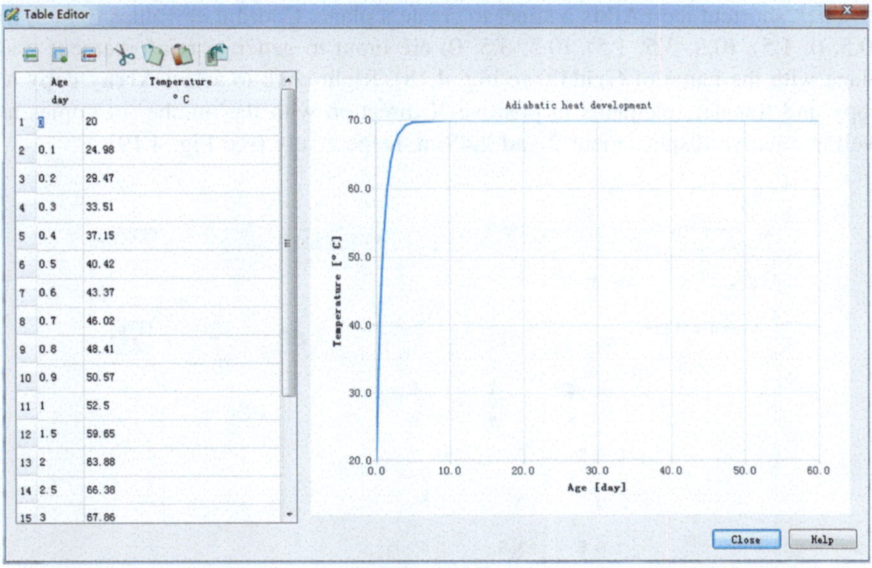

Fig. 4.16 Age–adiabatic temperature rise curve

Since solid elements are not required to specify cross-section geometric properties, after completion of material properties assignment shown as Fig. 4.17, clicking OK button directly, material properties are assigned to **Sheet1**.

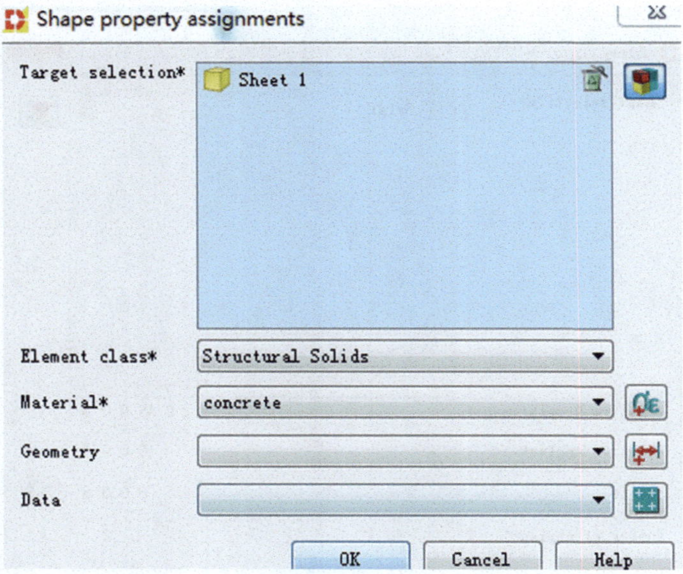

Fig. 4.17 Assignment of material properties

Click shortcut icon **Adds a sheet** to create a plane. Coordinate values (0.5, 0, 0) (0.5, 0, 1.5), (0.5, 3.5, 1.5), (0.5, 3.5, 0) are input to generate reinforcement grid plane with the name of **Grid1** (see Fig. 4.18). Right-click to select **Array copy** to copy and translate elements in positive X direction with the number of copies as well as relative displacement 2 and 2.45 m, respectively (see Fig. 4.19).

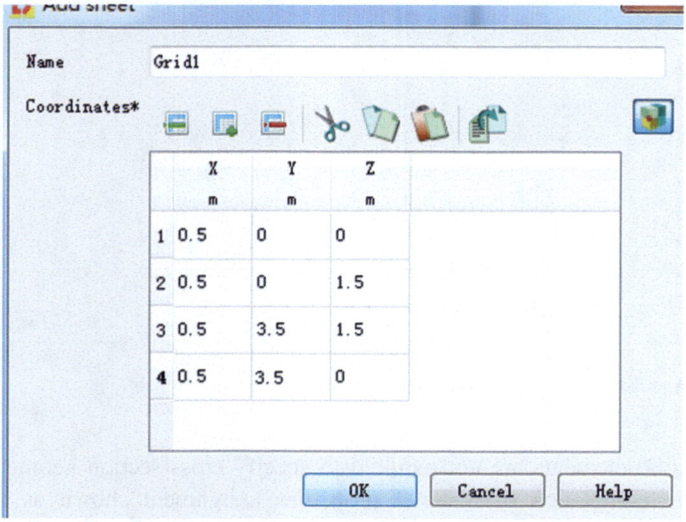

Fig. 4.18 Coordinate values of **Grid1**

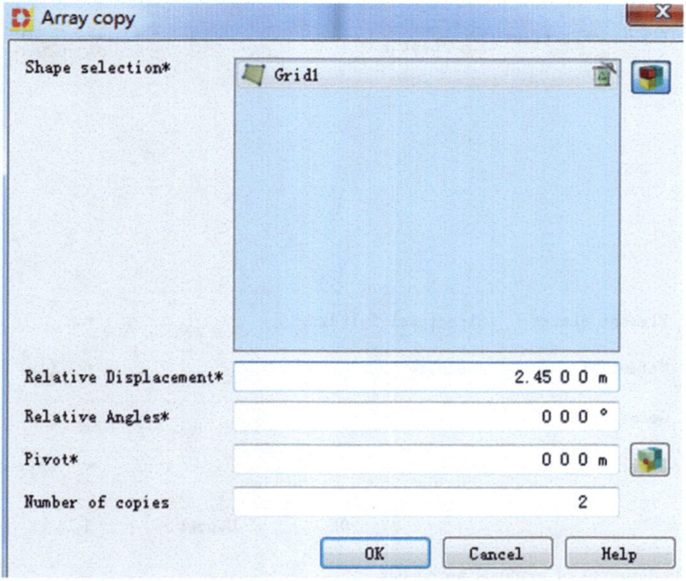

Fig. 4.19 Interface of **Array copy**

Input the coordinate values, as Table 4.2 displays, to generate lateral reinforcement grid planes with the name of **Grid4, Grid5, Grid6 and Grid7**. After generation of such grids, select **Grid4** and **Grid5** and right-click to select function of **Move shape** so as to translate them with the displacement 0.1 m in positive Y direction. Use the same manipulation for Grid6 and Grid7 in the negative Y direction with displacement 0.1 m, which are displayed in Figs. 4.20, 4.21, 4.22, 4.23 and 4.24.

Table 4.2 Coordinate values of lateral reinforcement grids

Grid4	(0.5, 0, 1.5) (2.95, 0, 1.5) (2.95, 0, 0), (0.5, 0, 0)
Grid5	(2.95, 0, 1.5), (5.4, 0, 1.5), (5.4, 0, 0), (2.95, 0, 0)
Grid6	(5.4, 3.5, 1.5) (5.4, 3.5, 0) (2.95, 3.5, 0) (2.95, 3.5, 1.5)
Grid7	(2.95, 3.5, 1.5) (2.95, 3.5, 0), (0.5, 3.5, 0), (0.5, 3.5, 1.5)

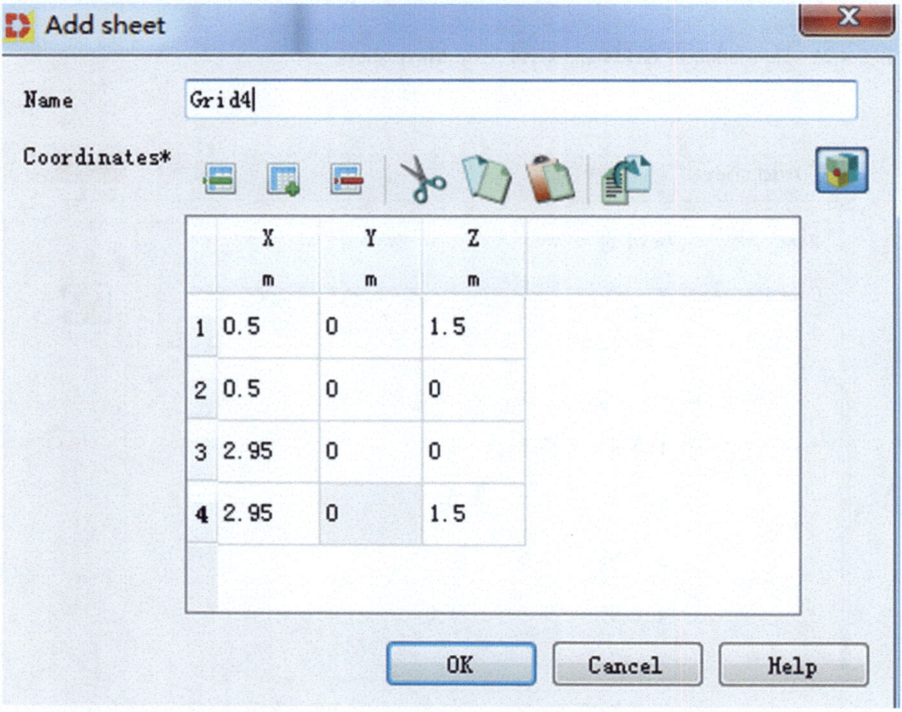

Fig. 4.20 Coordinate value of **Grid4**

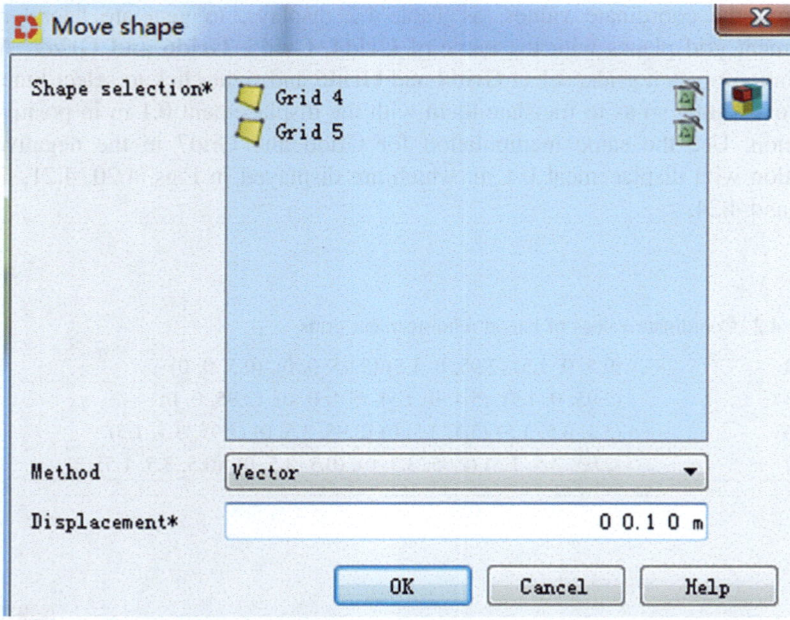

Fig. 4.21 Translation of **Grid4** and **Grid 5** via **Move shape**

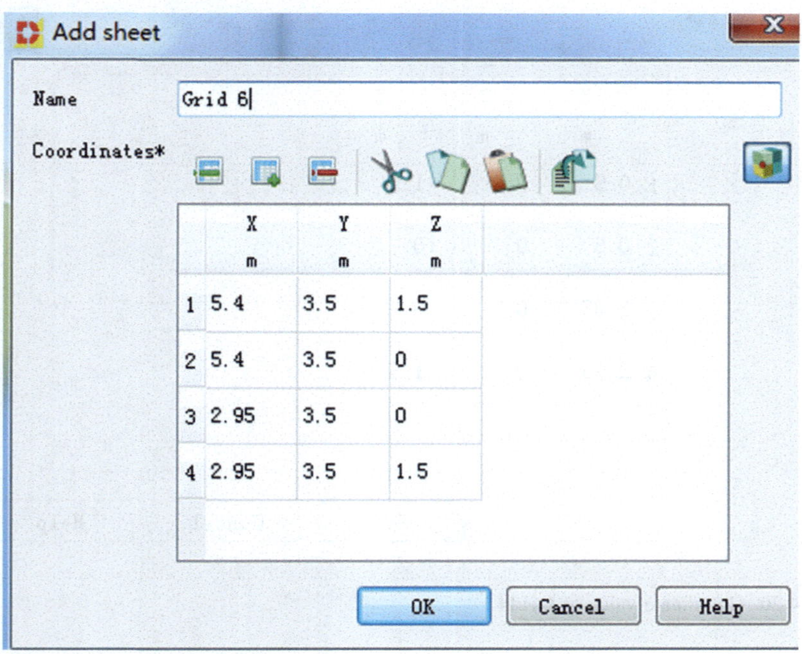

Fig. 4.22 Coordinate value of **Grid6**

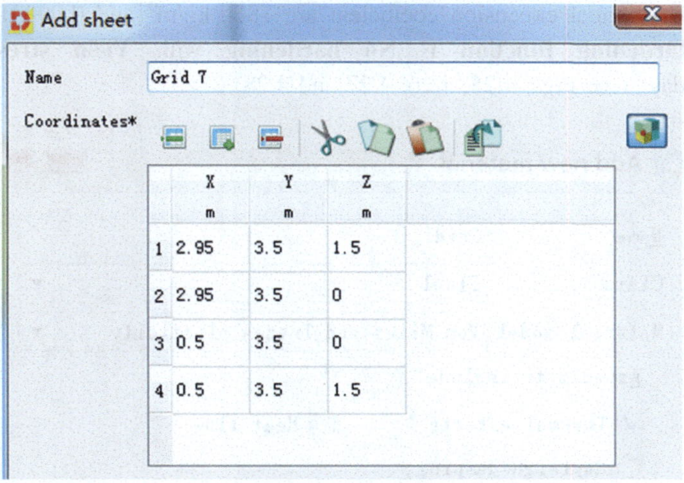

Fig. 4.23 Coordinate value of **Grid7**

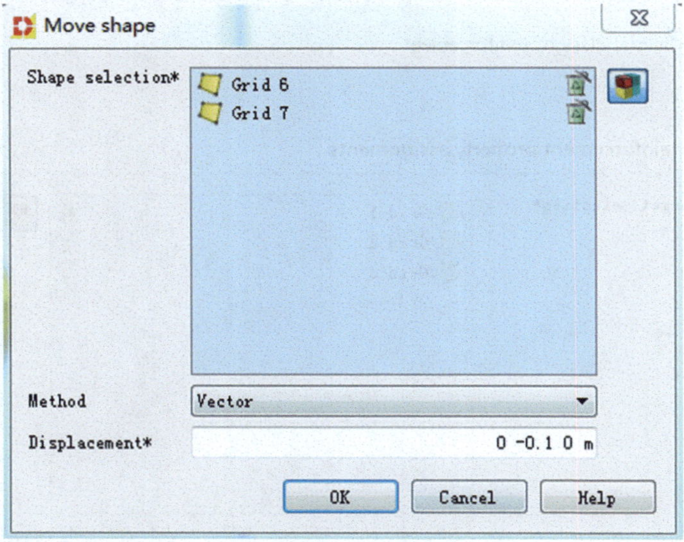

Fig. 4.24 Translation of **Grid4** and **Grid 5** via **Move shape**

The following procedure is an assignment for reinforcement grids. **Grid1, Gird2, Grid3** are selected and assigned in the same set with the set name **Grid1**. Clicking shortcut icon button **Edit reinforcement property assignments**, **von Mises and Tresca plasticity** model is chosen as **Material model** with elastic modulus and Poisson's ratio 2.1×10^{11} kg/mday^2 and 0.3, respectively. Mass density and thermal expansion coefficient are 7800 kg/m^3 and 0.00001, respectively. **Hardening function** is **No hardening** with **Yield stress** 4.4×10^8 kg/mday^2 (see Figs. 4.25, 4.26, 4.27 and 4.28).

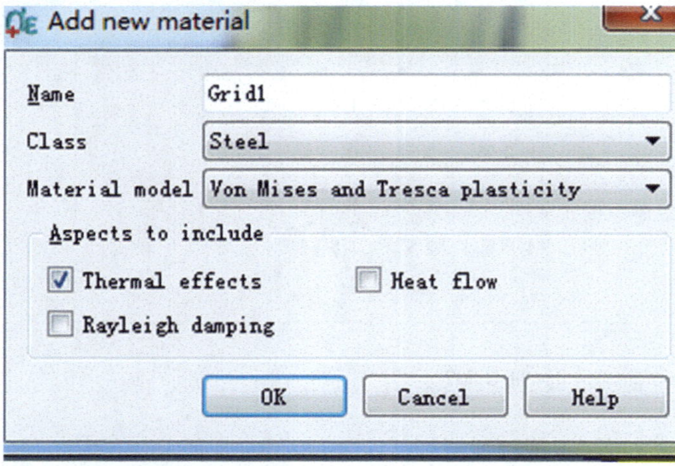

Fig. 4.25 Material class of reinforcement

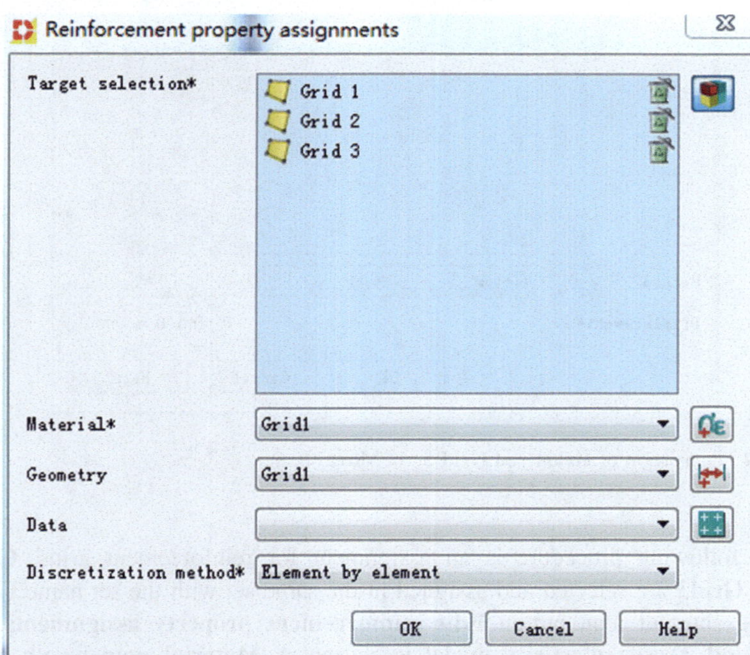

Fig. 4.26 **Grid1** set

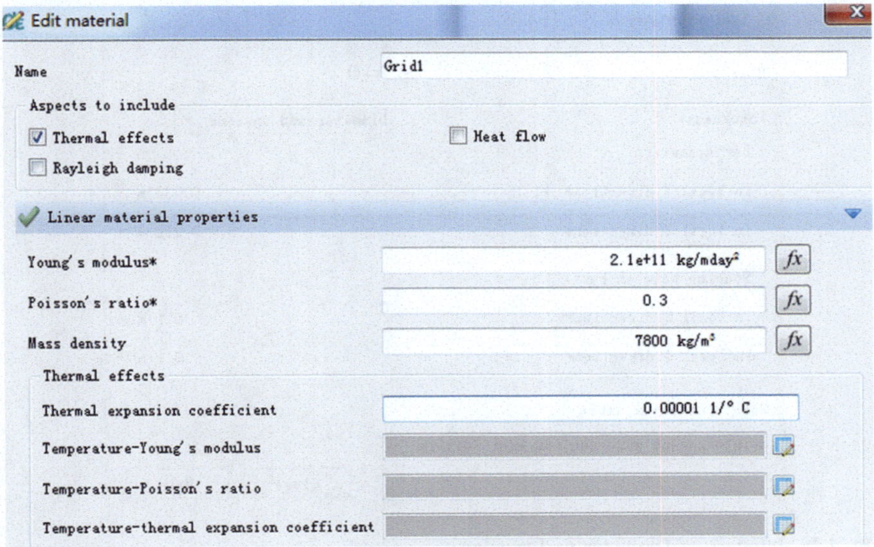

Fig. 4.27 Specifications for linear material properties

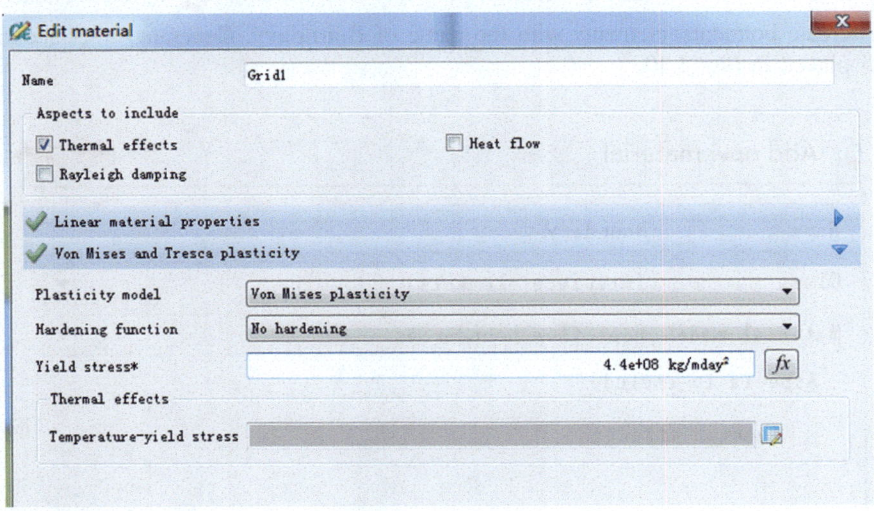

Fig. 4.28 Specifications for reinforcement **von Mises and Tresca plasticity** model

Thickness is assigned via **Diameter and spacing** way; diameters in local X and Y directions are both 0.032 m while spacing between bars are both 0.1 m (see Fig. 4.29). Reinforcement x-axis under local coordinate system corresponds to Y-axis under global coordinate system (see Fig. 4.29). **Grid4, Grid5, Grid6, Grid7** are assembled as set **Grid2**. Manipulations are the same as former and it is not repeated here.

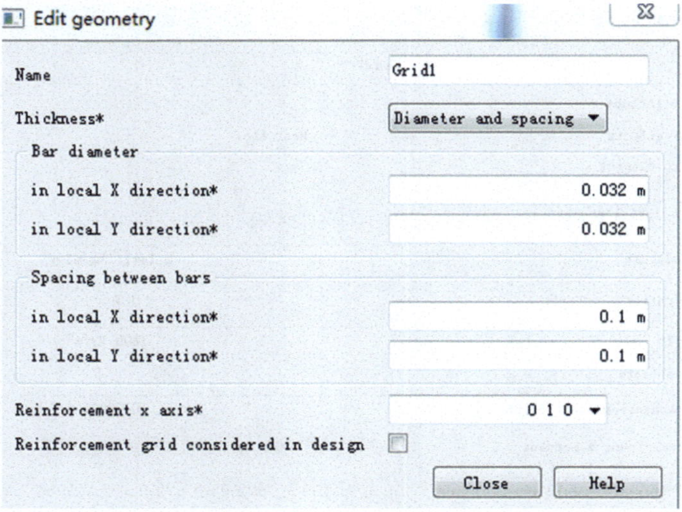

Fig. 4.29 Interface of **Diameter and spacing**

The shortcut icon **Edit connection property assignments**  in red is clicked to create boundary elements with the name of **Boundary**. Generated interface is displayed in Fig. 4.30.

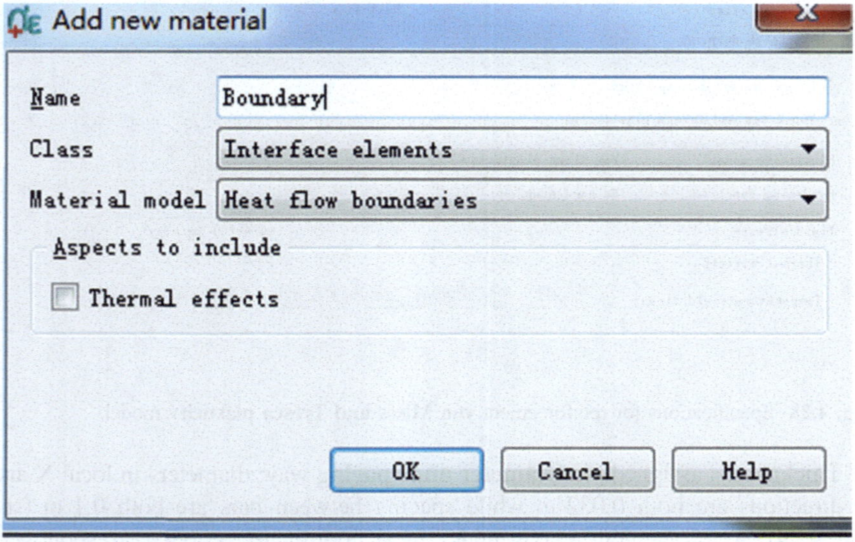

Fig. 4.30 Boundary connection

Heat flow boundaries is selected as **Material model**. Owing to the fact that concrete is solid element, thus there is no need to assign geometric properties. When defining boundaries of heat flow, there are four types: **Convection only**, **Radiation only**, **Convection and Radiation** and **None**, and **Convection only** is chosen as option, indicating that only heat convection rather than radiation is taken into account and convection coefficient is a constant shown as Fig. 4.31. Power exponent of heat convection function in this case is 1, which means that the function type is linear. Similar to former specification for concrete conductivity or capacity function assuming that they are both constant, **No dependency** is chosen for **Convection function**.

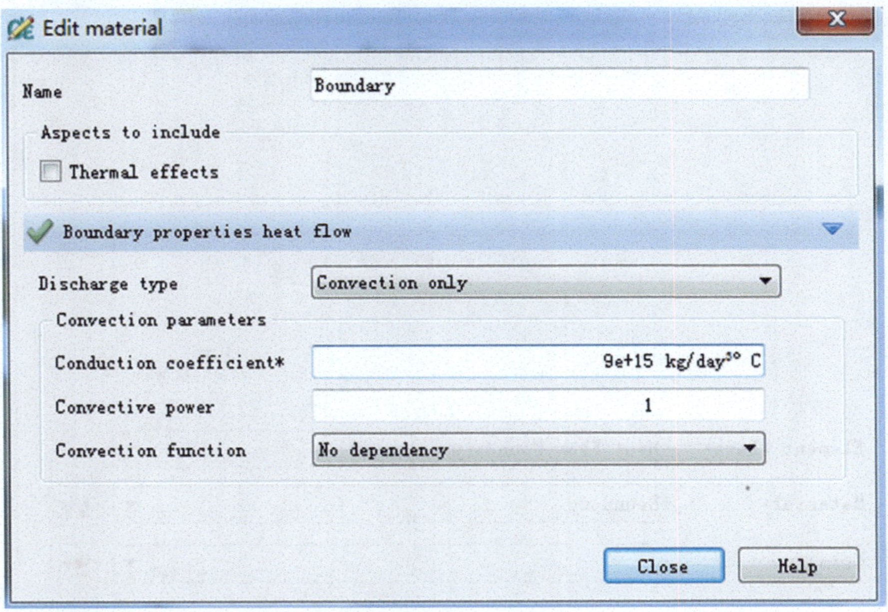

Fig. 4.31 Specifications for boundaries properties

Two lateral surfaces and the top surface of the **Sheet1** perpendicular to the X-axis are chosen as the thermal convection boundary surface and **Connection type** is **Boundary interface**. **Element class** is **Heat Flow Boundary** (see Fig. 4.32).

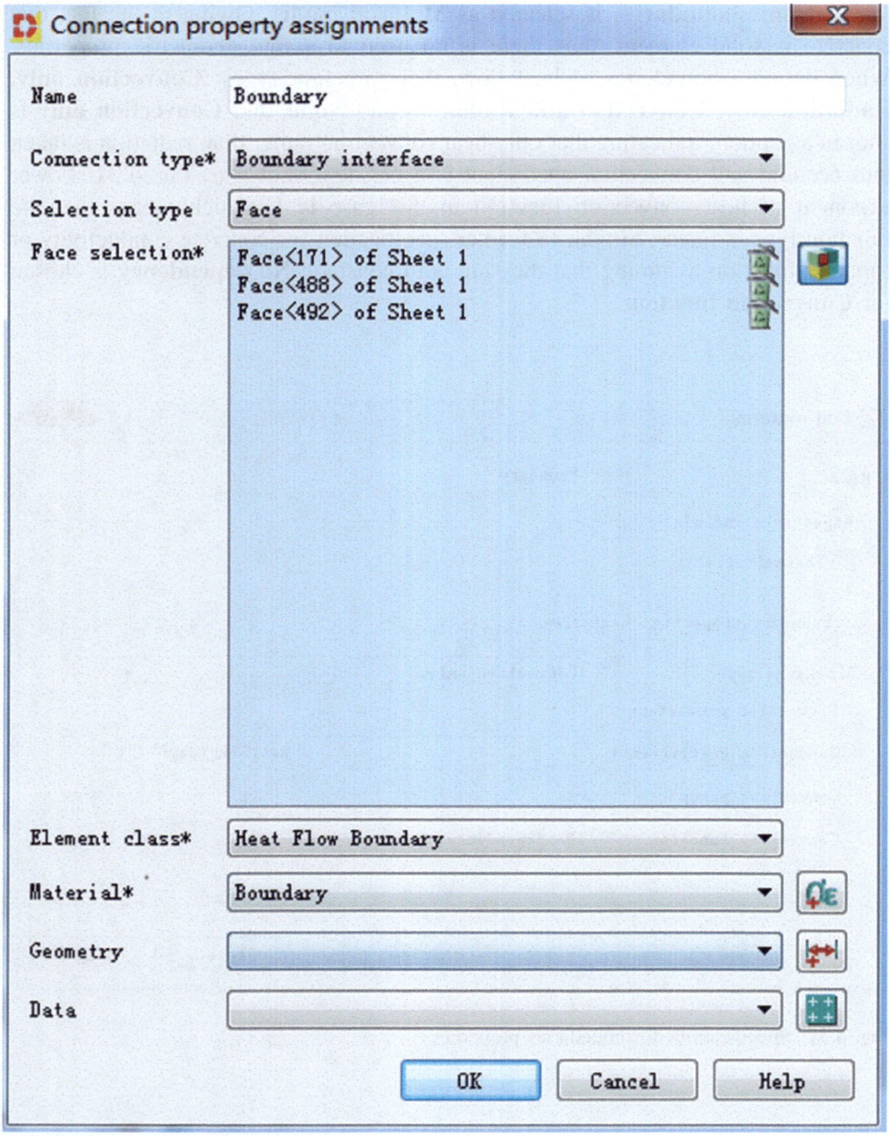

Fig. 4.32 Interface of defining heat flow boundaries

Specifying heat flow boundary conditions, **Boundary** is selected as **Boundary condition**. **Target type** is **Face** while **External temperature** is selected as **External temperature**, with the temperature 35 °C (see Fig. 4.33).

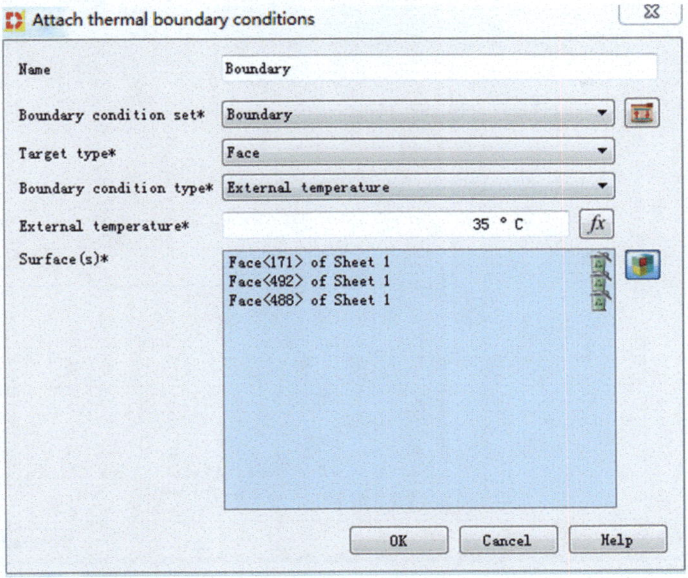

Fig. 4.33 Thermal boundary conditions

Click OK button to generate heat flow boundaries shown as in Fig. 4.34. Green zones represent successful definition for heat flow boundaries. Specifying time-dependent curve for boundary conditions, factors are always set as 1 with 60 days in this case (see Fig. 4.35).

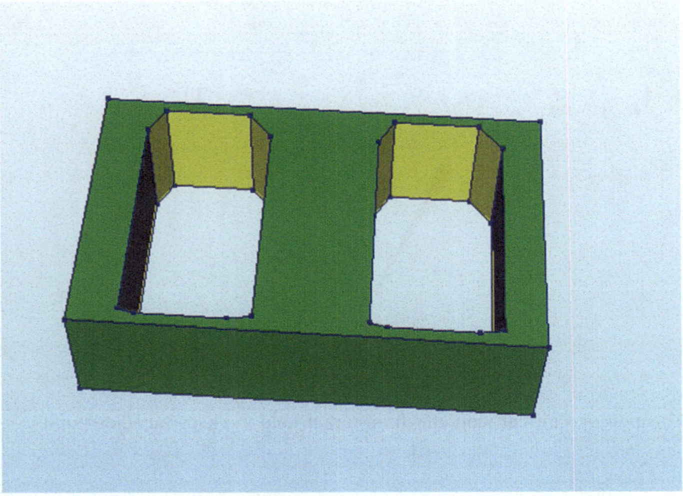

Fig. 4.34 Generation of heat flow boundaries

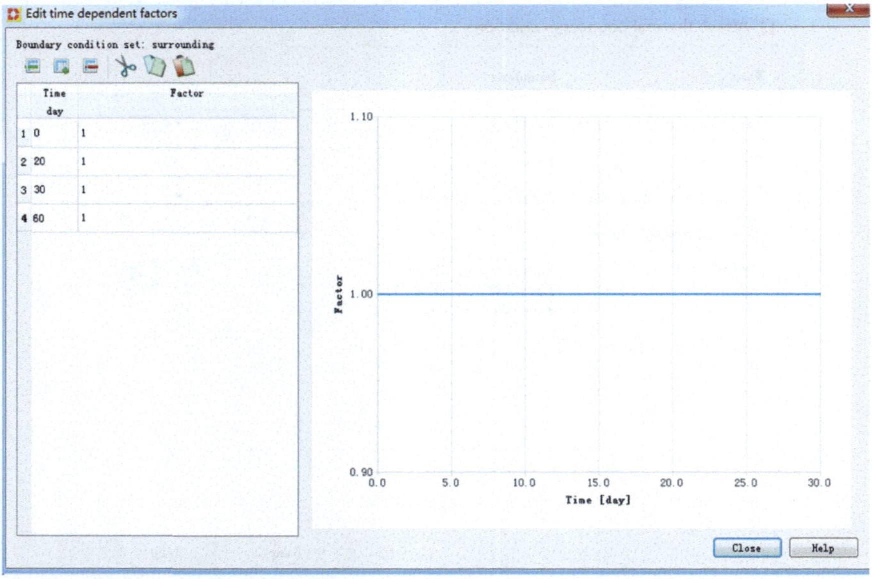

Fig. 4.35 Specification for time-dependent curve

Then initial temperature is defined. Click shortcut button **Attach an initial field to shape/face/line/point** (see Fig. 4.36) to generate an initial temperature field with the name of **initial** and select the whole solid model. **Initial field target type** is **Solid** with the initial temperature 20 °C (see Fig. 4.37).

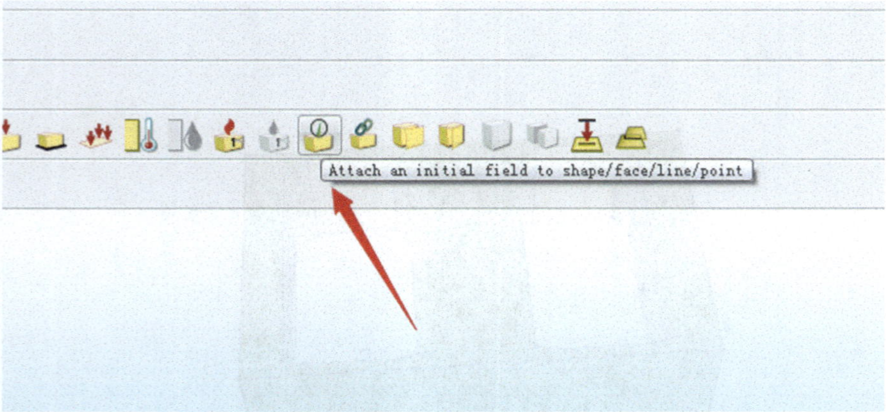

Fig. 4.36 Location of shortcut icon attach an initial field to shape/face/line/point

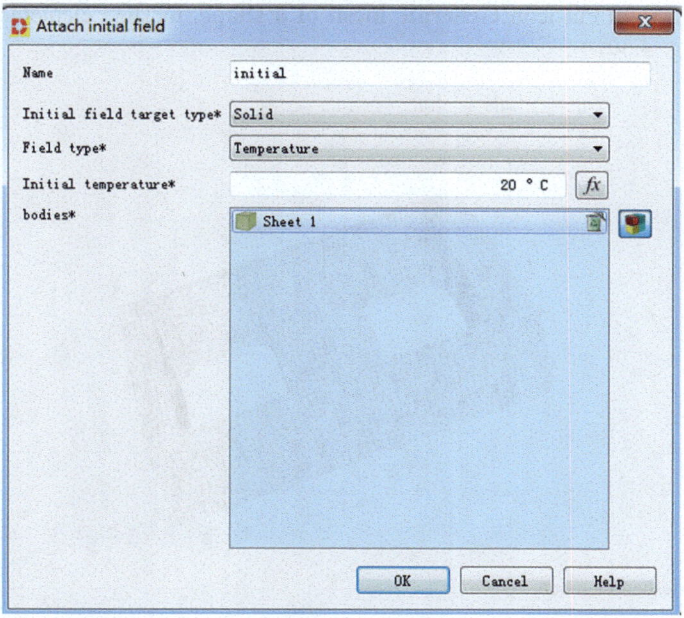

Fig. 4.37 Definition of initial temperature

After attachment of gravity, it is the time to mesh. All the geometric model is selected, then we right-click to set mesh properties. **Element size** is selected for **Seeding method** and **Desired size** is 0.1 m. Mesher type is **Hexa/Quad** (see Fig. 4.38).

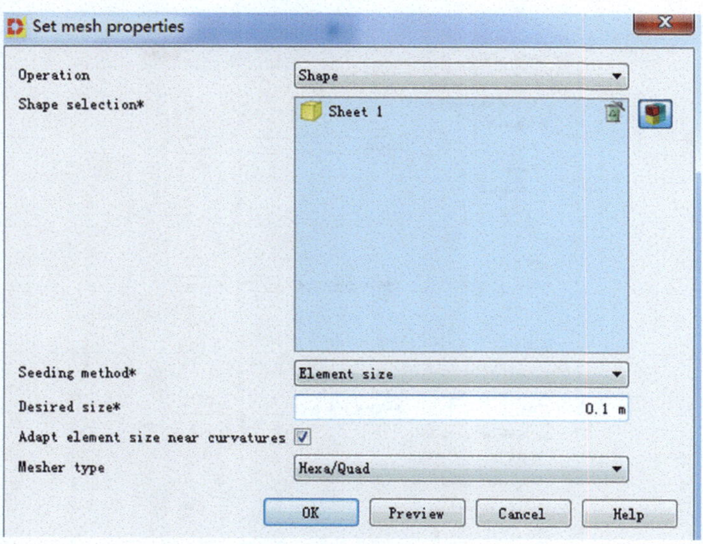

Fig. 4.38 Setting mesh properties

Clicking short cut icon **Generate mesh of a shape**, meshed elements are displayed (Fig. 4.39).

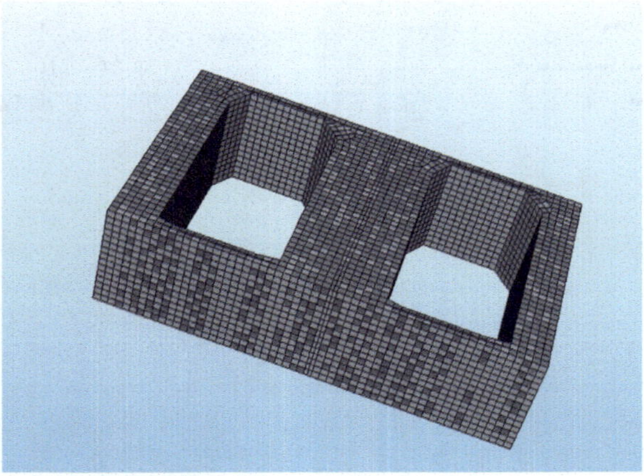

Fig. 4.39 Meshed elements

Setting analysis, right-clicking analysis to select **Transient heat transfer**, **Initial temperature field** is ticked, where the factor is 1. Analysis type is structural nonlinear. 1 day is input into **Start time**. **Hydration heat analysis** is ticked with default initial degree of reaction 0.01. **Calculate equivalent age** is also ticked with initial equivalent value 0 day (see Fig. 4.40). In **Execute analysis** block, step sizes in 35 days are specified as 0.500000(20) 1.00000(5) 10.0000(2) (as Fig. 4.41 shows).

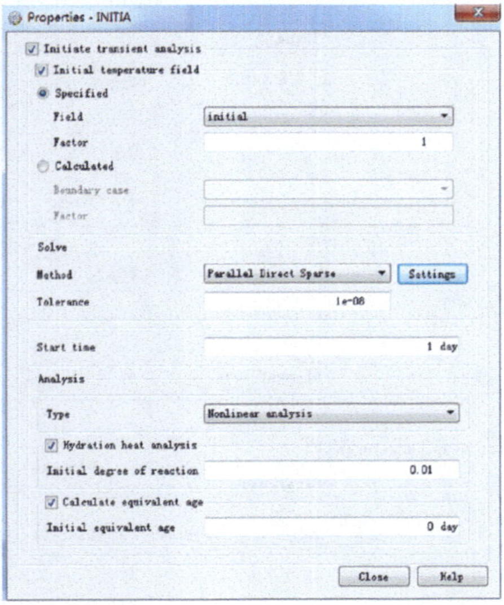

Fig. 4.40 Properties of **Transient heat transfer**

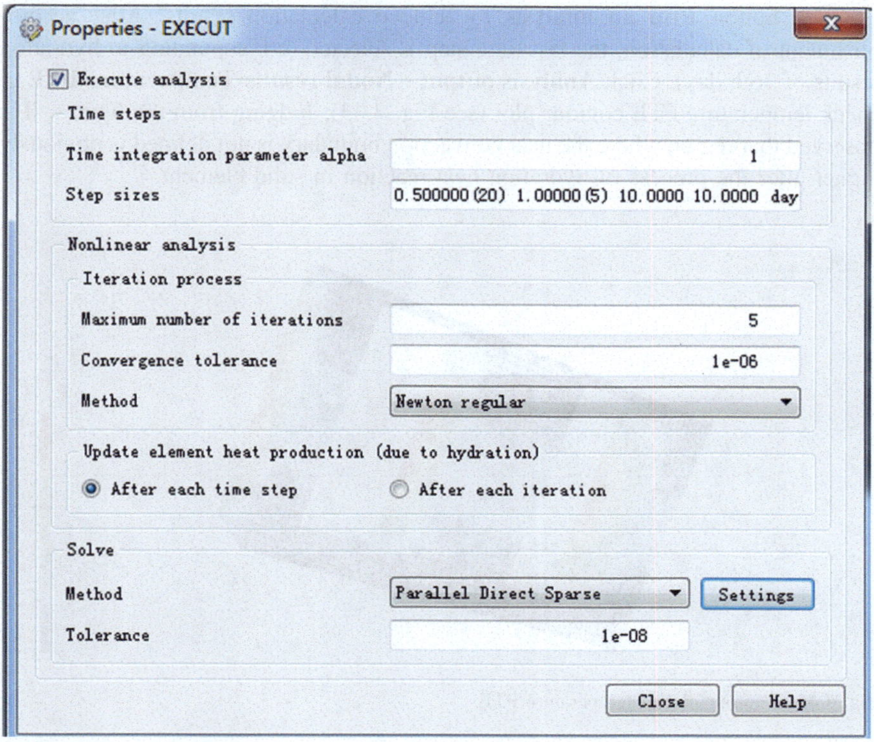

Fig. 4.41 Specification of **Step sizes**

Temperature (TEMPER), Reaction degree (REACTI TOTAL) and equivalent age (EQUAGE TOTAL) are selected as outputs in the **Result Selection** (see Fig. 4.42).

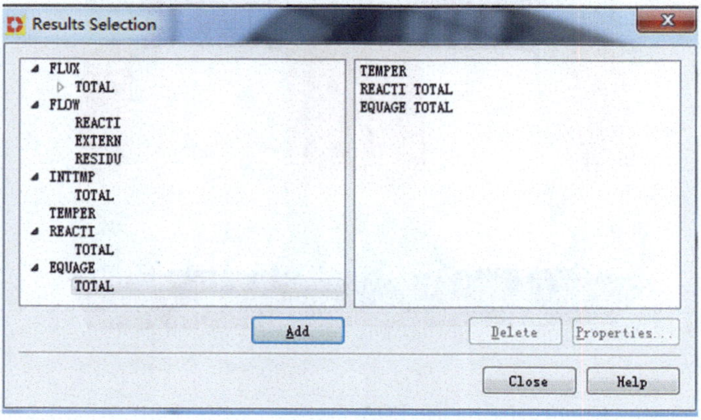

Fig. 4.42 OUTPUT

Click button **Run an analysis** to achieve calculation results. After accomplishment of calculation, the last time step is selected (corresponding to hydration results of 36th day). Click **Analysis output—Nodal results-Temperature-PTE** to check temperature PTE contour plot (see Fig. 4.43). Judging from the figures, it is observed that region where the heat convection boundary is not defined is obviously higher after the process of hydration heat reaction in solid element.

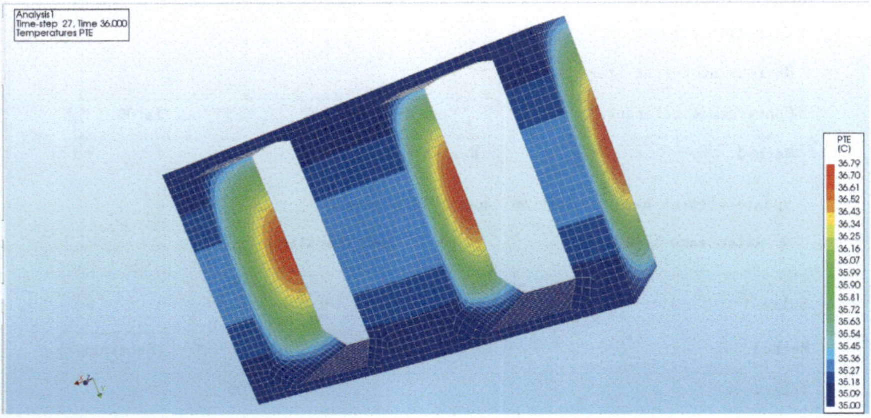

Fig. 4.43 Contour plot of temperature PTE

Contour plot of equivalent age (maturity) is shown in Fig. 4.44, from which it can be observed that the higher the exothermic temperature of hydration reaction, the larger is the equivalent age.

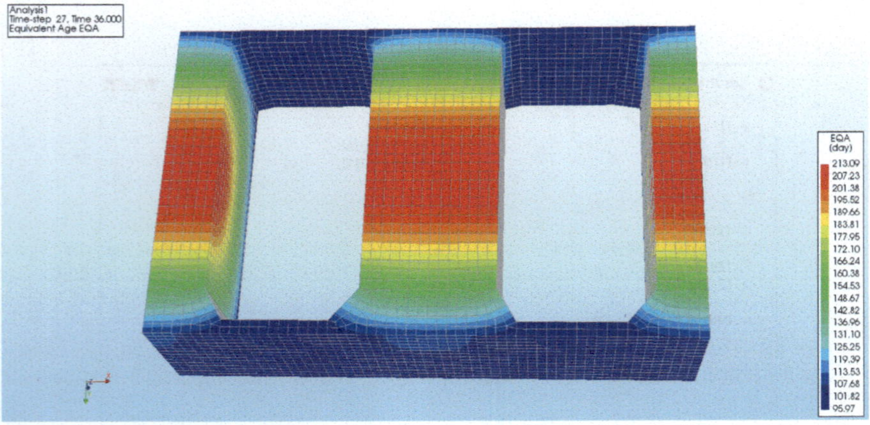

Fig. 4.44 Contour plot of equivalent age EQA

Command console is displayed as follows:

```
newProject( "guanlang", 100 )
setModelAnalysisAspects( [ "STRUCT", "HEATFL" ] )
setModelDimension( "3D" )
setDefaultMeshOrder( "LINEAR" )
setDefaultMesherType( "HEXQUAD" )
createSheet( "Sheet 1", [[ 0, 0, 0 ],[ 5.7, 0, 0 ],[ 5.7, 3.5, 0 ],[ 0, 3.5, 0 ]] )
createSheet( "Sheet 2", [[ 0.8, 0.15, 0 ],[ 0.6, 0.2, 0 ],[ 0.6, 3, 0 ],[ 0.8, 3.35, 0 ],[ 1.9, 3.35, 0 ],[ 2.2,
3, 0 ],[ 2.2, 0.2, 0 ],[ 1.9, 0.15, 0 ]] )
arrayCopy( [ "Sheet 2" ], [ 3, 0, 0 ], [ 0, 0, 0 ], [ 0, 0, 0 ], 1 )
saveProject(  )
subtract( "Sheet 1", [ "Sheet 3", "Sheet 2" ], False, True )
saveProject(  )
extrudeProfile( [ "Sheet 1" ], [ 0, 0, 1.5 ] )
addMaterial( "concrete", "CONCR", "TSCR", [ "HEATFL", "THERMA" ])
setParameter( "MATERIAL", "concrete", "LINEAR/ELASTI/YOUNG", 3.45e+10 )
setParameter( "MATERIAL", "concrete", "LINEAR/ELASTI/POISON", 0.15 )
setParameter( "MATERIAL", "concrete", "LINEAR/ELASTI/POISON", 0.15 )
setParameter( "MATERIAL", "concrete", "LINEAR/MASS/DENSIT", 2500 )
setParameter( "MATERIAL", "concrete", "LINEAR/THERMA/THERMX", 1e-05 )
setParameter( "MATERIAL", "concrete", "MODTYP/TOTCRK", "ROTATE" )
setParameter( "MATERIAL", "concrete", "TENSIL/TENSTR", 2600000 )
setParameter( "MATERIAL", "concrete", "TENSIL/TENSTR", 2600000 )
setParameter( "MATERIAL", "concrete", "TENSIL/TENCRV", "HORDYK" )
setParameter( "MATERIAL", "concrete", "TENSIL/GF1", 150 )
setParameter( "MATERIAL", "concrete", "TENSIL/RESTST", 0 )
setParameter( "MATERIAL", "concrete", "TENSIL/RESTST", 0 )
setParameter( "MATERIAL", "concrete", "COMPRS/COMCRV", "MC1990" )
setParameter( "MATERIAL", "concrete", "COMPRS/COMSTR", 58000000 )
setParameter( "MATERIAL", "concrete", "HEATFL/CONDUC", 1.73e+15 )
setParameter( "MATERIAL", "concrete", "HEATFL/CAPACI", 2e+16 )
setParameter( "MATERIAL", "concrete", "HEATFL/HEATHY/HYDRAT", "PREPRO" )
setParameter( "MATERIAL", "concrete", "HEATFL/HEATHY/ADIAB", [] )
setParameter( MATERIAL, "concrete", "HEATFL/HEATHY/ADIAB", [ 0, 20, 0.1, 24.98, 0.2,
29.47, 0.3, 33.51, 0.4, 37.15, 0.5, 40.42, 0.6, 43.37, 0.7, 46.02, 0.8, 48.41, 0.9, 50.57, 1, 52.5, 1.5,
59.65, 2, 63.88, 2.5, 66.38, 3, 67.86, 4, 69.25, 5, 69.73, 10, 70, 60, 70 ] )
setUnit( "TEMPER", "CELSIU" )
setUnit( "ANGLE", "DEGREE" )
setUnit( "TIME", "DAY" )
saveProject(  )
addGeometry( "Element geometry 1", "SOLID", "STRSOL", [] )
rename( "GEOMET", "Element geometry 1", "concrete" )
```

```
clearReinforcementAspects( [ "Sheet 1" ] )
setElementClassType( "SHAPE", [ "Sheet 1" ], "STRSOL" )
assignMaterial( "concrete", "SHAPE", [ "Sheet 1" ] )
resetGeometry( "SHAPE", [ "Sheet 1" ] )
resetElementData( "SHAPE", [ "Sheet 1" ] )
createSheet( "Grid1", [[ 0.5, 0, 0 ],[ 0.5, 0, 1.5 ],[ 0.5, 3.5, 1.5 ],[ 0.5, 3.5, 0 ]] )
saveProject(   )
arrayCopy( [ "Grid1" ], [ 2.45, 0, 0 ], [ 0, 0, 0 ], [ 0, 0, 0 ], 2 )
createSheet( "Grid5", [[ 0.5, 0, 1.5 ],[ 0.5, 0, 0 ],[ 2.95, 0, 0 ],[ 2.95, 0, 1.5 ]] )
createSheet( "Grid6", [[ 2.95, 0, 1.5 ],[ 2.95, 0, 0 ],[ 5.4, 0, 0 ],[ 5.4, 0, 1.5 ]] )
renameShape( "Grid5", "Grid4" )
renameShape( "Grid6", "Grid5" )
renameShape( "Grid4", "Grid 4" )
renameShape( "Grid5", "Grid 5" )
renameShape( "Grid3", "Grid 3" )
renameShape( "Grid2", "Grid 2" )
renameShape( "Grid1", "Grid 1" )
translate( [ "Grid 4", "Grid 5" ], [ 0, 0.1, 0 ] )
saveProject(   )
createSheet( "Grid 6", [[ 5.4, 3.5, 1.5 ],[ 5.4, 3.5, 0 ],[ 2.95, 3.5, 0 ],[ 2.95, 3.5, 1.5 ]] )
createSheet( "Grid 7", [[ 2.95, 3.5, 1.5 ],[ 2.95, 3.5, 0 ],[ 0.5, 3.5, 0 ],[ 0.5, 3.5, 1.5 ]] )
translate( [ "Grid 6", "Grid 7" ], [ 0, -0.1, 0 ] )
saveProject(   )
addMaterial( "Grid1", "MCSTEL", "TRESCA", [ "THERMA" ] )
setParameter( "MATERIAL", "Grid1", "LINEAR/ELASTI/YOUNG", 2.1e+11 )
setParameter( "MATERIAL", "Grid1", "LINEAR/ELASTI/POISON", 0.3 )
setParameter( "MATERIAL", "Grid1", "LINEAR/MASS/DENSIT", 7800 )
setParameter( "MATERIAL", "Grid1", "LINEAR/THERMA/THERMX", 1e-05 )
setParameter( "MATERIAL", "Grid1", "LINEAR/THERMA/THERMX", 1e-05 )
setReinforcementDiscretization( [ "Grid 1", "Grid 2", "Grid 3" ], "ELEMENT" )
addMaterial( "Grid2", "MCSTEL", "TRESCA", [ "THERMA" ] )
setParameter( "MATERIAL", "Grid2", "LINEAR/ELASTI/YOUNG", 2.1e+11 )
setParameter( "MATERIAL", "Grid2", "LINEAR/ELASTI/POISON", 0.3 )
setParameter( "MATERIAL", "Grid2", "LINEAR/MASS/DENSIT", 7800 )
setParameter( "MATERIAL", "Grid2", "LINEAR/THERMA/THERMX", 1e-05 )
setParameter( "MATERIAL", "Grid2", "TREPLA/YLDSTR", 4.4e+08 )
addGeometry( "Element geometry 2", "RSHEET", "REGRID", [] )
rename( "GEOMET", "Element geometry 2", "Grid 2" )
setParameter( "GEOMET", "Grid 2", "PHI", [ 0.032, 0.032 ] )
setParameter( "GEOMET", "Grid 2", "SPACIN", [ 0.1, 0.1 ] )
setParameter( "GEOMET", "Grid 2", "XAXIS", [ 1, 0, 0 ] )
```

```
setParameter( "GEOMET", "Grid 2", "XAXIS", [ 0, 1, 0 ] )
setParameter( "GEOMET", "Grid 2", "XAXIS", [ 0, 1, 0 ] )
saveProject(  )
setReinforcementAspects( [ "Grid 4", "Grid 5", "Grid 6", "Grid 7" ] )
assignMaterial( "Grid2", "SHAPE", [ "Grid 4", "Grid 5", "Grid 6", "Grid 7" ] )
assignGeometry( "Grid 2", "SHAPE", [ "Grid 4", "Grid 5", "Grid 6", "Grid 7" ] )
resetElementData( "SHAPE", [ "Grid 4", "Grid 5", "Grid 6", "Grid 7" ] )
setReinforcementDiscretization( [ "Grid 4", "Grid 5", "Grid 6", "Grid 7" ], "ELEMENT" )
saveProject(  )
addMaterial( "Boundary", "INTERF", "FLBOUN", [] )
setParameter( "MATERIAL", "Boundary", "HTBOUN/CONPAR/CONVEC", 9e+15 )
createSurfaceConnection( "Boundary" )
setParameter( "GEOMETRYCONNECTION", "Boundary", "CONTYP", "BOUNDA" )
attachTo( "GEOMETRYCONNECTION", "Boundary", "SOURCE", "Sheet 1", [[ 3.2693661, 3.5,
0.6396405 ],[ 2.4306339, 6.5822961e-33, 0.8603595 ],[ 0.2558562, 1.4924945, 1.5 ]] )
setElementClassType( "GEOMETRYCONNECTION", "Boundary", "HEABOU" )
assignMaterial( "Boundary", "GEOMETRYCONNECTION", "Boundary" )
resetGeometry( "GEOMETRYCONNECTION", "Boundary" )
resetElementData( "GEOMETRYCONNECTION", "Boundary" )
addSet( "GEOMETRYBCSET", "surrounding" )
createSurfaceBoundaryCondition( "THERMAL", "surrounding", "surrounding" )
setParameter( "GEOMETRYBC", "surrounding", "BOUTYP", "EXTEMP" )
setParameter( "GEOMETRYBC", "surrounding", "EXTEMP/VALUE", 35 )
attach( "GEOMETRYBC", "surrounding", "Sheet 1", [[ 3.2693661, 3.5, 0.6396405 ],[ 0.2558562,
1.4924945, 1.5 ],[ 2.4306339, 6.5822961e-33, 0.8603595 ]] )
saveProject(  )
setTimeDependentLoadFactors( "GEOMETRYBCSET", "surrounding", [ 0, 20, 30, 60 ], [ 1, 1, 1,
1 ] )
createBodyInitialField( "initial" )
setParameter( "GEOMETRYINIFIELD", "initial", "INITYP", "TEMPER" )
setParameter( "GEOMETRYINIFIELD", "initial", "TEMPER/VALUE", 20 )
attach( "GEOMETRYINIFIELD", "initial", [ "Sheet 1" ] )
addSet( "GEOMETRYLOADSET", "gravity" )
createModelLoad( "gravity", "gravity" )
setElementSize( [ "Sheet 1" ], 0.1, -1, True )
setMesherType( [ "Sheet 1" ], "HEXQUAD" )
setParameter( "MATERIAL", "concrete", "LINEAR/ELASTI/YOUNG", 3.45e+10 )
setParameter( "MATERIAL", "concrete", "TENSIL/TENSTR", 2600000 )
setParameter( "MATERIAL", "concrete", "TENSIL/GF1", 150 )
setParameter( "MATERIAL", "concrete", "COMPRS/COMSTR", 58000000 )
setParameter( "MATERIAL", "concrete", "HEATFL/CONDUC", 1.73e+15 )
```

```
setParameter( "MATERIAL", "concrete", "HEATFL/CONDUC", 1.73e+15 )
setParameter( "MATERIAL", "concrete", "HEATFL/CAPACI", 2e+16 )
setParameter( "MATERIAL", "concrete", "HEATFL/CAPACI", 2e+16 )
setParameter( MATERIAL, "concrete", "HEATFL/HEATHY/ADIAB", [ 0, 20, 0.1, 24.98, 0.2,
29.47, 0.3, 33.51, 0.4, 37.15, 0.5, 40.42, 0.6, 43.37, 0.7, 46.02, 0.8, 48.41, 0.9, 50.57, 1, 52.5, 1.5,
59.65, 2, 63.88, 2.5, 66.38, 3, 67.86, 4, 69.25, 5, 69.73, 10, 70, 60, 70 ] )
saveProject( )
generateMesh( [] )
hideView( "GEOM" )
showView( "MESH" )
addAnalysis( "Analysis1" )
addAnalysisCommand( "Analysis1", "HEATTR", "Transient heat transfer" )
addAnalysisCommandDetail( "Analysis1", "Transient heat transfer", "INITIA/TEMPER" )
setAnalysisCommandDetail( "Analysis1", "Transient heat transfer", "INITIA/TEMPER", True )
addAnalysisCommandDetail( "Analysis1", "Transient heat transfer", "INITIA/TIME0" )
setAnalysisCommandDetail( "Analysis1", "Transient heat transfer", "INITIA/TIME0", 1 )
setAnalysisCommandDetail( "Analysis1", "Transient heat transfer", "INITIA/ANATYP",
"NONLIN" )
addAnalysisCommandDetail( "Analysis1", "Transient heat transfer",
"INITIA/NONLIN/HYDRAT" )
setAnalysisCommandDetail( "Analysis1", "Transient heat transfer",
"INITIA/NONLIN/HYDRAT", True )
addAnalysisCommandDetail( "Analysis1", "Transient heat transfer",
"INITIA/NONLIN/EQUAGE" )
setAnalysisCommandDetail( "Analysis1", "Transient heat transfer",
"INITIA/NONLIN/EQUAGE", True )
saveProject( )
setAnalysisCommandDetail( "Analysis1", "Transient heat transfer", "OUTPUT(1)/SELTYP",
"USER" )
addAnalysisCommandDetail( "Analysis1", "Transient heat transfer", "OUTPUT(1)/USER" )
addAnalysisCommandDetail( "Analysis1", "Transient heat transfer",
"OUTPUT(1)/USER/TEMPER" )
addAnalysisCommandDetail( "Analysis1", "Transient heat transfer",
"OUTPUT(1)/USER/REACTI(1)/TOTAL" )
addAnalysisCommandDetail( "Analysis1", "Transient heat transfer",
"OUTPUT(1)/USER/EQUAGE(1)/TOTAL" )
setAnalysisCommandDetail( "Analysis1", "Transient heat transfer", "EXECUT/SIZES",
"0.500000(20) 1.00000(5) 10.0000 10.0000" )
runSolver( "Analysis1" )
showView( "RESULT" )
setResultCase( [ "Analysis1", "Analysis output", "Time-step 27, Time 36.000" ] )
setResultPlot( "contours", "Temperatures/node", "PTE" )
setResultPlot( "contours", "Degrees of Reaction/node", "DGR" )
setResultPlot( "contours", "Equivalent Age/node", "EQA" )
setResultPlot( "contours", "Temperatures/node", "PTE" )
```

4.2 Hydration Analysis for Mass Concrete Square Pile Block

Numerical model in this case is a mass plain concrete square pile block with the length of 1 m, and width and height are both 0.6 m (see Fig. 4.45). Concrete strength is C50 and solid element is applied. Parameters in heat flow are defined by three different codes: Japan Society of Civil Engineers (JSCE), European CEB-FIP Model Code 1990 (CEB-FIP1990) and AASHTO LRFD Highway Bridge Design Specifications (AASHTO), which are also compared in this chapter. Meanwhile, relative command console in Python language is attached in this part.

Essentials of learning

(1) *Learning to directly construct block* via *inputting dimensional sizes*
(2) *Learning to specify load case of nonlinear temperature calculation*
(3) *Learning to check contour plot of hydration cracking index.*

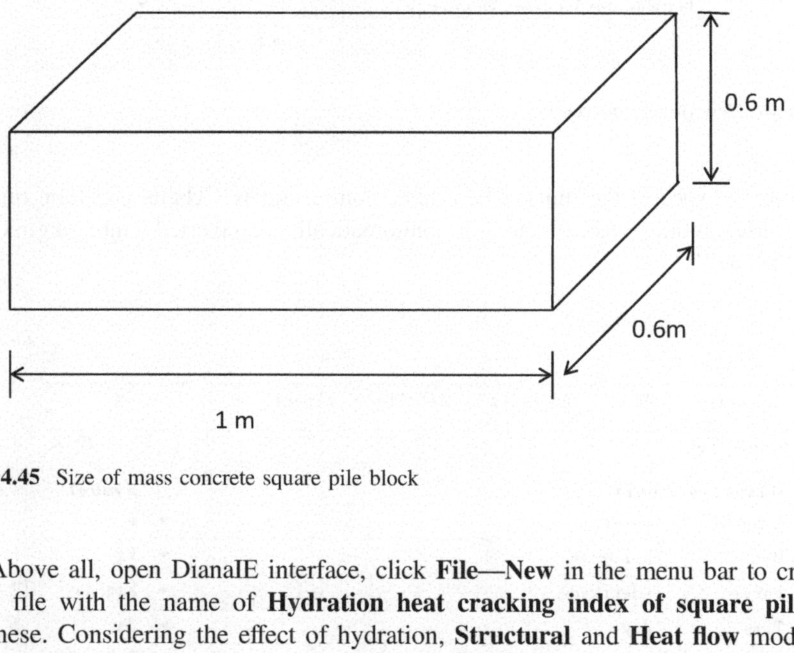

Fig. 4.45 Size of mass concrete square pile block

Above all, open DianaIE interface, click **File—New** in the menu bar to create new file with the name of **Hydration heat cracking index of square pile** in Chinese. Considering the effect of hydration, **Structural** and **Heat flow** modules are both selected. For further simulation of solid elements, dimensions are considered as three, with the maximum scope of model size 10 m. Default mesher type is **Hexa/Quad**, and **Default mesh order** is **Quadratic**. Determination of **Mid-side node location** is **On shape** (see Fig. 4.46).

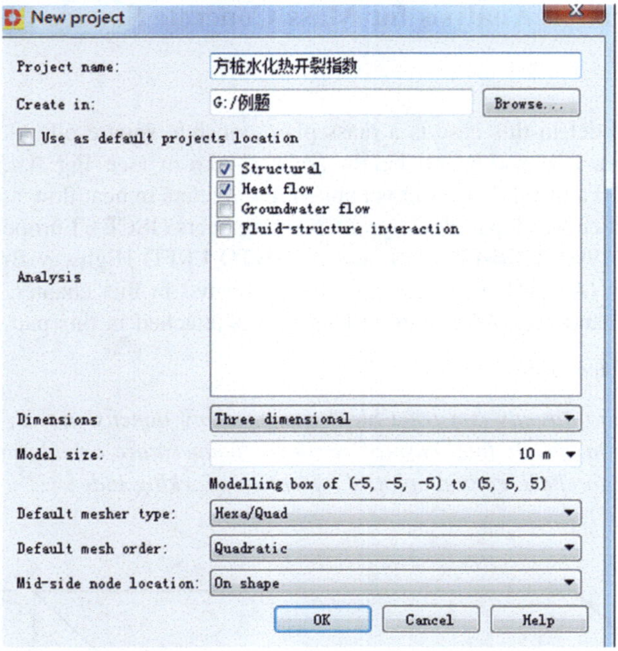

Fig. 4.46 New project interface

Then we specify the units, where temperature unit is **Celsius** and time unit is **day**. Meanwhile, force unit is automatically converted into kgm/day^2 (see Fig. 4.47).

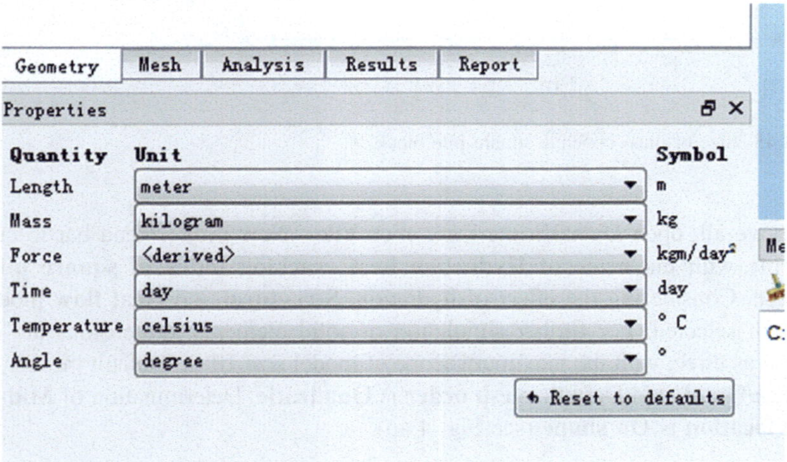

Fig. 4.47 Unit

To create geometric model of concrete clock, click shortcut icon button **Adds a block solid**; shape of solid element is brick, input starting point **Position** (0, 0, 0) and geometric dimensions of hexahedron input into **Size** according to length, width and height in turn, which are 1, 0.6 and 0.6 m, respectively. Click OK button to generate geometric block shape (see Figs. 4.48 and 4.49).

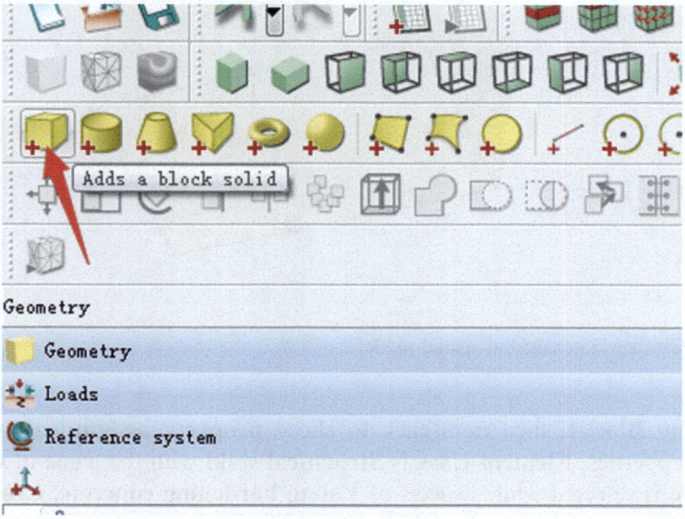

Fig. 4.48 Shortcut icon of **Adds a block solid**

Fig. 4.49 Creating block solid

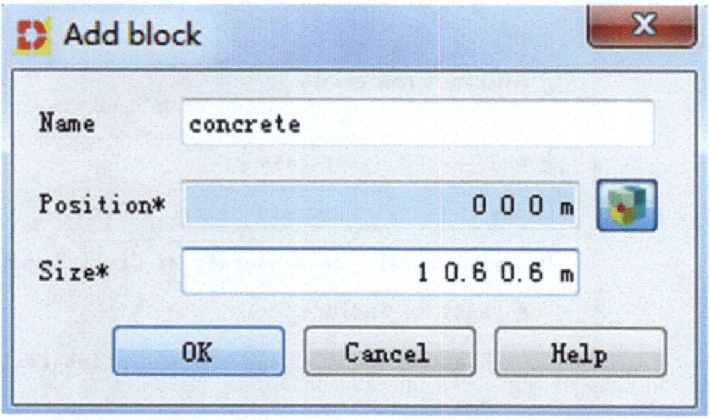

Click OK button to generate geometric hexahedral spatial model with the name of Block1, which is shown in Fig. 4.50.

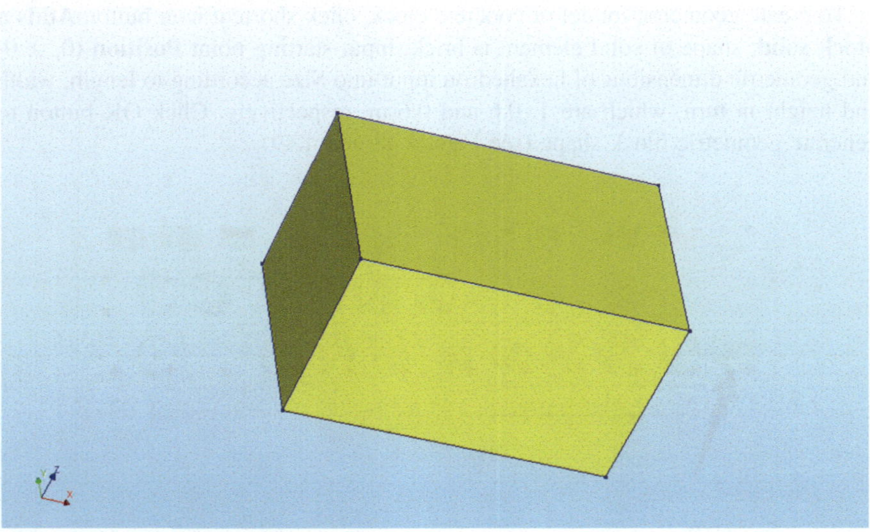

Fig. 4.50 Geometric hexahedral spatial model

Selecting **Block1**, then right-click to chose **property assignments** to assign material properties. **Element class** is **Structural solid** with the name of **concrete**. JSCE code is selected while aspects of **Young hardening concrete**, **Crack index** and **Heat flow** are ticked, as Fig. 4.51 shows. Then units are altered, where length, time and force are meter, day and Newton, respectively. Besides, temperature unit is Celsius and angle unit is radian, thus the unit of mass is automatically converted into Nday^2/m as derived one (see Fig. 4.52).

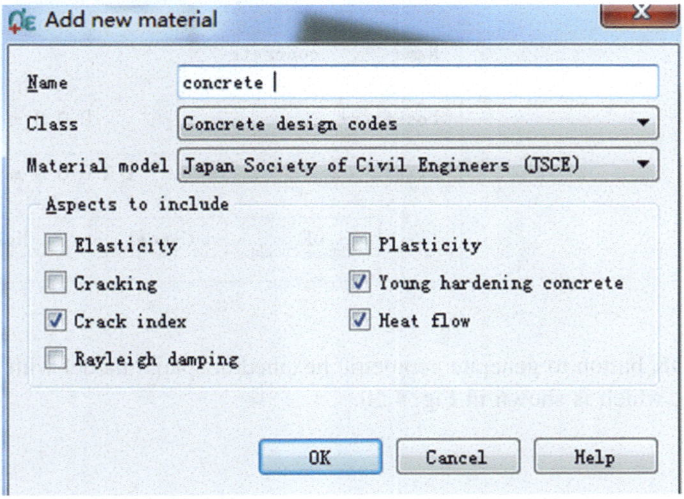

Fig. 4.51 Ticked aspects

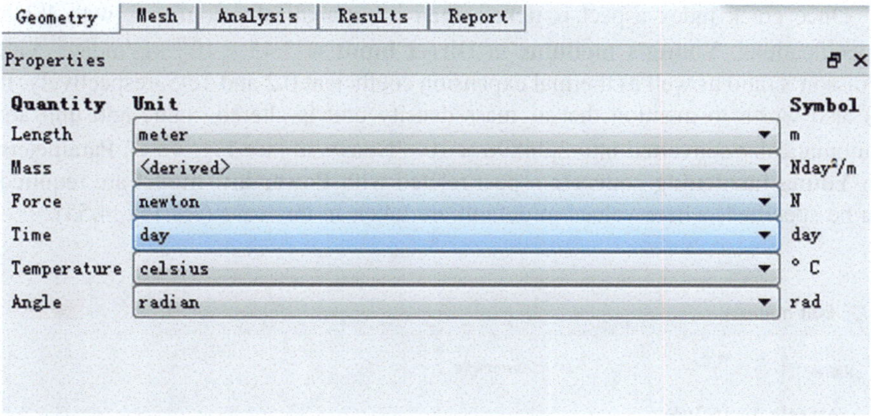

Fig. 4.52 Modification for unit

Specifying parameters in JSCE code, **Characteristic strength after 91 days** and **Modulus of elasticity at 91 days** are $3.24 \times 10^7\ \mathrm{N/m^2}$ and $3.45 \times 10^{10}\ \mathrm{N/m^2}$, respectively, while **Cement type** is **Normally and rapidly hardening** (see Fig. 4.53).

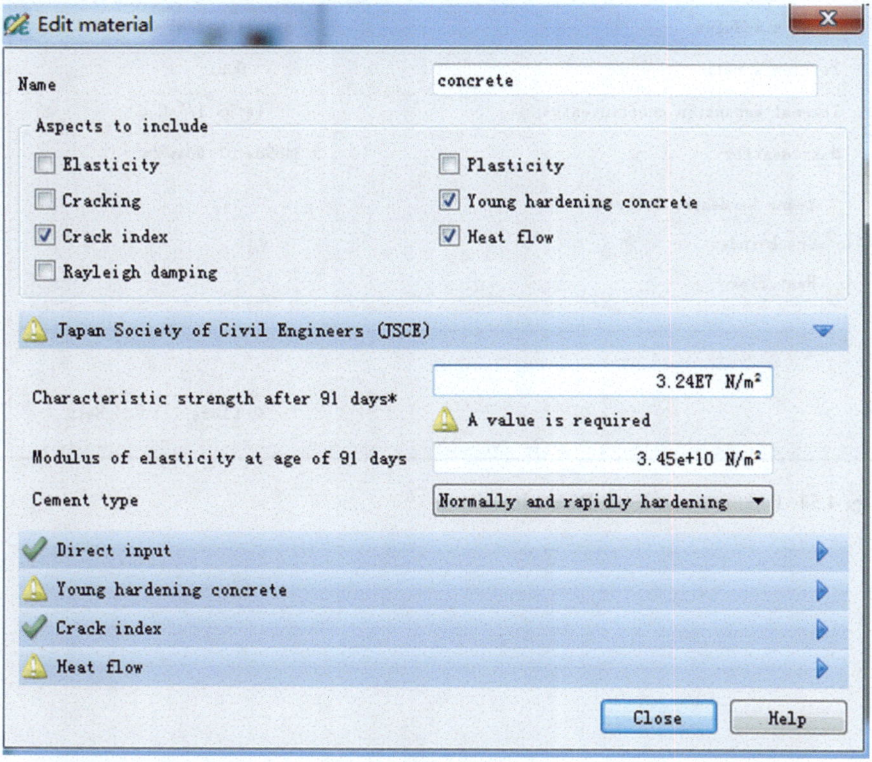

Fig. 4.53 Specifications for JSCE parameters

Once crack index aspect is ticked, then it automatically conforms with JSCE specifications. **Young's modulus** in **Direct input** is $3.45 \times 10^{10}\,\text{kg/mday}^2$ with Poisson's ratio as well as thermal expansion coefficient 0.2 and 1e-5, respectively. It is also worth to mention that as mass density unit is altered, value and unit are automatically converted into $3.34898 \times 10^{-7}\,\text{Nday}^2/\text{m}$ (see Fig. 4.54). Parameters in **Young hardening concrete** aspect related with **Power law model** are required to be specified, where values in default are taken in this case (see Fig. 4.55).

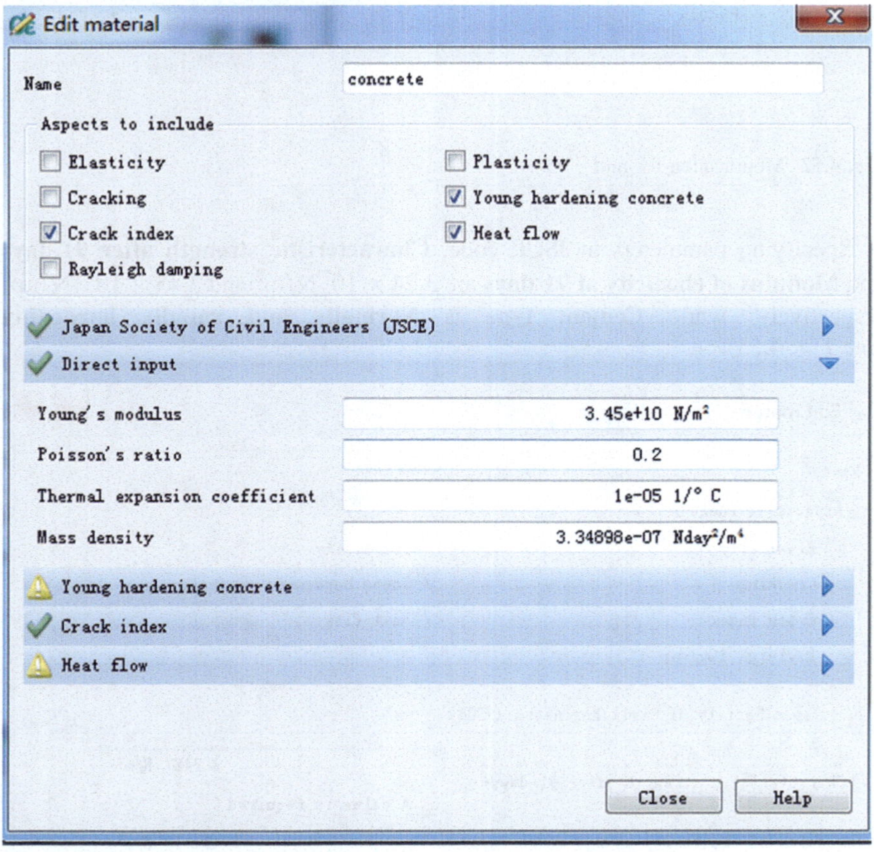

Fig. 4.54 Parameters in JSCE **Direct input**

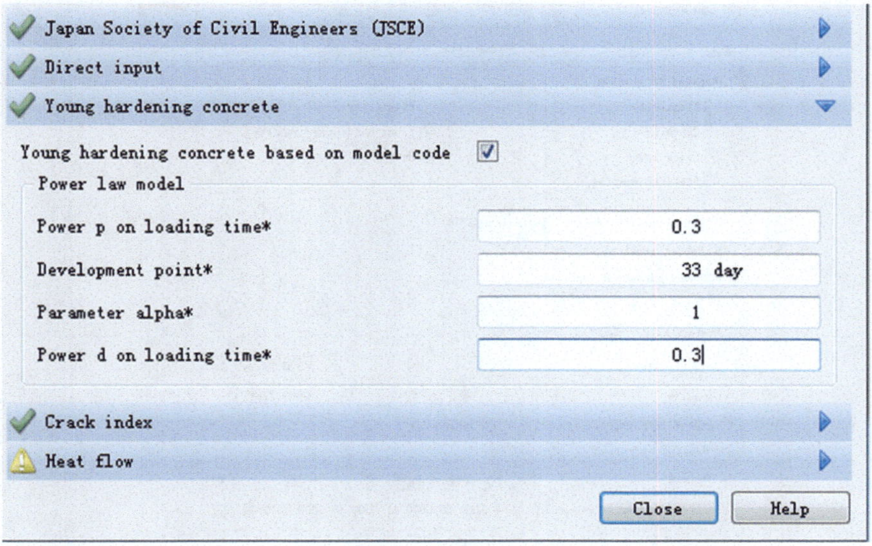

Fig. 4.55 Parameters in **Young hardening concrete** aspect

The adiabatic heating curve, heat conductivity and heat capacity can be defined by **Pre-processing** method. Both heat conductivity and heat capacity are specified as functions related to concrete element age, time and temperature. However, it is assumed that heat conductivity and heat capacity are constant here, so **No dependency** option is selected. There are three options in **Conductivity/capacity function**: **Preprocessing**, **Direct Input** as well as **User specified** concerning secondary development, where **Preprocessing** is selected in this case. **Conductivity** is 320 N/day °C while **Capacity** is 2660 J/m^3 °C. **Reference temperature** and **Arrhenius constant in Kelvin** are the same default value as in Sect. 4.1. In the **Adiabatic heat development** module, age–adiabatic temperature rise curve is specified (as Figs. 4.56 and 4.57 displays).

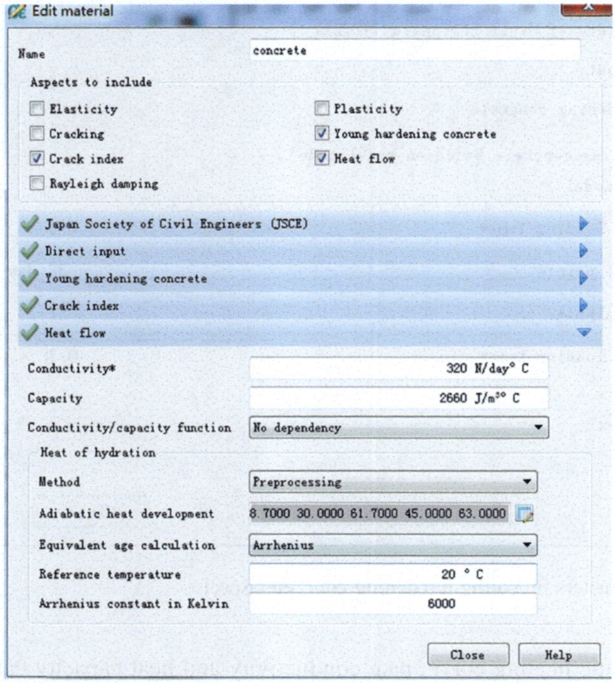

Fig. 4.56 Parameter specifications in Heat flow aspect

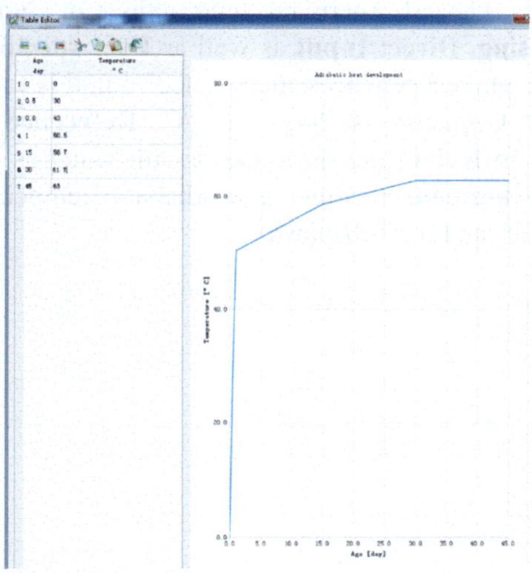

Fig. 4.57 Age–adiabatic temperature rise curve

Interface of property assignment is displayed as in Fig. 4.58. Directly click OK button to complete material properties assignment owing to the fact that there is no need to specify cross-section geometric properties in solid elements.

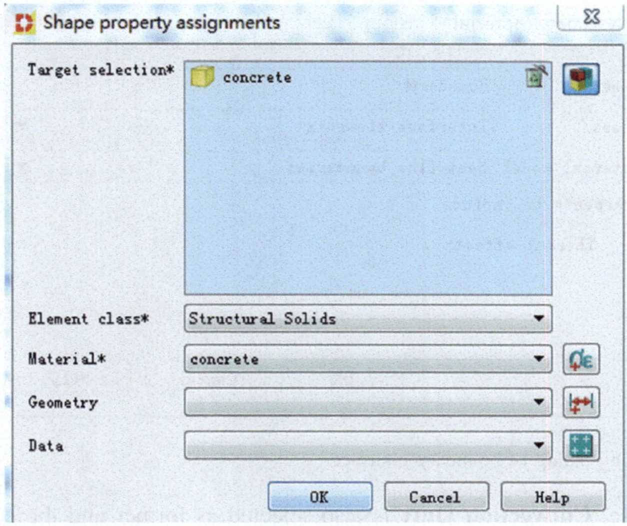

Fig. 4.58 Interface of property assignment

Clicking shortcut icon **Edit connection property assignments**  to create boundary elements with the name of **Boundary**, **Class** is **Interface elements** and **Material model** is still **Heat flow boundaries** (see Figs. 4.59 and 4.60, respectively).

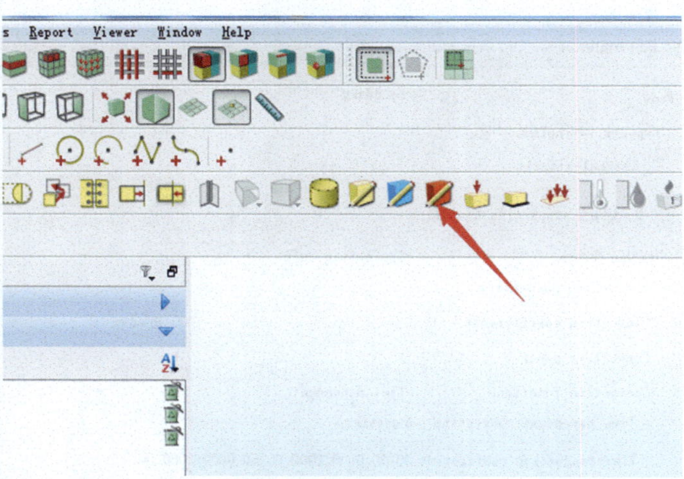

Fig. 4.59 Location of shortcut icon **Edit connection property assignments**

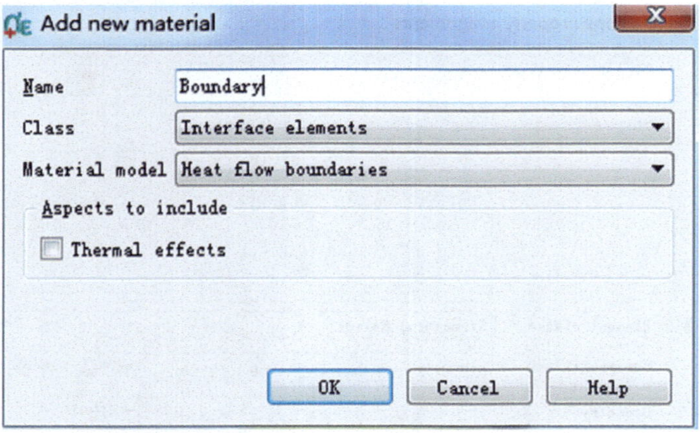

Fig. 4.60 Editing dialog of boundary interface

In this case, **Convection Only** is also selected as former and the coefficient is constant, which is 700 N/mday °C. Convective power is 1. Same as former part, it is assumed that convection function is related with time, thus **Time dependent** is selected as **Convection function** option. Time–heat conductivity curve is edited. Time is an independent variable and heat conductivity is a dependent one. The functional relationship between time and heat conductivity is considered. Total time is 60 days. The heat conductivity of the first two weeks (14 days) was 700 N/mday °C, rising to 2000 N/mday °C on 14th day and retaining such a value until the end of 60 days, as shown in Figs. 4.61 and 4.62.

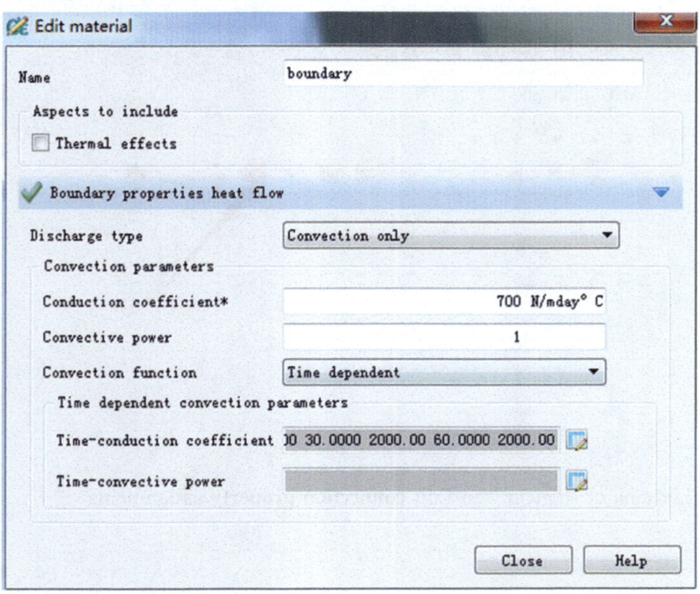

Fig. 4.61 Parameters in boundary properties

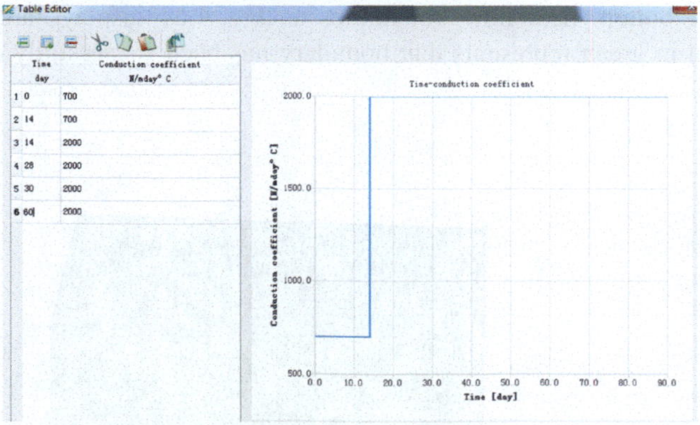

Fig. 4.62 Time–heat conductivity curve

The lateral side of the solid hexahedron perpendicular to the Y-axis and the top surface parallel to the Y-axis are selected as the thermal convection boundary surface and **Connection type** is **Boundary interface**. **Face** is selected as **Selection type** while **Element class** is **Heat Flow Boundary** (see Fig. 4.63).

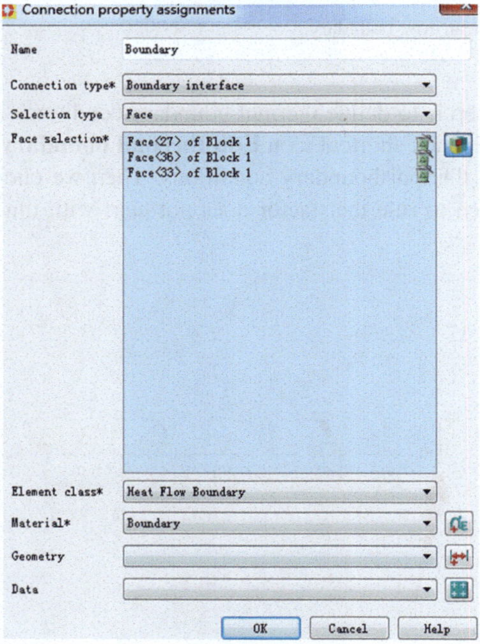

Fig. 4.63 Specification for boundary elements

Click OK button, heat flow boundaries as Fig. 4.64 displays are generated, where region in green represents that boundary has been successfully defined.

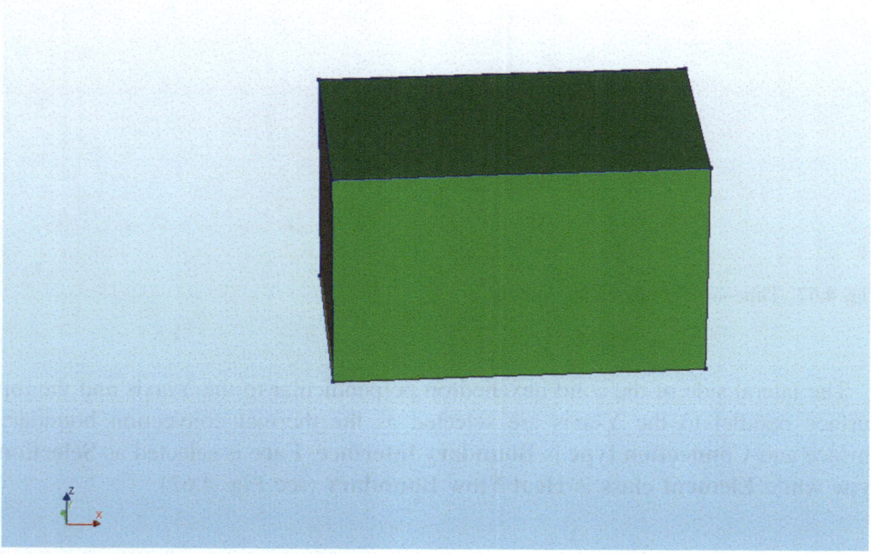

Fig. 4.64 Generated heat flow boundary

The following step is to define thermal boundary conditions after completion of heat flow boundary. Click shortcut icon **Edit thermal boundary conditions** shown in Fig. 4.65 to edit thermal boundary conditions. Then we click clock icon **Time dependency** in green to rule that factor does not alert with time with 60 days.

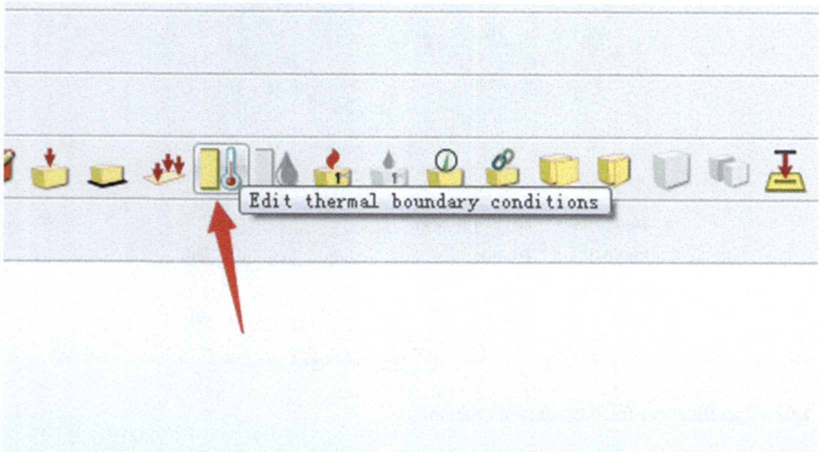

Fig. 4.65 Location of **Edit thermal boundary conditions** icon

Selecting all the heat flow boundary surface in green to define external temperature with the name of **surrounding**, **Face** is selected as **Target type**, and **Boundary condition type** is **External temperature** with the value 35 °C (see Fig. 4.66).

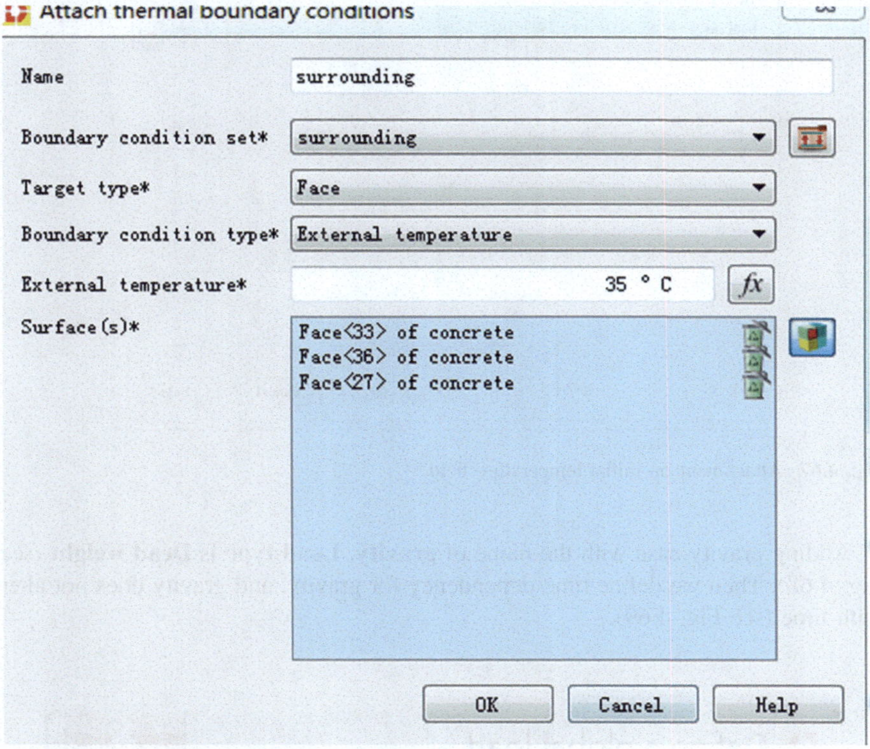

Fig. 4.66 External temperature specification

Defining initial temperature so as to attach initial temperature field to the whole model, clicking shortcut icon **Attach an initial field to shape/face/line/point**, selecting the whole model with the name of **initial**, **Initial field target type** is **Solid** and **Field type** is **Temperature** with the value of 25 °C (see Fig. 4.67).

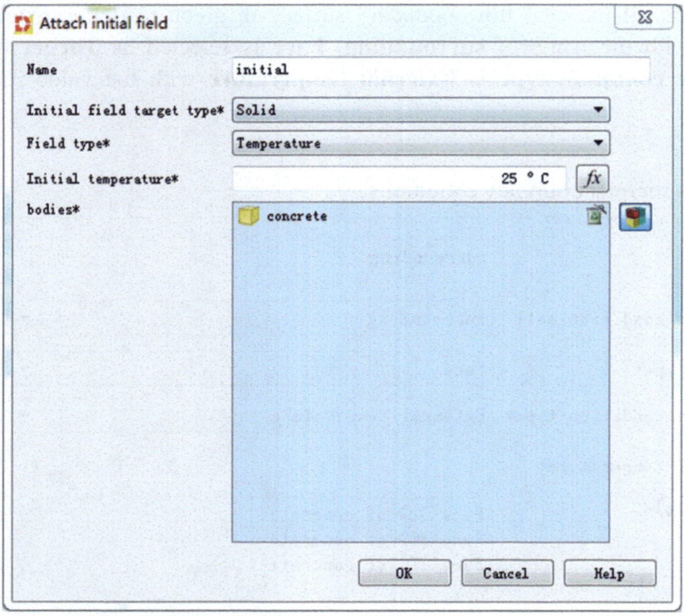

Fig. 4.67 Attachment on initial temperature field

Adding gravity case with the name of **gravity**, **Load type** is **Dead weight** (see Fig. 4.68). Then we define time-dependency for gravity, and gravity does not alter with time (see Fig. 4.69).

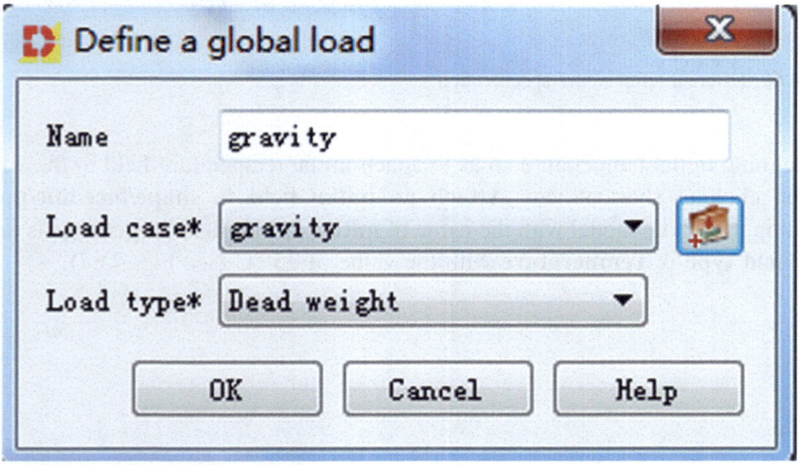

Fig. 4.68 Defining load case of gravity

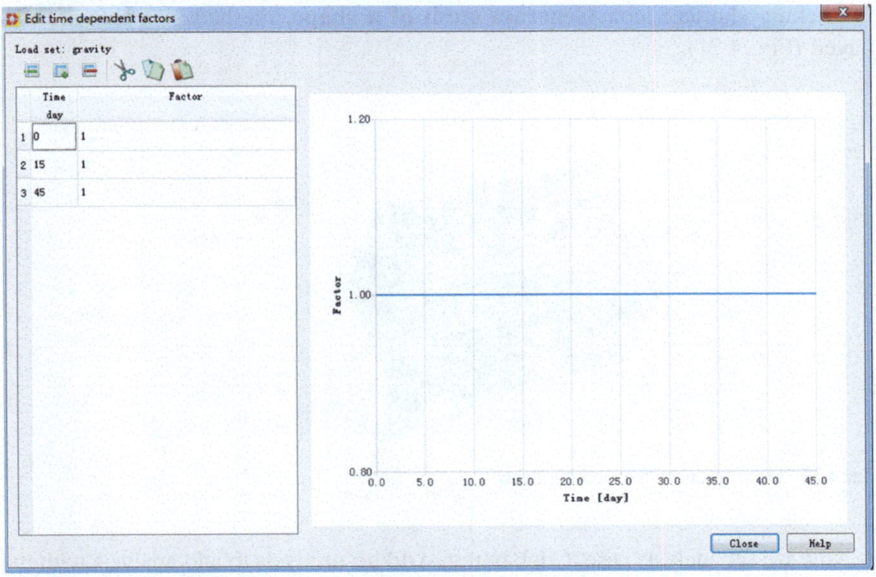

Fig. 4.69 Time-dependent relationship

Selecting all the solid model, right-clicking to select mesh properties, **Element size** is selected as Seeding method, where desired element size is 0.05 m. **Mesher type** is still **Hexa/Quad** and **Linear interpolation** is selected as **Mid-side node location** (see Fig. 4.70).

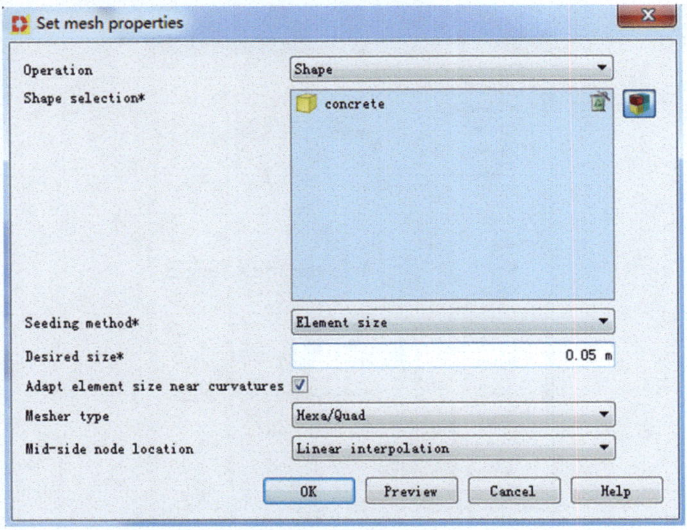

Fig. 4.70 Settings for mesh properties

Clicking shortcut icon **Generate mesh of a shape**, meshed elements are displayed (Fig. 4.71).

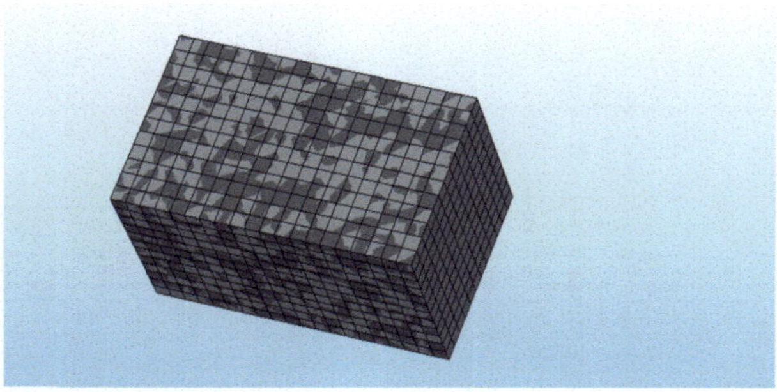

Fig. 4.71 Generation of meshed elements

Now we set analysis case. Click button **Add an analysis** to add analysis with the name of **Analysis**, right-click it to add **Transient heat transfer,** then **Initial temperature field** is ticked. **Nonlinear analysis** option and **Hydration heat analysis** as well as **Calculate equivalent age** under it are all selected while **Initial degree of reaction** and **Initial equivalent age** are 0.01 and 0 day, respectively (see Fig. 4.72). Total time for hydration reaction is 35 days with time step sizes **0.500000(20) 1.00000(5) 10.0000(2)** (see Fig. 4.73). **Maximum number of interactions** is 5 with convergence tolerance 1×10^{-6} by default. **Newton regular** is chosen as iterative method.

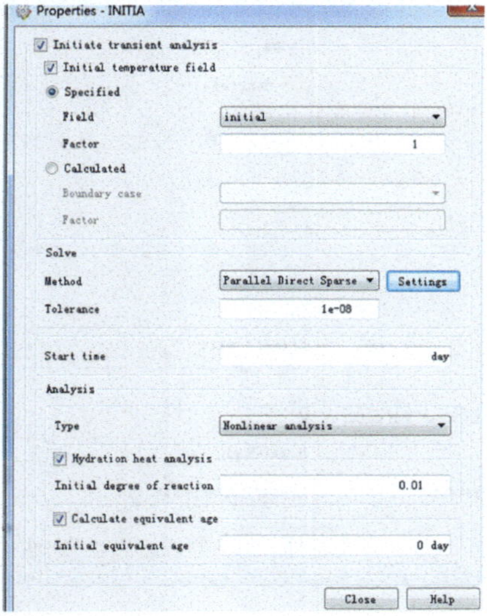

Fig. 4.72 Specification for **Transient heat transfer**

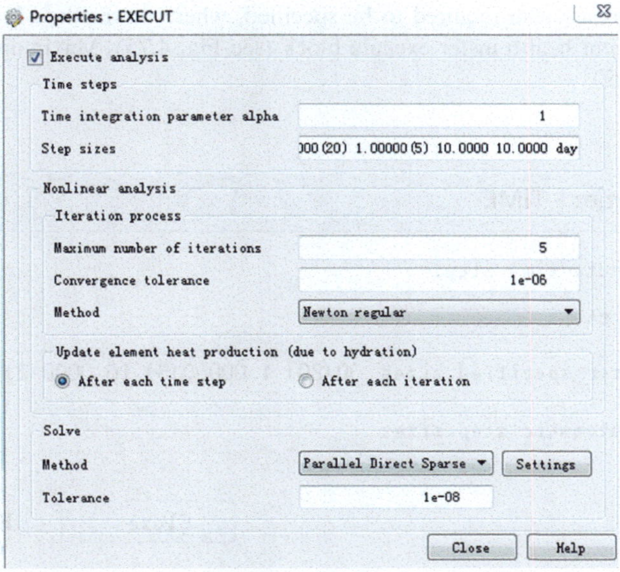

Fig. 4.73 Specification for time step sizes

Next, we set **Structural nonlinear** module. Initially, gravity is attached with the number of iterations 20. **Force** and **Displacement** are both ticked as convergence norm. **Load set** is gravity (see Fig. 4.74).

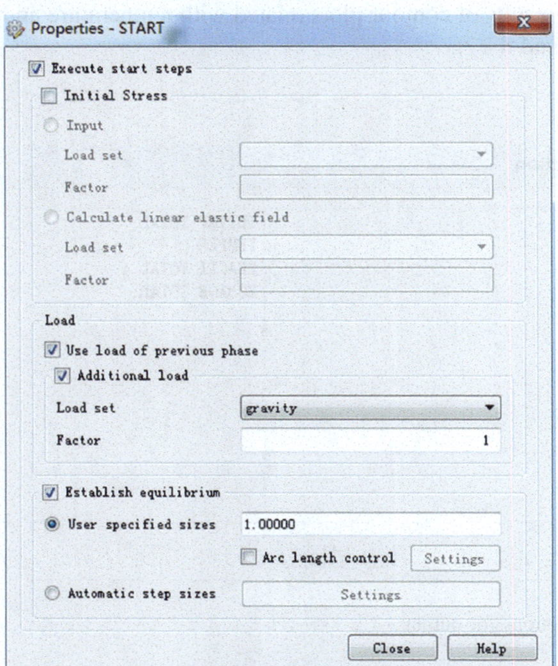

Fig. 4.74 Load set of gravity

Time steps are also required to be specified, where the method is the same as former transient heat transfer execute block (see Fig. 4.75). Maximum number of iterations is 50.

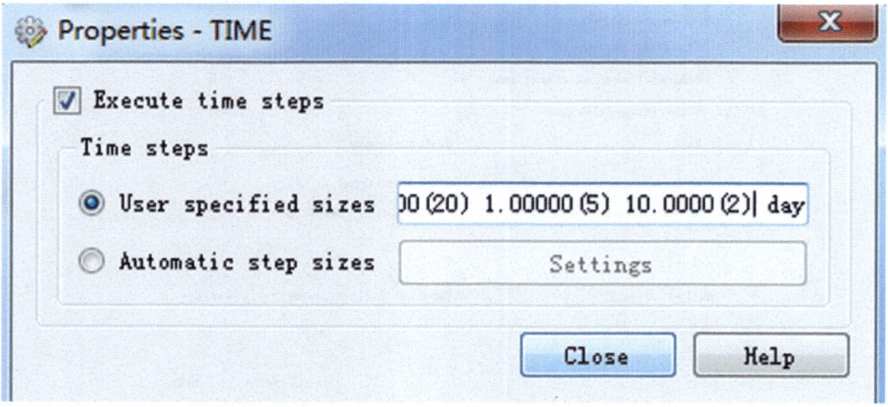

Fig. 4.75 User-specified seizes for time steps

User selection is chosen as output for **Transient heat transfer**, where **INITMP TOTAL**, **TEMPER**, **REACTI TOTAL**, **EQUAGE TOTAL** are chosen as output results to check results of contour plots related with temperature and equivalent age (see Figs. 4.76 and 4.77).

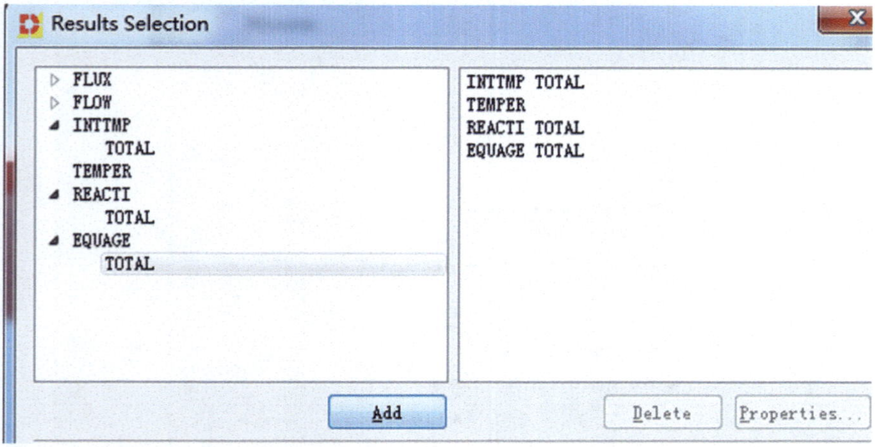

Fig. 4.76 Specifications for output

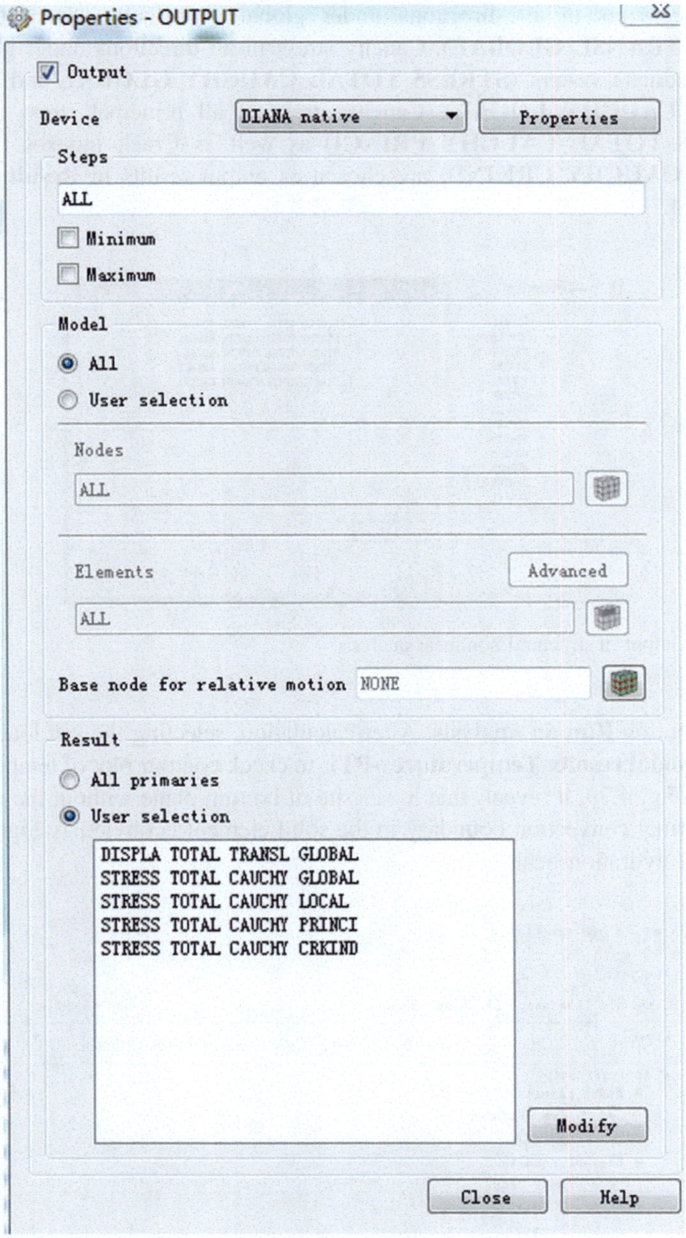

Fig. 4.77 User selection of OUTPUT

Displacements in all directions under global coordinate system **(DISPLA TOTAL TRANSL GLOBAL)**, Cauchy stress in all directions under global and local coordinate system (**STRESS TOTAL CAUCHY GLOBAL** and **STRESS TOTAL CAUCHY LOCAL)**, Cauchy stress in all principal stress directions **(STRESS TOTAL CAUCHY PRINCI)** as well as Crack indexes (**STRESS TOTAL CAUCHY CRKIND)** are chosen as output results in **Result selection** (see Fig. 4.78).

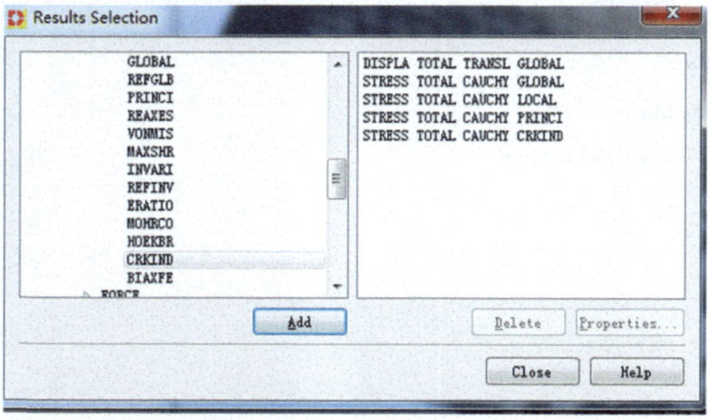

Fig. 4.78 Output in structural nonlinear analysis

Click button **Run an analysis**. After calculation, selecting the last load step via **Output-Nodal results-Temperatures-PTE** to check contour plot of temperature as shown in Fig. 4.79, it reveals that lower site of bottom plane without the definition of the thermal convection boundary in the solid element is obviously higher in the process of hydration heat.

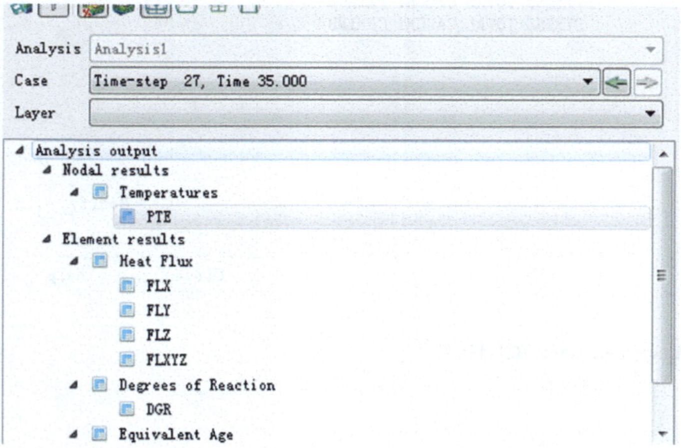

Fig. 4.79 PTE of Output

Select the last load step via **Output-Nodal results-Total displacements-TDtZ** to check ultimate contour plot of displacement in Z direction (see Fig. 4.80). It is observed that displacement in the Z direction is the largest in middle region.

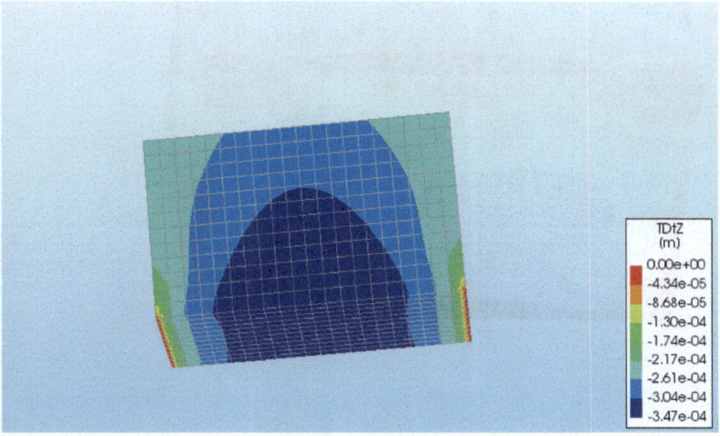

Fig. 4.80 Contour plot of displacement in Z direction

Contour plots of crack index representing characteristic of hydration heat reaction extent and probability of hydration crack under different days are displayed in Figs. 4.81, 4.82, 4.83, 4.84 and 4.85.

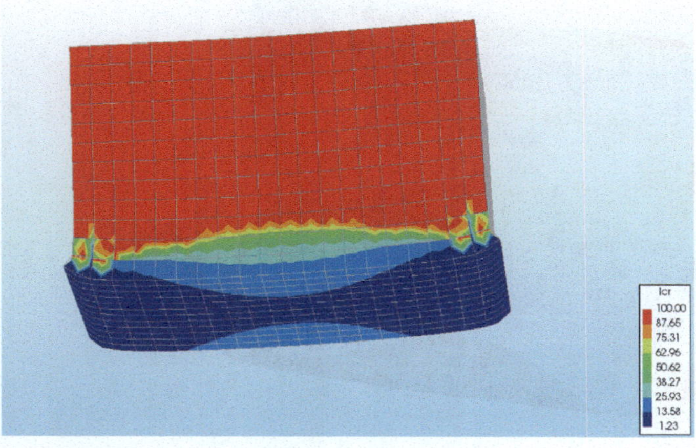

Fig. 4.81 Contour plot of initial crack index

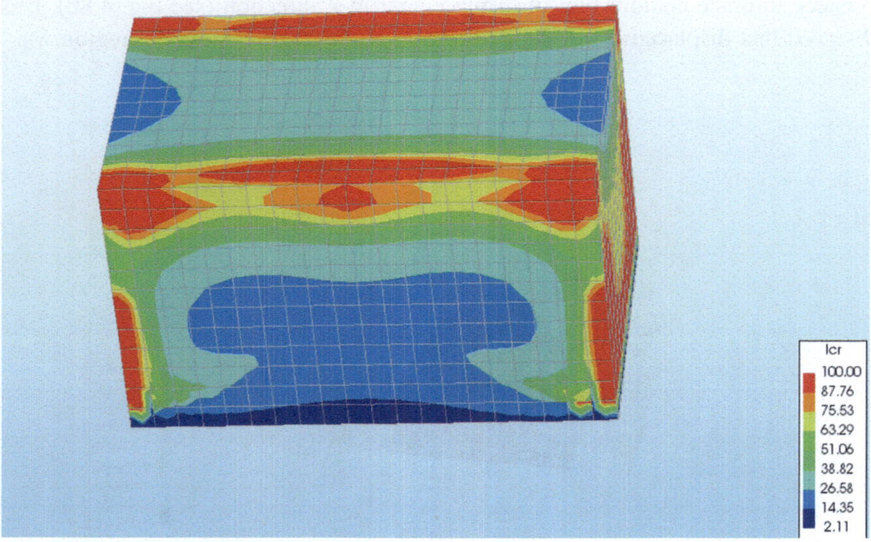

Fig. 4.82 Contour plot of crack index after 2 days

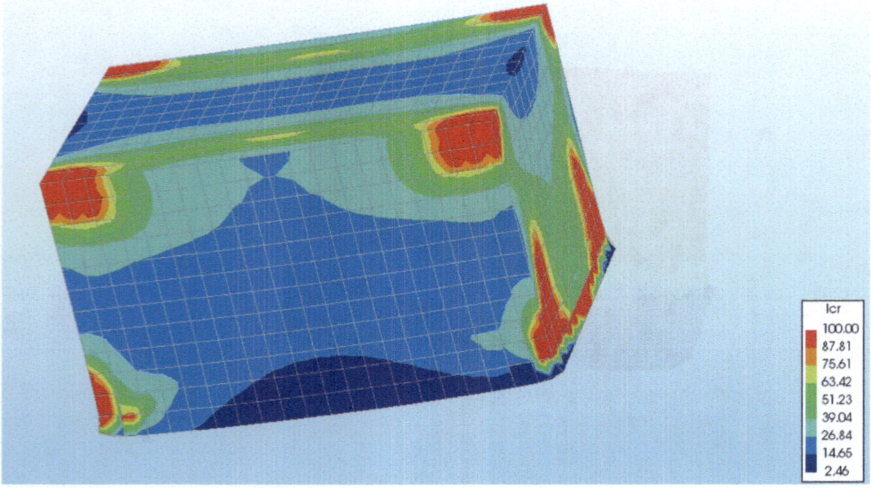

Fig. 4.83 Contour plot of crack index after 1 week

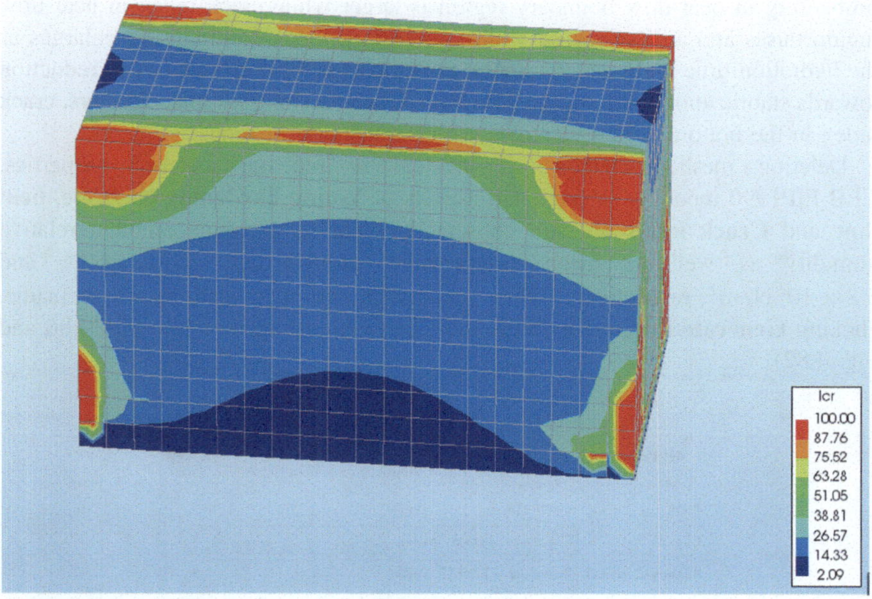

Fig. 4.84 Contour plot of crack index after 4 days

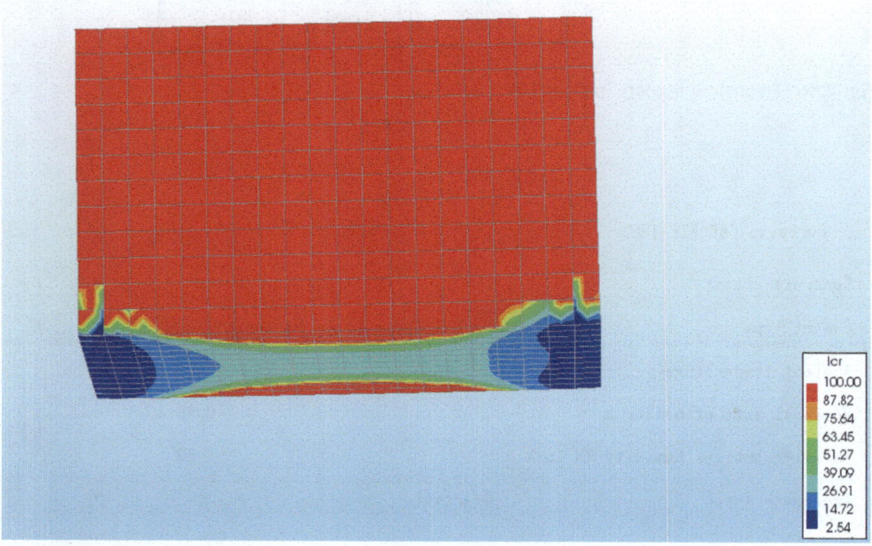

Fig. 4.85 Contour plot of crack index after 35 days

According to the contour plot of crack index, as hydration reaction starts, probability in heat flow boundary region is larger while crack index in heat flow region raises and area of region with higher crack index continuously enlarges as the hydration time increases gradually, thus the possibility of crack is in reduction towards stabilization until around 15th day. When reaching ultimate 35 days, crack index in the bottom region increases with the increase of time.

Deleting meshed elements in manual to redefine material properties, CEB-FIP1990 model is selected and aspects of **Young hardening concrete**, **heat flow** and **Crack index** are still chosen. Ambient temperature, ambient relative humidity as well as mean compressive strength are 20 °C, 69% and $5.8 \times 10^7\ \mathrm{N/m^2}$, respectively. Other parameters and specifications are the same, clicking **Generate mesh of a shape** then starts the calculation (see Fig. 4.86 and Fig. 4.87).

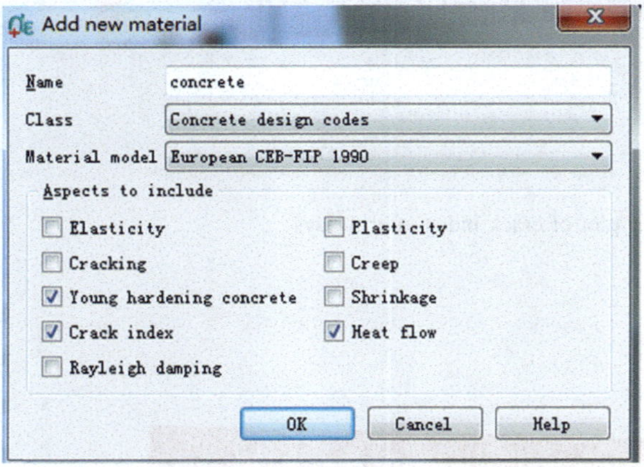

Fig. 4.86 Heat flow aspects in CEB-FIP 1990

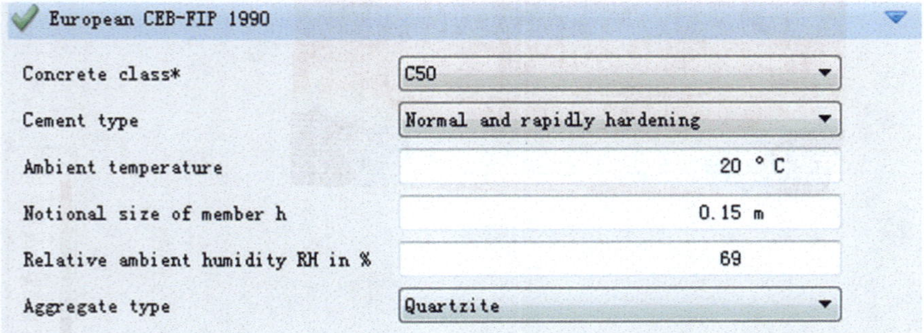

Fig. 4.87 Basic parameters in CEB-FIP 1990

After the calculation, results are checked, and displacement in Z direction is displayed (Fig. 4.88).

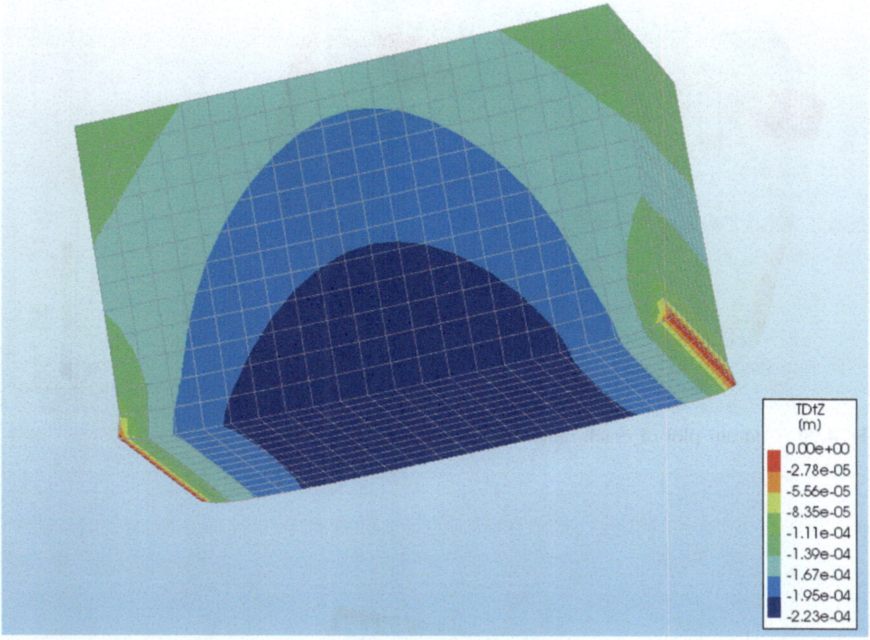

Fig. 4.88 Displacement contour plot in Z direction in CEB-FIP 1990

Selecting crack index under different days, figures are displayed as Figs. 4.89, 4.90, 4.91 and 4.92, respectively.

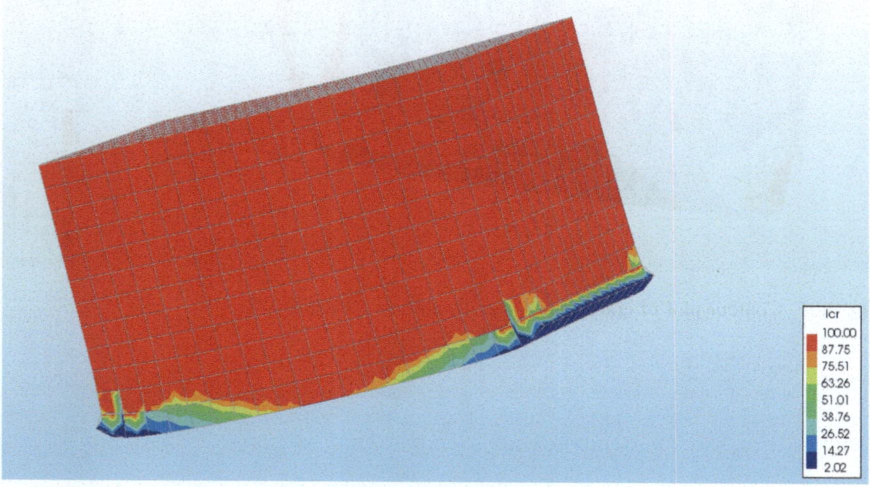

Fig. 4.89 Contour plot of initial crack index in CEB-FIP1990

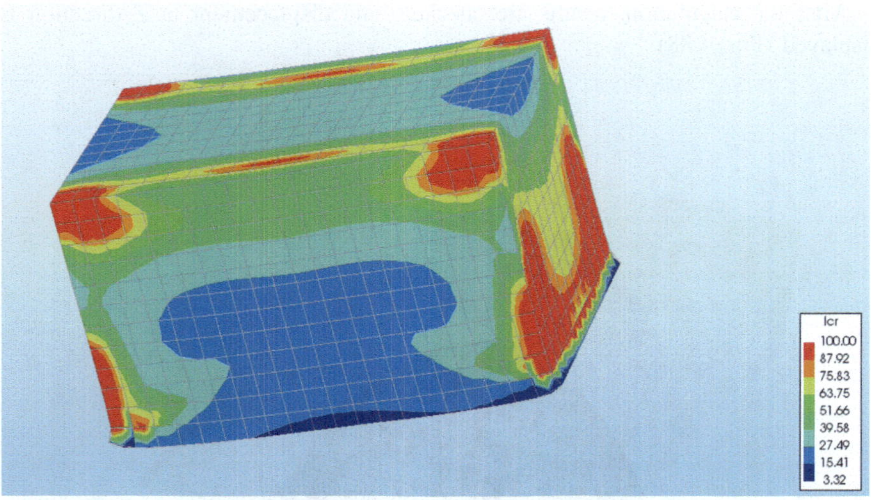

Fig. 4.90 Contour plot of crack index after 5 days in CEB-FIP1990

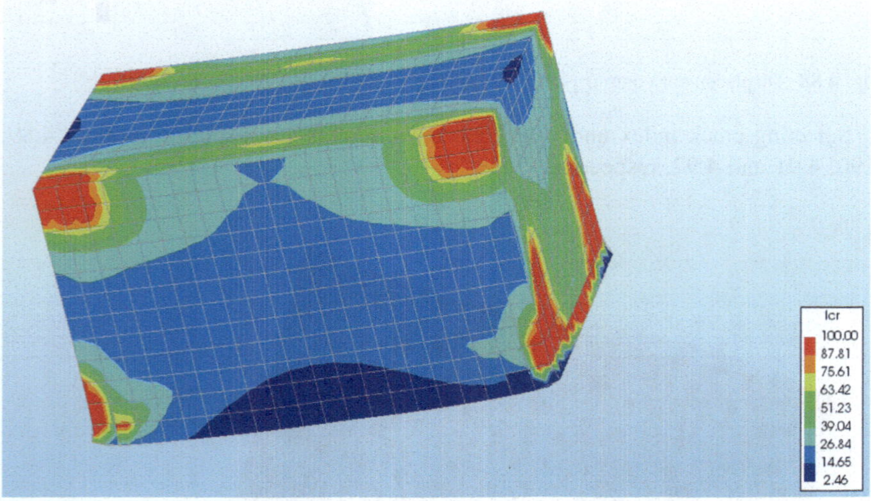

Fig. 4.91 Contour plot of crack index after 10 days in CEB-FIP1990

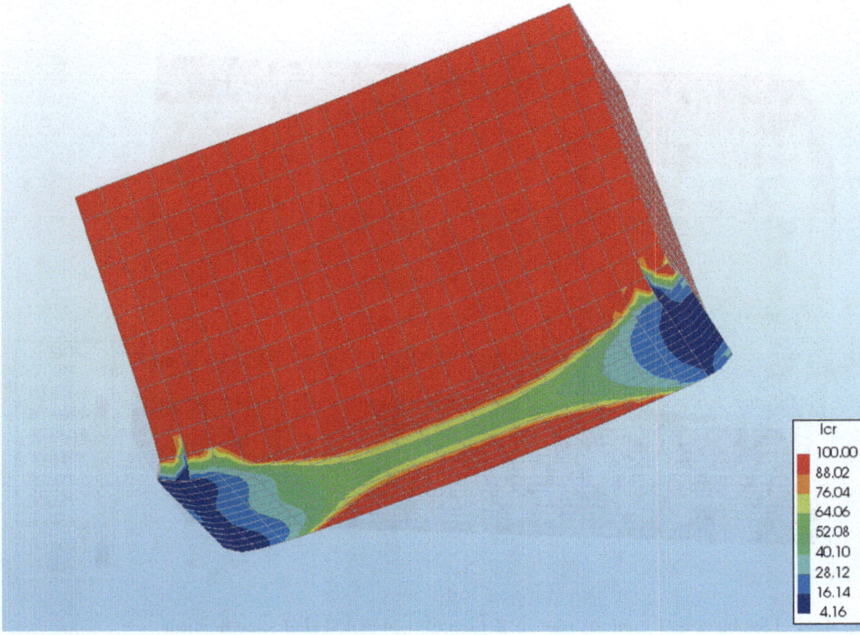

Fig. 4.92 Contour plot of crack index after 35 days in CEB-FIP1990

Similarly, replacing CEB-FIP 1990 code with AASHTO specification, corresponding constitutive parameters are input and other conditions are retained unchanged, Fig. 4.93. After calculation, generated displacement contour plot in the Z direction is displayed in Fig. 4.94.

Fig. 4.93 Basic parameters in AASHTO specification

Fig. 4.94 Displacement contour plot in Z direction in AASHTO specification

Crack index under different days calculated by AASHTO specification are displayed as Figs. 4.95, 4.96, 4.97 and 4.98.

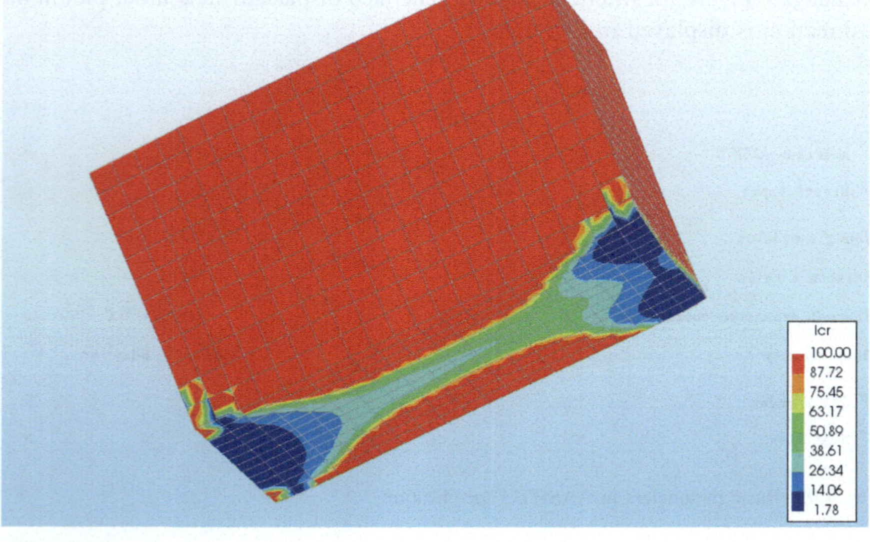

Fig. 4.95 Contour plot of initial crack index in AASHTO specification

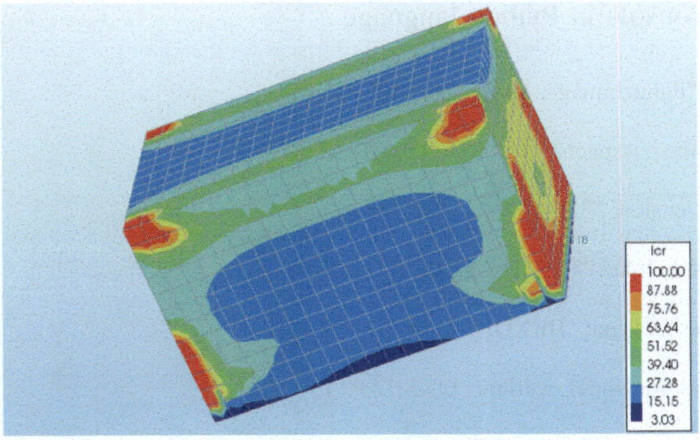

Fig. 4.96 Contour plot of crack index after 2 days in AASHTO specification

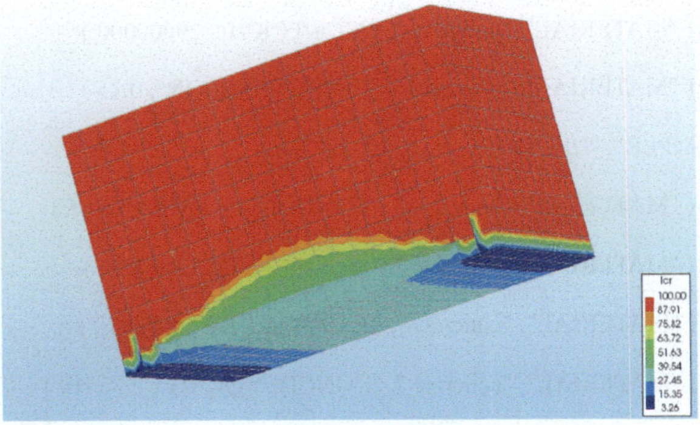

Fig. 4.97 Contour plot of crack index after 10 days in AASHTO specification

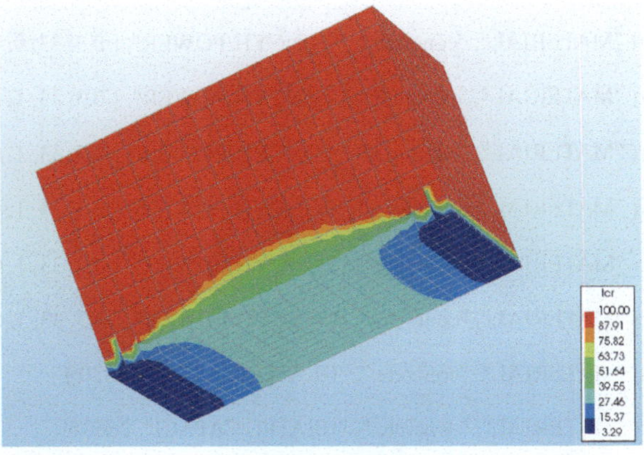

Fig. 4.98 Contour plot of crack index after 35 days in AASHTO specification

Command console in Python language

```
newProject( "Fangzhuang", 10 )
setModelAnalysisAspects( [ "STRUCT", "HEATFL" ] )
setModelDimension( "3D" )
setDefaultMeshOrder( "QUADRATIC" )
setDefaultMesherType( "HEXQUAD" )
setDefaultMidSideNodeLocation( "ONSHAP" )
createBlock( "concrete", [ 0, 0, 0 ], [ 1, 0.6, 0.6 ] )
addMaterial( "concrete ", "CONCDC", "JSCE", [ "CRKIDX", "HEATFL", "YOUNGH" ] )
setParameter( "MATERIAL", "concrete ", "JSCE/YOUN91", 2.7e+10 )
setParameter( "MATERIAL", "concrete ", "JSCE/FCK91", 29000000 )
setParameter( "MATERIAL", "concrete ", "CONCDI/POISON", 0.15 )
setUnit( "TEMPER", "CELSIU" )
setParameter( "MATERIAL", "concrete ", "CONCDI/THERMX", 1e-05 )
setParameter( "MATERIAL", "concrete ", "JSCE/FCK91", 32400000 )
setParameter( "MATERIAL", "concrete ", "JSCE/YOUN91", 2.7e+10 )
setParameter( "MATERIAL", "concrete ", "CONCDI/YOUNG", 2.7e+10 )
setUnit( "TIME", "DAY" )
setUnit( "FORCE", "N" )
setParameter( "MATERIAL", "concrete ", "CONCDI/DENSIT", 3.34898e-07 )
setParameter( "MATERIAL", "concrete ", "CONCYH/POWER", [ 0.3, 33, 0, 0 ] )
setParameter( "MATERIAL", "concrete ", "CONCYH/POWER", [ 0.3, 33, 1, 0 ] )
setParameter( "MATERIAL", "concrete ", "CONCYH/POWER", [ 0.3, 33, 1, 0 ] )
setParameter( "MATERIAL", "concrete ", "CONCYH/POWER", [ 0.3, 33, 1, 0.3 ] )
setParameter( "MATERIAL", "concrete ", "CONCYH/POWER", [ 0.3, 33, 1, 0.3 ] )
setParameter( "MATERIAL", "concrete ", "CONCYH/POWER", [ 0.3, 33, 1, 0.3 ] )
setParameter( "MATERIAL", "concrete ", "HEATFL/CONDUC", 320 )
setParameter( "MATERIAL", "concrete ", "HEATFL/CAPACI", 2660 )
setParameter( "MATERIAL", "concrete ", "HEATFL/HEATHY/HYDRAT", "PREPRO" )
```

```
setParameter( "MATERIAL", "concrete ", "HEATFL/HEATHY/ADIAB", [] )
setParameter( "MATERIAL", "concrete ", "HEATFL/HEATHY/ADIAB", [ 0, 0, 0.5, 30, 0.8,
43, 1, 50.5, 15, 58.7, 30, 61.7, 45, 63 ] )
clearReinforcementAspects( [ "concrete" ] )
setElementClassType( "SHAPE", [ "concrete" ], "STRSOL" )
assignMaterial( "concrete ", "SHAPE", [ "concrete" ] )
resetGeometry( "SHAPE", [ "concrete" ] )
resetElementData( "SHAPE", [ "concrete" ] )
saveProject( )
addMaterial( "boundary", "INTERF", "FLBOUN", [] )
setParameter( "MATERIAL", "boundary", "HTBOUN/CONPAR/CONVEC", 700 )
setParameter( "MATERIAL", "boundary", "HTBOUN/CONPAR/CONVEC", 700 )
setParameter( "MATERIAL", "boundary", "HTBOUN/CONPAR/CVTYPE", "TIMDEP" )
setParameter( "MATERIAL", "boundary", "HTBOUN/CONPAR/TIMDEP/TIMCNV", [] )
setParameter( "MATERIAL", "boundary", "HTBOUN/CONPAR/TIMDEP/TIMCNV", [ 0,
700, 14, 700, 14.1, 2000, 28, 2000, 30, 2000, 60, 2000 ] )
createSurfaceConnection( "boundary" )
setParameter( "GEOMETRYCONNECTION", "boundary", "CONTYP", "BOUNDA" )
attachTo( "GEOMETRYCONNECTION", "boundary", "SOURCE", "concrete", [[ 0.573573,
0.3441438, 0.6 ],[ 0.573573, 0.6, 0.3441438 ],[ 0.573573, 0, 0.2558562 ]] )
setElementClassType( "GEOMETRYCONNECTION", "boundary", "HEABOU" )
assignMaterial( "boundary", "GEOMETRYCONNECTION", "boundary" )
resetGeometry( "GEOMETRYCONNECTION", "boundary" )
resetElementData( "GEOMETRYCONNECTION", "boundary" )
saveProject( )
addSet( "GEOMETRYBCSET", "surrounding" )
createSurfaceBoundaryCondition( "THERMAL", "surrounding", "surrounding" )
setParameter( "GEOMETRYBC", "surrounding", "BOUTYP", "EXTEMP" )
```

```
setParameter( "GEOMETRYBC", "surrounding", "EXTEMP/VALUE", 35 )
attach( "GEOMETRYBC", "surrounding", "concrete", [[ 0.573573, 0,
0.2558562 ],[ 0.573573, 0.3441438, 0.6 ],[ 0.573573, 0.6, 0.3441438 ]] )
setTimeDependentLoadFactors( "GEOMETRYBCSET", "surrounding", [ 0, 15,45 ], [ 1, 1,1 ])
createBodyInitialField( "initial" )
setParameter( "GEOMETRYINIFIELD", "initial", "INITYP", "TEMPER" )
setParameter( "GEOMETRYINIFIELD", "initial", "TEMPER/VALUE", 25 )
attach( "GEOMETRYINIFIELD", "initial", [ "concrete" ] )
saveProject( )
addSet( "GEOMETRYLOADSET", "gravity" )
createModelLoad( "gravity", "gravity" )
setTimeDependentLoadFactors( "GEOMETRYLOADSET", "gravity", [ 0, 15,45 ], [ 1, 1,1 ] )
saveProject( )
addSet( "GEOMETRYSUPPORTSET", "co1" )
createLineSupport( "co1", "co1" )
setParameter( "GEOMETRYSUPPORT", "co1", "AXES", [ 1, 2 ] )
setParameter( "GEOMETRYSUPPORT", "co1", "TRANSL", [ 1, 0, 1 ] )
setParameter( "GEOMETRYSUPPORT", "co1", "ROTATI", [ 0, 0, 0 ] )
attach( "GEOMETRYSUPPORT", "co1", "concrete", [[ 0, 0.3, 0 ]] )
saveProject( )
addSet( "GEOMETRYSUPPORTSET", "Geometry support set 2" )
rename( "GEOMETRYSUPPORTSET", "Geometry support set 2", "co2" )
createLineSupport( "co2", "co2" )
setParameter( "GEOMETRYSUPPORT", "co2", "AXES", [ 1, 2 ] )
setParameter( "GEOMETRYSUPPORT", "co2", "TRANSL", [ 0, 0, 1 ] )
setParameter( "GEOMETRYSUPPORT", "co2", "ROTATI", [ 0, 0, 0 ] )
attach( "GEOMETRYSUPPORT", "co2", "concrete", [[ 1, 0.3, 0 ]] )
saveProject( )
```

```
setElementSize( [ "concrete" ], 0.05,-1, True )
setMesherType( [ "concrete" ], "HEXQUAD" )
setMidSideNodeLocation( [ "concrete" ], "LINEAR" )
saveProject( )
generateMesh( [] )
hideView( "GEOM" )
showView( "MESH" )
addAnalysis( "Analysis1" )
addAnalysisCommand( "Analysis1", "HEATTR", "Transient heat transfer" )
addAnalysisCommandDetail( "Analysis1", "Transient heat transfer", "INITIA/TEMPER" )
setAnalysisCommandDetail( "Analysis1", "Transient heat transfer", "INITIA/TEMPER",
True )
setAnalysisCommandDetail( "Analysis1", "Transient heat transfer", "INITIA/ANATYP",
"NONLIN" )
addAnalysisCommandDetail( "Analysis1", "Transient heat transfer",
"INITIA/NONLIN/HYDRAT" )
setAnalysisCommandDetail( "Analysis1", "Transient heat transfer",
"INITIA/NONLIN/HYDRAT", True )
addAnalysisCommandDetail( "Analysis1", "Transient heat transfer",
"INITIA/NONLIN/EQUAGE" )
setAnalysisCommandDetail( "Analysis1", "Transient heat transfer",
"INITIA/NONLIN/EQUAGE", True )
setAnalysisCommandDetail( "Analysis1", "Transient heat transfer", "EXECUT/SIZES",
"0.500000(20) 1.00000(5) 10.0000 10.0000" )
addAnalysisCommand( "Analysis1", "NONLIN", "Structural nonlinear" )
renameAnalysisCommandDetail( "Analysis1", "Structural nonlinear", "EXECUT(1)",
"gravity" )
```

```
renameAnalysisCommandDetail( "Analysis1", "Structural nonlinear", "EXECUT(1)",
"gravity" )
removeAnalysisCommandDetail( "Analysis1", "Structural nonlinear", "EXECUT(1)" )
setAnalysisCommandDetail( "Analysis1", "Structural nonlinear", "EXECUT/EXETYP",
"START" )
renameAnalysisCommandDetail( "Analysis1", "Structural nonlinear", "EXECUT(1)",
"gravity" )
addAnalysisCommandDetail( "Analysis1", "Structural nonlinear",
"EXECUT(1)/START/LOAD/ADD" )
setAnalysisCommandDetail( "Analysis1", "Structural nonlinear",
"EXECUT(1)/START/LOAD/ADD", True )
setAnalysisCommandDetail( "Analysis1", "Structural nonlinear",
"EXECUT(1)/ITERAT/MAXITE", 20 )
saveProject( )
setAnalysisCommandDetail( "Analysis1", "Structural nonlinear", "EXECUT/EXETYP",
"TIME" )
setAnalysisCommandDetail( "Analysis1", "Structural nonlinear",
"EXECUT(2)/TIME/STEPS/EXPLIC/SIZES", "0.500000(20) 1.00000(5) 10.0000(2)" )
setAnalysisCommandDetail( "Analysis1", "Structural nonlinear",
"EXECUT(2)/ITERAT/MAXITE", 50 )
saveProject( )
setAnalysisCommandDetail( "Analysis1", "Structural nonlinear", "OUTPUT(1)/SELTYP",
"USER" )
addAnalysisCommandDetail( "Analysis1", "Structural nonlinear", "OUTPUT(1)/USER" )
addAnalysisCommandDetail( "Analysis1", "Structural nonlinear",
"OUTPUT(1)/USER/DISPLA(1)/TOTAL/TRANSL/GLOBAL" )
addAnalysisCommandDetail( "Analysis1", "Structural nonlinear",
"OUTPUT(1)/USER/STRESS(1)/TOTAL/CAUCHY/GLOBAL" )
```

```
addAnalysisCommandDetail( "Analysis1", "Structural nonlinear",
"OUTPUT(1)/USER/STRESS(2)/TOTAL/CAUCHY/LOCAL" )
addAnalysisCommandDetail( "Analysis1", "Structural nonlinear",
"OUTPUT(1)/USER/STRESS(3)/TOTAL/CAUCHY/PRINCI" )
addAnalysisCommandDetail( "Analysis1 ", "Structural nonlinear",
"OUTPUT(1)/USER/STRESS(4)/TOTAL/CAUCHY/CRKIND" )
setAnalysisCommandDetail( "Analysis1", "Structural nonlinear",
"EXECUT(2)/TIME/STEPS/EXPLIC/SIZES", "0.500000(20) 1.00000(5) 10.0000 10.0000" )
saveProject( )
setAnalysisCommandDetail( "Analysis1", "Structural nonlinear", "OUTPUT(1)/SELTYP",
"USER" )
addAnalysisCommandDetail( "Analysis1", "Structural nonlinear", "OUTPUT(1)/USER" )
addAnalysisCommandDetail( "Analysis1", "Structural nonlinear",
"OUTPUT(1)/USER/DISPLA(1)/TOTAL/TRANSL/GLOBAL" )
addAnalysisCommandDetail( "Analysis1", "Structural nonlinear",
"OUTPUT(1)/USER/STRESS(1)/TOTAL/CAUCHY/GLOBAL" )
addAnalysisCommandDetail( "Analysis1", "Structural nonlinear",
"OUTPUT(1)/USER/STRESS(2)/TOTAL/CAUCHY/LOCAL" )
addAnalysisCommandDetail( "Analysis1", "Structural nonlinear",
"OUTPUT(1)/USER/STRESS(3)/TOTAL/CAUCHY/PRINCI" )
addAnalysisCommandDetail( "Analysis1", "Structural nonlinear",
"OUTPUT(1)/USER/STRESS(4)/TOTAL/CAUCHY/CRKIND" )
runSolver( "Analysis1" )
showView( "RESULT" )
setResultCase( [ "Analysis1", "Output", "Time-step 28, Time 35.000" ] )
setResultPlot( "contours", "Total Displacements/node", "TDtZ" )
setResultPlot( "contours", "Total Displacements/node", "TDtY" )
setResultPlot( "contours", "Total Displacements/node", "TDtX" )
```

```
setResultPlot( "contours", "Crack Indices/node", "Icr" )
setResultCase( [ "Analysis1", "Output", "Time-step 27, Time 25.000" ] )
setResultCase( [ "Analysis1", "Output", "Time-step 26, Time 15.000" ] )
setResultCase( [ "Analysis1", "Output", "Time-step 25, Time 14.000" ] )
setResultCase( [ "Analysis1", "Output", "Time-step 24, Time 13.000" ] )
setResultCase( [ "Analysis1", "Output", "Time-step 23, Time 12.000" ] )
setResultCase( [ "Analysis1", "Output", "Time-step 22, Time 11.000" ] )
setResultCase( [ "Analysis1", "Output", "Time-step 21, Time 10.000" ] )
setResultCase( [ "Analysis1", "Output", "Time-step 20, Time 9.5000" ] )
setResultCase( [ "Analysis1", "Output", "Time-step 19, Time 9.0000" ] )
setResultCase( [ "Analysis1", "Output", "Time-step 18, Time 8.5000" ] )
setResultCase( [ "Analysis1", "Output", "Time-step 17, Time 8.0000" ] )
setResultCase( [ "Analysis1", "Output", "Time-step 16, Time 7.5000" ] )
setResultCase( [ "Analysis1", "Output", "Time-step 15, Time 7.0000" ] )
setResultCase( [ "Analysis1", "Output", "Time-step 14, Time 6.5000" ] )
setResultCase( [ "Analysis1", "Output", "Time-step 13, Time 6.0000" ] )
setResultCase( [ "Analysis1", "Output", "Time-step 12, Time 5.5000" ] )
setResultCase( [ "Analysis1", "Output", "Time-step 11, Time 5.0000" ] )
setResultCase( [ "Analysis1", "Output", "Time-step 10, Time 4.5000" ] )
setResultCase( [ "Analysis1", "Output", "Time-step 9, Time 4.0000" ] )
setResultCase( [ "Analysis1", "Output", "Time-step 8, Time 3.5000" ] )
setResultCase( [ "Analysis1", "Output", "Time-step 7, Time 3.0000" ] )
setResultCase( [ "Analysis1", "Output", "Time-step 6, Time 2.5000" ] )
setResultCase( [ "Analysis1", "Output", "Time-step 5, Time 2.0000" ] )
setResultCase( [ "Analysis1", "Output", "Time-step 4, Time 1.5000" ] )
setResultCase( [ "Analysis1", "Output", "Time-step 3, Time 1.0000" ] )
setResultCase( [ "Analysis1", "Output", "Time-step 2, Time 0.50000" ] )
setResultCase( [ "Analysis1", "Output", "Start-step 1, Load-factor 1.0000" ] )
```

```
setResultCase( [ "Analysis1", "Output", "Time-step 2, Time 0.50000" ] )
setResultCase( [ "Analysis1", "Output", "Time-step 3, Time 1.0000" ] )
setResultCase( [ "Analysis1", "Output", "Time-step 4, Time 1.5000" ] )
setResultCase( [ "Analysis1", "Output", "Time-step 5, Time 2.0000" ] )
setResultCase( [ "Analysis1", "Output", "Time-step 6, Time 2.5000" ] )
setResultCase( [ "Analysis1", "Output", "Time-step 7, Time 3.0000" ] )
setResultCase( [ "Analysis1", "Output", "Time-step 8, Time 3.5000" ] )
setResultCase( [ "Analysis1", "Output", "Time-step 9, Time 4.0000" ] )
setResultCase( [ "Analysis1", "Output", "Time-step 10, Time 4.5000" ] )
setResultCase( [ "Analysis1", "Output", "Time-step 11, Time 5.0000" ] )
setResultCase( [ "Analysis1", "Output", "Time-step 12, Time 5.5000" ] )
setResultCase( [ "Analysis1", "Output", "Time-step 13, Time 6.0000" ] )
setResultCase( [ "Analysis1", "Output", "Time-step 14, Time 6.5000" ] )
setResultCase( [ "Analysis1", "Output", "Time-step 15, Time 7.0000" ] )
setResultCase( [ "Analysis1", "Output", "Time-step 16, Time 7.5000" ] )
setResultCase( [ "Analysis1", "Output", "Time-step 17, Time 8.0000" ] )
setResultCase( [ "Analysis1", "Output", "Time-step 18, Time 8.5000" ] )
setResultCase( [ "Analysis1", "Output", "Time-step 19, Time 9.0000" ] )
setResultCase( [ "Analysis1", "Output", "Time-step 20, Time 9.5000" ] )
setResultCase( [ "Analysis1", "Output", "Time-step 21, Time 10.000" ] )
setResultCase( [ "Analysis1", "Output", "Time-step 22, Time 11.000" ] )
setResultCase( [ "Analysis1", "Output", "Time-step 23, Time 12.000" ] )
setResultCase( [ "Analysis1", "Output", "Time-step 24, Time 13.000" ] )
setResultCase( [ "Analysis1", "Output", "Time-step 25, Time 14.000" ] )
setResultCase( [ "Analysis1", "Output", "Time-step 26, Time 15.000" ] )
setResultCase( [ "Analysis1", "Output", "Time-step 27, Time 25.000" ] )
setResultCase( [ "Analysis1", "Output", "Time-step 28, Time 35.000" ] )
```

The command console in the Python language of hydration heat reaction in European specification of CEB-FIP 1990 (**Note:** the following command console only lists the Python language that uses the CEB-FIP 1990 specification to replace the constitutive model defined in the JSCE specification, and the rest parts are identical, so the user can replace them by himself)

```
addMaterial( "concrete", "CONCDC", "MC1990", [ "CRKIDX", "HEATFL", "YOUNGH" ] )
setParameter( "MATERIAL", "concrete", "MC90CO/GRADE", "C50" )
setParameter( "MATERIAL", "concrete", "MC90CO/RH", 69 )
setParameter( "MATERIAL", "concrete", "MC90CO/RH", 69 )
setParameter( "MATERIAL", "concrete", "CONCDI/YOUNG", 3.8926e+10 )
setParameter( "MATERIAL", "concrete", "CONCDI/YOUN28", 3.45e+10 )
setParameter( "MATERIAL", "concrete", "CONCDI/POISON", 0.15 )
setParameter( "MATERIAL", "concrete", "CONCDI/THERMX", 1e-05 )
setParameter( "MATERIAL", "concrete", "CONCDI/DENSIT", 2500 )
setParameter( "MATERIAL", "concrete", "CONCDI/FCK28", 50000000 )
setParameter( "MATERIAL", "concrete", "CONCDI/FCM28", 58000000 )
setParameter( "MATERIAL", "concrete", "CONCDI/DENSIT", 3.34898e-07 )
setParameter( "MATERIAL", "concrete", "CONCYH/POWER", [ 0.3, 33, 1, 0.3 ] )
setParameter( "MATERIAL", "concrete", "HEATFL/CONDUC", 320 )
setParameter( "MATERIAL", "concrete", "HEATFL/CAPACI", 2660 )
setParameter( "MATERIAL", "concrete", "HEATFL/CNDTYP", "TIMDEP" )
setParameter( "MATERIAL", "concrete", "HEATFL/CNDTYP", "NONE" )
setParameter( "MATERIAL", "concrete", "HEATFL/CNDTYP", "NONE" )
setParameter( "MATERIAL", "concrete", "HEATFL/HEATHY/HYDRAT", "PREPRO" )
setParameter( "MATERIAL", "concrete", "HEATFL/HEATHY/ADIAB", [] )
setParameter( "MATERIAL", "concrete", "HEATFL/HEATHY/ADIAB", [ 0, 0, 0.5, 30, 0.8,
43, 1, 50.5, 15, 58.7, 30, 61.7, 45, 63 ] )
saveProject( )
clearReinforcementAspects( [ "concrete" ] )
setElementClassType( "SHAPE", [ "concrete" ], "STRSOL" )
assignMaterial( "concrete", "SHAPE", [ "concrete" ] )
resetGeometry( "SHAPE", [ "concrete" ] )
resetElementData( "SHAPE", [ "concrete" ] )
```

```
saveProject( )
generateMesh( [] )
hideView( "GEOM" )
showView( "MESH" )
runSolver( "Analysis1" )
showView( "RESULT" )
setResultCase( [ "Analysis1", "Output", "Time-step 9, Time 4.0000" ] )
setResultPlot( "contours", "Total Displacements/node", "TDtZ" )
setResultCase( [ "Analysis1", "Output", "Time-step 28, Time 35.000" ] )
setResultCase( [ "Analysis1", "Output", "Time-step 9, Time 4.0000" ] )
setResultCase( [ "Analysis1", "Output", "Time-step 28, Time 35.000" ] )
setParameter( "MATERIAL", "concrete", "MC90CO/CEMTYP", "RS" )
setParameter( "MATERIAL", "concrete", "MC90CO/CEMTYP", "NR" )
setResultPlot( "contours", "Crack Indices/node", "Icr" )
setResultCase( [ "Analysis1", "Output", "Start-step 1, Load-factor 1.0000" ] )
setResultCase( [ "Analysis1", "Output", "Time-step 2, Time 0.50000" ] )
setResultCase( [ "Analysis1", "Output", "Time-step 10, Time 4.5000" ] )
setResultCase( [ "Analysis1", "Output", "Time-step 9, Time 4.0000" ] )
setResultCase( [ "Analysis1", "Output", "Time-step 11, Time 5.0000" ] )
setResultCase( [ "Analysis1", "Output", "Time-step 5, Time 2.0000" ] )
setResultCase( [ "Analysis1", "Output", "Time-step 9, Time 4.0000" ] )
setResultCase( [ "Analysis1", "Output", "Time-step 8, Time 3.5000" ] )
setResultCase( [ "Analysis1", "Output", "Time-step 10, Time 4.5000" ] )
setResultCase( [ "Analysis1", "Output", "Time-step 13, Time 6.0000" ] )
setResultCase( [ "Analysis1", "Output", "Time-step 28, Time 35.000" ] )
setResultCase( [ "Analysis1", "Output", "Time-step 27, Time 25.000" ] )
setResultCase( [ "Analysis1", "Output", "Time-step 26, Time 15.000" ] )
setResultCase( [ "Analysis1", "Output", "Start-step 1, Load-factor 1.0000" ] )
```

```
setResultCase( [ "Analysis1", "Output", "Time-step 2, Time 0.50000" ] )

setResultCase( [ "Analysis1", "Output", "Start-step 1, Load-factor 1.0000" ] )

setResultCase( [ "Analysis1", "Output", "Time-step 2, Time 0.50000" ] )

setResultCase( [ "Analysis1", "Output", "Time-step 9, Time 4.0000" ] )

setResultCase( [ "Analysis1", "Output", "Time-step 26, Time 15.000" ] )

setResultCase( [ "Analysis1", "Output", "Time-step 16, Time 7.5000" ] )

setResultCase( [ "Analysis1", "Output", "Time-step 11, Time 5.0000" ] )
```

The command console in the Python language of hydration heat reaction in European specification of AASHTO (**Note**: the following command console only lists the python language applying the AASHTO specification to replace the constitutive model defined in the JSCE specification, and the rest parts are identical, so the user can replace them by himself)

```
setParameter( "MATERIAL", "CONCRETE", "HEATFL/HEATHY/HYDRAT", "PREPRO" )

setParameter( "MATERIAL", "CONCRETE", "HEATFL/HEATHY/ADIAB", [] )

setParameter( "MATERIAL", "CONCRETE", "HEATFL/HEATHY/ADIAB", [ 0, 0, 0.5, 30,
0.8, 43, 1, 50.5, 15, 58.7, 30, 61.7, 45, 63 ] )

setParameter( "MATERIAL", "CONCRETE", "HEATFL/CONDUC", 320 )

setParameter( "MATERIAL", "CONCRETE", "HEATFL/CONDUC", 320 )

saveProject(  )

setParameter( "MATERIAL", "CONCRETE", "HEATFL/CAPACI", 2660 )

setParameter( "MATERIAL", "CONCRETE", "HEATFL/CAPACI", 2660 )

rename( "MATERIAL", "CONCRETE", "concrete" )

setParameter( "MATERIAL", "concrete", "CONCDI/YOUNG", 0.3 )

removeParameter( "MATERIAL", "concrete", "CONCDI/YOUNG" )

setParameter( "MATERIAL", "concrete", "CONCDI/YOUNG", 3.45e+10 )
setParameter( "MATERIAL", "concrete", "CONCDI/POISON", 0.15 )

setParameter( "MATERIAL", "concrete", "CONCDI/THERMX", 1e-05 )

setParameter( "MATERIAL", "concrete", "CONCDI/DENSIT", 3.34898e-07 )

saveProject(  )

setParameter( "MATERIAL", "concrete", "MCAASH/FT28", 2640000 )
```

```
saveProject( )
setParameter( "MATERIAL", "concrete", "MCAASH/FT28", 2640000 )
generateMesh( [] )
hideView( "GEOM" )
showView( "MESH" )
runSolver( "Analysis1" )
showView( "RESULT" )
setResultPlot( "contours", "Total Displacements/node", "TDtZ" )
setResultCase( [ "Analysis1", "Output", "Time-step 28, Time 35.000" ] )
setResultPlot( "contours", "Crack Indices/node", "Icr" )
setResultCase( [ "Analysis1", "Output", "Time-step 2, Time 0.50000" ] )
setResultCase( [ "Analysis1", "Output", "Start-step 1, Load-factor 1.0000" ] )
setResultCase( [ "Analysis1", "Output", "Time-step 2, Time 0.50000" ] )
setResultPlot( "contours", "Total Displacements/node", "TDtZ" )
setParameter( "MATERIAL", "concrete", "MCAASH/RH", 69 )
saveProject( )
runSolver( "Analysis1" )
showView( "RESULT" )
setResultCase( [ "Analysis1", "Output", "Time-step 28, Time 35.000" ] )
setResultPlot( "contours", "Total Displacements/node", "TDtZ" )
show( "GEOMETRYSUPPORTSET", [ "co1" ] )
show( "GEOMETRYSUPPORTSET", [ "co2" ] )
setResultPlot( "contours", "Crack Indices/node", "Icr" )
setResultCase( [ "Analysis1", "Output", "Time-step 2, Time 0.50000" ] )
showIds( "NODE", [ 814, 118, 2910 ] )
setResultCase( [ "Analysis1", "Output", "Time-step 10, Time 4.5000" ] )
setResultCase( [ "Analysis1", "Output", "Time-step 5, Time 2.0000" ] )
setResultCase( [ "Analysis1", "Output", "Time-step 11, Time 5.0000" ] )
```

```
setResultCase( [ "Analysis1", "Output", "Time-step 21, Time 10.000" ] )
setResultCase( [ "Analysis1", "Output", "Time-step 28, Time 35.000" ] )
setResultCase( [ "Analysis1", "Output", "Time-step 21, Time 10.000" ] )
setResultCase( [ "Analysis1", "Output", "Time-step 28, Time 35.000" ] )
setResultCase( [ "Analysis1", "Output", "Time-step 11, Time 5.0000" ] )
setResultCase( [ "Analysis1", "Output", "Time-step 12, Time 5.5000" ] )
setResultCase( [ "Analysis1", "Output", "Time-step 28, Time 35.000" ] )
setResultCase( [ "Analysis1", "Output", "Time-step 5, Time 2.0000" ] )
```

Chapter 5
DIANA Modeling Cases for Precast Segmental Structures

Abstract Precast segmental structures are widely applied in current engineering owing to their rapid assembling efficiency, excellent quality control, low life-cycle cost and mitigated environmental disturbance. In this chapter, the focus of numerical simulation is put on precast segmental specimens. Targeting at the current emerging structures such as precast segmental bridges, various issues such as direct shear failure, bending failure and long-term deterioration are written as numerical engineering cases. Modeling methods for precast segmental structures in different joint shapes (shear keys as well as corbel joints) are also illustrated in this chapter. Moreover, as one of the unique and typical features for DIANA software, random field concerning forecast of precast segmental girders is specifically displayed in this chapter.

5.1 Direct Shear Failure of Shear Keys in Precast Segmental Concrete Specimens

Two precast segmental concrete specimens, which originated from ABAQUS model [1], are simulated under the effect of direct shear in this case, which is based on the platform of DIANA 10.1 release version. The height of contacted plane between blocks is 0.2 m and the whole height of each block is 0.42 m. Thickness of the specimen is 0.25 m, where the contacted part of shear keys height is 0.1 m. Additionally, both the ledge part and end length of shear keys are 0.05 m, where the thickness of shear keys is 0.25 m. The size of specimens is displayed in Fig. 5.1. Specimens are simulated by quadratic plane stress elements and the element size of mesh is 0.01 m. Hordijk type is selected as tension softening model. Vertical concentrated load and horizon pressure are applied on the specimens. Vertical displacement and crack distribution contours are investigated based on the multi-directional fixed crack model and total strain-based crack model, and load–displacement curve is plotted to illustrate the results of simulation. Concrete and steel parameters are shown in Tables 5.1 and 5.2, respectively.

S. Chai, *Finite Element Analysis for Civil Engineering with DIANA Software*,
https://doi.org/10.1007/978-981-15-2945-0_5

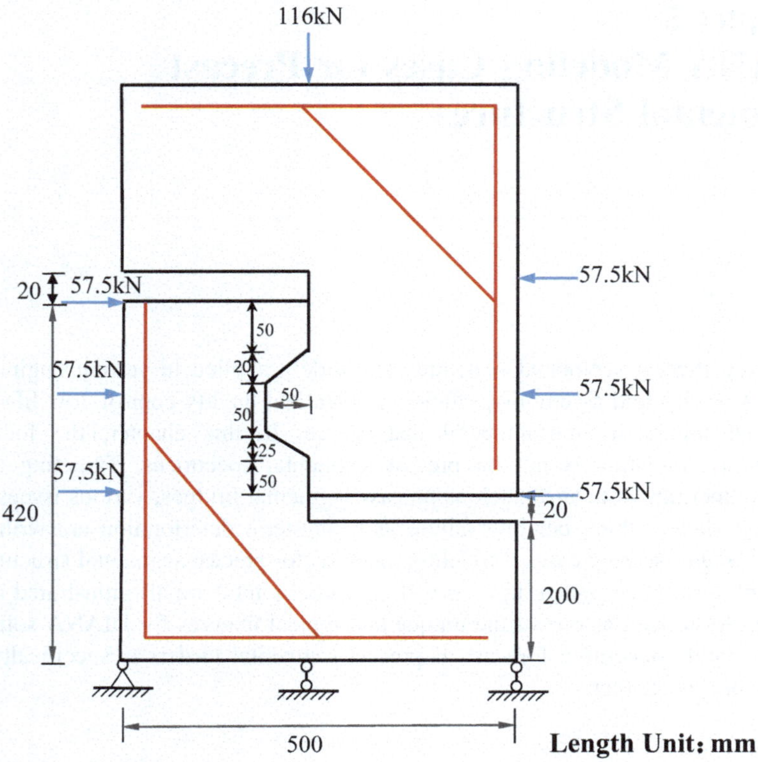

Fig. 5.1 Size of specimens

Table 5.1 Concrete parameters

Elastic modulus	3.45×10^{10} N/m^2
Compressive characteristic strength	3.24×10^7 N/m^2
Tensile characteristic strength	3.7×10^6 N/m^2
Poisson's ratio	0.15
Fracture energy	200 N/m
Mass density	2500 kg/m^3
Cracking model	Total strain-based crack model

Table 5.2 Steel parameters

Yield stress	400 MPa
Elastic modulus	210 GPa
Poisson's ratio	0.33

Above all, starting DianaIE to select 2D modeling plane, the maximum **Model size** is 10 m, and the coordinate points of specimens are displayed in Table 5.3.

Table 5.3 Coordinate points of specimens with shear keys

1	(0, 0)
2	(0.5, 0)
3	(0.5, 0.2)
4	(0.25, 0.2)
5	(0.25, 0.22)
6	(0.25, 0.27)
7	(0.2, 0.295)
8	(0.2, 0.345)
9	(0.25, 0.37)
10	(0.25, 0.42)
11	(0, 0.42)
12	(0, 0.22)
13	(0.5, 0.22)
14	(0.5, 0.62)
15	(0, 0.62)
16	(0, 0.44)
17	(0.25, 0.44)
18	(0.25, 0.42)
19	(0.25, 0.37)
20	(0.2, 0.345)
21	(0.2, 0.295)
22	(0.25, 0.27)

After the coordinate points are input, geometric model of specimens is displayed, as shown in Fig. 5.2.

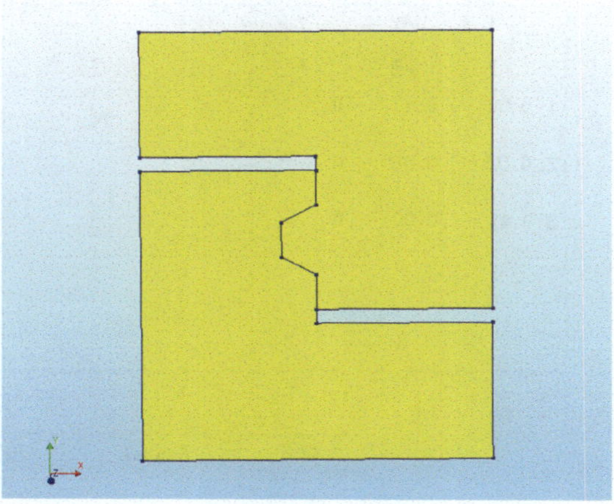

Fig. 5.2 Geometric model of specimens

Clicking shortcut icon as Fig. 5.3 demonstrates **Adds a polyline** to create polyline; untick the **Closed** function to complete the polygonal line part shape of reinforcement steel bars; the lower site of reinforcement steel is named as **bar1**, the coordinate of which is shown in Fig. 5.4. Consider the same method to construct polygonal line part shape of reinforcement steel in the upper specimen with the name of **bar2**, and the coordinate values of which are shown in Fig. 5.5.

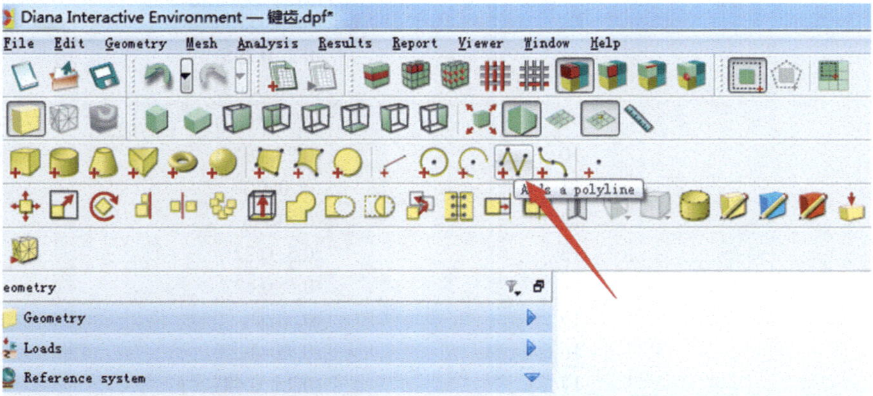

Fig. 5.3 Shortcut icon **Adds a polyline**

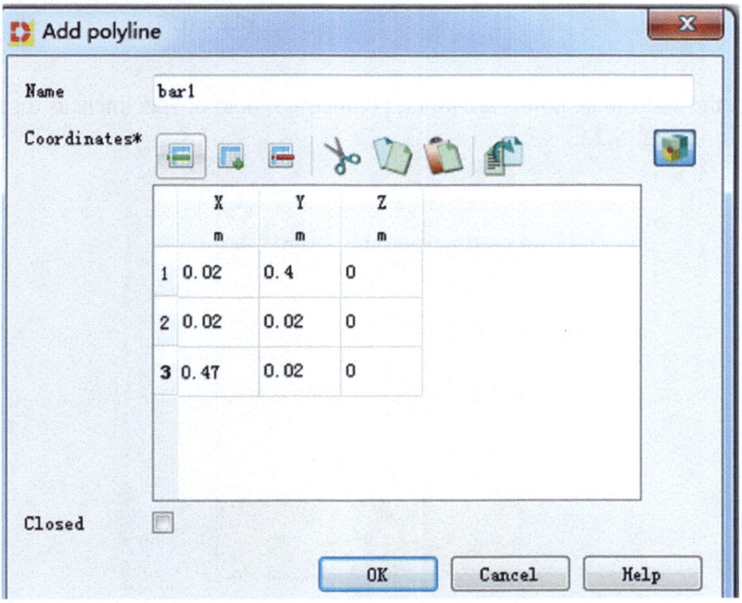

Fig. 5.4 Coordinate values of **bar1**

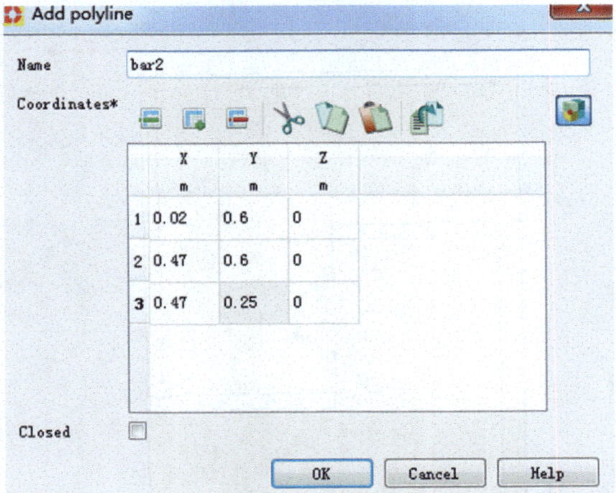

Fig. 5.5 Coordinate values of **bar2**

Construct reinforcement steel bars in line shape with the names **bar3** and **bar4,** respectively; the coordinate values of which are displayed in Figs. 5.6 and 5.7. On clicking OK button, complete geometric model of specimens is displayed (Fig. 5.8).

Fig. 5.6 Coordinate values of **bar3**

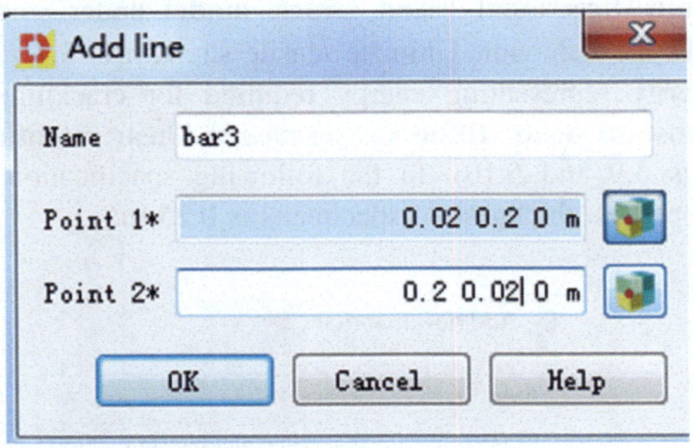

Fig. 5.7 Coordinate values of **bar4**

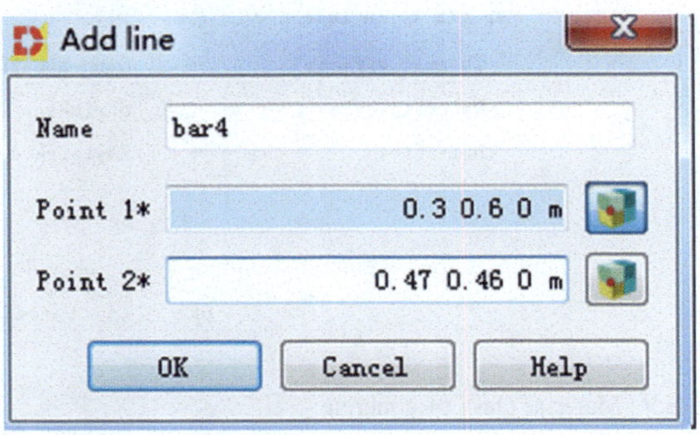

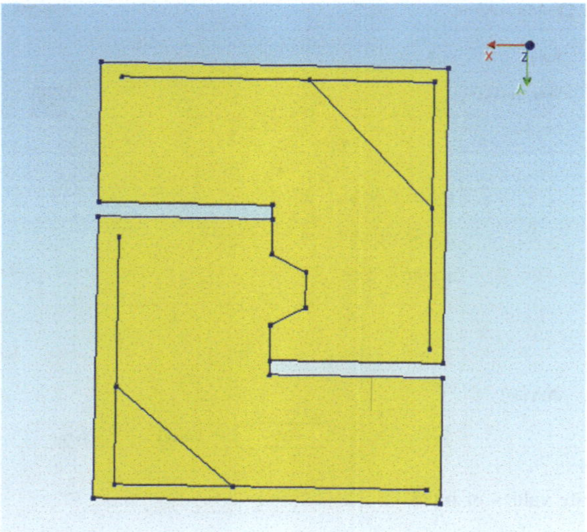

Fig. 5.8 Generated specimens

The following step is followed to assign material properties. Element type of concrete is quadratic plane stress element while **Hordijk softening** curve of **Multi-Directional Fixed Crack model** under smeared cracking is chosen as cracking behavior. Ultimate tensile strength is 3.7×10^6 N/m^2 and the **Fracture energy** representing energy required for cracking at unit width is 200 N/m. **Constant shear stiffness** is selected as **Shear retention** with the factor β 0.01 (see Figs. 5.9 and 5.10). In the following specification for cross-section geometric properties, thickness of specimens is 0.25 m.

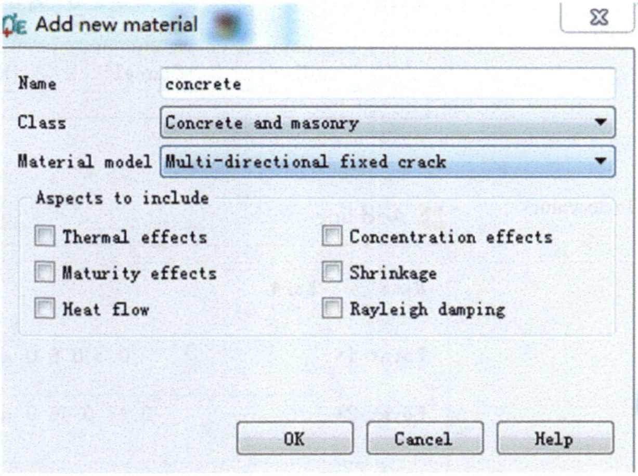

Fig. 5.9 Material class of concrete

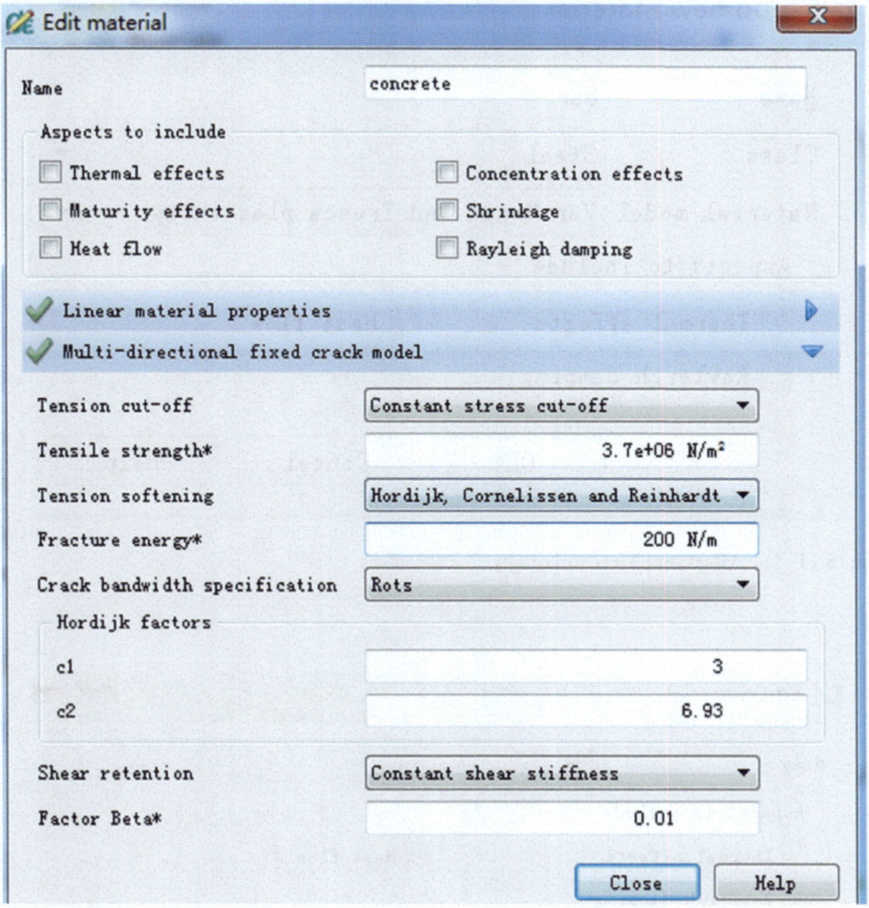

Fig. 5.10 Material specifications of concrete

Selecting all the reinforcement set (**bar1** to **bar4**) under the **Geometry** directory tree, right-clicking to select reinforcement material properties, **Steel** is selected as material **Class** and **von Mises and Tresca plasticity** is selected as **Material model. Elastic modulus** is 2.1×10^{11} N/m^2 with Poisson's ratio and mass density 0.33 and 7800 kg/m^3, respectively. **von Mises plasticity** is selected as **Plasticity model** while the **Hardening function** is **No hardening** with the **Yield stress** 4×10^8 N/m^2 (see Figs. 5.11, 5.12, 5.13).

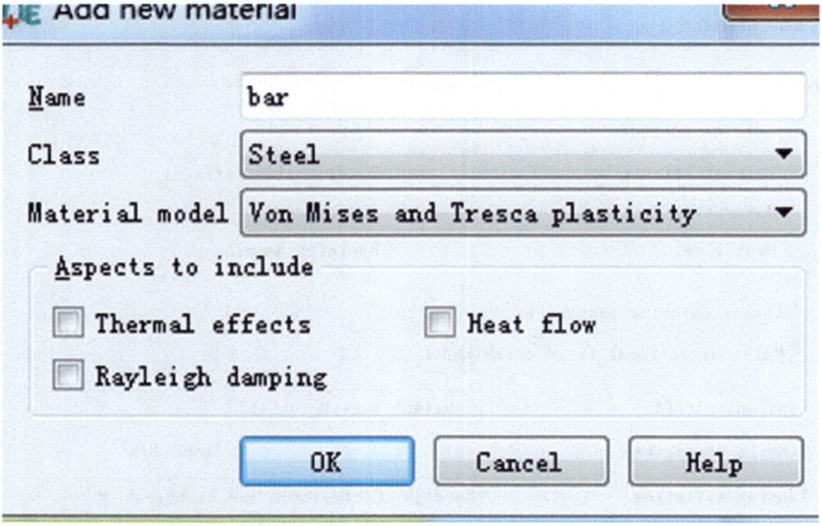

Fig. 5.11 von Mises and Tresca plasticity

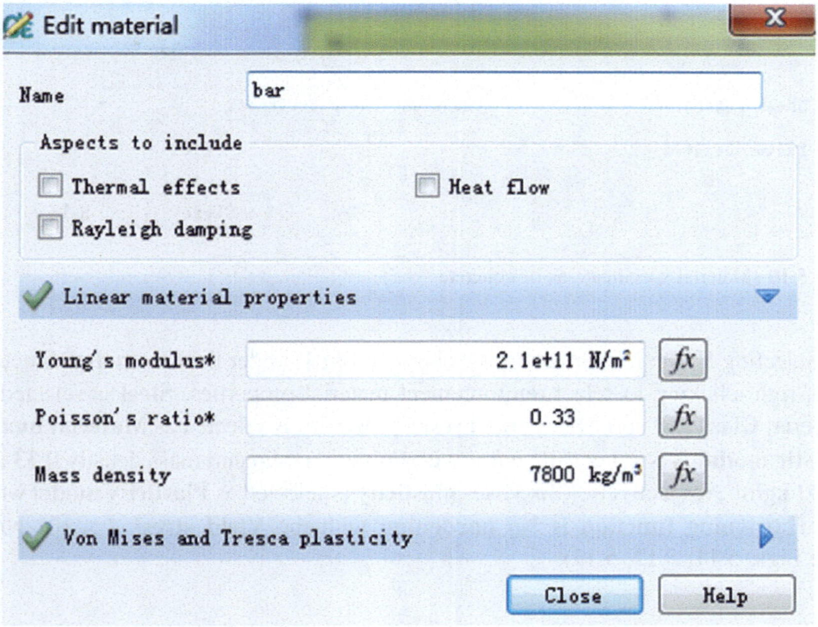

Fig. 5.12 Parameters of steel

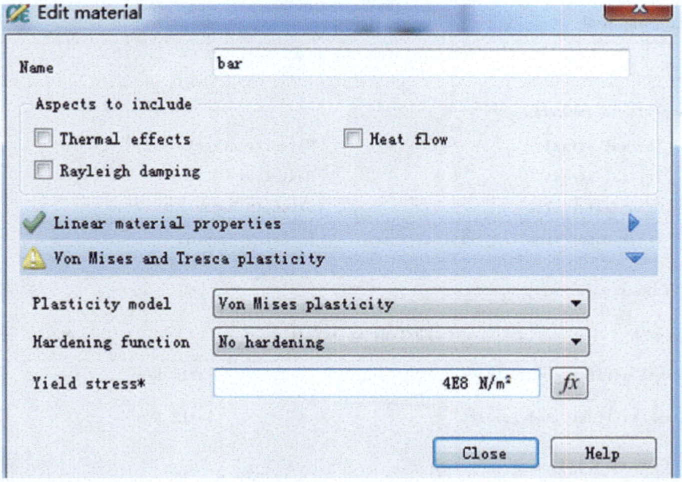

Fig. 5.13 Parameters under **von Mises plasticity** model

After the specification of material properties is accomplished, cross-section geometric properties are specified. Owing to the type of reinforcement steel is $2\phi 12$, cross-section area of reinforcement steel is 226 mm^2 after conversion (see Fig. 5.14).

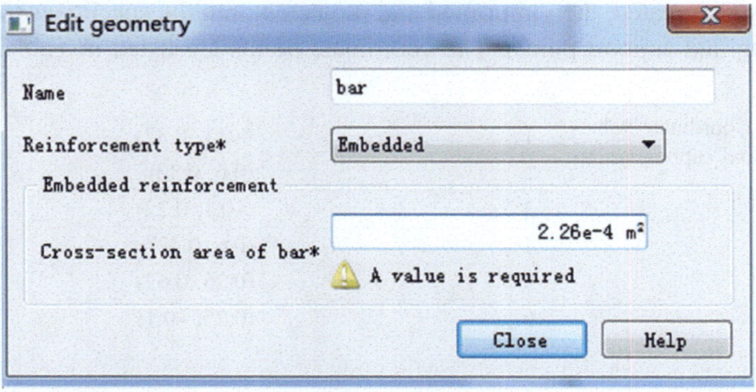

Fig. 5.14 Specification of cross-section area of steel

Adding structural interface element, see Fig. 5.15, the name of interface element is **int**, and 2D line to line connected interface element is chosen as element types. Considering the effects of relative dislocation and friction between shear keys, the resulting tangential friction in material constitutive model of interface element is **Coulomb friction** with normal and shear stiffness 1×10^{16} N/m^3 and 1×10^{12} N/m^3, respectively (see Fig. 5.15). In the **Coulomb friction** aspect, **Friction angle is 20°** and thickness of interface element in cross-section geometric properties is also 0.25 m.

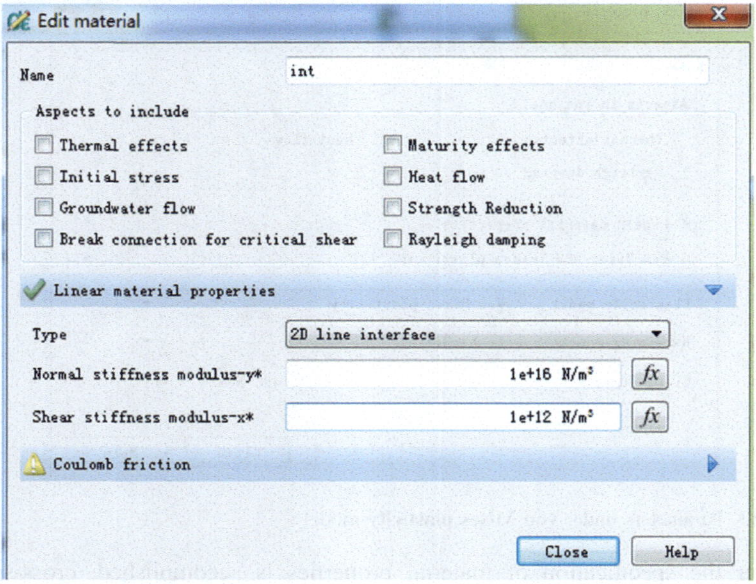

Fig. 5.15 Specification of normal and shear stiffness

In order to better attach load and for the success of meshing, vertexes outside of specimens are created, then imprinted and projected onto the specimens to generate the loading and support points. The coordinate points are listed in Table 5.4.

Table 5.4 Coordinate values of loading and support points

1	(–0.1, 0.32)
2	(0.6, 0.32)
3	(–0.1, 0.22)
4	(0.6, 0.42)
5	(0.25, 0.62)
6	(0.25, –0.1)

After geometric values are given as input, vertexes are imprinted and projected onto the specimens via the function of **Imprint projection**. It is required to note that there are projections along both the X and Y directions and the direction is either positive or negative; therefore, directions of **Imprint projection** should be in caution. Taking vertex 2 and vertex 4 for example, the whole procedure of **Imprint projection** in negative X direction is displayed in Figs. 5.16, 5.17 and 5.18. Other manipulations are completed with the same method.

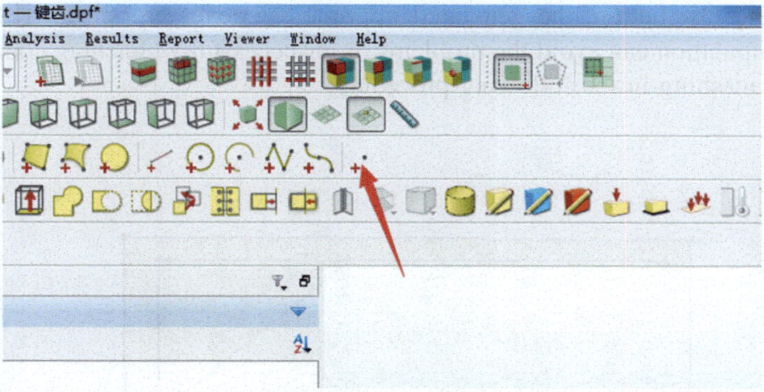

Fig. 5.16 Creating vertexes

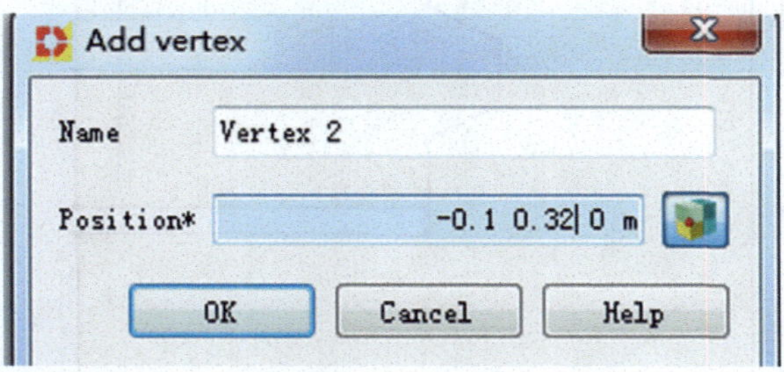

Fig. 5.17 Coordinate value of vertex 2

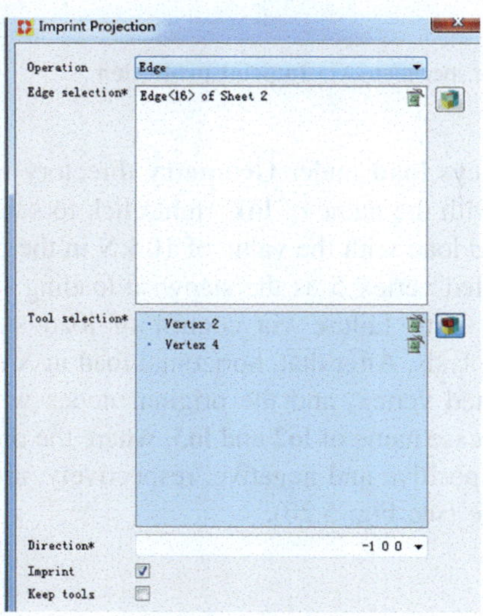

Fig. 5.18 Imprint projection of vertex 2 and vertex 4

Clicking OK button, the generated imprinted vertexes are shown (Fig. 5.19). This manipulation can avoid the problems of unsuccessfully attaching concentrated load and meshing in the following procedure.

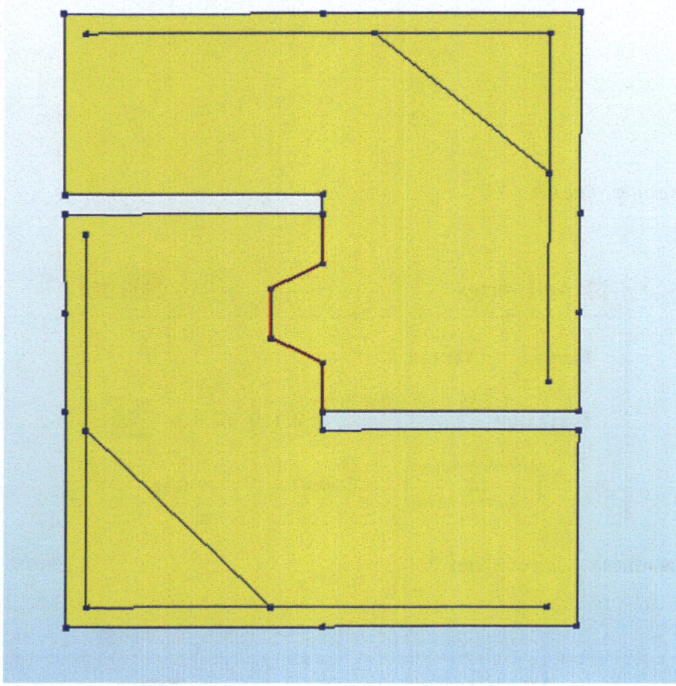

Fig. 5.19 Vertexes on the specimens via **Imprint projection**

Clicking shortcut keys **load** under **Geometry** directory tree creates **Geometry load combination 1** with the name of **lo1**; right-click to select function of **Attach load**, then concentrated load with the value of 10 kN in the negative Y direction is attached to the imprinted vertex 5 as the stepwise loading reference value, where specimens are loaded until failure via control of load steps, in the following structural nonlinear analysis. After that, horizontal load in X direction is attached to the rest of the imprinted vertex, and the original nodes with the common value 57.5 kN with the load case name of **lo2** and **lo3**, where the directions of load on the left and right side are positive and negative, respectively, are used to simulate the effect of prestress force (see Fig. 5.20).

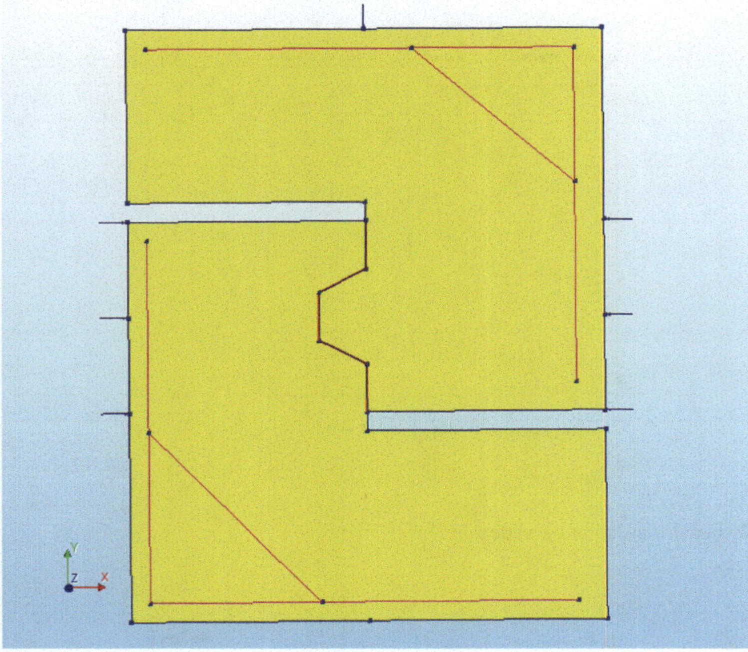

Fig. 5.20 Load attached on the specimens

The following step is the procedure to create geometry load combinations: horizontal **lo2** and **lo3** are set as initial **Geometry load combination 1** and the vertical concentrated load **lo1** is set as **Geometry load combination 2** (see Fig. 5.21).

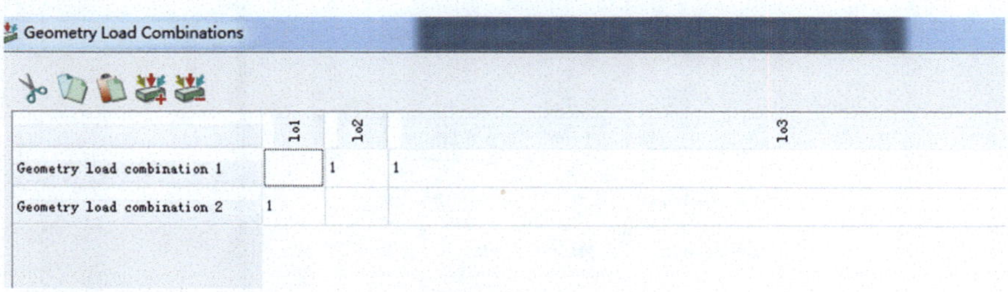

Fig. 5.21 Creating geometry load combinations

Then we attach supports. Fixed translation constraints in X and Y direction are attached to the vertex 11, which is located at the middle site of the edge belonging to the lower specimen. For the vertex 70 and vertex 15 located at the left and right site of the lower specimen, fixed translation constraints merely in Y direction are attached to them (see Figs. 5.22 and 5.23, respectively).

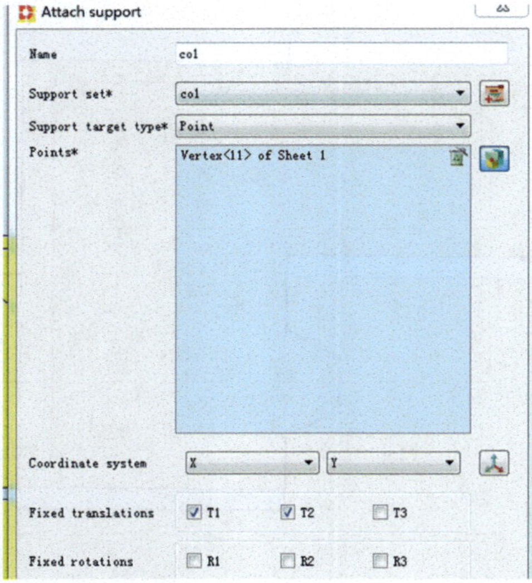

Fig. 5.22 Fixed translation on vertex 11

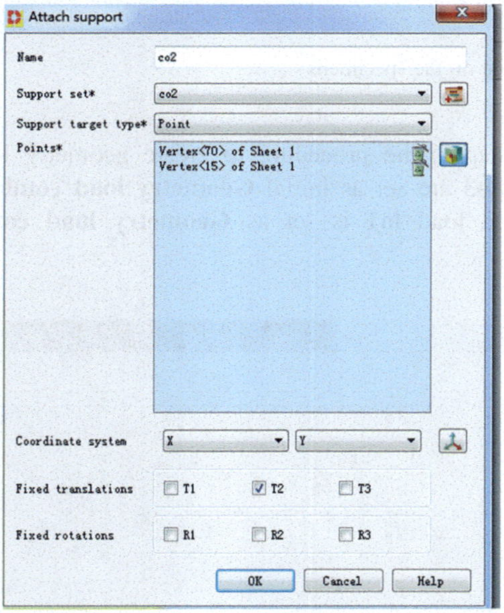

Fig. 5.23 Fixed translation on vertex 70 and vertex 15

Constraints after attachment are displayed in Fig. 5.24.

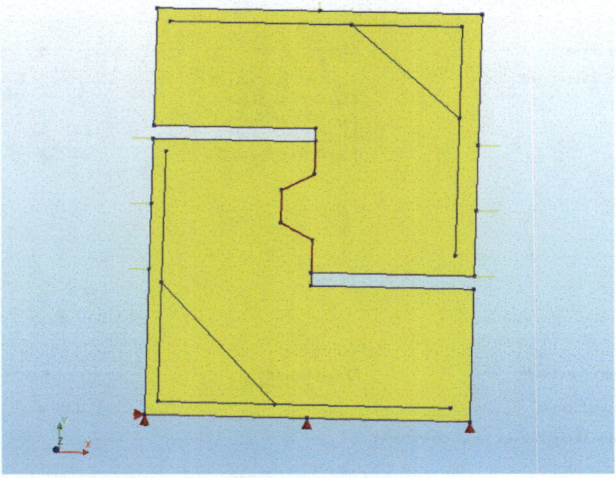

Fig. 5.24 Constraints after attachment

Selecting all the specimens, **Operation** is **Shape** while **Element size** is selected as **Seeding method** with the **Desired size** 0.01 m. **Mesher type** is **Hexa/Quad** and the way of determining **Mid-side node location** is **Linear interpolation** (see Figs. 5.25 and 5.26).

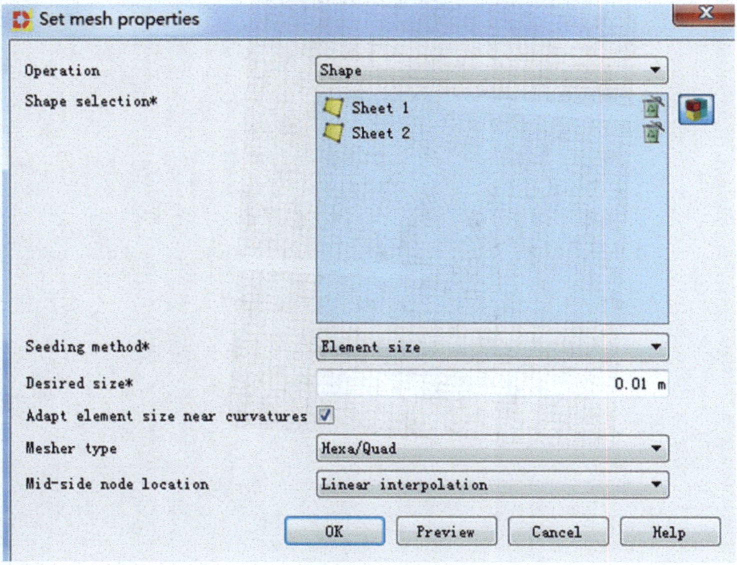

Fig. 5.25 Meshing interface

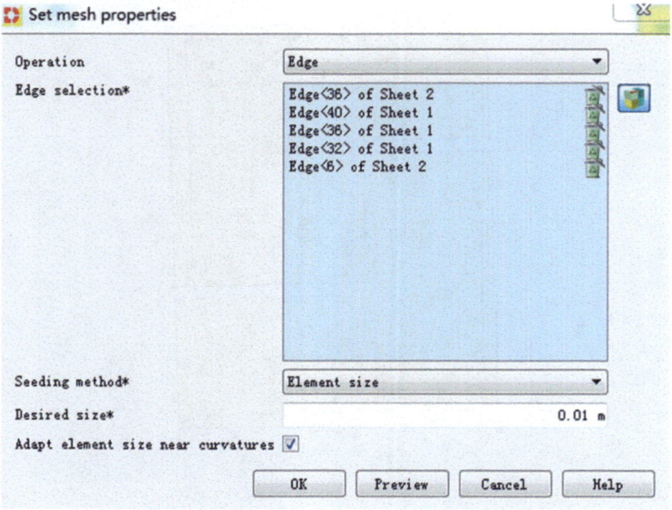

Fig. 5.26 Setting mesh properties

On clicking shortcut icon **Generate mesh of a shape**, the generated meshed elements are displayed (Fig. 5.27).

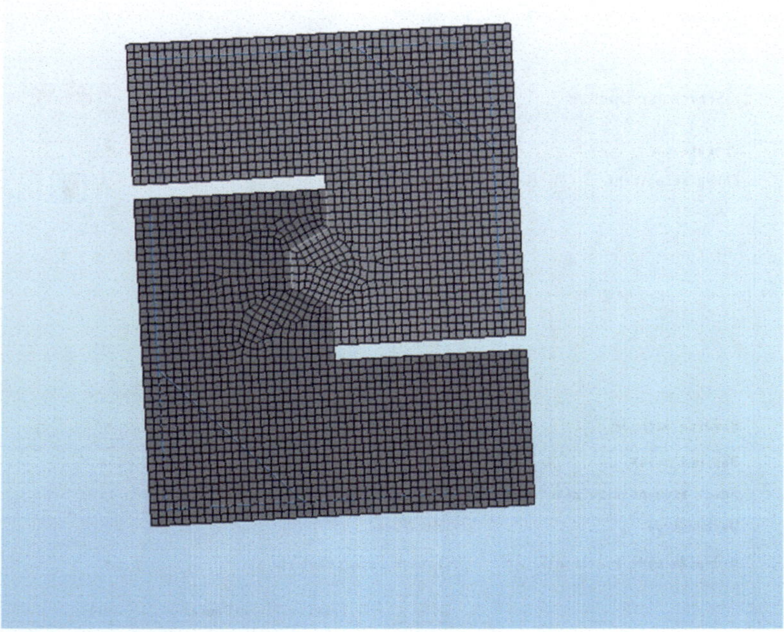

Fig. 5.27 Generated meshed elements

Adding analysis to generate analysis module1, right-clicking **Analysis-Add command**, **Structural nonlinear** is selected and load step block is added to it. **Geometry load combination 1** is chosen as **Load set** with both the number of load step and user-specified sizes 1. **Maximum number of iterations** is 20 and both the **Force and Displacement** options are selected as **Convergence norm**.

Next, we add **Geometry load combination 2**. Considering nonlinear analysis in the calculation, we found that the loading factors of the former substeps are larger than those of the latter ones. Basic loading reference value is 10 kN and the **User specified sizes** are 1.00000(10) 0.400000(5) (see Fig. 5.28). Other parameters in iteration calculation and convergence norm are the same as mentioned earlier.

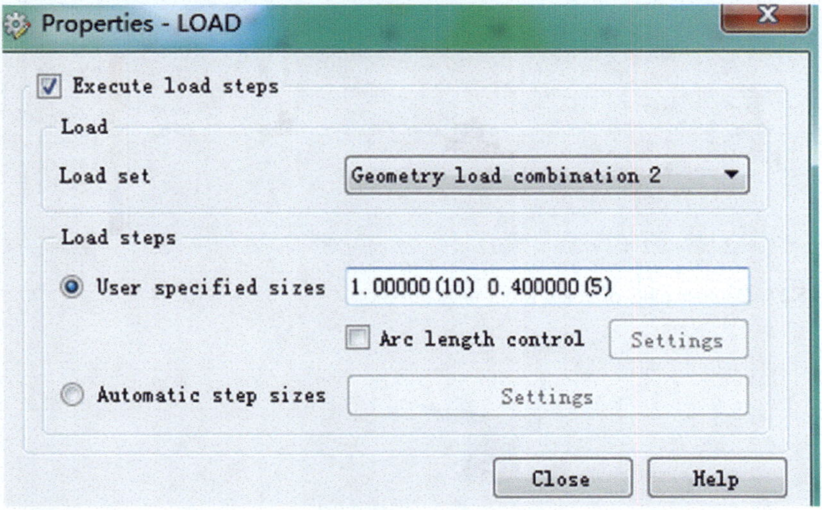

Fig. 5.28 User-specified sizes for geometry load combination 2

Click **Run analysis** to launch nonlinear calculation. After the calculation finishes, click **Output-Total displacement-TDtY** to check displacement in Y direction, which is 0.28 mm. Then the load value under ultimate state is 116 kN. Contour plots of crack width in X (EcwXX) and Y directions (EcwXX) under global coordinate system are displayed in Figs. 5.29 and 5.30. Normal cracking strain distribution contour after loading finished **Eknn** is shown in Fig. 5.31.

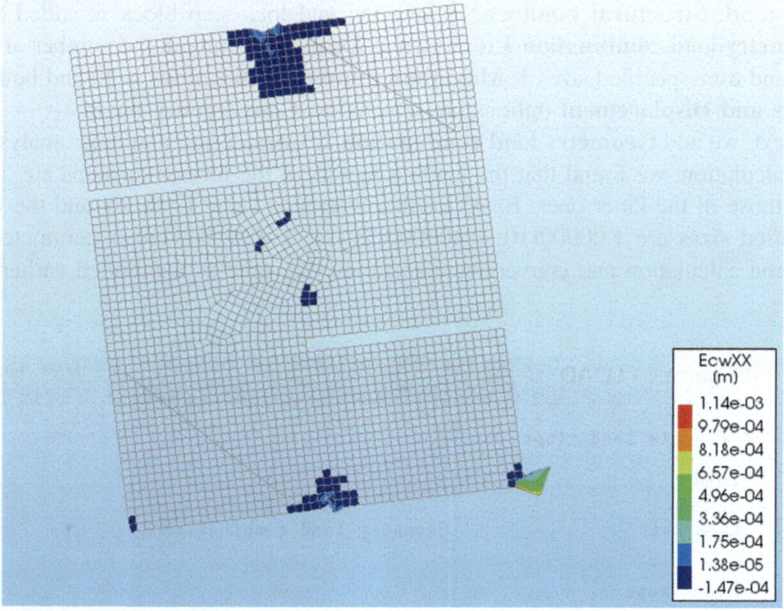

Fig. 5.29 Contour plots of crack width in X direction

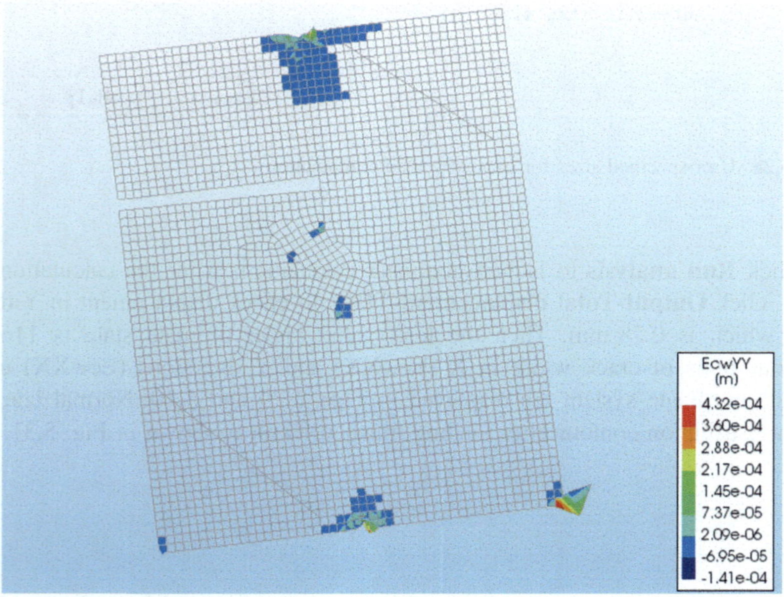

Fig. 5.30 Contour plots of crack width in Y direction

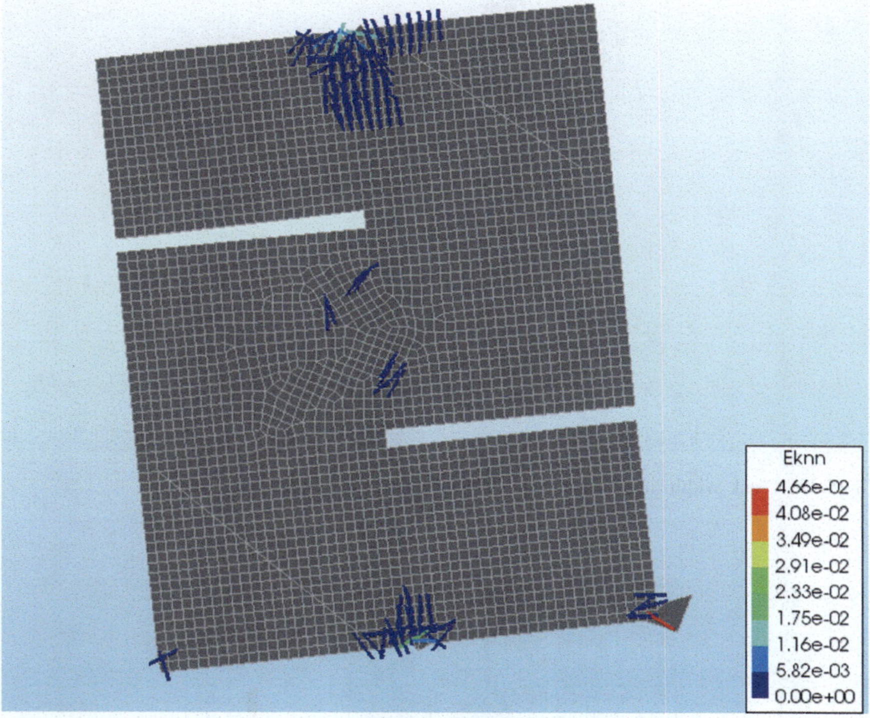

Fig. 5.31 Normal cracking strain distribution contour **Eknn**

Deleting the meshed elements in manual, **Multi-Directional Fixed Crack model** in material properties is replaced by the **Total Strain Based Crack Model**, keeping other specifications unchanged, and remeshing and calculating is performed. The ultimate load value is still 116 kN. Displacement in Y direction at the same site is 0.311 mm. Contour plots of crack width in X (EcwXX) under global coordinate system and normal cracking strain distribution contour are displayed in Figs. 5.32 and 5.33, respectively.

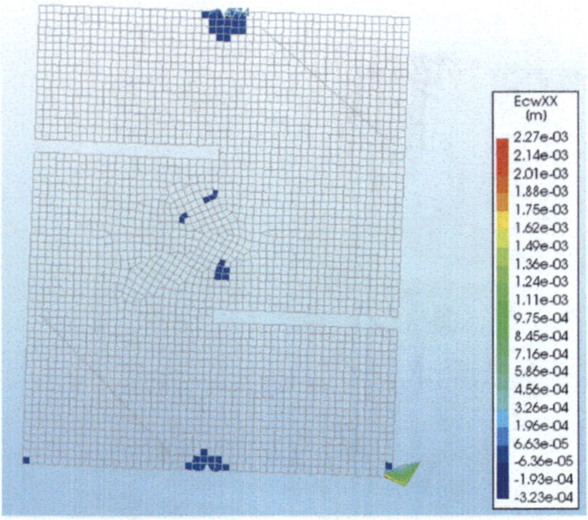

Fig. 5.32 Crack width in X direction (EcwXX) under global coordinate system

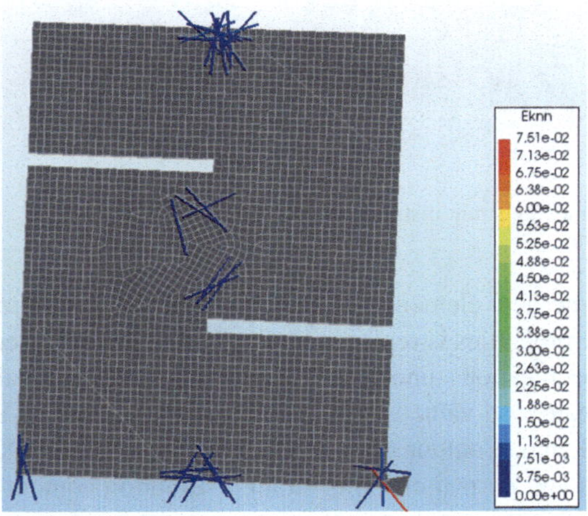

Fig. 5.33 Normal cracking strain distribution contour **Eknn**

To better illustrate the relationship between load and displacement, click **Viewer-node selection-Show ids-Output-Total displacement-TDtY-Show table** via right-clicking to check the node 6463. Extract the values of load and displacement under the process of stepwise loading from 0 to 116 kN, where these values under the two distinctive smeared cracking models are plotted with displacement and load values set as horizontal and longitudinal coordinates, respectively, and compare the curves. The curves processed by Origin are displayed in Fig. 5.34.

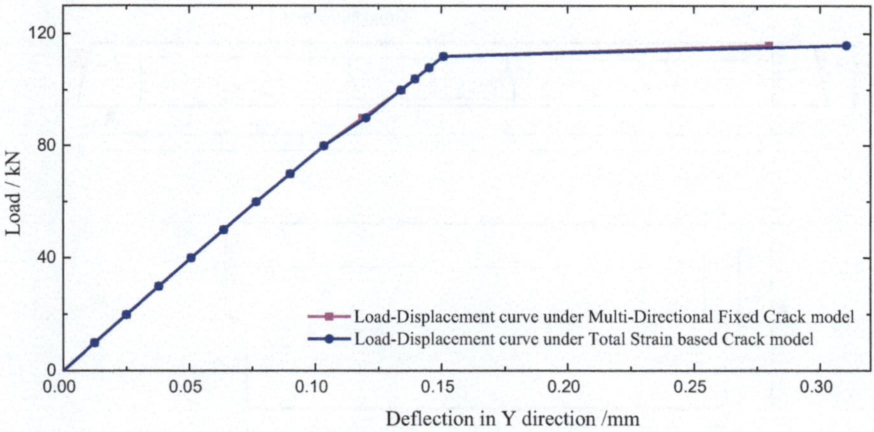

Fig. 5.34 Load–displacement curves under two different smeared cracking models

5.2 Time-Dependent Analysis of Precast Segmental Box Girders with Corbel Joints

A precast segmental concrete box girder with five segments in corbel joints is displayed as in Fig. 5.35. The longitudinal length is 4 m and the height is 0.39 m, with vertical symmetric distributed area value 200 kN/m^2 applied on top plate. Concrete grade is C50, and parameters conform with the European CEB-FIP 1990 code. Two internal bonded prestress tendons are in harp shape with elastic modulus and nominal strength 1.95×10^{11} N/m^2 and 1860 MPa, respectively. Time-dependent analysis for 3-year period is conducted in this numerical case and prestress loss as well as long-term deflection is investigated. Long-term prestress force loss along the path and between corbel segments is extracted and displayed in this part.

Note: It is worth to mention that in DIANA numerical cases, parameters such as prestress force and retention length in this book are assumed values. The appropriateness of the analysis results is another matter.

Essentials of learning

(1) *Learning to create geometric model of precast segmental box girder in corbel joints*
(2) *Learning to create discontinuous segmental stirrups* via *reinforcement grid elements*
(3) *Learning to define line to line connected interface elements between shells*
(4) *Learning to extract long-term prestress loss in corbel joints*
(5) *Learning to master fast parametric modeling of precast segmental box girders in editing and modifying Python command console*

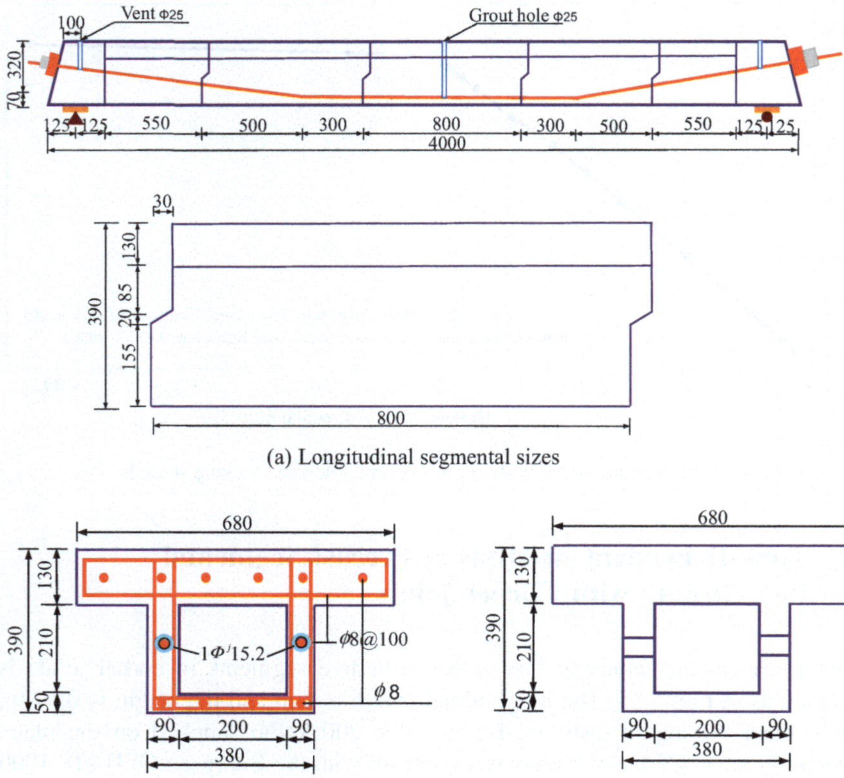

(a) Longitudinal segmental sizes

(b) Cross-section sizes

Fig. 5.35 Sizes of precast segmental bridges with corbel joints

Start DianaIE to open a new project with a file name of **Corbel** in computer E disk. **Structural** is selected as analysis type, and **Three dimensional** is chosen as **Dimensions** option. Maximum model size is 100 m, ranging from –50 to 50 m in the X, Y and Z directions. **Default mesher type** is **Hexa/Quad** while **Default mesh order** is **Quadratic,** where **Linear interpolation** is chosen as **Mid-side node location** (see Fig. 5.36).

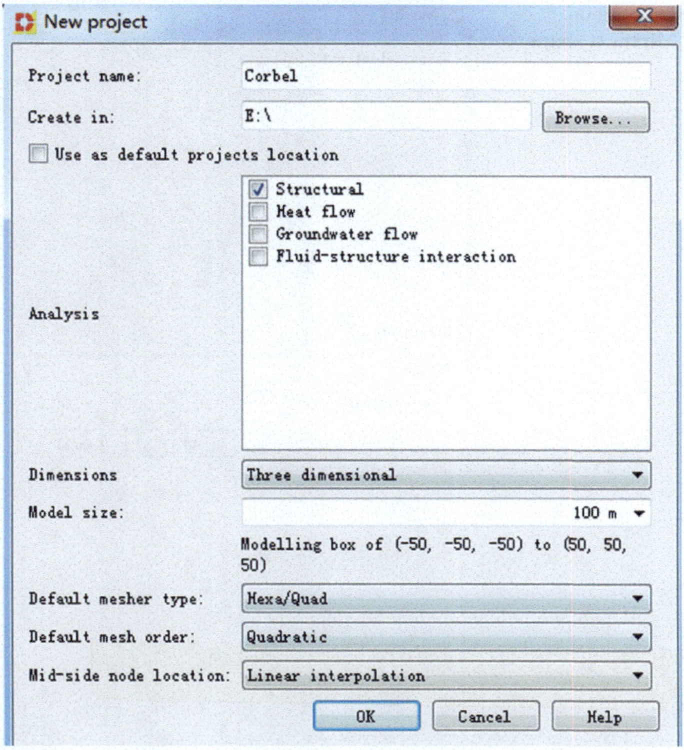

Fig. 5.36 New project interface

Clicking shortcut icon **Add a sheet**, input coordinate points [0, 0, 0], [0, 0.25, 0], [0, 0.25, 0.3], [0, 0.025, 0.3] to generate a new sheet with the name of **Sheet1**, then other sheets from **Sheet 2** to **Sheet 5** are established according to the coordinate values shown in Table 5.5.

Table 5.5 Coordinate values from **Sheet 2** to **Sheet 5**

Sheet 2	(0, 0.25, 0) (0, 0.8, 0) (0, 0.8, 0.13), (0, 0.83, 0.15) (0, 0.83, 0.3) (0, 0.25, 0.3)
Sheet 3	(0, 0.8, 0) (0, 1.6, 0) (0, 1.6, 0.13), (0, 1.63, 0.15) (0, 1.63, 0.3) (0, 0.83, 0.3) (0, 0.83, 0.15) (0, 0.8, 0.13)
Sheet 4	(0, 1.6, 0) (0, 2.4, 0) (0, 2.4, 0.13), (0, 2.43, 0.15) (0, 2.43, 0.3) (0, 1.63, 0.3) (0, 1.63, 0.15) (0, 1.6, 0.13)
Sheet 5	(0, 2.4, 0) (0, 3.2, 0) (0, 3.2, 0.13), (0, 3.23, 0.15) (0, 3.23, 0.3) (0, 2.43, 0.3) (0, 2.43, 0.15) (0, 2.4, 0.13)

After that, we select the **Sheet1**. Right-click to select **Mirror a shape** function; as Fig. 5.37 shows, mirror manipulation is along Y direction and the **Pivot** representing mirror symmetry axis is 2 m in Y direction. Clicking OK button generates Sheet6, as shown in Fig. 5.38.

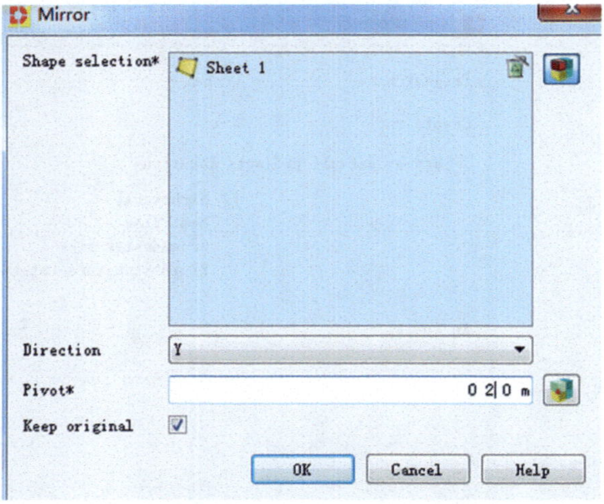

Fig. 5.37 Manipulation interface of **Mirror a shape**

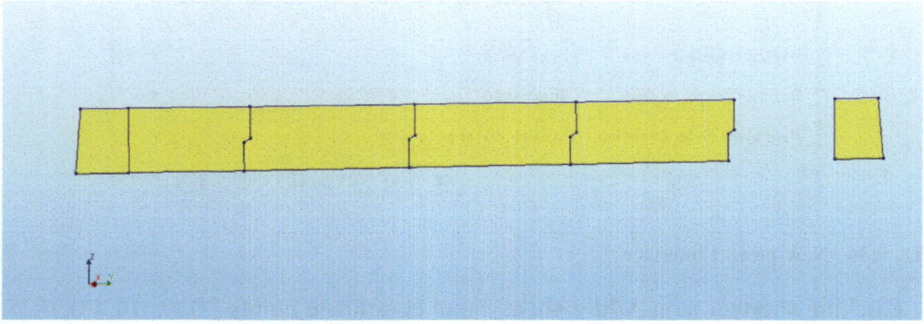

Fig. 5.38 Generation of **Sheet 6**

Then we create Sheet7 with the coordinate values [0, 3.2, 0], [0, 3.75, 0], [0, 3.75, 0.3], [0, 3.23, 0.3], [0, 3.23, 0.15], [0, 3.2, 0.13] to finish one side of web (see Fig. 5.39).

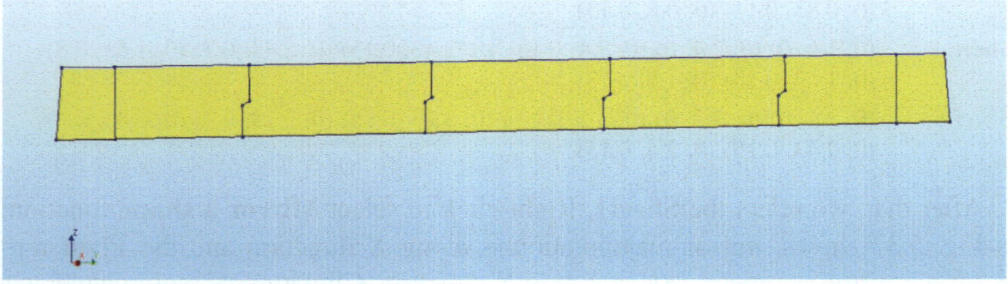

Fig. 5.39 One side of web

Select all the current sheets; right-click to select function of **Array copy. Relative Displacement** is input as 0.29 m in the positive X direction with the **Number of copies** 1, which is shown in Fig. 5.40.

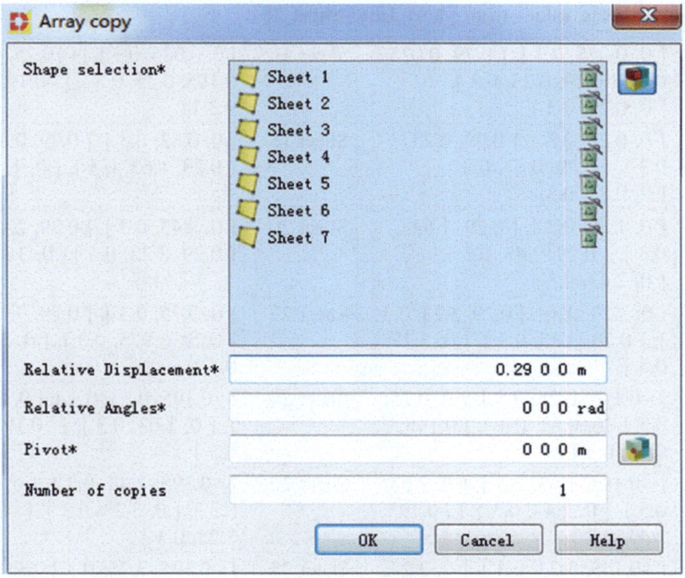

Fig. 5.40 Interface of **Array copy**

Clicking OK button, the generated webs on both sides are displayed (Fig. 5.41).

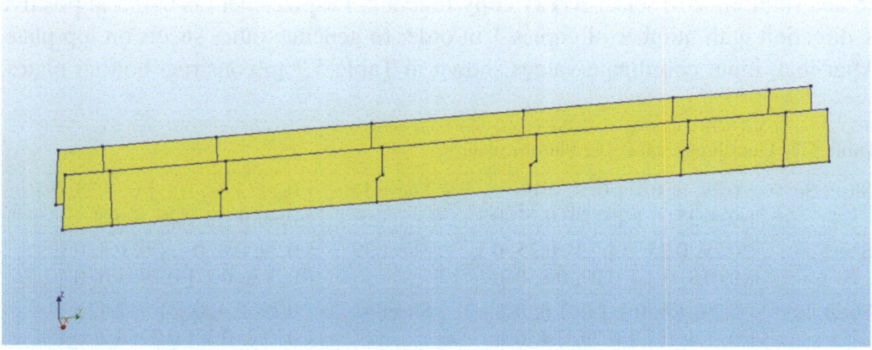

Fig. 5.41 Generated webs on both sides

Input coordinate values in Table 5.6 to create bottom plates and top plates, and name them Sheet15 to Sheet 42, respectively.

Table coordinate values of bottom and top plates.

Table 5.6 Coordinate values from **Sheet 15** to **Sheet 28**

Sheet 15	[0, 0.025, 0.3], [0.29, 0.025, 0.3], [0.29, 0.25, 0.3], [0, 0.25, 0.3]	Sheet 16	[0, 0.025, 0.3], [0, 0.25, 0.3], [–0.195, 0.25, 0.3], [–0.195, 0.025, 0.3]
Sheet 17	[0, 0.25, 0.3], [0.29, 0.25, 0.3], [0.29, 0.83, 0.3], [0, 0.83, 0.3]	Sheet 18	[0, 0.83, 0.3], [0.29, 0.83, 0.3], [0.29, 1.63, 0.3], [0, 1.63, 0.3]
Sheet 19	[0, 1.63, 0.3], [0.29, 1.63, 0.3], [0.29, 2.43, 0.3], [0, 2.43, 0.3]	Sheet 20	[0, 2.43, 0.3], [0.29, 2.43, 0.3], [0.29, 3.23, 0.3],[0, 3.23, 0.3]
Sheet 21	[0, 3.23, 0.3], [0.29, 3.23, 0.3], [0.29, 3.75, 0.3], [0, 3.75, 0.3]	Sheet 22	[0, 3.75, 0.3], [0.29, 3.75, 0.3], [0.29, 3.975, 0.3], [0, 3.975, 0.3]
Sheet 23	[–0.195, 0.25, 0.3], [0, 0.25, 0.3], [0, 0.83, 0.3], [-0.195, 0.83, 0.3]	Sheet 24	[–0.195, 0.83, 0.3], [0, 0.83, 0.3], [0, 1.63, 0.3], [–0.195, 1.63, 0.3]
Sheet 25	[–0.195, 1.63, 0.3], [0, 1.63, 0.3], [0, 2.43, 0.3], [–0.195, 2.43, 0.3]	Sheet 26	[–0.195, 2.43, 0.3], [0, 2.43, 0.3], [0, 3.23, 0.3], [–0.195, 3.23, 0.3]
Sheet 27	[–0.195, 3.23, 0.3], [0, 3.23, 0.3], [0, 3.75, 0.3], [–0.195, 3.75, 0.3]	Sheet 28	[–0.195, 3.75, 0.3], [0, 3.75, 0.3], [0, 3.975, 0.3], [–0.195, 3.975, 0.3]

Then select **Sheet 16, Sheet23, Sheet 24, Sheet 25, Sheet 26, Sheet 27, Sheet 28** and right-click to select **Array copy** function. Displacement is 0.485 in positive X direction with number of copies 1 in order to generate other sheets on top plate. After that, input coordinate values shown in Table 5.7 to construct bottom plates.

Table 5.7 Coordinate values of bottom plate

Sheet 36	[0.29, 0, 0], [0, 0, 0], [0, 0.25, 0], [0.29, 0.25, 0]	Sheet 37	[0.29, 3.75, 0], [0, 3.75, 0], [0, 4, 0], [0.29, 4, 0]
Sheet 38	[0.29, 0.25, 0], [0, 0.25, 0], [0, 0.8, 0], [0.29, 0.8, 0]	Sheet 39	[0.29, 0.8, 0], [0, 0.8, 0], [0, 1.6, 0], [0.29, 1.6, 0]
Sheet 40	[0.29, 1.6, 0], [0, 1.6, 0], [0, 2.4, 0], [0.29, 2.4, 0]	Sheet 41	[0.29, 2.4, 0], [0, 2.4, 0], [0, 3.2, 0], [0.29, 3.2, 0]
Sheet 42	[0.29, 3.2, 0], [0, 3.2, 0], [0, 3.75, 0], [0.29, 3.75, 0]		

When the **Sheet 42** is created, on clicking OK button, the geometric model of box girder is displayed (Fig. 5.42).

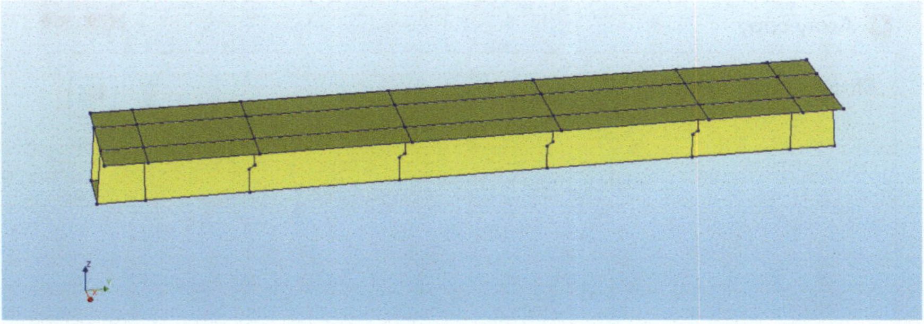

Fig. 5.42 Geometric model of box girder

The following step is to create geometric shape of internal harp tendon, with coordinate values displayed as follows. The name is **tenin** (see Fig. 5.43).

When geometric model of internal harp tendon is created, select it to copy and translate via the function of **Array Copy**. **Relative Displacement** is 0.29 m in the positive X direction while the **Number of copies** is 1 (see Fig. 5.44).

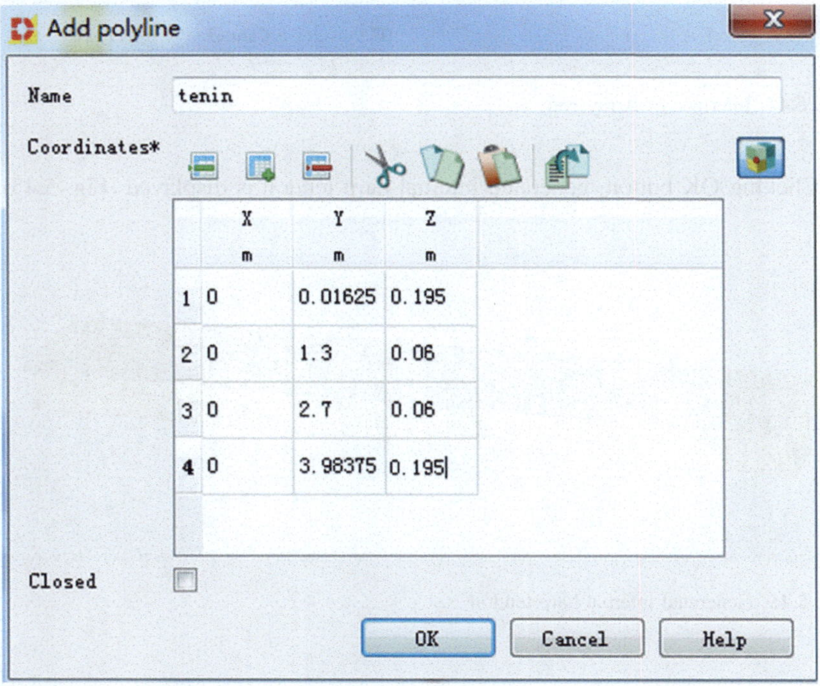

Fig. 5.43 Coordinate values of internal harp tendon

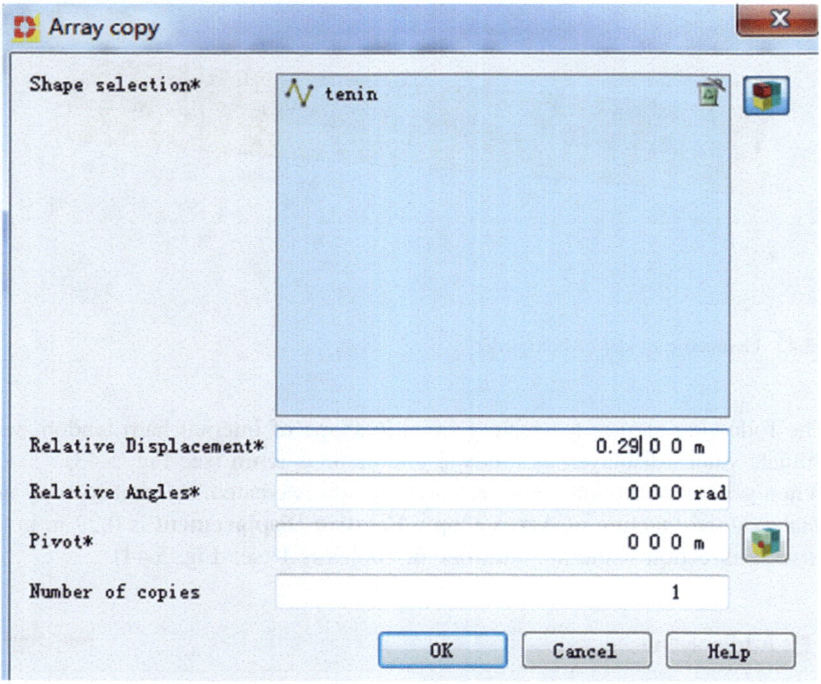

Fig. 5.44 Interface of Array copy

Clicking OK button, generated internal harp tendon is displayed (Fig. 5.45).

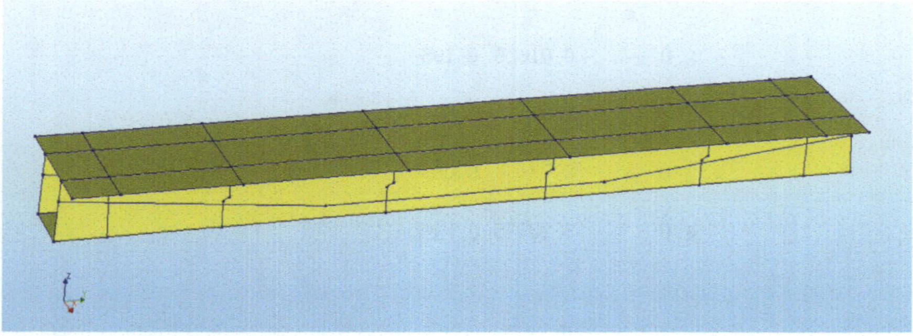

Fig. 5.45 Generated internal harp tendon

Units of temperature and angle are altered as Celsius and degree. Add new material dialog box for concrete material properties for top plate, bottom plate and webs on the end, as well as web in the middle region, with corresponding name **top, bot, mid1** and **mid2**, respectively. **Class** is Design codes while **Material model** for

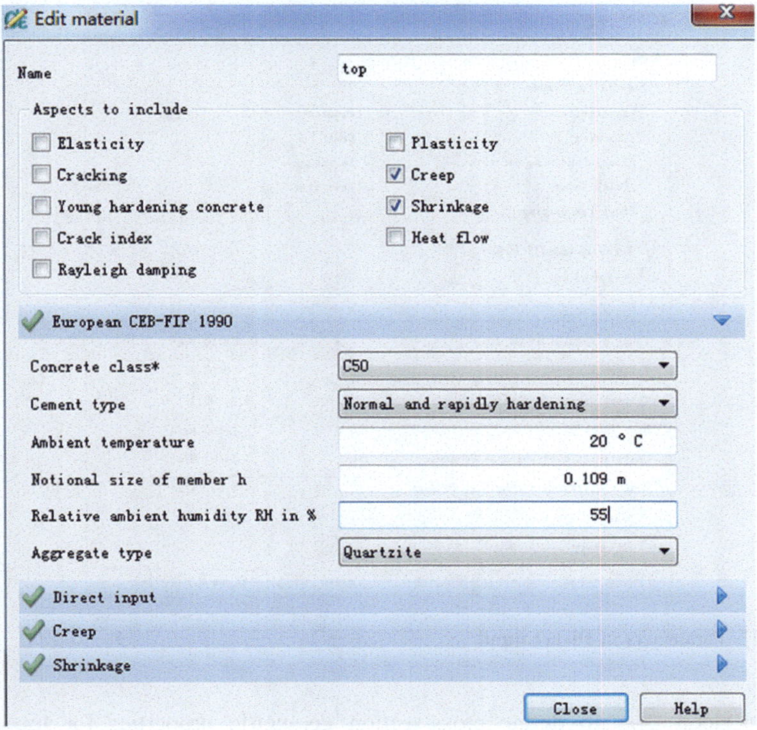

Fig. 5.46 Basic parameters in CEB-FIP 1990

time-dependent analysis is **CEB-FIP 1990**. **Creep** and **Shrinkage** are ticked. **Concrete class** is **C50** while **Cement type** is **Normal and rapidly hardening**. Ambient temperature is 20 °C and **Ambient relative humidity RH in %** is **55**. **Notional size of member** for top plate is 0.109 m according to the formula $\frac{0.68 \times 0.13}{(0.68+0.13)} = 0.109$ m. **Aggregate type** is chosen as **Quartzite** (see Fig. 5.46).

Then we enter **Direct input** module. **Young's modulus** and the value at 28 days are 3.8629×10^{10} N/m^2 and 3.45×10^{10} N/m^2, respectively. Poisson's ratio is 0.15 and thermal expansion coefficient for concrete is 1.2×10^{-5} 1/°C. Mass density is 2500 kg/m^3. Characteristics strength and mean compressive strength at 28 days are 5e7 and 5.8e7 N/m^2 (see Fig. 5.47).

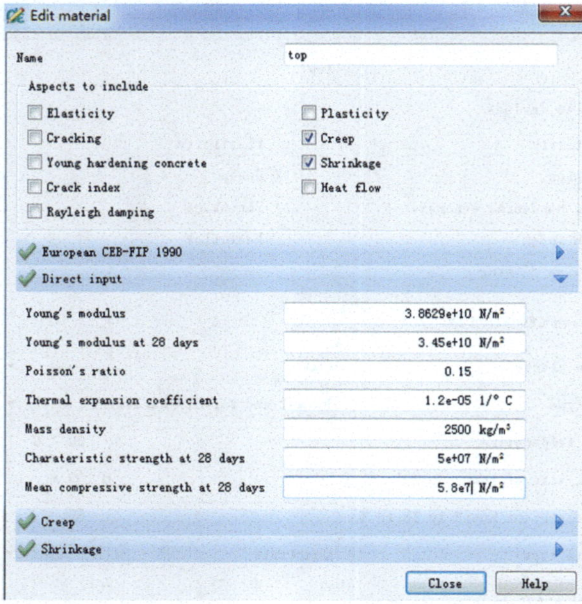

Fig. 5.47 Parameters in **Direct input**

Click icon to define cross-section geometric properties for top plate; thickness value is defined as 0.13 m, while **Element x-axis** in local coordinate corresponds to Y-axis in global coordinate system (see Fig. 5.48).

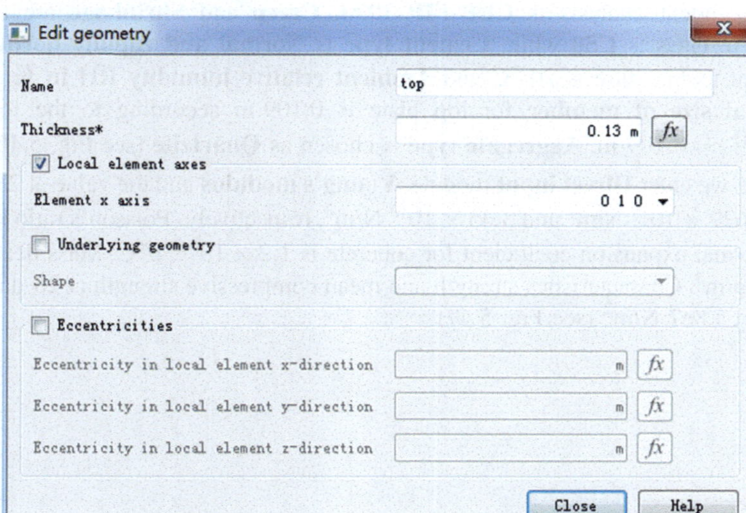

Fig. 5.48 Cross-section properties for top

Apply the same method to define material and cross-section for bottom plate, webs on the end and in the middle zone, with notional size of member and thickness 0.044, 0.226, 0.126 m as well as 0.05, 0.34, 0.09 m, respectively, and the following Figs. 5.49, 5.50, 5.51, 5.52, 5.53, 5.54 and 5.55 demonstrate part of specifications for them.

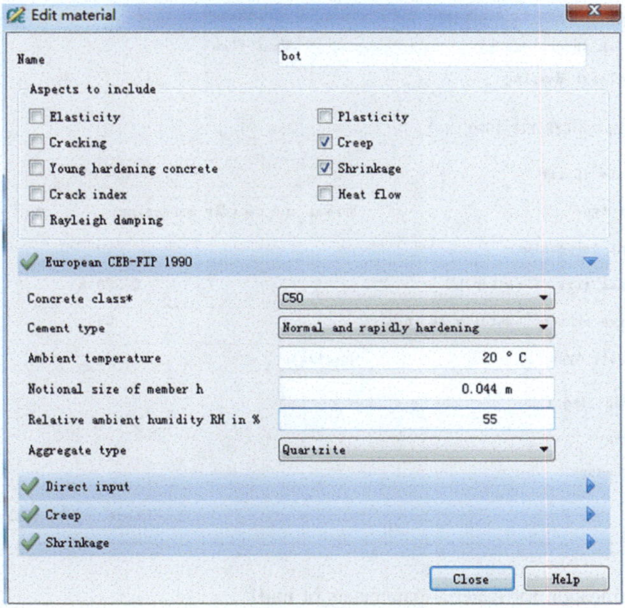

Fig. 5.49 Specification for material parameters of bottom plate

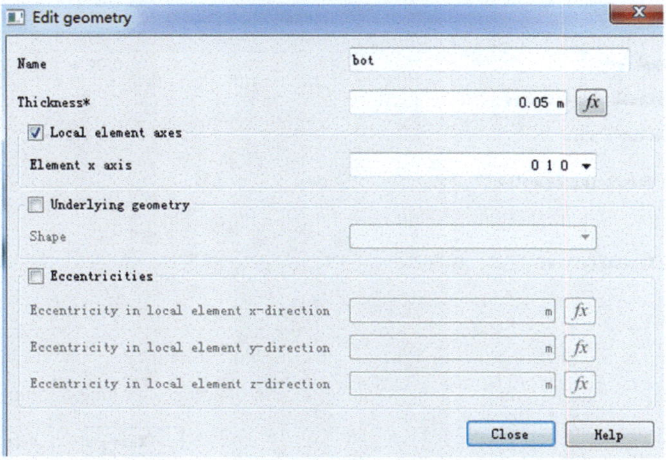

Fig. 5.50 Specification for cross-section geometric parameters of bottom plate

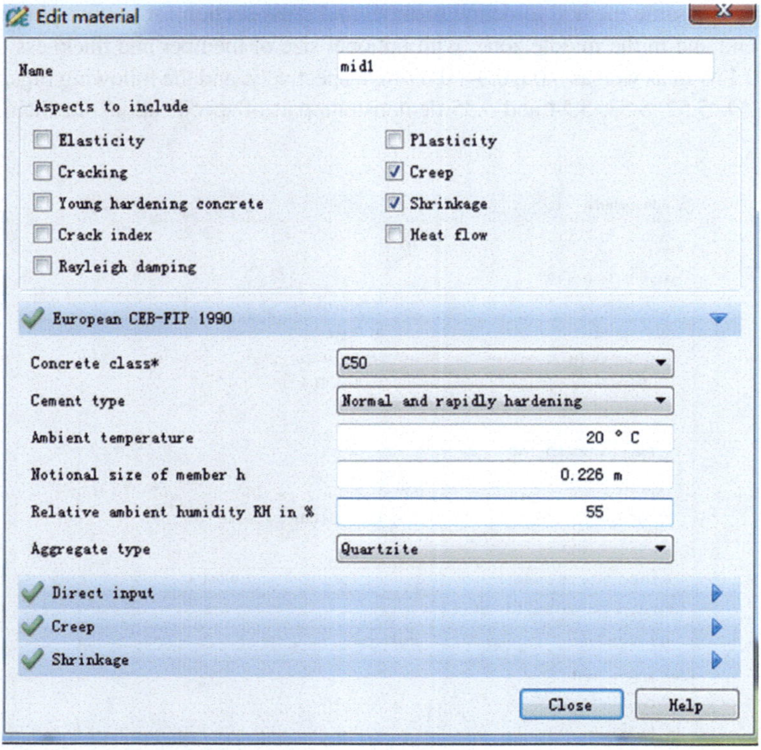

Fig. 5.51 Specification for material parameters of mid1

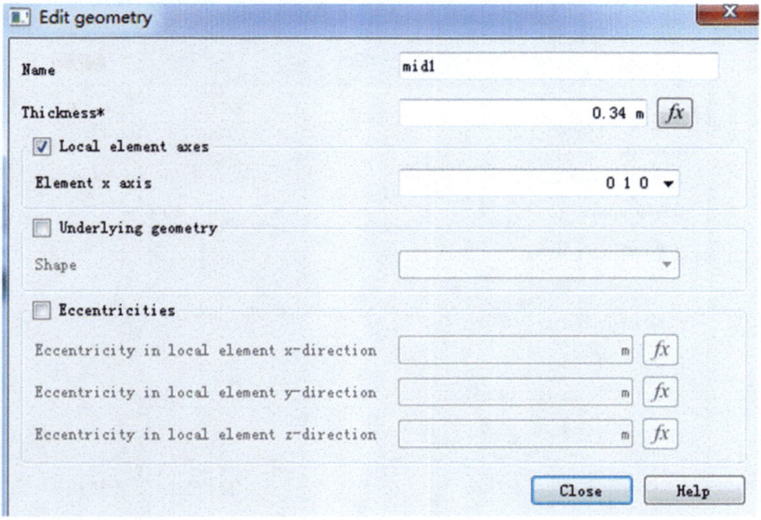

Fig. 5.52 Specification for cross-section geometric parameters of mid1

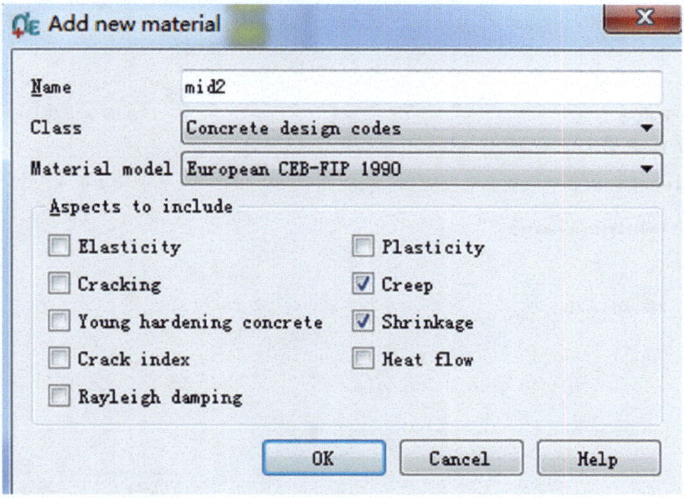

Fig. 5.53 Material class for web

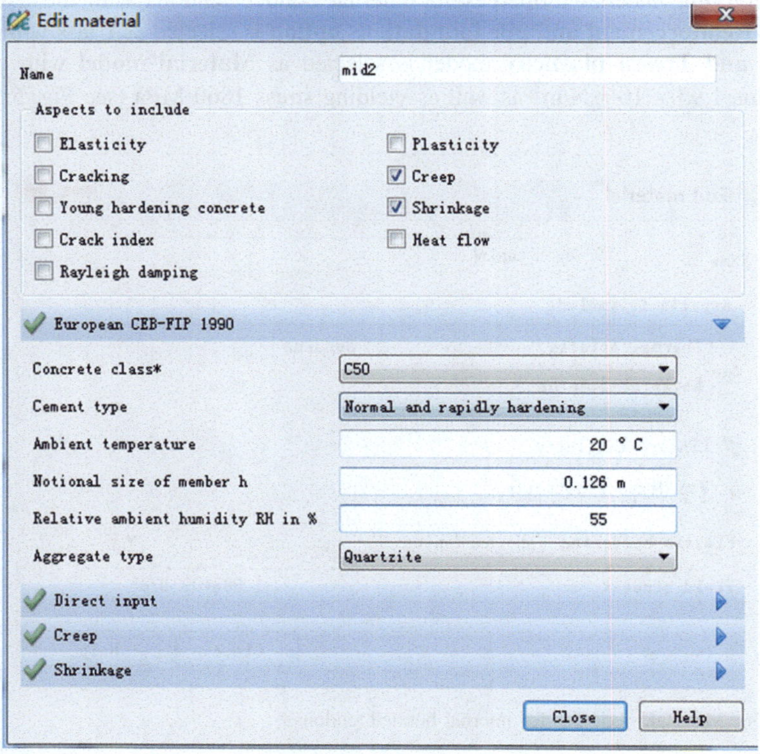

Fig. 5.54 Specification for material parameters of web

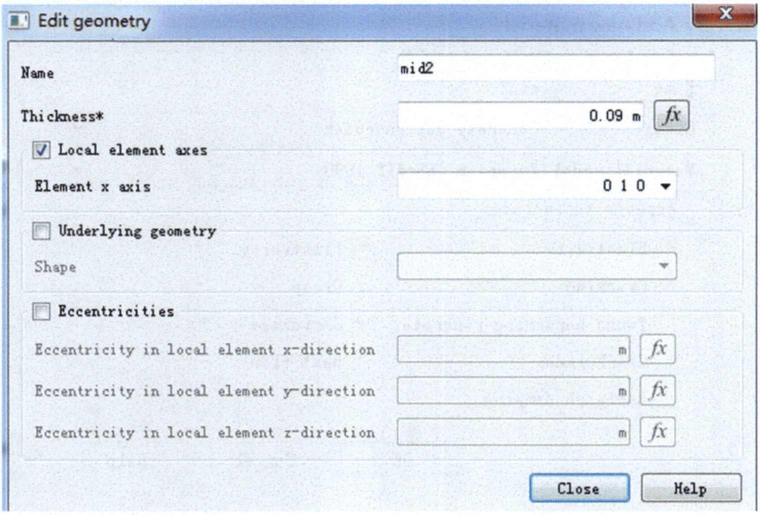

Fig. 5.55 Specification for cross-section geometric parameters of web

Specifying material properties for internal bonded tendons with the name of **tenin**, **Reinforcement and pile foundations option** is selected as **Class** while **von Mises and Tresca plasticity** model is selected as **Material model** with elastic modulus 1.95×10^{11} N/m^2 as well as yielding stress 1860 MPa (see Fig. 5.56).

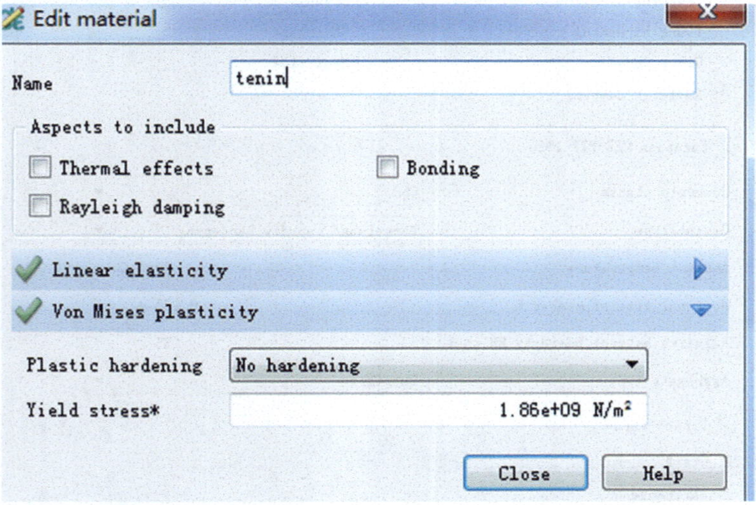

Fig. 5.56 Material properties for internal bonded tendons

Cross-section area of bar is 1.39e-4 and the **Reinforcement type** is **Embedded** (see Fig. 5.57).

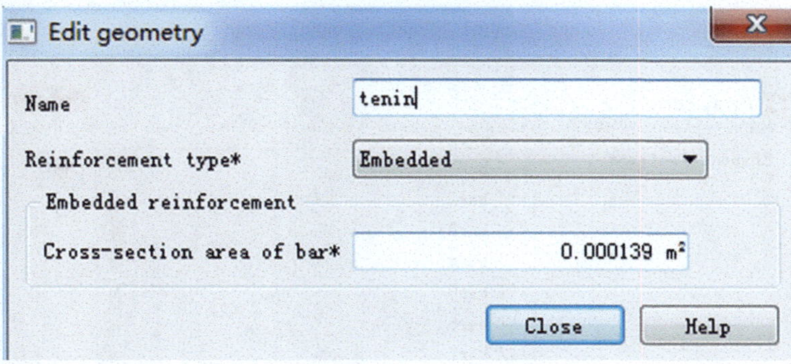

Fig. 5.57 Cross-section geometric properties for tendon

Now we start to create longitudinal bars. Adding shortcut icon **Add a line**  to create the first geometric model of longitudinal steel bar; the coordinate values are displayed in Fig. 5.58. Other longitudinal bars in the first segment are displayed in Tables 5.2, 5.3 and 5.4 (Table 5.8).

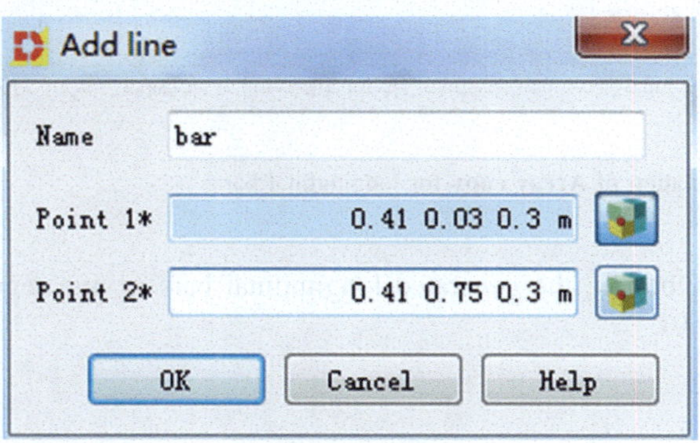

Fig. 5.58 Coordinate values of longitudinal bar

Table 5.8 Coordinate values of longitudinal bars

Bar1	[0.41, 0.03, 0.3], [0.41, 0.75, 0.3]	Bar2	[0.33, 0.03, 0.3], [0.33, 0.75, 0.3]
Bar3	[0.19, 0.03, 0.3], [0.19, 0.75, 0.3]	Bar4	[0.123, 0.03, 0.3], [0.123, 0.75, 0.3]
Bar5	[0.01, 0.03, 0.3], [0.01, 0.75, 0.3]	Bar6	[–0.14, 0.03, 0.3], [–0.14, 0.75, 0.3]
Bar7	[0.266, 0.03, 0], [0.266, 0.75, 0]	Bar8	[0.154, 0.03, 0], [0.154, 0.75, 0]
Bar9	[0.094, 0.03, 0], [0.094, 0.75, 0]	Bar10	[–0.007, 0.03, 0], [–0.007, 0.75, 0]

Selecting all the generated longitudinal bar, right-clicking to select function of **Array copy, Relative Displacement** is 0.8 m in the positive Y direction while **the number of copies** is 4 (see Fig. 5.59).

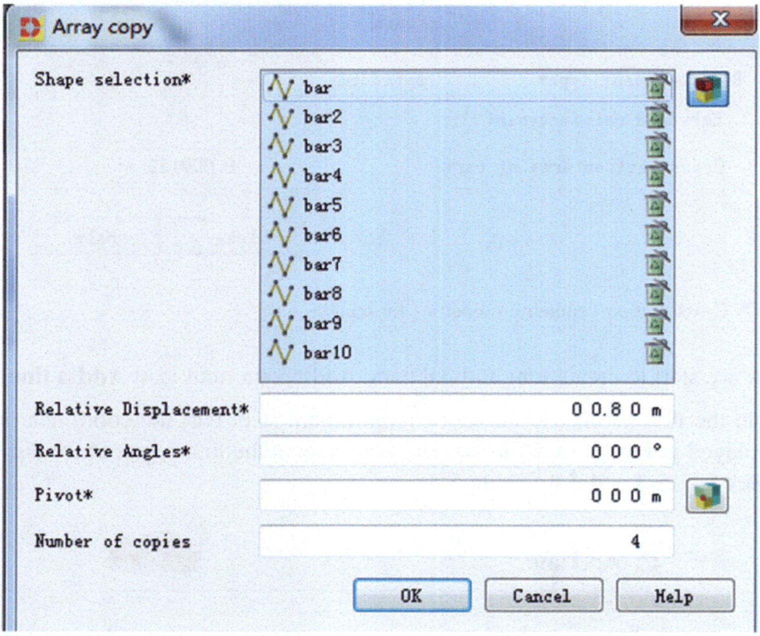

Fig. 5.59 Manipulation of **Array copy** for longitudinal bars

Clicking OK button, the generated longitudinal bars in five segments are displayed (Fig. 5.60).

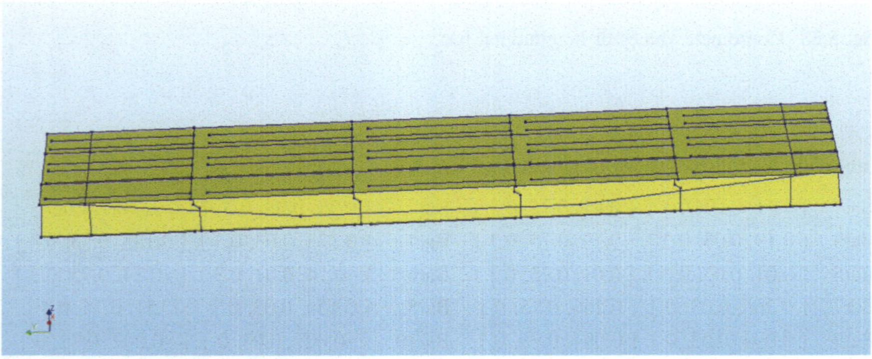

Fig. 5.60 Longitudinal bars in five segments

The following is to define cross-section properties for all the longitudinal bars with the name of **bar**. Similar to former, **Reinforcement type** is **Embedded** and **Cross-section area of bar** is 5.0265e – 5 m^2 (see Fig. 5.61).

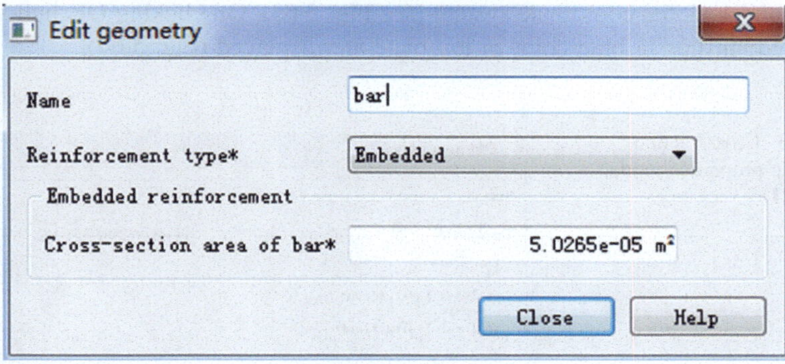

Fig. 5.61 Cross-section geometric properties for longitudinal bars

Select all the top plates; right-click to select **Array copy** to duplicate and translate them both in the negative and positive Z directions, respectively, with the relative displacement 0.041 and –0.041 m, respectively (see Fig. 5.62).

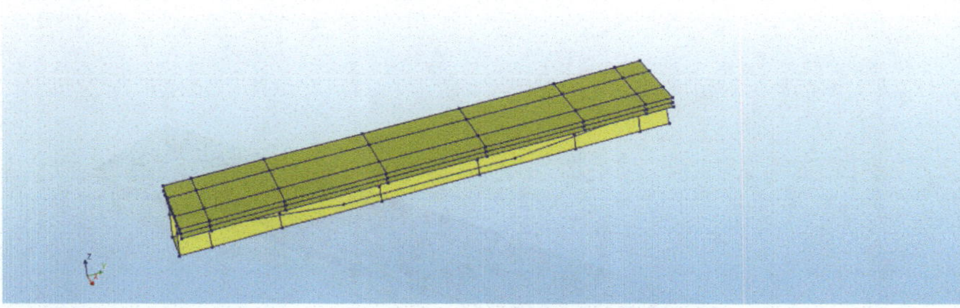

Fig. 5.62 Duplications and translations of top plate in the negative and positive Z directions

Note: Owing to the fact that both longitudinal reinforcement bars as well as stirrups in precast segmental girders are not in continuity, therefore, both longitudinal reinforcement bars and stirrups modeled via reinforcement grid elements in precast segmental girders should be created segment by segment!

Select all the new duplicated and translated sheets (from **Sheet 43** to **Sheet 84**); click shortcut icon button **Edit reinforcement property assignments** to assign material and geometric properties for stirrups. **Material model** is still **von Mises and Tresca plasticity** model, and **Class** is also **Reinforcement and pile foundations**. **Young's modulus** and **Yield stress** are 2.1×10^{11} N/m^2 and

400 MPa, respectively. In the module of cross-section geometric properties model, thickness is assigned via **Diameter and spacing** way, and diameters in local X and Y directions are 0 and 0.008 m, while spacing between bars are 0 and 0.1 m, respectively. Reinforcement x-axis under local coordinate system corresponds to Y-axis under global coordinate system (see Fig. 5.63). On clicking OK button, the generated stirrups in blue color in top plate are displayed (Fig. 5.64).

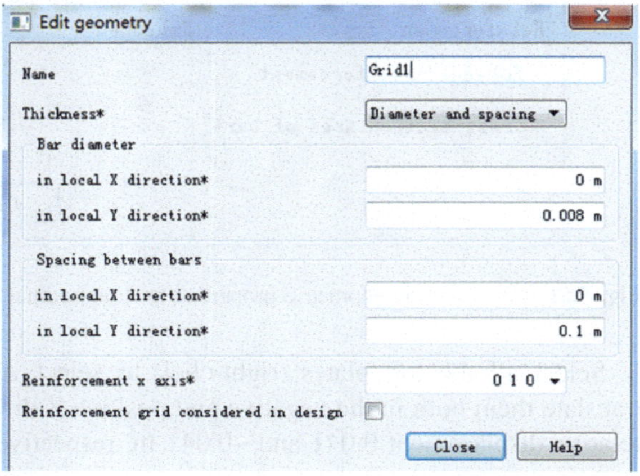

Fig. 5.63 Cross-section geometric properties model for **Grid1**

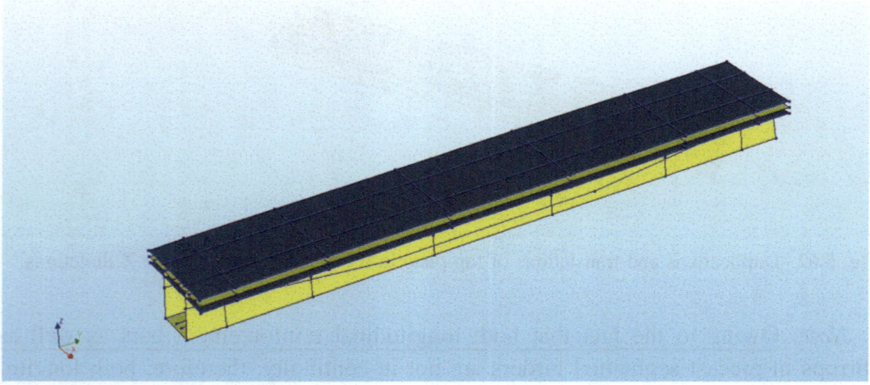

Fig. 5.64 Generated stirrups in top plate

Then we select all the web sheets to duplicate and translate via manipulation of **Array copy** in both positive as well as negative directions with relative displacement both 0.03 m and number of copies 1 to generate web stirrups in web region. After that, hide the reinforcement stirrup sheets on the flange, as shown in Fig. 5.65, and then input coordinate values, as given in Table 5.9, to create one-layer web stirrups in the top plate region on one side from **Sheet 113** to **Sheet 119**.

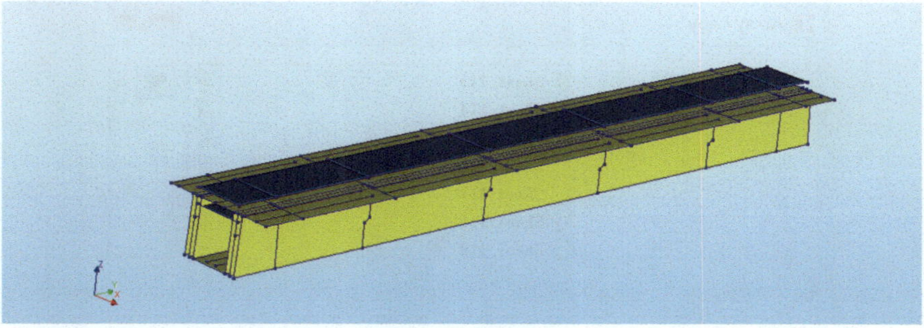

Fig. 5.65 Hiding the reinforcement stirrup sheets on the flange

Table 5.9 Coordinate values of stirrup webs on the top plate region

Sheet 113	[0.29, 0.025, 0.341], [0.29, 0.025, 0.3], [0.29, 0.25, 0.3], [0.29, 0.25, 0.341]	Sheet 114	[0.29, 0.25, 0.341], [0.29, 0.25, 0.3], [0.29, 0.83, 0.3], [0.29, 0.83, 0.341]
Sheet 115	[0.29, 0.83, 0.341], [0.29, 0.83, 0.3], [0.29, 1.63, 0.3], [0.29, 1.63, 0.341]	Sheet 116	[0.29, 1.63, 0.341], [0.29, 1.63, 0.3], [0.29, 2.43, 0.3], [0.29, 2.43, 0.341]
Sheet 117	[0.29, 2.43, 0.341], [0.29, 2.43, 0.3], [0.29, 3.23, 0.3], [0.29, 3.23, 0.341]	Sheet 118	[0.29, 3.23, 0.341], [0.29, 3.23, 0.3], [0.29, 3.75, 0.3], [0.29, 3.75, 0.341]
Sheet 119	[0.29, 3.75, 0.341], [0.29, 3.75, 0.3], [0.29, 3.975, 0.3], [0.29, 3.975, 0.341]		

When these sheets are generated, select all of them and right-click to select function of **Array copy** to duplicate and translate them in the negative X direction with the relative displacement 0.29 m in order to generate the other side of one layer web stirrups in the top plate region, and number of copies is 1, as shown in Fig. 5.66.

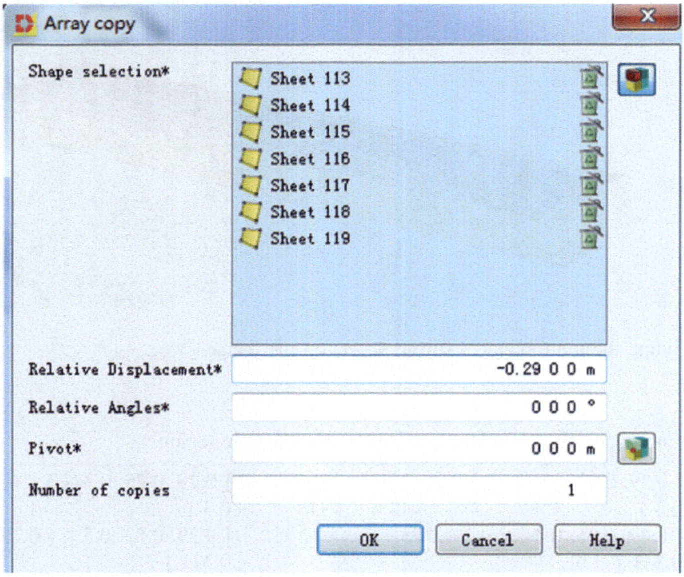

Fig. 5.66 Array copy for **Sheet 113** to **Sheet 119**

On clicking OK button, the generated other side of one layer web stirrups in the top plate region is displayed (Fig. 5.67).

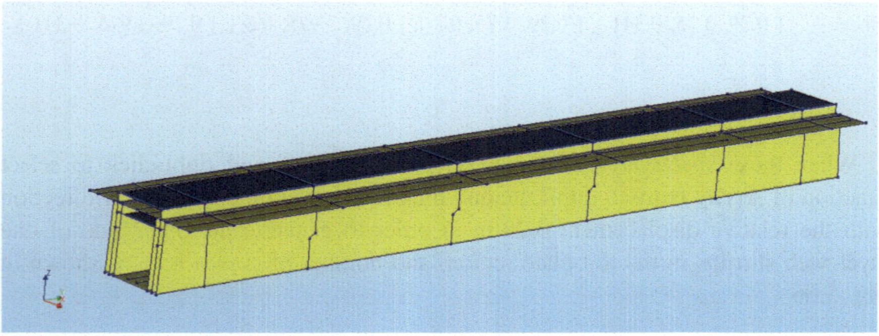

Fig. 5.67 The other side of web stirrups in the top plate region

Still selecting sheets **Sheet 113** to **Sheet 119**, right-click to select **Move a shape** function to move them to the correct reinforcement site connected with web reinforcement as an integral part in the negative X direction, where the movement is 0.03 m. Then use the same method to emerge reinforcement stirrups sheets in the top plate zone on the other side from **Sheet 120** to **Sheet 126**, as Fig. 5.68 shows.

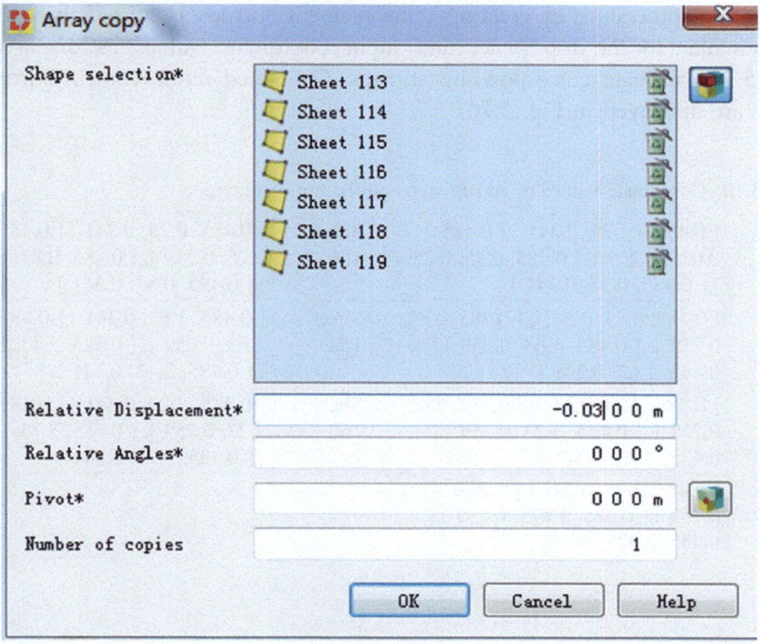

Fig. 5.68 Array copy for **Sheet 113** to **Sheet 119**

Select all the web stirrups (from **Sheet 85** to **Sheet 140**) to assign **Reinforcement and pile foundations material class** with **von Mises and Tresca plasticity** model with the material set name of **Grid2**, and the material and cross-section geometric properties parameters and specifications are the same as former. Clicking OK button, the generated web stirrups in blue are displayed (Fig. 5.69).

Use the same method to create reinforcement grids in bottom plate, with translation values –0.0085 and 0.0085 m, respectively, in the Z direction, and the material and cross-section geometric properties parameters and specifications are the same as former.

Fig. 5.69 Generated web stirrups in blue

The final procedure of creating reinforcement stirrups is to establish stirrups perpendicular to the top plate, and input coordinate values as displayed in Table 5.10 to create corresponding stirrups. Generated reinforcement stirrups in yellow are displayed in Fig. 5.70.

Table 5.10 Coordinate values of stirrup webs on the top plate region

Sheet 155	[0.485, 0.025, 0.341],[0.485, 0.025, 0.259], [0.485, 0.25, 0.259],[0.485, 0.25, 0.341]	Sheet 156	[0.485, 0.25, 0.341],[0.485, 0.25, 0.259], [0.485, 0.83, 0.259],[0.485, 0.83, 0.341]
Sheet 157	[0.485, 0.83, 0.341],[0.485, 0.83, 0.259], [0.485, 1.63, 0.259],[0.485, 1.63, 0.341]	Sheet 158	[0.485, 1.63, 0.341],[0.485, 1.63, 0.259], [0.485, 2.43, 0.259],[0.485, 2.43, 0.341]
Sheet 159	[0.485, 2.43, 0.341],[0.485, 2.43, 0.259], [0.485, 3.23, 0.259],[0.485, 3.23, 0.341]	Sheet 160	[0.485, 3.23, 0.341],[0.485, 3.23, 0.259], [0.485, 3.75, 0.259],[0.485, 3.75, 0.341]
Sheet 161	[0.485, 3.75, 0.341],[0.485, 3.75, 0.259],[0.485, 3.975, 0.259],[0.485, 3.975, 0.341]		

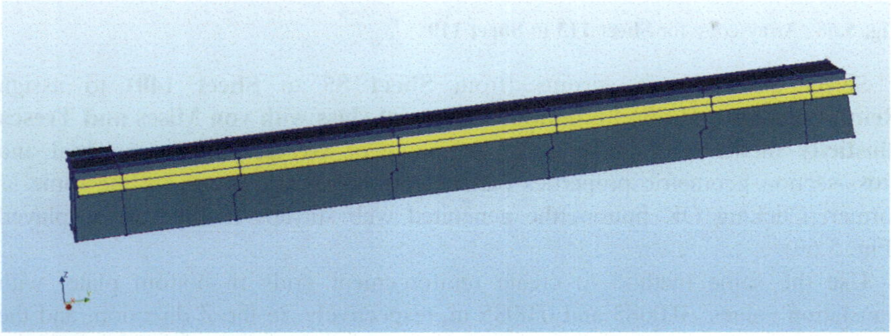

Fig. 5.70 Stirrups perpendicular to the top plate

Select Sheet 155 to Sheet 161; right-click to select **Move shape**. **Displacement** is 0.024 m in the negative X direction (see Fig. 5.71). Select them to duplicate and translate them 0.68 m along the negative X direction via the manipulation **Array copy** to generate stirrups perpendicular to the top plate on the other side, which are moved to correct location with the same method. Material and cross-section geometric properties parameters and specifications are the same as former and it is not repeated here.

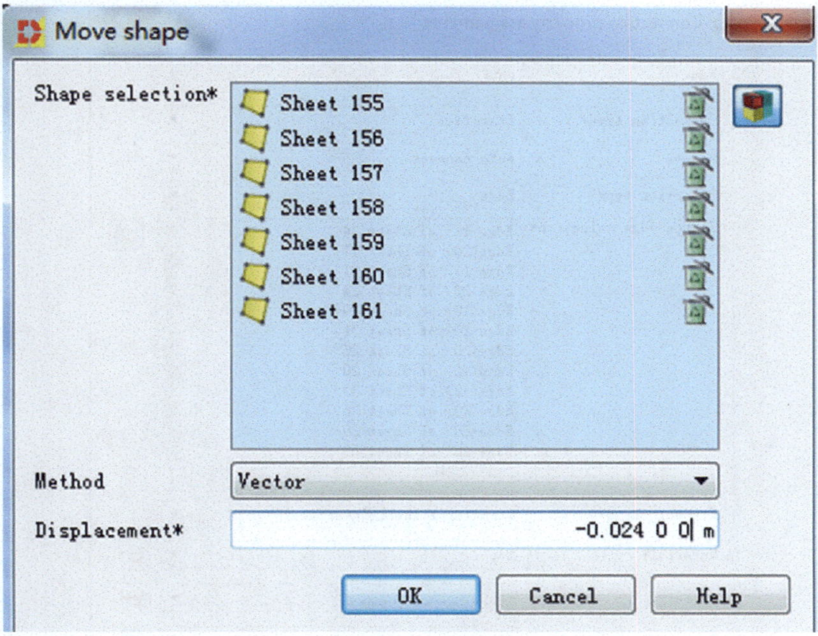

Fig. 5.71 Manipulation of **Move a shape**

Untick all the numerical reinforcement grids so as to render convenience for attaching interface elements. Click shortcut button  to add material properties for line to line connected interface elements on top plate with the name of **int1**. Edges between segments on the top plate are all selected and **Connection type** is **Interface** while **Selection type** is **Edge**. **Structural Shell Interfaces** is chosen as **Element class** (see Fig. 5.72). Click shortcut icon  to define interface material, where **Coulomb friction** is selected as material model. **3D line interface between shells** are selected with the normal stiffness modulus in y direction 3e16 N/m^3, while shear stiffness modulus in x and z directions is both 3e8 N/m^3. In the aspect of **Coulomb friction**, **Cohesion** is 0 N/m in order to simulate dry joints between segments. Friction angle and dilatancy angle are 25° and 35°, respectively (see Figs. 5.73 and 5.74).

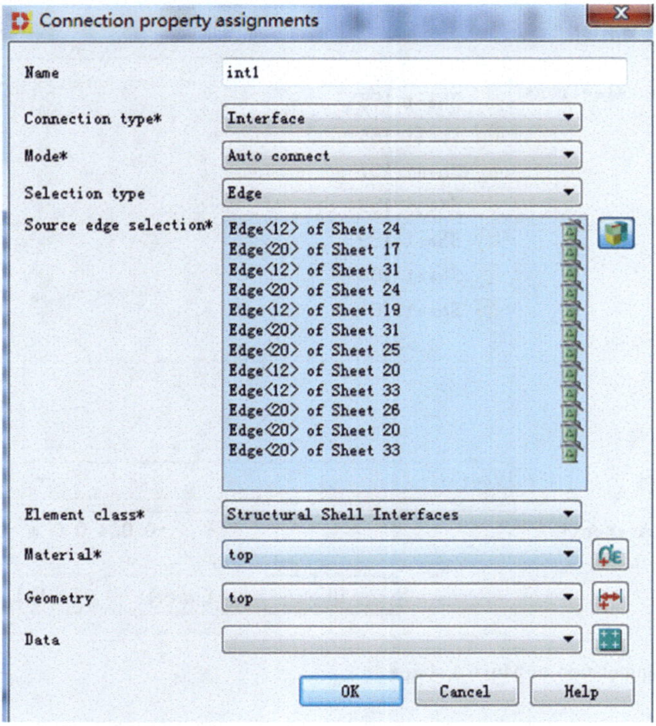

Fig. 5.72 Interface elements

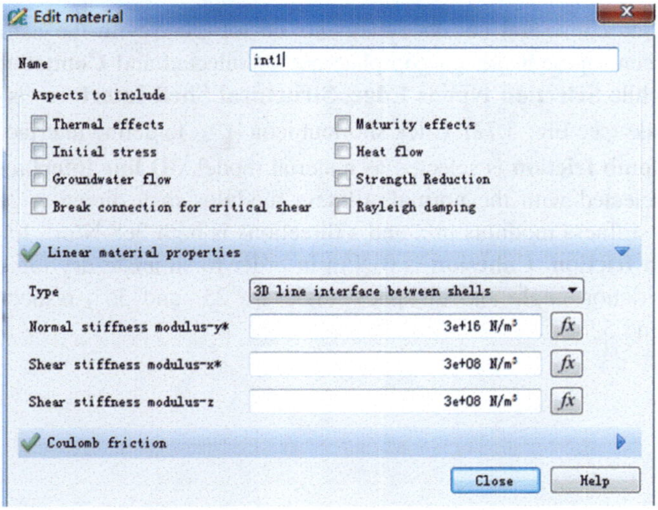

Fig. 5.73 Normal and shear stiffness modulus for interface elements

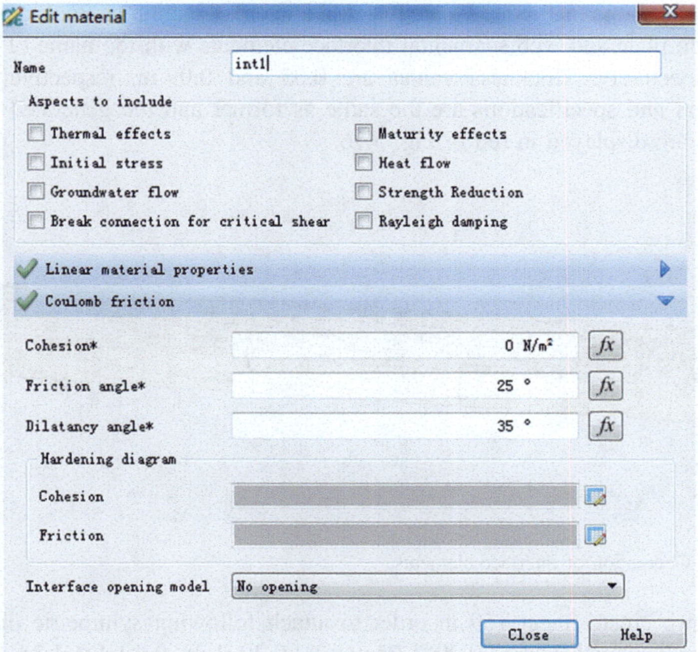

Fig. 5.74 Parameters in Coulomb friction

Then we specify cross-section properties for **int1**. Thickness value of top plate is 0.13 m, while **Element direction** is **Parallel to shell plane. Direction vector parallel** corresponds with negative Y direction (0, –1, 0), meaning that compressive mechanic behavior of interface element is in Y direction under global coordinate system (see Fig. 5.75).

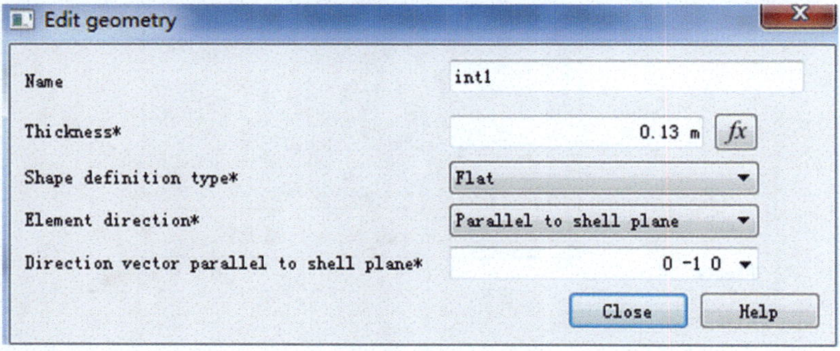

Fig. 5.75 Cross-section properties for **int1**

Apply the same method to create material and cross-section geometric properties for bottom plate and web segmental interface elements with the name of **int2** and **int3**, respectively. Thickness values are 0.05 and 0.09 m, respectively. Other parameters and specifications are the same as former and the generated interface elements are displayed in red in Fig. 5.76.

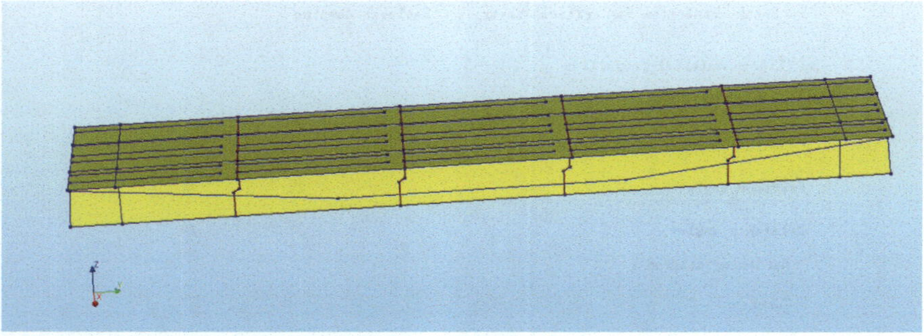

Fig. 5.76 Generation of interface elements

Then we create Sheet 169 in order to attach following symmetric distributed load. Coordinate values are [0.29, 1.05, 0.4], [0.29, 1.35, 0.4], [0, 1.35, 0.4], [0, 1.05, 0.4], which are then selected to be mirrored via the function **Mirror a shape** in Y direction and the **Pivot** representing symmetric axis of mirror symmetry is Y = 2 m, which is (0, 2, 0) (see Fig. 5.77). Clicking OK button, the generated surfaces are displayed (Fig. 5.78).

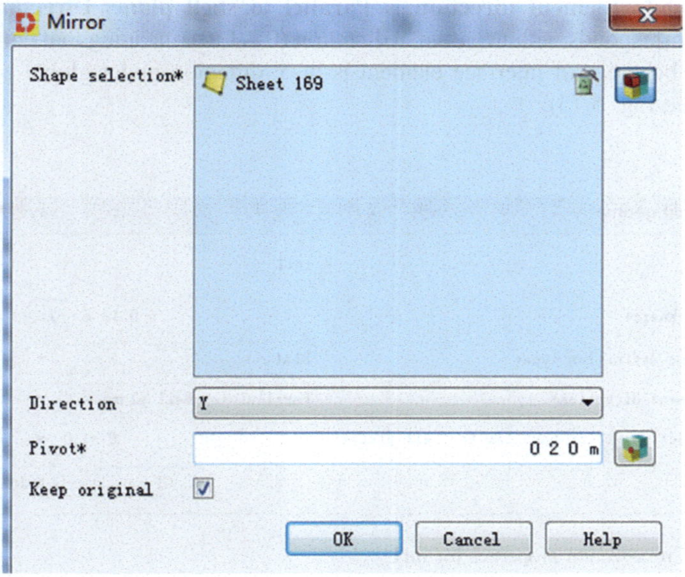

Fig. 5.77 Interface of **Mirror a shape**

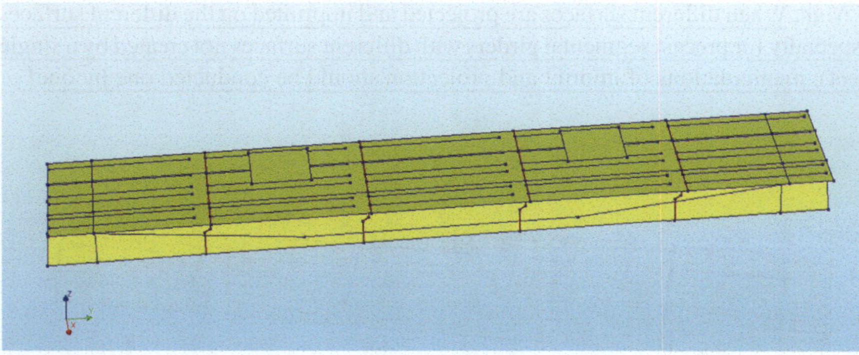

Fig. 5.78 Generation of surfaces after **Mirror a shape**

Then we imprint and project them to the top plate one by one. Above all, **Sheet 169** is selected, and then shortcut icon **Project edges, wires and points on solid, faces and edges** are clicked. **Operation** is **Face** and **Face selection** is top plate in the middle site of second segment while **Sheet 169** is selected as **Tool selection** with projection direction is in the negative Z direction, which is (0, 0, −1) (see Fig. 5.79).

Apply the same method to imprint and project **Sheet 170**. On clicking OK button, imprinted and projected numerical model is displayed (Fig. 5.80).

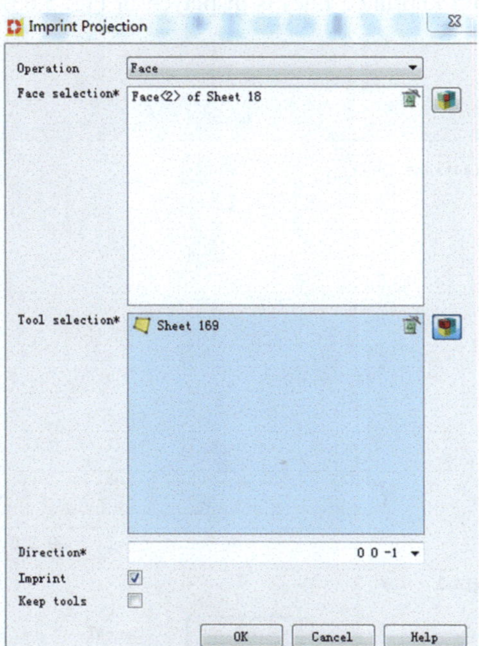

Fig. 5.79 Interface of imprint and projection

Note: When different surfaces are projected and imprinted on the different surfaces (especially for precast segmental girders with different surfaces not created by a single sheet), manipulations of imprint and projection should be conducted one by one!

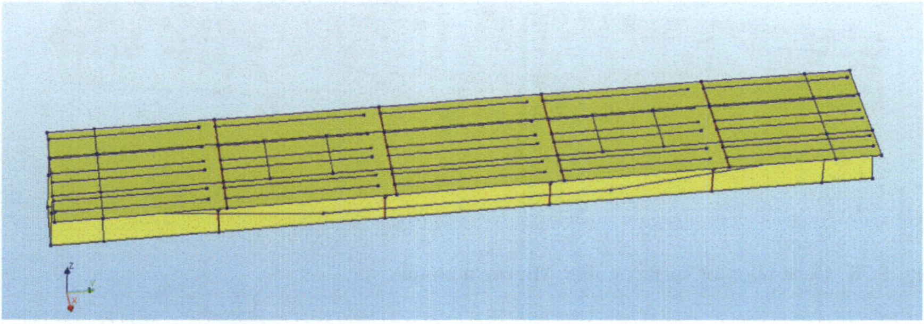

Fig. 5.80 Imprinted and projected numerical model

Now we begin to create supports. Above all, a line named **co1** with the coordinate values [0, 0.125, –0.1], [0.29, 0.125, –0.1] is created, which is further mirrored to symmetric site via the manipulation of **Mirror a shape** (see Fig. 5.81). The same method for imprint and projection is applied again to attach constraints on the correct site of bottom plate. It is worth to mention that since **co1** and **co2** are below the bottom plate, projection direction for them is in the positive Z direction, which is (0, 0, 1) (see Fig. 5.82). When this manipulation is accomplished, numerical model with imprinted lines is displayed in Fig. 5.83.

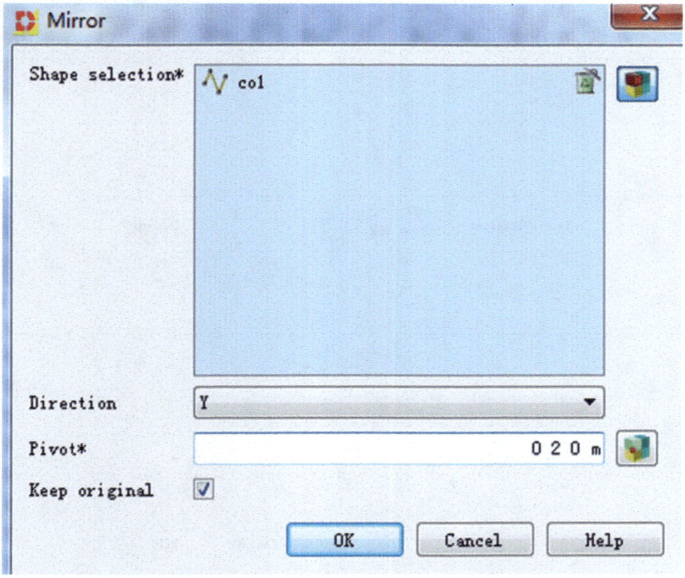

Fig. 5.81 Manipulation of **Mirror a shape** for **co1**

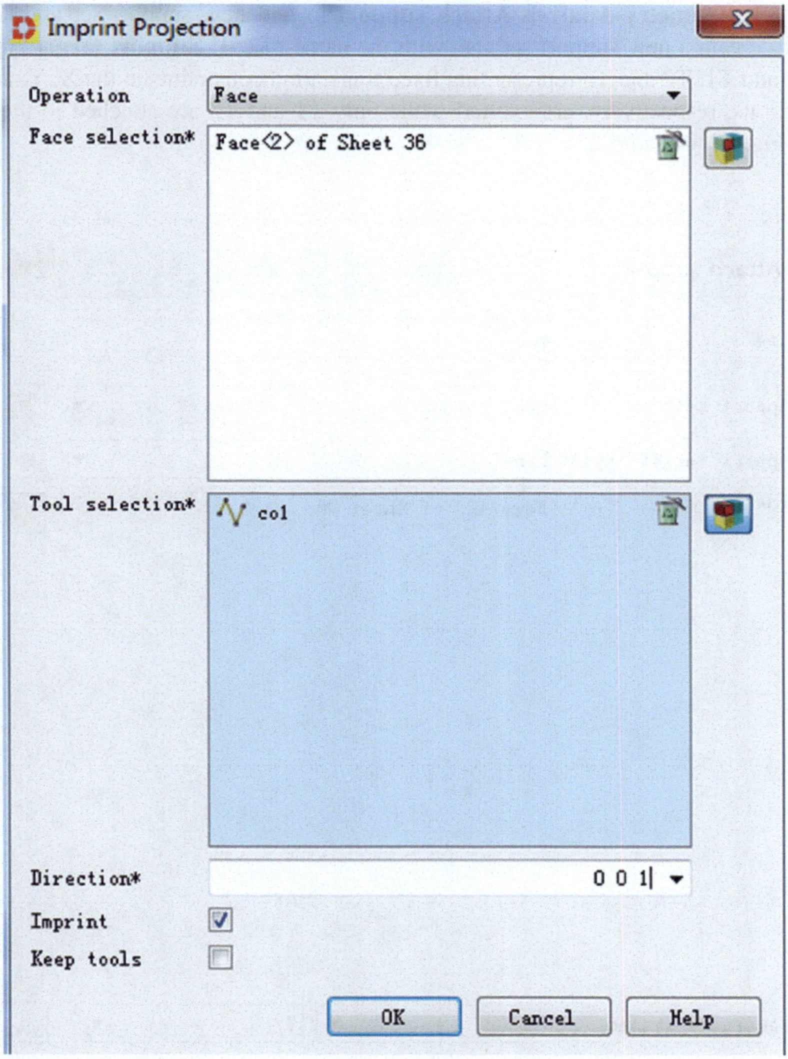

Fig. 5.82 Interface of imprint and projection for supports **col**

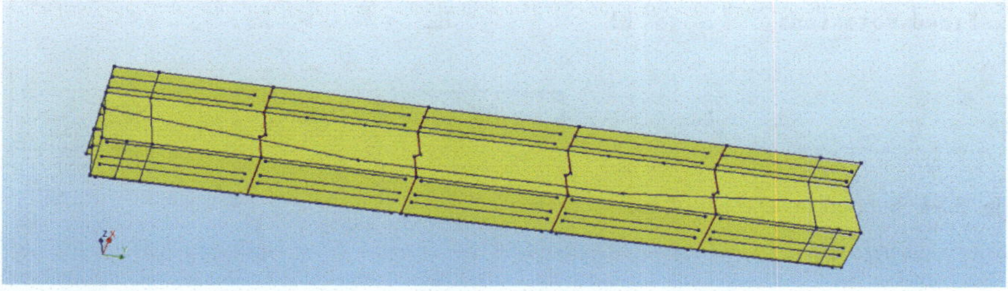

Fig. 5.83 Numerical model with imprinted lines after imprint and projection

Click **Geometry-Analysis-Attach support** to add new support set and select **co1** to create a new support set also with the name of co1. **Support target type** is **Line** and T1, T2 and T3 representing fixed translation constraints in the X, Y, and Z directions, respectively, are exerted while only T1 and T3 are attached to the **co2** (see Figs. 5.84 and 5.85).

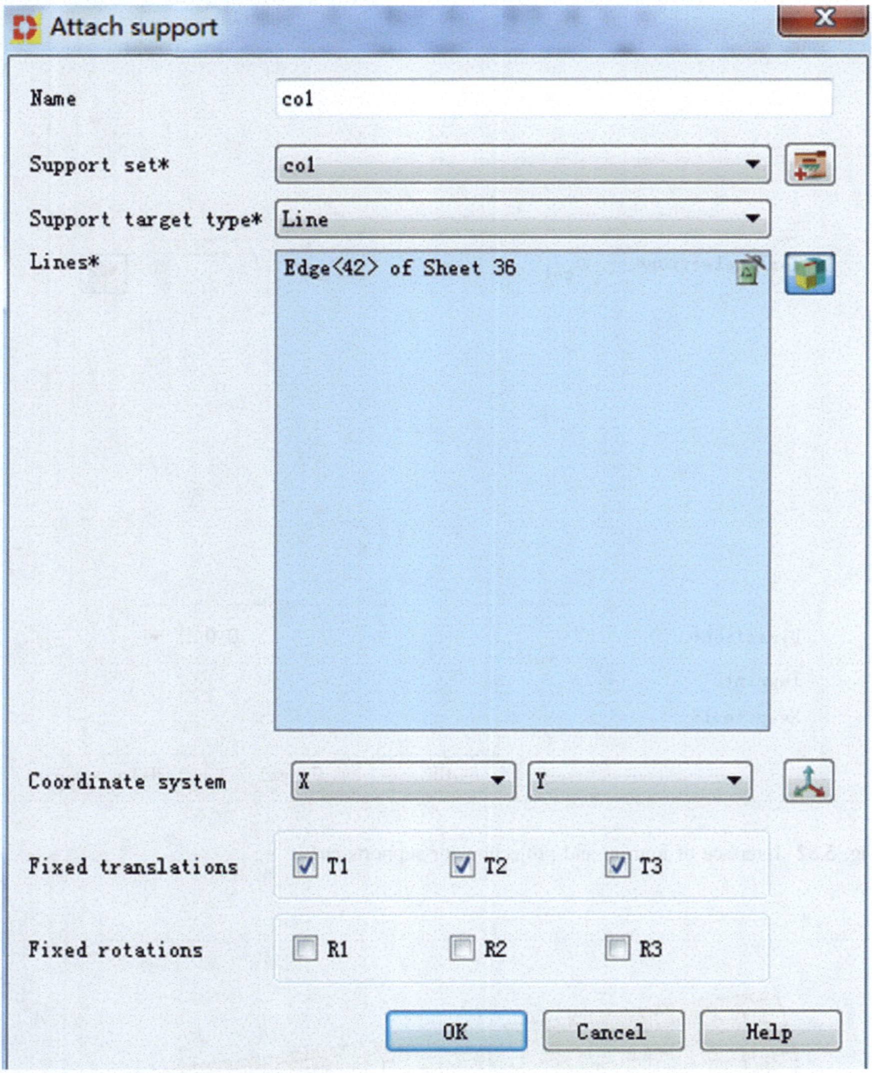

Fig. 5.84 Support attachment of **co1**

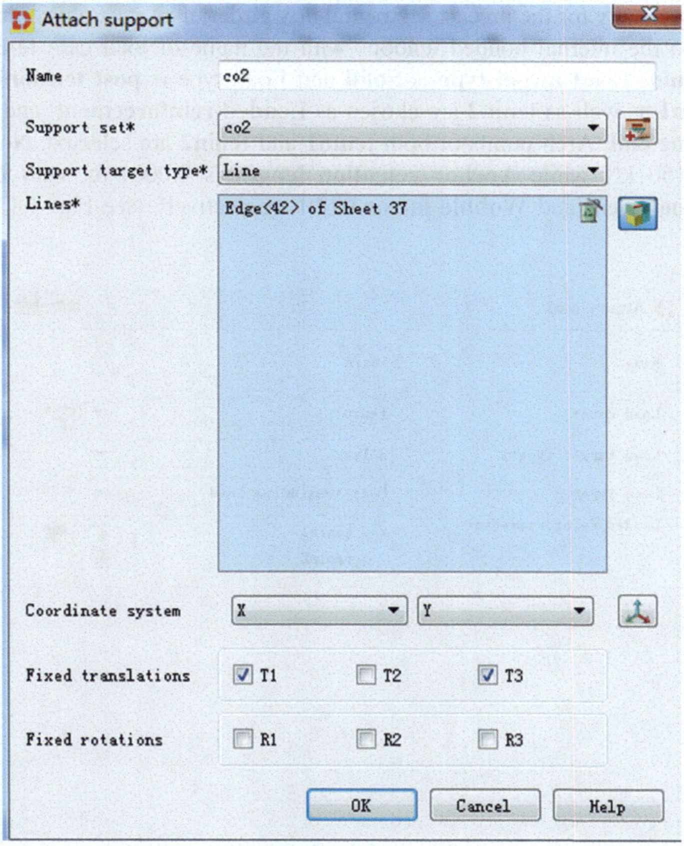

Fig. 5.85 Support attachment of **co2**

On clicking OK button, the generated numerical model with simply supported support is displayed (Fig. 5.86).

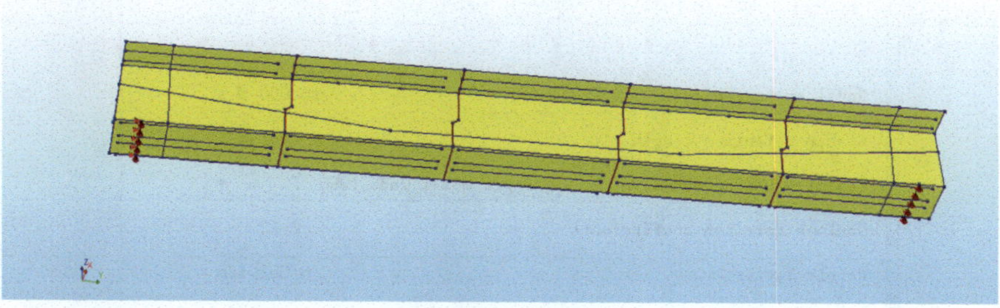

Fig. 5.86 Numerical model with simply supported support

Adding gravity for the precast segmental box girder, then post-tensioning load is attached to the internal bonded tendons with the name of load case **tenin**. **Load case** is **tenin**, **Load target type** is **Solid** and Load type is **post tensioning load**. Both **tenin1** as well as **tenin2** are chosen as **Loaded reinforcement**, and **Tension type** is **One end**. Arch points of both **tenin1** and **tenin2** are selected. **Nodal arch force** is 150 kN while **Anchor retention length** is 0.0001 m, with **Coulomb friction coefficient** and **Wobble factor** 0.001, respectively (see Fig. 5.87).

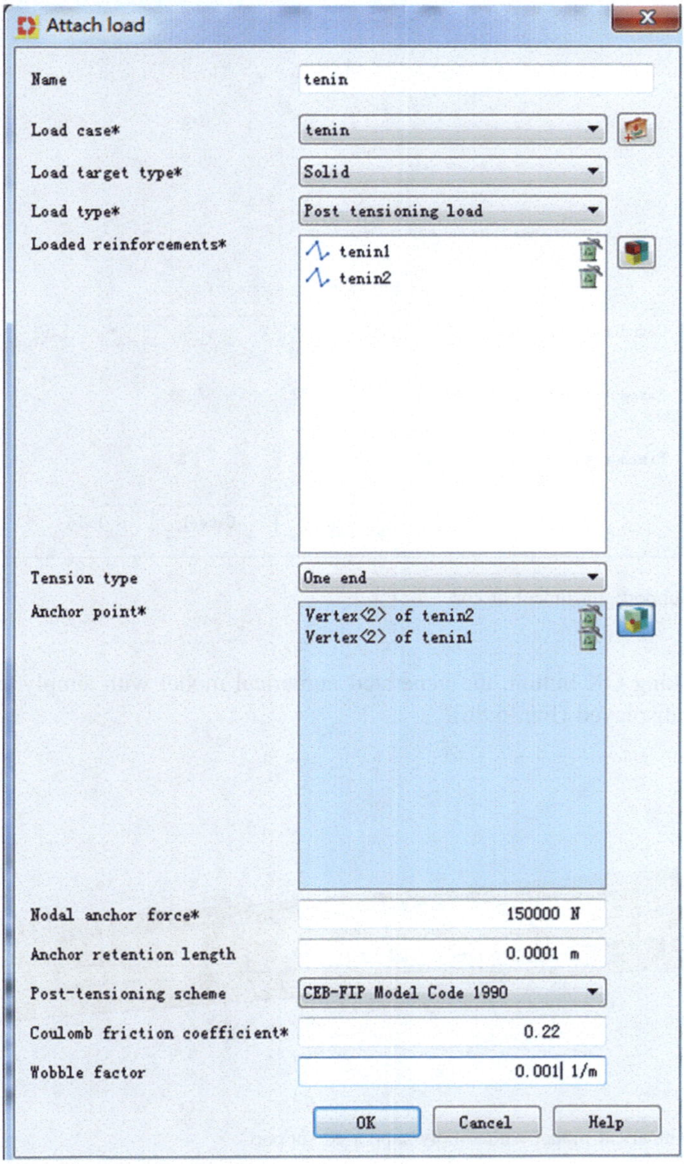

Fig. 5.87 Interface of attaching post-tensioning load

Holding down the mouse and scrolling the precast segmental box girder numerical model in the middle, distributed force is attached on the imprinted faces. The name of load case is **load** and **Load target type** is **Face. Surface force value** is 200 kN/m^2 in the negative Z direction, which is displayed in Fig. 5.88.

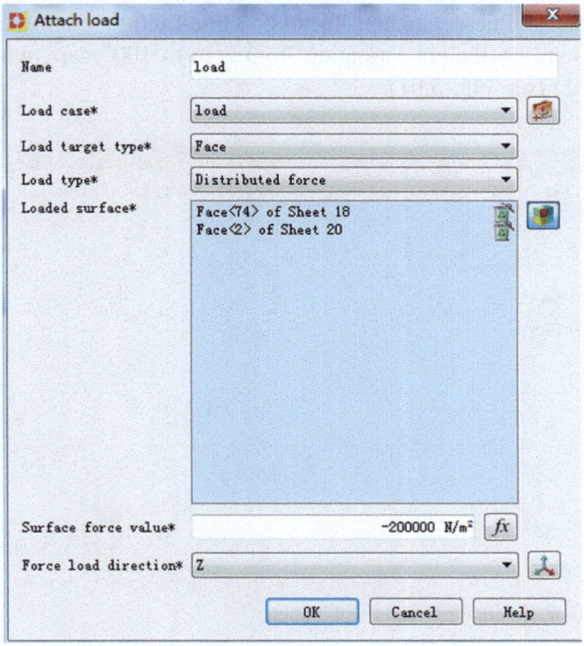

Fig. 5.88 Interface of attaching distributed force

Clicking OK button, the generated distributed force attached to the imprinted faces is displayed (Fig. 5.89).

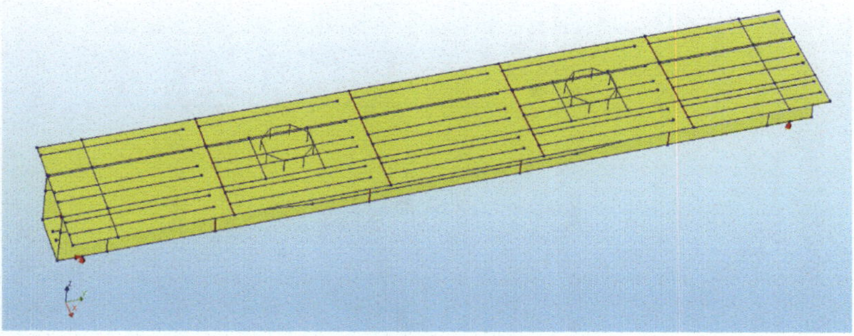

Fig. 5.89 Distributed force attached to the imprinted faces

Right-click **Combinations** to open **geometry load combination tables**, load cases of gravity and tendon are both added into the **Geometry load combination 1** and **load** case is added into **Geometry load combination 2** in solo with the factors all 1 (see Fig. 5.90). Then we specify time-dependent relationship for these two geometry load combinations. Click shortcut icon **Edit time dependency factors** to edit time-dependent factors; right-click **Edit time dependency** to open the dialog box of specifying time-dependent factors. **Time-factor** relationships are both defined as constant coefficient 1 ranging from zero to 100 years, which is (0 s, 1) (315360000 s, 1) (see Fig. 5.91).

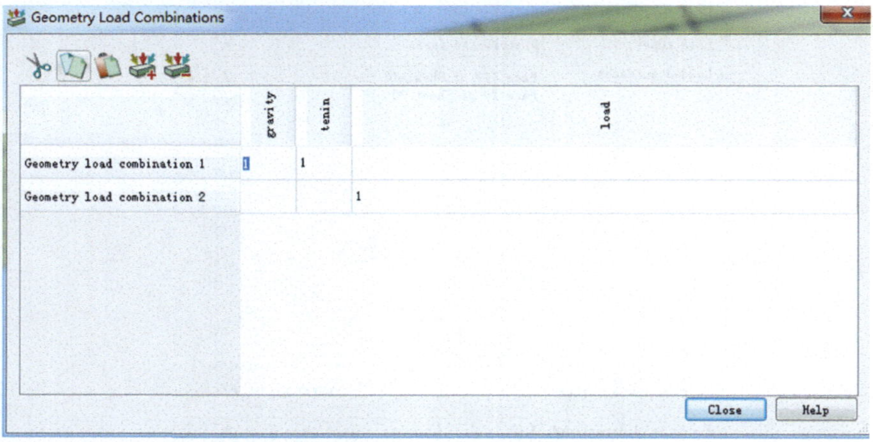

Fig. 5.90 Definition of geometry load combinations

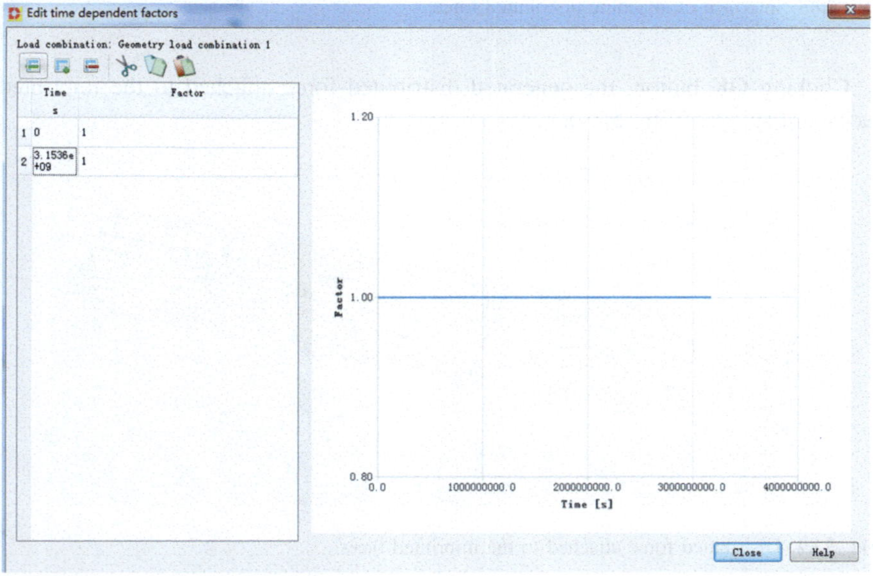

Fig. 5.91 Time–factorrelationships

Selecting all the sheets with concrete properties in the model to mesh them, **Seeding method** is **Element size** with **Desired size** 0.05 m. **Mesher type** is **Hexa/Quad** while determination of **Mid-side node location** is **Linear interpolation** (see Fig. 5.92). Similarly, selecting all the interface edges between segments in the edge selection, the **Operation** is **Edge** with desired size of element size 0.05 m (see Fig. 5.93).

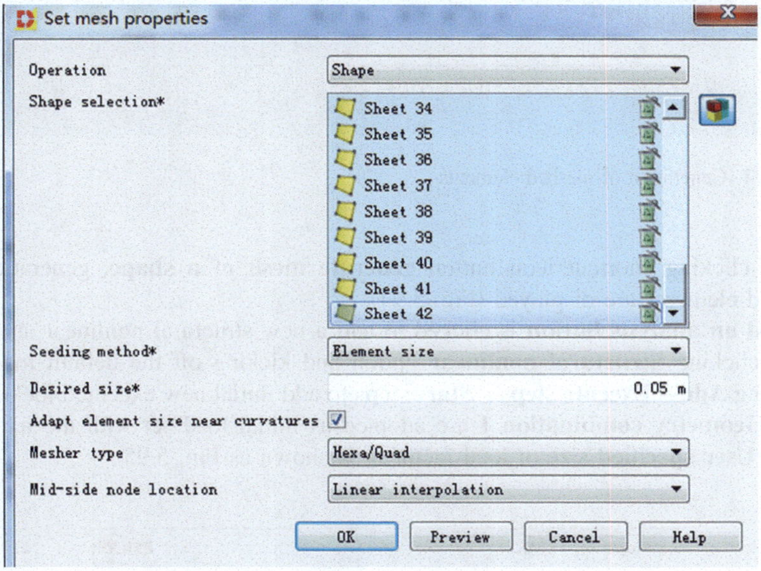

Fig. 5.92 Meshing properties for concrete sheets

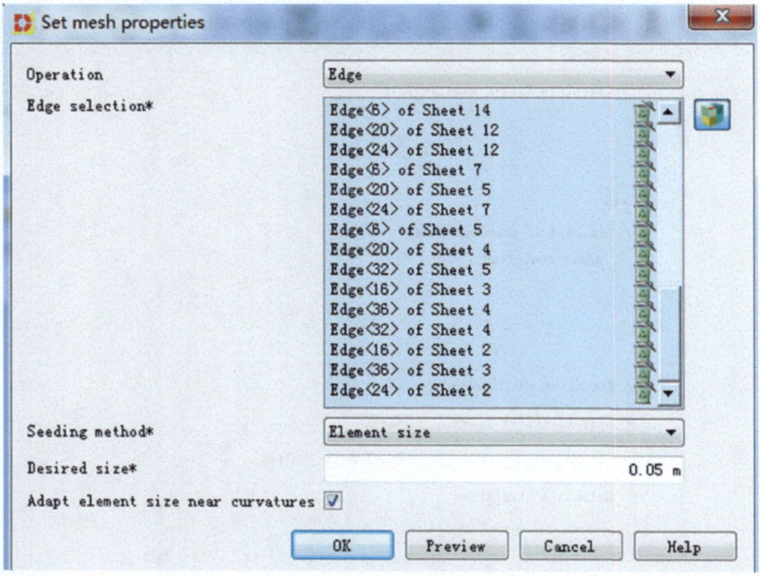

Fig. 5.93 Meshing properties for interface edges

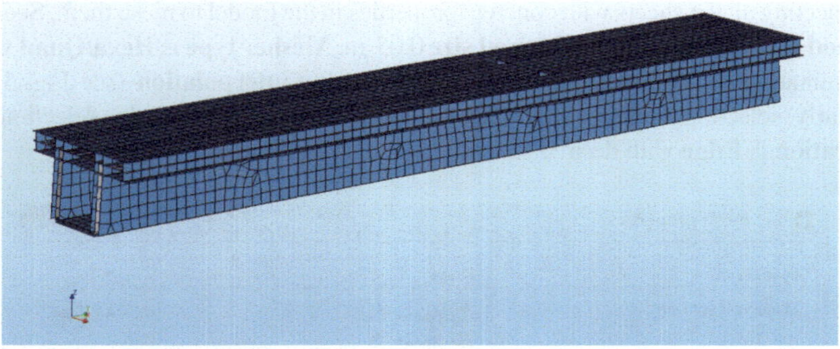

Fig. 5.94 Generation of meshed elements

On clicking shortcut icon button **generate mesh of a shape**, generation of meshed elements are displayed (Fig. 5.94).

Add an analysis button is clicked to add a new structural nonlinear analysis. Right-clicking **Structural nonlinear** option and kicking off the default load set, selecting **Add—Execute steps—Start steps** to add initial new execute block-**Start step**, **Geometry combination 1** are adopted as initial load set with the name of **tenin**. **User specified size** of load factor is 1, shown as Fig. 5.95.

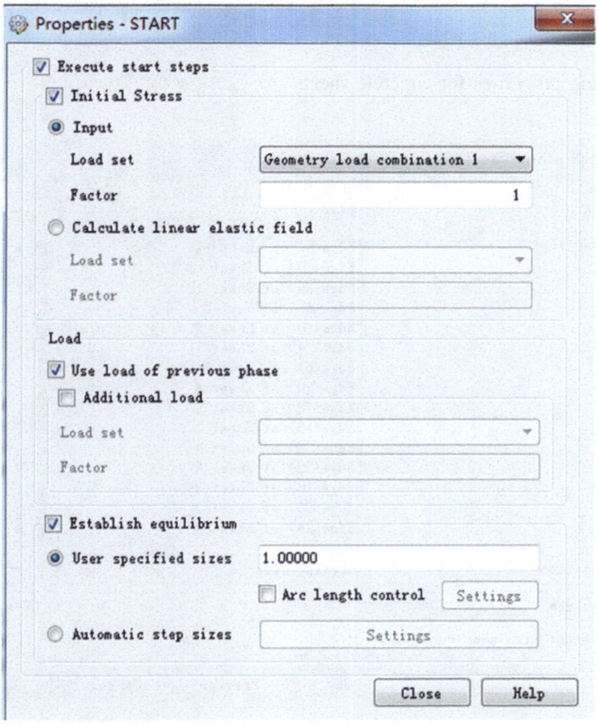

Fig. 5.95 Specification of initial step

Adding physical nonlinear properties and ticking fully bonded reinforcement at the same time, **Liquefaction** is unticked (see Fig. 5.96). Maximum number of iterations is 50 while both **Force** and **Displacement** are chosen as convergence norm (see Fig. 5.97).

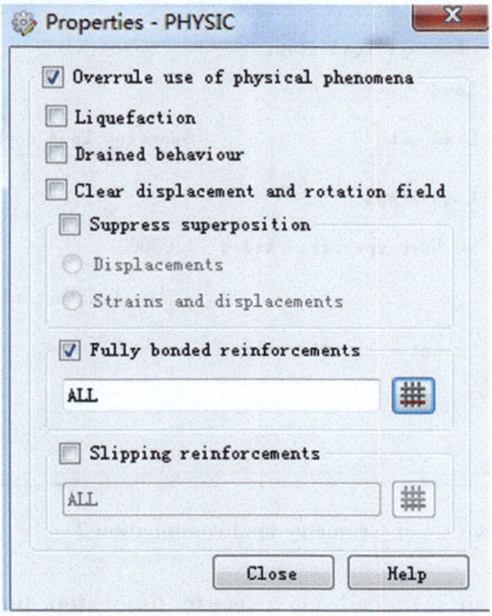

Fig. 5.96 Specification for physical nonlinear properties

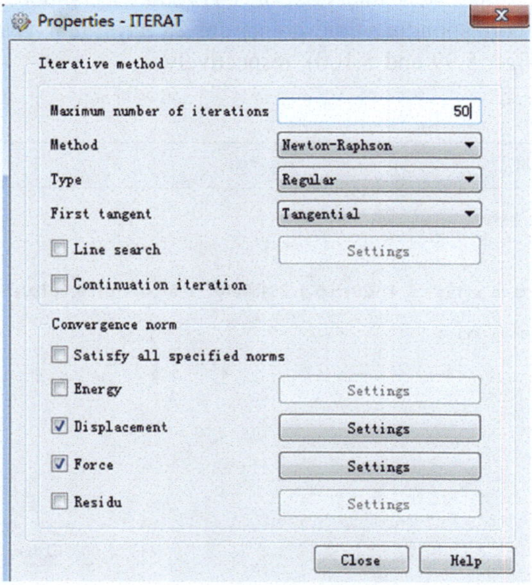

Fig. 5.97 Convergence norm

Geometry load combination 2 is added into load set of load step with the number of load step 1 as well as user-specified size 1.0000 (see Fig. 5.98). Other iteration parameters and specifications are the same as former.

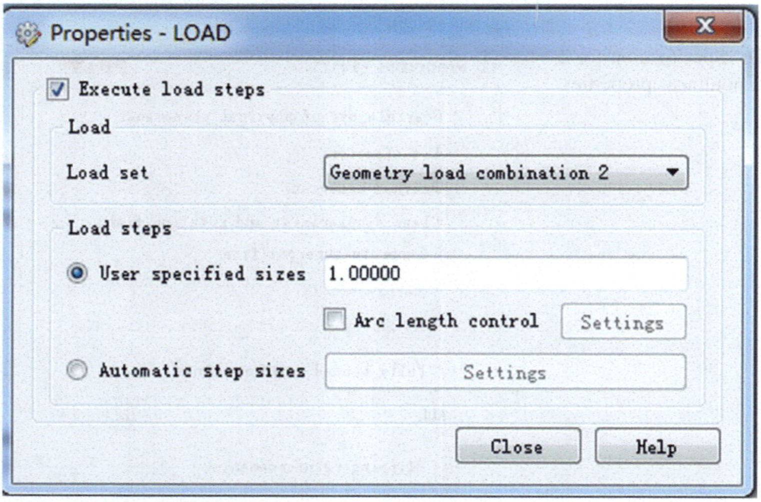

Fig. 5.98 Load set of **Geometry load combination 2**

Time step is added via **execute time step block** with the same iteration parameters as well as specifications except user-specified sizes, which are added according to the time step intervals of 2419200, 13348800, 15768000 and 63072000 s, and corresponding time points are 28 days, half year, one year and three years (see Figs. 5.99 and 5.100), respectively.

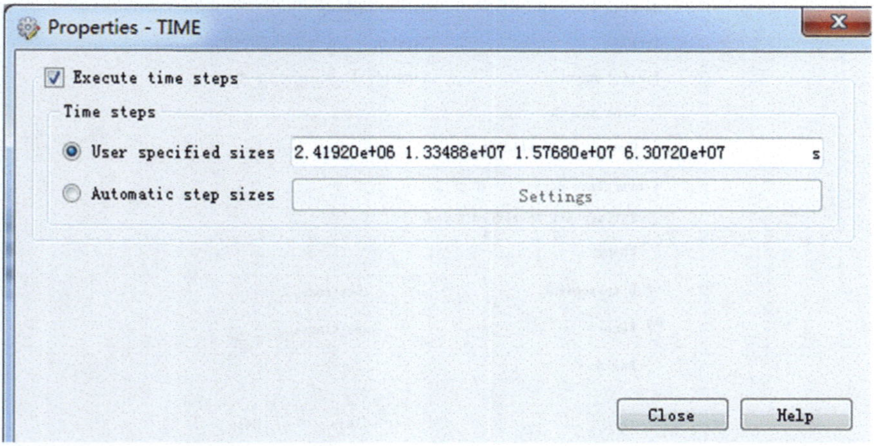

Fig. 5.99 Specification for time except user-specified sizes

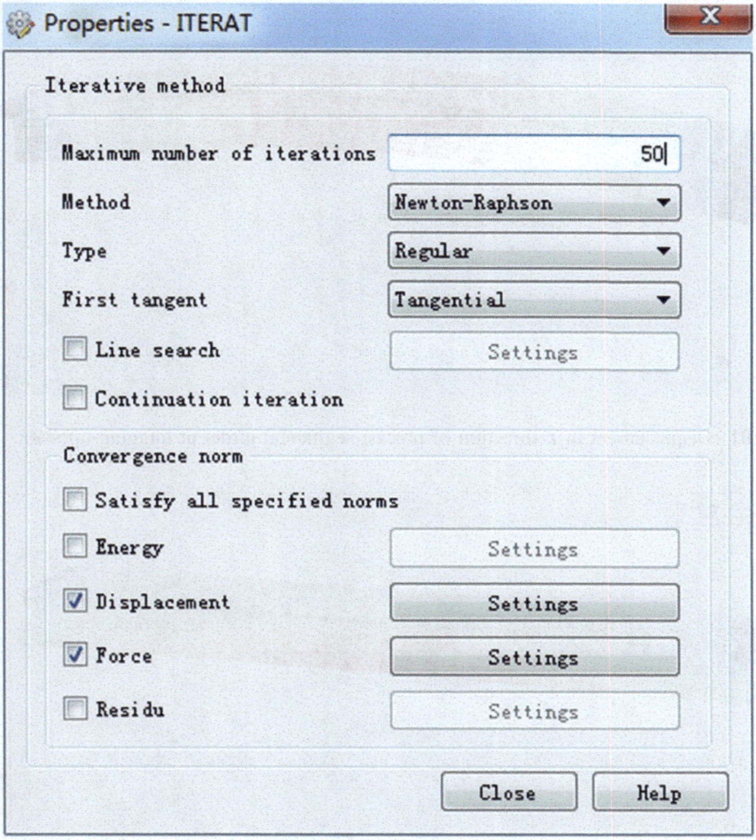

Fig. 5.100 Specifications for iteration parameters

Click **Run an analysis** button. After completion of calculation, contour plot of displacement in Z direction is checked via **Output-Nodal results-Total Displacements-TDtZ**, precast segmental girder at hugging-up state and ultimate state are displayed as Figs. 5.101 and 5.102, respectively. It is observed that hugging-up deflection is decreasing and ultimately reaching gentle as time goes on. We click **Viewer-node selection** to display all the nodes, then clicking **Reinforcement results-Reinforcement Cross-section Force-Nx**, selecting the node in the middle site of internal bonded tendon, where the node id is 16192, right-clicking to show table, outputs after loading and 3 years later are displayed as shown in Figs. 5.103 and 5.104. It can be calculated that prestress force loss in the three years is 7.095% according to the calculation $\frac{146151-135782}{146151} = 7.095\%$.

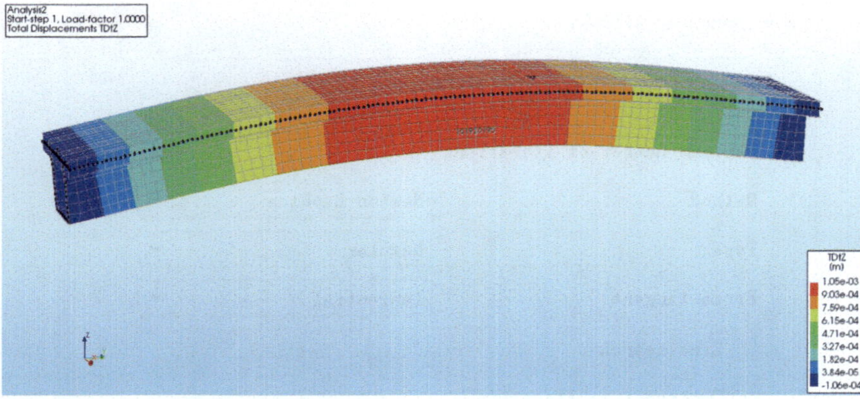

Fig. 5.101 Displacement in Z direction of precast segmental girder at hugging-up state

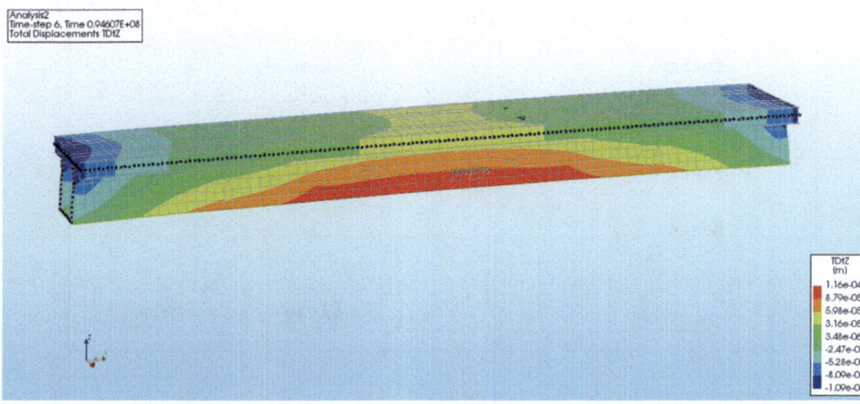

Fig. 5.102 Displacement in Z direction of precast segmental girder after 3 years

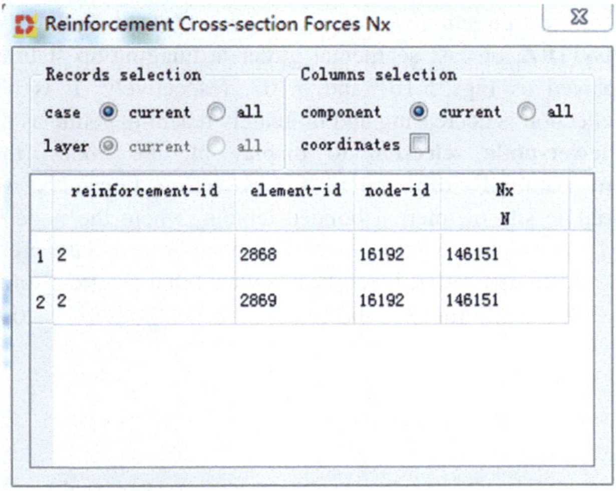

	reinforcement-id	element-id	node-id	Nx N
1	2	2868	16192	146151
2	2	2869	16192	146151

Fig. 5.103 Prestress force after loading

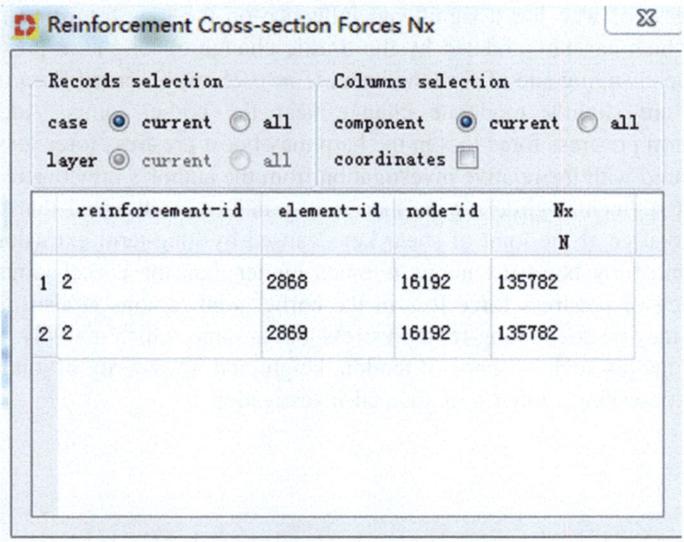

Fig. 5.104 Prestress force under ultimate state

Extracting prestress force loss at the length of 0.05, 0.15, 0.25, 0.3, 0.75, 0.8, 0.85, 1.3, 1.55, 1.6, 1.65 and 2 m, respectively, long-term prestress force loss during the 3 years is displayed as follows. Judging from the Fig. 5.105, conclusion can be drawn as follows:

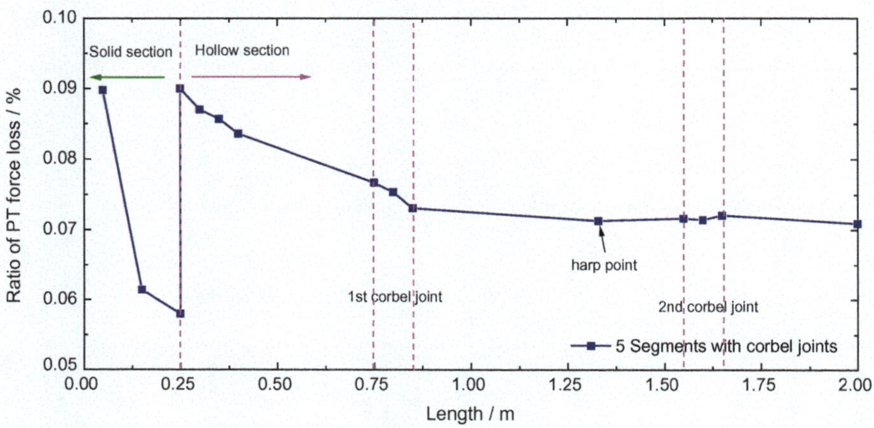

Fig. 5.105 Prestress force loss along the tendon path during the 3 years (in corbel joints)

1. Thickness of web has a significant influence on the long-term prestress force loss, which can be validated by the drastic change of long-term prestress loss near the changing site of web thickness (L = 0.25 m) observed from the figure.
2. There are slightly moderate change near the corbel joints. Additionally, long-term prestress force loss in the harp may boost prestress force loss slightly.
3. Compared with the relative investigation from the author's previous research on precast segmental girders with shear key joints, it is observed that long-term sudden surge at the joint of shear keys caused by long-term extrusion of concrete and fully bonded tendons is much higher than the corbel joints and the tendency of prestress force loss in the corbel joint regions is also distinctive, which may be due to long-term prestress loss at joints which is vastly influenced by the factors such as shape of tendon, height and irregularity of joint shape in reality deserving further systematical investigation.

Command console of this case in Python language is displayed as follows:

```
newProject( "Cobel", 100 )
setModelAnalysisAspects( [ "STRUCT" ] )
setModelDimension( "3D" )
setDefaultMeshOrder( "QUADRATIC" )
setDefaultMesherType( "HEXQUAD" )
setDefaultMidSideNodeLocation( "LINEAR" )
createSheet( "Sheet 1", [[ 0, 0, 0 ],[ 0, 0.25, 0 ],[ 0, 0.25, 0.3 ],[ 0, 0.025, 0.3 ]] )
createSheet( "Sheet 2", [[ 0, 0.25, 0 ],[ 0, 0.8, 0 ],[ 0, 0.8, 0.13 ],[ 0, 0.83, 0.15 ],[ 0, 0.83, 0.3 ],[ 0, 0.25, 0.3 ]] )
saveProject( )
createSheet( "Sheet 3", [[ 0, 0.8, 0 ],[ 0, 1.6, 0 ],[ 0, 1.6, 0.13 ],[ 0, 1.63, 0.15 ],[ 0, 1.63, 0.3 ],[ 0, 0.83, 0.3 ],[ 0, 0.83, 0.15 ],[ 0, 0.8, 0.13 ]] )
saveProject( )
createSheet( "Sheet 4", [[ 0, 1.6, 0 ],[ 0, 2.4, 0 ],[ 0, 2.4, 0.13 ],[ 0, 2.43, 0.15 ],[ 0, 2.43, 0.3 ],[ 0, 1.63, 0.3 ],[ 0, 1.63, 0.15 ],[ 0, 1.6, 0.13 ]] )
saveProject( )
createSheet( "Sheet 5", [[ 0, 2.4, 0 ],[ 0, 3.2, 0 ],[ 0, 3.2, 0.13 ],[ 0, 3.23, 0.15 ],[ 0, 3.23, 0.3 ],[ 0, 2.43, 0.3 ],[ 0, 2.43, 0.15 ],[ 0, 2.4, 0.13 ]] )
saveProject( )
mirror( [ "Sheet 1" ], [ 0, 2, 0 ], [ False, True, False ], True )
saveProject( )
createSheet( "Sheet 7", [[ 0, 3.2, 0 ],[ 0, 3.75, 0 ],[ 0, 3.75, 0.3 ],[ 0, 3.23, 0.3 ],[ 0, 3.23, 0.15 ],[ 0, 3.2, 0.13 ]] )
saveProject( )
arrayCopy( [ "Sheet 1", "Sheet 2", "Sheet 3", "Sheet 4", "Sheet 5", "Sheet 6", "Sheet 7" ], [ 0.29, 0, 0 ], [ 0, 0, 0 ], [ 0, 0, 0 ], 1 )
saveProject( )
createSheet( "Sheet 15", [[ 0, 0.025, 0.3 ],[ 0.29, 0.025, 0.3 ],[ 0.29, 0.25, 0.3 ],[ 0, 0.25, 0.3 ]] )
createSheet( "Sheet 16", [[ 0, 0.025, 0.3 ],[ 0, 0.25, 0.3 ],[ -0.195, 0.25, 0.3 ],[ -0.195, 0.025, 0.3 ]] )
saveProject( )
createSheet( "Sheet 17", [[ 0, 0.25, 0.3 ],[ 0.29, 0.25, 0.3 ],[ 0.29, 0.83, 0.3 ],[ 0, 0.83, 0.3 ]] )
createSheet( "Sheet 18", [[ 0, 0.83, 0.3 ],[ 0.29, 0.83, 0.3 ],[ 0.29, 1.63, 0.3 ],[ 0, 1.63, 0.3 ]] )
createSheet( "Sheet 19", [[ 0, 1.63, 0.3 ],[ 0.29, 1.63, 0.3 ],[ 0.29, 2.43, 0.3 ],[ 0, 2.43, 0.3 ]] )
createSheet( "Sheet 20", [[ 0, 2.43, 0.3 ],[ 0.29, 2.43, 0.3 ],[ 0.29, 3.23, 0.3 ],[ 0, 3.23, 0.3 ]] )
createSheet( "Sheet 21", [[ 0, 3.23, 0.3 ],[ 0.29, 3.23, 0.3 ],[ 0.29, 3.75, 0.3 ],[ 0, 3.75, 0.3 ]] )
createSheet( "Sheet 22", [[ 0, 3.75, 0.3 ],[ 0.29, 3.75, 0.3 ],[ 0.29, 3.975, 0.3 ],[ 0, 3.975, 0.3 ]] )
createSheet( "Sheet 23", [[ -0.195, 0.25, 0.3 ],[ 0, 0.25, 0.3 ],[ 0, 0.83, 0.3 ],[ -0.195, 0.83, 0.3 ]] )
```

```
createSheet( "Sheet 24", [[ -0.195, 0.83, 0.3 ],[ 0, 0.83, 0.3 ],[ 0, 1.63, 0.3 ],[ -0.195, 1.63, 0.3 ]] )
createSheet( "Sheet 25", [[ -0.195, 1.63, 0.3 ],[ 0, 1.63, 0.3 ],[ 0, 2.43, 0.3 ],[ -0.195, 2.43, 0.3 ]] )
createSheet( "Sheet 26", [[ -0.195, 2.43, 0.3 ],[ 0, 2.43, 0.3 ],[ 0, 3.23, 0.3 ],[ -0.195, 3.23, 0.3 ]] )
saveProject(  )
createSheet( "Sheet 27", [[ -0.195, 3.23, 0.3 ],[ 0, 3.23, 0.3 ],[ 0, 3.75, 0.3 ],[ -0.195, 3.75, 0.3 ]] )
saveProject(  )
createSheet( "Sheet 28", [[ -0.195, 3.75, 0.3 ],[ 0, 3.75, 0.3 ],[ 0, 3.975, 0.3 ],[ -0.195, 3.975, 0.3 ]] )
saveProject(  )
arrayCopy( [ "Sheet 16", "Sheet 23", "Sheet 24", "Sheet 25", "Sheet 26", "Sheet 27", "Sheet 28" ], [ 0.485, 0, 0 ], [ 0, 0, 0 ], [ 0, 0, 0 ], 1 )
saveProject(  )
createSheet( "Sheet 36", [[ 0.29, 0, 0 ],[ 0, 0, 0 ],[ 0, 0.25, 0 ],[ 0.29, 0.25, 0 ]] )
createSheet( "Sheet 37", [[ 0.29, 3.75, 0 ],[ 0, 3.75, 0 ],[ 0, 4, 0 ],[ 0.29, 4, 0 ]] )
saveProject(  )
createSheet( "Sheet 38", [[ 0.29, 0.25, 0 ],[ 0, 0.25, 0 ],[ 0, 0.8, 0 ],[ 0.29, 0.8, 0 ]] )
saveProject(  )
createSheet( "Sheet 39", [[ 0.29, 0.8, 0 ],[ 0, 0.8, 0 ],[ 0, 1.6, 0 ],[ 0.29, 1.6, 0 ]] )
createSheet( "Sheet 40", [[ 0.29, 1.6, 0 ],[ 0, 1.6, 0 ],[ 0, 2.4, 0 ],[ 0.29, 2.4, 0 ]] )
createSheet( "Sheet 41", [[ 0.29, 2.4, 0 ],[ 0, 2.4, 0 ],[ 0, 3.2, 0 ],[ 0.29, 3.2, 0 ]] )
createSheet( "Sheet 42", [[ 0.29, 3.2, 0 ],[ 0, 3.2, 0 ],[ 0, 3.75, 0 ],[ 0.29, 3.75, 0 ]] )
saveProject(  )
createPolyline( "Polyline 1", [[ 0, 0.01625, 0.195 ],[ 0, 1.3, 0.06 ],[ 0, 2.7, 0.06 ],[ 0, 3.98375, 0.195 ]], False )
renameShape( "Polyline 1", "tenin1" )
arrayCopy( [ "tenin1" ], [ 0.29, 0, 0 ], [ 0, 0, 0 ], [ 0, 0, 0 ], 1 )
saveProject(  )
setUnit( "TEMPER", "CELSIU" )
setUnit( "ANGLE", "DEGREE" )
addMaterial( "top", "CONCDC", "MC1990", [ "CREEP", "SHRINK" ] )
setParameter( MATERIAL, "top", "MC90CO/GRADE", "C50" )
setParameter( MATERIAL, "top", "MC90CO/H", 0.109 )
setParameter( MATERIAL, "top", "MC90CO/RH", 55 )
setParameter( MATERIAL, "top", "MC90CO/RH", 55 )
setParameter( MATERIAL, "top", "MC90CO/RH", 55 )
setParameter( MATERIAL, "top", "CONCDI/YOUNG", 3.8629e+10 )
setParameter( MATERIAL, "top", "CONCDI/YOUN28", 3.45e+10 )
```

```
setParameter( MATERIAL, "top", "CONCDI/POISON", 0.15 )
setParameter( MATERIAL, "top", "CONCDI/THERMX", 1.2e-05 )
setParameter( MATERIAL, "top", "CONCDI/DENSIT", 2500 )
setParameter( MATERIAL, "top", "CONCDI/FCK28", 50000000 )
setParameter( MATERIAL, "top", "CONCDI/FCM28", 58000000 )
setParameter( MATERIAL, "top", "CONCDI/FCM28", 58000000 )
setParameter( MATERIAL, "top", "CONCDI/FCM28", 58000000 )
setParameter( MATERIAL, "top", "CONCCP/AGETYP", "AGING" )
setParameter( MATERIAL, "top", "CONCCP/AGING", 2419200 )
setParameter( MATERIAL, "top", "CONCSH/CURAGE", 86400 )
addGeometry( "Element geometry 1", "SHEET", "CURSHL", [] )
rename( GEOMET, "Element geometry 1", "top" )
setParameter( GEOMET, "top", "THICK", 0.13 )
setParameter( GEOMET, "top", "LOCAXS", True )
setParameter( GEOMET, "top", "LOCAXS/XAXIS", [ 1, 0, 0 ] )
setParameter( GEOMET, "top", "LOCAXS/XAXIS", [ 0, 1, 0 ] )
setParameter( GEOMET, "top", "LOCAXS/XAXIS", [ 0, 1, 0 ] )
setParameter( GEOMET, "top", "LOCAXS/XAXIS", [ 0, 1, 0 ] )
setParameter( GEOMET, "top", "LOCAXS/XAXIS", [ 0, 1, 0 ] )
clearReinforcementAspects( [ "Sheet 15", "Sheet 16", "Sheet 17", "Sheet 18", "Sheet 19", "Sheet 20", "Sheet 21", "Sheet 22", "Sheet 23", "Sheet 24", "Sheet 25", "Sheet 26", "Sheet 27", "Sheet 28", "Sheet 29", "Sheet 30", "Sheet 31", "Sheet 32", "Sheet 33", "Sheet 34", "Sheet 35" ] )
setElementClassType( SHAPE, [ "Sheet 15", "Sheet 16", "Sheet 17", "Sheet 18", "Sheet 19", "Sheet 20", "Sheet 21", "Sheet 22", "Sheet 23", "Sheet 24", "Sheet 25", "Sheet 26", "Sheet 27", "Sheet 28", "Sheet 29", "Sheet 30", "Sheet 31", "Sheet 32", "Sheet 33", "Sheet 34", "Sheet 35" ], "CURSHL" )
assignMaterial( "top", SHAPE, [ "Sheet 15", "Sheet 16", "Sheet 17", "Sheet 18", "Sheet 19", "Sheet 20", "Sheet 21", "Sheet 22", "Sheet 23", "Sheet 24", "Sheet 25", "Sheet 26", "Sheet 27", "Sheet 28", "Sheet 29", "Sheet 30", "Sheet 31", "Sheet 32", "Sheet 33", "Sheet 34", "Sheet 35" ] )
assignGeometry( "top", SHAPE, [ "Sheet 15", "Sheet 16", "Sheet 17", "Sheet 18", "Sheet 19", "Sheet 20", "Sheet 21", "Sheet 22", "Sheet 23", "Sheet 24", "Sheet 25", "Sheet 26", "Sheet 27", "Sheet 28", "Sheet 29", "Sheet 30", "Sheet 31", "Sheet 32", "Sheet 33", "Sheet 34", "Sheet 35" ] )
resetElementData( SHAPE, [ "Sheet 15", "Sheet 16", "Sheet 17", "Sheet 18", "Sheet 19", "Sheet 20", "Sheet 21", "Sheet 22", "Sheet 23", "Sheet 24", "Sheet 25", "Sheet 26", "Sheet 27", "Sheet 28", "Sheet 29", "Sheet 30", "Sheet 31", "Sheet 32", "Sheet 33", "Sheet 34", "Sheet 35" ] )
saveProject(  )
addMaterial( "bot", "CONCDC", "MC1990", [ "CREEP", "SHRINK" ] )
setParameter( MATERIAL, "bot", "MC90CO/GRADE", "C50" )
setParameter( MATERIAL, "bot", "MC90CO/H", 0.15 )
setParameter( MATERIAL, "bot", "MC90CO/H", 0.044 )
setParameter( MATERIAL, "bot", "MC90CO/RH", 55 )
```

```
setParameter( MATERIAL, "bot", "MC90CO/RH", 55 )
setParameter( MATERIAL, "bot", "MC90CO/RH", 55 )
setParameter( MATERIAL, "bot", "CONCDI/YOUNG", 3.8629e+10 )
setParameter( MATERIAL, "bot", "CONCDI/YOUN28", 3.45e+10 )
setParameter( MATERIAL, "bot", "CONCDI/POISON", 0.15 )
setParameter( MATERIAL, "bot", "CONCDI/THERMX", 1.2e-05 )
setParameter( MATERIAL, "bot", "CONCDI/DENSIT", 2500 )
setParameter( MATERIAL, "bot", "CONCDI/FCK28", 50000000 )
setParameter( MATERIAL, "bot", "CONCDI/FCM28", 58000000 )
setParameter( MATERIAL, "bot", "CONCCP/AGETYP", "AGING" )
setParameter( MATERIAL, "bot", "CONCCP/AGING", 2419200 )
setParameter( MATERIAL, "bot", "CONCSH/CURAGE", 86400 )
addGeometry( "Element geometry 2", "SHEET", "CURSHL", [] )
rename( GEOMET, "Element geometry 2", "bot" )
setParameter( GEOMET, "bot", "THICK", 0.05 )
setParameter( GEOMET, "bot", "LOCAXS", True )
setParameter( GEOMET, "bot", "LOCAXS/XAXIS", [ 1, 0, 0 ] )
setParameter( GEOMET, "bot", "LOCAXS/XAXIS", [ 0, 1, 0 ] )
setParameter( GEOMET, "bot", "LOCAXS/XAXIS", [ 0, 1, 0 ] )
setParameter( GEOMET, "bot", "LOCAXS/XAXIS", [ 0, 1, 0 ] )
clearReinforcementAspects( [ "Sheet 36", "Sheet 37", "Sheet 38", "Sheet 39", "Sheet 40", 
"Sheet 41", "Sheet 42" ] )
setElementClassType( SHAPE, [ "Sheet 36", "Sheet 37", "Sheet 38", "Sheet 39", "Sheet 40", 
"Sheet 41", "Sheet 42" ], "CURSHL" )
assignMaterial( "bot", SHAPE, [ "Sheet 36", "Sheet 37", "Sheet 38", "Sheet 39", "Sheet 40", 
"Sheet 41", "Sheet 42" ] )
assignGeometry( "bot", SHAPE, [ "Sheet 36", "Sheet 37", "Sheet 38", "Sheet 39", "Sheet 40", 
"Sheet 41", "Sheet 42" ] )
resetElementData( SHAPE, [ "Sheet 36", "Sheet 37", "Sheet 38", "Sheet 39", "Sheet 40", 
"Sheet 41", "Sheet 42" ] )
saveProject(  )
addMaterial( "mid1", "CONCDC", "MC1990", [ "CREEP", "SHRINK" ] )
setParameter( MATERIAL, "mid1", "MC90CO/GRADE", "C50" )
setParameter( MATERIAL, "mid1", "MC90CO/H", 0.226 )
setParameter( MATERIAL, "mid1", "MC90CO/RH", 55 )
setParameter( MATERIAL, "mid1", "MC90CO/RH", 55 )
setParameter( MATERIAL, "mid1", "MC90CO/RH", 55 )
setParameter( MATERIAL, "mid1", "CONCDI/YOUNG", 3.8629e+10 )
setParameter( MATERIAL, "mid1", "CONCDI/YOUN28", 3.45e+10 )
setParameter( MATERIAL, "mid1", "CONCDI/POISON", 0.15 )
setParameter( MATERIAL, "mid1", "CONCDI/THERMX", 1.2e-05 )
```

```
setParameter( MATERIAL, "mid1", "CONCDI/DENSIT", 2500 )
setParameter( MATERIAL, "mid1", "CONCDI/FCK28", 50000000 )
setParameter( MATERIAL, "mid1", "CONCDI/FCM28", 58000000 )
setParameter( MATERIAL, "mid1", "CONCCP/AGETYP", "AGING" )
setParameter( MATERIAL, "mid1", "CONCCP/AGING", 2419200 )
setParameter( MATERIAL, "mid1", "CONCSH/CURAGE", 86400 )
addGeometry( "Element geometry 3", "SHEET", "CURSHL", [] )
rename( GEOMET, "Element geometry 3", "mid1" )
setParameter( GEOMET, "mid1", "THICK", 0.34 )
setParameter( GEOMET, "mid1", "LOCAXS", True )
setParameter( GEOMET, "mid1", "LOCAXS/XAXIS", [ 1, 0, 0 ] )
setParameter( GEOMET, "mid1", "LOCAXS/XAXIS", [ 0, 1, 0 ] )
setParameter( GEOMET, "mid1", "LOCAXS/XAXIS", [ 0, 1, 0 ] )
setParameter( GEOMET, "mid1", "LOCAXS/XAXIS", [ 0, 1, 0 ] )
setParameter( GEOMET, "mid1", "LOCAXS/XAXIS", [ 0, 1, 0 ] )
clearReinforcementAspects( [ "Sheet 1", "Sheet 6", "Sheet 8", "Sheet 13" ] )
setElementClassType( SHAPE, [ "Sheet 1", "Sheet 6", "Sheet 8", "Sheet 13" ], "CURSHL" )
assignMaterial( "mid1", SHAPE, [ "Sheet 1", "Sheet 6", "Sheet 8", "Sheet 13" ] )
assignGeometry( "mid1", SHAPE, [ "Sheet 1", "Sheet 6", "Sheet 8", "Sheet 13" ] )
resetElementData( SHAPE, [ "Sheet 1", "Sheet 6", "Sheet 8", "Sheet 13" ] )
saveProject(  )
addMaterial( "mid2", "CONCDC", "MC1990", [ "CREEP", "SHRINK" ] )
setParameter( MATERIAL, "mid2", "MC90CO/GRADE", "C50" )
setParameter( MATERIAL, "mid2", "MC90CO/AMBTEM", 20 )
setParameter( MATERIAL, "mid2", "MC90CO/H", 0.126 )
setParameter( MATERIAL, "mid2", "MC90CO/RH", 55 )
setParameter( MATERIAL, "mid2", "MC90CO/RH", 55 )
setParameter( MATERIAL, "mid2", "MC90CO/RH", 55 )
setParameter( MATERIAL, "mid2", "CONCDI/YOUNG", 3.8629e+10 )
setParameter( MATERIAL, "mid2", "CONCDI/YOUN28", 3.45e+10 )
setParameter( MATERIAL, "mid2", "CONCDI/POISON", 0.15 )
setParameter( MATERIAL, "mid2", "CONCDI/THERMX", 1.2e-05 )
setParameter( MATERIAL, "mid2", "CONCDI/DENSIT", 2500 )
setParameter( MATERIAL, "mid2", "CONCDI/FCK28", 50000000 )
setParameter( MATERIAL, "mid2", "CONCDI/FCM28", 58000000 )
setParameter( MATERIAL, "mid2", "CONCCP/AGETYP", "AGING" )
setParameter( MATERIAL, "mid2", "CONCCP/AGING", 2419200 )
setParameter( MATERIAL, "mid2", "CONCSH/CURAGE", 86400 )
addGeometry( "Element geometry 4", "SHEET", "CURSHL", [] )
rename( GEOMET, "Element geometry 4", "mid2" )
setParameter( GEOMET, "mid2", "THICK", 0.09 )
```

```
setParameter( GEOMET, "mid2", "LOCAXS", True )
setParameter( GEOMET, "mid2", "LOCAXS/XAXIS", [ 1, 0, 0 ] )
setParameter( GEOMET, "mid2", "LOCAXS/XAXIS", [ 0, 1, 0 ] )
setParameter( GEOMET, "mid2", "LOCAXS/XAXIS", [ 0, 1, 0 ] )
setParameter( GEOMET, "mid2", "LOCAXS/XAXIS", [ 0, 1, 0 ] )
clearReinforcementAspects( [ "Sheet 2", "Sheet 3", "Sheet 4", "Sheet 5", "Sheet 7", "Sheet 9", "Sheet 10", "Sheet 11", "Sheet 12", "Sheet 14" ] )
setElementClassType( SHAPE, [ "Sheet 2", "Sheet 3", "Sheet 4", "Sheet 5", "Sheet 7", "Sheet 9", "Sheet 10", "Sheet 11", "Sheet 12", "Sheet 14" ], "CURSHL" )
assignMaterial( "mid2", SHAPE, [ "Sheet 2", "Sheet 3", "Sheet 4", "Sheet 5", "Sheet 7", "Sheet 9", "Sheet 10", "Sheet 11", "Sheet 12", "Sheet 14" ] )
assignGeometry( "mid2", SHAPE, [ "Sheet 2", "Sheet 3", "Sheet 4", "Sheet 5", "Sheet 7", "Sheet 9", "Sheet 10", "Sheet 11", "Sheet 12", "Sheet 14" ] )
resetElementData( SHAPE, [ "Sheet 2", "Sheet 3", "Sheet 4", "Sheet 5", "Sheet 7", "Sheet 9", "Sheet 10", "Sheet 11", "Sheet 12", "Sheet 14" ] )
saveProject(  )
saveProject(  )
addMaterial( "tenin", "REINFO", "VMISES", [] )
setParameter( "MATERIAL", "tenin", "LINEAR/ELASTI/YOUNG", 1.95e+11 )
setParameter( "MATERIAL", "tenin", "PLASTI/HARDI1/YLDSTR", 1.86e+09 )
addGeometry( "Element geometry 5", "RELINE", "REBAR", [] )
rename( GEOMET, "Element geometry 5", "tenin" )
setParameter( "GEOMET", "tenin", "REIEMB/CROSSE", 0.000139 )
setReinforcementAspects( [ "tenin1", "tenin2" ] )
assignMaterial( "tenin", "SHAPE", [ "tenin1", "tenin2" ] )
assignGeometry( "tenin", "SHAPE", [ "tenin1", "tenin2" ] )
resetElementData( "SHAPE", [ "tenin1", "tenin2" ] )
setReinforcementDiscretization( [ "tenin1", "tenin2" ], "SECTION" )
saveProject(  )
createLine( "Line 1", [ 0.41, 0.03, 0.3 ], [ 0.41, 0.75, 0.3 ] )
renameShape( "Line 1", "bar1" )
createLine( "bar2", [ 0.33, 0.03, 0.3 ], [ 0.33, 0.75, 0.3 ] )
saveProject(  )
createLine( "bar3", [ 0.19, 0.03, 0.3 ], [ 0.19, 0.75, 0.3 ] )
saveProject(  )
createLine( "bar4", [ 0.123, 0.03, 0.3 ], [ 0.123, 0.75, 0.3 ] )
saveProject(  )
createLine( "bar5", [ 0.01, 0.03, 0.3 ], [ 0.01, 0.75, 0.3 ] )
saveProject(  )
createLine( "bar6", [ -0.14, 0.03, 0.3 ], [ -0.14, 0.75, 0.3 ] )
saveProject(  )
```

```
createLine( "bar7", [ 0.266, 0.03, 0 ], [ 0.266, 0.75, 0 ] )
createLine( "bar8", [ 0.154, 0.03, 0 ], [ 0.154, 0.75, 0 ] )
saveProject( )
createLine( "bar9", [ 0.094, 0.03, 0 ], [ 0.094, 0.75, 0 ] )
saveProject( )
createLine( "bar10", [ -0.007, 0.03, 0 ], [ -0.007, 0.75, 0 ] )
saveProject( )
arrayCopy( [ "bar1", "bar2", "bar3", "bar4", "bar5", "bar6", "bar7", "bar8", "bar9", "bar10" ],
[ 0, 0.8, 0 ], [ 0, 0, 0 ], [ 0, 0, 0 ], 4 )
saveProject( )
saveProject( )
addMaterial( "bar", "REINFO", "LINEAR", [] )
setParameter( "MATERIAL", "bar", "LINEAR/ELASTI/YOUNG", 2.1e+11 )
addGeometry( "Element geometry 6", "RELINE", "REBAR", [] )
rename( "GEOMET", "Element geometry 6", "bar" )
setParameter( "GEOMET", "bar", "REIEMB/CROSSE", 5.0625e-05 )
setReinforcementAspects( [ "bar1", "bar2", "bar3", "bar4", "bar5", "bar6", "bar7", "bar8",
"bar9", "bar10", "bar11", "bar12", "bar13", "bar14", "bar15", "bar16", "bar17", "bar18", "bar19",
"bar20", "bar21", "bar22", "bar23", "bar24", "bar25", "bar26", "bar27", "bar28", "bar29", "bar30",
"bar31", "bar32", "bar33", "bar34", "bar35", "bar36", "bar37", "bar38", "bar39", "bar40", "bar41",
"bar42", "bar43", "bar44", "bar45", "bar46", "bar47", "bar48", "bar49", "bar50" ] )
assignMaterial( "bar", "SHAPE", [ "bar1", "bar2", "bar3", "bar4", "bar5", "bar6", "bar7",
"bar8", "bar9", "bar10", "bar11", "bar12", "bar13", "bar14", "bar15", "bar16", "bar17", "bar18",
"bar19", "bar20", "bar21", "bar22", "bar23", "bar24", "bar25", "bar26", "bar27", "bar28", "bar29",
"bar30", "bar31", "bar32", "bar33", "bar34", "bar35", "bar36", "bar37", "bar38", "bar39", "bar40",
"bar41", "bar42", "bar43", "bar44", "bar45", "bar46", "bar47", "bar48", "bar49", "bar50" ] )
assignGeometry( "bar", "SHAPE", [ "bar1", "bar2", "bar3", "bar4", "bar5", "bar6", "bar7",
"bar8", "bar9", "bar10", "bar11", "bar12", "bar13", "bar14", "bar15", "bar16", "bar17", "bar18",
"bar19", "bar20", "bar21", "bar22", "bar23", "bar24", "bar25", "bar26", "bar27", "bar28", "bar29",
"bar30", "bar31", "bar32", "bar33", "bar34", "bar35", "bar36", "bar37", "bar38", "bar39", "bar40",
"bar41", "bar42", "bar43", "bar44", "bar45", "bar46", "bar47", "bar48", "bar49", "bar50" ] )
resetElementData( "SHAPE", [ "bar1", "bar2", "bar3", "bar4", "bar5", "bar6", "bar7", "bar8",
"bar9", "bar10", "bar11", "bar12", "bar13", "bar14", "bar15", "bar16", "bar17", "bar18", "bar19",
"bar20", "bar21", "bar22", "bar23", "bar24", "bar25", "bar26", "bar27", "bar28", "bar29", "bar30",
"bar31", "bar32", "bar33", "bar34", "bar35", "bar36", "bar37", "bar38", "bar39", "bar40", "bar41",
"bar42", "bar43", "bar44", "bar45", "bar46", "bar47", "bar48", "bar49", "bar50" ] )
setReinforcementDiscretization( [ "bar1", "bar2", "bar3", "bar4", "bar5", "bar6", "bar7",
"bar8", "bar9", "bar10", "bar11", "bar12", "bar13", "bar14", "bar15", "bar16", "bar17", "bar18",
"bar19", "bar20", "bar21", "bar22", "bar23", "bar24", "bar25", "bar26", "bar27", "bar28", "bar29",
"bar30", "bar31", "bar32", "bar33", "bar34", "bar35", "bar36", "bar37", "bar38", "bar39", "bar40",
"bar41", "bar42", "bar43", "bar44", "bar45", "bar46", "bar47", "bar48", "bar49", "bar50" ],
```

"SECTION")

saveProject( )

saveProject( )

arrayCopy(["Sheet 15", "Sheet 16", "Sheet 17", "Sheet 18", "Sheet 19", "Sheet 20", "Sheet 21", "Sheet 22", "Sheet 23", "Sheet 24", "Sheet 25", "Sheet 26", "Sheet 27", "Sheet 28", "Sheet 29", "Sheet 30", "Sheet 31", "Sheet 32", "Sheet 33", "Sheet 34", "Sheet 35"], [0, 0, 0.041], [0, 0, 0], [0, 0, 0], 1)

arrayCopy(["Sheet 15", "Sheet 16", "Sheet 17", "Sheet 18", "Sheet 19", "Sheet 20", "Sheet 21", "Sheet 22", "Sheet 23", "Sheet 24", "Sheet 25", "Sheet 26", "Sheet 27", "Sheet 28", "Sheet 29", "Sheet 30", "Sheet 31", "Sheet 32", "Sheet 33", "Sheet 34", "Sheet 35"], [0, 0, -0.041], [0, 0, 0], [0, 0, 0], 1)

saveProject( )

addMaterial("Grid1", "REINFO", "VMISES", [])

setParameter("MATERIAL", "Grid1", "LINEAR/ELASTI/YOUNG", 2.1e+11)

setParameter("MATERIAL", "Grid1", "PLASTI/HARDI1/YLDSTR", 4.4e+08)

addGeometry("Element geometry 7", "RSHEET", "REGRID", [])

rename("GEOMET", "Element geometry 7", "Grid1")

setParameter("GEOMET", "Grid1", "PHI", [0, 0.008])

setParameter("GEOMET", "Grid1", "SPACIN", [0, 0.1])

setParameter("GEOMET", "Grid1", "XAXIS", [1, 0, 0])

setParameter("GEOMET", "Grid1", "XAXIS", [0, 1, 0])

setParameter("GEOMET", "Grid1", "XAXIS", [0, 1, 0])

setReinforcementAspects(["Sheet 43", "Sheet 44", "Sheet 45", "Sheet 46", "Sheet 47", "Sheet 48", "Sheet 49", "Sheet 50", "Sheet 51", "Sheet 52", "Sheet 53", "Sheet 54", "Sheet 55", "Sheet 56", "Sheet 57", "Sheet 58", "Sheet 59", "Sheet 60", "Sheet 61", "Sheet 62", "Sheet 63", "Sheet 64", "Sheet 65", "Sheet 66", "Sheet 67", "Sheet 68", "Sheet 69", "Sheet 70", "Sheet 71", "Sheet 72", "Sheet 73", "Sheet 74", "Sheet 75", "Sheet 76", "Sheet 77", "Sheet 78", "Sheet 79", "Sheet 80", "Sheet 81", "Sheet 82", "Sheet 83", "Sheet 84"])

assignMaterial("Grid1", "SHAPE", ["Sheet 43", "Sheet 44", "Sheet 45", "Sheet 46", "Sheet 47", "Sheet 48", "Sheet 49", "Sheet 50", "Sheet 51", "Sheet 52", "Sheet 53", "Sheet 54", "Sheet 55", "Sheet 56", "Sheet 57", "Sheet 58", "Sheet 59", "Sheet 60", "Sheet 61", "Sheet 62", "Sheet 63", "Sheet 64", "Sheet 65", "Sheet 66", "Sheet 67", "Sheet 68", "Sheet 69", "Sheet 70", "Sheet 71", "Sheet 72", "Sheet 73", "Sheet 74", "Sheet 75", "Sheet 76", "Sheet 77", "Sheet 78", "Sheet 79", "Sheet 80", "Sheet 81", "Sheet 82", "Sheet 83", "Sheet 84"])

assignGeometry("Grid1", "SHAPE", ["Sheet 43", "Sheet 44", "Sheet 45", "Sheet 46", "Sheet 47", "Sheet 48", "Sheet 49", "Sheet 50", "Sheet 51", "Sheet 52", "Sheet 53", "Sheet 54", "Sheet 55", "Sheet 56", "Sheet 57", "Sheet 58", "Sheet 59", "Sheet 60", "Sheet 61", "Sheet 62", "Sheet 63", "Sheet 64", "Sheet 65", "Sheet 66", "Sheet 67", "Sheet 68", "Sheet 69", "Sheet 70", "Sheet 71", "Sheet 72", "Sheet 73", "Sheet 74", "Sheet 75", "Sheet 76", "Sheet 77", "Sheet 78", "Sheet 79", "Sheet 80", "Sheet 81", "Sheet 82", "Sheet 83", "Sheet 84"])

resetElementData("SHAPE", ["Sheet 43", "Sheet 44", "Sheet 45", "Sheet 46", "Sheet 47",

"Sheet 48", "Sheet 49", "Sheet 50", "Sheet 51", "Sheet 52", "Sheet 53", "Sheet 54", "Sheet 55", "Sheet 56", "Sheet 57", "Sheet 58", "Sheet 59", "Sheet 60", "Sheet 61", "Sheet 62", "Sheet 63", "Sheet 64", "Sheet 65", "Sheet 66", "Sheet 67", "Sheet 68", "Sheet 69", "Sheet 70", "Sheet 71", "Sheet 72", "Sheet 73", "Sheet 74", "Sheet 75", "Sheet 76", "Sheet 77", "Sheet 78", "Sheet 79", "Sheet 80", "Sheet 81", "Sheet 82", "Sheet 83", "Sheet 84"])

setReinforcementDiscretization(["Sheet 43", "Sheet 44", "Sheet 45", "Sheet 46", "Sheet 47", "Sheet 48", "Sheet 49", "Sheet 50", "Sheet 51", "Sheet 52", "Sheet 53", "Sheet 54", "Sheet 55", "Sheet 56", "Sheet 57", "Sheet 58", "Sheet 59", "Sheet 60", "Sheet 61", "Sheet 62", "Sheet 63", "Sheet 64", "Sheet 65", "Sheet 66", "Sheet 67", "Sheet 68", "Sheet 69", "Sheet 70", "Sheet 71", "Sheet 72", "Sheet 73", "Sheet 74", "Sheet 75", "Sheet 76", "Sheet 77", "Sheet 78", "Sheet 79", "Sheet 80", "Sheet 81", "Sheet 82", "Sheet 83", "Sheet 84"], "ELEMENT")

saveProject( )

hide("SHAPE", ["Sheet 63", "Sheet 62", "Sheet 61", "Sheet 60", "Sheet 59", "Sheet 58", "Sheet 57", "Sheet 44", "Sheet 51", "Sheet 52", "Sheet 53", "Sheet 54", "Sheet 55", "Sheet 56", "Sheet 78", "Sheet 79", "Sheet 80", "Sheet 81", "Sheet 82", "Sheet 84", "Sheet 83", "Sheet 77", "Sheet 76", "Sheet 75", "Sheet 74", "Sheet 72", "Sheet 73", "Sheet 65"])

arrayCopy(["Sheet 1", "Sheet 2", "Sheet 3", "Sheet 4", "Sheet 5", "Sheet 6", "Sheet 7", "Sheet 8", "Sheet 9", "Sheet 10", "Sheet 11", "Sheet 12", "Sheet 13", "Sheet 14"], [0.03, 0, 0], [0, 0, 0], [0, 0, 0], 1)

arrayCopy(["Sheet 1", "Sheet 2", "Sheet 3", "Sheet 4", "Sheet 5", "Sheet 6", "Sheet 7", "Sheet 8", "Sheet 9", "Sheet 10", "Sheet 11", "Sheet 12", "Sheet 13", "Sheet 14"], [-0.03, 0, 0], [0, 0, 0], [0, 0, 0], 1)

saveProject( )

createSheet("Sheet 113", [[0.29, 0.025, 0.341],[0.29, 0.025, 0.3],[0.29, 0.25, 0.3],[0.29, 0.25, 0.341]])

saveProject( )

createSheet("Sheet 114", [[0.29, 0.25, 0.341],[0.29, 0.25, 0.3],[0.29, 0.83, 0.3],[0.29, 0.83, 0.341]])

saveProject( )

createSheet("Sheet 115", [[0.29, 0.83, 0.341],[0.29, 0.83, 0.3],[0.29, 1.63, 0.3],[0.29, 1.63, 0.341]])

saveProject( )

createSheet("Sheet 116", [[0.29, 1.63, 0.341],[0.29, 1.63, 0.3],[0.29, 2.43, 0.3],[0.29, 2.43, 0.341]])

saveProject( )

createSheet("Sheet 117", [[0.29, 2.43, 0.341],[0.29, 2.43, 0.3],[0.29, 3.23, 0.3],[0.29, 3.23, 0.341]])

saveProject( )

createSheet("Sheet 118", [[0.29, 3.23, 0.341],[0.29, 3.23, 0.3],[0.29, 3.75, 0.3],[0.29, 3.75, 0.341]])

saveProject( )
createSheet("Sheet 119", [[0.29, 3.75, 0.341],[0.29, 3.75, 0.3],[0.29, 3.975, 0.3],[0.29, 3.975, 0.341]])
saveProject( )
arrayCopy(["Sheet 113", "Sheet 114", "Sheet 115", "Sheet 116", "Sheet 117", "Sheet 118", "Sheet 119"], [-0.29, 0, 0], [0, 0, 0], [0, 0, 0], 1)
saveProject( )
arrayCopy(["Sheet 113", "Sheet 114", "Sheet 115", "Sheet 116", "Sheet 117", "Sheet 118", "Sheet 119", "Sheet 120", "Sheet 121", "Sheet 122", "Sheet 123", "Sheet 124", "Sheet 125", "Sheet 126"], [0.03, 0, 0], [0, 0, 0], [0, 0, 0], 1)
saveProject( )
translate(["Sheet 113", "Sheet 114", "Sheet 115", "Sheet 116", "Sheet 117", "Sheet 118", "Sheet 119"], [-0.03, 0, 0])
saveProject( )
translate(["Sheet 120", "Sheet 121", "Sheet 122", "Sheet 123", "Sheet 124", "Sheet 125", "Sheet 126"], [-0.03, 0, 0])
saveProject( )
saveProject( )
addMaterial("Gird3", "REINFO", "VMISES", [])
setParameter("MATERIAL", "Gird3", "LINEAR/ELASTI/YOUNG", 2.1e+11)
setParameter("MATERIAL", "Gird3", "PLASTI/YLDTYP", "NONE")
setParameter("MATERIAL", "Gird3", "PLASTI/HARDI1/YLDSTR", 4e+08)
addGeometry("Element geometry 8", "RSHEET", "REGRID", [])
rename("GEOMET", "Element geometry 8", "Grid3")
setParameter("GEOMET", "Grid3", "PHI", [0, 0.008])
setParameter("GEOMET", "Grid3", "SPACIN", [0, 0.1])
setParameter("GEOMET", "Grid3", "XAXIS", [1, 0, 0])
setParameter("GEOMET", "Grid3", "XAXIS", [0, 1, 0])
setParameter("GEOMET", "Grid3", "XAXIS", [0, 1, 0])
setReinforcementAspects(["Sheet 85", "Sheet 86", "Sheet 87", "Sheet 88", "Sheet 89", "Sheet 90", "Sheet 91", "Sheet 92", "Sheet 93", "Sheet 94", "Sheet 95", "Sheet 96", "Sheet 97", "Sheet 98", "Sheet 99", "Sheet 100", "Sheet 101", "Sheet 102", "Sheet 103", "Sheet 104", "Sheet 105", "Sheet 106", "Sheet 107", "Sheet 108", "Sheet 109", "Sheet 110", "Sheet 111", "Sheet 112", "Sheet 113", "Sheet 114", "Sheet 115", "Sheet 116", "Sheet 117", "Sheet 118", "Sheet 119", "Sheet 120", "Sheet 121", "Sheet 122", "Sheet 123", "Sheet 124", "Sheet 125", "Sheet 126", "Sheet 127", "Sheet 128", "Sheet 129", "Sheet 130", "Sheet 131", "Sheet 132", "Sheet 133", "Sheet 134", "Sheet 135", "Sheet 136", "Sheet 137", "Sheet 138", "Sheet 139", "Sheet 140"])
assignMaterial("Gird3", "SHAPE", ["Sheet 85", "Sheet 86", "Sheet 87", "Sheet 88", "Sheet 89", "Sheet 90", "Sheet 91", "Sheet 92", "Sheet 93", "Sheet 94", "Sheet 95", "Sheet 96", "Sheet 97", "Sheet 98", "Sheet 99", "Sheet 100", "Sheet 101", "Sheet 102", "Sheet 103", "Sheet 104", "Sheet 105", "Sheet 106", "Sheet 107", "Sheet 108", "Sheet 109", "Sheet 110", "Sheet 111", "Sheet

112", "Sheet 113", "Sheet 114", "Sheet 115", "Sheet 116", "Sheet 117", "Sheet 118", "Sheet 119", "Sheet 120", "Sheet 121", "Sheet 122", "Sheet 123", "Sheet 124", "Sheet 125", "Sheet 126", "Sheet 127", "Sheet 128", "Sheet 129", "Sheet 130", "Sheet 131", "Sheet 132", "Sheet 133", "Sheet 134", "Sheet 135", "Sheet 136", "Sheet 137", "Sheet 138", "Sheet 139", "Sheet 140"])

assignGeometry("Grid3", "SHAPE", ["Sheet 85", "Sheet 86", "Sheet 87", "Sheet 88", "Sheet 89", "Sheet 90", "Sheet 91", "Sheet 92", "Sheet 93", "Sheet 94", "Sheet 95", "Sheet 96", "Sheet 97", "Sheet 98", "Sheet 99", "Sheet 100", "Sheet 101", "Sheet 102", "Sheet 103", "Sheet 104", "Sheet 105", "Sheet 106", "Sheet 107", "Sheet 108", "Sheet 109", "Sheet 110", "Sheet 111", "Sheet 112", "Sheet 113", "Sheet 114", "Sheet 115", "Sheet 116", "Sheet 117", "Sheet 118", "Sheet 119", "Sheet 120", "Sheet 121", "Sheet 122", "Sheet 123", "Sheet 124", "Sheet 125", "Sheet 126", "Sheet 127", "Sheet 128", "Sheet 129", "Sheet 130", "Sheet 131", "Sheet 132", "Sheet 133", "Sheet 134", "Sheet 135", "Sheet 136", "Sheet 137", "Sheet 138", "Sheet 139", "Sheet 140"])

resetElementData("SHAPE", ["Sheet 85", "Sheet 86", "Sheet 87", "Sheet 88", "Sheet 89", "Sheet 90", "Sheet 91", "Sheet 92", "Sheet 93", "Sheet 94", "Sheet 95", "Sheet 96", "Sheet 97", "Sheet 98", "Sheet 99", "Sheet 100", "Sheet 101", "Sheet 102", "Sheet 103", "Sheet 104", "Sheet 105", "Sheet 106", "Sheet 107", "Sheet 108", "Sheet 109", "Sheet 110", "Sheet 111", "Sheet 112", "Sheet 113", "Sheet 114", "Sheet 115", "Sheet 116", "Sheet 117", "Sheet 118", "Sheet 119", "Sheet 120", "Sheet 121", "Sheet 122", "Sheet 123", "Sheet 124", "Sheet 125", "Sheet 126", "Sheet 127", "Sheet 128", "Sheet 129", "Sheet 130", "Sheet 131", "Sheet 132", "Sheet 133", "Sheet 134", "Sheet 135", "Sheet 136", "Sheet 137", "Sheet 138", "Sheet 139", "Sheet 140"])

setReinforcementDiscretization(["Sheet 85", "Sheet 86", "Sheet 87", "Sheet 88", "Sheet 89", "Sheet 90", "Sheet 91", "Sheet 92", "Sheet 93", "Sheet 94", "Sheet 95", "Sheet 96", "Sheet 97", "Sheet 98", "Sheet 99", "Sheet 100", "Sheet 101", "Sheet 102", "Sheet 103", "Sheet 104", "Sheet 105", "Sheet 106", "Sheet 107", "Sheet 108", "Sheet 109", "Sheet 110", "Sheet 111", "Sheet 112", "Sheet 113", "Sheet 114", "Sheet 115", "Sheet 116", "Sheet 117", "Sheet 118", "Sheet 119", "Sheet 120", "Sheet 121", "Sheet 122", "Sheet 123", "Sheet 124", "Sheet 125", "Sheet 126", "Sheet 127", "Sheet 128", "Sheet 129", "Sheet 130", "Sheet 131", "Sheet 132", "Sheet 133", "Sheet 134", "Sheet 135", "Sheet 136", "Sheet 137", "Sheet 138", "Sheet 139", "Sheet 140"], "ELEMENT")

saveProject( )

arrayCopy(["Sheet 36", "Sheet 37", "Sheet 38", "Sheet 39", "Sheet 40", "Sheet 41", "Sheet 42"], [0, 0, 0.0085], [0, 0, 0], [0, 0, 0], 1)

arrayCopy(["Sheet 36", "Sheet 37", "Sheet 38", "Sheet 39", "Sheet 40", "Sheet 41", "Sheet 42"], [0, 0, -0.0085], [0, 0, 0], [0, 0, 0], 1)

saveProject( )

addMaterial("Grid4", "REINFO", "VMISES", [])

setParameter("MATERIAL", "Grid4", "LINEAR/ELASTI/YOUNG", 2.1e+11)

setParameter("MATERIAL", "Grid4", "PLASTI/YLDTYP", "NONE")

setParameter("MATERIAL", "Grid4", "PLASTI/HARDI1/YLDSTR", 4.4e+08)

setParameter("MATERIAL", "Grid1", "PLASTI/HARDI1/YLDSTR", 4e+08)

setParameter("MATERIAL", "Gird3", "PLASTI/HARDI1/YLDSTR", 4e+08)

setParameter("MATERIAL", "Grid4", "PLASTI/HARDI1/YLDSTR", 4e+08)

addGeometry("Element geometry 9", "RSHEET", "REGRID", [])
rename("GEOMET", "Element geometry 9", "Grid4")
setParameter("GEOMET", "Grid4", "PHI", [0, 0.008])
setParameter("GEOMET", "Grid4", "SPACIN", [0, 0.1])
setParameter("GEOMET", "Grid4", "XAXIS", [1, 0, 0])
setParameter("GEOMET", "Grid4", "XAXIS", [0, 1, 0])
setParameter("GEOMET", "Grid4", "XAXIS", [0, 1, 0])
setReinforcementAspects(["Sheet 141", "Sheet 142", "Sheet 143", "Sheet 144", "Sheet 145", "Sheet 146", "Sheet 147", "Sheet 148", "Sheet 149", "Sheet 150", "Sheet 151", "Sheet 152", "Sheet 153", "Sheet 154"])
assignMaterial("Grid4", "SHAPE", ["Sheet 141", "Sheet 142", "Sheet 143", "Sheet 144", "Sheet 145", "Sheet 146", "Sheet 147", "Sheet 148", "Sheet 149", "Sheet 150", "Sheet 151", "Sheet 152", "Sheet 153", "Sheet 154"])
assignGeometry("Grid4", "SHAPE", ["Sheet 141", "Sheet 142", "Sheet 143", "Sheet 144", "Sheet 145", "Sheet 146", "Sheet 147", "Sheet 148", "Sheet 149", "Sheet 150", "Sheet 151", "Sheet 152", "Sheet 153", "Sheet 154"])
resetElementData("SHAPE", ["Sheet 141", "Sheet 142", "Sheet 143", "Sheet 144", "Sheet 145", "Sheet 146", "Sheet 147", "Sheet 148", "Sheet 149", "Sheet 150", "Sheet 151", "Sheet 152", "Sheet 153", "Sheet 154"])
setReinforcementDiscretization(["Sheet 141", "Sheet 142", "Sheet 143", "Sheet 144", "Sheet 145", "Sheet 146", "Sheet 147", "Sheet 148", "Sheet 149", "Sheet 150", "Sheet 151", "Sheet 152", "Sheet 153", "Sheet 154"], "ELEMENT")
saveProject( )
createSheet("Sheet 155", [[0.485, 0.025, 0.341],[0.485, 0.025, 0.259],[0.485, 0.25, 0.259],[0.485, 0.25, 0.341]])
createSheet("Sheet 156", [[0.485, 0.25, 0.341],[0.485, 0.25, 0.259],[0.485, 0.83, 0.259],[0.485, 0.83, 0.341]])
show("SHAPE", ["Sheet 44"])
createSheet("Sheet 157", [[0.485, 0.83, 0.341],[0.485, 0.83, 0.259],[0.485, 1.63, 0.259],[0.485, 1.63, 0.341]])
createSheet("Sheet 158", [[0.485, 1.63, 0.341],[0.485, 1.63, 0.259],[0.485, 2.43, 0.259],[0.485, 2.43, 0.341]])
createSheet("Sheet 159", [[0.485, 2.43, 0.341],[0.485, 2.43, 0.259],[0.485, 3.23, 0.259],[0.485, 3.23, 0.341]])
createSheet("Sheet 160", [[0.485, 3.23, 0.341],[0.485, 3.23, 0.259],[0.485, 3.75, 0.259],[0.485, 3.75, 0.341]])
saveProject( )
createSheet("Sheet 161", [[0.485, 3.75, 0.341],[0.485, 3.75, 0.259],[0.485, 3.975, 0.259],[0.485, 3.975, 0.341]])
saveProject( )
arrayCopy(["Sheet 155", "Sheet 156", "Sheet 157", "Sheet 158", "Sheet 159", "Sheet 160",

"Sheet 161"], [-0.68, 0, 0], [0, 0, 0], [0, 0, 0], 1)

saveProject( )

translate(["Sheet 155", "Sheet 156", "Sheet 157", "Sheet 158", "Sheet 159", "Sheet 160", "Sheet 161"], [-0.024, 0, 0])

saveProject( )

translate(["Sheet 162", "Sheet 163", "Sheet 164", "Sheet 165", "Sheet 166", "Sheet 167", "Sheet 168"], [0.024, 0, 0])

saveProject( )

addMaterial("Grid3", "REINFO", "VMISES", [])

setParameter("MATERIAL", "Grid3", "LINEAR/ELASTI/YOUNG", 2.1e+11)

setParameter("MATERIAL", "Grid3", "PLASTI/HARDI1/YLDSTR", 4e+08)

addGeometry("Element geometry 10", "RSHEET", "REGRID", [])

setParameter("GEOMET", "Element geometry 10", "PHI", [0, 0.008])

setParameter("GEOMET", "Element geometry 10", "SPACIN", [0, 0.1])

setParameter("GEOMET", "Element geometry 10", "XAXIS", [1, 0, 0])

setParameter("GEOMET", "Element geometry 10", "XAXIS", [0, 1, 0])

setParameter("GEOMET", "Element geometry 10", "XAXIS", [0, 1, 0])

setReinforcementAspects(["Sheet 155", "Sheet 156", "Sheet 157", "Sheet 158", "Sheet 159", "Sheet 160", "Sheet 161", "Sheet 162", "Sheet 163", "Sheet 164", "Sheet 165", "Sheet 166", "Sheet 167", "Sheet 168"])

assignMaterial("Grid3", "SHAPE", ["Sheet 155", "Sheet 156", "Sheet 157", "Sheet 158", "Sheet 159", "Sheet 160", "Sheet 161", "Sheet 162", "Sheet 163", "Sheet 164", "Sheet 165", "Sheet 166", "Sheet 167", "Sheet 168"])

assignGeometry("Element geometry 10", "SHAPE", ["Sheet 155", "Sheet 156", "Sheet 157", "Sheet 158", "Sheet 159", "Sheet 160", "Sheet 161", "Sheet 162", "Sheet 163", "Sheet 164", "Sheet 165", "Sheet 166", "Sheet 167", "Sheet 168"])

resetElementData("SHAPE", ["Sheet 155", "Sheet 156", "Sheet 157", "Sheet 158", "Sheet 159", "Sheet 160", "Sheet 161", "Sheet 162", "Sheet 163", "Sheet 164", "Sheet 165", "Sheet 166", "Sheet 167", "Sheet 168"])

setReinforcementDiscretization(["Sheet 155", "Sheet 156", "Sheet 157", "Sheet 158", "Sheet 159", "Sheet 160", "Sheet 161", "Sheet 162", "Sheet 163", "Sheet 164", "Sheet 165", "Sheet 166", "Sheet 167", "Sheet 168"], "ELEMENT")

saveProject( )

saveProject( )

saveProject( )

rename("GEOMET", "Element geometry 10", "Grid2")

rename("MATERIAL", "Grid3", "Grid2")

rename("MATERIAL", "Gird3", "Grid3")

saveProject( )

addMaterial("int1", "INTERF", "FRICTI", [])

setParameter("MATERIAL", "int1", "LINEAR/IFTYP", "LIN3D")

```
setParameter( "MATERIAL", "int1", "LINEAR/ELAS4/DSNY", 3e+16 )
setParameter( "MATERIAL", "int1", "LINEAR/ELAS4/DSSX", 3e+8 )
setParameter( "MATERIAL", "int1", "LINEAR/ELAS4/DSSZ", 3e+8 )
setParameter( "MATERIAL", "int1", "COULOM/COHESI", 0 )
setParameter( "MATERIAL", "int1", "COULOM/COHESI", 0 )
setParameter( "MATERIAL", "int1", "COULOM/PHI", 25 )
setParameter( "MATERIAL", "int1", "COULOM/PSI", 35 )
addGeometry( "Element geometry 11", "LINE", "SHLLIF", [] )
rename( "GEOMET", "Element geometry 11", "int1" )
setParameter( "GEOMET", "int1", "THICK", 0.13 )
setParameter( "GEOMET", "int1", "XAXIS", [ 1, 0, 0 ] )
setParameter( "GEOMET", "int1", "XAXIS", [ 0, 1, 0 ] )
setParameter( "GEOMET", "int1", "XAXIS", [ 0, 1, 0 ] )
setParameter( "GEOMET", "int1", "THKDIR", "PARALL" )
setParameter( "GEOMET", "int1", "YAXIS", [ 1, 0, 0 ] )
setParameter( "GEOMET", "int1", "YAXIS", [ 0, 1, 0 ] )
setParameter( "GEOMET", "int1", "YAXIS", [ 0, -1, 0 ] )
createLineConnection( "int1" )
setParameter( "GEOMETRYCONNECTION", "int1", "CONTYP", "INTER" )
setParameter( "GEOMETRYCONNECTION", "int1", "MODE", "AUTO" )
attachTo( "GEOMETRYCONNECTION", "int1", "SOURCE", "Sheet 18", [[ 0.145, 0.83, 0.3 ],[ 0.145, 1.63, 0.3 ]] )
attachTo( "GEOMETRYCONNECTION", "int1", "SOURCE", "Sheet 19", [[ 0.145, 2.43, 0.3 ]] )
attachTo( "GEOMETRYCONNECTION", "int1", "SOURCE", "Sheet 20", [[ 0.145, 3.23, 0.3 ]] )
attachTo( "GEOMETRYCONNECTION", "int1", "SOURCE", "Sheet 24", [[ -0.0975, 0.83, 0.3 ]] )
attachTo( "GEOMETRYCONNECTION", "int1", "SOURCE", "Sheet 25", [[ -0.0975, 1.63, 0.3 ],[ -0.0975, 2.43, 0.3 ]] )
attachTo( "GEOMETRYCONNECTION", "int1", "SOURCE", "Sheet 26", [[ -0.0975, 3.23, 0.3 ]] )
attachTo( "GEOMETRYCONNECTION", "int1", "SOURCE", "Sheet 30", [[ 0.3875, 0.83, 0.3 ]] )
attachTo( "GEOMETRYCONNECTION", "int1", "SOURCE", "Sheet 31", [[ 0.3875, 1.63, 0.3 ]] )
attachTo( "GEOMETRYCONNECTION", "int1", "SOURCE", "Sheet 33", [[ 0.3875, 2.43, 0.3 ],[ 0.3875, 3.23, 0.3 ]] )
setElementClassType( "GEOMETRYCONNECTION", "int1", "SHLLIF" )
assignMaterial( "int1", "GEOMETRYCONNECTION", "int1" )
assignGeometry( "int1", "GEOMETRYCONNECTION", "int1" )
```

```
resetElementData( "GEOMETRYCONNECTION", "int1" )
saveProject( )
addMaterial( "int2", "INTERF", "FRICTI", [] )
setParameter( "MATERIAL", "int2", "LINEAR/IFTYP", "LIN3D" )
setParameter( "MATERIAL", "int2", "LINEAR/ELAS4/DSNY", 3e+16 )
setParameter( "MATERIAL", "int2", "LINEAR/ELAS4/DSSX", 3e+8 )
setParameter( "MATERIAL", "int2", "LINEAR/ELAS4/DSSZ", 3e+8 )
setParameter( "MATERIAL", "int2", "COULOM/COHESI", 0 )
setParameter( "MATERIAL", "int2", "COULOM/PHI", 25 )
setParameter( "MATERIAL", "int2", "COULOM/PSI", 35 )
setParameter( "MATERIAL", "int2", "COULOM/PSI", 35 )
addGeometry( "Element geometry 12", "LINE", "SHLLIF", [] )
rename( "GEOMET", "Element geometry 12", "int2" )
setParameter( "GEOMET", "int2", "THICK", 0.05 )
setParameter( "GEOMET", "int2", "THKDIR", "PARALL" )
setParameter( "GEOMET", "int2", "YAXIS", [ 1, 0, 0 ] )
setParameter( "GEOMET", "int2", "YAXIS", [ 0, 1, 0 ] )
setParameter( "GEOMET", "int2", "YAXIS", [ 0, -1, 0 ] )
createLineConnection( "int2" )
setParameter( "GEOMETRYCONNECTION", "int2", "CONTYP", "INTER" )
setParameter( "GEOMETRYCONNECTION", "int2", "MODE", "AUTO" )
attachTo( "GEOMETRYCONNECTION", "int2", "SOURCE", "Sheet 38", [[ 0.145, 0.8,
3.4345491e-19 ]] )
attachTo( "GEOMETRYCONNECTION", "int2", "SOURCE", "Sheet 39", [[ 0.145, 1.6,
4.9957078e-19 ]] )
attachTo( "GEOMETRYCONNECTION", "int2", "SOURCE", "Sheet 41", [[ 0.145, 2.4,
-4.9957078e-19 ],[ 0.145, 3.2, 4.9957078e-19 ]] )
setElementClassType( "GEOMETRYCONNECTION", "int2", "SHLLIF" )
assignMaterial( "int2", "GEOMETRYCONNECTION", "int2" )
assignGeometry( "int2", "GEOMETRYCONNECTION", "int2" )
resetElementData( "GEOMETRYCONNECTION", "int2" )
saveProject( )
addMaterial( "int3", "INTERF", "FRICTI", [] )
setParameter( "MATERIAL", "int3", "LINEAR/IFTYP", "LIN3D" )
setParameter( "MATERIAL", "int3", "LINEAR/ELAS4/DSNY", 3e+16 )
setParameter( "MATERIAL", "int3", "LINEAR/ELAS4/DSSX", 3e+8 )
setParameter( "MATERIAL", "int3", "LINEAR/ELAS4/DSSZ", 3e+8 )
setParameter( "MATERIAL", "int3", "LINEAR/ELAS4/DSSZ", 3e+8 )
setParameter( "MATERIAL", "int3", "COULOM/COHESI", 0 )
setParameter( "MATERIAL", "int3", "COULOM/PHI", 25 )
setParameter( "MATERIAL", "int3", "COULOM/PSI", 35 )
```

```
addGeometry( "Element geometry 13", "LINE", "SHLLIF", [] )
rename( "GEOMET", "Element geometry 13", "int3" )
setParameter( "GEOMET", "int3", "THICK", 0.09 )
setParameter( "GEOMET", "int3", "THKDIR", "PARALL" )
setParameter( "GEOMET", "int3", "YAXIS", [ 1, 0, 0 ] )
setParameter( "GEOMET", "int3", "YAXIS", [ 0, 1, 0 ] )
setParameter( "GEOMET", "int3", "YAXIS", [ 0, -1, 0 ] )
createLineConnection( "int3" )
setParameter( "GEOMETRYCONNECTION", "int3", "CONTYP", "INTER" )
setParameter( "GEOMETRYCONNECTION", "int3", "MODE", "AUTO" )
attachTo( "GEOMETRYCONNECTION", "int3", "SOURCE", "Sheet 2", [[ 0, 0.8, 0.065 ],[ 0, 0.815, 0.14 ],[ 0, 0.83, 0.225 ]] )
attachTo( "GEOMETRYCONNECTION", "int3", "SOURCE", "Sheet 3", [[ 0, 1.6, 0.065 ],[ 0, 1.615, 0.14 ]] )
attachTo( "GEOMETRYCONNECTION", "int3", "SOURCE", "Sheet 4", [[ 0, 2.4, 0.065 ],[ 0, 2.415, 0.14 ],[ 0, 2.43, 0.225 ],[ 0, 1.63, 0.225 ]] )
attachTo( "GEOMETRYCONNECTION", "int3", "SOURCE", "Sheet 5", [[ 0, 3.2, 0.065 ],[ 0, 3.215, 0.14 ],[ 0, 3.23, 0.225 ]] )
attachTo( "GEOMETRYCONNECTION", "int3", "SOURCE", "Sheet 9", [[ 0.29, 0.8, 0.065 ],[ 0.29, 0.815, 0.14 ],[ 0.29, 0.83, 0.225 ]] )
attachTo( "GEOMETRYCONNECTION", "int3", "SOURCE", "Sheet 10", [[ 0.29, 1.6, 0.065 ],[ 0.29, 1.615, 0.14 ],[ 0.29, 1.63, 0.225 ]] )
attachTo( "GEOMETRYCONNECTION", "int3", "SOURCE", "Sheet 11", [[ 0.29, 2.4, 0.065 ],[ 0.29, 2.415, 0.14 ],[ 0.29, 2.43, 0.225 ]] )
attachTo( "GEOMETRYCONNECTION", "int3", "SOURCE", "Sheet 12", [[ 0.29, 3.215, 0.14 ],[ 0.29, 3.23, 0.225 ]] )
attachTo( "GEOMETRYCONNECTION", "int3", "SOURCE", "Sheet 14", [[ 0.29, 3.2, 0.065 ]] )
setElementClassType( "GEOMETRYCONNECTION", "int3", "SHLLIF" )
assignMaterial( "int3", "GEOMETRYCONNECTION", "int3" )
assignGeometry( "int3", "GEOMETRYCONNECTION", "int3" )
resetElementData( "GEOMETRYCONNECTION", "int3" )
saveProject( )
saveProject( )
createSheet( "Sheet 169", [[ 0.29, 1.05, 0.4 ],[ 0.29, 1.35, 0.4 ],[ 0, 1.35, 0.4 ],[ 0, 1.05, 0.4 ]] )
mirror( [ "Sheet 169" ], [ 0, 2, 0 ], [ False, True, False ], True )
projection( SHAPEFACE, "Sheet 18", [[ 0.16633617, 1.2888584, 0.3 ]], [ "Sheet 169" ], [ 0, 0, -1 ], True )
removeShape( [ "Sheet 169" ] )
saveProject( )
```

```
projection( SHAPEFACE, "Sheet 20", [[ 0.16633617, 2.8888584, 0.3 ]], [ "Sheet 170" ], [ 0, 0, -1 ], True )
removeShape( [ "Sheet 170" ] )
saveProject( )
addSet( "GEOMETRYLOADSET", "gravituy" )
createModelLoad( "gravity", "gravituy" )
rename( "GEOMETRYLOADSET", "gravituy", "gravity" )
addSet( "GEOMETRYLOADSET", "Geometry load case 2" )
rename( "GEOMETRYLOADSET", "Geometry load case 2", "tenin" )
createBodyLoad( "tenin", "tenin" )
setParameter( "GEOMETRYLOAD", "tenin", "LODTYP", "POSTEN" )
setParameter( "GEOMETRYLOAD", "tenin", "POSTEN/TENTYP", "ONEEND" )
setParameter( "GEOMETRYLOAD", "tenin", "POSTEN/ONEEND/FORCE1", 150000 )
setParameter( "GEOMETRYLOAD", "tenin", "POSTEN/ONEEND/RETLE1", 0.0001 )
setParameter( "GEOMETRYLOAD", "tenin", "POSTEN/SHEAR", 0.22 )
setParameter( "GEOMETRYLOAD", "tenin", "POSTEN/WOBBLE", 0.001 )
attach( "GEOMETRYLOAD", "tenin", [ "tenin1", "tenin2" ] )
attachTo( "GEOMETRYLOAD", "tenin", "POSTEN/ONEEND/PNTS1", "tenin1", [[ 0, 0.01625, 0.195 ]] )
attachTo( "GEOMETRYLOAD", "tenin", "POSTEN/ONEEND/PNTS1", "tenin2", [[ 0.29, 0.01625, 0.195 ]] )
saveProject( )
createLine( "co1", [ 0, 0.125, -0.1 ], [ 0.29, 0.125, -0.1 ] )
mirror( [ "co1" ], [ 0, 2, 0 ], [ False, True, False ], True )
projection( SHAPEFACE, "Sheet 36", [[ 0.16633617, 0.14339325, -3.005904e-18 ]], [ "co1" ], [ 0, 0, 1 ], True )
removeShape( [ "co1" ] )
projection( SHAPEFACE, "Sheet 37", [[ 0.16633617, 3.8933932, -3.005904e-18 ]], [ "co2" ], [ 0, 0, 1 ], True )
removeShape( [ "co2" ] )
addSet( GEOMETRYSUPPORTSET, "Geometry support set 1" )
rename( GEOMETRYSUPPORTSET, "Geometry support set 1", "co1" )
createLineSupport( "co1", "co1" )
setParameter( GEOMETRYSUPPORT, "co1", "AXES", [ 1, 2 ] )
setParameter( GEOMETRYSUPPORT, "co1", "TRANSL", [ 1, 1, 1 ] )
setParameter( GEOMETRYSUPPORT, "co1", "ROTATI", [ 0, 0, 0 ] )
attach( GEOMETRYSUPPORT, "co1", "Sheet 36", [[ 0.145, 0.125, -3.8518599e-35 ]] )
saveProject( )
addSet( GEOMETRYSUPPORTSET, "Geometry support set 2" )
rename( GEOMETRYSUPPORTSET, "Geometry support set 2", "co2" )
createLineSupport( "co2", "co2" )
```

setParameter(GEOMETRYSUPPORT, "co2", "AXES", [1, 2])
setParameter(GEOMETRYSUPPORT, "co2", "TRANSL", [1, 0, 1])
setParameter(GEOMETRYSUPPORT, "co2", "ROTATI", [0, 0, 0])
attach(GEOMETRYSUPPORT, "co2", "Sheet 37", [[0.145, 3.875, 3.9404527e-32]])
addSet(GEOMETRYLOADSET, "Geometry load case 3")
rename(GEOMETRYLOADSET, "Geometry load case 3", "load")
createSurfaceLoad("load", "load")
setParameter(GEOMETRYLOAD, "load", "FORCE/VALUE", -200000)
setParameter(GEOMETRYLOAD, "load", "FORCE/DIRECT", 3)
attach(GEOMETRYLOAD, "load", "Sheet 18", [[0.16633617, 1.2220719, 0.3]])
attach(GEOMETRYLOAD, "load", "Sheet 20", [[0.16633617, 2.8220719, 0.3]])
saveProject( )
setDefaultGeometryLoadCombinations( )
setGeometryLoadCombinationFactor("Geometry load combination 1", "gravity", 1)
remove(GEOMETRYLOADCOMBINATION, "Geometry load combination 3")
remove(GEOMETRYLOADCOMBINATION, "Geometry load combination 2")
setGeometryLoadCombinationFactor("Geometry load combination 1", "gravity", 1)
setGeometryLoadCombinationFactor("Geometry load combination 1", "tenin", 1)
addGeometryLoadCombination("")
setGeometryLoadCombinationFactor("Geometry load combination 2", "load", 1)
setTimeDependentLoadFactors(GEOMETRYLOADCOMBINATION, "Geometry load combination 1", [0, 3.1536e+09], [1, 1])
setTimeDependentLoadFactors(GEOMETRYLOADCOMBINATION, "Geometry load combination 2", [0, 3.1536e+09], [1, 1])
saveProject( )

setElementSize(["Sheet 16", "Sheet 15", "Sheet 29", "Sheet 30", "Sheet 17", "Sheet 23", "Sheet 24", "Sheet 18", "Sheet 31", "Sheet 32", "Sheet 19", "Sheet 25", "Sheet 26", "Sheet 20", "Sheet 33", "Sheet 34", "Sheet 21", "Sheet 27", "Sheet 28", "Sheet 22", "Sheet 35", "Sheet 36", "Sheet 38", "Sheet 39", "Sheet 40", "Sheet 41", "Sheet 42", "Sheet 37", "Sheet 8", "Sheet 9", "Sheet 10", "Sheet 11", "Sheet 12", "Sheet 14", "Sheet 13", "Sheet 1", "Sheet 2", "Sheet 3", "Sheet 4", "Sheet 5", "Sheet 7", "Sheet 6"], 0.05, -1, True)

setMesherType(["Sheet 16", "Sheet 15", "Sheet 29", "Sheet 30", "Sheet 17", "Sheet 23", "Sheet 24", "Sheet 18", "Sheet 31", "Sheet 32", "Sheet 19", "Sheet 25", "Sheet 26", "Sheet 20", "Sheet 33", "Sheet 34", "Sheet 21", "Sheet 27", "Sheet 28", "Sheet 22", "Sheet 35", "Sheet 36", "Sheet 38", "Sheet 39", "Sheet 40", "Sheet 41", "Sheet 42", "Sheet 37", "Sheet 8", "Sheet 9", "Sheet 10", "Sheet 11", "Sheet 12", "Sheet 14", "Sheet 13", "Sheet 1", "Sheet 2", "Sheet 3", "Sheet 4", "Sheet 5", "Sheet 7", "Sheet 6"], "HEXQUAD")

setMidSideNodeLocation(["Sheet 16", "Sheet 15", "Sheet 29", "Sheet 30", "Sheet 17", "Sheet 23", "Sheet 24", "Sheet 18", "Sheet 31", "Sheet 32", "Sheet 19", "Sheet 25", "Sheet 26", "Sheet 20", "Sheet 33", "Sheet 34", "Sheet 21", "Sheet 27", "Sheet 28", "Sheet 22", "Sheet 35", "Sheet 36", "Sheet 38", "Sheet 39", "Sheet 40", "Sheet 41", "Sheet 42", "Sheet 37", "Sheet 8",

"Sheet 9", "Sheet 10", "Sheet 11", "Sheet 12", "Sheet 14", "Sheet 13", "Sheet 1", "Sheet 2", "Sheet 3", "Sheet 4", "Sheet 5", "Sheet 7", "Sheet 6"], "LINEAR")

saveProject( )

setElementSize("Sheet 2", 1, [[0, 0.8, 0.065],[0, 0.815, 0.14],[0, 0.83, 0.225]], 0.05, 0, True)

setElementSize("Sheet 3", 1, [[0, 1.6, 0.065],[0, 1.615, 0.14]], 0.05, 0, True)

setElementSize("Sheet 4", 1, [[0, 1.63, 0.225],[0, 2.4, 0.065],[0, 2.415, 0.14],[0, 2.43, 0.225]], 0.05, 0, True)

setElementSize("Sheet 5", 1, [[0, 3.2, 0.065],[0, 3.215, 0.14],[0, 3.23, 0.225]], 0.05, 0, True)

setElementSize("Sheet 9", 1, [[0.29, 0.8, 0.065],[0.29, 0.815, 0.14]], 0.05, 0, True)

setElementSize("Sheet 10", 1, [[0.29, 0.83, 0.225],[0.29, 1.6, 0.065],[0.29, 1.63, 0.225]], 0.05, 0, True)

setElementSize("Sheet 11", 1, [[0.29, 1.615, 0.14],[0.29, 2.415, 0.14],[0.29, 2.43, 0.225]], 0.05, 0, True)

setElementSize("Sheet 12", 1, [[0.29, 2.4, 0.065],[0.29, 3.215, 0.14],[0.29, 3.23, 0.225]], 0.05, 0, True)

setElementSize("Sheet 14", 1, [[0.29, 3.2, 0.065]], 0.05, 0, True)

setElementSize("Sheet 17", 1, [[0.145, 0.83, 0.3]], 0.05, 0, True)

setElementSize("Sheet 19", 1, [[0.145, 1.63, 0.3],[0.145, 2.43, 0.3]], 0.05, 0, True)

setElementSize("Sheet 20", 1, [[0.145, 3.23, 0.3]], 0.05, 0, True)

setElementSize("Sheet 23", 1, [[-0.0975, 0.83, 0.3]], 0.05, 0, True)

setElementSize("Sheet 25", 1, [[-0.0975, 1.63, 0.3]], 0.05, 0, True)

setElementSize("Sheet 26", 1, [[-0.0975, 2.43, 0.3],[-0.0975, 3.23, 0.3]], 0.05, 0, True)

setElementSize("Sheet 31", 1, [[0.3875, 0.83, 0.3],[0.3875, 1.63, 0.3]], 0.05, 0, True)

setElementSize("Sheet 33", 1, [[0.3875, 2.43, 0.3],[0.3875, 3.23, 0.3]], 0.05, 0, True)

setElementSize("Sheet 38", 1, [[0.145, 0.8, 3.4345491e-19]], 0.05, 0, True)

setElementSize("Sheet 39", 1, [[0.145, 1.6, 4.9957078e-19]], 0.05, 0, True)

setElementSize("Sheet 41", 1, [[0.145, 2.4, -4.9957078e-19],[0.145, 3.2, 4.9957078e-19]], 0.05, 0, True)

saveProject( )

generateMesh([])

hideView("GEOM")

showView("MESH")

addAnalysis("Analysis2")

addAnalysisCommand("Analysis2", "NONLIN", "Structural nonlinear")

renameAnalysis("Analysis2", "Analysis2")

removeAnalysisCommandDetail("Analysis2", "Structural nonlinear", "EXECUT(1)")

addAnalysisCommandDetail("Analysis2", "Structural nonlinear", "MODEL/EVALUA/REINFO/INTERF")

setAnalysisCommandDetail("Analysis2", "Structural nonlinear",

"MODEL/EVALUA/REINFO/INTERF", True)

setAnalysisCommandDetail("Analysis2", "Structural nonlinear", "EXECUT/EXETYP", "START")

setAnalysisCommandDetail("Analysis2", "Structural nonlinear", "EXECUT(1)/START/INITIA/STRESS/INPUT/LOAD", 1)

renameAnalysisCommand("Analysis2", "Structural nonlinear", "Structural nonlinear")

renameAnalysisCommandDetail("Analysis2", "Structural nonlinear", "EXECUT(1)", "tenin")

renameAnalysisCommandDetail("Analysis2", "Structural nonlinear", "EXECUT(1)", "tenin")

addAnalysisCommandDetail("Analysis2", "Structural nonlinear", "EXECUT(1)/START/INITIA/STRESS")

setAnalysisCommandDetail("Analysis2", "Structural nonlinear", "EXECUT(1)/START/INITIA/STRESS", True)

setAnalysisCommandDetail("Analysis2", "Structural nonlinear", "EXECUT(1)/ITERAT/MAXITE", 50)

setAnalysisCommandDetail("Analysis2", "Structural nonlinear", "EXECUT(1)/ITERAT/CONVER/DISPLA/NOCONV", "CONTIN")

setAnalysisCommandDetail("Analysis2", "Structural nonlinear", "EXECUT(1)/ITERAT/CONVER/FORCE/NOCONV", "CONTIN")

addAnalysisCommandDetail("Analysis2", "Structural nonlinear", "EXECUT(1)/PHYSIC")

renameAnalysisCommandDetail("Analysis2", "Structural nonlinear", "EXECUT(1)", "tenin")

setAnalysisCommandDetail("Analysis2", "Structural nonlinear", "EXECUT(1)/PHYSIC/LIQUEF", False)

setAnalysisCommandDetail("Analysis2", "Structural nonlinear", "EXECUT(1)/PHYSIC/BOND", True)

setAnalysisCommandDetail("Analysis2", "Structural nonlinear", "EXECUT/EXETYP", "LOAD")

renameAnalysisCommand("Analysis2", "Structural nonlinear", "Structural nonlinear")

renameAnalysisCommandDetail("Analysis2", "Structural nonlinear", "EXECUT(2)", "load")

setAnalysisCommandDetail("Analysis2", "Structural nonlinear", "EXECUT(2)/LOAD/LOADNR", 2)

setAnalysisCommandDetail("Analysis2", "Structural nonlinear", "EXECUT(2)/ITERAT/MAXITE", 50)

setAnalysisCommandDetail("Analysis2", "Structural nonlinear", "EXECUT(2)/ITERAT/CONVER/DISPLA/NOCONV", "CONTIN")

setAnalysisCommandDetail("Analysis2", "Structural nonlinear", "EXECUT(2)/ITERAT/CONVER/FORCE/NOCONV", "CONTIN")

setAnalysisCommandDetail("Analysis2", "Structural nonlinear",

```
"EXECUT(2)/LOAD/STEPS/EXPLIC/SIZES", "1.00000" )
    setAnalysisCommandDetail( "Analysis2", "Structural nonlinear", "EXECUT(1)/ITERAT/CONVER/DISPLA/TOLCON", 0.001 )
    setAnalysisCommandDetail( "Analysis2", "Structural nonlinear", "EXECUT(1)/ITERAT/CONVER/FORCE/TOLCON", 0.001 )
    setAnalysisCommandDetail( "Analysis2", "Structural nonlinear", "EXECUT(2)/ITERAT/CONVER/DISPLA/TOLCON", 0.001 )
    setAnalysisCommandDetail( "Analysis2", "Structural nonlinear", "EXECUT(2)/ITERAT/CONVER/FORCE/TOLCON", 0.001 )
    saveProject( )
    setAnalysisCommandDetail( "Analysis2", "Structural nonlinear", "EXECUT/EXETYP", "TIME" )
    renameAnalysisCommand( "Analysis2", "Structural nonlinear", "Structural nonlinear" )
    renameAnalysisCommandDetail( "Analysis2", "Structural nonlinear", "EXECUT(3)", "creep" )
    setAnalysisCommandDetail( "Analysis2", "Structural nonlinear", "EXECUT(3)/TIME/STEPS/EXPLIC/SIZES", "2419200 13348800 15768000 63072000" )
    saveProject( )
    setAnalysisCommandDetail( "Analysis2", "Structural nonlinear", "EXECUT(3)/ITERAT/MAXITE", 50 )
    setAnalysisCommandDetail( "Analysis2", "Structural nonlinear", "EXECUT(3)/ITERAT/CONVER/DISPLA/TOLCON", 0.001 )
    setAnalysisCommandDetail( "Analysis2", "Structural nonlinear", "EXECUT(3)/ITERAT/CONVER/DISPLA/NOCONV", "CONTIN" )
    setAnalysisCommandDetail( "Analysis2", "Structural nonlinear", "EXECUT(3)/ITERAT/CONVER/FORCE/NOCONV", "CONTIN" )
    setAnalysisCommandDetail( "Analysis2", "Structural nonlinear", "EXECUT(3)/ITERAT/CONVER/FORCE/TOLCON", 0.001 )
    saveProject( )
```

5.3 Random Field Numerical Case of Precast Segmental Box-Girder

This case is a random field model of a precast segmental thin-walled simply supported concrete box girder with the length of 32 m. Nominal strength of longitudinal prestress tendons with high strength and low-relaxation steel stranded wires is 1860 MPa and the nominal diameter is $\Phi^j 15.24$. Prestress tendons are in harp shape with the elastic modulus 1.95×10^{11} N/m^2. Curved shell element in DIANA software was applied to simulate the mechanical behavior of concrete and the concrete grade is C55. The whole random field numerical model is established based on JCSS probability model. The whole girder sustains stepwise loading. In view of initial distributed loading value 100 kN/m^2 symmetrically loaded on the girder, a positive symmetric half-structure model is established in this numerical case. The whole size of box-girder and reinforcement information is shown in Fig. 5.106a, b, respectively. Targeting at the influence of different correlation lengths on the simulation results, the numerical outcomes of thin-wall box girder under different correlation length values were also taken into consideration when correlation length was set at default value (L = 5 m), L = 0.1 m and L = 10 m, respectively, and the three outcomes are compared. Moreover, comparisons of different number of steps in sub-directions X and Z and number of steps in Y principal direction were also conducted in order to expect that these comparisons can render some suggestions for audience who have high requirement on numerical accuracy of random field. The whole numerical case is established based on the platform of DIANA release 10.1.

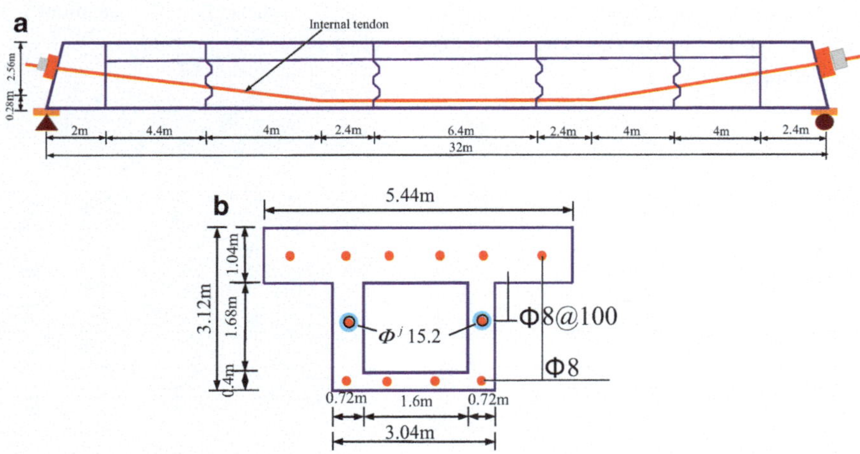

Fig. 5.106 **a** Longitudinal length and height of girder bridge, **b** Reinforcement on the end section

Essentials of Learning:

(1) ***Learning to create geometric model of thin-walled box girder* via *curved shell element of DIANA***
(2) ***Learning to use log-normal distribution type in JCSS random field definition to discretize material parameters in space.***
(3) ***Learning to establish 3D line interface elements between shells and nonlinear elasticity material constitutive model.***
(4) ***The influence of different correlation lengths on the results of random field simulation.***

Open DianaIE GUI interface environment; click the menu bar **"File—New"** and the New Project dialog box pops up. Then create a document in the directory of computer G-disk area named **"PSB-Random-32 m"** suffixed with the name of **. dpf**. **Structural** analysis is selected as the analysis type and the number of dimensions is **three.** In view of the whole size of numerical model in this case, the maximum **Model size** is 100 m, which means that scope of the whole graphical visualization zone ranges from –50 m to 50 m in all directions of the coordinate system. **Default mesher type** is chosen as **Hexa/Quad** element so that the geometric element shapes are all quadrilateral (2 dimensions) or hexahedron (3 dimensions), while the **Default mesh order** in this case is quadratic. Determination of mid-side node location is linear interpolation. The whole manipulations mentioned above are defined and determined by clicking **OK** button. Then the GUI graphical interface can be displayed immediately, as shown in Fig. 5.107.

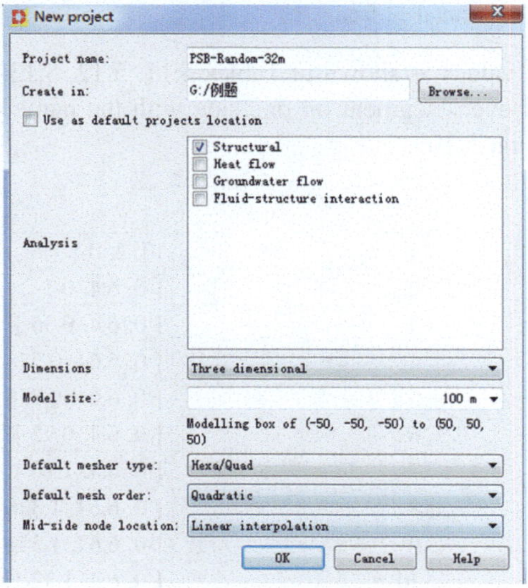

Fig. 5.107 New project operational interface

Create geometric model of precast segmental thin-wall box girder. Owing to the reason that curved shell element is applied to simulate concrete, geometric numerical model is established in the geometric neutral surface site according to the modeling method of shell element.

Clicking shortcut icon **Adds a sheet** and coordinate input dialog box pops up. Coordinate points (0, 0, 0), (0, 2, 0), (0, 2, 2.4) and (0, 0.2, 2.4) under Cartesian coordinate system are input to create geometric plane on end neutral surface-**Sheet1**, shown as Fig. 5.108

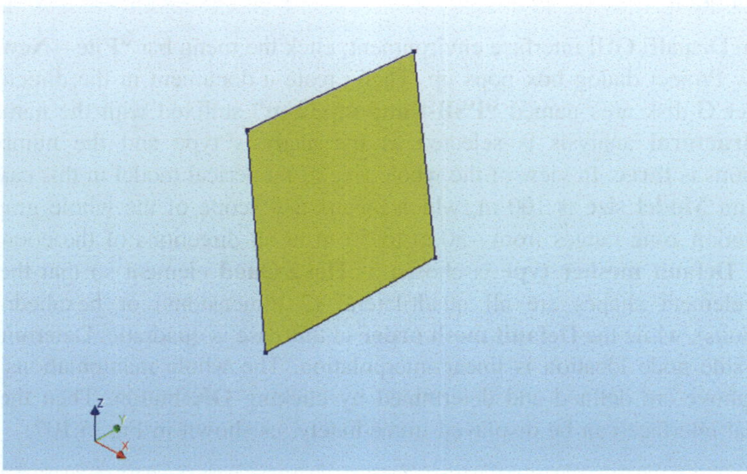

Fig. 5.108 Sheet1 on the neutral surface

Input coordinate values as shown in Tables 5.11, 5.12, 5.13 in turn to generate geometric model of every segment on one side with the name of **Sheet2, Sheet3, Sheet4,** shown as Fig. 5.109.

Table 5.11 Coordinate values of Sheet 2

1	[0, 2, 0]
2	[0, 6.4, 0]
3	[0, 6.4, 0.36]
4	[0, 6.64, 0.528]
5	[0, 6.64, 0.736]
6	[0, 6.4, 0.92]
7	[0, 6.4, 1.16]
8	[0, 6.64, 1.336]
9	[0, 6.64, 1.544]
10	[0, 6.4, 1.72]
11	[0, 6.4, 2.4]
12	[0, 2, 2.4]

Table 5.12 Coordinate values of Sheet 3

1	[0, 6.4, 0]
2	[0, 12.8, 0]
3	[0, 12.8, 0.36]
4	[0, 13.04, 0.528]
5	[0, 13.04, 0.736]
6	[0, 12.8, 0.92]
7	[0, 12.8, 1.16]
8	[0, 13.04, 1.336]
9	[0, 13.04, 1.544]
10	[0, 12.8, 1.72]
11	[0, 12.8, 2.4]
12	[0, 6.4, 2.4]
13	[0, 6.4, 1.72]
14	[0, 6.64, 1.544]
15	[0, 6.64, 1.336]
16	[0, 6.4, 1.16]
17	[0, 6.4, 0.92]
18	[0, 6.64, 0.736]
19	[0, 6.64, 0.528]
20	[0, 6.4, 0.36]

Table 5.13 Coordinate values of Sheet 4

1	[0, 12.8, 0]
2	[0, 16, 0]
3	[0, 16, 2.4]
4	[0, 12.8, 2.4]
5	[0, 12.8, 1.72]
6	[0, 13.04, 1.544]
7	[0, 13.04, 1.336]
8	[0, 12.8, 1.16]
9	[0, 12.8, 0.92]
10	[0, 13.04, 0.736]
11	[0, 13.04, 0.528]
12	[0, 12.8, 0.36]

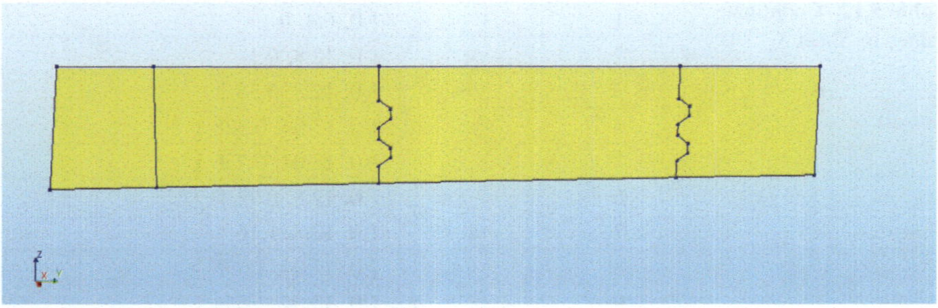

Fig. 5.109 Geometric model of box girder on one side

Selecting sheets from **Sheet1 to Sheet4** in the GUI graphical interface zone, then right-clicking the mouse and selecting function of **Array copy**, **Number of copies** is defined as 1 and the **Relative Displacement** is **2.32 m** in the positive X direction. **Pivot** and **Relative Angles** in the three directions are all defined as **zero** to demonstrate that this manipulation is parallel movement without any oblique movement or rotation, which is shown in Fig. 5.110.

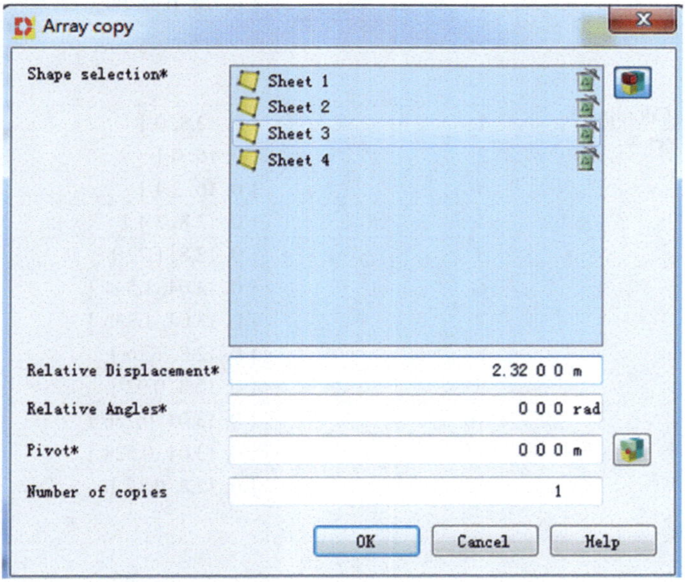

Fig. 5.110 Array copy editing interface

Clicking OK button, then the outcome of array copy is shown as Fig. 5.111 and the neutral surface of the web on the other side of the box girder has been established.

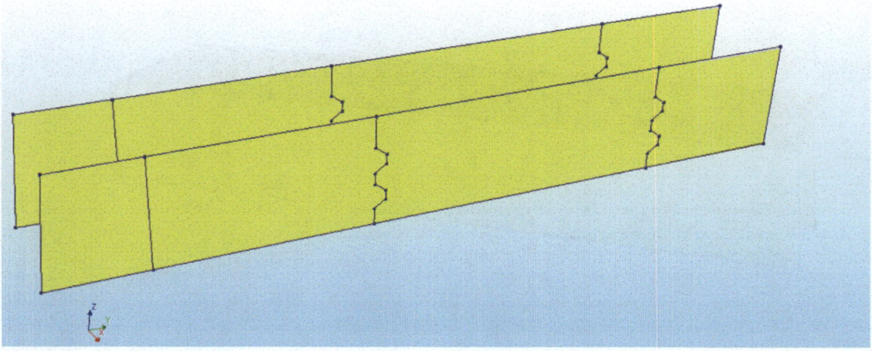

Fig. 5.111 Generation of neutral surfaces of webs on both sides

Next setting up the neutral surface of the top plate, connecting the points on the top edge of web in turn on the GUI, it is evident to see the geometric shape of the top plate from **Sheet 9** to **Sheet 16** and the coordinate values forging every sheet are collected and shown as Table 5.14.

Table 5.14 Coordinate values of Sheet 9 to Sheet 16

Sheet 9	1	[0, 0.2, 2.4]	Sheet 13	1	[0, 0.2, 2.4]
	2	[2.32, 0.2, 2.4]		2	[0, 2, 2.4]
	3	[2.32, 2, 2.4]		3	[−1.56, 2, 2.4]
	4	[0, 2, 2.4]		4	[−1.56,0.2,2.4]
Sheet 10	1	[0, 2, 2.4]	Sheet 14	1	[0, 2, 2.4]
	2	[2.32, 2, 2.4]		2	[0, 6.4, 2.4]
	3	[2.32, 6.4, 2.4]		3	[−1.56,6.4,2.4]
	4	[0, 6.4, 2.4]		4	[−1.56, 2, 2.4]
Sheet 11	1	[0, 6.4, 2.4]	Sheet 15	1	[−1.56,6.4,2.4]
	2	[2.32, 6.4, 2.4]		2	[0, 6.4, 2.4]
	3	[2.32, 12.8, 2.4]		3	[0, 12.8, 2.4]
	4	[0, 12.8, 2.4]		4	[−1.56,12.8,2.4]
Sheet 12	1	[0, 12.8, 2.4]	Sheet 16	1	[−1.56,12.8,2.4]
	2	[2.32, 12.8, 2.4]		2	[0, 12.8, 2.4]
	3	[2.32, 16, 2.4]		3	[0, 16, 2.4]
	4	[0, 16, 2.4]		4	[−1.56, 16, 2.4]

After inputting coordinate values of per point, the shape as Fig. 5.112 is displayed. Then the coordinate values (0, 0, 0), (2.32, 0, 0), (2.32, 2, 0), (0, 2, 0) are input to generate the **Sheet17** of the bottom plate.

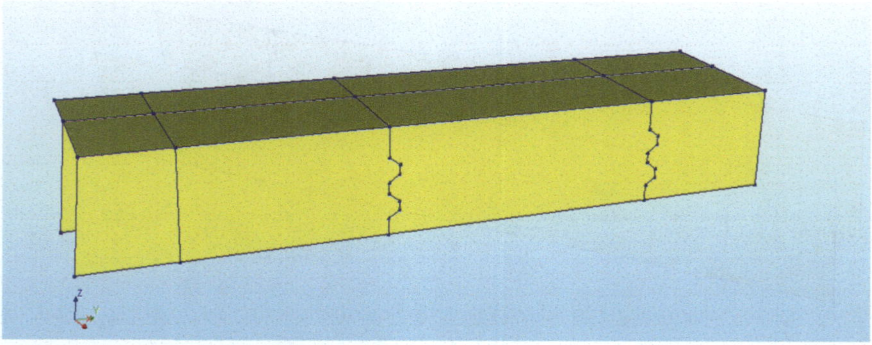

Fig. 5.112 Geometric model of Sheet1–Sheet16

Above all, Sheet10, Sheet11 as well as Sheet12 are selected on the top plate in the GUI zone, then **Array copy** function mentioned above is applied again along the negative Z direction. The **Number of copies** is still defined as 1 and the **Relative Displacement** is 2.4 m to generate the neutral surface of the bottom plate of the whole box girder. Similarly, Sheet 13 to Sheet 16 are selected and the **Number of copies** is 1. Displacement is 3.88 m in the X positive direction to create the neutral surface of the right flange plate so the whole semi-box girder on the geometric neutral surface site is shown in **Fig.** 5.113.

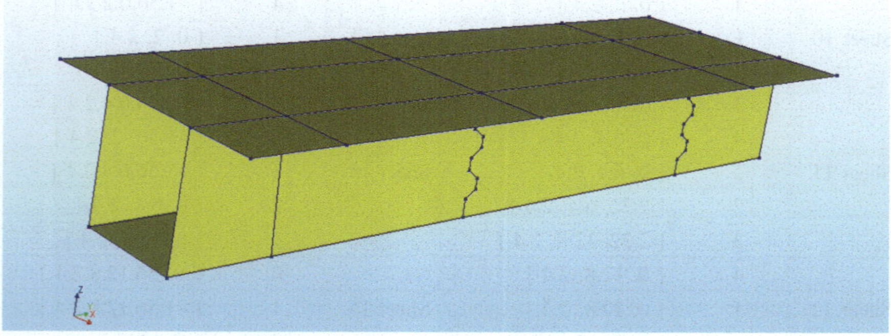

Fig. 5.113 The whole semi-box girder in the geometric neutral surface

Click shortcut icon of DianaIE **Add a line**  to create geometric model of longitudinal reinforcement bar; the coordinate values **(0, 0.16, 0)** and **(0, 6.24, 0)** are input and the generated line is named as **bar1**. Then select the bar and right-click **Array copy** to duplicate and translate other longitudinal steel bars on the bottom plate, the number of copies is 3 and the relative displacement is 0.778 m in the positive X direction, naming them **bar2** to **bar4**. Then selecting the **bar1–bar4**, duplicate and translate them in the positive Y direction. The number of copies is 1 and the relative displacement is 6.4 m to generate the longitudinal steel bars of the second segment.

After establishing geometric model of line longitudinal steel bars **bar9**, coordinate values of two terminal ends are **(0, 12.96, 0)** and **(0, 16, 0),** respectively. Similarly, the above **Array copy** function is still applied in the Y direction to generate four longitudinal steel bars on the bottom plate of third segment, which are named as **bar10–bar12**. Establish geometric model of **bar13** on the top plate, input coordinate values **(0, 0.24, 2.4)** and **(0, 6.24, 2.4)**, select **bar13,** then duplicating and translating it in the positive X direction, the number of copies is 4 while the displacement is 0.778 m. Still translating **bar13** 0.778 m in the negative X direction, the number of copies is 1 to complete the longitudinal steel bars on the top plate of first segment. Selecting **bar13–bar18,** repeating the same operation, the distance is 6.4 m and the number of copies is 1 to generate **bar19–bar24** on the top plate of the second segment. Repeating using **Adds a line** icon, inputting **(0, 12.96, 2.4)** and **(0, 16, 2.4)** to generate the line geometric model-**bar25**, it is duplicated **4** copies with the relative displacement **0.778 m** in the positive **X** direction and then we use the **Array copy** function one more time to duplicate **bar25** in 1 copy at the same translate distance in the negative X direction to generate the last steel bar on the top plate-**bar30**.

Note: Readers may take the method of creating lines in turn and inputting coordinate values directly one by one, but the modeling efficiency is far lower than the one that this case provides!

Click shortcut icon **Adds a polyline** to create geometric model of harp prestress tendon in the DianaIE with the name of **tenin1**. The values of three points in the coordinate system are used as input, as shown in Fig. 5.114.

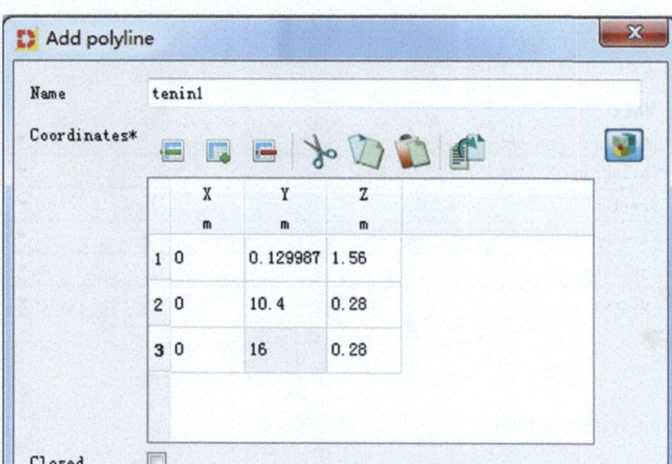

Fig. 5.114 Coordinate values of harp internal bonded tendon **tenin1**

Use the **Array copy** function again to create the geometric model of harp internal bonded tendon **tenin2** on the other side of the web, the number of copies is 1 and the relative displacement is 2.32 m in the positive X direction.

Clicking OK button, the reinforcement information of internal bonded tendons, longitudinal bar element and geometric model of semi-structural box girder with geometric neutral surface has been created and displayed (Fig. 5.115).

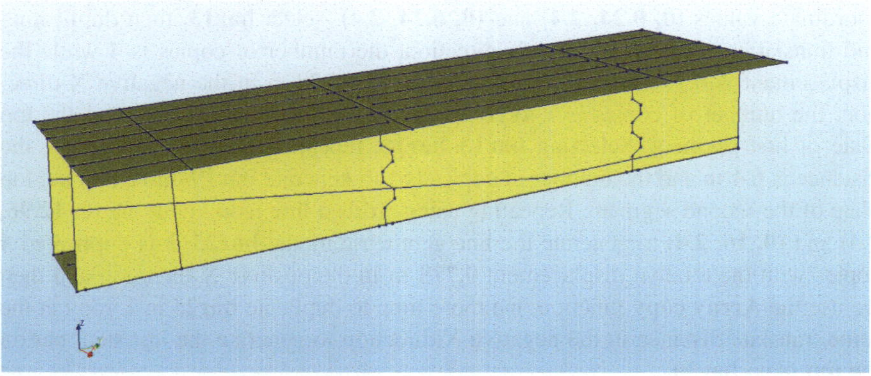

Fig. 5.115 Generation of numerical model of semi-structural box girder, reinforcement tendons and bar elements

The following step is to define material attribute. Before this work, click option **Unit** under **Reference system** and change the unit system properties of **Temperature** and **Angle** as **Celsius** and **degree** options, respectively, as Fig. 5.116 demonstrates.

Fig. 5.116 Unit system settings

Select the sheets forging neutral surface of the top plate of box girder in the GUI (Sheet 9, Sheet 10, Sheet 11, Sheet 12, Sheet 13, Sheet 14, Sheet 15, Sheet 16, Sheet 21, Sheet 22, Sheet 23, Sheet 24) and right-click to select **Property assignments,** the editing dialog box of material property ejects, the name of which is **top. Concrete design codes** option is selected in the material classification option —Class and **JCSS Probabilistic model code** is chosen as **Material model**, which is under the **Class** option. **JCSS Random field** option is ticked in the **Aspects to include,** as shown in Fig. 5.117. On clicking OK button, **Edit material** dialog box ejects and concrete constitutive parameters in the JCSS probabilistic model are edited. For the **Cement type** option, **Pre-case element** is selected and the concrete grade is **C55**. When the concrete grade is set, DIANA automatically generates mean value and standard deviation of compressive strength in the **Average basic concrete compressive strength** and **Standard deviation basic compressive strength** options respectively, which are 6.357e + 07 N/m^2 and 4.11e + 06 N/m^2. **Poisson ratio** is 0.15 while **Fracture energy** representing energy required for cracking at unit width is set as 500 N/m. Then concrete density is input as 2500 kg/m^3, as shown in Fig. 5.118.

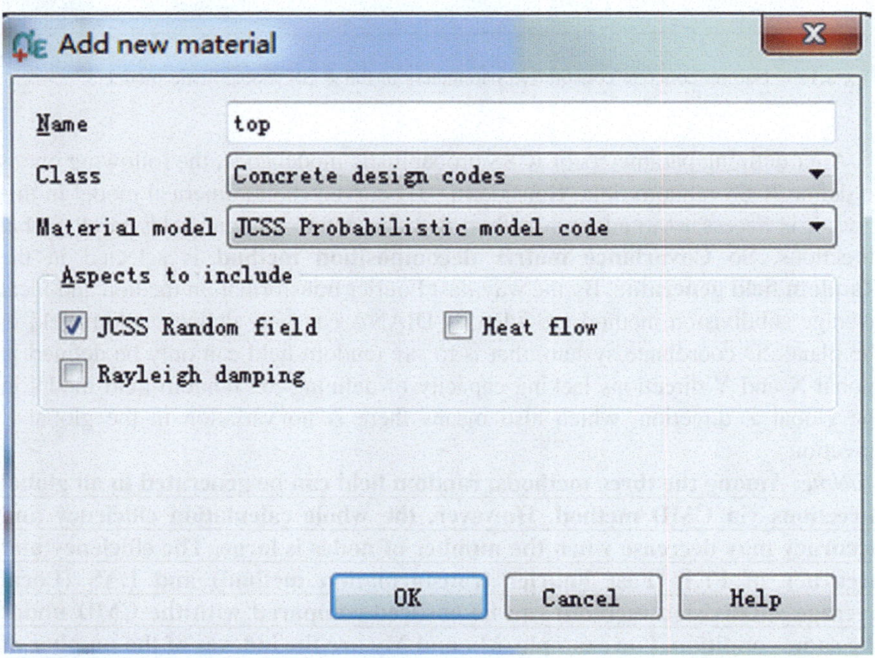

Fig. 5.117 Definition box of JCSS random field material attributes

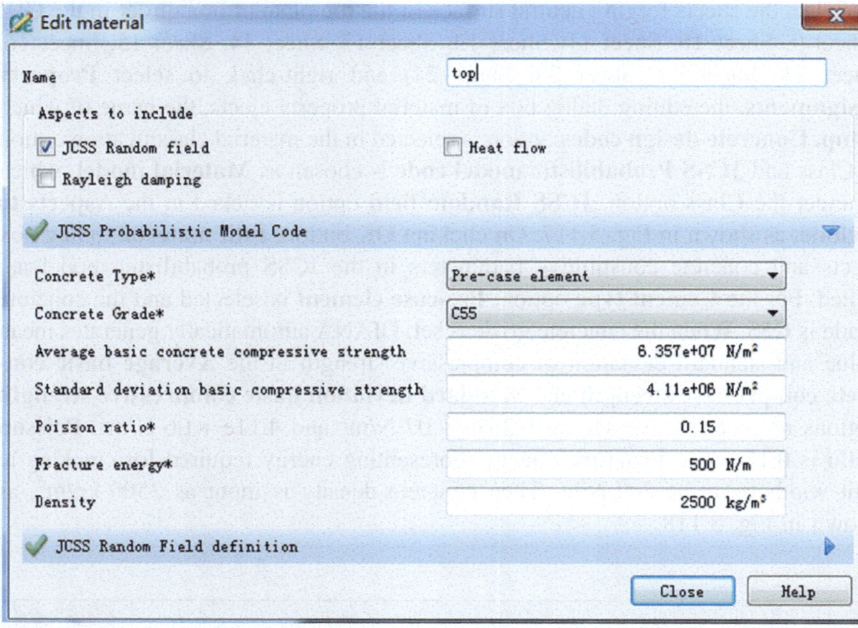

Fig. 5.118 Editing concrete constitutive parameters in the JCSS probabilistic model

After defining parameters of JCSS probabilistic model code, the following one is to define JCSS random field. Considering 3D curved shell numerical model in this case, it is necessary to adopt a method that can define random fields in all global directions. So **Covariance matrix decomposition method** is selected in the **Random field generator**. By the way, fast Fourier transformation method and local average subdivision method provided by DIANA can only define random field in the plane 2D coordinate system, that is to say random field can only be defined in global X and Y directions lacking capacity of defining 3D random field model in the global Z direction, which also means there is no variation in the global Z direction.

Note: **Among the three methods, random field can be generated in all global directions via CMD method. However, the whole calculation efficiency and accuracy may decrease when the number of nodes is large. The efficiency and accuracy of FFT (Fast Fourier transformation method) and LAS (Local average subdivision method) can be ensured compared with the CMD under the same condition. For example, Mx and My are the indexes of the number of mesh lines in the global X and Y directions of the random field generated by the FFT method. These indexed are for establishing number of mesh lines in the X and Y directions. That is to say, the following two conditions only suits 2D structural elements.**

Number of steps in global X/Z-direction are all set as 1 and **Number of steps in global Y-direction** is 10. **Decomposition method** is **Cholesky** while **Covariance type** is **Squared exponential** model, where exponential model usually applies to simulating modeling of spatial variability in soil properties. Type of distribution in this case is **Log-normal. Threshold value** is the default value **0.5 m** and **Correlation length** is 5 m, as Fig. 5.119 demonstrates.

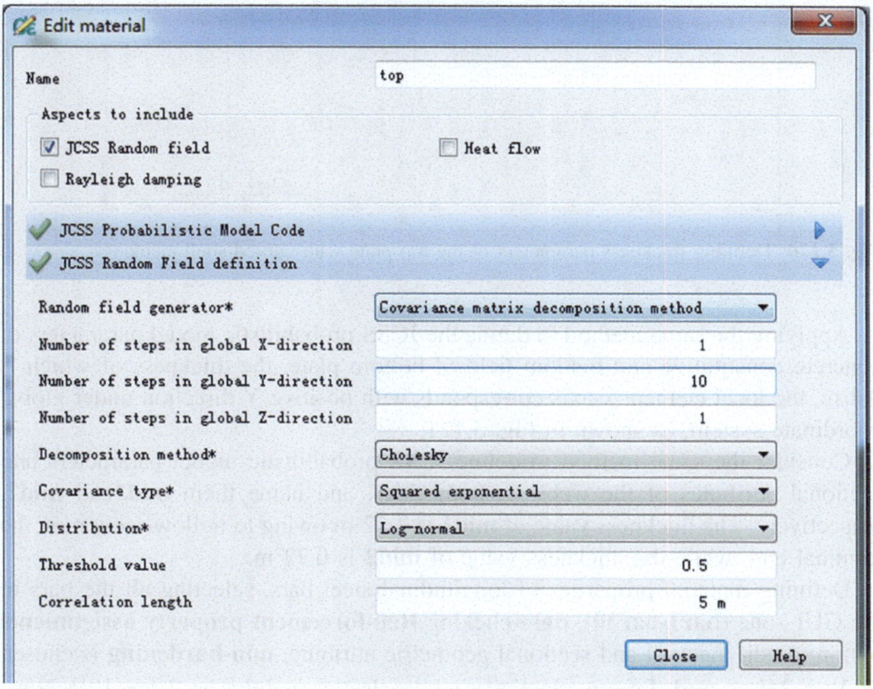

Fig. 5.119 Defining JCSS probabilistic model parameters of the top plate

Thickness of top plate is defined as 1.04 m while **the local element x axis** corresponds with positive Y direction (0, 1, 0) under global coordinate system, shown as Fig. 5.120.

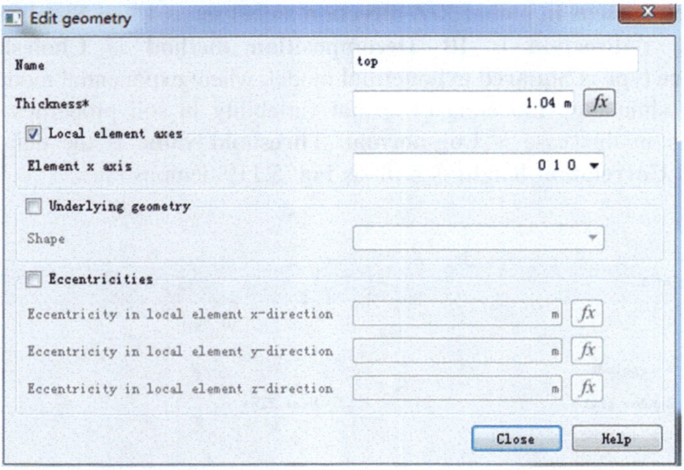

Fig. 5.120 Defining thickness parameters of the top plate with curved shell elements

Applying the same method to define the JCSS probabilistic model parameters of concrete constitutive and random field of bottom plate, the thickness of which is 0.4 m, the local element x-axis corresponds with positive Y direction under global coordinate system, as shown in Fig. 5.121.

Consider the same method to define JCSS probabilistic model parameters and sectional attributes of the webs on both sides, and name them **mid1**and **mid2,** respectively. The thickness value of **mid1** is 2.72 m owing to hollow section on the terminal end, while the thickness value of **mid2** is 0.72 m.

Defining material properties of longitudinal steel bars, selecting all the bars in the GUI zone (**bar1-bar30**), right-clicking **Reinforcement property assignments** option to edit material and sectional geometric attribute, **non-hardening** is chosen as **Von Mises and Tresca plasticity** while elastic modulus is 2.1×10^{11} N/m^2 coupled with density 7800 kg/m^{3}and Poisson ratio 0.3. The yield stress of steel bars is set as 4.4E8N/m^2.

Define internal bonded harp tendon with the same manipulation method above. Elastic modulus is 1.95×10^{11} N/m^2 while **Von Mises plasticity** is also non-hardening. **Yield stress** is set as 1.86×10^9 N/m^2.

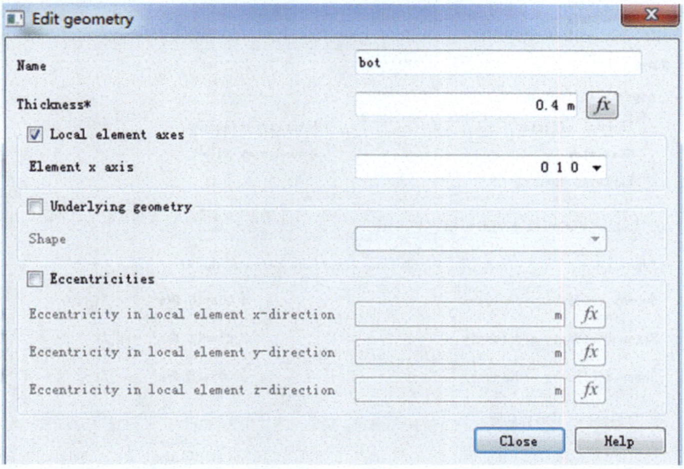

Fig. 5.121 Settings of geometric thickness parameters of the bottom plate

After defining material properties mentioned above, click shortcut icon **Edit connection property assignments**, constitutive attributes of interface elements on the top plate are defined with the name of **int1** in order to simulate mechanic behaviors between shear keys. In this case, **Structural Shell Interfaces** is selected as interface element in the **Element class** option so as to simulate line to line connected interface elements. Material model is **nonlinear elasticity,** shown as Fig. 5.122. Connection model of line to line interface elements in material nonlinear elastic properties editing box is **3D line interface between shells.** In specifying linear material properties, the **Normal stiffness modulus-y** is set as 3.65e16 N/m^3 while **Shear stiffness modulus-x/z** is all set as 3.65e12 N/m^3, as shown in Fig. 5.123. In the **Nonlinear elasticity** module, **No tension with constant shear stiffness** option is chosen in this case, meaning tensile stiffness is zero while shear stiffness is constant.

Fig. 5.122 Interface element

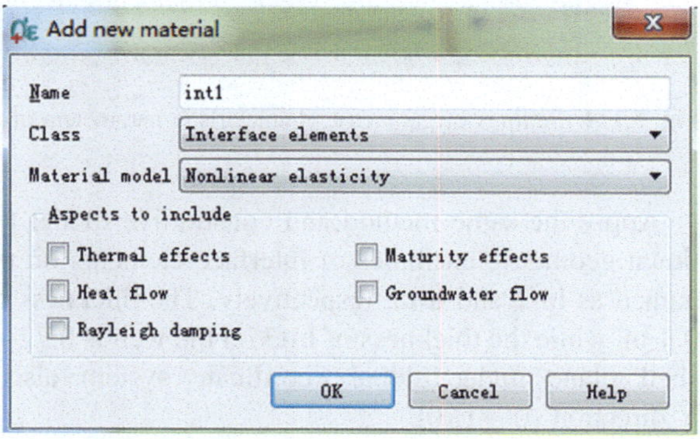

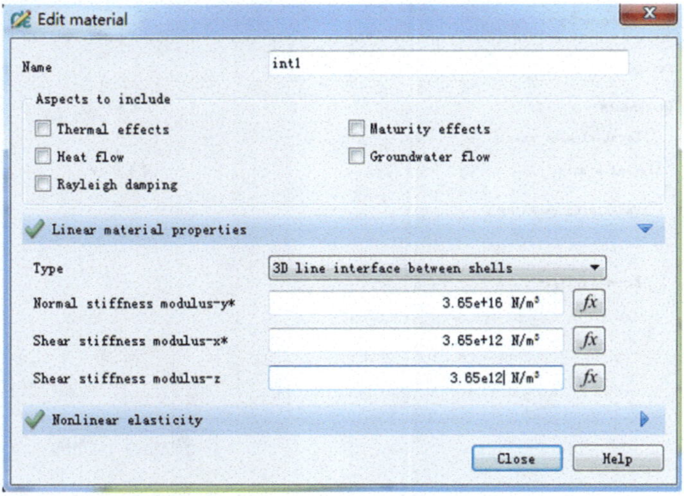

Fig. 5.123 Nonlinear elasticity constitutive model of interface element

Defining thickness value of interface element **int1** between shell elements of top plate is 1.04 m and the **Shape definition type** is flat, **Parallel to shell plane** is selected as **Element direction. Direction vector parallel to shell plane** corresponds with negative Y direction (0, –1, 0) under global coordinate system in order to achieve simulating normal compressive mechanic behavior of 3D line interface elements between shells. The operational interface is displayed in Fig. 5.124.

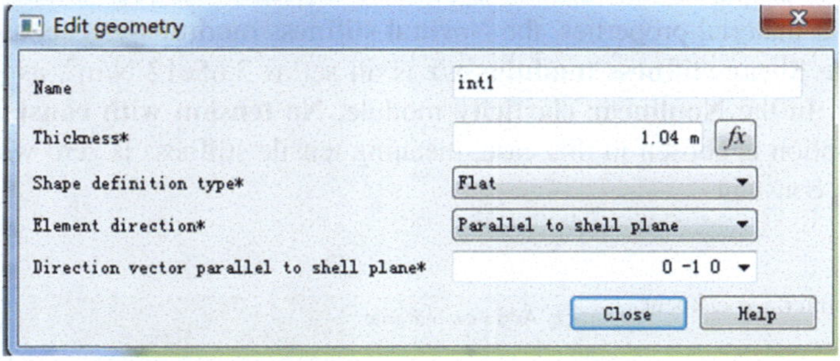

Fig. 5.124 Settings of geometric characteristic parameters of interface element section

Apply the same method and constitutive model to define constitutive and sectional geometric attributes of interface elements on the bottom plate and the web, named as **int2** and **int3**, respectively. The thickness of **int2** on the bottom plate is 0.4 m, while the thickness of **int3** on the web is 0.72 m. Direction vector parallel to shell plane under global coordinate system also corresponds with negative **Y** direction (0, –1, 0).

The following step is exerting constrains. Clicking menu bar **Geometry—Analysis—Attach support,** constraint definition dialog box ejects, the name of which is **co1**. **Support target type** is selected as **Line**. Select the bottom edge where the bottom plate on the neutral surface of box girder is located as the exerting object of line constraint (Edge 12 of Sheet 17) and T1, T2 and T3 representing fixed translation constraints in the X, Y and Z directions respectively are exerted, which is demonstrated in Fig. 5.125. Clicking **Add a new support set** icon, the name of the dialog box is co2.

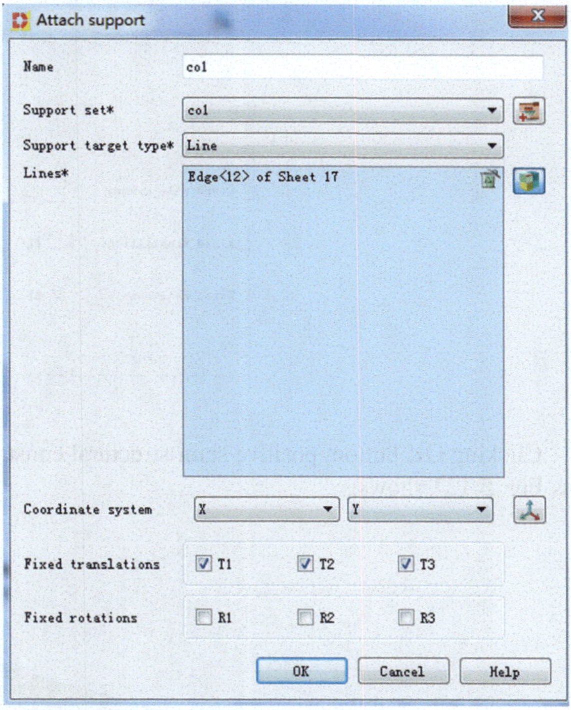

Fig. 5.125 Exerting co1 constrains

Clicking **Add a new support set** shortcut icon to create a constraint named **co2**, right-clicking to select **Attach support**, positive semi-structural constraints are exerted on the mid-span surface of edges in the semi-structure with the same method mentioned above. That means constraint **T2** is exerted to fix translation in the **Y** direction while **R1** and **R3** restrain rotation in the **X** and **Z** directions, respectively. Operations mentioned above are displayed in Fig. 5.126.

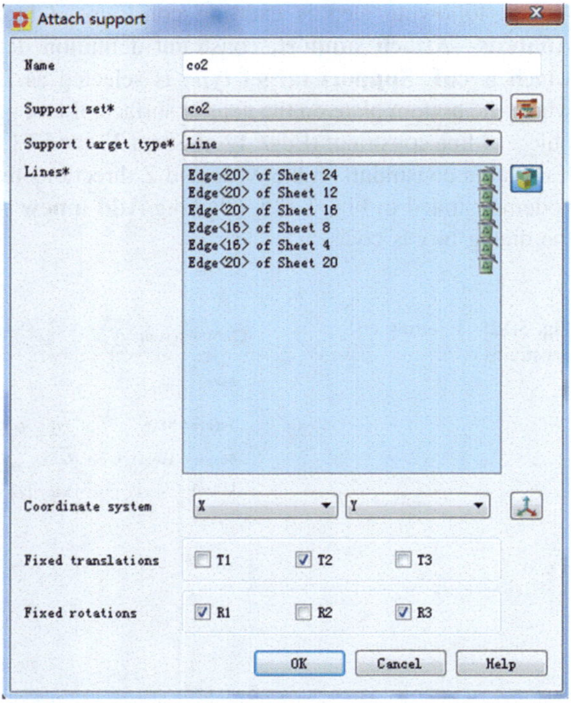

Fig. 5.126 Exerting co2 constrains

Clicking OK button, positive semi-structural constrained box-girder is generated as Fig. 5.127 shows.

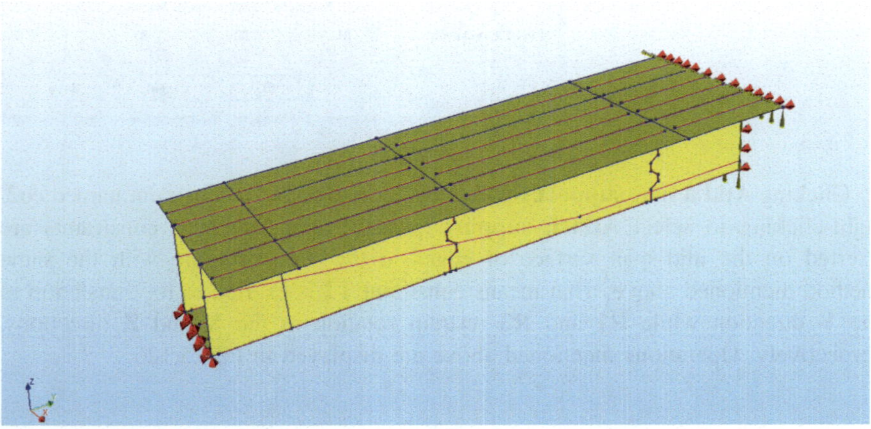

Fig. 5.127 Positive semi-structural constrained numerical model of box girder

Exert load on the numerical model. Click **Define a global load** to exert gravity under the **Load** bar then the icon **Add a new load case** is clicked so that the load definition dialog box pops up. Prestress force is named as **tenin** and **Solid** option is selected as **Load target type. Load type** is set as post-tensioning load while both **tenin1**and **tenin2** in the geometry bar are chosen as **Loaded reinforcements**. **Tension type** chooses option of **One end**. Initial tensioning force is set as 75% of the nominal stress of per strand so that **Nodal anchor force** is 1551.24 kN according to the calculation formula $\frac{1860\times10^6\times0.75\times1.39\times10^{-4}\times8}{1000} = 1551.24\,\text{kN}$. Anchor retention length is 0.0001 m with Coulomb friction coefficient 0.22, while Wobble curvature radius coefficient is 0.01 1/m, as shown in Fig. 5.128.

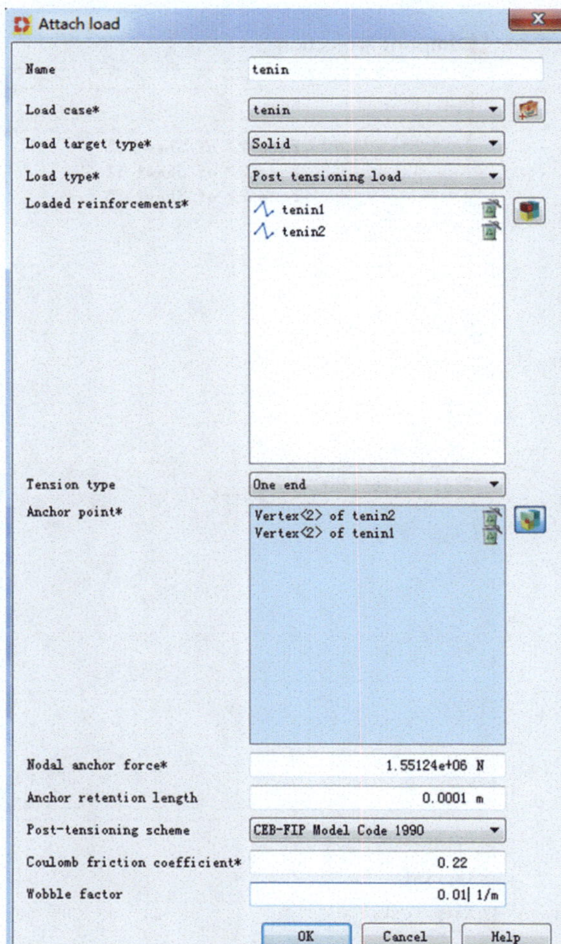

Fig. 5.128 Post-tensioning prestress force definition dialog box

Owing to the reason that distributed load cannot be exerted directly on the surface in DIANA, it is necessary to create loading plane and then imprint the projection onto the box girder. Above all, a loading plane is created above the box girder. Click **Add a sheet** icon and input the coordinate values (–1.56, 9.4, 2.5), (3.88, 9.4, 2.5), (3.88, 11.4, 2.5) and (–1.56, 11.4, 2.5) to create a geometric plane, naming it Sheet97, then icon **Project edges, wires and points on solid, faces and edges** are clicked and imprint editing interface ejects. **Operation** is **Face** while Sheet15, Sheet11, Sheet23 on the surface of box girder are selected in the **Face selection** operation. Sheet97 is set as the option of **Tool Selection**, then the whole procedure of imprint is completed via clicking OK button. The settings of **Imprint** editing interface and semi-structural model of box girder after imprinting are displayed in Figs. 5.129 and 5.130, respectively.

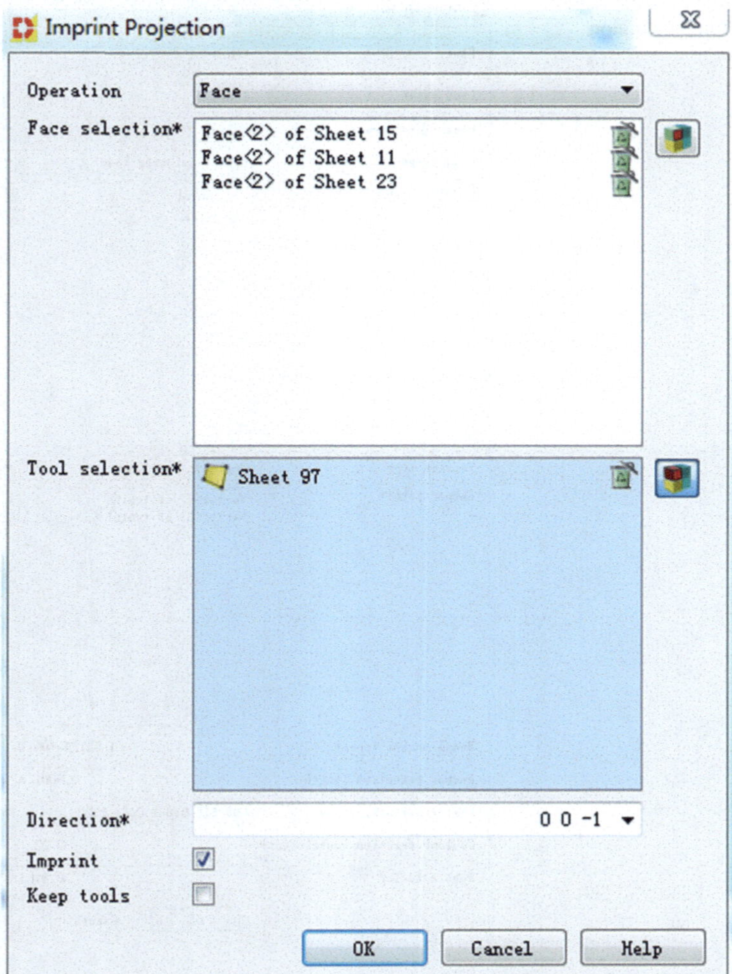

Fig. 5.129 Imprint operation

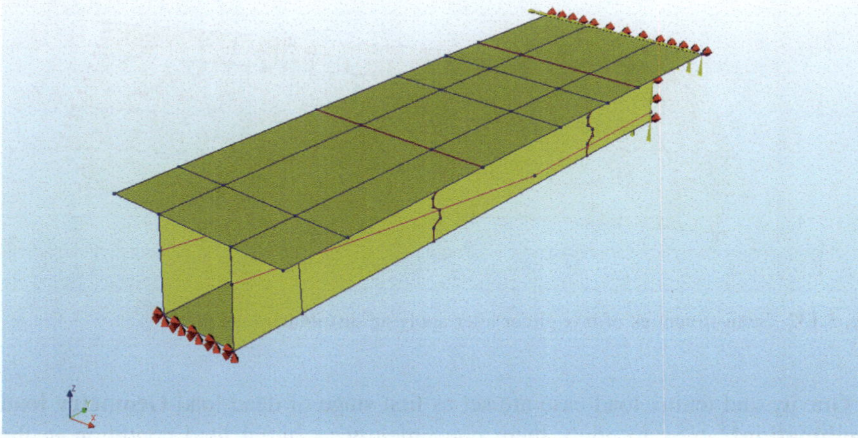

Fig. 5.130 Semi-structural model of box girder after imprinting

Loading is applied after imprinting and the name of load case is **load**. Right-clicking **Attach option**, the load editing dialog box pops up. **Face** is chosen as **Load target type** and **Load type** is **Distributed force**. **Loaded surface** contains three imprinted surfaces and the surface force value is 100 kN/m^2 in the vertical downward negative Z direction, as shown in Fig. 5.131. Click OK button to finish definition of distributed load. Now the distributed force applied on the structures is shown in Fig. 5.132.

Fig. 5.131 Settings of distributed load case

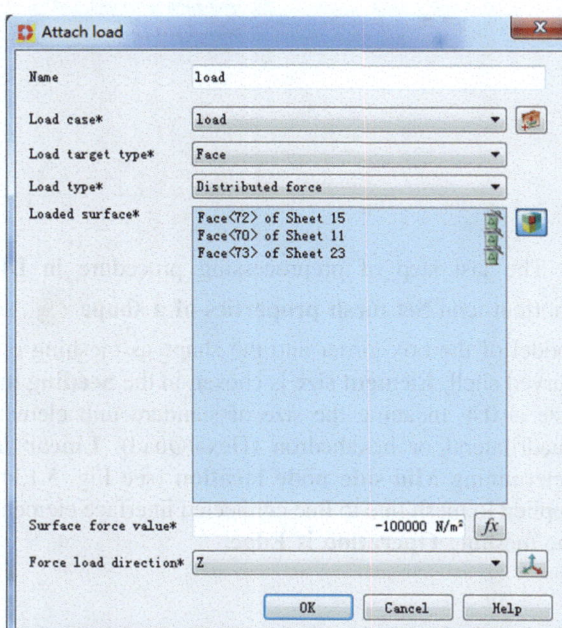

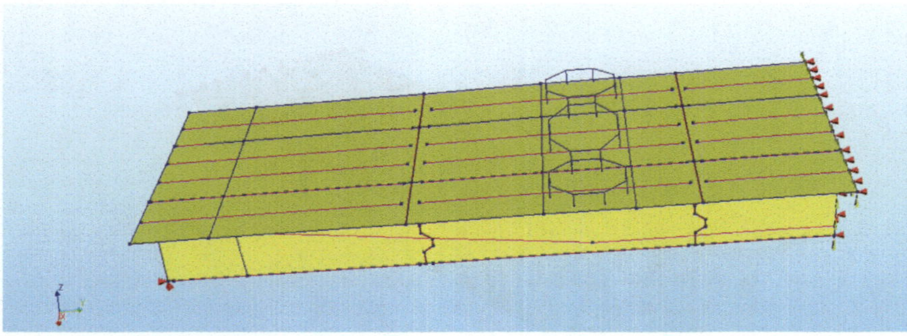

Fig. 5.132 Semi-structures of box girder after applying distributed load

Gravity and tenin1 load case are set as first stage of dead load-**Geometry load combination1** so as to apply them conveniently as initial load condition in the following structural nonlinear analysis. Load case is set as **Geometry load combination2** solo. Clicking close button, definition of load combination is finished, as shown in Fig. 5.133.

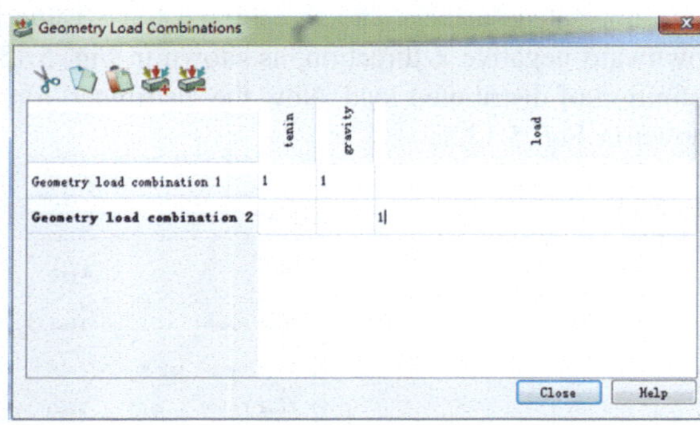

Fig. 5.133 Definition of geometric load combination

The last step of preprocessing procedure in DIANA is meshing. Clicking shortcut icon **Set mesh properties of a shape** , selecting the whole geometric model of the box girder and the shape as meshing operation in order to mesh the curved shell, **Element size** is chosen in the **Seeding method** option and the desired size is 0.4, meaning the size of standard unit element is 0.4 m. **Mesher type** is quadrilateral or hexahedron (Hexa/Quad). **Linear interpolation** is the way of determining **Mid-side node location** (see Fig. 5.134). Then the same method is applied to mesh line to line connected interface elements between curved shells, and the meshing **Operation** is **Edge.**

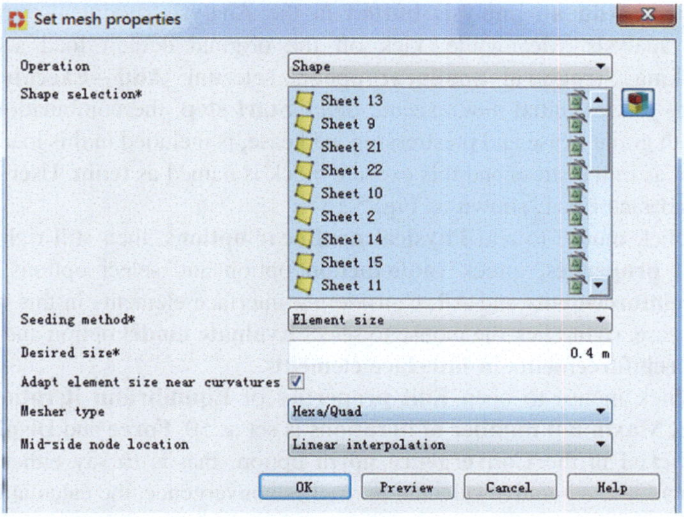

Fig. 5.134 Meshing settings

Click shortcut icon **Generate mesh of a shape**, meshed curved shell elements as Fig. 5.135 shows are generated. Click **Element types** option under mesh menu bar and the names of all generated meshed elements in the menu option of **Element types** can be displayed (Fig. 5.136).

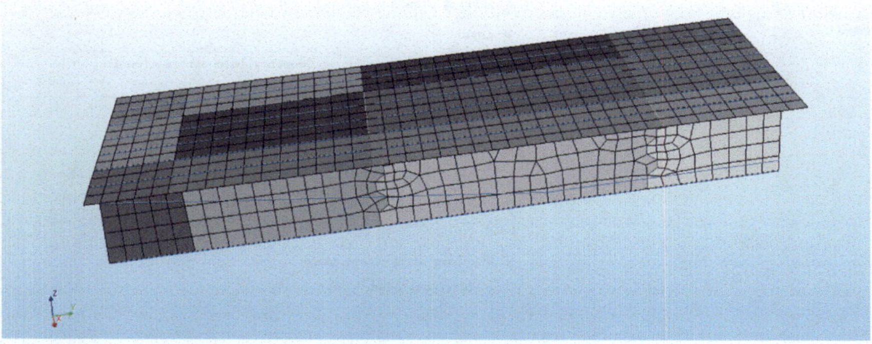

Fig. 5.135 Mesh generation

Fig. 5.136 Names of meshed elements

Click icon **Add an analysis button** in the Analysis module to create new nonlinear analysis. Meanwhile, kick off the original default load set setting. Right-clicking **Structural nonlinear** option, selecting **Add—Execute steps—Start steps** to add initial new execute block-**Start step**, the combination 1, containing both gravity case and prestressing load case, is included in this load set so as to be input as initial stress and this execute block is named as **tenin**. **User specified size** of load factor is 1, shown as Fig. 5.137.

Right-click mouse to add **Physical nonlinear options**, then still right-click to open **Edit properties,** untick **Liquefaction** option and select options of **Fully bonded reinforcements** and **All**. Considering interface elements in this numerical modeling case, right-click the mouse to select **Evaluate model** option and then tick **Evaluate reinforcements in interface elements**.

Right-click mouse to open **Edit properties** of **Equilibrium iterations** under **Start step, Maximum number of iterations** is set as 50. **Force** and **Displacement** are both ticked in the **Convergence norm** option, that is to say either force or displacement in the iterative calculation reaches convergence, the calculation step is deemed as convergence. Convergence tolerance in the **tenin** start step is set as 0.01 while **Abort criterion** is 10000.

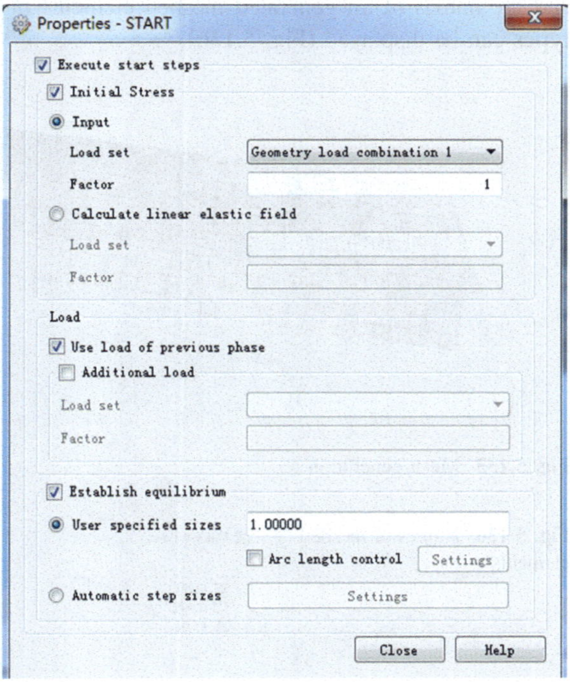

Fig. 5.137 Initial load case start step

Right-click **Structural nonlinear—Add—Execute steps—Load steps** to create another new execute block of load case. Adding **Geometry load combination 2** into **Load** set, altering the name of execute block as **load**, the **User specified size** is selected and set as (1.00000)(7), 0.200000(5) in the **Load steps**, as Fig. 5.138 shows.

Fig. 5.138 Adding **Geometry load combination 2** into **Load** set

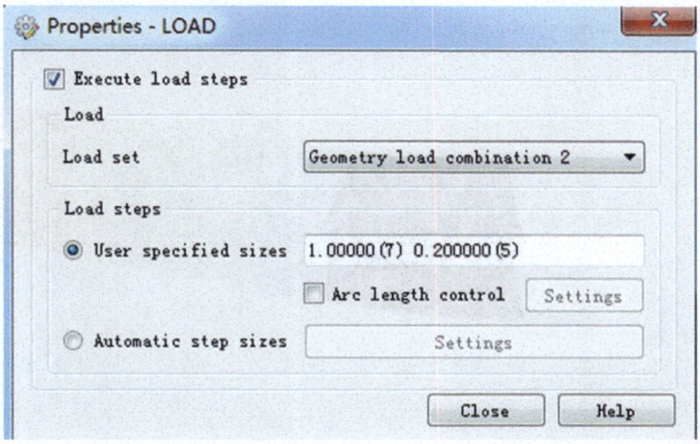

Under the load case-**load**, **Maximum number of iterations** is also set as 50 in the **Equilibrium iterations**. **Force** and **Displacement** are both selected as convergence norm. **Convergence tolerance** is still set as default value **0.01** while **Abort criterion** is also kept as **10000** unchanged.

Setting Output options, in view of checking JCSS probabilistic cracking condition under nonlinear calculation, translational displacement in all directions under global coordinate system (DISPLA TOTAL TRANSL GLOBAL), cracking strains in all directions (STRAIN CRACK GREEN), summed crack strains under local coordinate system (STRAIN CRKSUM GREEN LOCAL), summed crack strains under global coordinate system (STRAIN CRKSUM GREEN GLOBAL), summed crack strains in all principal stress directions (STRAIN CRKSUM GREEN PRINCI), crack width in all directions under global coordinate system (STRAIN CRKWDT GREEN GLOBAL) and crack width in all principal stress directions are selected as the outcomes of output, as shown in Fig. 5.139.

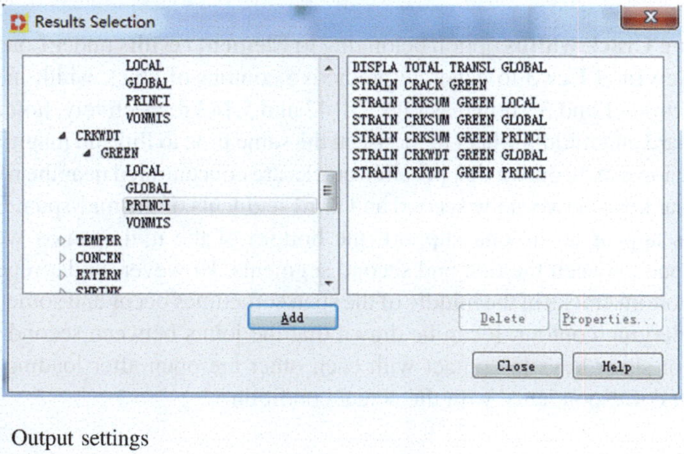

Fig. 5.139 Output settings

After Output is set, click Run analysis button to start calculation. Displacement contours of hugging-up and finished loading state (Load-step 13) in Z direction are shown in Figs. 5.140 and 5.141, respectively. It is evident to see that precast segmental box girder hugs up slightly when initial post-tensioning prestress force is applied.

Fig. 5.140 Displacement contour of hugging-up in Z direction

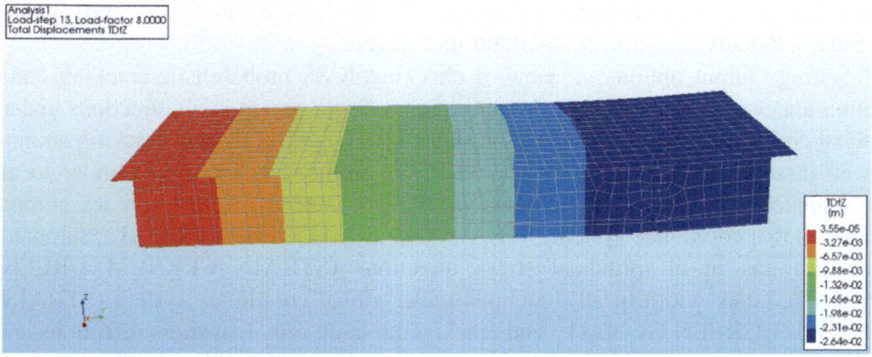

Fig. 5.141 Displacement contour of finished loading state in Z direction (load step 13)

Selecting **Crack widths** option belonging to **Element results** under Load-step 13, clicking **Ecw1**and **Ecw3** to generate predictive contour of crack width in principal stress directions 1 and 3, shown as Figs. 5.142 and 5.143 respectively, holding down Ctrl keyboard and middle wheel in mouse at the same time to flip and magnify model, the conclusion can be drawn that possible cracks are concentrated near the roots of the double shear keys between the second and third segments of the mid-span. Besides, a few cracks appear at the end support, the bottom of the mid-span as well as the occlusal zone between the first and second segments. However, under repeated calculation, bottom cracks in the middle of the span sometimes occur and sometimes not. Further enlarging contour, it can be drawn that the joints between second and third segments of shear keys in contact with each other are open after loading finished, which is in correspondence with the actual condition.

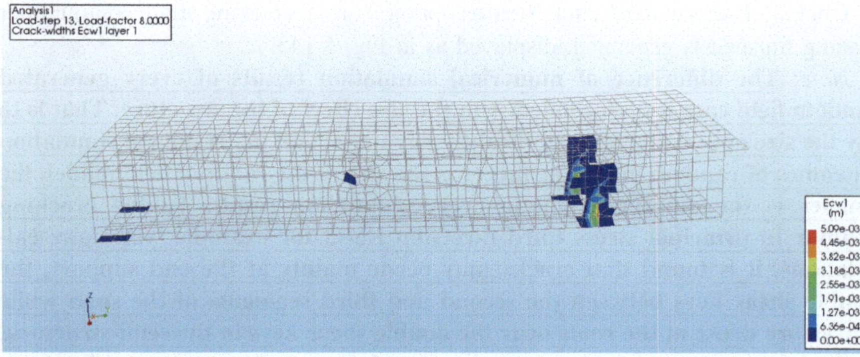

Fig. 5.142 Crack width in principal stress 1st direction Ecw1

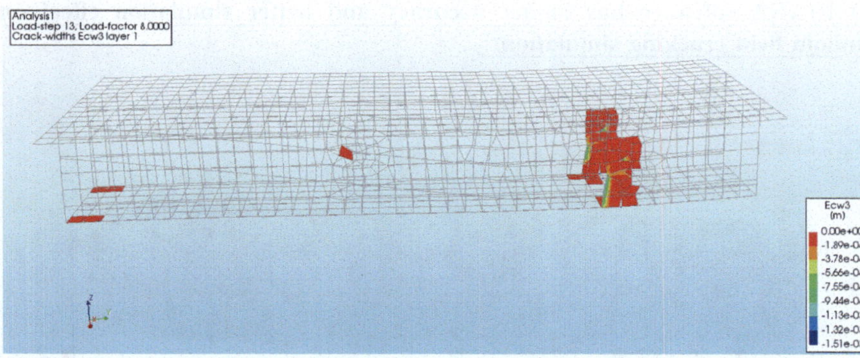

Fig. 5.143 Crack width contour in principal stress 3rd direction Ecw3

Then we click **EcwXX** under **Crack widths** option to generate crack width contour in global X direction, as shown in Fig. 5.144.

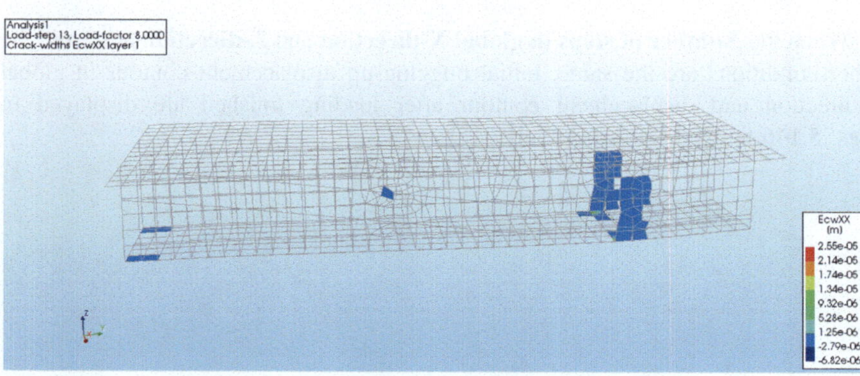

Fig. 5.144 Crack width contour in global X direction

Clicking **Eknn** under **Crack Strains** option, normal cracking strain contour after loading finished is generated, displayed as in Fig. 5.145.

Note: **The difference of numerical simulation results of every generated random field case is high related with the size effect of the structure. That is to say the size of a model has significantly key impact on the results of simulation outcomes of random field. The larger is the size of the random field model, the smaller is the difference among every simulation result. Taking cracking contour in principal stress third direction Ecw3 for example, via many calculations, it is found that cracks may occur mainly at the end support, the double shear keys between the second and third segments of the span while cracks are dense at the roots near the double shear keys in the semi-structural numerical model, which is in accordance with the cracking features of precast segmental girders. This illustrates that:(1) the uncertainty state of JCSS probabilistic model tends to stabilize with the increase of model size. (2) DIANA software has higher accuracy and better simulation effect on random field cracking simulation.**

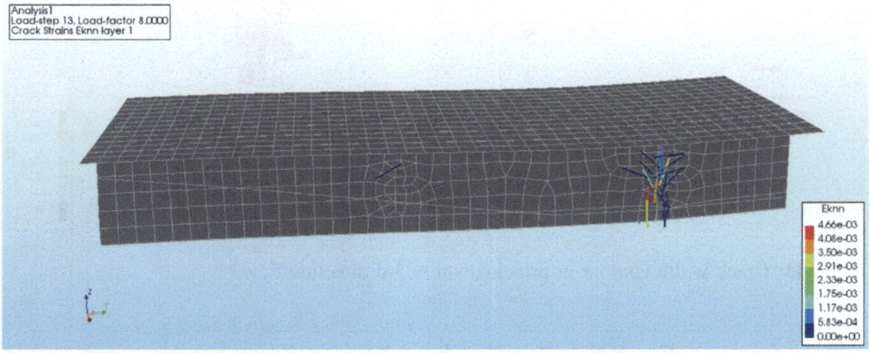

Fig. 5.145 Normal cracking strain contour after loading finished Eknn (Load-step 13)

When the **Number of steps in global X-direction** and **Z-direction** are 3 and the other conditions are the same, initial hugging-up displacement contour in global Z-direction and displacement contour after loading finished are displayed in Figs. 5.146 and 5.147, respectively.

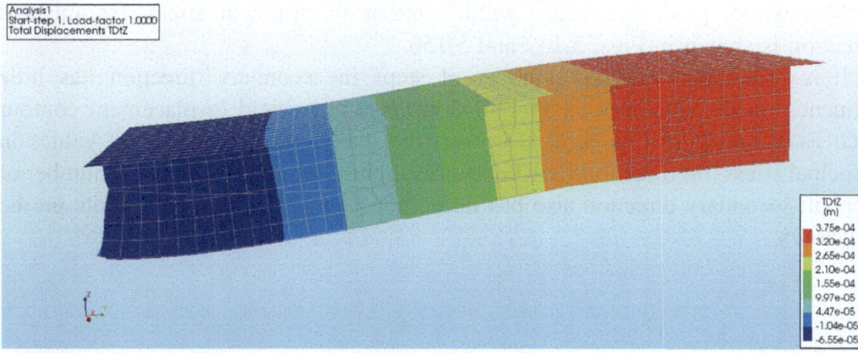

Fig. 5.146 Initial hugging-up displacement contour in global Z direction (**Number of steps in global X direction** and **Z direction** is 3)

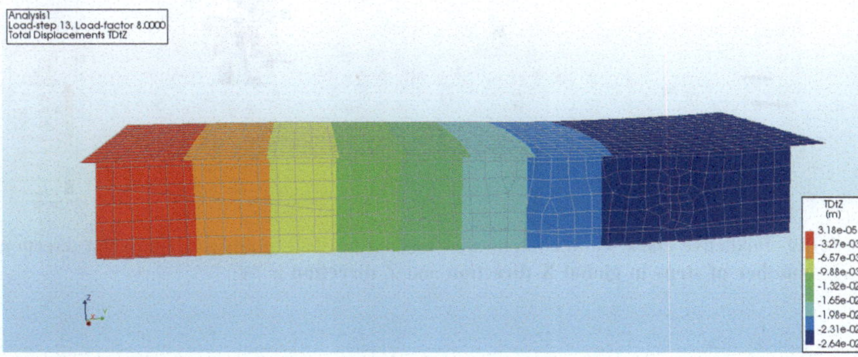

Fig. 5.147 Displacement contour in global Z direction after loading finished (**Number of steps in global X direction** and **Z direction** is 3)

Now the predictive normal cracking strain contour **Eknn** is shown in Fig. 5.148.

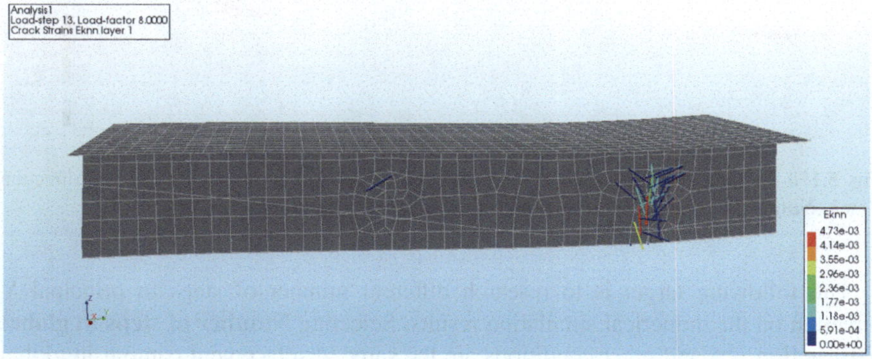

Fig. 5.148 Predictive normal cracking strain contour **Eknn** after loading finished (**Number of steps in global X direction** and **Z direction** is 3)

Meanwhile, predictive crack width contour in principal stress 1st and 3rd direction is shown in Figs. 5.149 and 5.150.

It is found that different number of steps in secondary direction has little influence on the results of JCSS probabilistic random field displacement contour such as X and Z in this case. However, the maximum and minimum values in principal stress 3rd direction **Ecw3** decrease slightly. Moreover, different number of steps in secondary direction also has little impact on potential cracking site on the structures.

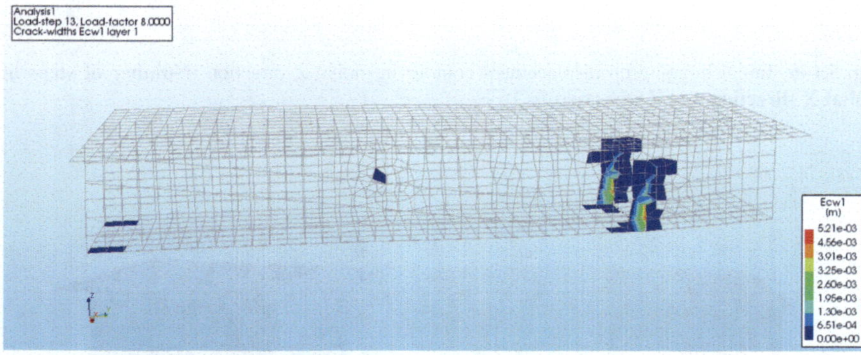

Fig. 5.149 Predictive crack width contour after loading finished in principal stress 1st direction **Ecw1** (**Number of steps in global X direction** and **Z direction** is 3)

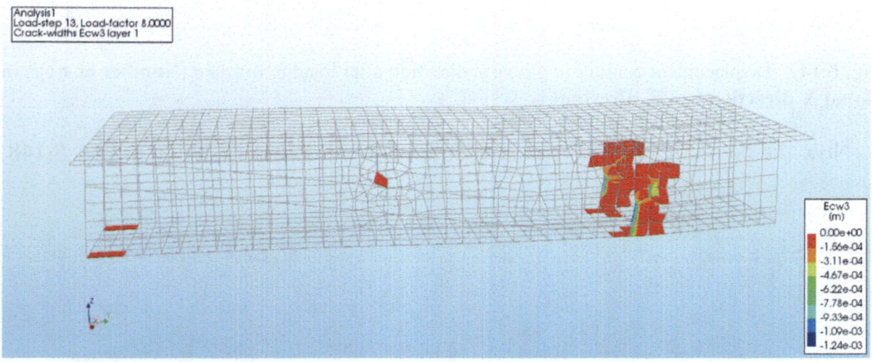

Fig. 5.150 Predictive crack width contour after loading finished in principal stress 3rd direction **Ecw3** (**Number of steps in global X direction** and **Z direction** is 3)

The following target is to research different number of steps in principal Y direction on the numerical simulation results. Selecting **Number of steps in global Y-direction** is 5 and other conditions are the same, displacement contour in global Z direction under recalculation is shown in Fig. 5.151.

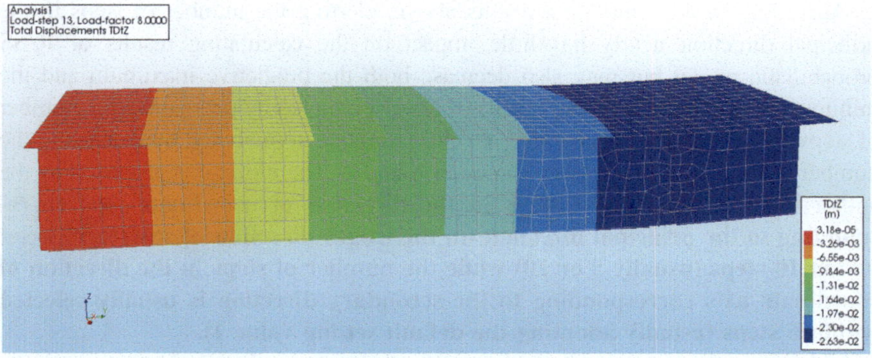

Fig. 5.151 Displacement contour in global Z direction (**Number of steps in global Y direction** is 5)

Crack width contour of Ecw1 and Ecw3 after loading finished is shown in Figs. 5.152 and 5.153, respectively.

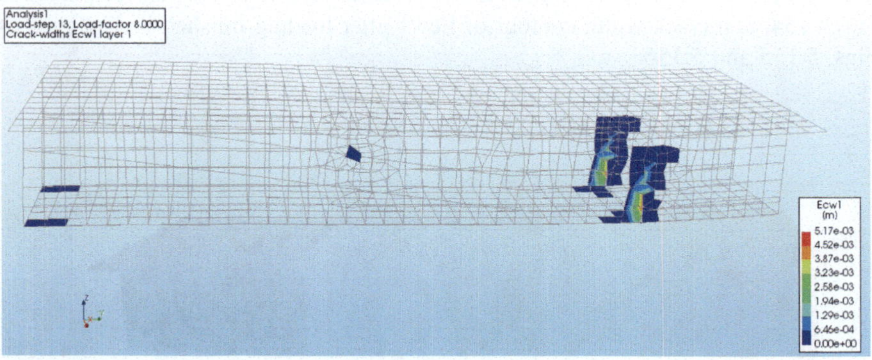

Fig. 5.152 Crack width contour of Ecw1 (**Number of steps in global Y direction** is 5)

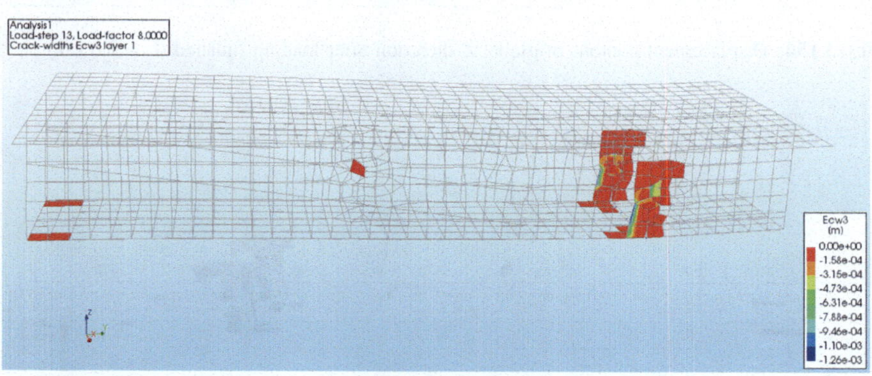

Fig. 5.153 Crack width contour of Ecw3 (**Number of steps in global Y direction** is 5)

According to the simulation results above, alerting the number of steps in the principal direction nearly has little impact on the calculating results of JCSS probabilistic model but may also decrease both the predictive maximum and the minimum values of **Ecw3**. The results of Ecw3 predicted by decreasing the number of steps in the principal direction Y are close to those of predicted by increasing the number of steps in the secondary direction X and Z. Therefore, it is proposed by the author that **the number of steps in the direction of coordinate axis corresponding to the principal direction (in the length direction of structure) is set within 10 steps (usually 5 or 10) while the number of steps in the direction of coordinate axis corresponding to the secondary direction is usually selected within 5 steps (usually adopting the default setting value 1).**

Ultimately, the impacts of correlation length on the nonlinear calculation results in this case are worth discussing under three conditions: L = 5 m, L = 0.05 m and L = 10 m. As is known to all, concrete is a kind of inhomogeneous material and the inhomogeneity of aggregate strength is directly affected by the correlation length. The shorter the correlation length, the more uniform is the strength distribution of concrete and vice versa. So the displacement contour in global Z direction after loading finished under the condition that correlation length is 0.05 m is displayed in Fig. 5.154, and crack width contour of Ecw1 after loading finished is displayed as Figs. 5.155 and 5.156.

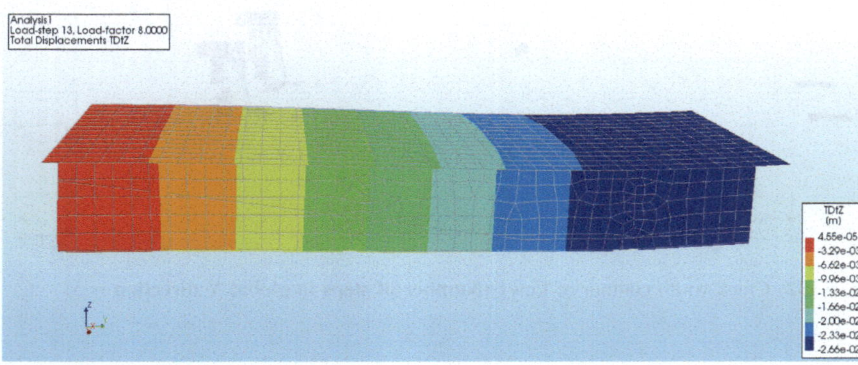

Fig. 5.154 Displacement contour in global Z direction after loading finished (L = 0.05 m)

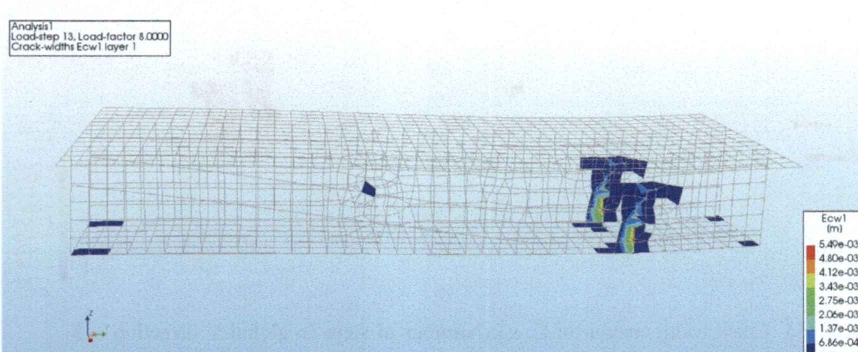

Fig. 5.155 Crack width contour of Ecw1 after loading finished (L = 0.05 m)

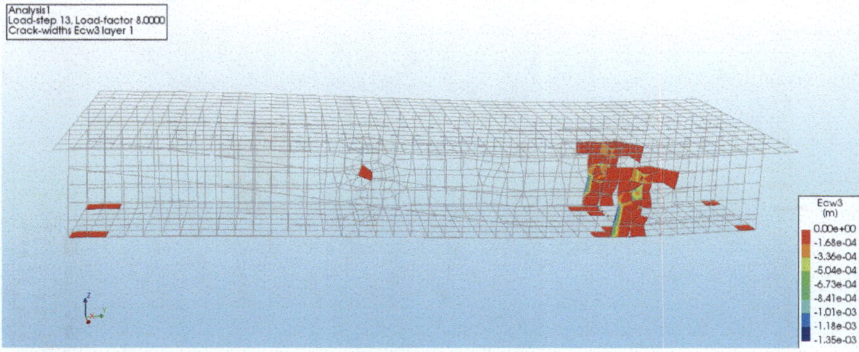

Fig. 5.156 Crack width contour of Ecw3 after loading finished (L = 0.05 m)

According to the contour results, it is evident to know that potential cracking site may have little alteration when decreasing correlation length, but the maximum vertical displacement value in Z direction diminishes. Meanwhile, the predictive maximum and minimum values of Ecw3 in the contour are also decreasing.

When the correlation length is 10 m and other conditions are the same, the predictive contours of JCSS probabilistic values are displayed in Figs. 5.157, 5.158, and 5.159.

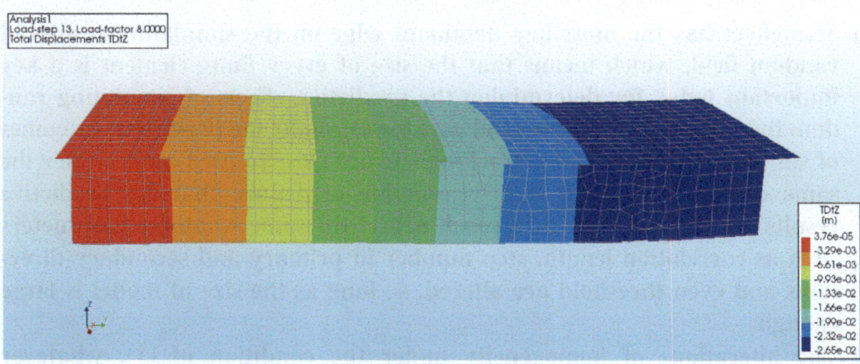

Fig. 5.157 Displacement contour in global Z direction after loading finished (L = 10 m)

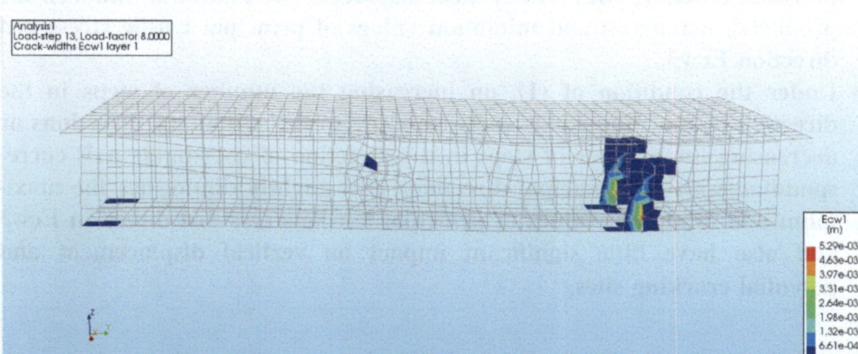

Fig. 5.158 Crack width contour of Ecw1 after loading finished (L = 10 m)

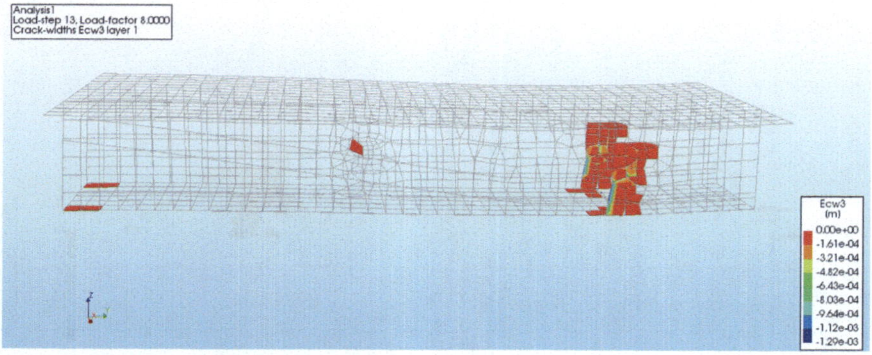

Fig. 5.159 Crack width contour of Ecw3 after loading finished (L = 10 m)

According to the contour results, potential cracking site may have little alteration when increasing correlation length, but the maximum vertical displacement value in Z direction diminishes. Meanwhile, the predictive maximum and minimum values of Ecw3 in the contour are also further decreasing.

It is easy to draw the following conclusions via the calculation and comparison mentioned above:

(1) **Size effect has the most key dominant edge on the simulating results of random field, which means that the size of every finite element is a key important index for determining the predictive effects of simulating random field. Taking this numerical case for example, the predictive outcomes of simulation and potential cracking site are close to each other under the same calculation model. It is reasonable to deduce that the predictive results of JCSS random field reach nearly the same whatever parameters such as correlation length, step number in primary and secondary directions and even threshold are altered, so long as the size of model is large enough.**

(2) **The alterations of homogeneity under the condition of (1), whatever increasing or decreasing correlation length, have little influence on potential cracking sites and vertical displacement values but diminish the predictive maximum and minimum values of principal tensile stress 3rd direction Ecw3.**

(3) **Under the condition of (1), on increasing the number of steps in the direction of coordinate axis corresponding to the secondary directions or decreasing the number of steps in the direction of coordinate axis corresponding with the principal directions only slightly diminishes the maximum and minimum values of principal tensile stress 3rd direction Ecw3 and also have little significant impact on vertical displacement and potential cracking sites.**

Python command console is as follows:

```
newProject( "PSB-Random-32m", 100 )
setModelAnalysisAspects( [ "STRUCT" ] )
setModelDimension( "3D" )
setDefaultMeshOrder( "QUADRATIC" )
setDefaultMesherType( "HEXQUAD" )
setDefaultMidSideNodeLocation( "LINEAR" )
createSheet( "Sheet 1", [[ 0, 0, 0 ],[ 0, 2, 0 ],[ 0, 2, 2.4 ],[ 0, 0.2, 2.4 ]] )
createSheet( "Sheet 2", [[ 0, 2, 0 ],[ 0, 6.4, 0 ],[ 0, 6.4, 0.36 ],[ 0, 6.64, 0.528 ],[ 0, 6.64,
0.736 ],[ 0, 6.4, 0.92 ],[ 0, 6.4, 1.16 ],[ 0, 6.64, 1.336 ],[ 0, 6.64, 1.544 ],[ 0, 6.4, 1.72 ],[ 0, 6.4,
2.4 ],[ 0, 2, 2.4 ]] )
createSheet( "Sheet 3", [[ 0, 6.4, 0 ],[ 0, 12.8, 0 ],[ 0, 12.8, 0.36 ],[ 0, 13.04, 0.528 ],[ 0, 13.04,
0.736 ],[ 0, 12.8, 0.92 ],[ 0, 12.8, 1.16 ],[ 0, 13.04, 1.336 ],[ 0, 13.04, 1.544 ],[ 0, 12.8, 1.72 ],[ 0,
12.8, 2.4 ],[ 0, 6.4, 2.4 ],[ 0, 6.4, 1.72 ],[ 0, 6.64, 1.544 ],[ 0, 6.64, 1.336 ],[ 0, 6.4, 1.16 ],[ 0, 6.4,
0.92 ],[ 0, 6.64, 0.736 ],[ 0, 6.64, 0.528 ],[ 0, 6.4, 0.36 ]] )
saveProject(  )
createSheet( "Sheet 4", [[ 0, 12.8, 0 ],[ 0, 16, 0 ],[ 0, 16, 2.4 ],[ 0, 12.8, 2.4 ],[ 0, 12.8,
1.72 ],[ 0, 13.04, 1.544 ],[ 0, 13.04, 1.336 ],[ 0, 12.8, 1.16 ],[ 0, 12.8, 0.92 ],[ 0, 13.04, 0.736 ],[ 0,
13.04, 0.528 ],[ 0, 12.8, 0.36 ]] )
saveProject(  )
arrayCopy( [ "Sheet 1", "Sheet 2", "Sheet 3", "Sheet 4" ], [ 2.32, 0, 0 ], [ 0, 0, 0 ], [ 0, 0, 0 ],
1 )
createSheet( "Sheet 9", [[ 0, 0.2, 2.4 ],[ 2.32, 0.2, 2.4 ],[ 2.32, 2, 2.4 ],[ 0, 2, 2.4 ]] )
createSheet( "Sheet 10", [[ 0, 2, 2.4 ],[ 2.32, 2, 2.4 ],[ 2.32, 6.4, 2.4 ],[ 0, 6.4, 2.4 ]] )
createSheet( "Sheet 11", [[ 0, 6.4, 2.4 ],[ 2.32, 6.4, 2.4 ],[ 2.32, 12.8, 2.4 ],[ 0, 12.8, 2.4 ]] )
createSheet( "Sheet 12", [[ 0, 12.8, 2.4 ],[ 2.32, 12.8, 2.4 ],[ 2.32, 16, 2.4 ],[ 0, 16, 2.4 ]] )
createSheet( "Sheet 13", [[ 0, 0.2, 2.4 ],[ 0, 2, 2.4 ],[ -1.56, 2, 2.4 ],[ -1.56, 0.2, 2.4 ]] )
createSheet( "Sheet 14", [[ 0, 2, 2.4 ],[ 0, 6.4, 2.4 ],[ -1.56, 6.4, 2.4 ],[ -1.56, 2, 2.4 ]] )
createSheet( "Sheet 15", [[ -1.56, 6.4, 2.4 ],[ 0, 6.4, 2.4 ],[ 0, 12.8, 2.4 ],[ -1.56, 12.8, 2.4 ]] )
createSheet( "Sheet 16", [[ -1.56, 12.8, 2.4 ],[ 0, 12.8, 2.4 ],[ 0, 16, 2.4 ],[ -1.56, 16, 2.4 ]] )
createSheet( "Sheet 17", [[ 0, 0, 0 ],[ 2.32, 0, 0 ],[ 2.32, 2, 0 ],[ 0, 2, 0 ]] )
arrayCopy( [ "Sheet 10", "Sheet 11", "Sheet 12" ], [ 0, 0, -2.4 ], [ 0, 0, 0 ], [ 0, 0, 0 ], 1 )
arrayCopy( [ "Sheet 13", "Sheet 14", "Sheet 15", "Sheet 16" ], [ 3.88, 0, 0 ], [ 0, 0, 0 ], [ 0, 0,
0 ], 1 )
createLine( "bar1", [ 0, 0.16, 0 ], [ 0, 6.24, 0 ] )
arrayCopy( [ "bar1" ], [ 0.778, 0, 0 ], [ 0, 0, 0 ], [ 0, 0, 0 ], 3 )
saveProject(  )
arrayCopy( [ "bar1", "bar2", "bar3", "bar4" ], [ 0, 6.4, 0 ], [ 0, 0, 0 ], [ 0, 0, 0 ], 1 )
saveProject(  )
createLine( "bar9", [ 0, 12.96, 0 ], [ 0, 16, 0 ] )
arrayCopy( [ "bar9" ], [ 0.778, 0, 0 ], [ 0, 0, 0 ], [ 0, 0, 0 ], 3 )
saveProject(  )
```

```
createLine( "bar13", [ 0, 0.24, 2.4 ], [ 0, 6.24, 2.4 ] )
arrayCopy( [ "bar13" ], [ 0.778, 0, 0 ], [ 0, 0, 0 ], [ 0, 0, 0 ], 4 )
arrayCopy( [ "bar13" ], [ -0.778, 0, 0 ], [ 0, 0, 0 ], [ 0, 0, 0 ], 1 )
arrayCopy( [ "bar13", "bar14", "bar15", "bar16", "bar17", "bar18" ], [ 0, 6.4, 0 ], [ 0, 0, 0 ], [ 0, 0, 0 ], 1 )
createLine( "bar25", [ 0, 12.96, 2.4 ], [ 0, 16, 2.4 ] )
arrayCopy( [ "bar25" ], [ 0.78, 0, 0 ], [ 0, 0, 0 ], [ 0, 0, 0 ], 4 )
saveProject(   )
arrayCopy( [ "bar25" ], [ -0.78, 0, 0 ], [ 0, 0, 0 ], [ 0, 0, 0 ], 1 )
createPolyline( "tenin1", [[ 0, 0.129987, 1.56 ],[ 0, 10.4, 0.28 ],[ 0, 16, 0.28 ]], False )
arrayCopy( [ "tenin1" ], [ 2.32, 0, 0 ], [ 0, 0, 0 ], [ 0, 0, 0 ], 1 )
setUnit( "TEMPER", "CELSIU" )
setUnit( "ANGLE", "DEGREE" )
addMaterial( "top", "CONCDC", "JCSSPR", [ "JCSSRF" ] )
setParameter( MATERIAL, "top", "JCSSMC/JCSSTP", "PRECST" )
setParameter( MATERIAL, "top", "JCSSMC/JCSSG2/JCSSGR", "C55" )
setParameter( MATERIAL, "top", "JCSSMC/DENSIT", 2500 )
setParameter( MATERIAL, "top", "JCSSMC/DENSIT", 2500 )
setParameter( MATERIAL, "top", "JCSSMC/DENSIT", 2500 )
setParameter( MATERIAL, "top", "JCSSMC/GF1", 500 )
setParameter( MATERIAL, "top", "JCSSRF/COVARI/NX", 1 )
setParameter( MATERIAL, "top", "JCSSRF/COVARI/NY", 10 )
setParameter( MATERIAL, "top", "JCSSRF/COVARI/NZ", 1 )
setParameter( MATERIAL, "top", "JCSSMC/JCSSTP", "RMIXED" )
setParameter( MATERIAL, "top", "JCSSMC/JCSSTP", "PRECST" )
addGeometry( "Element geometry 3", "SHEET", "CURSHL", [] )
rename( GEOMET, "Element geometry 3", "top" )
setParameter( GEOMET, "top", "THICK", 1.04 )
setParameter( GEOMET, "top", "LOCAXS", True )
setParameter( GEOMET, "top", "LOCAXS/XAXIS", [ 1, 0, 0 ] )
setParameter( GEOMET, "top", "LOCAXS/XAXIS", [ 0, 1, 0 ] )
setParameter( GEOMET, "top", "LOCAXS/XAXIS", [ 0, 1, 0 ] )
setParameter( GEOMET, "top", "LOCAXS/XAXIS", [ 0, 1, 0 ] )
clearReinforcementAspects( [ "Sheet 9", "Sheet 10", "Sheet 11", "Sheet 12", "Sheet 13", "Sheet 14", "Sheet 15", "Sheet 16", "Sheet 21", "Sheet 22", "Sheet 23", "Sheet 24" ] )
setElementClassType( SHAPE, [ "Sheet 9", "Sheet 10", "Sheet 11", "Sheet 12", "Sheet 13", "Sheet 14", "Sheet 15", "Sheet 16", "Sheet 21", "Sheet 22", "Sheet 23", "Sheet 24" ], "CURSHL" )
assignMaterial( "top", SHAPE, [ "Sheet 9", "Sheet 10", "Sheet 11", "Sheet 12", "Sheet 13", "Sheet 14", "Sheet 15", "Sheet 16", "Sheet 21", "Sheet 22", "Sheet 23", "Sheet 24" ] )
assignGeometry( "top", SHAPE, [ "Sheet 9", "Sheet 10", "Sheet 11", "Sheet 12", "Sheet 13", "Sheet 14", "Sheet 15", "Sheet 16", "Sheet 21", "Sheet 22", "Sheet 23", "Sheet 24" ] )
resetElementData( SHAPE, [ "Sheet 9", "Sheet 10", "Sheet 11", "Sheet 12", "Sheet 13", "Sheet 14", "Sheet 15", "Sheet 16", "Sheet 21", "Sheet 22", "Sheet 23", "Sheet 24" ] )
addMaterial( "bot", "CONCDC", "JCSSPR", [ "JCSSRF" ] )
```

```
setParameter( MATERIAL, "bot", "JCSSMC/JCSSTP", "PRECST" )
setParameter( MATERIAL, "bot", "JCSSMC/JCSSG2/JCSSGR", "C55" )
setParameter( MATERIAL, "bot", "JCSSMC/DENSIT", 2500 )
setParameter( MATERIAL, "bot", "JCSSMC/GF1", 500 )
setParameter( MATERIAL, "bot", "JCSSRF/COVARI/NX", 1 )
setParameter( MATERIAL, "bot", "JCSSRF/COVARI/NY", 10 )
setParameter( MATERIAL, "bot", "JCSSRF/COVARI/NZ", 1 )
addGeometry( "Element geometry 4", "SHEET", "CURSHL", [] )
rename( GEOMET, "Element geometry 4", "bot" )
setParameter( GEOMET, "bot", "THICK", 0.4 )
setParameter( GEOMET, "bot", "LOCAXS", True )
setParameter( GEOMET, "bot", "LOCAXS/XAXIS", [ 1, 0, 0 ] )
setParameter( GEOMET, "bot", "LOCAXS/XAXIS", [ 0, 1, 0 ] )
setParameter( GEOMET, "bot", "LOCAXS/XAXIS", [ 0, 1, 0 ] )
setParameter( GEOMET, "bot", "LOCAXS/XAXIS", [ 0, 1, 0 ] )
setParameter( GEOMET, "bot", "LOCAXS/XAXIS", [ 0, 1, 0 ] )
clearReinforcementAspects( [ "Sheet 17", "Sheet 18", "Sheet 19", "Sheet 20" ] )
setElementClassType( SHAPE, [ "Sheet 17", "Sheet 18", "Sheet 19", "Sheet 20" ],
"CURSHL" )
assignMaterial( "bot", SHAPE, [ "Sheet 17", "Sheet 18", "Sheet 19", "Sheet 20" ] )
assignGeometry( "bot", SHAPE, [ "Sheet 17", "Sheet 18", "Sheet 19", "Sheet 20" ] )
resetElementData( SHAPE, [ "Sheet 17", "Sheet 18", "Sheet 19", "Sheet 20" ] )
saveProject(  )
addMaterial( "mid1", "CONCDC", "JCSSPR", [ "JCSSRF" ] )
setParameter( MATERIAL, "mid1", "JCSSMC/JCSSTP", "PRECST" )
setParameter( MATERIAL, "mid1", "JCSSMC/JCSSG2/JCSSGR", "C55" )
setParameter( MATERIAL, "mid1", "JCSSMC/DENSIT", 2500 )
setParameter( MATERIAL, "mid1", "JCSSMC/GF1", 500 )
setParameter( MATERIAL, "mid1", "JCSSRF/COVARI/NX", 1 )
setParameter( MATERIAL, "mid1", "JCSSRF/COVARI/NY", 10 )
setParameter( MATERIAL, "mid1", "JCSSRF/COVARI/NZ", 1 )
addGeometry( "Element geometry 5", "SHEET", "CURSHL", [] )
rename( GEOMET, "Element geometry 5", "mid1" )
setParameter( GEOMET, "mid1", "THICK", 2.72 )
setParameter( GEOMET, "mid1", "LOCAXS", True )
setParameter( GEOMET, "mid1", "LOCAXS/XAXIS", [ 1, 0, 0 ] )
setParameter( GEOMET, "mid1", "LOCAXS/XAXIS", [ 0, 1, 0 ] )
setParameter( GEOMET, "mid1", "LOCAXS/XAXIS", [ 0, 1, 0 ] )
clearReinforcementAspects( [ "Sheet 1", "Sheet 5" ] )
setElementClassType( SHAPE, [ "Sheet 1", "Sheet 5" ], "CURSHL" )
assignMaterial( "mid1", SHAPE, [ "Sheet 1", "Sheet 5" ] )
```

```
assignGeometry( "mid1", SHAPE, [ "Sheet 1", "Sheet 5" ] )
resetElementData( SHAPE, [ "Sheet 1", "Sheet 5" ] )
saveProject( )
addMaterial( "mid2", "CONCDC", "JCSSPR", [ "JCSSRF" ] )
setParameter( MATERIAL, "mid2", "JCSSMC/JCSSTP", "PRECST" )
setParameter( MATERIAL, "mid2", "JCSSMC/JCSSG2/JCSSGR", "C55" )
setParameter( MATERIAL, "mid2", "JCSSMC/DENSIT", 2500 )
setParameter( MATERIAL, "mid2", "JCSSMC/GF1", 500 )
setParameter( MATERIAL, "mid2", "JCSSRF/COVARI/NX", 1 )
setParameter( MATERIAL, "mid2", "JCSSRF/COVARI/NY", 10 )
setParameter( MATERIAL, "mid2", "JCSSRF/COVARI/NZ", 1 )
addGeometry( "Element geometry 6", "SHEET", "CURSHL", [] )
rename( GEOMET, "Element geometry 6", "mid2" )
setParameter( GEOMET, "mid2", "THICK", 0.72 )
setParameter( GEOMET, "mid2", "LOCAXS", True )
setParameter( GEOMET, "mid2", "LOCAXS/XAXIS", [ 1, 0, 0 ] )
setParameter( GEOMET, "mid2", "LOCAXS/XAXIS", [ 0, 1, 0 ] )
setParameter( GEOMET, "mid2", "LOCAXS/XAXIS", [ 0, 1, 0 ] )
clearReinforcementAspects( [ "Sheet 2", "Sheet 3", "Sheet 4", "Sheet 6", "Sheet 7", "Sheet 8" ] )
setElementClassType( SHAPE, [ "Sheet 2", "Sheet 3", "Sheet 4", "Sheet 6", "Sheet 7", "Sheet 8" ], "CURSHL" )
assignMaterial( "mid2", SHAPE, [ "Sheet 2", "Sheet 3", "Sheet 4", "Sheet 6", "Sheet 7", "Sheet 8" ] )
assignGeometry( "mid2", SHAPE, [ "Sheet 2", "Sheet 3", "Sheet 4", "Sheet 6", "Sheet 7", "Sheet 8" ] )
resetElementData( SHAPE, [ "Sheet 2", "Sheet 3", "Sheet 4", "Sheet 6", "Sheet 7", "Sheet 8" ] )
saveProject( )
addMaterial( "tenin", "REINFO", "VMISES", [] )
setMaterialAspects( "tenin", [ "FRLGTH" ] )
setMaterialAspects( "tenin", [] )
setParameter( "MATERIAL", "tenin", "PLASTI/HARDI1/YLDSTR", 1.86e+09 )
setParameter( "MATERIAL", "tenin", "LINEAR/ELASTI/YOUNG", 1.95e+11 )
setMaterialAspects( "tenin", [ "FRLGTH" ] )
setParameter( "MATERIAL", "tenin", "FREELE/FRLGTH", 1E-5)
addGeometry( "Element geometry 6", "RELINE", "REBAR", [] )
rename( "GEOMET", "Element geometry 6", "tenin" )
setParameter( "GEOMET", "tenin", "REIEMB/CROSSE", 0.001112 )
setParameter( "GEOMET", "tenin", "REIEMB/CROSSE", 0.001112 )
setReinforcementAspects( [ "tenin1", "tenin2" ] )
```

```
assignMaterial( "tenin", "SHAPE", [ "tenin1", "tenin2" ] )
assignGeometry( "tenin", "SHAPE", [ "tenin1", "tenin2" ] )
resetElementData( "SHAPE", [ "tenin1", "tenin2" ] )
setReinforcementDiscretization( [ "tenin1", "tenin2" ], "SECTION" )
saveProject(  )
addMaterial( "BAR", "MCSTEL", "TRESCA", [] )
setParameter( "MATERIAL", "BAR", "LINEAR/ELASTI/YOUNG", 2.1e+11 )
setParameter( "MATERIAL", "BAR", "LINEAR/ELASTI/POISON", 0.3 )
setParameter( "MATERIAL", "BAR", "LINEAR/MASS/DENSIT", 7800 )
setParameter( "MATERIAL", "BAR", "TREPLA/TRESSH", "NONE" )
setParameter( "MATERIAL", "BAR", "TREPLA/YLDSTR", 4.4e+08 )
setParameter( "MATERIAL", "BAR", "TREPLA/TRESSH", "KAPSIG" )
setParameter( "MATERIAL", "BAR", "TREPLA/YIELD", "TRESCA" )
setParameter( "MATERIAL", "BAR", "TREPLA/YIELD", "VMISES" )
setParameter( "MATERIAL", "BAR", "TREPLA/TRESSH", "NONE" )
addGeometry( "Element geometry 7", "RELINE", "REBAR", [] )
rename( "GEOMET", "Element geometry 7", "BAR" )
setParameter( "GEOMET", "BAR", "REIEMB/CROSSE", 5.0625e-05 )
setReinforcementAspects( [ "bar1", "bar2", "bar3", "bar4", "bar5", "bar6", "bar7", "bar8",
"bar9", "bar10", "bar11", "bar12", "bar13", "bar14", "bar15", "bar16", "bar17", "bar18", "bar19",
"bar20", "bar21", "bar22", "bar23", "bar24", "bar25", "bar26", "bar27", "bar28", "bar29",
"bar30" ] )
assignMaterial( "BAR", "SHAPE", [ "bar1", "bar2", "bar3", "bar4", "bar5", "bar6", "bar7",
"bar8", "bar9", "bar10", "bar11", "bar12", "bar13", "bar14", "bar15", "bar16", "bar17", "bar18",
"bar19", "bar20", "bar21", "bar22", "bar23", "bar24", "bar25", "bar26", "bar27", "bar28", "bar29",
"bar30" ] )
assignGeometry( "BAR", "SHAPE", [ "bar1", "bar2", "bar3", "bar4", "bar5", "bar6", "bar7",
"bar8", "bar9", "bar10", "bar11", "bar12", "bar13", "bar14", "bar15", "bar16", "bar17", "bar18",
"bar19", "bar20", "bar21", "bar22", "bar23", "bar24", "bar25", "bar26", "bar27", "bar28", "bar29",
"bar30" ] )
resetElementData( "SHAPE", [ "bar1", "bar2", "bar3", "bar4", "bar5", "bar6", "bar7", "bar8",
"bar9", "bar10", "bar11", "bar12", "bar13", "bar14", "bar15", "bar16", "bar17", "bar18", "bar19",
"bar20", "bar21", "bar22", "bar23", "bar24", "bar25", "bar26", "bar27", "bar28", "bar29",
"bar30" ] )
setReinforcementDiscretization( [ "bar1", "bar2", "bar3", "bar4", "bar5", "bar6", "bar7",
"bar8", "bar9", "bar10", "bar11", "bar12", "bar13", "bar14", "bar15", "bar16", "bar17", "bar18",
"bar19", "bar20", "bar21", "bar22", "bar23", "bar24", "bar25", "bar26", "bar27", "bar28", "bar29",
"bar30" ], "SECTION" )
saveProject(  )
addMaterial( "int1", "INTERF", "NONLIF", [] )
setParameter( MATERIAL, "int1", "LINEAR/IFTYP", "LIN3D" )
```

```
setParameter( MATERIAL, "int1", "LINEAR/ELAS4/DSNY", 3.65e+16 )
setParameter( MATERIAL, "int1", "LINEAR/ELAS4/DSSX", 3.65e+12 )
setParameter( MATERIAL, "int1", "LINEAR/ELAS4/DSSZ", 3.65e+12 )
setParameter( MATERIAL, "int1", "NONLIN/IFNOTE", "NOTENS" )
addGeometry( "Element geometry 12", "LINE", "SHLLIF", [] )
rename( GEOMET, "Element geometry 12", "int1" )
setParameter( GEOMET, "int1", "THICK", 1.04 )
setParameter( GEOMET, "int1", "THKDIR", "PARALL" )
setParameter( GEOMET, "int1", "YAXIS", [ 1, 0, 0 ] )
setParameter( GEOMET, "int1", "YAXIS", [ 0, 1, 0 ] )
saveProject(   )
setParameter( GEOMET, "int1", "YAXIS", [ 0, -1, 0 ] )
setParameter( GEOMET, "int1", "YAXIS", [ 0, -1, 0 ] )
createLineConnection( "int1" )
setParameter( GEOMETRYCONNECTION, "int1", "CONTYP", "INTER" )
setParameter( GEOMETRYCONNECTION, "int1", "MODE", "AUTO" )
attachTo( GEOMETRYCONNECTION, "int1", "SOURCE", "Sheet 10", [[ 1.16, 6.4, 2.4 ]] )
attachTo( GEOMETRYCONNECTION, "int1", "SOURCE", "Sheet 12", [[ 1.16, 12.8,
2.4 ]] )
attachTo( GEOMETRYCONNECTION, "int1", "SOURCE", "Sheet 14", [[ -0.78, 6.4, 2.4 ]] )
attachTo( GEOMETRYCONNECTION, "int1", "SOURCE", "Sheet 15", [[ -0.78, 12.8,
2.4 ]] )
attachTo( GEOMETRYCONNECTION, "int1", "SOURCE", "Sheet 22", [[ 3.1, 6.4, 2.4 ]] )
attachTo( GEOMETRYCONNECTION, "int1", "SOURCE", "Sheet 23", [[ 3.1, 12.8, 2.4 ]] )
setElementClassType( GEOMETRYCONNECTION, "int1", "SHLLIF" )
assignMaterial( "int1", GEOMETRYCONNECTION, "int1" )
assignGeometry( "int1", GEOMETRYCONNECTION, "int1" )
resetElementData( GEOMETRYCONNECTION, "int1" )
addMaterial( "int2", "INTERF", "NONLIF", [] )
setParameter( MATERIAL, "int2", "LINEAR/IFTYP", "LIN3D" )
setParameter( MATERIAL, "int2", "LINEAR/ELAS4/DSNY", 3.65e+16 )
setParameter( MATERIAL, "int2", "LINEAR/ELAS4/DSSX", 3.65e+12 )
setParameter( MATERIAL, "int2", "LINEAR/ELAS4/DSSZ", 3.65e+12 )
setParameter( MATERIAL, "int2", "NONLIN/IFNOTE", "NOTENS" )
addGeometry( "Element geometry 13", "LINE", "SHLLIF", [] )
rename( GEOMET, "Element geometry 13", "int2" )
setParameter( GEOMET, "int2", "THICK", 0.4 )
setParameter( GEOMET, "int2", "THKDIR", "PARALL" )
setParameter( GEOMET, "int2", "YAXIS", [ 1, 0, 0 ] )
setParameter( GEOMET, "int2", "YAXIS", [ 0, 1, 0 ] )
setParameter( GEOMET, "int2", "YAXIS", [ 0, -1, 0 ] )
```

```
createLineConnection( "int2" )
setParameter( GEOMETRYCONNECTION, "int2", "CONTYP", "INTER" )
setParameter( GEOMETRYCONNECTION, "int2", "MODE", "AUTO" )
attachTo( GEOMETRYCONNECTION, "int2", "SOURCE", "Sheet 18", [[ 1.16, 6.4, 0 ]] )
attachTo( GEOMETRYCONNECTION, "int2", "SOURCE", "Sheet 19", [[ 1.16, 12.8, 0 ]] )
setElementClassType( GEOMETRYCONNECTION, "int2", "SHLLIF" )
assignMaterial( "int2", GEOMETRYCONNECTION, "int2" )
assignGeometry( "int2", GEOMETRYCONNECTION, "int2" )
resetElementData( GEOMETRYCONNECTION, "int2" )
saveProject(  )
addMaterial( "int3", "INTERF", "NONLIF", [] )
setParameter( MATERIAL, "int3", "LINEAR/IFTYP", "LIN3D" )
setParameter( MATERIAL, "int3", "LINEAR/ELAS4/DSNY", 3.65e+16 )
setParameter( MATERIAL, "int3", "LINEAR/ELAS4/DSSX", 3.65e+12 )
setParameter( MATERIAL, "int3", "LINEAR/ELAS4/DSSZ", 3.65e+12 )
setParameter( MATERIAL, "int3", "NONLIN/IFNOTE", "NOTENS" )
addGeometry( "Element geometry 14", "LINE", "SHLLIF", [] )
rename( GEOMET, "Element geometry 14", "int3" )
setParameter( GEOMET, "int3", "THICK", 0.72 )
setParameter( GEOMET, "int3", "THKDIR", "PARALL" )
setParameter( GEOMET, "int3", "YAXIS", [ 1, 0, 0 ] )
setParameter( GEOMET, "int3", "YAXIS", [ 0, 1, 0 ] )
setParameter( GEOMET, "int3", "YAXIS", [ 0, -1, 0 ] )
setParameter( GEOMET, "int3", "THICK", 0.72 )
createLineConnection( "int3" )
setParameter( GEOMETRYCONNECTION, "int3", "CONTYP", "INTER" )
setParameter( GEOMETRYCONNECTION, "int3", "MODE", "AUTO" )
attachTo( GEOMETRYCONNECTION, "int3", "SOURCE", "Sheet 2", [[ 0, 6.4, 0.18 ],[ 0, 6.52, 0.444 ],[ 0, 6.64, 0.632 ],[ 0, 6.4, 1.04 ],[ 0, 6.52, 1.248 ],[ 0, 6.64, 1.44 ],[ 0, 6.4, 2.06 ]] )
attachTo( GEOMETRYCONNECTION, "int3", "SOURCE", "Sheet 3", [[ 0, 12.92, 0.828 ],[ 0, 12.8, 1.04 ],[ 0, 12.92, 1.248 ],[ 0, 12.92, 1.632 ],[ 0, 12.8, 2.06 ],[ 0, 6.52, 0.828 ],[ 0, 6.52, 1.632 ]] )
attachTo( GEOMETRYCONNECTION, "int3", "SOURCE", "Sheet 4", [[ 0, 12.8, 0.18 ],[ 0, 12.92, 0.444 ],[ 0, 13.04, 0.632 ],[ 0, 13.04, 1.44 ]] )
attachTo( GEOMETRYCONNECTION, "int3", "SOURCE", "Sheet 6", [[ 2.32, 6.4, 2.06 ],[ 2.32, 6.52, 1.632 ],[ 2.32, 6.4, 1.04 ],[ 2.32, 6.52, 0.828 ],[ 2.32, 6.64, 0.632 ],[ 2.32, 6.52, 0.444 ],[ 2.32, 6.4, 0.18 ]] )
attachTo( GEOMETRYCONNECTION, "int3", "SOURCE", "Sheet 7", [[ 2.32, 6.64, 1.44 ],[ 2.32, 6.52, 1.248 ],[ 2.32, 12.8, 0.18 ],[ 2.32, 12.92, 0.444 ],[ 2.32, 12.92, 0.828 ],[ 2.32, 12.8, 1.04 ],[ 2.32, 12.92, 1.248 ],[ 2.32, 13.04, 1.44 ],[ 2.32, 12.92, 1.632 ]] )
attachTo( GEOMETRYCONNECTION, "int3", "SOURCE", "Sheet 8", [[ 2.32, 13.04,
```

0.632],[2.32, 12.8, 2.06]])

setElementClassType(GEOMETRYCONNECTION, "int3", "SHLLIF")
assignMaterial("int3", GEOMETRYCONNECTION, "int3")
assignGeometry("int3", GEOMETRYCONNECTION, "int3")
resetElementData(GEOMETRYCONNECTION, "int3")
saveProject( )
addSet(GEOMETRYSUPPORTSET, "co1")
createLineSupport("co1", "co1")
setParameter(GEOMETRYSUPPORT, "co1", "AXES", [1, 2])
setParameter(GEOMETRYSUPPORT, "co1", "TRANSL", [1, 1, 1])
setParameter(GEOMETRYSUPPORT, "co1", "ROTATI", [0, 0, 0])
attach(GEOMETRYSUPPORT, "co1", "Sheet 17", [[1.16, 0, 0]])
saveProject( )
addSet(GEOMETRYSUPPORTSET, "co2")
createLineSupport("co2", "co2")
setParameter(GEOMETRYSUPPORT, "co2", "AXES", [1, 2])
setParameter(GEOMETRYSUPPORT, "co2", "TRANSL", [0, 1, 0])
setParameter(GEOMETRYSUPPORT, "co2", "ROTATI", [1, 0, 1])
attach(GEOMETRYSUPPORT, "co2", "Sheet 4", [[0, 16, 1.2]])
attach(GEOMETRYSUPPORT, "co2", "Sheet 8", [[2.32, 16, 1.2]])
attach(GEOMETRYSUPPORT, "co2", "Sheet 12", [[1.16, 16, 2.4]])
attach(GEOMETRYSUPPORT, "co2", "Sheet 16", [[-0.78, 16, 2.4]])
attach(GEOMETRYSUPPORT, "co2", "Sheet 20", [[1.16, 16, 0]])
attach(GEOMETRYSUPPORT, "co2", "Sheet 24", [[3.1, 16, 2.4]])
saveProject( )
addSet(GEOMETRYLOADSET, "gravity")
createModelLoad("gravity", "gravity")
saveProject( )
addSet(GEOMETRYLOADSET, "tenin")
createBodyLoad("tenin", "tenin")
setParameter(GEOMETRYLOAD, "tenin", "LODTYP", "POSTEN")
setParameter(GEOMETRYLOAD, "tenin", "POSTEN/TENTYP", "ONEEND")
setParameter(GEOMETRYLOAD, "tenin", "POSTEN/ONEEND/FORCE1", 1551240)
setParameter(GEOMETRYLOAD, "tenin", "POSTEN/ONEEND/RETLE1", 0.0001)
setParameter(GEOMETRYLOAD, "tenin", "POSTEN/SHEAR", 0.22)
setParameter(GEOMETRYLOAD, "tenin", "POSTEN/WOBBLE", 0.01)
attachTo(GEOMETRYLOAD, "tenin", "POSTEN/ONEEND/PNTS1", "tenin1", [[0, 0.129987, 1.56]])
attachTo(GEOMETRYLOAD, "tenin", "POSTEN/ONEEND/PNTS1", "tenin2", [[2.32, 0.129987, 1.56]])
attach(GEOMETRYLOAD, "tenin", ["tenin1", "tenin2"])

```
saveProject( )
addSet( GEOMETRYLOADSET, "Geometry load case 1" )
rename( GEOMETRYLOADSET, "Geometry load case 1", "load" )
createSheet( "Sheet 97", [[ -1.56, 9.4, 2.5 ],[ 3.88, 9.4, 2.5 ],[ 3.88, 11.4, 2.5 ],[ -1.56, 11.4, 2.5 ]] )
saveProject( )
projection( SHAPEFACE, "Sheet 11", [[ 1.3306894, 10.070867, 2.4 ]], [ "Sheet 97" ], [ 0, 0, -1 ], True )
projection( SHAPEFACE, "Sheet 15", [[ -0.66522612, 10.070867, 2.4 ]], [ "Sheet 97" ], [ 0, 0, -1 ], True )
projection( SHAPEFACE, "Sheet 23", [[ 3.2147739, 10.070867, 2.4 ]], [ "Sheet 97" ], [ 0, 0, -1 ], True )
removeShape( [ "Sheet 97" ] )
saveProject( )
createSurfaceLoad( "load", "load" )
setParameter( GEOMETRYLOAD, "load", "FORCE/VALUE", -100000 )
setParameter( GEOMETRYLOAD, "load", "FORCE/DIRECT", 3 )
attach( GEOMETRYLOAD, "load", "Sheet 11", [[ 1.3306894, 10.547146, 2.4 ]] )
attach( GEOMETRYLOAD, "load", "Sheet 15", [[ -0.66522612, 10.547146, 2.4 ]] )
attach( GEOMETRYLOAD, "load", "Sheet 23", [[ 3.2147739, 10.547146, 2.4 ]] )
setDefaultGeometryLoadCombinations( )
setGeometryLoadCombinationFactor( "Geometry load combination 1", "tenin", 1 )
remove( GEOMETRYLOADCOMBINATION, "Geometry load combination 3" )
remove( GEOMETRYLOADCOMBINATION, "Geometry load combination 2" )
setGeometryLoadCombinationFactor( "Geometry load combination 1", "tenin", 1 )
setGeometryLoadCombinationFactor( "Geometry load combination 1", "gravity", 1 )
addGeometryLoadCombination( "" )
setGeometryLoadCombinationFactor( "Geometry load combination 2", "load", 1 )
setElementSize( [ "Sheet 1", "Sheet 2", "Sheet 3", "Sheet 4", "Sheet 5", "Sheet 6", "Sheet 7", "Sheet 8", "Sheet 9", "Sheet 10", "Sheet 11", "Sheet 12", "Sheet 13", "Sheet 14", "Sheet 15", "Sheet 16", "Sheet 17", "Sheet 18", "Sheet 19", "Sheet 20", "Sheet 21", "Sheet 22", "Sheet 23", "Sheet 24" ], 0.4, -1, True )
setMesherType( [ "Sheet 1", "Sheet 2", "Sheet 3", "Sheet 4", "Sheet 5", "Sheet 6", "Sheet 7", "Sheet 8", "Sheet 9", "Sheet 10", "Sheet 11", "Sheet 12", "Sheet 13", "Sheet 14", "Sheet 15", "Sheet 16", "Sheet 17", "Sheet 18", "Sheet 19", "Sheet 20", "Sheet 21", "Sheet 22", "Sheet 23", "Sheet 24" ], "HEXQUAD" )
setMidSideNodeLocation( [ "Sheet 1", "Sheet 2", "Sheet 3", "Sheet 4", "Sheet 5", "Sheet 6", "Sheet 7", "Sheet 8", "Sheet 9", "Sheet 10", "Sheet 11", "Sheet 12", "Sheet 13", "Sheet 14", "Sheet 15", "Sheet 16", "Sheet 17", "Sheet 18", "Sheet 19", "Sheet 20", "Sheet 21", "Sheet 22", "Sheet 23", "Sheet 24" ], "LINEAR" )
setElementSize( "Sheet 2", 1, [[ 0, 6.4, 0.18 ],[ 0, 6.64, 0.632 ],[ 0, 6.52, 0.828 ],[ 0, 6.4,
```

1.04],[0, 6.52, 1.248],[0, 6.64, 1.44],[0, 6.4, 2.06]], 0.4, 0, True)

setElementSize("Sheet 3", 1, [[0, 12.92, 0.444],[0, 13.04, 0.632],[0, 12.92, 0.828],[0, 12.8, 1.04],[0, 12.92, 1.248],[0, 12.92, 1.632],[0, 6.52, 0.444],[0, 6.52, 1.632]], 0.4, 0, True)

setElementSize("Sheet 4", 1, [[0, 12.8, 0.18],[0, 13.04, 1.44],[0, 12.8, 2.06]], 0.4, 0, True)

setElementSize("Sheet 6", 1, [[2.32, 6.4, 0.18],[2.32, 6.52, 0.444],[2.32, 6.52, 0.828],[2.32, 6.52, 1.248],[2.32, 6.64, 1.44],[2.32, 6.52, 1.632],[2.32, 6.4, 2.06]], 0.4, 0, True)

setElementSize("Sheet 7", 1, [[2.32, 6.64, 0.632],[2.32, 6.4, 1.04],[2.32, 12.8, 0.18],[2.32, 12.92, 0.444],[2.32, 13.04, 0.632],[2.32, 12.92, 0.828],[2.32, 12.8, 1.04],[2.32, 12.92, 1.248],[2.32, 13.04, 1.44],[2.32, 12.8, 2.06]], 0.4, 0, True)

setElementSize("Sheet 8", 1, [[2.32, 12.92, 1.632]], 0.4, 0, True)

setElementSize("Sheet 11", 1, [[1.16, 6.4, 2.4],[1.16, 12.8, 2.4]], 0.4, 0, True)

setElementSize("Sheet 14", 1, [[-0.78, 6.4, 2.4]], 0.4, 0, True)

setElementSize("Sheet 15", 1, [[-0.78, 12.8, 2.4]], 0.4, 0, True)

setElementSize("Sheet 19", 1, [[1.16, 6.4, 0],[1.16, 12.8, 0]], 0.4, 0, True)

setElementSize("Sheet 22", 1, [[3.1, 6.4, 2.4]], 0.4, 0, True)

setElementSize("Sheet 24", 1, [[3.1, 12.8, 2.4]], 0.4, 0, True)

generateMesh([])

hideView("GEOM")

showView("MESH")

addAnalysis("Analysis1")

addAnalysisCommand("Analysis1", "NONLIN", "Structural nonlinear")

renameAnalysis("Analysis1", "Analysis1")

removeAnalysisCommandDetail("Analysis1", "Structural nonlinear", "EXECUT(1)")

addAnalysisCommandDetail("Analysis1", "Structural nonlinear", "MODEL/EVALUA/REINFO/INTERF")

setAnalysisCommandDetail("Analysis1", "Structural nonlinear", "MODEL/EVALUA/REINFO/INTERF", True)

setAnalysisCommandDetail("Analysis1", "Structural nonlinear", "EXECUT/EXETYP", "START")

renameAnalysisCommandDetail("Analysis1", "Structural nonlinear", "EXECUT(1)", "tenin")

addAnalysisCommandDetail("Analysis1", "Structural nonlinear", "EXECUT(1)/START/INITIA/STRESS")

setAnalysisCommandDetail("Analysis1", "Structural nonlinear", "EXECUT(1)/START/INITIA/STRESS", True)

addAnalysisCommandDetail("Analysis1", "Structural nonlinear", "EXECUT(1)/PHYSIC")

renameAnalysisCommandDetail("Analysis1", "Structural nonlinear", "EXECUT(1)", "tenin")

setAnalysisCommandDetail("Analysis1", "Structural nonlinear",

```
"EXECUT(1)/PHYSIC/BOND", True )
    setAnalysisCommandDetail( "Analysis1", "Structural nonlinear",
"EXECUT(1)/PHYSIC/LIQUEF", False )
    saveProject( )
    setAnalysisCommandDetail( "Analysis1", "Structural nonlinear", "EXECUT/EXETYP",
"LOAD" )
    renameAnalysisCommand( "Analysis1", "Structural nonlinear", "Structural nonlinear" )
    renameAnalysisCommandDetail( "Analysis1", "Structural nonlinear", "EXECUT(2)",
"load" )
    setAnalysisCommandDetail( "Analysis1", "Structural nonlinear",
"EXECUT(2)/LOAD/STEPS/EXPLIC/SIZES", "1.00000" )
    setAnalysisCommandDetail( "Analysis1", "Structural nonlinear",
"EXECUT(2)/LOAD/LOADNR", 2 )
    setAnalysisCommandDetail( "Analysis1", "Structural nonlinear",
"EXECUT(2)/LOAD/STEPS/EXPLIC/SIZES", "1.00000(7) 0.2(5)" )
    setAnalysisCommandDetail( "Analysis1", "Structural nonlinear",
"EXECUT(2)/ITERAT/MAXITE", 50 )
    setAnalysisCommandDetail( "Analysis1", "Structural nonlinear",
"EXECUT(2)/ITERAT/CONVER/DISPLA/NOCONV", "CONTIN" )
    setAnalysisCommandDetail( "Analysis1", "Structural nonlinear",
"EXECUT(2)/ITERAT/CONVER/FORCE/NOCONV", "CONTIN" )
    setAnalysisCommandDetail( "Analysis1", "Structural nonlinear",
"EXECUT(1)/ITERAT/MAXITE", 50 )
    setAnalysisCommandDetail( "Analysis1", "Structural nonlinear",
"EXECUT(1)/ITERAT/CONVER/DISPLA/NOCONV", "CONTIN" )
    setAnalysisCommandDetail( "Analysis1", "Structural nonlinear",
"EXECUT(1)/ITERAT/CONVER/FORCE/NOCONV", "CONTIN" )
    setAnalysisCommandDetail( "Analysis1", "Structural nonlinear", "OUTPUT(1)/SELTYP",
"PRIMAR" )
    setAnalysisCommandDetail( "Analysis1", "Structural nonlinear", "OUTPUT(1)/SELTYP",
"USER" )
    addAnalysisCommandDetail( "Analysis1", "Structural nonlinear", "OUTPUT(1)/USER" )
    addAnalysisCommandDetail( "Analysis1", "Structural nonlinear",
"OUTPUT(1)/USER/DISPLA(1)/TOTAL/TRANSL/GLOBAL" )
    addAnalysisCommandDetail( "Analysis1", "Structural nonlinear",
"OUTPUT(1)/USER/STRAIN(7)/CRACK/GREEN" )
    addAnalysisCommandDetail( "Analysis1", "Structural nonlinear",
"OUTPUT(1)/USER/STRAIN(2)/CRKSUM/GREEN/LOCAL" )
    addAnalysisCommandDetail( "Analysis1", "Structural nonlinear",
"OUTPUT(1)/USER/STRAIN(3)/CRKSUM/GREEN/GLOBAL" )
    addAnalysisCommandDetail( "Analysis1", "Structural nonlinear",
```

```
"OUTPUT(1)/USER/STRAIN(4)/CRKSUM/GREEN/PRINCI" )
    addAnalysisCommandDetail( "Analysis1", "Structural nonlinear",
"OUTPUT(1)/USER/STRAIN(5)/CRKWDT/GREEN/GLOBAL" )
    addAnalysisCommandDetail( "Analysis1", "Structural nonlinear",
"OUTPUT(1)/USER/STRAIN(6)/CRKWDT/GREEN/PRINCI" )
    runSolver( "Analysis1" )
    showView( "RESULT" )
    setResultPlot( "contours", "Total Displacements/node", "TDtZ" )
    setResultCase( [ "Analysis1", "Output", "Load-step 2, Load-factor 1.0000" ] )
    setResultCase( [ "Analysis1", "Output", "Load-step 3, Load-factor 2.0000" ] )
    setResultPlot( "contours", "Summed Crack Strains/node", "Ekxx" )
    setResultPlot( "contours", "Summed Crack Strains/node", "Ekyy" )
    setResultPlot( "contours", "Summed Crack Strains/node", "Ekzz" )
    setResultPlot( "contours", "Summed Crack Strains/node", "Ek1" )
    setResultCase( [ "Analysis1", "Output", "Load-step 4, Load-factor 3.0000" ] )
    setResultCase( [ "Analysis1", "Output", "Load-step 5, Load-factor 4.0000" ] )
    setResultCase( [ "Analysis1", "Output", "Load-step 6, Load-factor 5.0000" ] )
    setResultCase( [ "Analysis1", "Output", "Load-step 7, Load-factor 6.0000" ] )
    setResultCase( [ "Analysis1", "Output", "Load-step 8, Load-factor 7.0000" ] )
    setResultCase( [ "Analysis1", "Output", "Load-step 9, Load-factor 7.2000" ] )
    setResultCase( [ "Analysis1", "Output", "Load-step 10, Load-factor 7.4000" ] )
    setResultCase( [ "Analysis1", "Output", "Load-step 11, Load-factor 7.6000" ] )
    setResultCase( [ "Analysis1", "Output", "Load-step 12, Load-factor 7.8000" ] )
    setResultCase( [ "Analysis1", "Output", "Load-step 13, Load-factor 8.0000" ] )
    setResultPlot( "contours", "Summed Crack Strains/node", "Ek3" )
    setResultPlot( "contours", "Summed Crack Strains/node", "Ek2" )
    setResultPlot( "contours", "Crack-widths/node", "EcwXX" )
    setResultPlot( "contours", "Crack-widths/node", "EcwYY" )
    setResultPlot( "contours", "Crack-widths/node", "EcwZZ" )
    setResultPlot( "contours", "Crack-widths/node", "Ecw2" )
    setResultPlot( "contours", "Crack-widths/node", "Ecw1" )
    setResultPlot( "contours", "Crack-widths/node", "Ecw2" )
    setResultPlot( "cracks", "Crack Strains/mappedcrack", "Eknn" )
    setResultPlot( "cracks", "Crack Strains/mappedcrack", "Gknt" )
    setResultPlot( "cracks", "Crack Strains/mappedcrack", "Gkns" )
    setResultPlot( "contours", "Crack-widths/node", "Ecw3" )
    saveProject( )
```

Reference

1. Song ST (2015) Experimental study and theoretical analysis on bending and joint shear resistance of precast segmental box girder bridges of high-speed railway. PhD Dissertation, Southeast University, Nan Jing (In Chinese)

Chapter 6
Proposals for Further Improvements

Abstract During the process of manipulating DIANA software, author can feel that DIANA is a powerful finite-element analysis software typical for civil engineering, and there are continuous improvement in all the current and emerging release versions. Integrated with author's experience, however, there are still issues that deserve further improvement although vast performance enhancement is achieved in these release versions.

(1) It is advisable to add relevant module on frost resistance and impermeability in the durability investigation, and fiber-reinforced composite materials (FRP) can also be added in the subsequent upgrade versions. According to the mechanical properties of different FRP composite material types, users can choose constitutive options such as CFRP, AFRP and GFRP and they are not bothered to take the method specified by the user.

(2) It is observed that the nonlinear calculation situation is very complex when the concrete constitutive material aspects of shrinkage, creep and cracking model are ticked at the same time in the European CEB-FIP1990 code, JSCE code and the European fib 2010 code. Users always confront poor convergence or even non-convergence. It is expected to be improved in the future upgrading research and development of Diana advanced version.

(3) It is also recommended that DIANA should track the mechanical properties of current new civil engineering materials such as UHPC concrete and FRP material and set new specific material properties, aspects and blocks for them. Meanwhile, emphasis should also be attached on the time-dependent block concerning UHPC concrete international codes.

(4) Meshing procedure should be simplified and the memory footprint as well as consuming time should be further decreased in order to adapt ordinary computer CPU. Additionally, deletion and modification of meshed elements should be conducted in manual, which is adverse for users, especially for beginners.

(5) DIANA errors sometimes occur and it is sometimes hard to open under repeated recalculation in all kinds of versions. Moreover, two different .dpf files cannot be opened at the same time. Therefore, the reliability needs further improvement.

S. Chai, *Finite Element Analysis for Civil Engineering with DIANA Software*,
https://doi.org/10.1007/978-981-15-2945-0_6

(6) For the crack propagation part, it may be considered that which point crack width is the largest while which point crack width is the smallest should be monitored in the post-processing interface of the finite element, so that the user can compare the experimental results with the numerical simulation at a glance.

As a powerful software in the finite-element nonlinear analysis, the author expects that the suggestions and shortcomings mentioned above can be further improved in the new version of DIANA software while applying this superior software. However, one flaw cannot obscure the splendor of the jade. Although there are also minor problems above-mentioned needing improvement or some suggestions that can be adopted in higher-level versions, the author believes that in the next higher-level version, DIANA software will dominate a higher level in the original excellent human-machine operation, compatibility and reliability, thus truly makes its own in the non-linear field analysis, becoming a high-level structural analysis software highly praised by more and more users and civil engineering experts!

Attachment: DIANA Shortcut Manipulations and Default Terms

Enlarge	Mouse wheeling up
Narrow	Mouse wheeling down.
Rotation	Pressing and holding the mouse wheeling key to rotate the model.
Translation	Pressing and holding the mouse wheeling key and moving the mouse left and right.
Saving .doff model and .pie file	Ctrl+S.
Duplication of files	Ctrl+C.
Undo former manipulation	Ctrl+Z.
Running a model	F5.
Windows+arrow in left	Moving model to the left screen region.
Windows+arrow in left	Moving model to the right screen region.
1, 0, 0	X direction in the global coordinate system.
0, 1, 0	Y direction in the global coordinate system.
0, 0, 1	Z direction in the global coordinate system

The manufacturer's authorised representative in the EU is Springer Nature Customer Service Centre GmbH, Europaplatz 3, 69115 Heidelberg, Germany. If you have any concerns regarding our products, please contact ProductSafety@springernature.com

Printed and bound by CPI Group (UK) Ltd, Croydon, CR0 4YY
23/09/2026
02221532-0017